T0155985

Handbook of Meta-Analysis

Handbooks of Modern Statistical Methods

Series Editor

Garrett Fitzmaurice

Department of Biostatistics, Harvard School of Public Health, Boston, MA, USA

The objective of the series is to provide high-quality volumes covering the state-of-the-art in the theory and applications of statistical methodology. The books in the series are thoroughly edited and present comprehensive, coherent, and unified summaries of specific methodological topics from statistics. The chapters are written by the leading researchers in the field and present a good balance of theory and application through a synthesis of the key methodological developments and examples and case studies using real data.

Published Titles

Handbook of Methods for Designing, Monitoring, and Analyzing Dose-Finding Trials
John O'Quigley, Alexia Iasonos, and Björn Bornkamp

Handbook of Quantile Regression
Roger Koenker, Victor Chernozhukov, Xuming He, and Limin Peng

Handbook of Statistical Methods for Case-Control Studies
Ørnulf Borgan, Norman Breslow, Nilanjan Chatterjee, Mitchell H. Gail, Alastair Scott, and Chris J. Wild

Handbook of Environmental and Ecological Statistics
Alan E. Gelfand, Montserrat Fuentes, Jennifer A. Hoeting, and Richard L. Smith

Handbook of Approximate Bayesian Computation
Scott A. Sisson, Yanan Fan, and Mark Beaumont

Handbook of Graphical Models
Marloes Maathuis, Mathias Drton, Steffen Lauritzen, and Martin Wainwright

Handbook of Mixture Analysis
Sylvia Frühwirth-Schnatter, Gilles Celeux, and Christian P. Robert

Handbook of Infectious Disease Data Analysis
Leonhard Held, Niel Hens, Philip O'Neill, and Jacco Walllinga

Handbook of Forensic Statistics
David L. Banks, Karen Kafadar, David H. Kaye, and Maria Tackett

Handbook of Meta-Analysis
Christopher H. Schmid, Theo Stijnen, and Ian R. White

For more information about this series, please visit: https://www.crcpress.com/Chapman--HallCRC-Handbooks-of-Modern-Statistical-Methods/book-series/CHHANMODSTA

Handbook of Meta-Analysis

Edited by
Christopher H. Schmid, Theo Stijnen,
and Ian R. White

CRC Press is an imprint of the
Taylor & Francis Group, an **informa** business
A CHAPMAN & HALL BOOK

First edition published 2021
by CRC Press
6000 Broken Sound Parkway NW, Suite 300, Boca Raton, FL 33487-2742

and by CRC Press
2 Park Square, Milton Park, Abingdon, Oxon, OX14 4RN

CRC Press is an imprint of Taylor & Francis Group, LLC

Library of Congress Cataloging-in-Publication Data

Names: Schmid, Christopher H., editor.
Title: Handbook of meta-analysis / [edited by] Christopher H. Schmid, Ian White, Theo Stijnen.
Description: First edition. | Boca Raton : Taylor and Francis, [2020] | Series: Chapman & Hall/CRC handbooks of modern statistical methods | Includes bibliographical references and index. | Summary: "Meta-analysis is the application of statistics to combine results from multiple studies and draw appropriate inferences. Its use and importance have exploded over the years as the need for a robust evidence base has become clear in many scientific areas like medicine and health, social sciences, education, psychology, ecology and economics"-- Provided by publisher.
Identifiers: LCCN 2020014279 (print) | LCCN 2020014280 (ebook) | ISBN 9781498703987 (hardback) | ISBN 9780367539689 (paperback) | ISBN 9781315119403 (ebook)
Subjects: LCSH: Meta-analysis.
Classification: LCC R853.M48 H35 2020 (print) | LCC R853.M48 (ebook) | DDC 610.72--dc23
LC record available at https://lccn.loc.gov/2020014279
LC ebook record available at https://lccn.loc.gov/2020014280

ISBN: 9781498703987 (hbk)
ISBN: 9781315119403 (ebk)

Typeset in Palatino
by Deanta Global Publishing Services, Chennai, India

Contents

Preface

Why Did We Write This Book?

Meta-analysis is the statistical combination of results from multiple studies in order to yield results which make the best use of all available evidence. It has become increasingly important over the last 25 years as the need for a robust evidence base has become clear in many scientific areas including medicine and health, social sciences, education, psychology, ecology, and economics.

Alongside the explosion of use of meta-analysis has come an explosion of methods for handling complexities in meta-analysis, including explained and unexplained heterogeneity between studies, publication bias, and sparse data. At the same time, meta-analysis has been extended beyond simple two-group comparisons to multiple comparisons and complex observational studies, and beyond continuous and binary outcomes to survival and multivariate outcomes. Many of these methods are statistically complex.

Many books overview the role of meta-analysis in the broader research synthesis process or cover particular aspects of meta-analysis in more statistical detail. This book, by contrast, aims to cover the full range of statistical methodology used in meta-analysis, in a statistically rigorous and up-to-date way. It provides a comprehensive, coherent, and unified overview of the statistical foundations behind meta-analysis as well as a detailed description of the primary methods and their application to specific types of data.

Who Is This Book For?

In editing the book, we have kept in mind a broad audience of graduate students, researchers, and practitioners interested in the theory and application of statistical methods for meta-analysis. Our target audience consists of statisticians involved in the development of new methods for meta-analysis, as well as scientists applying established methods in their research. The book is written at the level of graduate courses in statistics, but will be of interest to and readable for quantitative scientists from a range of disciplines. The book can be used as a graduate level textbook, as a general reference for methods, or as an introduction to specialized topics using state-of-the art methods.

What Scientific Fields Are Covered?

Different scientific areas sometimes address different types of meta-analysis questions and hence require different methods of meta-analysis. However, they often tackle the same types of meta-analysis questions in different ways, and in this case, different scientific areas can learn from each other. Most of the statistical methods described in this book are appropriate to all scientific areas. In fact, most of the authors of this book come from a biomedical background and so most examples are from the biomedical sciences, but the majority of the methods discussed are equally relevant for other life, natural, and social sciences.

How Should I Use the Book?

The book is designed to be read from beginning to end, though readers can also use it as a reference book.

Chapters 1–6 are the core material. We first give the background to performing a meta-analysis: Chapter 1 describes the broader systematic review process; Chapter 2 discusses general issues in meta-analysis that recur throughout the book; and Chapter 3 describes the extraction of data for meta-analysis. We then present the fundamental statistical tools for meta-analysis: Chapter 4 covers meta-analysis of study-level results ("two-stage methods") while Chapter 5 covers meta-analysis where more detailed data are available from each study ("one-stage methods"); these both follow a frequentist approach, and Chapter 6 covers the alternative Bayesian approach.

Chapters 7–13 present key extensions to these basic methods: meta-regression to explore whether results relate to study-level characteristics (Chapter 7); use of individual participant data (Chapter 8); handling studies that estimate multiple quantities (Chapter 9) or that compare multiple treatments (Chapter 10); checking meta-analysis models (Chapter 11); and handling bias within studies (Chapter 12) or in study reporting (Chapter 13). By this point, the reader has covered all the techniques used in standard published meta-analyses.

Chapters 14–22 discuss a number of extensions to particular fields of biomedical and social research, showing how the methods described in Chapters 4–13 are applied or modified for these settings: relating treatment effects to overall risk (Chapter 14); meta-analysis of survival data (Chapter 15), correlation matrices (Chapter 16), genetic data (Chapter 17), dose-response relationships (Chapter 18), and diagnostic test data (Chapter 19); using meta-analysis to evaluate surrogate endpoints (Chapter 20) and to combine complex observational data (Chapter 21) and prognostic models (Chapter 22). We end in Chapter 23 with a discussion of the uses of meta-analysis in planning future research.

Alongside the theory, the book covers a number of practical examples. The data for most of these examples (except those using individual participant data) are available on the book's website (https://www.ctu.mrc.ac.uk/books/handbook-of-meta-analysis). Meta-analysis software is discussed in each chapter, and the website also provides syntax in various statistical packages by which the results of the examples can be reproduced.

Comments

Terminology in the meta-analysis field has grown up over many years, and while most is good, some is unhelpful. We have chosen to avoid using the term "fixed-effect model" or "fixed-effects model" to mean the meta-analysis model with no heterogeneity, and instead we call this the "common-effect" model. This allows us to use the term "fixed effect" in its standard statistical sense (see Chapter 2). We hope that the meta-analysis community will follow this lead.

There has been much discussion recently about whether the concept of statistical significance is outdated, for example, Wasserstein et al. (2019). We have taken the view in this book that, though statistical significance is overused, it remains a useful concept.

Bibliography

Wasserstein, R. L., Schirm, A. L. and Lazar, N. A. (2019). Moving to a World Beyond "p<0.05." *The American Statistician*, 73(sup1), 1–19.

Editors

Christopher H. Schmid is Professor of Biostatistics at Brown University. He received his BA in Mathematics from Haverford College in 1983 and his PhD in Statistics from Harvard University in 1991. In 1991, he joined the Institute for Clinical Research and Health Policy Studies at Tufts Medical Center and joined the medical faculty at Tufts University in 1992. He became the director of the Biostatistics Research Center in 2006 and Associate Director of the Tufts Clinical and Translational Research training program in 2009. In 2012, he moved to Brown University to co-found the Center for Evidence Synthesis in Health. In 2016, he became Director of the Clinical Study Design, Epidemiology and Biostatistics Core of the Rhode Island Center to Advance Translational Science and in 2018 became Chair of Biostatistics in the School of Public Health.

Dr. Schmid has a long record of collaborative research and training activities in many different clinical and public health research areas. His research focuses on Bayesian methods for meta-analysis, including networks of treatments and N-of-1 designs, as well as open-source software tools and methods for developing and assessing predictive models using data from multiple databases, for example, the current standard biomarker prediction tool for the glomerular filtration rate (GFR). He is the author of nearly 300 publications, including coauthored consensus CONSORT reporting guidelines for N-of-1 trials and single-case designs, and PRISMA guideline extensions for meta-analysis of individual participant studies and for network meta-analyses, as well as the Institute of Medicine report that established US standards for systematic reviews.

Dr. Schmid is an elected member of the Society for Research Synthesis Methodology and co-founding editor of its journal, *Research Synthesis Methods*. He is a Fellow of the American Statistical Association and long-time Statistical Editor of the *American Journal of Kidney Diseases*.

Ian R. White is Professor of Statistical Methods for Medicine at the Medical Research Council Clinical Trials Unit at University College London, UK. He originally studied Mathematics at Cambridge University, and his first career was as a teacher of mathematics in The Gambia, Cambridge, and London. He obtained his MSc in Statistics from University College London, where he subsequently worked in the Department of Epidemiology and Public Health. He was then Senior Lecturer in the Medical Statistics Unit at the London School of Hygiene and Tropical Medicine and for 16 years Program Leader at the Medical Research Council Biostatistics Unit in Cambridge. He received his PhD by publication in 2011.

His research interests are in statistical methods for the design and analysis of clinical trials, observational studies, and meta-analyses. He is particularly interested in developing methods for handling missing data, correcting for departures from randomized treatment, novel trial designs, simulation studies, and network meta-analysis. He runs courses on various topics and has written a range of Stata software.

Theo Stijnen is Emeritus Professor of Medical Statistics at the Leiden University Medical Center, The Netherlands. He obtained his MSc in Mathematics at Leiden University in 1973 and received his PhD in Mathematical Statistics at the University of Utrecht in 1980. Then he decided to leave mathematical statistics and to specialize in applied medical statistics, a choice he has never regretted. In 1981, he was appointed Assistant Professor of Medical

Statistics at the Leiden University Medical Faculty. In 1987, he became Associate Professor of Biostatistics at the Erasmus University Medical Center in Rotterdam, where he was appointed Full Professor in 1998. In 2007, he returned to Leiden again to become the Head of the Department of Medical Statistics and Bioinformatics, which was recently renamed the Department of Biomedical Data Sciences. He has broad experience in teaching statistics to various audiences and his teaching specialties include mixed modeling, survival analysis, epidemiological modeling, and meta-analysis. In 2009, he was a co-founder of the MSc program Statistical Science for the Life and Behavioral Sciences, the first MSc program in this field in The Netherlands. He has extensive experience in statistical consultancy for medical researchers, resulting in more than 400 co-authorships in the medical scientific literature, of which about 25 are on medical meta-analyses. His biostatistical research interests include clinical trials methodology, epidemiological methods, mixed modeling, and meta-analysis. He is (co-)author of over 70 methodological articles, of which about 25 are on meta-analysis. He retired on December 14, 2016. He now works part-time as an independent Biostatistical Consultant and continues doing research.

Contributors

Ariel Aloe
Educational Measurement and Statistics
University of Iowa
Iowa City, IA

Ariel Alonso
Interuniversity Institute for Biostatistics and statistical Bioinformatics (I-BioStat)
KU Leuven
Leuven, Belgium

Betsy Jane Becker
Synthesis Research Group, College of Education
Florida State University
Tallahassee, FL

Tomasz Burzykowski
Interuniversity Institute for Biostatistics and statistical Bioinformatics (I-BioStat)
Hasselt University
Hasselt, Belgium
and
International Drug Development Institute (IDDI)
Louvain-la-Neuve, Belgium

Marc Buyse
International Drug Development Institute (IDDI)
Louvain-la-Neuve, Belgium
and
Interuniversity Institute for Biostatistics and statistical Bioinformatics (I-BioStat)
Hasselt University
Hasselt, Belgium

Bradley Carlin
Counterpoint Statistical Consulting, LLC
Minneapolis, MN

Yong Chen
Department of Biostatistics, Epidemiology and Informatics, Perelman School of Medicine
University of Pennsylvania
Philadelphia, PA

Michael Cheung
Department of Psychology
National University of Singapore
Singapore

Haitao Chu
Division of Biostatistics, School of Public Health
University of Minnesota Twin Cities
Minneapolis, MN

Karl Claxton
Professor of Economics, Department of Economics and Centre for Health Economics
University of York
York, UK

Thomas Debray
Julius Center for Health Sciences and Primary Care
University Medical Center Utrecht
Utrecht, the Netherlands

Rui Duan
Department of Biostatistics, Epidemiology and Informatics
University of Pennsylvania
Philadelphia, PA

Marta Fiocco
Mathematical Institute Leiden University
Leiden, the Netherlands

Susan Griffin
Centre for Health Economics
University of York
York, UK

Annamaria Guolo
Department of Statistical Sciences
University of Padua
Padua, Italy

Julian Higgins
Population Health Sciences
Bristol Medical School
University of Bristol
Bristol, UK

Dan Jackson
Statistical Innovation Group
AstraZeneca
Cambridge, UK

Hayley E. Jones
Population Health Sciences
Bristol Medical School
University of Bristol
Bristol, UK

Stephen Kaptoge
Department of Public Health and Primary Care
University of Cambridge
Cambridge, UK

Hendrik Koffijberg
Department Health Technology & Services Research
Faculty of Behavioural, Management and Social Sciences
Technical Medical Centre
University of Twente
Enschede, the Netherlands

Martin Law
Medical Research Council Biostatistics Unit
Cambridge, UK

Yulun Liu
Department of Population and Data Sciences
UT Southwestern Medical Center
Dallas, TX

Jose López-López
Department of Basic Psychology & Methodology, Faculty of Psychology
University of Murcia
Murcia, Spain

Xiaoye Ma
Genentech
San Francisco, CA

Arielle Marks-Anglin
Department of Biostatistics, Epidemiology and Informatics
Perelman School of Medicine
University of Pennsylvania
Philadelphia, PA

Cosetta Minelli
National Heart and Lung Institute
Imperial College London
London, UK

Geert Molenberghs
Interuniversity Institute for Biostatistics and statistical Bioinformatics (I-BioStat)
Hasselt University
Diepenbeek, Belgium
and
Interuniversity Institute for Biostatistics and statistical Bioinformatics (I-BioStat)
KU Leuven
Leuven, Belgium

Daan Nieboer
Erasmus MC
Rotterdan, the Netherlands

Adriani Nikolakopoulou
Institute of Social and Preventive Medicine
University of Bern
Bern, Switzerland

Clare Oliver-Williams
Department of Public Health and Primary Care
University of Cambridge
Cambridge, UK

Nicola Orsini
Department of Global Public Health
Karolinska Institutet
Stockholm, Sweden

Orestis Panagiotou
Department of Health Services, Policy & Practice
Brown University School of Public Health
Providence, RI

Richard D. Riley
Centre for Prognosis Research, School of Primary, Community and Social Care
Keele University
Newcastle, UK

Claire Rothery
Senior Research Fellow in Health Economics, Centre for Health Economics
University of York
York, UK

Georgia Salanti
Institute of Social and Preventive Medicine (ISPM)
University of Bern
Bern, Switzerland

Jelena Savović
Population Health Sciences
Bristol Medical School
University of Bristol
Bristol, UK
and
NIHR Applied Research Collaboration (ARC) West
University Hospitals Bristol NHS Foundation Trust
Bristol, UK

Christopher H. Schmid
Department of Biostatistics and Center for Evidence Synthesis in Health
Brown University School of Public Health
Providence, RI

Ziv Shkedy
Interuniversity Institute for Biostatistics and statistical Bioinformatics (I-BioStat)
Hasselt University
Hasselt, Belgium

Mark Simmonds
Centre for Reviews and Dissemination
University of York
York, UK

Donna Spiegelman
Susan Dwight Bliss Professor of Biostatistics and Director
Center on Methods for Implementation and Prevention Science (CMIPS)
Yale School of Public Health
Newhaven, CT

Lesley Stewart
Centre for Reviews and Dissemination
University of York
York, UK

Ewout Steyerberg
Leiden University Medical Center
Leiden, the Netherlands

Theo Stijnen
Leiden University Medical Center
Leiden, the Netherlands

Michael Sweeting
Department of Health Sciences
University of Leicester
Leicester, UK

John Thompson
Department of Health Sciences
University of Leicester
Leicester, UK

Rebecca Turner
MRC Clinical Trials Unit
University College London
London, UK

Wim Van der Elst
Janssen Pharmaceutica
Beerse, Belgium
and
Interuniversity Institute for Biostatistics and statistical Bioinformatics (I-BioStat)
Hasselt University
Hasselt, Belgium

Hans van Houwelingen
Leiden University Medical Center
Leiden, the Netherlands

Wolfgang Viechtbauer
Department of Psychiatry and Neuropsychology
Maastricht University
Maastricht, the Netherlands

Nicky J. Welton
Population Health Sciences
Bristol Medical School
University of Bristol
Bristol, UK

Ian R. White
MRC Clinical Trials Unit
University College London
London, UK

Angela Wood
Department of Public Health and Primary Care
University of Cambridge
Cambridge, UK

1

Introduction to Systematic Review and Meta-Analysis

Christopher H. Schmid, Ian R. White, and Theo Stijnen

CONTENTS

1.1 Introduction

The growth of science depends on accumulating knowledge building on the past work of others. In health and medicine, such knowledge translates into developing treatments for diseases and determining the risks of exposures to harmful substances or environments. Other disciplines benefit from new research that finds better ways to teach students, more effective ways to rehabilitate criminals, and better ways to protect fragile environments. Because the effects of treatments and exposures often vary with the conditions under which they are evaluated, multiple studies are usually required to ascertain their true extent. As the pace of scientific development quickens and the amount of information in the literature continues to explode (for example, about 500,000 new articles are added to the National Library of Medicine's PubMed database each year), scientists struggle to keep up with the latest research and recommended practices. It is impossible to read all the studies in even a specialized subfield and even more difficult to reconcile the often-conflicting messages that they present. Traditionally, practitioners relied on experts to summarize the literature and make recommendations in articles that became known as narrative reviews.

Over time, however, researchers began to investigate the accuracy of such review articles and found that the evidence often did not support the recommendations (Antman et al., 1992). They began to advocate a more scientific approach to such reviews that did not rely on one expert's idiosyncratic review and subjective opinion. This approach required documented evidence to back claims and a systematic process carried out by a multidisciplinary team to ensure that all the evidence was reviewed.

This process is now called a *systematic review*, especially in the healthcare literature. Systematic reviews use a scientific approach that carefully searches for and reviews all evidence using accepted and pre-specified analytic techniques (Committee on Standards, 2011). A systematic review encompasses a structured search of the literature in order to combine information across studies using a defined protocol to answer a focused research question. The process seeks to find and use all available evidence, both published and unpublished, evaluate it carefully and summarize it objectively to reach defensible recommendations. The synthesis may be qualitative or quantitative, but the key feature is its adherence to a set of rules that enable it to be replicated. The widespread acceptance of systematic reviews has led to a revolution in the way practices are evaluated and practitioners get information on which interventions to apply. Table 1.1 outlines some of the fundamental differences between narrative reviews and systematic reviews.

Systematic reviews are now common in many scientific areas. The modern systematic review originated in psychology in a 1976 paper by Gene Glass that quantitatively summarized all the studies evaluating the effectiveness of psychotherapy (Glass, 1976). Glass called the technique *meta-analysis* and the method quickly spread into diverse fields such as education, criminal justice, industrial organization, and economics (Shadish and Lecy, 2015). It also eventually reached the physical and life sciences, particularly policy-intensive areas like ecology (Järvinen, 1991; Gurevitch et al., 1992). It entered the medical literature in the 1980s with one of the earliest influential papers being a review of the effectiveness of beta blockers for patients suffering heart attacks (Yusuf et al., 1985) and soon grew very popular. But over time, especially in healthcare, the term *meta-analysis* came to refer primarily to the quantitative analysis of the data from a systematic review. In other words, systematic reviews without a quantitative analysis in health studies are not called a meta-analysis, although this distinction is not yet firmly established in other fields. We will maintain the distinct terms in this book, however, using meta-analysis to refer to the statistical analysis of the data collected in a systematic review. Before exploring the techniques available for meta-analysis in the following chapters, it will be useful first to discuss the parts of the systematic review process in this chapter. This will enable us to understand

TABLE 1.1

Key Differences between Narrative and Systematic Review

Narrative review	Systematic review
Broad overview of topic	Focus on well-formulated questions
Content experts	Multidisciplinary team
Not guided by a protocol	*A priori* defined protocol
No systematic literature search	Comprehensive, reproducible literature search
Unspecified selection of studies	Study selection by eligibility criteria
No critical appraisal of studies	Quality assessment of individual studies
Formal quantitative synthesis unlikely	Meta-analysis often performed when data available
Conclusions based on opinion	Conclusions follow analytic plan and protocol
Direction for future research rarely given	States gaps in current evidence

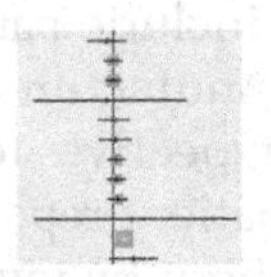

FIGURE 1.1
Systematic review process.

the sources of the data and how the nature of those sources affects the subsequent analysis of the data and interpretation of the results.

Systematic reviews generally involve six major components: topic preparation, literature search, study screening, data extraction, analysis, and preparation of a report (Figure 1.1). Each involves multiple steps and a well-conducted review should carefully attend to all of them (Wallace et al., 2013). The entire process is an extended one and a large, funded review may take over a year and cost hundreds of thousands of dollars. Fortunately, several organizations have written standards and manuals describing proper ways to carry out a review. Excellent references are the Institute of Medicine's *Standards for Systematic Reviews of Comparative Effectiveness Research* (Committee on Standards, 2011), the Cochrane Collaboration's *Cochrane Handbook for Systematic Reviews of Interventions* (Higgins et al., 2019) and *Handbook for Diagnostic Test Accuracy Reviews* (Cochrane, https://methods.cochrane.org/sdt/handbook-dta-reviews), and the Agency for Healthcare Research and Quality (AHRQ) *Methods Guide for Effectiveness and Comparative Effectiveness Reviews* (Agency for Healthcare Research and Quality, 2014). We briefly describe each component and reference additional sources for readers wanting more detail. Since the process is most fully developed and codified in health areas, we will discuss the process in that area. However, translating the ideas and techniques into any scientific field is straightforward.

1.2 Topic Preparation

The Institute of Medicine's *Standards for Systematic Review* (Committee on Standards, 2011) lists four steps to take when preparing a topic: establishing a review team, consulting with stakeholders, formulating the review topic, and writing a review protocol.

The review team should have appropriate expertise to carry out all phases of the review. This includes not only statisticians and systematic review experts, but librarians, science writers, and a wide array of experts in various aspects of the subject matter (e.g., clinicians, nurses, social workers, epidemiologists).

Next, for both the scientific validity and the impact of the review, the research team must consult with and involve the review's stakeholders, those individuals to whom the endeavor is most important and who will be the primary users of the review's conclusions.

Stakeholders may include patients, clinicians, caregivers, policy makers, insurance companies, product manufacturers, and regulators. Each of these groups of individuals will bring different perspectives to ensure that the review answers the most important questions. The use of patient-reported outcomes provides an excellent example of the change in focus brought about by involvement of all stakeholders. Many older studies and meta-analyses focused only on laboratory measurements or clinical outcomes but failed to answer questions related to patient quality of life. When treatments cannot reduce pain, improve sleep, or increase energy, patients may perceive them to be of little benefit even if they do improve biological processes. It is also important to address potential financial, professional, and intellectual conflicts of interest of stakeholders and team members in order to ensure an unbiased assessment (Committee on Standards, 2011).

Thoroughly framing the topic to be studied and constructing the right testable questions forms the foundation of a good systematic review. The foundation underlies all the later steps, especially analysis for which the proper approach depends on addressing the right question. Scope is often motivated by available resources (time, money, personnel), prior knowledge about the problem and evidence. Questions must carefully balance the tradeoff between breadth and depth. Very broadly defined questions may be criticized for not providing a precise answer to a question. Very narrowly focused questions have limited applicability and may be misleading if interpreted broadly; there may also be little or no evidence to answer them.

An analytic framework is often helpful when developing this formulation. An analytic framework is a graphical representation that presents the chain of logic that links the intervention to outcomes and helps define the key questions of interest, including their rationale (Anderson et al., 2011). The rationale should address both research and decision-making perspectives. Each link relating test, intervention, or outcome represents a potential key question. Stakeholders can provide important perspectives. Figure 1.2 provides an example from an AHRQ evidence report on the relationship between cardiovascular disease and omega-3 fatty acids (Balk et al., 2016).

For each question, it is important to identify the PICOS elements: Populations (participants and settings), Interventions (treatments and doses), Comparators (e.g., placebo, standard of care or an active comparator), Outcomes (scales and metrics), and Study designs (e.g., randomized and observational) to be included in the review. Reviews of studies of diagnostic test accuracy modify these components slightly to reflect a focus on tests, rather than treatments. Instead of interventions and comparators, they examine index tests and gold standards (see Chapter 19). Of course, some reviews may have non-comparative outcomes (e.g., prevalence of disease) and so would not have a comparator. Table 1.2 shows potential PICOS components for this study to answer the question posed in the omega-3 review "Are omega-3 fatty acids beneficial in reducing cardiovascular disease?"

As with primary studies, it is also important to construct a thorough protocol that defines all of the review's inclusion and exclusion criteria and also carefully describes how the study will carry out the remaining components of the systematic review: searching, screening, extraction, analysis, and reporting (Committee on Standards, 2011; Moher et al., 2015).

Because the PICOS elements comprise a major part of the protocol that informs the whole study design, it is useful to discuss each element of PICOS in turn. Defining the appropriate populations is crucial for ensuring that the review applies to the contexts for which it is intended to apply. Often, inferences are intended to apply widely, but studies in the review may only focus on narrow settings and groups of individuals. For example,

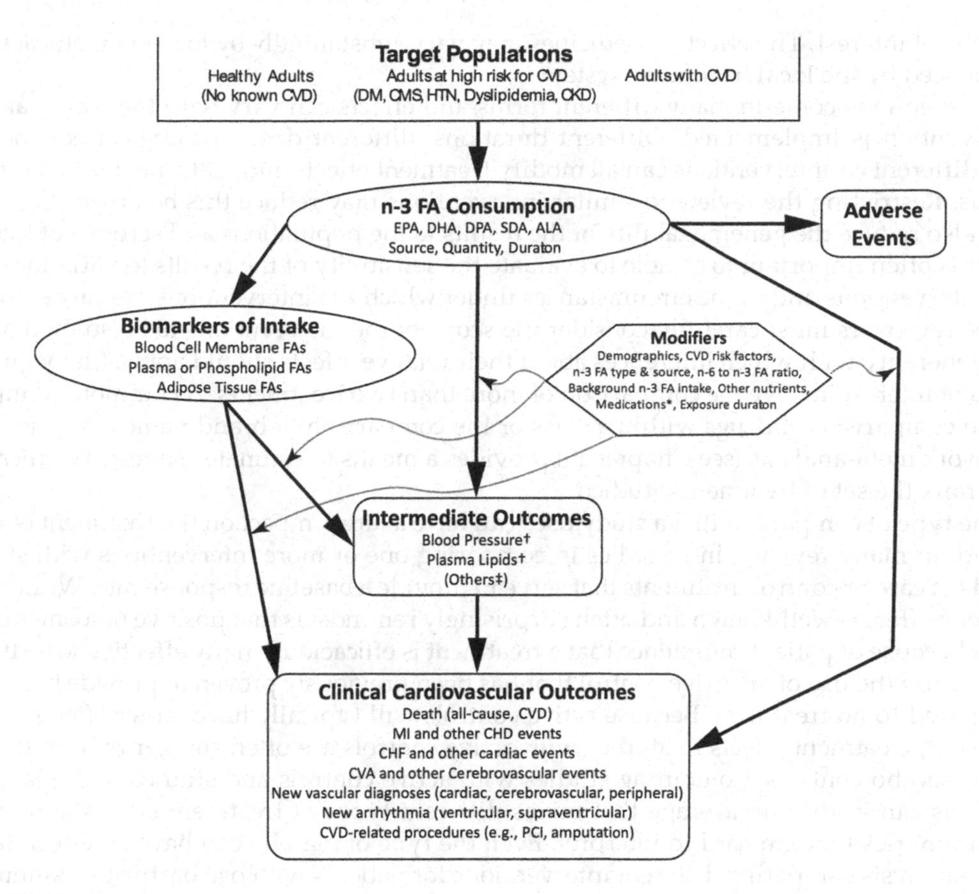

FIGURE 1.2
Analytic framework for omega-3 fatty acid intake and cardiovascular disease (Balk et al., 2016).

TABLE 1.2
Potential PICOS Criteria for Addressing the Question: "Are Omega-3 Fatty Acids Beneficial in Reducing Cardiovascular Disease?"

Participants	Interventions	Comparator	Outcomes	Study Design
Primary prevention	Fish, EPA, DHA, ALA	Placebo	Overall mortality	RCTs
Secondary prevention	Dosage	No control	Sudden death	Observational studies
	Background intake	Isocaloric control	Revascularization	Follow-up duration
	Duration		Stroke	Sample size
			Blood pressure	

many studies exclude the elderly and so do not contribute information about the complete age spectrum. Even when some studies do include all subpopulations, an analysis must be carefully designed in order to evaluate whether benefits and harms apply differently to each subpopulation. Homogeneity of effects across geographic, economic, cultural, or other units may also be difficult to test if most studies are conducted in the same environment. For some problems, inferences may be desired in a particular well-defined area, such as effects in a single country, but in others variation in wider geographic regions may be of interest. The effect of medicines may vary substantially by location if efficacy is influenced by the local healthcare system.

Interventions come in many different forms and effects can vary with the way that an intervention is implemented. Different durations, different doses, different frequencies and different co-interventions can all modify treatment effects and make results heterogeneous. Restricting the review to similar interventions may reduce this heterogeneity but will also reduce the generalizability of the results to the populations and settings of interest. It is often important to be able to evaluate the sensitivity of the results to variations in the interventions and to the circumstances under which the interventions are carried out. Thus, reviewers must carefully consider the scope of the interventions to be studied and the generality with which inferences about their relative effects should apply. Many questions of interest involve the comparison of more than two treatments. A common example is the comparison of drugs within a class or the comparison of brand names to generics. Network meta-analysis (see Chapter 10) provides a means to estimate the relative efficacy and rank the set of treatments studied.

The type of comparator that a study uses can have a large impact on the treatment effect found. In many reviews, interest lies in comparing one or more interventions with standards of care or control treatments that serve to provide a baseline response rate. While the placebo effect is well-known and often surprisingly reminds us that positive outcomes can arise because of patient confidence that a treatment is efficacious, many effectiveness studies require the use of an active control that has been previously proven to provide benefits compared to no treatment. Because active controls will typically have larger effects than placebos, treatment effects in studies with active controls are often smaller than in those with placebo controls. Combining studies with active controls and studies with placebo controls can lead to an average that mixes different types of treatment effects and lead to summaries that are hard to interpret. Even the type of placebo can have an effect. In a meta-analysis comparing different interventions for patients with osteoarthritis, Bannuru et al. found that placebos given intravenously worked better than those given orally, thus distorting the comparison between oral and intravenous treatments which were typically compared with different placebos (Bannuru et al., 2015). Regression analyses may be needed when comparators differ in these ways (see Chapter 7).

The studies included in a review may report many different types of outcomes and the choice of which to summarize can be daunting. Outcomes selected should be meaningful and useful, based on sound scientific principles. Some outcomes correspond to well-defined events such as death or passing a test. Other outcomes are more subjective: amount of depression, energy levels, or ability to do daily activities. Outcomes can be self-reported or reported by a trained evaluator; they can be extracted from a registry or measured by an instrument. Some are more important to the clinician, teacher, or policymaker; others are more important to the research participant, patient, or student. Some are more completely recorded than others; some are primary and some secondary; some relate to benefits, others relate to harms. All of these variations affect the way in which analyses are carried out and interpreted. They change the impact that review conclusions have on different stakeholders and the degree of confidence they inspire. All of these considerations play a major role in the choice of methods used for meta-analysis.

Reviews can summarize studies with many different types of designs. Studies may be randomized or observational, parallel or crossover, cohort or case-control, prospective or retrospective, longitudinal or cross-sectional, single or multi-site. Different techniques are needed for each. Study quality can also vary. Not all randomized trials use proper randomization techniques, appropriate allocation concealment, and double blinding.

Not all studies use standardized protocols, appropriately follow up participants, record reasons for withdrawal, and monitor compliance to treatment. All of these design differences among studies can introduce heterogeneity into a meta-analysis and require careful consideration of whether the results will make sense when combined. Careful consideration of the types and quality of studies to be synthesized in a review can help to either limit this heterogeneity or expand the review's generalizability, depending on the aims of the review's authors. In many cases, different sensitivity analyses will enable reviewers to judge the impact of this heterogeneity on conclusions.

1.3 Literature Search

The PICOS elements motivate the strategy for searching the relevant literature using a variety of sources to address research questions. Bibliographic databases such as Medline or PsycINFO are updated continually and are freely available to the public. Medline, maintained by the US National Library of Medicine since 1964, indexes more than 5500 biomedical journals and more than 20 million items with thousands added each day. A large majority are English language publications. Other databases are available through an annual paid subscription. EMBASE, published by Elsevier, indexes 7500 journals and more than 20 million items in healthcare. Although it overlaps substantially with Medline, it includes more European journals. Other databases provide registries of specific types of publications. The Cochrane Controlled Trials Registry is part of the Cochrane Library and indexes more than 500,000 controlled trials identified through manual searches by volunteers in Cochrane review groups. Many other databases are more specific to subject matter areas. CINAHL covers nursing and allied health fields in more than 1600 journals; PsycINFO covers more than 2000 journals related to psychology; and CAB (Commonwealth Agricultural Bureau) indexes nearly 10,000 journals, books, and proceedings in applied life sciences and agriculture. Sources like Google Scholar are broader but less well-defined making structured, reproducible searches more difficult to carry out and complicating the capture of all relevant articles.

To ensure that searches capture studies missing from databases, researchers should search the so-called gray literature for results not published as full text papers in journals (Balshem et al., 2013). These include sources such as dissertations, company reports, regulatory filings at government agencies such as the US Food and Drug Administration, online registries such as clinicaltrials.gov, and conference proceedings that contain abstracts presented. Some of these items may be available through databases that index gray literature, others may be identified through contact with colleagues and others may require manual searching of key journals and reference lists of identified publications. Preliminary reports like abstracts often present data that will change with final publication, so it is a good idea to continue to check the literature for the final report.

Searches are often restricted to specific languages, especially English, for expediency. Research has not been completely consistent on the potential impact of this language bias (Morrison et al., 2012; Jüni et al., 2002; Pham et al., 2005), but its impact for studying certain treatments is undeniable. For example, reviews of Chinese medical treatments such as acupuncture that ignore the Chinese literature will be incomplete (Wang et al., 2008). Other considerations in choosing sources to search include the quality of the studies, the accessibility of journals, the cost of accessing articles, and the presence of peer review.

Accurate and comprehensive searches require knowledge of the structure of the databases and the syntax needed to search them (e.g., Boolean combinations using AND, OR, and NOT). Medline citations, for example, include the article's title, authors, journal, publication date, language, publication type (e.g., article, letter), and 5–15 controlled vocabulary Medical Subject Heading (MeSH) search terms (i.e., keywords) chosen from a structured vocabulary that ensures uniformity and consistency of indexing and so greatly facilitates searching. The MeSH terms are divided into headings (e.g., disease category or body region), specialized subheadings (e.g., diagnosis, therapy, epidemiology, human, animal), publication types (e.g., journal article, randomized controlled trial), and a large list of supplementary concepts related to the specific article (e.g., type of intervention). Information specialists like librarians trained in systematic review can help construct efficient algorithms of keywords, headings, and subclassifications that best use the search tools in order to optimize sensitivity (not missing relevant items) and specificity (not capturing irrelevant items) during database searches.

Searching is an iterative process. The scope of the search can vary greatly depending on the topic and the questions asked, as well as on the time and manpower resources available. Reviews must balance completeness of the search with the costs incurred. Each database will require its own search strategy to take advantage of its unique features, but general features will remain the same. Some generic search filters that have been developed for specific types of searches (Glanville et al., 2006) can be easily modified to provide a convenient starting strategy for a specific problem. A search that returns too many citations may indicate that the questions being asked are too broad and that the topic should be reformulated in a more focused manner. Manual searches that identify items missed by the database search may suggest improved strategies. It is important to document the exact search strategy, including its date, and the disposition of each report identified including reasons for exclusion, so that the search and the final collection of documents can be reproduced (Liberati et al., 2009).

1.4 Study Screening

Once potential articles are identified, they must be screened to determine which are relevant to the review based on the protocol's pre-specified criteria. Because a review may address several different questions, each article is not necessarily relevant to all questions. For instance, a review addressing the benefits and harms of an intervention may include only randomized trials in assessing benefits, but both trials and observational studies in assessing harms.

Traditionally, screening has been a laborious process, poring through a large stack of printed abstracts. Recently, computerized systems have been developed to facilitate the process (Wallace et al., 2012). These systems organize and store the abstracts recovered by the search and enable the screener to read, comment on, and highlight text and make decisions electronically. Systems for computer-aided searching using text mining and machine learning have also been developed (Wallace et al., 2010; Marshall et al., 2016; Marshall et al., 2018).

Experts recommend independent screening by at least two members of the research team in order to minimize errors (Committee on Standards, 2011), although many teams do not have resources for such an effort. Computerized screening offers a possible solution to this problem by allowing the computer to check the human extractor (Jap et al., 2019). Often, one screener is a subject matter expert and the other a methodologist so that both aspects of inclusion criteria are covered. Teams also often have a senior researcher work with a junior member to ensure accuracy and for training purposes. Screening is a tedious process and requires careful attention. Sometimes, it is possible to screen using only a title, but in other cases, careful reading of the abstract is required. Articles are often screened in batches to optimize effort. If duplicate screening is used, the pair of screeners meet at the end to reconcile any differences. Once the initial screening of abstracts is completed, the articles identified for further review are retrieved and examined more closely in a second screening phase. Because sensitivity is so important, reviewers screen abstracts conservatively and may end up retrieving many articles that ultimately do not meet criteria.

1.5 Data Extraction

After screening, the review team must extract data from the studies identified as relevant. In most cases, the report itself will provide the information; in some cases, investigators may have access to the study dataset or may need to contact investigators for additional information. Tables, figures, and text from the report provide quantitative and qualitative summary information including bibliographic information and the PICOS elements relating to the demographics, disease characteristics, comorbidities, enrollments, baseline measurements, exposures and interventions, outcomes, and design elements. Outcomes are usually reported by treatment group; participant characteristics are usually aggregated as averages or proportions across or by study groups (e.g., mean age or proportion female). It is also important to extract how each study has defined and ascertained the outcome for use in assessing study quality. Combined with study design features such as location or treatment dosage, these study-level variables aid in assessing the relevance of the study for answering the research questions and for making inferences to the population of interest, including assessing how the effect of treatments might change across studies that enroll different types of participants or use different study designs (see Chapter 7). Extracted items should also include those necessary to assess study quality and the potential for bias (see Chapter 12).

As with screening, extraction should follow a carefully pre-specified process using a structured form developed for the specific systematic review. Independent extraction by two team members, often a subject matter expert and a methodologist who meet to reconcile discrepancies, reduces errors as does the use of structured items using precise operational definitions of items to extract. For example, one might specify whether the study's location is defined by where it was conducted or where its authors are employed. If full duplicate extraction is too costly, duplicate extraction of a random sample of studies or of specific important items may help to determine whether extraction is consistent and whether further training may be necessary. It is often helpful to categorize variables into pre-defined levels in order to harmonize extraction and reduce free text items. This can be especially useful with non-numeric items like drug classes or racial groups. Pilot testing of the form on a few studies using all extractors can identify inadvertently omitted and ill-defined items and help reduce the need to re-categorize variables or re-extract data upon reconciliation. The pilot testing phase is also useful for training inexperienced extractors.

Advances in natural language processing are beginning to facilitate computerized data extraction (Marshall et al., 2016).

Quite often, meta-analysts identify errors and missing information in the published reports and may contact the authors for corrections. In other cases, it may be possible to infer missing information from other sources by back calculation (e.g., using a confidence interval to determine a standard error) or by digitizing software from a graph (Shadish et al., 2009).

Many software tools such as spreadsheets, databases, and dedicated systematic review packages (e.g., the Cochrane Collaboration's RevMan) aid in collection of data extracted from papers. The Systematic Review Data Repository is one example of a tool that provides many facilities for extracting and storing information from studies (Ip et al., 2012). These tools can be evaluated based on their cost, ease of setup, ease of use, versatility, portability, accessibility, data management capability, and ability to store and retrieve data.

1.6 Critical Appraisal of Study and Assessment of Risk of Bias

Many elements extracted from a study help to assess its quality and validity. These assess the relevance of the study's populations, interventions, and outcome measures to the systematic review criteria; the fidelity of the implementation of interventions; and potential risk of bias that study elements pose to each study's conclusions and to the overall synthesis. Elements that inform potential risk of bias in a study include adequacy of randomization, allocation concealment and blinding in experiments (Schulz et al., 1995), and proper adjustment for confounding in observational studies (Sterne et al., 2016). Across studies, biases may arise from missing information in the evidence base caused by missing or incompletely reported studies that lack full documentation of all outcomes collected or all patients enrolled (e.g., from withdrawals and loss to follow-up) (Deo et al., 2011).

One way to get a sense of potential bias is to compare the study as actually conducted with the study as it should ideally have been conducted. Some trials fail to assign participants in a truly random fashion or fail to ensure that the randomized treatment for a given participant is concealed. In other cases, participants may be properly randomized but may find out their treatment during the study. If this knowledge changes their response, then their response is no longer that intended by the study. Or participants in the control group may decide to seek out the treatment on their own, leading to an outcome that no longer represents that under the assigned control. This would bias a study of efficacy but might actually provide a better estimate in a study of effectiveness (i.e., an estimate of how a treatment works when applied in a real-world setting). In other cases, individuals may drop out of a study because of adverse effects. Other study performance issues that may cause bias include co-interventions given unequally across study arms and outcome assessments made inconsistently. These issues are more problematic when studies are not blinded and those conducting the study are influenced by the assignment or exposure to treat groups differently.

Several quality assessment tools are available depending on whether the study is experimental or observational (Whiting et al., 2016; Sterne et al., 2016). These provide excellent checklists of items to check for bias and provide guidelines for defining the degree of bias. Low risk of bias studies should not lead to large differences between the study estimate and the intended estimand; high risk of bias studies may lead to large differences. Chapter 12 examines strategies for assessing and dealing with such bias, including analytic strategies such as sensitivity analyses that can be used to assess the impact on conclusions.

It is important to bear in mind that study conduct and study reporting are different things. A poor study report does not necessarily imply that the study was poorly done and therefore has biased results. Conversely, a poorly done study may be reported well. In the end, the report must provide sufficient information for its readers to be confident enough to know how to proceed to use it.

1.7 Analysis

Analysis of the extracted data can take many forms depending on the questions asked and data collected but can be generally categorized as either qualitative or quantitative. Systematic reviews may consist solely of qualitative assessments of individual studies when insufficient data are available for a full quantitative synthesis or they may involve both qualitative and quantitative components.

A qualitative synthesis typically summarizes the scientific and methodological characteristics of the included studies (e.g., size, population, interventions, quality of execution); the strengths and limitations of their design and execution and the impact of these on study conclusions; their relevance to the populations, comparisons, co-interventions, settings, and outcomes or measures of interest defined by the research questions; and patterns in the relationships between study characteristics and study findings. Such qualitative summaries help answer questions not amenable to statistical analysis.

Meta-analysis is the quantitative synthesis of information from a systematic review. It employs statistical analyses to summarize outcomes across studies using either aggregated summary data from trial reports (e.g., trial group summary statistics like means and standard deviations) or complete data from individual participants. When comparing the effectiveness and safety of treatments between groups of individuals that receive different treatments or exposures, the meta-analysis summarizes the differences as treatment effects where size and corresponding uncertainty estimates are expressed by standard metrics that depend on the scale of the outcome measured such as continuous, categorical, count, or time-to-event. Examples include differences in means of continuous outcomes and differences in proportions for binary outcomes. When comparison between treatments is not the object of the analysis, the summaries may take the form of means or proportions of a single group (see Chapter 3 for discussion of the different types of effect measures in meta-analysis).

Combining estimates across studies not only provides an overall estimate of how a treatment is working in populations and subgroups but can overcome the lack of power that leads many studies to non-statistically significant conclusions because of insufficient sample sizes. Meta-analysis also helps to explore heterogeneity of results across studies and helps identify research gaps that can be filled by future studies. Synthesis can help uncover differential treatment effects according to patient subgroups, form of intervention delivery, study setting, and method of measuring outcomes. It can also detect bias that may arise from poor research such as lack of blinding in randomized studies or failure to follow all individuals enrolled in a study. As with any statistical analysis, it is also important to assess the sensitivity of conclusions to changes in the protocol, study selection, and analytic assumptions. Findings that are sensitive to small changes in these elements are less trustworthy.

The extent to which a meta-analysis captures the truth about treatment effects depends on how accurately the studies included represent the populations and settings for which inferences must be made. Research gaps represent studies that need to be done. Publication bias and reporting bias relate to studies that have been done but that have been incompletely reported. Publication bias refers to the incorrect estimation of a summary treatment effect from the loss of information resulting from studies that are not published because they had uninteresting, negative, or non-statistically significant findings. Failure to include such studies in a review leads to an overly optimistic view of treatment effects, biased toward positive results (Dickersin, 2005).

Reporting bias arises when studies report only a subset of their findings, often a subset of all outcomes examined (Schmid, 2017). Bias is not introduced if the studies fail to collect the outcomes for reasons unrelated to the outcome values. For example, a study on blood pressure treatments may collect data on cardiovascular outcomes, but not on kidney outcomes. A subsequent meta-analysis of the effect of blood pressure treatments on kidney outcomes can omit those studies without issues of bias. However, if the outcomes were collected but were not reported because they were negative, then bias will be introduced if the analysis omits those studies without adjusting for the missing outcomes. Adjustment for selective outcome reporting is difficult because the missing data mechanism is usually not known and thus hard to incorporate into analysis. Comparing study protocols with published reports can often detect potential reporting bias. Sometimes, authors have good reasons for reporting only some outcomes as when the full report is too long to publish or outcomes relate to different concepts that might be reported in different publications. In other cases, it is useful to contact authors to find out why outcomes were unreported and whether these results were negative. Chapter 13 discusses statistical and non-statistical methods for handling publication and reporting bias.

1.8 Reporting

Generation of a report summarizing the findings of the meta-analysis is the final step in the systematic review process. The most important part of the report contains its conclusions regarding the evidence found to answer the review's research questions. In addition to stating whether the evidence does or does not favor the research hypotheses, the report needs to assess the strength of that evidence in order that proper decisions be drawn from the review (Agency for Healthcare Research and Quality). Strength of evidence involves both the size of the effect found and the confidence in the stability and validity of that effect. A meta-analysis that finds a large effect based on studies of low quality is weaker than one that finds a small effect based on studies of high quality. Likewise, an effect that disappears with small changes in model assumptions or that is sensitive to leaving out one study is not very reliable. Thus, a report must summarize not only the analyses leading to a summary effect estimate, but also the analyses assessing study quality and their potential for bias. Randomized studies have more internal validity than observational studies; studies that ignore dropouts are more biased than studies that perform proper missing data adjustments; analyses that only include studies with individual participant data available and that ignore the results from known studies with only summary data available may be both inefficient and biased. It is important to bear in mind that study conduct and study reporting are different things. A poor report does not necessarily imply that the study was poorly done and therefore has biased results. Conversely, a poorly done study may be reported well. In the end, the report must provide sufficient information for its readers to be confident enough to know how to proceed to use it.

The Preferred Reporting Items for Systematic Reviews and Meta-Analyses (PRISMA) statement (Moher et al., 2009) is the primary guideline to follow for reporting a meta-analysis of randomized studies. It lists a set of 27 items that should be included in each report. These include wording for the title and elements in the introduction, methods, results, and discussion. Slightly different items are needed for observational studies and these can be found in the Meta-analysis of Observational Studies in Epidemiology (MOOSE) statement (Stroup et al., 2000). Modified guidelines have been written for particular types of meta-analyses such as those for individual participant data (Stewart et al., 2015), networks of treatments (Hutton et al., 2015), diagnostic tests (Bossuyt et al., 2003), and N-of-1 studies (Vohra et al., 2015; Shamseer et al., 2015). These are now required by most journals that publish meta-analyses. Later chapters discuss these and other types of meta-analyses and explain the need for these additional items.

1.9 Using a Systematic Review

Once the systematic review is complete, a variety of different stakeholders may use it for a variety of different purposes. Many reviews are commissioned by organizations seeking to set evidence-based policy or guidelines. For example, many clinical societies publish guidelines for their members to follow in treating patients. Such guidelines ensure that best practices are used but may also protect members from malpractice claims when proper treatment fails to produce a desired outcome. Government and private health insurance coverage decisions make use of systematic reviews to estimate the safety and effectiveness of new treatments. Government agencies regulating the approval or funding of new drug treatments or medical devices such as the Food and Drug Administration in the United States, the European Medicines Agency, or the National Institute for Clinical Excellence (NICE) in the United Kingdom now often require applicants to present the results of a meta-analysis summarizing all studies related to the product under review in order to provide context to the application. Many educational policy decisions are motivated by reviews of the evidence deposited in the Institute of Education's What Works Clearinghouse (https://ies.ed.gov/ncee/wwc). Many national environmental policies rely on systematic reviews of chemical exposures, ecological interventions, and natural resources. Often, the impact and cost-effectiveness of decisions is modeled using inputs derived from systematic reviews. Other stakeholders include businesses making decisions about marketing strategies and consumer advocacy groups pursuing legal action.

In addition to the What Works Clearinghouse, two other prominent repositories of systematic reviews are maintained to support these efforts. The Cochrane Collaboration maintains a large and growing database of nearly 8000 reviews of all types of healthcare interventions and diagnostic modalities in the Cochrane Database of Systematic Reviews (www.cochranelibrary.com). The Campbell Collaboration's repository is smaller, but of a similar structure that covers reviews in the social sciences (www.campbellcollaboration.org/library.html). These repositories have become quite influential, used by many researchers and the public and quoted frequently in the popular press. They speak to the growing influence of systematic reviews which are highly referenced in the scientific literature.

In the United States and Canada, the Agency for Healthcare Research and Quality (AHRQ) has supported 10–15 Evidence-Based Practice Centers since 1997. These centers carry out large reviews of questions nominated by stakeholders and refined through a consensus process. Stakeholders for AHRQ reports include clinical societies, payers, the United States Congress, and consumers. Like NICE and the Cochrane Collaboration, AHRQ has published guidance documents for its review teams that emphasize not only review methods, but review processes and necessary components of final reports in order to ensure uniformity and adherence to standards (AHRQ, 2008).

Systematic reviews also serve an important role in planning future research. A review often identifies areas where further research is needed either because the overall evidence is inconclusive or because uncertainty remains about outcomes in certain circumstances such as in specific subpopulations, settings, or under treatment variations. Ideally, decision models should incorporate systematic review evidence and be able to identify which new studies would best inform the model. Systematic review results can also provide important sources for inputs such as effect sizes and variances needed in sample size calculations. These can be explicitly incorporated into calculations in a Bayesian framework (Sutton et al., 2007; Schmid et al., 2004). Chapter 23 discusses these issues.

1.10 Summary

Systematic reviews have become a standard approach for summarizing the existing scientific evidence in many different fields. They rely on a set of techniques for framing the proper question, identifying the relevant studies, extracting the relevant information from those studies, and synthesizing that information into a report that interprets findings for its audience. This chapter has summarized the basic principles of these steps as background for the main focus of this book which is on the statistical analysis of data using meta-analysis. Readers interested in further information about the non-statistical aspects of systematic reviews are urged to consult the many excellent references on these essential preliminaries. The guidance documents referenced in this chapter provide a good starting point and contain many more pointers in their bibliographies. Kohl et al. (2018) provide a detailed comparison of computerized systems to aid in the review process.

References

Agency for Healthcare Research and Quality, 2008-. Methods Guide for Effectiveness and Comparative Effectiveness Reviews [Internet]. Rockville, MD: Agency for Healthcare Research and Quality (US). https://www.ahrq.gov/research/findings/evidence-based-reports/technical/methodology/index.html and https://www.ncbi.nlm.nih.gov/books/NBK47095.

Anderson LM, Petticrew M, Rehfuess E, Armstrong R, Ueffing E, Baker P, Francis D and Tugwell P, 2011. Using logic models to capture complexity in systematic reviews. *Research Synthesis Methods* **2**(1): 33–42.

Antman EM, Lau J, Kupelnick B, Mosteller F and Chalmers TC, 1992. A comparison of results of meta-analyses of randomized control trials and recommendations of clinical experts. Treatments for myocardial infarction. *JAMA* **268**(2): 240–248.

Balk EM, Adam GP, Langberg V, Halladay C, Chung M, Lin L, Robertson S, Yip A, Steele D, Smith BT, Lau J, Lichtenstein AH and Trikalinos TA, 2016. Omega-3 Fatty Acids and Cardiovascular Disease: An Updated Systematic Review. Evidence Report/Technology Assessment No. 223. (Prepared by the Brown Evidence-based Practice Center under Contract No. 290-2012-00012-I.) AHRQ Publication No. 16-E002-EF. Rockville, MD: Agency for Healthcare Research and Quality, August 2016. www.effectivehealthcare.ahrq.gov/reports/final.cfm.

Balshem H, Stevens A, Ansari M, Norris S, Kansagara D, Shamliyan T, Chou R, Chung M, Moher D and Dickersin K, 2013. *Finding Grey Literature Evidence and Assessing for Outcome and Analysis Reporting Biases When Comparing Medical Interventions: AHRQ and the Effective Health Care Program. Methods Guide for Comparative Effectiveness Reviews.* AHRQ Publication No. 13(14)-EHC096-EF. Rockville, MD: Agency for Healthcare Research and Quality. www.effectivehealthcare.ahrq.gov/reports/final.cfm.

Bannuru RR, Schmid CH, Kent D, Wong J and McAlindon T, 2015. Comparative effectiveness of pharmacological interventions for knee osteoarthritis: A systematic review and network meta-analysis. *Annals of Internal Medicine* **162**(1): 46–54.

Bossuyt PM, Reitsma JB, Bruns DE, Gatsonis CA, Glasziou PP, Irwig LM, Lijmer JG, Moher D, Rennie D and de Vet HC, 2003. Towards complete and accurate reporting of studies of diagnostic accuracy: The STARD initiative. *Clinical Chemistry* **49**(1): 1–6.

Cochrane Handbook for Systematic Reviews of Diagnostic Test Accuracy. https://methods.cochrane.org/sdt/handbook-dta-reviews.

Committee on Standards for Systematic Review of Comparative Effectiveness Research. 2011. In Eden J, Levit L, Berg A and Morton S (Eds). *Finding What Works in Health Care: Standards for Systematic Reviews.* Washington, DC: Institute of Medicine of the National Academies.

Deo A, Schmid CH, Earley A, Lau J and Uhlig K, 2011. Loss to analysis in randomized controlled trials in chronic kidney disease. *American Journal of Kidney Diseases* **58**(3): 349–355.

Dickersin K, 2005. Publication bias: recognizing the problem, understanding its origins and scope, and preventing harm. In Rothstein HR, Sutton AJ and Borenstein M (Eds). *Publication Bias in Meta-Analysis: Prevention, Assessment and Adjustments.* John Wiley & Sons.

Glanville JM, Lefebvre C, Miles JN and Camosso-Stefinovic J, 2006. How to identify randomized controlled trials in Medline: Ten years on. *Journal of the Medical Library Association* **94**(2): 130–136.

Glass GV, 1976. Primary, secondary and meta-analysis of research. *Educational Researcher* **5**(10): 3–8.

Gurevitch J, Morrow LL, Wallace A and Walsh JS, 1992. A meta-analysis of competition in field experiments. *The American Naturalist* **140**(4): 539–572.

Higgins JPT, Thomas J, Chandler J, Cumpston M, Li T, Page MJ, Welch VA (editors), 2019. *Cochrane Handbook for Systematic Reviews of Interventions.* 2nd Edition. Chichester, UK: John Wiley & Sons. Also online version 6.0 (updated July 2019) available from www.training.cochrane.org/handbook.

Hutton T, Salanti G, Caldwell DM, Chaimani A, Schmid CH, Cameron C, Ioannidis JPA, Straus S, Shing LK, Thorlund K, Jansen J, Mulrow C, Catala-Lopez F, Gotzsche PC, Dickersin K, Altman D and Moher D, 2015. The PRISMA extension statement for reporting of systematic reviews incorporating network meta-analyses of healthcare interventions: Checklist and explanations. *Annals of Internal Medicine* **162**(11): 777–784.

Ip S, Hadar N, Keefe S, Parkin C, Iovin R, Balk EM and Lau J, 2012. A Web-based archive of systematic review data. *Systematic Reviews* **1**: 15. https://srdr.ahrq.gov/.

Jap J, Saldanha I, Smith BT, Lau J, Schmid CH, Li T on behalf of the Data Abstraction Assistant Investigators, 2019. Features and functioning of Data Abstraction Assistant, a software application for data abstraction during systematic reviews. *Research Synthesis Methods* **10**(1): 2–14.

Järvinen A, 1991. A meta-analytic study of the effects of female age on laying- date and clutch-size in the great tit Parus major and the pied flycatcher Ficedula hypoleuca. *Ibis* **133**(1): 62–67.

Jüni P, Holenstein F, Sterne J, Bartlett C and Egger M, 2002. Direction and impact of language bias in meta-analyses of controlled trials: Empirical study. *International Journal of Epidemiology* **31**(1): 115–123.

Kohl C, McIntosh EJ, Unger S, Haddaway NR, Kecke S, Schiemann J and Wilhelm R, 2018. Online tools supporting the conduct and reporting of systematic reviews and systematic maps: A case study on CADIMA and review of existing tools. *Environmental Evidence* **7**(1): 8.

Liberati A, Altman DG, Tetzlaff J, Mulrow C, Gøtzsche PC, Ioannidis JP, Clarke M, Devereaux PJ, Kleijnen J and Moher D, 2009. The PRISMA statement for reporting systematic reviews and meta-analyses of studies that evaluate health care interventions: Explanation and elaboration.*Annals of Internal Medicine* **151**(4): W65–94.

Marshall IJ, Kuiper J and Wallace BC, 2016. RobotReviewer: Evaluation of a system for automatically assessing bias in clinical trials. *Journal of the American Medical Informatics Association JAMIA* **23**(1): 193–201.

Marshall IJ, Noel-Storr A, Kuiper J, Thomas J and Wallace BC, 2018. Machine learning for identifying randomized controlled trials: An evaluation and practitioner's guide. *Research Synthesis Methods* **9**(4): 602–614.

Moher D, Liberati A, Tetzlaff J, Altman DG and the PRISMA Group, 2009. Preferred reporting items for systematic reviews and meta-analyses: The PRISMA statement. *BMJ* **339**: b2535.

Moher D, Shamseer L, Clarke M, Ghersi D, Liberati A, Petticrew M, Shekelle P, Stewart LA and the PRISMA-P Group, 2015. Preferred reporting items for systematic review and meta-analysis protocols (PRISMA-P) 2015 statement. *Systematic Reviews* **4**: 1.

Morrison A, Polisena J, Husereau D, Moulton K, Clark M, Fiander M, Mierzwinski-Urban M, Clifford T, Hutton B and Rabb D, 2012. The effect of English-language restriction on systematic review-based meta-analyses: A systematic review of empirical studies. *International Journal of Technology Assessment in Health Care* **28**(2): 138–144.

Pham B, Klassen TP, Lawson ML and Moher D, 2005. Language of publication restrictions in systematic reviews gave different results depending on whether the intervention was conventional or complementary. *Journal of Clinical Epidemiology* **58**(8): 769–776. Erratum *Journal of Clinical Epidemiology*, 2006 **59**(2): 216.

Schmid CH, 2017. Outcome reporting bias: A pervasive problem in published meta-analyses. *American Journal of Kidney Diseases* **69**(2): 172–174.

Schmid CH, Cappelleri JC and Lau J, 2004. Bayesian methods to improve sample size approximations. In Johnson ML and Brand L (Eds). *Methods in Enzymology Volume 383: Numerical Computer Methods Part D*. New York: Elsevier, 406–427.

Schulz KF, Chalmers I, Hayes RJ and Altman DG, 1995. Empirical evidence of bias. Dimensions of methodological quality associated with estimates of treatment effects in controlled trials. *JAMA* **273**(5): 408–412.

Shadish WR, Brasil ICC, Illingworth DA, White KD, Galindo R, Nagler ED and Rindskopf DM, 2009. Using UnGraph to extract data from image files: Verification of reliability and validity. *Behavior Research Methods* **41**(1): 177–183.

Shadish WR and Lecy JD, 2015.The meta-analytic big bang. *Research Synthesis Methods* **6**(3): 246–264.

Shamseer L, Sampson M, Bukutu C, Schmid CH, Nikles J, Tate R, Johnson BC, Zucker DR, Shadish W, Kravitz R, Guyatt G, Altman DG, Moher D, Vohra S and the CENT Group, 2015. CONSORT extension for N-of-1 trials (CENT): Explanation and elaboration. *BMJ* **350**: h1793.

Sterne JA, Hernán MA, Reeves BC, Savović J, Berkman ND, Viswanathan M, Henry D, Altman DG, Ansari MT, Boutron I, Carpenter JR, Chan AW, Churchill R, Deeks JJ, Hróbjartsson A, Kirkham J, Jüni P, Loke YK, Pigott TD, Ramsay CR, Regidor D, Rothstein HR, Sandhu L, Santaguida PL, Schünemann HJ, Shea B, Shrier I, Tugwell P, Turner L, Valentine JC, Waddington H, Waters E, Wells GA, Whiting PF and Higgins JP, 2016. Robins-I: A tool for assessing risk of bias in non-randomised studies of interventions. *BMJ* **355**: i4919.

Stewart LA, Clarke M, Rovers M, Riley RD, Simmonds M, Stewart G, Tierney JF and the PRISMA-IPD Development Group, 2015. Preferred reporting items for systematic review and meta-analyses of individual participant data: The PRISMA-IPD Statement. *JAMA* **313**(16): 1657–1665.

Stroup DF, Berlin JA, Morton SC, Olkin I, Williamson GD, Rennie D, Moher D, Becker BJ, Sipe TA and Thacker SB, 2000. Meta-analysis of observational studies in epidemiology: A proposal for reporting. *JAMA* **283**(15): 2008–2012.

Sutton AJ, Cooper NJ, Jones DR, Lambert PC, Thompson JR and Abrams KR, 2007. Evidence-based sample size calculations based upon updated meta-analysis. *Statistics in Medicine* **26**(12): 2479–2500.

Vohra S, Shamseer L, Sampson M, Bukutu C, Schmid CH, Tate R, Nikles J, Zucker DR, Kravitz R, Guyatt G, Altman DG, Moher D and the CENT Group, 2015. CONSORT statement: An extension for N-of-1 trials (CENT). *BMJ* **350**: h1738.

Wallace BC, Dahabreh IJ, Schmid CH, Lau J and Trikalinos TA, 2013. Modernizing the systematic review process to inform comparative effectiveness: Tools and methods. *Journal of Comparative Effectiveness Research* **2**(3): 273–282.

Wallace BC, Small K, Brodley CE, Lau J and Trikalinos TA, 2012. Deploying an interactive machine learning system in an evidence-based practice center: Abstrackr. In *Proceedings of the 2nd ACM SIGHIT International Health Informatics Symposium*, Miami, Florida, 28–30 January 2012. New York Association for Computing Machinery, 819–824.

Wallace BC, Trikalinos TA, Lau J, Brodley C and Schmid CH, 2010. Semi-automated screening of biomedical citations for systematic reviews. *BMC Bioinformatics* **11**: 55.

Wang C, De Pablo P, Chen X, Schmid C and McAlindon T, 2008. Acupuncture for pain relief in patients with rheumatoid arthritis: A systematic review. *Arthritis and Rheumatism (Arthritis Care and Research)* **59**(9): 1249–1256.

Whiting P, Savović J, Higgins JP, Caldwell DM, Reeves BC, Shea B, Davies P, Kleijnen J, Churchill R and ROBIS group, 2016. ROBIS: A new tool to assess risk of bias in systematic reviews was developed. *Journal of Clinical Epidemiology* **69**: 225–234.

Yusuf S, Peto R, Lewis J, Collins R and Sleight P, 1985. Beta blockade during and after myocardial infarction: An overview of the randomized trials. *Progress in Cardiovascular Diseases* **27**(5): 335–371.

2

General Themes in Meta-Analysis

Christopher H. Schmid, Theo Stijnen, and Ian R. White

CONTENTS

2.1 Introduction

Chapter 1 reviewed the parts of a systematic review and noted that they often include a quantitative synthesis or meta-analysis, when sufficient data are available. Meta-analysis uses statistical methods to combine data across studies in order to estimate parameters of interest. In general, meta-analysis is used to address four types of questions. The first type is descriptive, summarizing some characteristic of a distribution such as the prevalence of a disease, the mean of a population characteristic, or the sensitivity of a diagnostic test. The second type of question is comparative: how does one treatment compare with another in terms of reducing the risk of a stroke; does a new method of teaching improve student test scores compared with the current method; or does exposure to warmer water change the number of fish caught? Some of these questions involve specific interventions, other relate to prevalent exposures and others are related to diagnostic tests. We will use the general term "treatment" to apply to all of them unless a specific need arises to differentiate them. This comparative type is the most common. A third type of question involves non-comparative associations such as correlations between outcomes or the structure of an underlying pathway (Chapter 16) and associations between variables in a regression model (Chapter 18). A fourth type of question involves developing a prognostic or predictive model for an outcome. Frequently different studies report different models or parts of models that involve predictive factors. Chapter 22 explores methods for combining such data. Typically, meta-analysis estimates the size and uncertainty of the parameters of interest expressed by standard metrics that depend on the scale of the outcome. Chapter 3 discusses these metrics in detail for different types of outcomes.

Using meta-analysis to combine information across studies offers a variety of benefits. It provides an estimate of the average size of a characteristic of a population or of the effectiveness or harm of a treatment (exposure) as well as a sense of the variation of these quantities across different study settings. To the extent that the variation is not large or can be understood, meta-analysis can increase the generalizability of the research findings and determine their effects in subgroups. By combining small studies, meta-analysis can also increase the precision with which key parameters are estimated and help to explain inconsistent results that arise when underpowered studies report non-statistically significant conclusions because of insufficient sample sizes. Meta-analysis can also focus attention on discrepancies in study findings that might argue against combining their results or might argue for more subtle interpretations of parameters whose true values might vary with characteristics of the populations studied or with the manner in which interventions are undertaken. In certain cases, exploring the causes of such heterogeneity can lead to important conclusions in their own right or might point to the need for further studies to fill in research gaps. Integration of meta-analysis with study of risk of bias (Chapter 12) can also pinpoint weaknesses in the data and evaluate the sensitivity of the conclusions to poor study processes.

This chapter introduces the general themes that motivate the methods used for carrying out meta-analysis and provides a map with which the reader can navigate the rest of the book. Meta-analysis comes in a variety of flavors motivated by the types of data available, the research questions to be answered and the inferences desired. Section 2.2 discusses various types of data that may be available and what they imply about the types of questions that may be asked. Section 2.3 explores the types of models that may be fit with these data.

Before embarking on this overview, however, it is important to bear in mind that meta-analysis is not appropriate for all systematic reviews. Reviews with few studies, or having studies with dissimilar outcomes or outcome scales, studying diverse interventions or interventions that may have evolved over the course of time covered by the review, or that combine different study designs (e.g., observational vs. experimental) may not fit into a single statistical framework and may be better handled qualitatively or by separating into separate meta-analyses.

2.2 Data Structures

2.2.1 Study-Level Data

The data extracted from published reports of each study in a systematic review typically include the PICOS elements (see Chapter 1) comprising elements of the study design, information on the study participants such as demographics and personal history, their exposures, and their outcomes. Usually, this information will come from study reports and protocols, either published or unpublished. If a study is comparative, the *study-level data* are often reported by treatment group (or arm) and not at the level of the individual participant. Data items include the number of individuals who received each treatment and the number of outcomes these individuals had (for categorical outcomes) or the mean and standard deviation (for continuous outcomes). Additional characteristics of the participants in each group (e.g., the number who were female, their average age, what proportion went to school, or who had a certain type of disease) or of the treatments or exposures which they had are also recorded for descriptive purposes and perhaps to evaluate reasons why treatment effects might differ between studies.

If the meta-analysis is focused on a comparison of two treatment groups, the key statistic will often be the difference in their means or in a function of their means (e.g., the log odds ratio is a difference between the logits of two proportions). Representing the group mean for group j of study i as y_{ij}, the study treatment effect is $y_i = y_{i2} - y_{i1}$. Studies often only report this mean contrast and its resulting standard error s_i. Sometimes the standard errors will need to be backcalculated from a reported confidence interval as discussed in Chapter 3.

In rare cases, studies may also report the treatment effects within subgroups such as male and female and one can investigate treatment by covariate interactions. More commonly, results may be reported by dose levels (see Chapter 18). Usually, though, effects within subgroups are unavailable and heterogeneity of treatment effects across studies can only be studied through the correlation of the study effects with a summary measure of the subgroup in each study such as the proportion of females or the average age of participants. Within-study heterogeneity due to factors such as participant age or education level that vary within-study by participant cannot be investigated at all. Chapter 7 discusses meta-regression methods for assessing heterogeneity that test for interactions of treatment with study-level risk factors such as design characteristics that apply to all individuals in a given study or to summary measures like average age of a study's participants.

2.2.2 Individual Participant Data

Sometimes, however, it may be possible to obtain *individual participant data* (IPD) on each individual in each study (see Chapter 8). IPD substantially increase the capacity to model within- and between-study variation. With IPD, one can investigate within-study heterogeneity of the associations between the outcome and treatments that manifest between subgroups. For example, IPD can compare treatment efficacy between younger and older individuals, rather than just between populations with different average ages. IPD permits inferences at the individual participant level; study-level data only allow inference to study populations. IPD are necessary for meta-analyses of prognostic studies (see Chapter 22) where one needs to combine predicted outcomes from models that vary by study. IPD can also facilitate modeling time-to-event outcomes (Chapter 15) because the IPD provide censoring and follow-up information for individual participants. This can be especially helpful when ongoing longitudinal studies have additional follow-up information that earlier reports do not capture. IPD have other advantages too, including the ability to discover outcomes and do analyses that may have been left out of the study reports, to identify errors and fill in missing values, and to harmonize variable definitions to reduce potential sources of heterogeneity (Stewart and Clarke, 1995). These benefits come with a cost, however, because the study databases usually require considerable work to harmonize with each other and investigators may be unwilling to share data (Schmid et al., 2003).

2.2.3 Randomized versus Observational Data

Interpreting meta-analyses that combine studies in which participants have been randomized is more straightforward than when the data are non-randomized. Experimental designs that use randomization to assign participants give unconfounded estimates of treatment comparisons, with distributions of all potential measured and unmeasured confounders balanced by treatment group on average. If the studies are free of bias, each of their estimates of treatment effect y_i provides an independent contribution to the meta-analytic average, $\hat{\theta}$. Many scientific research studies compare exposures that cannot be randomized either practically or ethically, though. When studies are not randomized, meta-analysis is potentially complicated by confounding. To remove this type of bias, many studies will report estimates adjusted for potential confounders using either multivariable regression or special methods such as propensity scores or matching. Because the meta-analyst must rely on these reported study analyses unless IPD are available, meta-analysis of non-randomized studies can be biased if the studies do not properly adjust for confounders or if they adjust in different ways. Determining whether appropriate adjustment has taken place can be difficult to determine from a published report alone since it is not always clear which factors were considered for adjustment and whether those missing were unavailable or rejected as not significant. Chapter 21 discusses this issue.

2.2.4 Multivariate Data

Systematic reviews typically evaluate several outcomes related to different review questions. These might involve a combination of benefits and harms or efficacy and resource use. Traditionally, each outcome is addressed in a separate meta-analysis, but methods are available to assess outcomes simultaneously in multivariate models (Jackson et al., 2011). These allow incorporation of correlations among outcomes and can lead to more precise estimates of effects (Riley et al., 2008; Trikalinos et al., 2014). Chapter 9 lays out general models for multivariate data. Certain types of meta-analysis use particular multivariate data structures. In meta-analyses of tests to diagnose a particular disease or condition, both the sensitivity and the specificity of the test are of interest. Since the use of a threshold to define the test result as positive or negative introduces correlation between the sensitivity and specificity, it is usual to model these simultaneously with a bivariate model (see Chapter 19). When combining correlations between the results of educational tests, it is common to meta-analyze all the correlations together, leading to a multivariate meta-analysis (Chapter 16).

A different type of multivariate data structure manifests when meta-analyzing a network of more than two treatments. In such networks, interest lies in the comparison of all pairs of treatments and in their rank ordering by outcomes using network meta-analysis. Complexity arises because each study compares only a subset of the entire set of treatments and some pairs of treatments are never compared in any study. The missing treatment effects can be recovered indirectly by making assumptions that the comparative difference between any two treatments is the difference of their respective comparisons to a third treatment with which they have both been compared. Further assumptions about the studies being exchangeable and about the correlations among the treatment effects are required to undertake the modeling process. Chapter 10 provides an overview of network meta-analysis.

2.3 Models

The observed group-level and study-level summaries, y_{ij} and y_i, respectively, depend on parameters of the data-generating process. We can construct a model of this measurement process by assuming that each observed effect is a realization of a stochastic process with group-level means θ_{ij} or contrast-level means θ_i and associated variances σ_{ij} and σ_i. Typically, $\theta_i = \theta_{i2} - \theta_{i1}$. Although the data may be collected at the group-level, the mean effect θ and perhaps the individual study effects θ_i are the parameters of interest in a meta-analysis focused on comparing two groups; θ represents the best single number summary of the treatment effects in the group of studies. It is informed by all of the studies together. But sometimes it may be helpful to determine whether the estimate of the treatment effect, θ_i in a single study, particularly a small one, could be improved by information from other studies similar to it. We will find that certain models can do this.

2.3.1 Homogeneous or Heterogeneous Effects

In order to draw inferences about the study effects θ_i, we must consider whether we are only interested in the particular studies in the meta-analysis or whether we wish to extrapolate to a larger population of similar studies. There are three basic structural models for the treatment effects. The most common is the *random-effects model* which treats the θ_i as following a distribution, usually a $N(\theta, \tau^2)$ distribution but other random-effects distributions are sometimes used too, such as beta and gamma distributions. Because it is more interpretable on the scale of the data, it is often more useful to report the standard deviation τ rather than the variance τ^2. τ is often considered a nuisance parameter for estimating θ, although τ is important when estimating the uncertainty associated with predicting the effects in new studies and assessing the amount of heterogeneity in the data. Large values of τ indicate that study effects vary considerably between studies and suggest searching for factors associated with this heterogeneity. A common hypothesis of interest is whether the studies exhibit any heterogeneity. The absence of heterogeneity implies that $\tau = 0$ or, equivalently that $\theta_i = \theta$ for all i. In other words, the studies are all estimating a common treatment effect θ. We call such a model a *common-effect* (CE) model. The third model, which we shall call a *separate-effects* model, treats the θ_i as fixed effects (Rice et al., 2017). In statistics, a fixed effect traditionally refers to an unknown parameter that is not randomly drawn from a larger population. The separate-effects model is appropriate if we do not wish to assume any relationship among the different treatment effects, but rather consider them separately. In this case, it may still be of interest to estimate an average of the separate effects. Laird and Mosteller (1990) suggested using an unweighted mean since there would be no reason to treat any of the studies differently from each other if one believed them equally important.

This raises an important comment about terminology. Many authors use the term *fixed-effect* model instead of *common-effect* model to describe the model in which each study is estimating the same effect. We avoid this terminology because this is not a fixed effect in the common statistical parlance. Furthermore, some authors such as Rice et al. (2017) use the term *fixed-effects* model to refer to what we have called the *separate-effects* model. Although the use of the plural for the word "effects" technically avoids the improper statistical terminology, it is confusing both because of the similarity in wording to "fixed effect" and the confusion with the colloquial use of the term for the common-effect model. We prefer distinct terms to distinguish distinct models.

The three models also correspond to different points on a continuum with respect to how much knowledge about one study should be influenced by information from other studies. The separate-effects model corresponds to a no pooling model in which the estimate of the treatment effect in a given study is determined only by the data in that study. The common-effect model describes the opposite end of the spectrum, a complete pooling model in which the information from all studies is combined into a single estimate believed to describe each study effect. The random-effects model serves as a compromise between the two with the treatment effect θ_i in a given study estimated by a weighted average of the estimated effect y_i in that study and the overall mean θ from all the studies. This is commonly termed a *shrinkage estimate* and reflects the belief that each study is informed by the other studies. The weights correspond to the relative within- and between-study variability. While we say little more about the separate-effects model, common-effect and random-effects models are discussed throughout the book and in particular in Chapters 4 and 5.

2.3.2 One- and Two-Stage Models

Traditionally, meta-analysts have assumed the true σ_i to be known and equal to the sample standard error s_i, although this is a strong assumption (Jackson and White, 2018). In this case, we may speak of a two-stage approach to modeling in which the first stage encompasses the separate analysis of each study's results to obtain y_i and s_i and the second stage focuses on using these to estimate θ (and perhaps also the θ_i). Chapter 4 discusses the two-stage approach in detail. The two-stage approach focuses on the study treatment effects and relies heavily on the asymptotic normality of the contrast estimates and on the assumption of known within-study variances. This approach works particularly well with larger randomized studies where the balancing of covariates across arms avoids the need to use estimated effects adjusted for confounding and when the average effect is of primary importance. However, many problems involve studies that are non-randomized, small, involve rare events, or for which the normality and known variance assumptions fail to hold. In such cases, one may need to model the summary data from study arms directly using their exact likelihoods. This leads to the one-stage models of Chapter 5.

One-stage models are commonly applied to binary outcomes. In such cases, the group-level summary statistics are the counts in 2×2 tables. Using separate binomial distributions for each study group, the fundamental parameters are the two event proportions in each group. In this case, the binomial mean and variance are functions of the same single parameter and one can construct a multilevel model to simultaneously model the group-level outcomes and their study-level parameters. The four counts in each study are the sufficient statistics and can be used to construct the exact likelihood.

One-stage models are also useful when fitting survival models (Chapter 15) and if one wants to model variances instead of assuming them known when the effects are continuous. Chapter 5 discusses four different models for meta-analysis of continuous outcomes that treat the group-level variances as separate from each other or constrains them to be equal across arms, across studies, or across both arms and studies.

2.3.3 Fixed or Random Study Effects

It is often helpful to reparameterize the group-level parameters of the one-stage model in terms of the effect in the reference or control group, γ_i, and the differences between groups, θ_i. In problems focusing only on comparative treatment effects, the γ_i can be considered nuisance parameters. If we want to estimate the relationship between treatment effects and the underlying control risk, the γ_i are parameters of interest themselves (see Chapter 14). Two different models have been used for these two scenarios.

The first model treats each γ_i as a random effect drawn from a distribution of potential control risks with unknown mean and variance. The γ_i and the θ_i are then estimated together. Such a formulation is necessary if one wants to form a regression model relating the treatment effects to the control risks or one wants to make inference about results in each treatment group in a new study. Because the γ_i and the θ_i are being simultaneously estimated, their effects can leak into each other. This is a basic consequence of parameter shrinkage from multilevel models. As an example, a study with a high observed rate of events in the control group will tend to have an estimated true study control risk rate shrunk to the average control risk. This lowering of the control risk in the study will tend to make the estimated study treatment effect (difference between treated and control) bigger and may therefore change the overall treatment effect estimate.

To avoid leakage, one could use a second model that treats each γ_i as a fixed effect that applies only to that study and has no relationship to any other study. Assuming that one has correctly estimated the γ_i, one might hope that the estimate of the θ_i and of θ are independent of the γ_i and so one has solved the leakage problem. However, this model introduces a new problem because the number of γ_i parameters is directly associated with the number of studies in the meta-analysis, so that the number of parameters is now increasing at the same rate as the number of data points. This has some poor theoretical properties as discussed in Chapter 5.

Whether to treat control risk parameters as fixed or random remains controversial in the literature (Jackson et al., 2018). One side argues, particularly when combining randomized studies, that the parameters must be treated as fixed effects in order to not potentially change the treatment effects estimates which are unbiased in each study as a result of the randomization. The other side argues that the asymptotic properties of the maximum likelihood estimates of treatment effect (and perhaps also of Bayesian estimates under a non-informative prior) are compromised when the number of parameters is increasing at the same rate as the number of studies. If one wants to make generalized inferences to larger populations, however, a complete model of all the parameters using a random-effect formulation is needed (see Chapter 5).

2.3.4 Model Checking

It is important to emphasize the importance of verifying model assumptions, no matter what model is being used. Both formal model checks as described in Chapter 11 and sensitivity analyses in which the model is refit under different scenarios (such as leaving one study out in turn) are important to carry out in order to ensure that results are not driven by outliers and that assumptions are met. As many of the most common meta-analysis models are types of linear and generalized linear models, these diagnostics will be familiar. Assumptions about missing data are particularly important. These can involve participants excluded from studies, unreported outcomes, and unpublished studies. Chapters 12 and 13 discuss these issues.

2.3.5 Bayesian or Frequentist

The book discusses both frequentist and Bayesian approaches to estimation. Chapters 4 and 5 outline the basic frequentist approaches focusing on normal-distribution based models and likelihood-based models, respectively. Chapter 6 introduces the Bayesian framework. Later chapters choose one or the other or sometimes both approaches, depending on the methods most common in each particular area.

2.4 Conclusion

This chapter has summarized the basic data structures and models encountered in meta-analysis. As indicated, many specific problems implement specific variations of these structures and models. Before proceeding to discussing models in more detail, we discuss in Chapter 3 how to choose an appropriate effect measure and how to extract the data necessary for analyzing these effect measures.

References

Jackson D and White IR, 2018. When should meta-analysis avoid making hidden normality assumptions? *Biometrical Journal* **60**(6): 1040–1058.

Jackson D, Law M, Stijnen T, Viechtbauer W and White IR, 2018. A comparison of 7 random-effects models for meta-analyses that estimate the summary odds ratio. *Statistics in Medicine* **37**(7): 1059–1085.

Jackson D, Riley R and White IR, 2011. Multivariate meta-analysis: Potential and promise (with discussion). *Statistics in Medicine* **30**(20): 2481–2510.

Laird NM and Mosteller F, 1990 Some statistical methods for combining experimental results. *International Journal of Technology Assessment in Health Care* **6**(1): 5–30.

Rice K, Higgins JPT and Lumley T, 2017. A re-evaluation of fixed effect(s) meta-analysis. *Journal of the Royal Statistical Society: Series A* **181**(1): 205–227.

Riley RD, Thompson JR and Abrams KR, 2008. An alternative model for bivariate random-effects meta-analysis when the within-study correlations are unknown. *Biostatistics* **9**(1): 172–186.

Schmid CH, Landa M, Jafar TH, Giatras I, Karim T, Reddy M, Stark PC Levey AS and Angiotensin-Converting Enzyme Inhibition in Progressive Renal Disease (AIPRD) Study Group, 2003. Constructing a database of individual clinical trials for longitudinal analysis. *Controlled Clinical Trials* **24**(3): 324–340.

Stewart LA, Clarke MJ on behalf of the Cochrane Working Group on Meta-Analysis Using Individual Patient Data, 1995. Practical methodology of meta-analyses (overviews) using updated individual patient data. *Statistics in Medicine* **14**(19): 2057–2079.

Trikalinos TA, Hoaglin DC and Schmid CH, 2014. An empirical comparison of univariate and multivariate meta-analyses for categorical outcomes. *Statistics in Medicine* **33**(9): 1441–1459.

3

Choice of Effect Measure and Issues in Extracting Outcome Data

Ian R. White, Christopher H. Schmid, and Theo Stijnen

CONTENTS

3.1 Introduction

Chapter 1 described the basic concepts of systematic reviews, and Chapter 2 introduced meta-analysis and different ways in which meta-analysis data might be structured. In this chapter, we assume that a systematic review has been performed and has identified a set of studies to be included in a meta-analysis. We discuss two related issues that must precede the meta-analysis: choice of effect measure and outcome data extraction.

The "effect measure" or "metric" describes the particular way in which the meta-analysis aims are quantified and defines the parameter (or parameters) whose combination across studies is the main focus of the meta-analysis. For example, in a descriptive study, the effect measure could be the mean of a quantitative outcome or the proportion experiencing a binary outcome; in a comparative study, the effect measure could be the difference of means or the odds ratio between groups. In Section 3.2, we discuss how to choose the effect measure.

"Outcome data extraction" is the process by which summary outcome data suitable for estimation of the effect measure are extracted from the primary studies. We may extract the data at various levels: individual participant data (IPD), group-level summaries (summary statistics that are ideally both commonly reported and informative), or a study-level summary (an estimate of the effect measure with its standard error). In Section 3.3 we explain how to perform the outcome data extraction at group level and study level. Meta-analysis of study-level summaries is flexible and popular (Chapter 4), so Section 3.4 describes how to convert group-level summaries to study-level summaries. Finally, complications often arise in extracting study-level summaries from published reports, and some pitfalls and tricks are discussed in Section 3.5.

3.2 Choosing the Effect Measure

The choice of effect measure is usually driven by the aim of the meta-analysis and the type of outcome data. By "outcome data", we mean the data in the individual studies and not any summaries that are constructed for meta-analysis.

3.2.1 Descriptive Studies

We start by considering meta-analysis of single-group studies, for which Table 3.1 lists possible choices of effect measure and statistical notation. These are usually studies with a descriptive aim: for example, estimating the prevalence of a disease, estimating the sensitivity or specificity of a diagnostic test, describing educational attainment in a population, or predicting five-year survival in a particular condition. In most cases, the type of data directly determines the effect measure: for example, if the outcome is the prevalence of a disease, then the only available effect measure is the prevalence. By count data we refer to numbers of events within a group with known total person-time of follow-up, not to summaries of binary or ordinal data which would count events within a group with a known number of individuals. Quantitative data are often called continuous data, and time-to-event data are often called survival data.

Time-to-event data can be summarized in a number of ways. Survival probability at a specific time point is easily understood but does not usually make full use of the data; multiple time points may be used instead (Chapter 15). Mean survival requires unlimited follow-up, whereas restricted mean survival refers to a limited follow-up period. Note that the formula in Table 3.1 for the mean survival, $\int_0^\infty S_i(t)dt$, is equivalent to the more standard formula $\int_0^\infty tf_i(t)dt$ where $f_i(t) = -S_i'(t)$ is the density function.

3.2.2 Comparative Studies

We now turn to meta-analysis of comparative studies, such as randomized trials. In a randomized trial, we usually call the randomized groups "treatment" and "control", as a shorthand for "experimental treatment" and "standard treatment". In an observational

TABLE 3.1

Choice of Effect Measure for Meta-Analysis of Descriptive Studies

Type of data	Effect measure	Notation in study i
Quantitative	Mean	μ_i
Binary	Proportion (risk, prevalence)	π_i
Categorical	Often dichotomized and handled as binary	
	Ordered categorical data may be regarded as quantitative	
Count	Rate	λ_i
Time to event	Survival at fixed time t_0	$S_i(t_0)$
	Median survival	$S_i^{-1}(0.5)$
	Mean survival	$\int_0^\infty S_i(t)dt$
	Restricted mean survival to time t_0	$\int_0^{t_0} S_i(t)dt$

study, the groups might be "exposed" and "unexposed". Because the aim here is to compare groups, there are more possible effect measures and the choice is more complex. Table 3.2 lists possible choices of effect measure for meta-analysis of comparative studies.

A key criterion in choosing the effect measure is that it should be estimable from the primary studies. It should also be interpretable and scientifically relevant, or convertible into an interpretable and scientifically relevant form. Two more technical criteria are also important. First, if effect measures might differ in their degree of heterogeneity between studies, then one would prefer an effect measure that is likely to be less heterogeneous. Second, one would wish to avoid an effect measure with poor statistical properties, for example, possibility of bias or large variance in the meta-analysis.

With a quantitative outcome, the preferred effect measure in biomedical studies is usually the mean difference or (occasionally) the ratio of means. When the outcome variable is well understood, this yields easily interpreted results: for example, that a treatment reduces blood pressure by 10 millimeters of mercury. However, using the mean difference requires all trials to report the outcome measure on the same scale. It is very common in social sciences for different studies to measure the same concept in different ways—for example, different psychology studies often use different rating scales to assess the same construct. In this case, it would be inappropriate to combine mean differences, since one-unit changes on different scales would not be comparable. Instead, the standardized mean difference allows outcomes measured in different ways to be compared by dividing the mean difference by the population standard deviation, which may be estimated from the control group or pooled within groups. Use of the standardized mean difference yields effect sizes that can be related to an absolute scale, for example, 0.2 is "small", 0.5 is "medium", and 0.8 is "large" (Cohen, 1988). However, use of the standardized mean difference is sensitive to noise that increases the standard deviation (Lu et al., 2014).

With a binary outcome, the three common choices of effect measure (risk difference, risk ratio, and odds ratio) all behave similarly when studies have similar overall outcome proportions. However, they may behave very differently when studies have markedly different overall outcome proportions: for example, studies with outcome proportion near 0.5 tend to be down-weighted by the risk difference and up-weighted by the odds ratio. Studies of many meta-analyses show that the risk difference tends to have larger heterogeneity and is hence less desirable in general, but that the risk ratio and odds ratio have

TABLE 3.2

Choice of Effect Measure for Meta-Analysis of Comparative Studies

Type of data	Effect measure	Definition in study *i*	Comment
Quantitative	Mean difference	$\mu_{i2}-\mu_{i1}$	Appropriate when all studies report the same outcome (e.g., systolic blood pressure)
	Ratio of means	μ_{i2}/μ_{i1}	Appropriate when studies report the same or different outcomes and are likely to have similar proportionate changes in outcome
	Standardized mean difference	$(\mu_{i2}-\mu_{i1})/\sigma_i$	May be appropriate when different studies report different outcomes; σ_i is a suitable standard deviation
Binary	Risk difference	$\pi_{i2}-\pi_{i1}$	Not usually appropriate: tends to be affected by the overall level of risk in the study, and hence more heterogeneous than alternatives (Deeks, 2002)
	Risk ratio	π_{i2}/π_{i1}	Usually appropriate: less likely to be affected by the overall risk
	Odds ratio	$\frac{\pi_{i2}}{1-\pi_{i2}}\Big/\frac{\pi_{i1}}{1-\pi_{i1}}$	Usually appropriate: less likely to be affected by the overall risk
Categorical	Binary measures[a] after dichotomizing outcome		Makes incomplete use of data
	Mean difference, etc.		Treats outcome as quantitative; may be appropriate if categories are ordered
	Odds ratio for ordinal outcome		Estimated from proportional odds model if categories are ordered
	Odds ratios for multinomial outcome		Estimated from multinomial logit model; multiple parameters
Count	Rate difference	$\lambda_{i2}-\lambda_{i1}$	Not usually appropriate: tends to be affected by the overall rate
	Rate ratio	$\lambda_{i2}/\lambda_{i1}$	Usually appropriate: less likely to be affected by the overall rate
Time to event	Binary measures[a] for event before a fixed time t_0	$S_{i2}(t_0)-S_{i1}(t_0)$, etc.	Appropriate, but ignores data after the fixed period
	Binary measures[a] for event occurrence at any time		Inappropriately ignores censoring and length of follow-up, thus involving approximation and risking bias
	Hazard ratio[b]	$h_{i2}(t)/h_{i1}(t)$	Assumed constant over time; allows for censoring and length of follow-up
	Comparisons of other measures, e.g., restricted mean survival or median survival		Appropriate but less often used

[a] Binary measures are risk difference, risk ratio, and odds ratio.

[b] Not estimable from group-level summaries, unless a constant hazard is assumed.

similar levels of heterogeneity (Deeks, 2002). With the risk ratio there is the further choice of whether to consider the beneficial or harmful outcome (Deeks, 2002).

Two other effect measures for binary outcomes deserve mention. The number needed to treat, used in medicine, is the number of individuals that would need to be treated to reduce the expected count by one, and equals the reciprocal of the risk difference. It has poor statistical properties and should therefore not be used as the effect measure in the meta-analysis, but can be derived after meta-analysis of another effect measure (Furukawa et al., 2002). The arcsine difference is defined as $\arcsin\sqrt{p_{i2}} - \arcsin\sqrt{p_{i1}}$. Its main advantage is that its asymptotic variance $\frac{1}{4n_{i1}} + \frac{1}{4n_{i2}}$ depends only on the study sample size. However, its lack of easy interpretation has limited its wider use as an effect measure.

Adjustment for baseline covariates in randomized trials can increase power. Any of the effect measures in Table 3.2 can be adjusted for baseline covariates. For a quantitative outcome, covariate adjustment in randomized trials does not change the effect measure being estimated: for example, the true covariate-adjusted mean difference in a randomized trial is the same as the unadjusted mean difference (although their estimates differ in any given dataset). The increase in power due to covariate adjustment therefore corresponds to a reduction (on average) in standard errors, and randomized trials analyzed with different degrees of covariate adjustment can reasonably be combined. For other outcomes, however, covariate adjustment does change the effect measure being estimated: for example, the true covariate-adjusted odds ratio in a randomized trial is further from one than the unadjusted odds ratio (Robinson and Jewell, 1991). As a result, the increase in power due to covariate adjustment can accompany an increase (on average) in standard errors. It could be argued that randomized trials analyzed with different degrees of covariate adjustment should not be combined since they do not estimate the same effect measure: however, the differences tend to be small in practice. The case of observational studies is different, since covariate adjustment is often required to control for confounding, and studies with different levels of covariate adjustment should not be directly combined unless their control of confounding can be shown to be similar; alternative approaches are discussed in Chapter 21. Specific issues when the baseline covariate is the baseline level of the outcome are discussed in Section 3.5.2.

3.2.3 General Association Studies

A third type of study explores more general associations between variables. Suitable effect measures here are regression coefficients from a suitable regression model (e.g., Chapter 18) or correlation coefficients (Chapter 16).

3.2.4 Avoiding Effect Measures

It is possible to avoid the use of an effect measure by performing a meta-analysis of p-values, yielding just a summary p-value. For example, given p-values p_i in the ith study (either all one-tailed or all two-tailed), Fisher's method uses the statistic $\sum_{i=1}^{I} -2\log p_i$: if the null hypothesis is true for all studies then this statistic follows the χ^2_{2I} distribution (Fisher, 1934). Meta-analysis of p-values is usually not desirable because it fails to quantify the magnitude of effect, because there is no assessment of heterogeneity, and because p-values are insufficient to tell whether results are scientifically significant—for example, whether treatment reduces risk by a large enough amount to justify its increased burden on the patient. Meta-analysis of p-values is sometimes used in genetics (Chapter 17) and forms part of the "fail-safe N" method for publication bias (Chapter 13).

3.2.5 Transformation

Many of the above effect measures (including the odds ratio, risk ratio, and hazard ratio in Table 3.2) would be transformed onto the log scale before analysis. Other transformations can be used to make the modeling assumptions described in later chapters more plausible: for example, a binary proportion might be transformed onto the logit scale and the correlation coefficient might be transformed by the Fisher z-transformation.

3.3 How to Perform Outcome Data Extraction

As noted in Chapter 2, data may be extracted at three levels (Figure 3.1): individual participant level (yielding IPD), group level (yielding simple and informative summary statistics), or study level (yielding estimates of the effect measure with associated standard errors). Chapter 2 introduced the advantages and disadvantages of collecting IPD. The meta-analyst who chooses this approach needs to negotiate with study investigators to assemble the data required to perform the required analyses. This is discussed further in Chapter 8. Other meta-analysts need to extract suitable outcome data from published reports, which is the topic of this section.

3.3.1 Extracting Group-Level Summaries

Group-level summaries are usually taken to be simple summaries of the raw data, as summarized for each type of data in Table 3.3. We now discuss several pitfalls that may arise in extracting such summaries from publications.

For quantitative data, care must be taken to distinguish the standard deviation from the standard error, since these can be confused in publications. In the presence of missing data, the numbers of individuals should exclude any individuals who did not contribute to the mean and standard deviation. Often the group-level summary statistics presented are not those wanted—for example, the median and interquartile range and not the mean and SD. Methods for inferring the mean and SD in this case are given by Follmann et al. (1992), Hozo et al. (2005), Higgins et al. (2019), and Wan et al. (2014).

Survival data also present challenges. It is common to extract the number of participants and the number of events at any time and hence to treat the data as a proportion, although

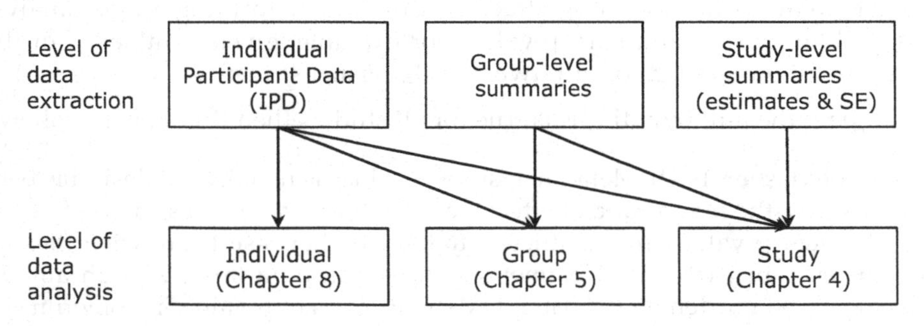

FIGURE 3.1
Levels of outcome data extraction and levels of data analysis in a meta-analysis.

TABLE 3.3

Group-Level Summaries for Different Types of Outcome Data

Type of outcome data	Parameters	Group-level summary in group j of study i
Quantitative	Mean μ_{ij} Standard deviation σ_{ij}	Mean m_{ij} Standard deviation s_{ij} Number of individuals n_{ij}
Binary	Probability of event π_{ij}	Number of events y_{ij} Number of individuals n_{ij}
Ordinal	Probability of outcome level l π_{ijl}	Number y_{ijl} at outcome level l Number of individuals n_{ij}
Count	Rate λ_{ij}	Number of events y_{ij} Number of person-years T_{ij}
Time to event	Survival $S_{ij}(t)$ or hazard $h_{ij}(t)$	Can be difficult: see text

this does not account for different follow-up and potential censoring. For example, the MATRIX trial was a randomized trial comparing radial with femoral access in patients undergoing invasive management for acute coronary syndromes (Valgimigli et al., 2015). The primary outcome event occurred in 369 of 4197 patients randomized to radial access and in 429 of 4207 randomized to femoral access. A meta-analysis used these four summary statistics, approximating the time-to-event outcome with a binary outcome (Andò and Capodanno, 2015).

Alternative group-level approaches are more faithful to the time-to-event nature of the data. If the hazard is roughly constant, then the total person-time and total number of events are a useful summary (as for count data). When hazards may be non-constant, usually no group-level summaries contain enough information to reproduce a full survival analysis such as estimation of a hazard ratio under a Cox proportional hazards model. In this case, it is best to attempt to reconstruct the IPD from group-specific Kaplan–Meier curves (Parmar et al., 1998; Tierney et al., 2007).

3.3.2 Extracting Study-Level Summaries

A study-level summary is an estimate of the effect measure, together with its standard error, that is either reported by authors using a statistical model or computed by the meta-analyst.

Often the authors report a contrast such as a rate ratio with a confidence interval; in this case, it is necessary to work backwards to derive the standard error on the appropriate scale. For example, in the MATRIX trial discussed above, the authors reported a rate ratio from survival analysis of 0.85 (95% CI, 0.74 to 0.99). We can deduce a log rate ratio of approximately $\log 0.85 = -0.163$ and a standard error of approximately $(\log 0.99 - \log 0.74) / 2z_{0.975} = 0.074$. If instead the authors had reported the rate ratio of 0.85 with a p-value of 0.03, then we could have deduced a z-statistic of 2.16 and hence a standard error of approximately $|\log 0.85| / 2.16 = 0.075$.

Group-level summaries and study-level summaries need not agree. For example, in the MATRIX trial, the group-level summaries cannot be used to estimate a rate ratio. Instead of using the study authors' estimate of the rate ratio, the meta-analysis authors used the group-level summaries to estimate a risk ratio of 0.86 (95% CI, 0.76 to 0.98). This is not numerically identical to the reported rate ratio but is very similar because the event proportion was very low (under 10%). Because estimation of the risk ratio does not account for censoring and wrongly assumes that all participants were followed up for the same period, we would prefer to use the author-reported rate ratio here.

3.3.3 Which Summaries Are Appropriate?

Extracting data at the group level enables a one-stage analysis (Chapter 5). This is especially useful for small studies where the approximation involved in two-stage analysis is poor. Further, data extracted at the group level can later be converted into study-level summaries for two-stage analysis if required. On the other hand, extracting data at the study level forces a two-stage analysis.

In some cases, study authors report complex statistical analyses. Study-level data extraction allows these to be used. This can have various advantages: for example, complex analyses in randomized trials might improve power by adjusting for covariates; they might avoid bias in the standard error by allowing for a complex study design such as a cluster-randomized trial; or they might reduce the risk of bias in the point estimate through an appropriate handling of missing data. In observational epidemiology, the need to adjust for confounders usually implies the use of study-level summaries rather than group-level summaries, though IPD are preferable (Chapter 21).

The flexibility of complex statistical analyses in study publications brings with it the potential for selective reporting (Chapter 13): for example, one might be concerned that a study report adjusted for a covariate because doing so improved the statistical significance of the result. If multiple analyses are reported, then the meta-analyst needs to formulate rules for choosing between them. For these reasons, the Cochrane Collaboration tends to prefer group-level summaries for systematic reviews of randomized evaluations (Higgins et al., 2019).

An intermediate option between group-level and study-level summaries as described above is to summarize data at group level by estimating the group-level parameters (e.g., the log odds) together with their standard errors. This may be useful for time-to-event data when the effect measure relates to survival to a fixed time point. Otherwise it is rarely used, because suitable data are rarely available and because it brings the cost of two-stage analysis (requiring a normal approximation in analysis) without gaining its flexibility.

3.4 Calculating Study-Level Summaries from Group-Level Summaries

Sometimes it is necessary to calculate study-level summaries (estimated effect measures and their variances) from group-level summaries or from published reports. This section describes some tricks and pitfalls in doing so. If all studies provide group-level summaries, then a one-stage analysis avoiding the use of study-level summaries is possible, but a two-stage analysis using study-level summaries is always possible (Figure 3.1). Formulae for computing study-level summaries (after transformation where appropriate) are given in Table 3.4. Useful general references are Greenland (1987) and Lipsey and Wilson (2000).

Some of the study-level summaries in Table 3.4 suffer from small-sample bias. For example, the observed proportion y_{i1}/n_{i1} is an unbiased estimate of the true probability in group 1, but the observed log odds, $\log \frac{y_{i1}}{n_{i1} - y_{i1}}$, is not an unbiased estimate of the true log odds,

TABLE 3.4

Extracting Study-Level Summaries from Group-Level Summaries

Type of data	Effect measure	Study-level summaries: Estimate	Study-level summaries: Estimated variance
Single-group studies			
Quantitative	Mean	m_i	s_i^2 / n_i
Binary	Proportion	y_i/n_i	$y_i(n_i - y_i) / n_i^3$
Count	Rate	y_i/T_i	y_i / T_i^2
Time to event	(See text)		
Comparative studies			
Quantitative	Mean difference	$m_{i2} - m_{i1}$	$\sum_j s_{ij}^2 / n_{ij}$
	SMD[a]	$(m_{i2} - m_{i1})J(v_i)/s_i$	$\sum_j 1/n_{ij} + SMD^2 K(v_i)$
Binary	Risk difference	$y_{i2}/n_{i2} - y_{i1}/n_{i1}$	$\sum_j y_{ij}(n_{ij} - y_{ij}) / n_{ij}^3$
	Log risk ratio	$\log(y_{i2}/n_{i2}) - \log(y_{i1}/n_{i1})$	$\sum_j (1/y_{ij} - 1/n_{ij})$
	Log odds ratio	$\log\frac{y_{i2}}{n_{i2} - y_{i2}} - \log\frac{y_{i1}}{n_{i1} - y_{i1}}$	$\sum_j \left(\frac{1}{y_{ij}} + \frac{1}{n_{ij} - y_{ij}}\right)$
Count	Log rate ratio	$\log(y_{i2}/T_{i2}) - \log(y_{i1}/T_{i1})$	$\sum_j 1/y_{ij}$
Time to event	(see text)		
Association studies			
Two quantitative variables	Fisher's transformation of the correlation coefficient r_i	$\frac{1}{2}\log\frac{1+r_i}{1-r_i}$	$1/(n_i - 3)$

Notation is defined in Table 3.3, with subscripts i for study and (in the case of comparative studies) j for group. All sums are over $j = 1,2$.

[a] Standardized mean difference. Here s_i is an estimate of the population standard deviation, $v_i = n_{i1} + n_{i2} - 2$ is its degrees of freedom, and $J(v) \approx 1 - 3/(4v - 1)$ and $K(v) \approx 1/2(v - 1.94)$ with exact expressions given by White and Thomas (2005).

since the log odds is a non-linear transformation of the observed proportion. Cox and Snell (1989) showed that this bias can be largely removed by the estimate $\log\frac{y_{i1} + 0.5}{n_{i1} - y_{i1} + 0.5}$: that is, by adding 0.5 to all cell counts in the 2×2 table. The quantity 0.5 is called the continuity correction and is discussed in detail by Sweeting et al. (2004). It is particularly important in the case of zero events or non-events ($y_{i1} = 0$ or $y_{i1} = n_{i1}$), since otherwise the log odds is undefined. In practice, the continuity correction is only applied in studies with a zero cell count, even though the bias reduction would apply in all studies. The continuity correction also avoids problems in estimating the sampling variance.

Another problem with the summaries in Table 3.4 is that estimated variances in small studies can be correlated with point estimates, leading to bias in the meta-analysis (Jackson and White, 2018). Specific solutions to remove this correlation have been proposed

(Emerson et al., 1993; Rücker et al., 2009) but if this correlation is a problem then a one-stage analysis is usually the best solution (Chapter 5).

3.5 Calculating Study-Level Summaries from Published Results

This final section describes some difficulties that may arise in outcome data extraction of study-level summaries.

3.5.1 Missing Information

A common theme is that some information required for a valid analysis may be missing: for example, the standard deviation s_{ij} of a quantitative outcome may not be reported. The standard deviation is not the primary focus of analysis, but it is required to estimate standard errors validly. Missing standard deviations may be tackled in a simple way by assuming they are equal to standard deviations observed in another study—preferably one chosen to have similar characteristics. Ideally, allowance would be made for the standard deviations varying between studies via a hierarchical model, but this is usually not done. It would not be appropriate to assume missing standard errors are equal to standard errors observed in another study, unless sample sizes are equal.

Other examples of missing information arise in the next three subsections and may be handled in a similar way.

3.5.2 Before-and-After Data

Before-and-after data arise when the same variable is observed in two groups before and after some intervention. A particular difficulty arises when before-and-after data are extracted at group level (Riley et al., 2013). Assume that the group level summaries in the jth group of the ith study are presented as the count n_{ij}, means m_{ijB} before and m_{ijA} after intervention, and standard deviations s_{ijB} before and s_{ijA} after intervention. For simplicity, we pool the variances across groups to give $s_{i,B}$ and $s_{i,A}$, though group-specific variances can also be used in the formulae below. Then the intervention effect for group $j=2$ compared with group $j=1$ may be estimated in three different ways:

1. A difference of after-intervention means, $m_{i2A}-m_{i1A}$, with variance $s_{i.A}^2(1/n_{i1}+1/n_{i2})$.
2. A difference of change scores, $(m_{i2A}-m_{i2B})-(m_{i1A}-m_{i1B})$ with variance $(s_{i.A}^2-2\rho_i s_{i.A}s_{i.B}+s_{i.B}^2)(1/n_{i1}+1/n_{i2})$, where ρ_i is the within-group correlation of the before-intervention and after-intervention values, assumed equal across groups.
3. An adjusted difference, $(m_{i2A}-m_{i1A})-\lambda_i(m_{i2B}-m_{i1B})$ with variance $s_{i.A}^2(1-\rho_i^2)(1/n_{i1}+1/n_{i2})$, where $\lambda_i=\rho_i s_{i.A}/s_{i.B}$ is the within-group regression of after-intervention values on before-intervention values.

In a randomized trial, where before-intervention means m_{i1B} and m_{i2B} are equal in expectation, all three estimators have the same expectation, but the adjusted difference has the lowest variance. Thus, if ρ_i is reported then the adjusted difference should be computed.

If ρ_i is not reported, it may be possible to infer it from other reported results (e.g., from the standard deviation of change scores as well as s_{ijB} and s_{ijA}) or to assume it is equal to an available value in another study as in Section 3.5.1. Otherwise, it may instead be necessary to use a less statistically efficient estimate, the difference of after-intervention means, or change scores. Once these decisions are made, it is reasonable to meta-analyze studies where intervention effects have been estimated in different ways.

In an observational study, the three estimators are different. Subject-matter arguments should be used to choose between them, and it is rarely reasonable to meta-analyze studies where intervention effects have been estimated in different ways.

3.5.3 Cluster-Randomized Trials

In a cluster-randomized trial, participants are grouped in clusters, and the resulting correlation ("clustering") must be allowed for in order to give correct standard errors. Clustering is measured by the intra-class correlation (ICC) (Kerry and Bland, 1998).

If study-level summary statistics are reported using an analysis that allows for the clustering, then there is no problem. However, commonly reported study-level summaries do not adjust for the clustering, and therefore the outcome data extraction must estimate this adjustment (Donner and Klar, 2002). The relevant adjustment multiplies the variance by the "design effect" $1+(m-1)\text{ICC}$, where m is the cluster size (Kerry and Bland, 1998). The formula can be extended for unequal cluster sizes (Kerry and Bland, 2001). The ICC is sometimes reported; if not, then it may be necessary to assume it equal to an observed value from another study as in Section 3.5.1. Once this is done, cluster-randomized trials may be combined with other trial designs in a two-stage meta-analysis.

The simple group-level summary statistics described in Table 3.3 are also not sufficient to give correct standard errors in cluster-randomized trials. A common approach is to convert them to study-level summary statistics and apply the design effect.

3.5.4 Cross-Over Trials

In a cross-over trial, participants are randomized to groups representing different sequences of treatments, and outcomes are observed after each treatment. Again, group-level summaries often prove inadequate and a correlation is also needed.

Assuming a two-treatment two-period design with a quantitative outcome, the data may be reported as the outcome means m_{ijk} for group $j=1,2$ of the ith study when receiving treatment $k=T,C$, with corresponding sample sizes $n_{ijk}=n_{ij}$ and standard deviations s_{ijk}. For example, suppose in study i that group $j=1$ represents the sequence T then C. Then m_{i1T} is the mean of group 1 in period 1 receiving T, and m_{i1C} is the mean of group 1 in period 2 receiving C. To eliminate the effect of any trend in outcome over time (a period effect), the treatment effect is estimated as $\frac{m_{i1T}-m_{i1C}}{2}+\frac{m_{i2T}-m_{i2C}}{2}$, with estimated variance $\frac{\sigma_{i1T}^2+\sigma_{i1C}^2-2\rho_{i1TC}\sigma_{i1T}\sigma_{i1C}}{4n_{i1}}+\frac{\sigma_{i2T}^2+\sigma_{i2C}^2-2\rho_{i2TC}\sigma_{i2T}\sigma_{i2C}}{4n_{i2}}$. Often a correlation ρ_{ijTC} between outcomes on treatment and control in the jth group of the ith study is not reported. It may then be necessary to assume it is equal to an observed value from another study as in Section 3.5.1. Once this is done, however, cross-over trials may be combined with other trial designs in a two-stage meta-analysis (Curtin et al., 2002).

3.5.5 Comparative Studies with More Than Two Groups

With more than two groups, there are multiple effect measures; such studies usually include one group which is a control group, so the effect measures are all defined as above relative to the control group. A complication arises when study-level summaries are used, since then the estimated effect measures are correlated due to their common control group. The correlation is easily estimated and is needed for many multivariate analyses such as those discussed in Chapters 9 and 10.

3.5.6 Studies Report Different Effect Measures

Often, different studies included in the meta-analysis report study-level summaries for different effect measures. In these cases, it is useful to be able to make conversions. Here we only give some examples without formulae; more information is given by Lipsey and Wilson (2000) and Wilson (2001). For quantitative data, one can convert between mean difference and standardized mean difference using the formulae in Table 3.4. Where a quantitative outcome is dichotomized in some studies, it is possible to convert between an odds ratio and a standardized mean difference (Chinn, 2000). For binary data, if the event is rare then the risk, odds, and rate are very similar, and hence so also are the risk ratio, odds ratio, and rate ratio. For commoner events, some conversions are available, for example, odds ratio to risk ratio (Zhang and Yu, 1998). For survival data, it is useful to know that rate and hazard are the same quantity.

3.5.7 What Next?

Once an effect measure has been chosen and data have been extracted, we are ready to perform the meta-analysis. This will be covered in future chapters: in Chapter 4, we meta-analyze study-level summaries; in Chapter 5, we meta-analyze group-level summaries; and in Chapter 8, we meta-analyze IPD.

References

Andò G and Capodanno D, 2015. Radial versus femoral access in invasively managed patients with acute coronary syndrome. *Annals of Internal Medicine* **163**(12): 932.

Chinn S, Nov 2000. A simple method for converting an odds ratio to effect size for use in meta-analysis. *Statistics in Medicine* **19**(22): 3127–3131.

Cohen J, 1988. *Statistical Power Analysis for the Behavioral Sciences*. Lawrence Erlbaum Associates, Inc.

Cox DR and Snell EJ, 1989. *Analysis of Binary Data*. Chapman & Hall/CRC.

Deeks JJ, 2002. Issues in the selection of a summary statistic for meta-analysis of clinical trials with binary outcomes. *Statistics in Medicine* **21**(11): 1575–1600.

Donner A and Klar N, 2002. Issues in the meta-analysis of cluster randomized trials. *Statistics in Medicine* **21**(19): 2971–2980.

Emerson JD, Hoaglin DC and Mosteller F, 1993. A modified random-effect procedure for combining risk difference in sets of 2×2 tables from clinical trials. *Journal of the Italian Statistical Society* **2**(3): 269–290.

Fisher RA, 1934. *Statistical Methods for Research Workers*, 5th edition. Oliver and Boyd.

Follmann D, Elliott P, Suh I and Cutler J, 1992. Variance imputation for overviews of clinical trials with continuous response. *Journal of Clinical Epidemiology* **45**(7): 769–773.

Furukawa TA, Guyatt GH and Griffith LE, 2002. Can we individualize the 'number needed to treat'? An empirical study of summary effect measures in meta-analyses. *International Journal of Epidemiology* **31**(1): 72–76.

Greenland S, 1987. Quantitative methods in the review of epidemiologic literature. *Epidemiologic Reviews* **9**: 1–30.

Higgins JPT, Thomas J, Chandler J, Cumpston M, Li T, Page MJ, Welch VA (editors), 2019. *Cochrane Handbook for Systematic Reviews of Interventions*. 2nd Edition. Chichester, UK: John Wiley & Sons. Also online version 6.0 (updated July 2019) available from www.training.cochrane.org/handbook.

Hozo SP, Djulbegovic B and Hozo I, 2005. Estimating the mean and variance from the median, range, and the size of a sample. *BMC Medical Research Methodology* **5**(1): 13.

Jackson D and White IR, 2018. When should meta-analysis avoid making hidden normality assumptions? *Biometrical Journal* **60**(6): 1040–1058.

Kerry SM and Bland JM, 1998. Statistics notes: the intracluster correlation coefficient in cluster randomisation. *British Medical Journal*, **316**(7142): 1455–1460.

Kerry SM and Bland JM, 2001. Unequal cluster sizes for trials in English and Welsh general practice: Implications for sample size calculations. *Statistics in Medicine* **20**(3): 377–390.

Lipsey MW and Wilson DB, 2000. *Practical Meta Analysis*. Thousand Oaks, CA: Sage.

Lu G, Kounali D and Ades AE, 2014. Simultaneous multioutcome synthesis and mapping of treatment effects to a common scale. *Value in Health* **17**(2): 280–287.

Parmar MKB, Torri V and Stewart L, 1998. Extracting summary statistics to perform meta-analyses of the published literature for survival endpoints. *Statistics in Medicine* **17**(24): 2815–2834.

Riley RD, Kauser I, Bland M, Thijs L, Staessen JA, Wang J, Gueyffier F and Deeks JJ, 2013. Meta-analysis of randomised trials with a continuous outcome according to baseline imbalance and availability of individual participant data. *Statistics in Medicine* **32**(16): 2747–2766.

Robinson LD and Jewell NP, 1991. Some surprising results about covariate adjustment in logistic regression models. *International Statistical Review / Revue Internationale de Statistique* **59**(2): 227.

Rucker G, Schwarzer G, Carpenter J and Olkin I, 2009. Why add anything to nothing? The arcsine difference as a measure of treatment effect in meta-analysis with zero cells. *Statistics in Medicine* **28**(5): 721–738.

Sweeting MJ, Sutton AJ and Lambert PC, 2004. What to add to nothing? Use and avoidance of continuity corrections in meta-analysis of sparse data. *Statistics in Medicine* **23**(9): 1351–1375.

Tierney JF, Stewart LA, Ghersi D, Burdett S and Sydes MR, 2007. Practical methods for incorporating summary time-to-event data into meta-analysis. *Trials* **8**(1): 16.

Valgimigli M, Gagnor A, Calabro P, Frigoli E, Leonardi S, Zaro T, Rubartelli P, Briguori C, Ando G, Repetto A, Limbruno U, Cortese B, Sganzerla P, Lupi A, Galli M, Colangelo S, Ierna S, Ausiello A, Presbitero P, Sardella G, Varbella F, Esposito G, Santarelli A, Tresoldi S, Nazzaro M, Zingarelli A, de Cesare N, Rigattieri S, Tosi P, Palmieri C, Brugaletta S, Rao SV, Heg D, Rothenbuhler M, Vranckx P and Jüni P, 2015. Radial versus femoral access in patients with acute coronary syndromes undergoing invasive management: A randomised multicentre trial. *Lancet* **385**: 2465–2476.

Wan X, Wang W, Liu J and Tong T, 2014. Estimating the sample mean and standard deviation from the sample size, median, range and / or interquartile range. *BMC Medical Research Methodology* **14**: 135.

White IR and Thomas J, 2005. Standardised mean differences in individually-randomised and cluster-randomised trials, with applications to metaanalysis. *Clinical Trials* **2**(2): 141–151.

Wilson DB, 2001. Practical meta-analysis effect size calculator. http://www.campbellcollaboration.org/escalc/html/EffectSizeCalculator-R3.php.

Zhang J and Yu KF, 1998. What's the relative risk? A method of correcting the odds ratio in cohort studies of common outcomes. *JAMA* **280**(19): 1690.

4

Analysis of Univariate Study-Level Summary Data Using Normal Models

Theo Stijnen, Ian R. White, and Christopher H. Schmid

CONTENTS

4.1 Introduction

Chapters 2 and 3 noted that univariate meta-analysis of published results can be performed on study-level summaries or arm-level summaries. In this chapter, we consider the very common situation where study-level summaries—effect measures and their standard errors—are used with a normal approximation to their distribution. The effect measure might be, for instance, a (log) odds ratio, a (standardized) difference between two means, a correlation coefficient, a single proportion or a rate, or any other effect measure of interest. In this chapter, we will restrict to data in the form of a single (univariate) effect measure per study. Many other chapters in this book, for instance, Chapters 9, 10, 14, 15, 16, and 19, consider two or more effect measures at the same time, leading to different forms of multivariate meta-analysis.

The normal approximation has a number of limitations which we make clear in this chapter. Chapter 5 will instead consider meta-analysis of arm-level summaries using the correct likelihood for the form of the data (e.g., normal, proportions, or counts) which avoids these limitations.

Throughout this chapter we will assume that each of $i = 1,\ldots,I$ studies yields an estimate y_i of the effect measure, together with its standard error s_i. The unknown parameter that is estimated by y_i in the ith study is called the true effect measure in study i, denoted by θ_i. We will assume that y_i follows a normal distribution with standard error estimated by s_i: this is called the within-studies model. The summary data y_i and s_i may have been provided in the study report, calculated from other data ingredients given in the report by methods such as in Chapter 3, or calculated from IPD. The methods of this chapter are therefore also called two-stage methods, where the first stage is estimating y_i and s_i, and the second stage is the meta-analysis.

4.1.1 Common-Effect versus Random-Effects versus Separate-Effects

The aim of a meta-analysis is to combine or aggregate the study-specific estimates in order to reach overall conclusions. On the aggregated level, different target parameters (estimands) could be defined, depending on the choices and assumptions that one is willing to make.

If one is willing to make the assumption that all θ_i are equal to one common value θ, then this parameter will be the estimand of the meta-analysis. This leads to a common-effect (CE) meta-analysis, frequently referred to as a fixed-effect meta-analysis (see Chapter 2 for discussion of the choice of nomenclature).

If one wants to allow that the θ_i vary between studies and one is willing to assume that they can be regarded as a random sample from an unknown distribution, then the mean of that distribution, θ, will be a straightforward choice for the main estimand of the meta-analysis. A secondary estimand will then be a parameter characterizing the variability or heterogeneity of that distribution, usually the standard deviation. This leads to what is called random-effects (RE) meta-analysis.

We discuss the within-studies model in Section 4.2, the common-effect model in Section 4.3, and the random-effects model in Section 4.4, with estimation of the heterogeneity variance in Section 4.5. We end the chapter with a short section on meta-analysis software (Section 4.6).

4.2 The Within-Studies Model

Standard assumptions in two-stage modeling (both for CE and RE modeling) are first that y_i has a normal distribution and second that y_i has variance s_i^2, given the true effect θ_i:

$$y_i \sim N\left(\theta_i, s_i^2\right). \tag{4.1}$$

These assumptions will at best only be approximately true. In practice, y_i is asymptotically normally distributed, so the larger the sample size, the better the normal approximation will be. The accuracy of the approximation also depends on the choice of the effect measure. For instance, the logarithm of ratio measures like the odds ratio approaches the normal distribution with much smaller sample sizes than the odds ratio itself. This is the main reason why it is customary to logarithmically transform ratio measures like the odds ratio and risk ratio. The same holds for Fisher's Z-transformed correlation coefficient (Chapter 3). If sample sizes are too small for (4.1) to be a reasonable approximation, a one-stage method as discussed in Chapter 5 might be better. Although the standard errors s_i are estimated, they are always treated as known in the two-stage meta-analysis approach.

4.3 The Common-Effect Model

The simplest model for incorporating all the studies assumes that all θ_i are equal, $\theta_i = \theta$. Combining this with the within-studies model (4.1) gives the common-effect (CE) model

$$y_i \sim N\left(\theta, s_i^2\right). \tag{4.2}$$

4.3.1 Estimation

The maximum likelihood estimate of θ, or equivalently the weighted least squares estimate, is the average of the study-specific estimated effects weighted by the inverse squared standard errors:

$$\hat{\theta}_{CE} = \frac{\sum_i w_{i,CE} y_i}{\sum_i w_{i,CE}} \text{ with } w_{i,CE} = \frac{1}{s_i^2} \text{ and } se\left(\hat{\theta}_{CE}\right) = \frac{1}{\sqrt{\sum_i w_{i,CE}}}. \tag{4.3}$$

This is also the minimum variance unbiased estimator of θ (Viechtbauer, 2005). (The normality assumption in (4.2) is in fact not needed for this property.) Under (4.2), $\hat{\theta}_{CE}$ is normally distributed, and thus a test and confidence interval for θ may be based on the normal distribution. The test statistic $Z_{CE} = \hat{\theta}_{CE} / se(\hat{\theta}_{CE})$ has a standard normal distribution under $H_0\colon \theta = 0$ and the $(1-\alpha)$ 100% confidence interval is $\hat{\theta}_{CE} \pm z_{1-\alpha/2} se(\hat{\theta}_{CE})$, where $z_{1-\alpha/2}$ denotes the $(1-\alpha)$ quantile of a standard normal distribution.

Alternatively, model (4.2) can be written as a linear model

$$y_i = \theta + \varepsilon_i \text{ with } \varepsilon_i \sim N\left(0, s_i^2\right).$$

The model is usually fitted in specialist meta-analysis software. However, it can also be fitted with a weighted linear regression program with the $w_{i,CE}$ as weights and the residual variance set equal to 1.

4.3.2 Testing Heterogeneity

The null hypothesis of a common effect $\theta_i = \theta$ for all i may be tested with Cochran's Q-statistic (Cochran, 1954, 1973):

$$Q_{CE} = \sum_i w_{i,CE}\left(y_i - \hat{\theta}_{CE}\right)^2. \tag{4.4}$$

Under the model, Q_{CE} follows a χ^2-distribution with df$=I$-1 under the null hypothesis. Cochran's Q-test has low power (Hardy and Thompson, 1998), especially if the number of studies is small. Furthermore, the true study effects are nearly always likely to differ since the individual studies are never identical with respect to study populations and other factors that can cause differences between studies. Therefore, the assumption of a common effect remains questionable even if Cochran's Q-test is not significant.

4.3.3 Critique

What is a CE analysis worth if the common-effect assumption is not fulfilled? In that case, the interpretation of the overall effect $\hat{\theta}$ is a weighted average of the trial specific effects, where the weights are determined by the standard errors, which are mainly determined by the sample sizes. Since the sample sizes are arbitrary, the overall estimated effect lacks a meaningful interpretation. It might be better to choose the weights ourselves in a way that results in a meaningful summary parameter, for instance, the unweighted mean of the θ_i. Then (4.3) can be used with the chosen weights filled in. In fact, this would be a separate-effects analysis as introduced in Chapter 2.

If the CE assumption is not fulfilled, the p-value of the test for H_0: $\theta = 0$ is still a valid test of the strong null hypothesis that $\theta_i = 0$ for all i. If it is rejected, the conclusion is that there is an effect in at least one study.

If the CE assumption is fulfilled, not only for the populations where the study samples in the meta-analysis come from but also for other populations, the result of the meta-analysis can be considered to be directly applicable to other populations. However, if this assumption is not reasonable, applying the results of the CE meta-analysis to other populations may be problematic.

A second criticism of the two-stage CE model is that the within-studies model may not hold. If the normality in (4.2) is not satisfied, approximate normality of $\hat{\theta}_{CE}$ could be justified if the number of studies is large by the central limit theorem. For a small number of studies in which approximate normality of the y_i does not hold, other methods such as those in Chapter 5 should be employed. Assuming the s_i to be known usually has little impact on estimation of θ (Hardy and Thompson, 1996), but this is not always true (e.g., Jackson and White, 2018). See chapter 5 for more on this. Similarly, the χ^2-distribution of Q_{CE} is only approximate when the s_i are estimated: see, for instance, Kulinskaya and Dollinger (2015).

A silent assumption for the unbiasedness of the weighted estimate $\hat{\theta}$ is that the weights are uncorrelated with the y_i. There might be correlation in practice, for statistical or non-statistical reasons. An example of the latter is where the larger studies have been conducted in somewhat different patient populations with a smaller effect. That causes a positive correlation between θ_i and s_i^2, and a negative correlation between θ_i and w_i, leading to negative bias in the estimate of θ (Chapter 13). Correlation due to statistical reasons arises if the outcome is non-normal and the mean and the variance of the distribution of the outcome are functions of the same parameter. Then y_i and s_i^2 will in general be correlated. For instance, if θ_i is a binomial proportion estimated by y_i, one has $s_i^2 = y_i(1-y_i)/n_i$, which is positively or negatively correlated with y_i depending on whether $\theta_i < 0.5$ or > 0.5. See Chapter 5 for more examples and a more extensive discussion of this. In practice, this kind of correlation is almost always negligible, except with rare events (Stijnen et al., 2010; Chapter 5).

4.3.4 Examples

Throughout this chapter we use two examples to illustrate the methods. The first example is the meta-analysis of Colditz et al. (1994), who report data from 13 clinical trials on the efficacy of the BCG vaccine in the prevention of tuberculosis (TB). The data are reproduced in Table 4.1.

We choose the log odds ratio as the effect measure θ_i. The 95% confidence intervals for the odds ratios of the individual studies are depicted in a forest plot (Lewis and Clarke, 2001), which gives a graphical description of the meta-analysis data (Figure 4.1). For each study, the estimated effect measure y_i and the corresponding 95% confidence interval

TABLE 4.1
BCG Data

	Vaccinated		Non-vaccinated		ln(OR)	Var(ln(OR))
Trial	TB	No TB	TB	No TB	y_i	s_i^2
1	4	119	11	128	−0.9387	0.3571
2	6	300	29	274	−1.6662	0.2081
3	3	228	11	209	−1.3863	0.4334
4	62	13536	248	12619	−1.4564	0.0203
5	33	5036	47	5761	−0.2191	0.0520
6	180	1361	372	1079	−0.9581	0.0099
7	8	2537	10	619	−1.6338	0.2270
8	505	87886	499	87892	0.0120	0.0040
9	29	7470	45	7232	−0.4717	0.0570
10	17	1699	65	1600	−1.4012	0.0754
11	186	50448	141	27197	−0.3408	0.0125
12	5	2493	3	2338	0.4466	0.5342
13	27	16886	29	17825	−0.0173	0.0716

Data from 13 clinical trials comparing a vaccinated with a non-vaccinated group, showing numbers with and without incident tuberculosis (TB) in each group. Log odds ratios (ln(OR) and corresponding variances (var(ln(OR)) are calculated using formulas given in Table 3.4.

$y_i \pm 1.96 \cdot s_i$ is given. The forest plot gives a nice overview of the available data and the accuracy of the estimates of the different studies and gives a quick impression of the variability between the studies. In a forest plot, the eye tends to be drawn to the studies with the biggest confidence interval, but in fact these are the least important. The studies with the smaller confidence intervals are most important since these are the most precise studies. To emphasize the more precise studies, the point estimates are denoted by squares with sizes proportional to the inverse variances. Notice that the odds ratios are graphed on a logarithmic scale.

The last line in Figure 4.1 gives the results of fitting the CE model (4.3) using the inverse variances as weights. The calculations were of course done on the log scale. We find $\hat{\theta}_{CE} = -0.436$ with standard error $se(\theta_{CE}) = 0.042$, $Z_{CE} = -10.32$ ($p < 0.0001$), and 95% CI $-0.436 \pm 1.96 \times 0.042 = (-0.519, -0.353)$. Exponentiation gives the estimate and the confidence interval for the common odds ratio given in Figure 4.1. However, the horizontal separation of some of the 95% confidence intervals makes it clear that there is huge variability across trials and that a CE estimate is unrealistic in this example. The Q_{CE}-statistic (4.4) is 163 with df = 12, so $p < 0.0001$ for the test of H_0: $\theta_i = \theta$ for all i.

As the second example in this chapter, with a smaller number of studies, we consider a meta-analysis of Lopes da Silva et al. (2014), comparing mean plasma iron levels between Alzheimer's disease (AD) patients and controls. The data are shown in Table 4.2.

Figure 4.2 gives a forest plot. From this it is clear that the between-studies heterogeneity in this example is much less than in the BCG example. In this case, Cochran's Q-test is not statistically significant ($Q_{CE} = 7.73$ on df = 4; $p = 0.10$). Thus, there is no strong statistical indication of heterogeneity. Nevertheless, given the low power of Cochran's Q-test, the common-effect assumption remains questionable.

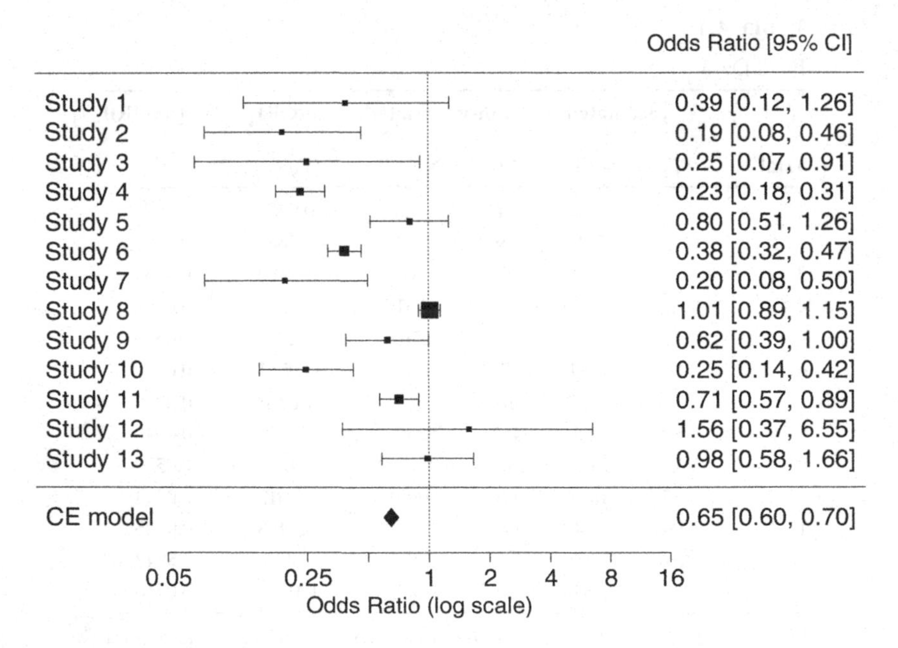

FIGURE 4.1
Forest plot and results of the common-effect model (inverse variances) for the BCG data of Table 4.1 made in the R-package `metafor` (Viechtbauer, 2010).

TABLE 4.2
Iron Data

	Controls			Alzheimer's disease			Mean difference	Variance
Study	Number	Mean	SD	Number	Mean	SD	y_i	s_i^2
1	26	114	25	20	100	39	−14	100.08
2	20	89	32	26	90	33	1	93.08
3	421[a]	63	30	31	56	22	−7	17.75
4	28	101	31	26	114	35	13	81.43
5	50	81	31	50	67	23	−14	29.80

Iron levels (μg/dL) in Alzheimer's disease (AD) patients compared with controls. Mean difference (mean of AD group minus mean of control group) and its variance are calculated as indicated in Table 3.4.

[a] This is not an error. This observational study recruited many more controls than cases.

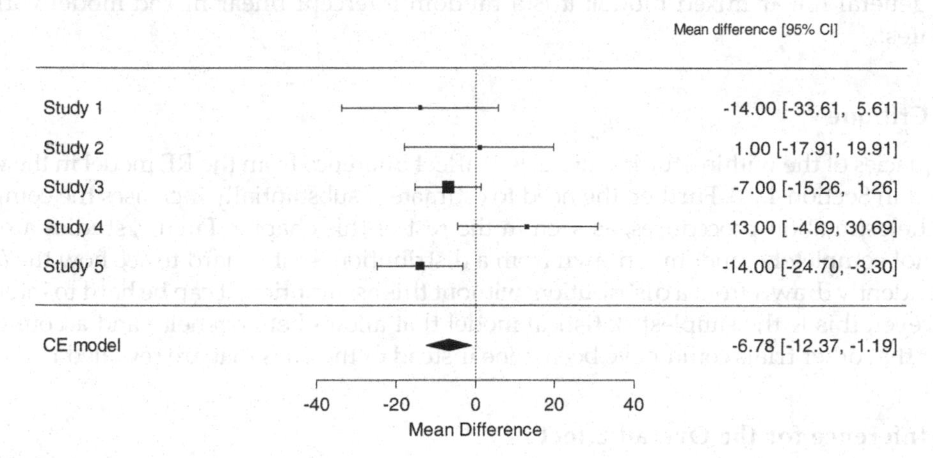

FIGURE 4.2
Forest plot and results of the common-effect model (inverse variances) for the iron data of Table 4.2.

4.4 The Random-Effects Model

4.4.1 The Model

In a random-effects approach, one allows that the θ_i may be different. More specifically, it is assumed that the θ_i are randomly drawn from some unknown distribution. The typical assumption is that this distribution is normal:

$$\theta_i \sim N\left(\theta, \tau^2\right). \tag{4.5}$$

In addition, the within-studies model (4.1) is still assumed to hold. Together, model (4.1) and the between-studies model (4.5) define the random-effects (RE) meta-analysis model as presented by Hedges and Olkin (1985) and DerSimonian and Laird (1986). The choice between the CE and the RE model should not be based on (the significance of) the Q_{CE}-statistic (although this is still unfortunately common in practice) since this leads to choosing the CE model too often, and hence to too small p-values and too narrow confidence intervals for θ (e.g., Brockwell and Gordon, 2001).

Both parameters in (4.5) are unknown and have to be estimated. Usually the main parameter of interest is θ, the mean of the distribution from which the study-specific effects θ_i are drawn. The standard deviation τ is the natural parameter to describe the heterogeneity of the true effects across studies. Note that if $\tau = 0$ there is no heterogeneity and the model reduces to the CE model.

Alternative formulations of the RE model are

$$y_i \sim N\left(\theta, \tau^2 + s_i^2\right) \tag{4.6}$$

(the "marginal form") or

$$y_i = \theta + \delta_i + \varepsilon_i \text{ with } \delta_i \sim N\left(0, \tau^2\right) \text{and } \varepsilon_i \sim N\left(0, s_i^2\right) \tag{4.7}$$

(the "hierarchical form" or normal-normal model). The model is just a particular case of the general linear mixed model: it is a random intercept linear mixed model without covariates.

4.4.2 Critique

Inadequacies of the within-studies model (4.1) affect inference from the RE model in the ways outlined in Section 4.3.3. Further, the need to estimate τ^2 substantially increases the complexity of the estimation procedures, as seen in the rest of this chapter. Finally, studies are certainly not completely randomly drawn from a distribution, so it is hard to see how the θ_i are independently drawn from a distribution; without this assumption, it can be hard to interpret θ. However, this is the simplest statistical model that allows heterogeneity and accounts for the fact that other trials could have been done instead of the ones that are revealed to us.

4.4.3 Inference for the Overall Effect

As the RE model (4.6) or (4.7) is a particular case of the nowadays standard linear mixed model (LMM), standard (restricted) maximum likelihood ((RE)ML) methods for the LMM may be used for inference on θ and τ^2. However, at the time the RE model was proposed, the LMM was much less developed than in the last two decades, and no software was available in the general statistical packages. Moreover, the RE model is a special kind of LMM in that the residual variances are assumed known, and also meta-analysis datasets are typically very small, in particular for the estimation of τ^2, making good performance of the asymptotic (RE)ML not guaranteed. Therefore, a variety of inference methods was developed specially for the RE model since the early 1990s. In the rest of this chapter, we discuss the most important of these methods, more or less following historical lines.

If τ were known (as well as the s_i^2), inference for θ would be straightforward along the lines of Section 4.3.1. Given τ^2, the maximum likelihood estimate of θ (also the minimum variance unbiased estimate) is

$$\hat{\theta}\left(\tau^2\right) = \frac{\sum_i w_{i,RE}\, y_i}{\sum_i w_{i,RE}} \text{ with } w_{i,RE} = \frac{1}{\tau^2 + s_i^2} \text{ and } se\left(\hat{\theta}\left(\tau^2\right)\right) = \frac{1}{\sqrt{\sum_i w_{i,RE}}}, \tag{4.8}$$

and the test and confidence interval for θ are based on the normal distribution. Note that the standard error of $\hat{\theta}(\tau^2)$ from the RE model is always larger than from the CE model, unless τ^2 is zero. That leads to wider confidence intervals and often to less significant p-values, but it is the price to pay for more realistic and broader inference. From (4.8), it is also seen that the studies are weighted more equally compared with the CE model. The larger the between-studies variance and the larger the sample sizes, the more the overall estimate will resemble the unweighted mean of the study-specific effect estimates.

In practice, τ^2 is unknown, and an estimate needs to be substituted into (4.8). Most if not all (frequentist) methods for estimating θ are weighted means of the study-specific estimates y_i, usually of the form (4.8) with some plug-in estimate for τ^2. Some authors have proposed to use fixed, non-estimated, weights, for instance, equal weights or weights proportional to the sample size, and argue that these may be more robust to deviations from the model than estimated weights (Shuster, 2010). However, under the model, weights of the form (4.8) are most efficient.

The literature contains many different estimators of τ^2, the most common method being the one due to DerSimonian and Laird (1986). We postpone discussing these until Section 4.5. Until then, we assume that an estimate $\hat{\tau}$ has been obtained, and we focus on estimation of θ. In practice, for different estimates of τ^2, the resulting estimates $\hat{\theta}(\hat{\tau}^2)$ are mostly very similar. This is not surprising, since each y_i is an unbiased estimate of θ and every weighted mean of the y_i as well. However, different estimates for τ^2 can greatly influence the standard error, as is clear from (4.8). In the sequel, we will often use $\hat{\theta}$ as shorthand notation for $\hat{\theta}(\hat{\tau}^2)$.

4.4.3.1 Confidence Interval and Test

The straightforward method as implemented in most software is to calculate $se(\hat{\theta}) = \frac{1}{\sqrt{\sum_i 1/(\hat{\tau}^2 + s_i^2)}}$ as the standard error, $\hat{\theta} \pm 1.96 \cdot se(\hat{\theta})$ as the approximate 95% (Wald) confidence interval, and $Z = \hat{\theta}/se(\hat{\theta})$ as the standard (Wald) approximate normal test statistic for H_0: $\theta = 0$. This is based on asymptotic arguments and will only work well for a large number of studies, while most meta-analyses tend to have few studies. By using the standard normal distribution, the method does not take into account the uncertainty in the estimate of the standard error $se(\hat{\theta})$, which arises from uncertainty in the estimate $\hat{\tau}$. Therefore, some authors recommend using a *t*-distribution, for instance, with $I-1$ df (Follmann and Proschan, 1999; Higgins et al., 2009). If a linear mixed model program is used, one can chose from a variety of methods to determine the number of degrees of freedom, such as the Satterthwaite and Kenward–Roger methods.

Another shortcoming of the above method is that the standard error is estimated with bias, even if $\hat{\tau}^2$ is unbiased, since the standard error depends on $\hat{\tau}^2$ in a non-linear way. After an example, we will describe two methods to address these shortcomings.

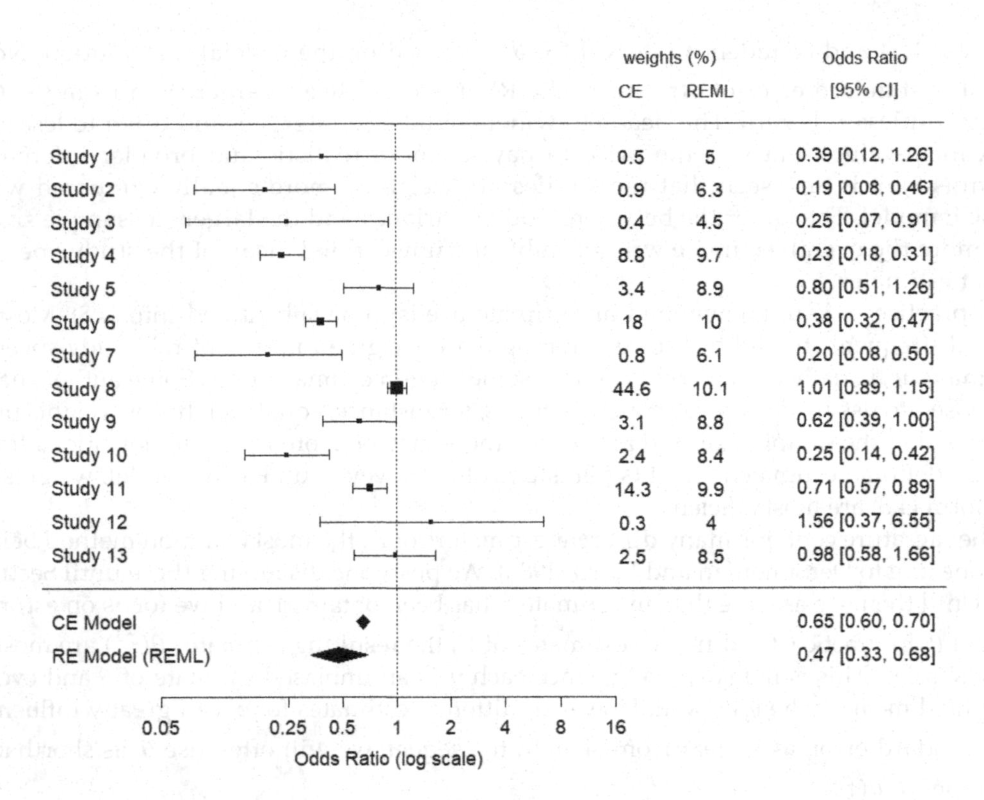

FIGURE 4.3
CE and RE models fitted to the BCG data of Table 4.1.

4.4.3.2 Example

We fit the RE model to the BCG data with the log odds ratio as effect measure. Estimating τ^2 by the REML method described below gives $\hat{\tau}^2 = 0.338$, $\hat{\theta} = -0.745$ with $se(\hat{\theta}) = 0.186$ and 95% CI equal to (−1.11, −0.38). The comparison between the RE and CE models is illustrated in Figure 4.3.

Notice that the RE confidence interval is much wider than the CE confidence interval. Also, the weights are considerably different. The CE results are dominated by study 8, with a weight of almost 45%, while five studies have a weight less than 1%. In the RE model, study 8 is down-weighted to 10%, and the small studies get much more weight. Since in this meta-analysis the small studies tend to show a larger treatment effect, the estimated overall effect of the RE model in this dataset is considerably larger.

4.4.3.3 The Hartung–Knapp–Sidik–Jonkman Correction

Hartung and Knapp (2001) and Sidik and Jonkman (2002) proposed the same improved confidence interval and test for θ; we follow the development of Sidik and Jonkman (2002). If w_i are the true weights (using the true unknown τ^2) from (4.8), one can show that under the RE model (4.6) or (4.7)

$$\frac{\hat{\theta} - \theta}{1/\sqrt{\Sigma w_i}} \sim N(0,1) \text{ and } \Sigma w_i (y_i - \hat{\theta})^2 \sim \chi^2_{I-1} \tag{4.9}$$

and that these two statistics are independent. Hence

$$\frac{\left(\hat{\theta}-\theta\right)}{\sqrt{\Sigma w_i(y_i-\hat{\theta})^2/\left((I-1)\Sigma w_i\right)}} \sim t_{I-1}. \tag{4.10}$$

Notice that the square of the denominator in (4.10) is an unbiased estimate of the variance of $\hat{\theta}$ if the weights w_i are known. Plugging in the estimated weights (denoted by $\hat{w}_i = 1/(\hat{\tau}^2 + s_i^2)$) leads to the Hartung–Knapp–Sidik–Jonkman (HKSJ) standard error

$$se_{HKSJ}\left(\hat{\theta}\right) = \sqrt{\frac{\Sigma \hat{w}_i\left(y_i-\hat{\theta}\right)^2}{(I-1)\Sigma \hat{w}_i}}. \tag{4.11}$$

The test and confidence interval for θ use the t-distribution with $I-1$ degrees of freedom in combination with this standard error. Thus, the HKSJ method addresses both the bias and the uncertainty in the estimation of the standard error. Van Aert and Jackson (2019) show that the HKSJ method corresponds to a LMM with within-study variances proportional to, rather than equal to, the s_i^2. In simulation studies, see, for instance, IntHout et al. (2014) or Langan et al. (2018), the HKSJ method works much better than the approximate standard Wald type method given above, with the smaller numbers of studies encountered in practice. Published meta-analysis results can be easily converted into HKSJ results. IntHout et al. (2014), in an empirical study using 689 meta-analyses from the Cochrane Database of Systematic Reviews, showed that 25% of the significant findings for the DL method were non-significant with the HKSJ method. Note that the HKSJ standard error and confidence interval can be used with any estimator $\hat{\tau}^2$ inserted into the weights and are easily calculated. In extensive simulation studies, Langan et al. (2018) observed that results from the HKSJ method are remarkably robust against the choice of the between-studies variance estimation method.

As an example, we apply the HKSJ correction in our two data examples, using different methods (to be introduced in Section 4.5) for estimating the between-studies variance. The results are shown in Table 4.3.

It is seen that the HKSJ correction affects the standard error in the BCG meta-analysis less than in the iron data example, probably due to the fact that the BCG example has many more studies. In these examples, the largest effect from the HKSJ correction comes from using the t_{I-1} distribution instead of the normal distribution for the confidence interval and the p-value.

4.4.3.4 The Profile Likelihood Method

Another method that takes into account the uncertainty in the estimate of τ^2 is based on the profile likelihood (PL) (Hardy and Thompson, 1996). The log-likelihood of model (4.6) or (4.7) is, apart from a constant term,

$$\ln L\left(\theta,\tau^2\right) = -\frac{1}{2}\Sigma \ln(s_i^2+\tau^2) - \frac{1}{2}\Sigma \frac{(y_i-\theta)^2}{\left(s_i^2+\tau^2\right)}. \tag{4.12}$$

TABLE 4.3
The HKSJ Correction Applied to the BCG and Iron Data Using Various Estimation Methods for τ^2

		BCG data: log OR					Iron data: mean difference				
Method for $\hat{\tau}^2$	**HKSJ**	$\hat{\theta}$	$\hat{\tau}$	**se**	**95% CI**	**p**	$\hat{\theta}$	$\hat{\tau}$	**se**	**95% CI**	**p**
DL	No	−0.747	0.61	0.192	−1.12, −0.37	0.0001	−5.57	6.62	4.38	−14.16, 3.02	0.20
	Yes			0.187	−1.16, −0.34	0.0018			4.68	−18.56, 7.42	0.30
REML	No	−0.745	0.58	0.186	−1.11, −0.38	< .0001	−5.52	6.87	4.47	−14.28, 3.24	0.22
	Yes			0.187	−1.15, −0.34	0.0018			4.70	−18.57, 7.53	0.31
Paule–Mandel	No	−0.746	0.58	0.187	−1.11, −0.38	< .0001	−5.37	7.70	4.76	−14.70, 3.97	0.26
	Yes			0.187	−1.15, −0.34	0.0018			4.76	−18.58, 7.85	0.33

The profile log-likelihood for θ is the log-likelihood maximized over τ^2 given θ:

$$pl(\theta) = \max_{\tau^2} \ln L(\theta, \tau^2). \tag{4.13}$$

To test H_0: $\theta = 0$ we use

$$-2\left(pl(0) - pl\left(\hat{\theta}_{ML}\right)\right) \sim \chi_1^2, \tag{4.14}$$

where $\hat{\theta}_{ML}$ is the maximum likelihood estimator maximizing (4.12). A (1-α)100% confidence interval is given by all θ that satisfy

$$pl(\theta) > pl\left(\hat{\theta}_{ML}\right) - \chi_{1,1-\alpha}^2 / 2. \tag{4.15}$$

An iterative numerical procedure is needed to calculate the test and confidence interval. In a simulation study, Brockwell and Gordon (2001) compared the PL method with the standard DL method and the Wald test based on the ordinary likelihood. They concluded that the PL method performed best, with coverage slightly below the nominal level. The resulting PL 95% confidence intervals for our examples are (–1.13, –0.37) for the BCG data and (–14.59, 4.82) for the iron data. Comparing these with the confidence intervals of Table 4.3, we see that they are less wide than the HKSJ corrected confidence intervals, in particular for the iron data.

4.4.3.5 Method Comparisons

Veroniki et al. (2016) reviewed the many published comparisons of methods to estimate the between-studies variance. They conclude that comprehensive simulation studies where all methods are compared under the same scenarios are lacking. They tentatively suggest the REML estimator for continuous outcomes and the Paule–Mandel estimator for both dichotomous and continuous outcomes. Recently, Langan et al. (2018), in an extensive simulation study, compared eight different methods to estimate τ^2. Their results show that the DL method has negative bias in meta-analyses with large heterogeneity and small studies, while the Paule–Mandel method has considerable positive bias with large differences in study size, and they recommend the REML method over the other methods. These biases appear to arise from failures of the within-studies model. Different estimates of τ^2 do not have much impact on estimates of θ unless there is funnel plot asymmetry, but they do have impact on the standard error and hence the CI width.

Table 4.3 gives the results of the leading methods for estimation of the between-studies standard deviation for the two data examples. For the BCG data, the differences are relatively small, possibly because this meta-analysis is relatively large. In contrast, the iron data which include only five studies show considerable differences between the different methods.

4.4.4 Describing Heterogeneity

The key results of a meta-analysis that should be reported in practice should at least comprise the estimate of the overall effect together with a confidence interval and the estimate of the between studies standard deviation. In practice, the importance of reporting the estimate of τ is under-appreciated. However, without that the implications for practice of the results of the meta-analysis cannot be satisfactorily assessed (e.g., Higgins et al., 2009). We describe two other ways to describe heterogeneity in the RE model: the I^2 measure and the prediction interval.

4.4.4.1 I^2 Measure

A popular method to characterize the between-studies variability is the I^2 measure proposed by Higgins and Thompson (2002). It is defined as

$$I^2 = \frac{Q_{CE} - (I-1)}{Q_{CE}}. \tag{4.16}$$

The I on the right-hand side denotes the number of studies and Q_{CE} is Cochran's Q given by (4.4). I^2 measures the relative excess of Q_{CE} over its expected value $I-1$ under $\tau^2=0$. If the right-hand side of (4.16) is negative, I^2 is set to zero. Alternatively, I^2 can be written as

$$I^2 = \frac{\hat{\tau}^2_{DL}}{\hat{\tau}^2_{DL} + s^2} \text{ with } s^2 = \frac{(I-1)\Sigma w_{i,CE}}{\left(\Sigma w_{i,CE}\right)^2 - \Sigma w^2_{i,CE}}. \tag{4.17}$$

Since s^2 can be considered as a typical value for the within-study variances (Higgins and Thompson, 2002), I^2 can be interpreted as the proportion of the total (within+between studies) variance due to between-studies variability. Thus, I^2 is not a direct measure of the between-studies heterogeneity. Higgins and Thompson describe it as measuring the *impact* of the heterogeneity.

A confidence interval for I^2 can be obtained by calculating first a confidence interval for τ^2 and then substituting the endpoints of that confidence interval into (4.17). Higgins and Thompson (2002) also provide easy to calculate confidence intervals only based on the value of Q_{CE} and the number of studies I. Note that these intervals do not take account of the uncertainty in the standard errors s_i, since these are still assumed fixed and known.

An advantage of I^2 is that it is a unitless measure, independent of the metric of the effect sizes. This is in contrast to the between-studies standard deviation τ, which is expressed in the metric of the effect sizes. However, at the same time, this means that I^2 only indirectly tells us by how much the true effect sizes differ across studies. Though the interpretation of I^2 by itself is very clear, it is much less straightforward to assess what it means for the interpretation of the results of a meta-analysis. *The Cochrane Handbook for Systematic Reviews of Interventions* (Higgins et al., 2019) gives a rough guide to assess the size of I^2.

Published applications of meta-analysis often omit the estimate of τ and instead report only I^2. However, as noted above, this is a relative measure and cannot be used to describe how the true effect can vary from one study to another. Much more valuable in that respect is the prediction interval, which is discussed next.

4.4.4.2 Prediction Interval

An undervalued aid in the interpretation of the results of a meta-analysis is the prediction interval, which explicitly uses the estimate of τ^2 (Van Houwelingen et al., 2002; Higgins et al., 2009). Suppose a physician has read the results of a meta-analysis of a new treatment and is considering introduction of the new treatment in her center. The question is then: What would be the (true) treatment effect if she conducted a study in her center? A

$(1-\alpha)100\%$ prediction interval, constructed such that the probability is approximately 95% that the true effect θ_{new} in a new study lies in the prediction interval, answers this question.

If τ^2 were known, $\hat{\theta} \sim N(\theta, se(\hat{\theta})^2)$ with $se(\hat{\theta})$ given by (4.8). It is assumed that the new patient population is comparable with and independent of the patient populations covered by the meta-analysis: more precisely, it is assumed that θ_{new} is independently drawn from the same distribution, $\theta_{new} \sim N(\theta, \tau^2)$. Therefore, $\theta_{new} - \hat{\theta} \sim N(0, \tau^2 + se(\hat{\theta})^2)$ which implies that $(\theta_{new} - \hat{\theta})/\sqrt{\tau^2 + se(\hat{\theta})^2}$ has a standard normal distribution. In practice, τ^2 is estimated and, to account for that, the standard normal distribution is approximated by a t-distribution with the approximate, but somewhat arbitrary, choice of $(I$-$2)$ degrees of freedom:

$$\frac{\theta_{new} - \hat{\theta}}{\sqrt{\hat{\tau}^2 + se(\hat{\theta})^2}} \sim t_{I-2}. \tag{4.18}$$

Therefore, an approximate $(1-\alpha)100\%$ prediction interval for θ_{new} is given by

$$\hat{\theta} \pm t_{\alpha,I-2}\sqrt{\hat{\tau}^2 + se(\hat{\theta})^2} \tag{4.19}$$

where $t_{\alpha/2,I-2}$ is the $(1-\alpha/2)$ quantile of the t-distribution with I–2 degrees of freedom.

For the BCG example (using the REML estimate of τ^2 from Section 4.5), the mean log odds ratio was −0.745 with standard error 0.186, and the estimate of $\hat{\tau}^2_{DL}$ is 0.336. The 95% prediction interval for the true effect in a new study as calculated from (4.19) is therefore (−2.09, 0.60) for the log odds ratio and (0.12, 1.82) for the odds ratio: see Figure 4.4. Note that

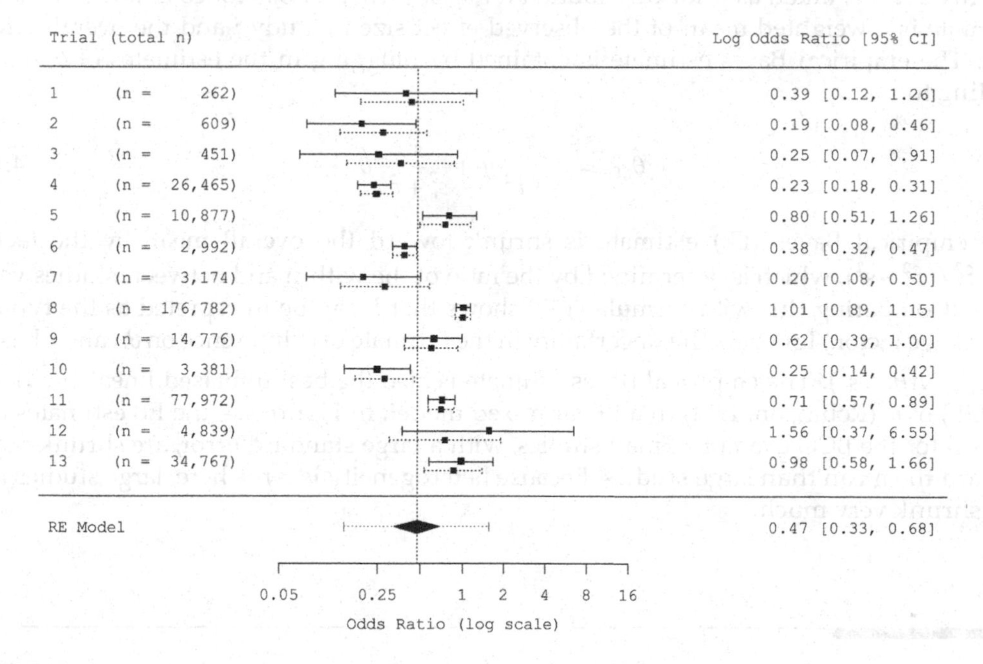

FIGURE 4.4
Forest plot for BCG data with empirical Bayes estimates and confidence intervals (dotted lines) for the effects in the individual studies (Section 4.4.5). The dotted line accompanying the overall effect estimate is the 95% prediction interval for the true effect of a new study (Section 4.4.4).

this interval contains values larger than 1, meaning an adverse effect of the BCG vaccination. Based on (4.18), the estimated probability of a new study having an adverse effect is 0.12. If only odds ratios of 0.80 or smaller would be considered worthwhile, the probability of a vaccination effect that is not worthwhile in a new study is 0.21. This shows that the overall effect estimate and corresponding confidence interval, even if clinically relevant and statistically very significant, do not necessarily provide enough information to confidently make a decision to introduce the treatment in a new setting. If the prediction interval is large, or if it overlaps considerably with the range of not clinically relevant treatment effect values, then a well-informed decision requires more insight into the factors causing the variability in the treatment effects.

4.4.5 Empirical Bayes Estimates of Study-Specific Effects

Under the random-effects model, the true effect θ_i in an individual trial is assumed to be randomly drawn from a normal distribution with mean θ and variance τ^2. This can be employed to improve the estimate of an individual θ_i; this might be useful if the physician considering introducing the new treatment in her center is working in a setting very similar to that of study i. If θ and τ^2 were known, the estimator for θ_i that minimizes the mean squared error is

$$\frac{\tau^2}{s_i^2+\tau^2}y_i+\frac{s_i^2}{s_i^2+\tau^2}\theta.$$

This is the Bayesian estimator of θ_i under the model $y_i \approx N(\theta_i, s_i^2)$ (with known s_i^2) when$N(\theta,\tau^2)$ is taken as prior distribution for θ_i (see Chapter 6). Notice that the Bayesian estimate is a weighted mean of the observed effect size in study i and the overall effect size. The empirical Bayes estimate is obtained by plugging in the estimates of θ and τ^2 leading to

$$\hat{\theta}_{i,EB}=\frac{\hat{\tau}^2}{s_i^2+\hat{\tau}^2}y_i+\frac{s_i^2}{s_i^2+\hat{\tau}^2}\hat{\theta}. \tag{4.20}$$

The empirical Bayes (EB) estimate is shrunk toward the overall mean by the factor $B_i=\hat{\tau}^2/(\hat{\tau}^2+s_i^2)$, which is determined by the ratio of the within and between-studies variance. (Comparing this with formula (4.17) shows that I^2 can be interpreted as the typical shrinkage factor.) Ignoring the uncertainty in the estimate of θ, inference on θ_i can be based on $\hat{\theta}_i \sim N(\theta_i, B_i s_i^2)$. The empirical Bayes estimate is also the best unbiased linear predictor (BLUP) of θ_i (Robinson, 1991) in a linear mixed model. In Figure 4.4, the EB estimates are shown for the BCG example. Small studies, with a large standard error, are shrunk more toward the mean than large studies. Because heterogeneity is large here, large studies are not shrunk very much.

4.5 Estimating the Heterogeneity Variance

We now turn to estimation of the heterogeneity variance for use in the methods above. The most commonly used estimator in practice is due to DerSimonian and Laird (DL) (1986).

Numerous authors, including DerSimonian and Kacker (2007), Hardy and Thompson (1996), Hartung and Knapp (2001), IntHout et al. (2014), and Cornell et al. (2014), have pointed out its shortcomings. Nevertheless, it is still implemented as the default method in much meta-analysis software and is used in the large majority of applications. We start by introducing in Section 4.5.1 the DL estimator because it has formed the basis of many later developments. In Section 4.5.2, we present other methods for estimation and further inference on τ^2.

4.5.1 The Method of DerSimonian and Laird

In a seminal paper, DerSimonian and Laird (1986) proposed a method-of-moments estimate derived by equating the observed Cochran's Q_{CE} statistic (4.4) to its expected value. First Q_{CE} is written as

$$\begin{aligned} Q_{CE} &= \sum w_{i,CE}\left[(y_i-\theta)-\left(\hat{\theta}_{CE}-\theta\right)\right]^2 \\ &= \sum w_{i,CE}(y_i-\theta)^2-(\Sigma w_{i,CE})(\hat{\theta}_{CE}-\theta)^2. \end{aligned} \tag{4.21}$$

Thus, its expected value is

$$\begin{aligned} E(Q_{CE}) &= \sum w_{i,CE}\,\mathrm{var}(y_i)-\left(\sum w_{i,CE}\right)\mathrm{var}(\hat{\theta}_{CE}) \\ &= \sum w_{i,CE}(w_{i,CE}^{-1}+\tau^2)-\left(\sum w_{i,CE}\right)\frac{\sum w_{i,CE}^2(w_{i,CE}^{-1}+\tau^2)}{\left(\sum w_{i,CE}\right)^2} \\ &= I-1+\left(\sum w_{i,CE}-\frac{\sum w_{i,CE}^2}{\sum w_{i,CE}}\right)\tau^2. \end{aligned} \tag{4.22}$$

Note that $\mathrm{var}(\hat{\theta}_{CE})$ is the variance of $\hat{\theta}_{CE}$ under the RE model for θ. Equating the right-hand side of (4.22) to Q_{CE}, solving for τ^2, and setting the result to zero if it is negative yields

$$\hat{\tau}_{DL}^2 = \max\left\{0, \frac{\sum_i w_{i,CE}(y_i-\hat{\theta}_{CE})^2-(I-1)}{\sum_i w_{i,CE}-\sum_i w_{i,CE}^2/\sum_i w_{i,CE}}\right\}. \tag{4.23}$$

Thus, the standard method of DerSimonian and Laird (DL) as implemented in most meta-analysis software is

$$\hat{\theta}_{DL} = \frac{\sum_i w_{i,DL}y_i}{\sum_i w_{i,DL}} \text{ with } w_{i,DL} = \frac{1}{\hat{\tau}_{DL}^2+s_i^2} \text{ and } se\left(\hat{\theta}_{DL}\right) = \frac{1}{\sqrt{\sum_i w_{i,DL}}}, \tag{4.24}$$

with (Wald) confidence interval and p-value for the overall effect θ based on the standard normal distribution. This method works reasonably well if the number of studies is relatively large, typically 20 or more (Jackson et al., 2010). However, most meta-analyses in practice are smaller (in healthcare, not necessarily in social science). The DL method can behave quite badly for smaller numbers of studies, with inflated type I error probabilities and coverage probabilities that are much too small (see, e.g., Hartung and Knapp, 2001; Sidik and Jonkman, 2002; Jackson et al., 2010; IntHout et al., 2014; Langan et al., 2018). One reason for its poor performance in smaller meta-analyses is that the standard error in (4.24) does not take account of the uncertainty in the weights due to the uncertainty in the estimate of τ^2. Another reason is that the variance estimate (4.23) is often seriously biased, as for instance shown by Langan et al. (2018) in a wide range of simulation scenarios. Nevertheless, the DL method is popular because it can be computed non-iteratively by a simple formula.

Notice that the normality (4.5) of the random-effects distribution is not used in the estimation of τ^2 and therefore also in that of θ, implying robustness against this assumption.

4.5.2 Other Methods for Inference on the Between-Studies Variance

After the initial proposal of DerSimonian and Laird, many alternative methods have been proposed in the literature to estimate τ^2. Veroniki et al. (2016) give a comprehensive overview. They discuss 16 different methods including eight method-of-moment estimators, three estimators based on the likelihood, a model error variance estimator, several Bayesian estimators, and a non-parametric bootstrap method. Recently, Langan et al. (2018) also compared eight different heterogeneity estimators in an extensive simulation study. Here we discuss method-of-moments and likelihood-based estimators.

4.5.2.1 Method-of-Moment Estimators

The DL method (4.23) is an example of a method-of-moments estimator. General method-of-moment estimators can be derived by looking at estimators of θ of type $\hat{\theta}_a = \sum a_i y_i / \sum a_i$, for an arbitrary set of positive weights a_i, and the corresponding Q-statistic $Q_a = \sum a_i (y_i - \hat{\theta}_a)^2$. By straightforward calculations (cf. (4.21) and (4.22)) it can be shown that the expected value of Q_a under the RE model is equal to

$$E(Q_a) = \sum_i a_i \left(\tau^2 + s_i^2\right) - \sum_i a_i^2 \left(\tau^2 + s_i^2\right) / \sum_i a_i. \tag{4.25}$$

Equating (4.25) to the observed value of Q_a leads to the general method-of-moments estimate as introduced by Kacker (2004)

$$\hat{\tau}_a^2 = \max\left\{0, \frac{\left(\sum_i a_i (y_i - \hat{\theta}_a)\right)^2 - \left(\sum_i a_i s_i^2 - \sum_i a_i^2 s_i^2 / \sum_i a_i\right)}{\left(\sum_i a_i - \sum_i a_i^2 / \sum_i a_i\right)}\right\}. \tag{4.26}$$

Choosing $a_i = 1/s_i^2$ yields the DL estimate (4.23). Other choices might be $a_i = 1/I$ (Hedges and Olkin, 1985; Cochran, 1954) yielding estimator $\hat{\tau}_{HO}^2$; $a_i = 1/n_i$ (where n_i is the sample size of study i); the two-step DL method with $a_i = 1/\left(\hat{\tau}_{DL}^2 + s_i^2\right)$ (DerSimonian and Kacker, 2007);

the two-step Hedges and Olkin (HO) method with $a_i = 1/\left(\hat{\tau}_{HO}^2 + s_i^2\right)$ (DerSimonian and Kacker, 2007); and $a_i = 1/(s_i^2 + x)^p$ for different choices of x and p (Jackson, 2013).

An estimate of τ^2 recommended by many authors (e.g., Veroniki et al., 2016; DerSimonian and Kacker, 2007) is due to Paule and Mandel (1982). The idea of their method is to use the optimal weights $a_i = 1/(\tau^2 + s_i^2)$, which are unknown. For this special choice of the weights, we use the notation $Q(\tau^2)$. From (4.25) it is seen that the expected value of $Q(\tau^2)$ is equal to $(I-1)$. By equating $Q(\tau^2)$ to $(I-1)$ we get the Paule–Mandel estimating equation

$$Q(\tau^2) - (I-1) = \sum_i \left(\frac{y_i - \hat{\theta}(\tau^2)}{\tau^2 + s_i^2} \right)^2 - (I-1) = 0, \tag{4.27}$$

which can be shown to have at most one solution for $\tau^2 \geq 0$. If there is no solution, the estimate is set to zero. In general, there is no analytical solution, but a numerical solution can be obtained by simple iterative algorithms, for instance, the one given by DerSimonian and Kacker (2007), which is easily implemented in a spreadsheet.

4.5.2.2 Likelihood-Based Estimators

Likelihood-based estimation methods are attractive because of their optimal large sample properties. They explicitly use the normal distribution assumption for the random effects, and therefore might in principle lack the robustness of the method-of-moment estimators that do not need this assumption.

The log-likelihood of model (4.6) or (4.7) is given in (4.12). Van Houwelingen et al. (2002) and Stijnen et al. (2010) call this the approximate likelihood, since the within-study variances are treated as known; the exact likelihood (Chapter 5) also involves the exact likelihoods of y_i given θ_i at the study level.

Setting the partial derivatives to zero leads to an estimating equation that can be written as (Hardy and Thompson, 1996)

$$\tau^2 = \frac{\sum \left[\left\{ y_i - \hat{\theta}(\tau^2) \right\}^2 - s_i^2 \right] / (\tau^2 + s_i^2)}{\sum 1/(\tau^2 + s_i^2)^2}. \tag{4.28}$$

Simple iterative numerical methods such as Newton–Raphson can be used to compute the ML estimators $\hat{\tau}^2_{ML}$ and $\hat{\theta}_{ML} = \hat{\theta}\left(\hat{\tau}^2_{ML}\right)$.

The ML estimator of τ^2 is biased downwards since it does not take into account the loss of the degrees of freedom due to estimation of θ. A better alternative therefore is the restricted maximum likelihood (REML) method, which yields nearly unbiased estimates. The REML log-likelihood is

$$\ln L(\theta, \tau^2) = -\frac{1}{2} \sum \ln(s_i^2 + \tau^2) - \frac{1}{2} \sum \frac{(y_i - \theta)^2}{s_i^2 + \tau^2} - \frac{1}{2} \ln\left(\sum \frac{1}{s_i^2 + \tau^2} \right) \tag{4.29}$$

and the REML estimates $\hat{\tau}^2_{REML}$ and $\hat{\theta}_{REML}$ can be calculated with iterative numerical methods such as Newton–Raphson.

4.5.2.3 Methods for Calculation of Confidence Intervals for τ^2

First, we remark that the asymptotic normality needed for a Wald type CI, such as

$$\hat{\tau}^2_{ML} \pm 1.96 \cdot se\left(\hat{\tau}^2_{ML}\right), \tag{4.30}$$

is only satisfactory for impractically high sample sizes. Moreover, the method does not account for the standard error in (4.30) being estimated. Veroniki et al. (2016) discuss seven alternative methods to construct confidence intervals (CIs) for the between-studies variance. Here we present the three that seem to be the most widely used.

Hardy and Thompson (1996) proposed the profile likelihood (PL) method based on the likelihood (4.12). Alternatively, one could choose the REML profile likelihood based on (4.29). This approach takes into account that θ has to be estimated as well. In contrast to a Wald type CI, the PL CI is invariant under re-parametrization and asymmetric, which is more natural for a parameter that cannot be negative. The PL method is based on the likelihood ratio test for $H_0: \tau^2 = \tau_0^2$ for an arbitrary positive value τ_0^2. Under H_0, the test statistic asymptotically has a χ_1^2 distribution,

$$-2\ln\left(\frac{L\left(\hat{\theta}\left(\tau_0^2\right),\tau_0^2\right)}{L\left(\hat{\theta}_{ML},\hat{\tau}_{ML}^2\right)}\right) \sim \chi_1^2, \tag{4.31}$$

where $L\left(\hat{\theta}\left(\tau_0^2\right),\tau_0^2\right)$ means the likelihood maximized over θ for fixed τ_0^2. The null hypothesis is rejected if the statistic is larger than the $(1-\alpha)$ percentile $\chi_{1,1-\alpha}^2$. The $(1-\alpha)$100% confidence interval then consists of all values of τ_0^2 that are not rejected, i.e., all values for which

$$\ln L\left(\hat{\theta}\left(\tau_0^2\right),\tau_0^2\right) > \ln L\left(\hat{\theta}_{ML},\hat{\tau}_{ML}^2\right) - \chi_{1,1-\alpha}^2/2. \tag{4.32}$$

The inequality has to be iteratively solved. For small or zero values of τ^2, the method behaves poorly with overly wide CIs and excessive coverage probabilities (Viechtbauer, 2007). The problem is that for small values of τ_0^2 the approximate chi-square distribution of (4.31) only holds for very large numbers of studies, and for $\tau_0^2 = 0$ the approximation is wrong for any number of studies, the true null distribution then being a fifty-fifty mixture of a point mass at zero and a χ_1^2-distribution. The PL method is based on asymptotic likelihood arguments and possesses the optimality properties of likelihood methods in large samples. Though efficient in large meta-analyses, the method is not expected to work well in the more typical small meta-analyses. There is a need for more precise confidence intervals appropriate for smaller numbers of studies.

Our next method is an exact method, the *Q*-profile (QP) method of Viechtbauer (2007), which is based on the generalized *Q*-statistic $Q(\tau^2)$ used in the Paule–Mandel method above. Because $Q(\tau^2)$ follows an exact χ_{I-1}^2distribution, we have

$$\Pr(\chi_{I-1,\alpha/2}^2 \le Q\left(\tau^2\right) \le \chi_{I-1,1-\alpha/2}^2) = 1-\alpha. \tag{4.33}$$

Then a $(1-\alpha)$100% CI for τ^2 is given by $\left(\tau_{lo}^2,\tau_{hi}^2\right)$ through solving $Q\left(\tau_{lo}^2\right) = \chi_{I-1,1-\alpha/2}^2$ and $Q\left(\tau_{hi}^2\right) = \chi_{I-1,\alpha/2}^2$. It can, for instance, be computed by calculating $Q(\tau^2)$ on a grid of non-negative σ^2 values ("profiling the generalized *Q*-statistic"). If $Q(0) \le \chi_{I-1,1-\alpha/2}^2$ then the CI is set to [0,0]. The method is exact under the assumptions of the RE model, but not necessarily optimal in the sense of being the shortest possible interval. Note that there exists no generally accepted estimator for τ^2 in natural conjunction with the QP method. Viechtbauer (2007) gives the DL estimate alongside his QP confidence interval, while the R package `metafor` (Viechtbauer, 2010) gives the REML estimate. The only way to ensure that the estimator lies within the QP confidence interval is to report the value of τ^2 for which $Q(\tau^2)$ is equal to the median $\chi^2_{I-1,0.5}$, which is a median unbiased estimator and can be calculated as the QP confidence interval with $\alpha = 1$.

While the exact distribution of $Q(\tau^2)$ has a simple χ^2 form, the generalized Q-statistics with known weights have complex distributions. Jackson (2013) investigated alternative CIs for τ^2 based on the generalized Cochran heterogeneity statistic Q_a for other known choices of the weights a_i. For arbitrary fixed a_i he showed that Q_a is distributed as a linear combination of χ^2 distributed random variables with coefficients depending on the s_i^2, a_i, and τ^2. He also showed that the distribution function of Q_a is a strictly monotonically decreasing function of τ^2. It follows that Q_a can be used as pivotal quantity in the same way as for the QP method, leading to a "Generalized Q" method. However, Jackson's method is somewhat more demanding computationally since the distribution of Q_a is not available in closed form. Jackson (2013) looked at weights of the form $a_i = 1/(s_i^2 + x)^p$ for different choices of x and p. As a special case, $x = 0$ and $p = 1$ yields the CE weights and leads to the CI as proposed by Biggerstaff and Jackson (2008). In a simulation study, Jackson (2013) compared his method for different choices of x and p with the QP method. Though the QP method performed well in general, it was outperformed for small and moderate values of τ^2 by $x = 0$ and $p = 1$ (Biggerstaff and Jackson method) and $x = 0$ and $p = 1/2$ (weights equal to inverse standard errors).

We applied the REML profile likelihood method, the QP method (Viechtbauer, 2007), the Generalized Q method with $x = 0$ and $p = 1$ (Jackson, 2013), and the Generalized Q method with $x = 0$ and $p = 0.5$ (Biggerstaff and Jackson, 2008) to the BCG data (Table 4.1) and iron data (Table 4.2). The results are shown in Table 4.4. As software usually gives the standard error of τ^2 instead of τ, the confidence interval was calculated by (4.30) and taking square roots afterwards. As mentioned before, the Wald method is not appropriate for τ, as is, for instance, illustrated by the value zero being included in the confidence interval while the evidence for heterogeneity between studies is overwhelming. The Generalized Q estimators are the method-of-moments estimators as given by (4.23) with the appropriate weights, i.e., $w_i = 1/s_i^2$ and $w_i = 1/s_i$, respectively. Note that in the former case, the estimate is identical to the DL estimate.

TABLE 4.4
Point Estimates and Confidence Intervals for τ for the BCG and Iron Data

Between studies standard deviation estimate			**95% confidence interval**		
Method	**BCG data**	**Iron data**	**Method**	**BCG data**	**Iron data**
DL	0.61	6.6	Wald	(0, 0.94)	(0, 13)
REML	0.58	6.9	Profile REML	(0.36, 1.10)	(0, 25)
Paule–Mandel	0.58	7.7	QP method	(0.36, 1.09)	(0, 32)
Generalized Q (x = 0, p = 1)	0.61	6.6	Generalized Q (x = 0, p = 1)	(0.35, 1.37)	(0, 28)
Generalized Q (x = 0, p = 0.5)	0.59	7.5	Generalized Q (x = 0, p = 0.5)	(0.37, 1.12)	(0, 29)

4.6 Software

A popular program for meta-analysis, used by many medical researchers, is RevMan, which is provided by the Cochrane Collaboration. At present, RevMan only provides the inverse-variance common-effect, the Mantel–Haenszel common-effect (Chapter 5), and the DL random-effects methods. Many more possibilities are offered by the user-written meta-analysis programs in Stata (Palmer and Sterne, 2016). Even more methods are implemented in the R-package `metafor` from Viechtbauer (2010). There are several other R-packages for meta-analysis, mostly for specific methods. A well-known special purpose meta-analysis software package is Comprehensive Meta-Analysis (www.meta-analysis.com). However, many of the methods discussed in this chapter are not implemented in this package. The likelihood-based methods can be performed in many mixed model programs of standard statistical packages, see Van Houwelingen et al. (2002) for SAS and Bagos (2015) for Stata. All the results in the examples of this chapter were produced using `metafor`, except for the PL confidence intervals for the overall effect, which were made using SAS (Van Houwelingen et al., 2002).

4.7 Conclusions

The concepts introduced in this chapter—the common-effect and random-effects models, heterogeneity, I^2, prediction intervals—are of enormous importance in meta-analysis and will recur throughout the book. On the other hand, the specific methods of analysis proposed are hampered by the simple form of data assumed and the possible failure of the within-study model assumptions. Later chapters explore improvements to the estimation procedures and extensions to the model.

References

Bagos P, 2015. Meta-analysis in Stata using gllamm. *Research Synthesis Methods* **6**: 310–332.

Biggerstaff BJ and Jackson D, 2008. The exact distribution of Cochran's heterogeneity statistic in one-way random-effects meta-analysis. *Statistics in Medicine* **27**: 6093–6110.

Brockwell SE and Gordon IR, 2001. A comparison of statistical methods for meta-analysis. *Statistics in Medicine* **20**: 825–840.

Cochran WG, 1954. The combination of estimates from different experiments. *Biometrics* **10**: 101–129.

Cochran WG, 1973. Problems arising in the analysis of a series of similar experiments. *Journal of the Royal Statistical Society (Supplement)* **4**: 102–118.

Colditz GA, Brewer FB, Berkey CS, Wilson EM, Burdick E, Fineberg HV and Mosteller F, 1994. Efficacy of BCG vaccine in the prevention of tuberculosis. *Journal of the American Medical Association* **271**: 698–702.

Cornell JE, Mulrow CD, Localio R, Stack CB, Meibohm AR, Guallar E, and Goodman SN, 2014. Random-effects meta-analysis of inconsistent effects: A time for change. *Annals of Internal Medicine* **160**: 267–270.

DerSimonian R and Laird N, 1986. Meta-analysis in clinical trials. *Controlled Clinical Trials* **7**: 177–188.

DerSimonian R and Kacker R, 2007. Random-effects model for meta-analysis of clinical trials: An update. *Contemporary Clinical Trials* **28**: 105–114.

Follmann DA and Proschan MA, 1999. Valid inference in random-effects meta-analysis. *Biometrics* **55**: 732–737.

Hardy RJ and Thompson SG, 1996. A likelihood approach to meta-analysis with random-effects. *Statistics in Medicine* **15**: 619–629.

Hardy RJ and Thompson SG, 1998. Detecting and describing heterogeneity in meta-analysis. *Statistics in Medicine* **17**: 841–856.

Hartung J and Knapp G, 2001. A refined method for the meta-analysis of controlled clinical trials with binary outcome. *Statistics in Medicine* **20**: 3875–3889.

Hedges LV and Olkin I, 1985. *Statistical Methods for Meta-Analysis*. Orlando: Academic Press.

Higgins JPT and Thompson SG, 2002. Quantifying heterogeneity in a meta-analysis. *Statistics in Medicine* **21**: 1539–1558.

Higgins JPT, Thomas J, Chandler J, Cumpston M, Li T, Page MJ and Welch VA (editors), 2019. *Cochrane Handbook for Systematic Reviews of Interventions*. 2nd Edition. Chichester, UK: John Wiley & Sons. Also online version 6.0 (updated July 2019) available from www.training.cochrane.org/handbook.

Higgins JPT, Thompson SG and Spiegelhalter DJ, 2009. A re-evaluation of random-effects meta-analysis. *Journal of the Royal Statistical Society: Series A* **172**(Part 1): 137–159.

IntHout J, Ioannidis JP and Borm GF (2014). The Hartung-Knapp-Sidik-Jonkman method for random-effects meta-analysis is straightforward and considerably outperforms the standard DerSimonian-Laird method. *BMC Medical Research Methodology*. DOI: 10.1186/1471-2288-14-25.

Jackson D, Bowden J and Baker R (2010). How does the DerSimonian and Laird procedure for random effects meta-analysis compare with its more efficient but harder to compute counterparts? *Journal of Statistical Planning and Inference* **140**: 961–970.

Jackson D, 2013. Confidence intervals for the between-study variance in random-effects meta-analysis using generalised Cochran heterogeneity statistics. *Research Synthesis Methods* **4**: 220–229.

Jackson D and White IR (2018). When should meta-analysis avoid making hidden normality assumptions? *Biometrical Journal* **60**: 1040–1058.

Kacker RN, 2004. Combining information from interlaboratory evaluations using a random-effects model. *Metrologia* **41**: 132–136.

Kulinskaya E and Dollinger MB, 2015. An accurate test for homogeneity of odds ratios based on Cochran's *Q*-statistic. *BMC Medical Research Methodology* **15**: 49.

Laird NM and Mosteller F, 1990. Some statistical methods for combining experimental results. *International Journal of Technology Assessment in Health Care* **6**: 5–30.

Langan D, Higgins JPT, Jackson D, Bowden J, Veroniki AA, Kontopantelis E, Viechtbauer W and Simmonds M, 2019. A comparison of heterogeneity variance estimators in simulated random-effects meta-analyses. *Research Synthesis Methods* **10**: 83–98.

Lewis S and Clarke M (2001). Forest plots: Trying to see the wood and the trees. *British Medical Journal* **322**: 1479–1480.

Lopes da Silva S, Vellas B, Elemans S, Luchsingerd J, Kamphuis P, Yaffee K, Sijbena J, Groenendijk M and Stijnen T, 2014. Plasma nutrient status of patients with Alzheimer's disease: Systematic review and meta-analysis. *Alzheimer's & Dementia* **10**: 485–502.

Palmer T and Sterne JAC (editors), 2016. *Meta-Analysis in Stata: An Updated Collection from the Stata Journal*. Stata Press. ISBN 9781597181471.

Paule RC and Mandel J, 1982. Consensus values and weighting factors. *Journal of Research of the National Bureau of Standards* **87**: 377–385.

Robinson GK, 1991. That BLUP is a good thing: The estimation of random-effects. *Statistical Science* **6**: 15–32.

Shuster JJ, 2010. Empirical vs natural weighting in random-effects meta-analysis. *Statistics in Medicine* **29**: 1259–1265.

Sidik K and Jonkman JN, 2002. A simple confidence interval for meta-analysis. *Statistics in Medicine* **21**: 3153–3159.

Stijnen T, Hamza TH and Özdemir P, 2010. Random effects meta-analysis of event outcome in the framework of the generalized linear mixed model with applications in sparse data. *Statistics in Medicine* **29**: 3046–3067.

Van Aert RCM and Jackson D, 2019. A new justification of the Hartung-Knapp method for random effects meta-analysis based on least squares regression. *Research Synthesis Methods*. DOI: 10.1002/jrsm/term.1356.

Van Houwelingen HC, Arends LR and Stijnen T, 2002. Advanced methods in meta-analysis: Multivariate approach and meta-regression. *Statistics in Medicine* **21**: 589–624.

Viechtbauer W, 2005. Bias and efficiency of meta-analytic variance estimators in the random-effects model. *Journal of Educational and Behavioral Statistics* **30**: 261–293. doi:10.3102/10769986030003261

Viechtbauer W, 2007. Confidence intervals for the amount of heterogeneity in meta-analysis. *Statistics in Medicine* **26**: 37–52.

Viechtbauer W, 2010. Conducting meta-analyses in R with the `metafor` package. *Journal of Statistical Software* **36**: 1–48.

Veroniki AA, Li Ka S, Jackson D, Viechtbauer W, Bender R, Bowden J, Knapp G, Kuss O, Higgins JPT, Langan D and Salanti G, 2016. Methods to estimate the between-study variance and its uncertainty in meta-analysis. *Research Synthesis Methods* **7**: 55–79.

5

Exact Likelihood Methods for Group-Based Summaries

Theo Stijnen, Christopher H. Schmid, Martin Law, Dan Jackson, and Ian R. White

CONTENTS

5.1 Introduction

In Chapter 4, we discussed approaches that were based on an estimate y_i of the true treatment effect measure θ_i. The estimate y_i and its standard error s_i were either directly taken from the study report or calculated by the meta-analyst based on group-level summary data given in the study report. The methods of Chapter 4 relied on the assumption of approximate normality for y_i, and the estimated standard errors s_i were treated as fixed

TABLE 5.1

Potential Drawbacks of Two-Stage Methods

Potential drawbacks of two-stage methods (Chapter 4)
1. Violation of normality assumption for y_i given θ_i.
2. Failure to account for uncertainty in estimation of s_i, leading to underestimation of the uncertainty in the overall effect estimate.
3. Failure to account for possible correlation between y_i and s_i (when the outcome is not normal), potentially leading to bias in the estimates of the overall effect and between-studies variance.

and known. In this chapter, we consider methods that are directly based on group-level summary data. They avoid the normality assumption and take the uncertainty in the estimated standard errors into account. This is done by using the exact likelihood of the group-level summary data. Sometimes these methods are referred to as one-stage methods, in contrast to the methods of Chapter 4, which are referred to as two-stage methods.

The normal within-study model of Chapter 4 is usually justified by sufficiently large study sample sizes and an appropriate choice of the effect measure, such as the log odds ratio rather than the odds ratio. However, this approach is not always appropriate and may have some disadvantages, as listed in Table 5.1. In meta-analysis of sparse count data, all these drawbacks apply at the same time.

In this chapter, we will discuss methods that aim to replace the approximate normal within-study likelihood by the true or at least a better likelihood of the within-study data. Whether and how that can be done depends on the choice of the effect measure and the nature of the data. These methods avoid the explicit calculation of study-specific effect sizes y_i. The idea is that when publications report the sufficient statistics per treatment group (such as mean, standard deviation, and sample size for normal data or number of events and sample size for dichotomous data), it is possible to write down the likelihood of the original (unobserved) individual outcomes. Practically, the analysis is then the same as if one had the individual patient data and the same software can be used.

In this chapter, we will cover the three most commonly occurring outcome types: continuous, dichotomous, and count outcomes which are modeled by normal, binomial, and Poisson distributions, respectively. We will discuss a number of examples, covering the most frequently occurring situations in practice. It will turn out that the models are linear mixed models (LMMs) in the case of normal outcomes, and generalized linear mixed models (GLMMs) in the case of other outcome types, in particular, mixed effects (conditional) logistic or Poisson regression models. These models can all be fitted with the LMM and GLMM programs available in most statistical packages.

5.2 Continuous Outcome

In this section, we consider the case where the outcome variable is continuous, and two groups are compared using the difference between the two means as the effect measure. We assume that the mean, standard deviation, and sample size are available for each group in a study. These are the sufficient statistics if the outcome variable has a normal distribution. Standard software packages cannot account for the standard deviations being estimated in analyzing data in this format. However, one can easily create a dataset of pseudo individual participant data (IPD) that matches the observed sufficient statistics and therefore has the same likelihood as the true but unknown IPD (Papadimitropoulou et al., 2019). Analyzing any such pseudo IPD dataset with a likelihood-based method leads to identical results as if one had the true IPD. This dataset has the same structure as data from a multi-center clinical trial, and the same methods as for multi-center trials can be employed, with standard statistical linear (mixed) model software.

5.2.1 Common-Effect and Separate-Effects Models

If y_{ijk} is the outcome for patient k in group j of study i, one could fit the following linear model

$$Y_{ijk} = \beta_{0i} + \beta_1 X_{ijk} + \varepsilon_{ijk} \quad \text{with} \quad \varepsilon_{ijk} \sim N(0, \sigma_{ij}^2). \tag{5.1}$$

The covariate $X_{ijk} = j$ denotes the treatment group (0 = control, 1 = treated). The parameter β_1 is the common treatment effect, assumed the same in all studies; β_{0i} is a fixed study-specific intercept representing the mean in group 0 of study i, and the residual variance σ_{ij}^2 is the group and study-specific within-study variance of the outcome variable. We call this the common-effect (CE) model. Note that, to be in line with standard linear regression notation, we use β_1 here instead of θ for the overall effect. In this model, every study has its own fixed intercept and every group in every study has its own residual variance. We call this variance model V1. Other variance models are listed in Table 5.2.

Note that under V4, the model reduces to an ordinary linear regression model. The two-stage approach implicitly uses one of variance models V1–V4. For instance, `metafor` calculates the standard errors s_i using separate variances for the groups within a study and thus uses model V1. Other programs might employ the pooled standard deviation within studies, thus implicitly using V2. Models V3 and V4 are rarely applied in the two-stage approach, though they might be entirely reasonable in certain applications. The results of the CE two-stage approach using different variance models can be reproduced through fitting model (5.1) keeping the σ_{ij} fixed to their appropriate values. Notice the advantage of the one-stage approach of circumventing the assumption of fixed and known standard deviations.

As an alternative to the common treatment effects in (5.1), one could assume separate, unrelated treatment effects:

$$Y_{ijk} = \beta_{0i} + \beta_{1j} X_{ijk} + \varepsilon_{ijk} \quad \text{with} \quad \varepsilon_{ijk} \sim N(0, \sigma_{ij}^2). \tag{5.2}$$

This is a one-stage analogue of the separate-effects model introduced in Chapter 2. This model is equivalent to adding the interaction between study and treatment to model (5.1). The test of the interaction could serve as a test of the common-effect assumption of model (5.1).

TABLE 5.2

Models for the Residual Standard Deviation

Models for the residual variance
• V1: σ_{ij} may vary by study and group
• V2: σ_{ij} may vary by study, but not by group: $\sigma_{ij} = \sigma_i$
• V3: σ_{ij} may vary by group, but not by study: $\sigma_{ij} = \sigma_j$
• V4: σ_{ij} does not vary: $\sigma_{ij} = \sigma$

As an alternative to the fixed intercepts in (5.1), one could assume random study-specific intercepts b_i drawn from a common distribution (i.e., a random-intercepts model):

$$Y_{ijk} = \beta_0 + b_i + \beta_1 X_{ijk} + \varepsilon_{ijk} \quad \text{with } b_i \sim N(0,\tau_0^2) \text{ and } \varepsilon_{ijk} \sim N(0,\sigma_{ij}^2). \tag{5.3}$$

We will show later (in 5.3.2) that this permits between-study information on the control groups leaking into the estimate of the treatment effect (Senn, 2010), as discussed in Section 2.3.3. Models (5.1) and (5.3) can be fitted with standard (ordinary (ML) or restricted (REML)) likelihood methods using linear mixed model software.

5.2.1.1 Example

As an example, we consider the meta-analysis of Lopes da Silva et al. (2014) comparing mean plasma iron levels between Alzheimer's disease patients and controls. The data were introduced in Chapter 4 and are shown in Table 4.2. We choose the (unstandardized) mean difference as effect measure and create a dataset with pseudo individual patient data such that in all groups the mean, standard deviation, and sample size are as in Table 4.2. The pseudo IPD outcomes can be constructed in infinitely many ways, but that does not matter since the resulting likelihood is always identical to the likelihood of the true (unknown) individual patient data (Papadimitropoulou et al., 2019). We make the pseudo data by generating for each study i and each study group j a sample of size n_{ij} from a standard normal distribution and scaling them to have the required mean and standard deviation.

Following a commonly used strategy for model selection in linear mixed models (see, for instance, Fitzmaurice et al. [2011]), we start by searching for a good model for the residual variance. This is best done in the model with the richest specification of the fixed parameters, model (5.2). We compare models V1–V4 defined above for the residual variances using the REML likelihood ratio test. Standard linear regression programs do not provide the option of different residual variances in subgroups. However, one can fit these models in standard linear mixed model programs, even though the models are not really mixed models. Since models V2–V4 all fit significantly worse than model V1 (results not shown), we adopt model V1 in the subsequent analyses.

Lines 1, 2, and 3 of Table 5.4 give the results of models (5.1)–(5.3). The first line reproduces the results of the two-stage CE method of Section 4.2 through fitting model (5.1) with the σ_{ij}^2 set fixed to the observed variances, giving results that are indeed identical to the results of Figure 4.2. The results of fitting model (5.1), but now with the σ_{ij}^2 estimated, turn out to be very similar to the results of the two-stage CE method. The results of the random-intercept CE model (5.3) are almost identical to the results of the model with study as fixed factor. The common-effect assumption might be tested by adding the interaction between study and treatment to model (5.1). The interaction is not significant, $p=0.103$, very comparable with the p-value of the Q_{CE} test from the two-stage CE analysis: $p=0.102$. We also fit the separate-effects model (5.2) and estimate the unweighted mean of the effects of the five trials in the meta-analysis. The results from the separate-effects model differ from those of the other models because the estimand and therefore the weights are different.

5.2.2 Random-Effects Models

The separate-effects model (5.2) allows heterogeneity in the treatment effects but does not yield a clear estimand. We now turn to a more popular way to allow heterogeneity in the treatment effects, which does yield a clear estimand. Adding the random effect for group in (5.1) gives the following LMM:

$$Y_{ijk} = \beta_{0i} + \beta_1 X_{ijk} + b_i X_{ijk} + \varepsilon_{ijk} \quad \text{with } b_i \sim N(0,\tau^2) \text{ and } \varepsilon_{ijk} \sim N(0,\sigma_{ij}^2). \tag{5.4}$$

The parameter β_1 is the overall treatment effect, $\beta_1 + b_i$ represents the (random) treatment effect in study i, τ^2 is the between-studies variance of the treatment effects and β_{0i} is a study-specific intercept.

Again, different alternative models V1–V4 are possible for the residual variance. If the σ_{ij} are forced to be fixed at the observed standard deviations in (5.4), and the model is fitted with REML/ML, the results will be almost identical to the two-stage approach REML/ML method of Chapter 4.

As an alternative to (5.4) one could assume a model with random intercepts:

$$\begin{gathered} Y_{ijk} = \beta_0 + \beta_1 X_{ijk} + b_{0i} + b_{1i} X_{ijk} + \varepsilon_{ijk} \\ \text{with} \begin{pmatrix} b_{0i} \\ b_{1i} \end{pmatrix} \sim N\left(\begin{pmatrix} 0 \\ 0 \end{pmatrix}, \begin{pmatrix} \tau_0^2 & \tau_{01} \\ \tau_{01} & \tau^2 \end{pmatrix} \right) \text{ and } \varepsilon_{ijk} \sim N(0,\sigma_{ij}^2). \end{gathered} \tag{5.5}$$

This model allows correlation between the study effect and treatment effect, but one could also assume independence ($\tau_{01} = 0$). If independence is assumed, one should realize that the model becomes sensitive to the coding of the treatment covariate. For instance, coding $X_{ijk} = 1$ for treatment and $X_{ijk} = -1$ for control would change the model.

Like model (5.3), this model permits recovering between-studies information. In a very extensive simulation study, Legha et al. (2018) compare the performance of the stratified model (5.4) with the random-intercept model (5.5) in IPD meta-analyses. Though the authors did not mention it, the results also apply without restriction to the pseudo IPD approach for summary data meta-analysis. The simulations were restricted to meta-analysis of randomized clinical trials with a 1:1 treatment:control allocation ratio, and a zero covariance τ_{01} was assumed in model (5.5). Their main conclusions were: (1) bias and coverage of the summary treatment effect estimate are very similar when using fixed- or random-intercepts models fitted with REML, and thus either approach could be taken in practice; (2) confidence intervals are generally best derived using either a Kenward-Roger or Satterthwaite correction, although these are occasionally overly conservative; and (3) if maximum likelihood estimation is used, a random-intercept model performs better than a stratified-intercept model. However, bias can arise when allocation ratios vary across studies (Jackson et al., 2018).

5.2.2.1 Advantages and Disadvantages of the One-Stage Approach

In general, this one-stage approach using LMMs gives much freedom to the meta-analyst in the statistical modeling, though a disadvantage may be that knowledge of mixed modeling is required to avoid errors when using mixed model programs. Nevertheless, use of the linear mixed model can have several advantages, as listed in Table 5.3.

5.2.2.1.1 Example

We applied this approach to the Iron data example of Table 4.2, the same dataset as used in the previous section. In the common-effect modeling considered in the previous

TABLE 5.3

Advantages of the One-Stage Linear Mixed Modeling Approach to Random-Effect Modeling of Continuous Outcome

1. It avoids the assumption of fixed and known standard deviations.
2. It allows inference on the residual variances. For instance, it might be interesting to test the equality of the variances of the treatment and control groups: H_0: $\sigma_{i0}^2 = \sigma_{i1}^2$ for all studies i, which is easily performed through the likelihood ratio test. If treatment effects are the same for all individuals within a study then variances in the treatment and control groups must be equal: so different variances indicate heterogeneity of the treatment effect among individuals within a study which might be of scientific interest.
3. The Wald type test and confidence intervals are adjusted for the covariance parameters being estimated. Most LMM programs offer methods for adjusting the denominator degrees of freedom, such as Satterthwaite's or Kenward and Roger's method.
4. It easily allows (REML) likelihood ratio-based tests and confidence intervals for the parameters of interest.
5. It easily allows testing the strong null hypothesis that there is no effect in any study, i.e., H_0: $\beta_1 = \tau^2 = 0$, through the likelihood ratio test. Testing this hypothesis might sometimes be of greater interest than testing H_0: $\beta_1 = 0$, for instance, in genetic meta-analyses (Lebrec et al., 2010).
6. Model (5.5) allows investigation of the possible association between the size of the treatment effect and the level in the control group, through the τ_{01} parameter: this is discussed fully in Chapter 14.
7. This modeling approach allows straightforward and immediate extension to meta-regression (Chapter 7), multivariate meta-analysis (Chapter 9), and network meta-analysis (Chapter 10).

subsection, we saw that in this example the best model for the patient-level variance was the model that leaves all σ_{ij} free. Using this model for the residual variances, the results of fitting models (5.4) and (5.5) (with and without correlation between study and treatment effects) using REML are shown in Table 5.4. For comparison, the result of the two-stage REML method from Chapter 4 is also given. Compared with the CE models, the RE models give a slightly less negative estimate of the overall group effect β_1, because they down-weight the larger studies (3 and 5) which show more strongly negative effects. Also note that the two RE models with random study effects show a less negative effect than the RE model with fixed study effects. Probably this is an illustration of the random study effects models allowing the use of between-studies information. The three smallest studies

TABLE 5.4

Common-Effect, Separate-Effects, and Random-Effects Analysis of the Iron Data

Model	Overall effect	se	P	95% CI	$\hat{\tau}^2$
Common-effect model					
Study-level (two-stage) CE model	−6.78	2.85	0.017	(−12.37, −1.19)	n.a.
CE with fixed study effects (model [5.1])	−6.91	2.86	0.016	(−12.52, −1.30)	n.a.
CE with random study effects (model [5.3])	−6.89	2.94	0.020	(−12.70, −1.09)	n.a.
Separate-effects model					
Separate-effects model (model [5.2])	−4.20	3.59	0.24	(−11.25, 2.85)	n.a.
Random-effects model					
Study-level (two-stage) RE model (REML)	−5.52	4.47	0.22	(−14.28, 3.24)	47.3
RE with fixed study effects (model [5.4])	−5.60	4.89	0.36	(−25.1, 13.9)	45.1
RE with random study effects (model [5.5])	−4.55	5.04	0.45	(−24.1, 15.0)	44.2
RE with random study effects (model [5.5] with zero correlation)	−5.19	5.06	0.39	(−22.7,12.4)	60.0

The overall effect of the separate-effects model was chosen to be the unweighted mean over the studies in the meta-analysis.

(1, 2, and 4) happen to have the largest control means, and, because of the random study effect, these are shrunk downwards, making their group effects less negative. The group effects of the two larger studies, especially the largest study which has a very large control sample size, are much less affected by this shrinkage effect. Therefore, the overall group effect becomes less negative. This phenomenon is hardly seen for the CE models, probably because the two largest studies dominate much more in the CE than in the RE models. According to the advice of Legha et al. (2018), models (5.4) and (5.5) were fitted using the Kenward-Roger correction.

5.3 Dichotomous Outcomes

In this section, we focus on the situation where two groups are compared with respect to the probability of some outcome event of interest. We suppose that the data for each of I studies consist of a 2×2 table (numbers of subjects with/without event for two groups). The most popular choice for the effect measure is the (log) odds ratio and we will use that choice. Other effect measures such as the risk ratio or the risk difference lead to models that are very difficult to fit. The number of events in a group follows a binomial distribution and in Section 5.3.1 we will use the binomial likelihood in order to avoid the normal within-study distribution assumption of Chapter 4. All problems with the two-stage approach mentioned in Table 5.1 apply for dichotomous outcomes, including number 3, which did not apply for the continuous outcome case discussed in the previous section. These problems get worse as the numbers of events get smaller.

5.3.1 Problems with the Two-Stage Approach

We first look in more detail at problems arising in the two-stage analysis due to the correlation between the estimated log odds ratio $\hat{\theta}_i$ and its variance estimate s_i^2. Since the variance is in the denominator of the weights used in the weighted mean of the two-stage methods, a negative correlation means that relatively small $\hat{\theta}_i$ get too little weight and relatively large $\hat{\theta}_i$ get too much weight. Thus, a negative correlation leads to positive bias in $\hat{\theta}$, and a positive correlation leads to negative bias. Some more insight into the correlation between $\hat{\theta}_i$ and s_i^2 can be obtained by writing $\hat{\theta}_i = \hat{\theta}_{i1} - \hat{\theta}_{i0}$, with $\hat{\theta}_{ij} = \log(y_{ij} / (n_i - y_{ij}))$ the estimated log odds in group j, and writing $s_i^2 = s_{i0}^2 + s_{i1}^2$, with $s_{ij}^2 = 1/y_{ij} + 1/(n_i - y_{ij})$ the estimated variance of $\hat{\theta}_{ij}$. Then, the covariance can be written as

$$\mathrm{cov}(\hat{\theta}_i, s_i^2) = \mathrm{cov}(\hat{\theta}_{i1}, s_{i1}^2) - \mathrm{cov}(\hat{\theta}_{i0}, s_{i0}^2). \tag{5.6}$$

The covariances on the right-hand side depend on whether the true θ_{ij} is positive or negative. If it is negative, the dominating term in s_{ij}^2 is $1/y_{ij}$ and the covariance, and thus the correlation, will be negative. If it is positive, the dominating term is $1/(n_i - y_{ij})$ and the correlation will be positive. In practice, most of the pairs $(\theta_{i0}, \theta_{i1})$ lie on the same side of $\theta = 0$, and the covariances on the right-hand side will therefore at least partly cancel out, since they have the same sign. However, this is not always the case. For instance, in diagnostic test meta-analysis (Chapter 19), the proportions of patients with a positive test are almost always larger than 0.5 in patients with the disease, and smaller than 0.5 in patients without the disease. Thus, one of the θ_{i0} and θ_{i1} will be positive and the other negative. In that case, the two covariances in (5.6) have opposite signs and reinforce each other. Also, in cases where the effect size is large or when there is a large imbalance in the sample sizes, the two covariances can differ considerably and may not cancel out.

Another, minor, problem in two-stage meta-analysis of dichotomous outcomes is that one must decide on how to estimate the study-specific log odds ratio and corresponding standard error, especially if the number of events or non-events is zero. See Section 3.4 for a discussion of this.

5.3.2 Common-Effect Models

5.3.2.1 Logistic Regression

Since the number of subjects with and without the event are available for each treatment group, we have the individual outcomes per subject. As in the previous section, Y_{ijk} denotes the individual level outcome (equal to 0/1 for an individual without/with the event), and covariate X_{ijk} denotes the treatment group. Note that in contrast to the continuous outcome case of Section 5.2, where we had pseudo individual outcomes, we now have the actual observed outcomes. If a common effect is assumed, one could, analogously to model (5.1), fit the following logistic regression model

$$\text{logit}\left(\text{P}(Y_{ijk}=1)\right)=\log\left(\frac{\text{P}(Y_{ijk}=1)}{\text{P}(Y_{ijk}=0)}\right)=\beta_{0i}+\beta_1 X_{ijk}. \tag{5.7}$$

The parameter β_1 is the common-effect measure expressed as a log odds ratio, and β_{0i} is a fixed study-specific intercept interpreted as the log odds of the event in group 0 of study *i*. Logistic regression is available in all standard statistical packages and standard likelihood methods are used for the inference. Notice that the number of parameters in this model is large, $I+1$. Hence, as the sample size (the number of studies) becomes large, the number of parameters to estimate also becomes large. The theory of likelihood inference requires that the number of parameters is fixed as the sample size goes to infinity, and that is not the case here. Therefore, model (5.7) does not fulfil the regularity conditions required for maximum likelihood estimation to possess its usual good properties. (In the continuous outcome case of Section 5.2 this causes no problem, as REML is used.) Breslow and Day (1980) show that indeed the estimate of the common log odds ratio in principle can be highly biased as the number of studies goes to infinity. However, this does not mean that the performance of this model in practical meta-analyses is necessarily bad. We do not expect problems with this model when the numbers of control group events are reasonably large. As far as we know, the performance of model (5.7) has not been studied extensively in the context of meta-analysis. Given that, one should be careful in using this model in case of a large number of small trials.

We applied model (5.7) to the data of the BCG example of Table 4.1, giving OR = 0.620 with 95% CI (0.572, 0.673), similar to the results of the two-stage method of Section 4.2. The latter is not surprising since in this example the trials are not small and the approximate normality assumption for the estimated study-specific log odds ratios is well fulfilled.

5.3.2.2 Random-Intercept Logistic Regression

Model (5.7) gives each study its own control log odds. Analogously to model (5.3), we could also model the control log odds by a random effect for study, assuming the following random-intercept logistic regression model:

$$\text{logit}\left(P(Y_{ijk} = 1)\right) = \beta_0 + b_i + \beta_1 X_{ij} \quad \text{with } b_i \sim N(0, \sigma_b^2). \tag{5.8}$$

As in model (5.2), we could allow different fixed treatment effects in this model, as well as in model (5.7), leading to separate-effects models. Model (5.8) is in fact a simple particular case of a GLMM. It can be fitted in the GLMM modules of almost all general statistical packages, such as SAS, Stata, R, and SPSS.

An advantage of model (5.8) is that it prevents the potential problem associated with the many parameters in model (5.7) in case of rare events. A disadvantage is that an extra normality assumption is needed for the distribution of the control log odds. Moreover, it might have the disadvantage that some between-study information may be used to estimate the overall effect (Senn, 2010). However, Jackson et al. (2018) found that the impact of between-study information is very small. We applied model (5.8) to the BCG example, giving OR = 0.620 with 95% CI (0.567, 0.679), almost identical to the results of model (5.7).

5.3.2.3 Conditional Logistic Regression

Another way to prevent the possible problem associated with the many parameters in model (5.7) is to switch to conditional logistic regression (CLR), see, for instance, the books by Breslow and Day (1980) and Clayton and Hills (1993). CLR is a variant of logistic regression meant for stratified data, as we have in meta-analysis, where the study is the stratification factor. The motivation for CLR is to get rid of the stratum-specific intercepts, which in a meta-analysis are the study-specific baseline log odds. In contrast to ordinary logistic regression, which is based on the usual (unconditional) likelihood of the data, CLR uses the likelihood of the data conditioned on the total number of events per stratum (= study in the case of meta-analysis). The likelihood changes from a binomial likelihood to a non-central hypergeometric likelihood, see also the next subsection. Because this conditional likelihood does not contain the stratum-specific intercepts (the β_{0i} in model (5.7)) and these are therefore no longer estimated, one avoids the possible statistical difficulty associated with too many parameters in the ordinary likelihood. The option of CLR is not always provided by the logistic regression procedures of the general statistical packages. In the BCG example, the results were OR = 0.621 with 95% CI (0.572, 0.673), almost identical to the results of unconditional logistic regression (there are differences in the fourth decimal place).

5.3.2.4 Mantel–Haenszel Method

Both conditional and unconditional logistic regression require iterative numerical methods to estimate the parameters. One popular non-iterative method for estimating the common odds ratio across 2×2 tables is the Mantel–Haenszel procedure (Mantel and Haenszel, 1959), which is implemented in most general statistical packages and special purpose meta-analysis software. If we denote Y_{ij}, n_{ij} and n_i, respectively, as the number of events in group j of study i, the sample size of group j of study i, and the total sample size of study i, respectively, then the estimated common odds ratio is given by

$$OR_{MH} = \frac{\sum_{i=1}^{I} y_{i1}(n_{i0} - y_{i0}) / n_i}{\sum_{i=1}^{I} y_{i0}(n_{i1} - y_{i1}) / n_i}. \tag{5.9}$$

It can also be written as a weighted mean of the observed study-specific odds ratios OR_i:

$$OR_{MH} = \sum_i w_i OR_i / \sum_i w_i \quad \text{with} \quad w_i = Y_{i0}(n_{i1} - Y_{i1}) / n_i.$$

There are several published estimators for the logarithmic standard error of OR_{MH}. Here we give the one due to Robins et al. (1986) which has the advantage that it is consistent both under a large-strata limiting model (where the cell sizes go to infinity for a fixed number of studies) and under a sparse-data limiting model (where the number of studies goes to infinity but the cell sizes are bounded):

$$se\left(\ln(OR_{MH})^2\right) = \frac{S_3}{2S_1^2} + \frac{S_5}{2S_1S_2} + \frac{S_4}{2S_2^2},$$

where

$$S_1 = \sum_{i=1}^{I} \frac{Y_{i1}(n_{i0} - Y_{i0})}{n_i},$$

$$S_2 = \sum_{i=1}^{I} \frac{Y_{i0}(n_{i1} - Y_{i1})}{n_i},$$

$$S_3 = \sum_{i=1}^{I} \frac{Y_{i1}(n_{i0} - Y_{i0})(Y_{i1} + n_{i0} - Y_{i0})}{n_i^2},$$

$$S_4 = \sum_{i=1}^{I} \frac{Y_{i0}(n_{i1} - Y_{i1})(Y_{i0} + n_{i1} - Y_{i1})}{n_i^2}$$

and

$$S_5 = \sum_{i=1}^{I} \frac{Y_{i0}(n_{i1} - Y_{i1})(Y_{i1} + n_{i0} - Y_{i0}) + Y_{i1}(n_{i0} - Y_{i0})(Y_{i0} + n_{i1} - Y_{i1})}{n_i^2}.$$

The 100(1−α)% CI for the common OR then is $\left(OR_{MH}e^{-z_{\alpha/2}se(\ln(OR_{MH}))}, OR_{MH}e^{z_{\alpha/2}se(\ln(OR_{MH}))}\right)$.

The null hypothesis H_0: $OR = 1$ is tested using the Cochran–Mantel–Haenszel (CMH) statistic $Z = (O - E)/\sqrt{V}$, with $O = \sum_i Y_{i1}$ the total number of observed events in the treatment groups, $E = \sum_i E_i = \sum_i n_{i1} Y_i / n_i$ the total number of expected events in the treatment groups under the null hypothesis conditional on the total numbers of events Y_i per study, and V is the conditional variance of O given by $V = \sum_i V_i$ with $V_i = n_{i0} n_{i1} Y_{i1}(n_i - Y_{i1}) / \left(n_i^2(n_i - 1)\right)$ the conditional variance of Y_{i1} according to the hypergeometric distribution. The test statistic has approximately a standard normal distribution under the null hypothesis. The CMH test is very closely related to CLR. In fact, it is identical to the score test in CLR. In the BCG example, the Mantel–Haenszel method yields results very similar to those of CLR: OR=0.623 with 95% CI=(0.575, 0.675). For the case of very rare events, some statistical packages (e.g., SAS) provide an exact p-value and confidence interval for the Mantel–Haenszel and/or the CLR method.

5.3.2.5 Method of Peto

Another, very simple, method to estimate the common odds ratio avoiding the iterative method of (C)LR is due to Peto (Yusuf et al., 1985). The estimated log odds ratio is given by

$$\ln(OR_{\text{PETO}}) = \frac{O - E}{V}, \tag{5.10}$$

where O, E and V are defined as above for the Mantel–Haenszel method. Peto's estimate (5.10) can be seen as a "one-step estimate", i.e., it is the estimate of the common log odds ratio on the first step of a Newton–Raphson procedure to maximize the conditional likelihood. The standard error is $se(\ln(OR_{\text{PETO}})) = 1/V$. Thus the Peto $(1-\alpha)100\%$ CI of the common odds ratio is $\left(OR_{\text{PETO}}\, e^{-z_{\alpha/2} se(\ln(OR_{\text{PETO}}))}, OR_{\text{PETO}}\, e^{z_{\alpha/2} se(\ln(OR_{\text{PETO}}))}\right)$. The null hypothesis H_0: $OR=1$ is tested by $Z = \ln(OR_{\text{PETO}}) / se(\ln(OR_{\text{PETO}}))$, which follows approximately a standard normal distribution under the null hypothesis. Greenland and Salvan (1990) have shown that Peto's common odds ratio estimator is not consistent. It can yield extremely biased results when applied to unbalanced data, but also the bias may be unacceptable for balanced data. Therefore the (exact) Mantel–Haenszel method is preferred if one wishes to avoid CLR.

We used `metafor` to obtain the results of Peto's method for the BCG data. The common log odds ratio estimate is −0.474 with se=0.041. The estimated common odds ratio is 0.622 with 95% CI (0.575, 674), very close to the results of the Mantel-Haenszel and (C)LR methods.

5.3.2.6 Studies with Zero Events in One or Both Groups

Except for model (5.8), all methods discussed in this subsection do not make use of any study with zero events in both groups: such a study is not by itself informative about the odds ratio and can only contribute if between-studies information is used as in model (5.8). However, all methods do use the studies with zero events in just one group. Further, they use these studies without modification, whereas the two-stage approach of Chapter 4 must modify these studies' data, for instance, by adding a constant such as ½ to all the numbers in the 2×2 table.

5.3.3 Random-Effects Approaches

5.3.3.1 Meta-Analysis of Proportions

Although in this chapter we focus on the comparison between two groups, we start for didactical reasons with the simple but very instructive case where one wishes to meta-analyze a single series of proportions.* The proportions to be meta-analyzed may, for instance, be incidences of an adverse event, or sensitivities or specificities of a diagnostic test. In general, it is the probability that a certain event occurs, and in each study we observe the number of patients with the event Y_i and the sample size n_i. The true value of the probability of the event in study i is denoted by π_i and the corresponding log odds by $\theta_i = \log(\pi_i / (1-\pi_i))$. Again, all potential problems of Table 5.1 apply, in particular number 3. As explained above, the two-stage random-effect estimate $\hat{\theta}$ is positively biased if $\theta<0$ and negatively biased if $\theta>0$. On the probability scale, the bias is toward $\pi=1/2$ (Chang et al., 2001). For larger numbers of both events and non-events, the bias will be negligible, but the bias can be serious when events or non-events are relatively rare.

The potential difficulties associated with the two-stage methods can be circumvented by using the following random-intercept logistic regression model (Hamza et al., 2008a) without covariates:

$$\text{logit}\left(P(Y_{ik}=1)\right) = \theta + b_i \quad \text{with } b_i \sim N(0,\tau^2). \tag{5.11}$$

Here Y_{ik} denotes the outcome for patient k in study i, i.e., $Y_{ik}=1$ if the event occurred and $Y_{ik}=0$ if the event did not occur. The Y_{ik} are independently Bernoulli(π_i) distributed, thus the normal within-study model assumed in the two-stage method is now replaced by a binomial model. Sometimes the resulting model is called the binomial-normal model (Stijnen et al., 2010). It can be fitted using GLMM software, which is available in almost all general statistical packages. The method is also implemented in the R meta-analysis package `metafor` (Viechtbauer, 2010). Hamza et al. (2008a) and Stijnen et al. (2010) give SAS syntax examples, using SAS PROC NLMIXED, but an easier approach would be to use SAS PROC GLIMMIX.

Hamza et al. (2008a) performed an extensive simulation study to compare the random-intercept logistic regression model with the two-stage approach of Chapter 4. The random-intercept logistic regression always performed better than the two-stage approach, which behaved quite poorly in many scenarios, in particular, when the proportions were close to zero and the events were relatively rare. The random-intercept logistic regression gave unbiased estimates of θ and the confidence intervals gave reasonable coverage probabilities. GLMM programs often use the t-distribution with df $=I$-1 (with I the number of studies) for a test and confidence interval of θ, to allow for the fact that $\hat{\tau}^2$ is estimated. This kind of adjustment is standard in LMM programs and it is also recommended by some authors (Follmann and Proschan, 1999; Higgins et al., 2009) for the two-stage approach. Alternatively, one could use the likelihood ratio test for testing θ and the profile likelihood method for the corresponding confidence interval.

As well as the logistic link function used above, most GLMM programs allow the choice of another link function, such as the inverse standard normal cumulative distribution function (leading to a random-intercept probit model), or the complementary log-log link. However, link functions that do not map probabilities onto the whole real line, such as the identity or log, are not appropriate in GLMMs. Zhou et al. (1999) consider a model where the random effect is put directly on the π_i by assuming them to be a random sample from a beta distribution. This leads to a relatively simple expression for the marginal likelihood

* In the previous subsection on the common-effect approaches, we did not consider this case. The reason is that if it is assumed that the true proportions are equal, the data just reduce to the total number of events and the total sample size. The meta-analysis is then trivial: just apply exact or approximate statistical methods for estimating a binomial proportion.

of Y_i, not involving an integral that has to be numerically solved, as is the case for the above logistic model (5.11).

5.3.3.1.1 Example

Niël-Weise et al. (2007) reported a meta-analysis on the effect of anti-infectively treated central venous catheters on the probability of catheter related bloodstream infections (CRBI). The data are shown in Table 5.5.

For this example, we analyzed the standard catheter and anti-infective catheter groups separately, to get an overall incidence of CRBIs for each kind of catheter. (This is done for the sake of the example; in the next section, both groups will be meta-analyzed together as one will be interested in the comparison.) Notice that we have relatively small numbers of events, in particular in the anti-infective catheter groups, so the approximate normality of the log odds estimates is questionable. The results are given in Table 5.6.

For the standard catheter, the difference in estimated mean log odds between the two models is 0.20, meaning that the odds of getting a CRBI is estimated to be 22% higher (exp(0.20)-1) by the logistic regression compared with the standard approach. For the anti-infective catheter, the odds are as much as 73% higher by the logistic regression approach. Also, the estimated between-studies variance is considerably larger for the logistic regression method, which leads to a larger standard error for θ. A confidence interval for θ can be calculated using the standard error and the t-distribution with df = 17. The median incidence of CRBI and its confidence interval were calculated by back transformation to the

TABLE 5.5

Catheter Infection Risk Data: Number of Patients with a CRBI and Sample Sizes for Patients Using an Anti-Infective Catheter Compared with Patients Using a Standard Catheter

	Anti-infective catheter		Standard catheter	
Study i	No. of CRBIs Y_{i1}	No. of patients n_{i1}	No. of CRBIs Y_{i0}	No. of patients n_{i0}
1	0	116	3	117
2	1	44	3	35
3	2	208	9	195
4	0	130	7	136
5	5	151	6	157
6	1	98	4	139
7	1	174	3	177
8	1	74	2	39
9	1	97	19	103
10	1	113	2	122
11	0	66	7	64
12	0	70	1	58
13	3	188	5	175
14	6	187	11	180
15	0	118	0	105
16	0	252	1	262
17	1	345	3	362
18	4	64	1	69

TABLE 5.6

Catheter Infection Risk Data: Results of the Standard Two-Stage Model of Chapter 4 Fitted with REML and the Random-Intercept Logistic Regression Model (5.11), Separately Applied to the Data for the Standard and Anti-Infective Catheters

	Model	
	Standard two-stage (REML)	**Random-intercept logistic regression**
Standard catheter		
Mean log odds, θ (se)	−3.30 (0.24)	−3.50 (0.26)
Between-studies standard deviation of log odds, τ	0.81	0.90
Median incidence, π (95% CI)	0.035 (0.022, 0.057)	0.029 (0.018, 0.050)
Prediction interval for π_i	(0.006, 0.18)	(0.004, 0.18)
Anti-infective catheter		
Mean log odds, θ (se)	−4.26 (0.26)	−4.81 (0.36)
Between-studies standard deviation of log odds, τ	0.62	0.91
Median incidence, π (95% CI)	0.014 (0.008, 0.024)	0.008 (0.004, 0.017)
Prediction interval for π_i	(0.003,0.056)	(0.001, 0.060)

probability scale via $\pi = (1 + \exp(-\theta))^{-1}$. Notice that since this is a non-linear transformation, the mean on the log odds scale is not transformed into the mean but into the median on the probability scale. Therefore, the between-studies variation is quite skewed on the probability scale. Also notice that the confidence interval of the mean log odds for the random-intercept logistic regression is larger than for the standard method (as seen from the standard errors), while the confidence interval for the median incidence is shorter. The explanation is that the mean log odds estimate for the logistic regression method is closer to zero, and the log odds transformation stretches the probability scale more for probabilities close to zero. In this example, the difference between the two models is especially large, at least relatively, since the events are relatively rare. However, even if the events are not that rare, the difference is not necessarily negligible.

The phenomenon that the two-stage modeling approach of Chapter 4 for meta-analysis of proportions leads to bias toward 0.5 is well known in diagnostic test meta-analysis (e.g., Chu and Cole [2006]; Hamza et al. [2008a]; Chapter 19). The overall sensitivity and specificity tend to lie above 0.5 so are biased downwards, and the biases reinforce each other in the calculation of diagnostic odds ratios and summary receiver operating curves (Hamza et al., 2008b; Chapter 19). The prediction intervals for the true probability π_i of a new study were calculated by first using formula (4.19) to get a prediction interval for θ_i, and transforming the result back to the probability scale.

5.3.3.2 Meta-Analysis of Odds Ratios

Now, we consider the most frequently occurring case in medical meta-analysis, where we have a dichotomous outcome for two groups per study. As an alternative to the two-stage approach, the mixed effects logistic regression of the previous subsection can be straightforwardly extended to model the odds ratio. We now have two binomial distributions, one for each group. Thus, we just add the group as a fixed and random covariate to model (5.11).

We have a choice for what to do with the intercept: either we specify a fixed intercept for each study or specify a random intercept. This mixed effects logistic regression modeling approach has been discussed by several authors, see, for instance, Van Houwelingen et al. (1993), Turner et al. (2000), Stijnen et al. (2010), Simmonds and Higgins (2016), and Jackson et al. (2018).

Analogously to model (5.4) we can assume the following mixed effects logistic regression model

$$\text{logit}\,\Pr(Y_{ijk}=1)=\beta_{0i}+\beta_1 X_{ijk}+b_{1i}X_{ijk} \quad \text{with } b_{1i}\sim N(0,\tau^2). \tag{5.12}$$

In this model, every study has its own fixed intercept β_{0i}, interpreted as the log odds in the control group, β_1 is the mean treatment effect over studies, and b_{1i} is the deviation of the true effect of study i from the mean effect β_1. The b_{1i} are assumed to be independent zero mean normally distributed with between-studies variance τ^2.

Alternatively, analogously to model (5.5), instead of modeling the control log odds as fixed parameters, we can assume these are random:

$$\begin{gathered}\text{logit}\,\Pr(Y_{ijk}=1)=\beta_0+b_{0i}+\beta_1 X_{ijk}+b_{1i}X_{ijk}\\ \text{with } \begin{bmatrix} b_{0i}\\ b_{1i}\end{bmatrix}\sim N\left(\begin{bmatrix}0\\0\end{bmatrix},\begin{bmatrix}\tau_0^2 & \tau_{01}\\ \tau_{01} & \tau^2\end{bmatrix}\right).\end{gathered} \tag{5.13}$$

In this model, we assume that the true log odds in the control groups follow a normal distribution and β_0 represents the mean log odds in the control groups; in other words, we assume model (5.11) for the control groups. In fact, model (5.13) assumes that (5.11) holds for both the control and experimental groups, and additionally allows the control log odds and the experimental log odds to be correlated. Therefore, it can be used to study the relation between control rate and treatment effect (Chapter 14; Arends et al., 2000; Schmid et al. 1998; Schmid et al., 2004; Sharp et al., 1996).

Model (5.13) was introduced for meta-analysis in 1993 by Van Houwelingen et al. A practical hindrance for its use in practice at that time was that software to fit the model was not yet generally available. However, as powerful GLMM programs have become available in almost all general statistical packages, fitting models (5.12) and (5.13) is nowadays very easy in practice. Model (5.13) has become very popular in diagnostic test meta-analysis (Chapter 19; Chu and Cole, 2006; Arends et al., 2008). It is also available in special purpose meta-analysis software, for example, in the Stata collection of meta-analysis programs (`metandi`) (Harbord and Whiting, 2009). Model (5.12) is very common in Bayesian meta-analysis (Chapter 6; Smith et al., 1995), though (5.13), especially with $\tau_{01}=0$, is sometimes used as well.

GLMM programs permit giving structure to the covariance matrix of the random effects, though we consider it most natural to leave it unstructured. Nevertheless, one possibility might be to assume b_{0i} and b_{1i} to be independent. In addition, one could code $X_{ijk}=-0.5$ for the control and $X_{ijk}=0.5$ for the treatment. This has the advantage that the model is symmetric in the labeling of the groups, and this model has been implemented in `metafor` (Viechtbauer, 2010; model="UM.FS" in function `rma.glmm`).

Model (5.12) has the advantage that it assumes less than model (5.13). However, as for model (5.7), the common-effect equivalent of (5.12), the latter has the potential disadvantage that the number of parameters becomes larger as the number of studies becomes larger, violating a regularity condition needed for the likelihood method to possess its good properties (Van Houwelingen and Senn, 1999). Jackson et al. (2018), in an extensive simulation study, did not find evidence of bias in β_1, though only scenarios were simulated where the treatment effect was independent of the control log odds. However, they also found that the estimate of the between-studies variance τ^2 was seriously downwards biased. Therefore, they recommended to not use model (5.12). However, they show that model (5.12) with −0.5/+0.5 coding and independent random effects performs better.

Senn (2010) pointed out that using a random effect for the control group risk enables the model to make use of between-study information, which could bias the log odds ratio. This bias arises because by assuming the control group risks are exchangeable, the control group event rate in each study provides information about the control rates in other studies, and so the treatment effects in those studies. The simulations done by Jackson et al. (2018) indicate that the potential bias from the use of between-study information is probably negligible.

An alternative to models (5.12) and (5.13), avoiding the problems mentioned, is a random-effects conditional logistic regression model, which employs the likelihood of model (5.12) conditional on the total number of events, say Y_i, in a study. Since the number of events in a group given Y_i follows a non-central hypergeometric distribution, the within-study likelihood is

$$L_i(\beta_1, b_{1i}) = \frac{\binom{n_{i1}}{Y_{i1}}\binom{n_{i0}}{Y_i - Y_{i1}}\exp\left((\beta_1 + b_{1i})Y_{i1}\right)}{\sum_m \binom{n_{i1}}{m}\binom{n_{i0}}{Y_i - m}\exp\left((\beta_1 + b_{1i})m\right)}, \tag{5.14}$$

where n_{i0}, n_{i1}, and Y_{i0}, Y_{i1} are the sample sizes and numbers of events in group 0 and 1, respectively. The summation in the denominator of (5.14) is over the admissible values $\max(0, Y_{i1} - n_{i0}) \le m \le \min(Y_i, n_{i1})$. Unfortunately, CLR with random effects is not implemented in GLMM programs. However, the conditional likelihood can be maximized in SAS PROC NLMIXED, since that program allows a user specified likelihood. See Stijnen et al. (2010) for syntax examples. The method is also implemented in `metafor` (Viechtbauer, 2010).

Because of the sum in the denominator of (5.14), the likelihood can be computationally cumbersome, sometimes leading to numerical problems such as non-convergence or unreliable standard errors (Jackson et al., 2018). For that reason, CLR programs often use Breslow's approximation (Breslow, 1974). The right-hand side of (5.14) is the probability of observing Y_{i1} events with $X=1$ if Y_i events are randomly sampled without replacement from study i and the odds of an event with $X=1$ is $\exp(\beta_1 + b_{1i})$ times as large as with $X=0$. Breslow's approximation replaces sampling without replacement by sampling with replacement. Then the non-central hypergeometric likelihood changes into the much simpler likelihood of the following binomial distribution

$$Y_{i1} \sim \text{Binomial}\left(Y_i, \frac{\exp\left(\log(n_{1i}/n_{0i}) + \beta_1 + b_{1i}\right)}{1 + \exp\left(\log(n_{1i}/n_{0i}) + \beta_1 + b_{1i}\right)}\right). \tag{5.15}$$

This model can be fitted using the following model in the subset of patients with an event:

$$\text{logit}\, P(X_{ijk} = 1 \mid Y_{ijk} = 1) = \beta_1 + \log(n_{i1} / n_{i0}) + b_{1i} \quad \text{with } b_{1i} \sim N(0, \tau^2). \tag{5.16}$$

This is a simple random-intercept logistic regression model without covariates and with offset variable $\log(n_{i1}/n_{i0})$ and can be fitted in standard GLMM programs. Model (5.16) is fitted on the subset of individuals with the event in the individual patient dataset, and $\log(n_{i1}/n_{i0})$ should be available as a variable in this dataset in order to use it as an offset. If the number of events in a study is small compared with the group sizes, sampling with replacement will hardly differ from sampling without replacement, and the likelihood of (5.15) will be an excellent approximation of the exact conditional likelihood (5.14).

As for the CE model, studies with zero events in both groups are not included in the conditional likelihood. In the stratified-intercept model (5.12), these studies are included in the likelihood, but they do not contribute to the estimation of the overall treatment effect. In the random-intercept model (5.13) double-zero studies are included and can contribute to the estimation of the overall treatment effect through the use of between-study information.

5.3.3.2.1 *Example*

We continue the previous example, the data of which are given in Table 5.5, but now we focus on estimation of the treatment effect. We apply the above described mixed effects logistic regression methods and compare them with the approximate two-stage method (with the REML estimate of the between-studies variance). The results are given in Table 5.7. The logistic regression methods were fitted with SAS PROC GLIMMIX, except for the CLR, which was fitted with SAS PROC NLMIXED. The two-stage method was carried out with SAS PROC MIXED.

The estimates of the overall log odds ratio β_1 from the four logistic regression methods do not differ much, but the between-studies variance from the model with fixed study effects is much smaller than from the other three, in agreement with the simulation results of Jackson et al. (2018) which suggest that model (12) underestimates the between-studies

TABLE 5.7

Catheter Infection Risk Data: Estimated Mean Log Odds Ratio (Standard Error) and Between-Studies Variance (Standard Error) for Four Different Random-Effects Logistic Regression Methods Compared with the Two-Stage Random-Effects Model Fitted with REML

Model	$\hat{\theta}$ *(se)*	$\hat{\tau}^2$ *(se)*
Two-stage with REML (model [4.7])	−0.960 (0.238)	0.0140 (0.330)
Two-stage with REML, omitting double-zero study (model [4.7])	−0.980 (0.244)	0.036 (0.345)
LR with fixed study and random treatment effects (model [5.12])	−1.348 (0.316)	0.381 (0.452)
LR with random study and treatment effects (model [5.13])	−1.274 (0.373)	0.711 (0.631)
CLR with random treatment effects (model [5.14], exact likelihood)	−1.353 (0.351)	0.694 (0.640)
CLR with random treatment effects (model [5.16], approximate likelihood)	−1.303 (0.339)	0.601 (0.592)

For the two-stage method ½ was added to all cells in 2×2 tables with at least one zero count.

variance. The two-stage model fitted with REML is considerably different, as may be expected with such small counts: the estimated overall log odds ratio and the between-studies variance are much closer to 0 than in the LR models, probably because the large treatment effect means that the covariances in (5.6) do not cancel out, but also the other two problems mentioned in Table 5.1 apply.

5.4 Count Outcome

5.4.1 Common-Effect Approaches

In cohort studies, participants are followed until some event of interest occurs, while the duration of follow-up may differ among the participants. In meta-analysis of cohort studies, the summary statistics available per group are often the number of events and the total time of follow-up. Let the total number of events in group j of study i be denoted by Y_{ij}, the total time of follow-up by T_{ij} and the treatment by X_{ij}. In the two-stage approach of Chapter 4, the log incidence rate in a group is estimated as $\log(Y_{ij}/T_{ij})$ with standard error $1/Y_{ij}$. Two groups ($j=0,1$) are compared using the log incidence rate ratio as effect measure. The estimate of the effect measure and corresponding standard error are

$$\hat{\theta}_i = \log(Y_{i1}/T_{i1}) - \log(Y_{i0}/T_{i0}) \text{ and } s_i^2 = 1/Y_{i1} + 1/Y_{i0}, \tag{5.17}$$

and a normal distribution is assumed. The estimate and standard error from (5.17) form the input for the two-stage approaches of Chapter 4. If there are zero events, one usually takes $Y_{ij}=0.5$. Arguments against this use of the approximate normal within-study likelihood apply as before (Table 5.1). Therefore, we show how to replace it by the exact likelihood.

Model (5.17) is based on an underlying model that assumes a constant incidence rate over time across individuals in group j of study i, say $\exp(\theta_{ij})$. This model is also known as the exponential survival model and the corresponding method is sometimes referred to as the "person years" method. See, for instance, Breslow and Day (1980) and Clayton and Hills (1993) for an extensive discussion. The exact within-study likelihood can be derived as follows. Let Y_{ijk} be 1 or 0 if individual k in group j of study i experienced the event or not, and let T_{ijk} be the follow-up time of this individual, so that $Y_{ij} = \sum_k Y_{ijk}$ and $T_{ij} = \sum_k T_{ijk}$. According to the negative exponential distribution, the within-study likelihood contribution of an individual without the event is $\exp\left(-\exp(\theta_{ij})T_{ijk}\right)$ and of an individual with the event $\exp\left(-\exp(\theta_{ij})T_{ijk}\right)\exp(\theta_{ij})$. The contribution to the within-study likelihood of all individuals in group ij together then becomes

$$\prod_k \exp\left(-\exp(\theta_{ij})T_{ijk}\right)\exp(\theta_{ij})^{Y_{ijk}} = \exp\left(-\exp(\theta_{ij})T_{ij} + \theta_{ij}Y_{ij}\right). \tag{5.18}$$

Notice that this is the likelihood of a Poisson distribution (up to a factor not depending on the parameters θ_{ij}) with parameter $\exp(\theta_{ij})T_{ij}$. Thus, although Y_{ij} does not really have a Poisson distribution, we can use Poisson regression modeling with Y_{ij} as dependent variable. Therefore, assuming an equal incidence rate ratio across studies, we can adopt the following common-effect model

$$Y_{ij} \sim \text{Poisson}(\mu_{ij}) \text{ with } \log(\mu_{ij}) = \beta_{0i} + \beta_1 X_{ij} + \log(T_{ij}), \tag{5.19}$$

where $\log(T_{ij})$ is an offset variable. The β_{0i} are study-specific fixed intercepts, interpreted as the log incidence rates in the control groups, and β_1 is the common-effect measure expressed as the log incidence rate ratio. Note the analogy with model (5.1) for continuous and model (5.7) for dichotomous outcome, though a difference is that the dependent variable in (5.19) is at group-level, while in models (5.1) and (5.7) the dependent variable is the individual patient outcome or pseudo outcome. However, we can also formulate model (5.19) in terms of pseudo IPD outcomes, using $T_{ijk} = T_{ij}/n_{ij}$ (where n_{ij} is the group size) as the pseudo length of follow-up for patient k in group j of study i. Then model (5.19) becomes

$$Y_{ijk} \sim \text{Poisson}(\mu_{ijk}) \text{ with } \log(\mu_{ijk}) = \beta_{0i} + \beta_1 X_{ijk} + \log(T_{ijk}). \tag{5.20}$$

Under this model, the total contribution of all patients in group j of study i to the within-study likelihood is equal to the right-hand side of (5.18), thus (5.19) and (5.20) are indeed effectively identical models.

As an alternative for model (5.19), analogous to model (5.3) for continuous outcome and model (5.8) for dichotomous outcome, one could adopt a random-intercept model:

$$Y_{ij} \sim \text{Poisson}(\mu_{ij}) \text{ with } \log(\mu_{ij}) = \beta_0 + b_i + \beta_1 X_{ij} + \log(T_{ij}) \text{ and } b_i \sim N(0, \tau_0^2). \tag{5.21}$$

As an alternative to the normal distribution of the random effect in (5.21), one could assume a mean-zero log gamma distribution, as is standard practice in mixed effects survival analysis. Then (5.21) changes into a negative binomial regression model.

Comparing models (5.19) and (5.21), the same remarks apply as for the analogous unconditional logistic regression models discussed in Sections 5.3.1 and 5.3.2. Model (5.19) has the disadvantage that the number of parameters goes to infinity as the number of trials goes to infinity, violating a regularity condition of the theory of making maximum likelihood estimation. Model (5.21) has the disadvantage that in principle it uses between-study information (Senn, 2010). Analogous to the conditional regression approach of Section 5.3.2.3, we can avoid modeling the control group incidence rates by conditioning on the total number of events in a study $Y_i = Y_{i0} + Y_{i1}$. Given Y_i, Y_{i1} follows the following binomial distribution:

$$Y_{i1} \text{ given } Y_i \sim \text{Bin}\left(Y_i, \frac{\exp(\beta_1 + \log(T_{i1}/T_{i0}))}{1 + \exp(\beta_1 + \log(T_{i1}/T_{i0}))}\right). \tag{5.22}$$

Thus, the model reduces to a simple intercept-only logistic regression model with an offset variable. In practice, the model can be fitted using the individual patient dataset as indicated above. After constructing $\log(T_{i1}/T_{i0})$ as a variable in the dataset as the offset variable, one fits the following logistic model with X as the dependent variable in the subset of individuals with the event:

$$\text{logit}\Pr(X_{i1k} = 1 \mid Y_{i1k} = 1) = \beta_1 + \log(T_{i1}/T_{i0}). \tag{5.23}$$

In this subsection, we have shown that one-stage common-effect meta-analysis of incidence rates can be performed by fitting relatively simple Poisson, negative binomial, or logistic models. All models can be fitted in the generalized linear model programs available in all general statistical packages, except model (5.21) which requires a generalized linear mixed model program. In the next subsection, we will consider random-effects approaches by extending these models with a random group effect.

5.4.2 Random-Effects Approaches

5.4.2.1 Meta-Analysis of Incidence Rates

We first consider the situation where we have one group per study, for which we have the number of events Y_i and the total time of follow-up T_i. In the previous subsection, for the common-effect approaches, we did not consider this case explicitly. The reason is that if it is assumed that the true incidence rates are equal across studies, the data just reduce to the total number of events and the total time of follow-up. The meta-analysis is then trivial: just apply exact or approximate statistical methods for estimating the parameter of a Poisson distribution.

To estimate an overall incidence rate in the presence of heterogeneity across studies, we can adopt the following model.

$$Y_i \sim \text{Poisson}(\mu_i) \text{ with } \log(\mu_i) = \beta_0 + b_i + \log(T_i) \text{ and } b_i \sim N(0, \tau^2). \tag{5.24}$$

In fact, we have written down model (5.19) for only one group per study and extended it with a random effect. The result is a random-intercept Poisson regression model without covariates and with offset variable $\log(T_i)$. It can be fitted in the GLMM modules available in many popular statistical packages such as SAS, R, and Stata.

5.4.2.1.1 Example

Niël-Weise et al. (2008) reported a second meta-analysis on the effect of anti-infectively treated central venous catheters (CVCs) compared with standard catheters on the prevention of CRBIs. This meta-analysis considered the use of CVCs for total parenteral or chemotherapy, while the previous meta-analysis considered the use of CVCs in the acute care setting. The data are shown in Table 5.8. The structure of this meta-analysis is very similar

TABLE 5.8

Catheter Infection Rate Data: Numbers of CRBIs and Total Number of Catheter Days Per Group

	Anti-infective catheter		Standard catheter	
Study i	No. of CRBIs Y_{i1}	No. of catheter days T_{i1}	No. of CRBIs Y_{i0}	No. of catheter days T_{i0}
1	7	1491	11	1988
2	8	1638	8	1460
3	3	11484	14	10962
4	6	1446	10	1502
5	1	370	1	482
6	1	786	8	913
7	17	6840	15	6840
8	3	1156	7	1015
9	2	400	3	440

to that of Table 5.5, but in the present example, the follow-up lengths (number of days that a catheter was applied to a patient) of the patients varied, and thus we cannot assume a binomial distribution as we did in the example of Table 5.5, where the follow-up times were constant among all patients in a group. Therefore, the total number of catheter days is given as a group summary alongside the number of CRBIs, which makes it possible to compare the two different catheters in terms of the CRBI incidence rates. We call these data the catheter infection *rate* data to distinguish them from the catheter infection *risk* data. Once again, we begin by considering each treatment group separately in order to estimate an overall incidence of CRBI for each the two catheters, before estimating their relative efficacy.

The results of model (5.24) fitted to the two groups separately are given in Table 5.9. For comparison, the results of the standard two-stage approach with REML estimation are also given. The standard method uses $\log(Y_{ij}/T_{ij})$ and $1/Y_{ij}$ as effect measure and standard error; because these are respectively increasing and decreasing in Y_{ij}, they are negatively correlated.

In this example, the differences between the Poisson regression results and the two-stage approach are not that dramatic. As expected, because of the negative correlation between the estimated incidence and its standard error, the estimated incidence rates from the two-stage approach are somewhat higher; the increases (7% for the standard catheter and 14% for the anti-infective catheter) are somewhat more in the anti-infective catheter groups, probably because the events are more rare in these groups. The estimates of the between-studies variance are comparable in this example.

5.4.2.2 Meta-Analysis of Incidence Rate Ratios

Now we consider the situation that we have two groups per study. The standard two-stage meta-analysis model uses (5.17) as study level summary and standard error, and applies (4.12) with the DL or another estimate for the between-studies variance. To estimate the overall incidence rate ratio in the presence of heterogeneity across studies using the one-stage approach, we extend models (5.19), (5.21), or (5.23) with a random group effect, leading to the following models:

TABLE 5.9

Catheter Infection Rate Data: Results of Fitting the Standard Two-Stage Model and the Random-Intercept Poisson Regression Model (5.24)

	Model	
	Two-stage (REML)	**Random-intercept Poisson regression**
Standard catheter		
Mean log incidence rate, $\hat{\theta}$ (se)	1.47 (0.24)	1.40 (0.23)
Between-studies variance log incidence rate, $\hat{\tau}^2$ (se)	0.19 (0.19)	0.32 (0.20)
Median incidence rate, $\exp(\hat{\theta})$ (95% CI)	4.35 (2.72, 6.96)	4.06 (2.58, 6.36)
Anti-infective catheter		
Mean log incidence rate, $\hat{\theta}$ (se)	0.95 (0.31)	0.82 (0.32)
Between-studies variance log incidence rate, $\hat{\tau}^2$ (se)	0.60 (0.45)	0.60 (0.39)
Median incidence rate, $\exp(\hat{\theta})$ (95% CI)	2.59 (1.41, 4.75)	2.27 (1.21, 4.25)

The incidence rates are per year.

$$\log(\mu_{ij}) = \beta_{0i} + \beta_1 X_{ij} + b_{1i} X_{ij} + \log(T_{ij}) \text{ with } b_{1i} \sim N(0,\tau^2); \tag{5.25}$$

$$\log(\mu_{ij}) = \beta_0 + b_{0i} + \beta_1 X_{ij} + b_{1i} X_{ij} + \log(T_{ij})$$
$$\text{with } \begin{bmatrix} b_{0i} \\ b_{1i} \end{bmatrix} \sim MVN\left(\begin{bmatrix} 0 \\ 0 \end{bmatrix}, \begin{bmatrix} \tau_0^2 & \tau_{01} \\ \tau_{01} & \tau^2 \end{bmatrix}\right); \tag{5.26}$$

$$\text{logit}\Pr(X_{ijk} = 1 \mid Y_{ijk} = 1) = \beta_1 + b_{1i} + \log(T_{i1}/T_{i0}) \text{ with } b_{1i} \sim N(0,\tau^2). \tag{5.27}$$

Models (5.25) and (5.26) are mixed effects Poisson regression models for the group level summary data. Model (5.27) is a mixed effects logistic regression model for an individual patient dataset as described in the previous subsection, where the condition $Y_{ijk}=1$ indicates that only patients with the event are used. These models can be fitted in the GLMM programs of many general statistical packages. Model (5.25) is also implemented in `metafor` using an alternative coding $X_{ij}=-0.5$ and $X_{ij}=0.5$ for the control and treatment group, respectively. Model (5.26) is implemented in `metafor`, but only in the special case with the alternative coding for X and assuming independence between the random effects.

5.4.2.2.1 Example

We continue with the catheter infection rate data example of Table 5.8 and focus now on the estimation of the incidence rate ratio comparing the anti-infective catheter with the standard catheter. The results are given in Table 5.10.

The three GLMMs were fitted using SAS PROC GLIMMIX. As expected, the median incidence rate ratio as estimated by the GLMMs is somewhat more extreme than that of the two-stage approach, but not by much, probably because the covariance between the first term of $\hat{\theta}_i$ and the first term of s_i^2 in (5.6) largely cancels out against the covariance between the second terms. The conditional Poisson regression is the simplest model with the fewest assumptions and gives the largest treatment effect, a median rate ratio 8% lower than that of the two-stage method, and also the largest between-studies variance. The between-studies variance is estimated to be zero in the Poisson regression model with the fixed study effects: GLMM programs do not report a standard error in these cases.

TABLE 5.10

Catheter Infection Rate Data: Three One-Stage GLMM Approaches Compared with the Standard Two-Stage Approach

Model	$\hat{\theta}$ (se)	$\hat{\tau}^2$ (se)
Two-stage with REML	–0.42 (0.22)	0.08 (0.20)
Poisson regression with fixed study and random treatment effects (model [5.25])	–0.44 (0.18)	0
Poisson regression with random study and treatment effects (model [5.26])	–0.47 (0.24)	0.10 (0.17)
Conditional Poisson regression (model [5.27])	–0.50 (0.23)	0.11 (0.19)

5.5 Concluding Remarks

Traditionally, meta-analysis is performed using the standard two-stage approach, which makes use of an approximate within-study likelihood. However, often information is available that allows the use of the exact within-study likelihood. This chapter has shown how to do this in meta-analysis with normal, binary, or event count/follow-up time outcomes. The one-stage approach is identical or very similar to an IPD meta-analysis (Chapter 8).

The one-stage approach discussed in this chapter has several statistical advantages. It avoids the problems of the two-stage methods listed in Table 5.1. The advantages of the one-stage approach as listed in Table 5.3 apply to all methods introduced in this chapter, except for advantages 1 and 2, which only apply in the case of continuous outcome. All one-stage methods of this chapter can be carried out in the LMM and GLMM modules of the popular statistical packages. In fact, the one-stage approach brings meta-analysis largely back into mainstream statistics. The extension to meta-regression and multivariate meta-analysis including network meta-analysis is conceptually and practically straightforward by adding a fixed and random term to the (G)LMM for every extra treatment or extra outcome. If the number of parameters in the covariance matrix of the random effects becomes too large, then one of the simplified structures that are offered by GLMM programs can be used. Chapters 9, 10, and 15 as well as the papers by Arends et al. (2000) and Stijnen et al. (2010) give examples of multivariate meta-analysis and network meta-analysis.

When studies are large, with many events in case of non-continuous outcome, the statistical advantages of the one-stage models become less important. Two-stage methods have advantages too. One is that the two-stage approach estimates the overall effect explicitly as a weighted mean of the study-specific effect measures, which is simple, straightforward, and very easy for non-statisticians to understand. They also give explicit study weights which are instructive. In addition, the two-stage approach can be applied to any effect measure, such as the risk difference, the rate difference, and the risk ratio. Extension of the GLMM approach of this chapter to measures such as the risk difference and ratio is difficult, if not impossible, because these models allow group-specific probabilities to lie below 0 or above 1. In the Bayesian approach this is easier, since parameter values can be forced to be within their permissible regions (Warn et al., 2002).

All models discussed in this chapter are (G)LMM models, and the inference has been performed by applying standard likelihood methods (or REML in the LMM case). However, all models can be fitted by Bayesian methods too. These are described in Chapter 6.

References

Arends LR, Hoes AW, Lubsen J, Grobbee DE and Stijnen T, 2000. Baseline risk as predictor of treatment benefit: Three clinical meta-re-analyses. *Statistics in Medicine* **19**(24): 3497–3518.

Arends LR, Hamza TH, van Houwelingen JC, Heijenbrok-Kal MH, Hunink MGM and Stijnen T, 2008. Bivariate random effects meta-analysis of ROC curves. *Medical Decision Making* **28**(5): 621–638.

Arends LR, Voko Z and Stijnen T, 2003. Combining multiple outcome measures in a meta-analysis: An application. *Statistics in Medicine* **22**(8): 1335–1353.

Breslow NE, 1974. Covariance analysis of censored survival data. *Biometrics* **30**(1): 89–99.

Breslow NE and Day NE, 1980. *Statistical Methods in Cancer Research. Volume 1: The Analysis of Case-Control Studies*. Lyon, France: International Agency on the Research of Cancer.

Chang BH, Waternaux C and Lipsitz S, 2001. Meta-analysis of binary data: Which study variance estimate to use? *Statistics in Medicine* **20**(13): 1947–1956.

Chu H and Cole SR, 2006. Bivariate meta-analysis for sensitivity and specificity with sparse data: A generalized linear mixed model approach (letter to the Editor). *Journal of Clinical Epidemiology* **59**(12): 1331–1331.

Clayton DG and Hills M, 1993. *Statistical Models in Epidemiology*. Oxford, UK: Oxford University Press.

Fitzmaurice GM, Laird NM and Ware JH, 2011. *Applied Longitudinal Analysis*. 2nd Edition. John Wiley & Sons. ISBN 978-0-470-38027-7

Follmann DA and Proschan MA, 1999. Valid inference in random-effects meta-analysis. *Biometrics* **55**(3): 732–737.

Greenland S and Salvan A, 1990. Bias in the one-step method for pooling study results. *Statistics in Medicine* **9**(3): 247–252.

Hamza TH, Reitsma JB and Stijnen T, 2008b. Meta-analysis of diagnostic studies: A comparison of random intercept, normal-normal, and binomial-normal bivariate summary ROC approaches. *Medical Decision Making* **28**(5): 639–649.

Hamza TH, Van Houwelingen HC and Stijnen T, 2008a. The binomial distribution of meta-analysis was preferred to model within-study variability. *Journal of Clinical Epidemiology* **61**(1): 41–51.

Harbord RM and Whiting P, 2009. metandi: Meta-analysis of diagnostic accuracy using hierarchical logistic regression. *The STATA Journal* **9**(2): 211–229.

Higgins JPT, Thompson SG and Spiegelhalter DJ, 2009. A re-evaluation of random-effects meta-analysis. *Journal of the Royal Statistical Society: Series A* **172**(1): 137–159.

Jackson D, Law M, Stijnen T, Viechtbauer W and White IR, 2018. A comparison of 7 random-effects models for meta-analyses that estimate the summary odds ratio. *Statistics in Medicine* **37**(7): 1059–1085.

Lebrec JJ, Stijnen T and van Houwelingen JC, 2010. Dealing with heterogeneity between cohorts in Genomewide SNP Association Studies. *Statistical Applications in Genetics and Molecular Biology* **9**(1): Article 8.

Legha A, Riley RD, Ensor J, Snell KIE, Morris TP, Danielle L and Burke DL, 2018. Individual participant data meta-analysis of continuous outcomes: A comparison of approaches for specifying and estimating one-stage models. *Statistics in Medicine* **37**(29): 4404–4420.

Lopes da Silva S, Vellas B, Elemans S, Luchsingerd J, Kamphuis P, Yaffee K, Sijbena J, Groenendijk M and Stijnen T, 2014. Plasma nutrient status of patients with Alzheimer's disease: Systematic review and meta-analysis. *Alzheimer's & Dementia* **10**(4): 485–502.

Mantel N and Haenszel W, 1959. Statistical aspects of the analysis of data from retrospective studies of disease. *Journal of the National Cancer Institute* **22**(4): 719–748.

Niël-Weise BS, Stijnen T and van den Broek PJ, 2007. Anti-infective-treated central venous catheters: A systematic review of randomized controlled trials. *Intensive Care Medicine* **33**(12): 2058–2068.

Niël-Weise BS, Stijnen T and Van den Broek PJ, 2008. Anti-infective treated central venous catheters for total parenteral nutrition or chemotherapy? A systematic review. *The Journal of Hospital Infection* **69**(2): 114–123.

Papadimitropoulou K, Stijnen T, Dekkers OM and le Cessie S, 2019. One-stage random effects meta-analysis using linear mixed models for aggregate continuous outcome data. *Research Synthesis Methods* **10**(3): 360–375.

Robins J, Breslow N and Greenland S, 1986. Estimators of the Mantel-Haenszel variance consistent in both sparse data and large-strata limiting models. *Biometrics* **42**(2): 311–323.

Schmid CH, Lau J, McIntosh MW and Cappelleri JC, 1998. An empirical study of the effect of the control rate as a predictor of treatment efficacy in meta-analysis of clinical trials. *Statistics in Medicine* **17**(17): 1923–1942.

Schmid CH, Stark PC, Berlin JA, Landais P and Lau J, 2004. Meta-regression detected associations between heterogeneous treatment effects and study-level, but not patient-level, factors. *Journal of Clinical Epidemiology* **57**(7): 683–697.

Senn S, 2010. Hans van Houwelingen and the art of summing up. *Biometrical Journal* **52**(1): 85–94.
Sharp SJ, Thompson SG and Altman DG, 1996. The relation between treatment benefit and underlying risk in meta-analysis. *British Medical Journal* **313**(7059): 735–738.
Simmonds MC and Higgins JP, 2016. A general framework for the use of logistic regression models in meta-analysis. *Statistical Methods in Medical Research* **25**(6): 2858–2877.
Smith TC, Spiegelhalter DJ and Thomas A, 1995. Bayesian approaches to random effects meta-analysis: A comparative study. *Statistics in Medicine* **14**(24): 2685–2699.
Stijnen T, Hamza TH and Özdemir P, 2010. Random effects meta-analysis of event outcome in the framework of the generalized linear mixed model with applications in sparse data. *Statistics in Medicine* **29**(29): 3046–3067.
Turner RM, Omar RZ, Yang M, Goldstein H and Thompson SG, 2000. A multilevel model framework for meta-analysis of clinical trials with binary outcomes. *Statistics in Medicine* **19**(24): 3417–3432.
Van Houwelingen HC, Zwinderman KH and Stijnen T, 1993. A bivariate approach to meta-analysis. *Statistics in Medicine* **12**(24): 2273–2284.
Van Houwelingen JC and Senn S, 1999. Investigating underlying risk as a source of heterogeneity in meta-analysis. *Statistics in Medicine* **18**: 107–115.
Viechtbauer W, 2010. Conducting meta-analyses in R with the metafor package. *Journal of Statistical Software* **36**(3): 1–48.
Warn DE, Thompson SG and Spiegelhalter DJ, 2002. Bayesian random effects meta-analysis of trials with binary outcomes: Methods for the absolute risk difference and relative risk scales. *Statistics in Medicine* **21**(11): 1601–1623.
Yusuf S, Peto R, Lewis J, Collins R and Sleight P, 1985. Beta blockade during and after myocardial infarction: An overview of the randomized trials. *Progress in Cardiovascular Diseases* **27**(5): 335–371.
Zhou XH, Brizendine EJ and Pritz MB, 1999. Methods for combining rates from several studies. *Statistics in Medicine* **18**(5): 557–566.

6

Bayesian Methods for Meta-Analysis

Christopher H. Schmid, Bradley P. Carlin, and Nicky J. Welton

CONTENTS

6.1 Introduction

Chapters 4 and 5 introduced approaches to estimate an average effect across studies providing data either as summaries of a contrast between groups or as treatment arm-level summary observations. All the methods relied on a frequentist approach that seeks to estimate model parameters either by maximizing the likelihood or by method of moments.

This chapter describes the Bayesian approach to meta-analysis, wherein inferences are based on a posterior distribution that describes knowledge about model parameters conditional on the data and prior or external information (Schmid, 2001; Schmid and Mengersen, 2013). As data accrue, the posterior can be continually updated to reflect information coming into the likelihood. Statements about the probability of scientific hypotheses follow directly from the posterior distribution. For example, the hypothesis that cleaning with an antibacterial product reduces the probability of infection by at least 10% can be evaluated by the posterior probability that a parameter representing the change in the probability exceeds 10%.

We begin in Section 6.2 by reviewing how the likelihood and prior distribution are combined into a posterior distribution and how prior distributions may be chosen. Section 6.3 discusses various point and interval summary estimates from the posterior distribution. Sections 6.4 and 6.5 then introduce the common and random meta-analysis models from previous chapters in a Bayesian context, discussing how the hierarchical structure of data from individuals nested within studies that are accumulating over time fits naturally into a Bayesian framework. Section 6.6 reviews Bayesian computation using Markov chain Monte Carlo (MCMC); Section 6.7 discusses models for discrete outcomes; Section 6.8 covers prediction; Section 6.9 covers model fit and assessment of model assumptions; Section 6.10 discusses potential model generalizations, Section 6.11 discusses software; and finally Section 6.12 summarizes the advantages and disadvantages of the Bayesian approach.

6.2 Bayesian Inference

Scientific knowledge commonly increases by incorporating new evidence to update our beliefs based on a model of previous evidence. The Bayesian approach reflects this learning process. In the Bayesian framework, the parameters Θ of a model are treated as random variables about which we have uncertainty. Bayesian inference seeks to describe our knowledge about the parameters given information available to us coming from the observed data $\mathbf{y}$ through the process that generated the data, the likelihood $p(\mathbf{y}|\Theta)$, and from our prior beliefs about the parameters described by a prior distribution $p(\Theta)$. For example, if we have data from a randomized controlled trial (RCT) comparing a new antibacterial product with standard care and reporting the number of patients developing an infection, we might assume that the data are generated through a binomial model with parameters Θ representing the probability of infection in each treatment group. The parameter of interest is usually the difference in these probabilities or some function of them such as their logits for a log odds ratio. Once generated, the data observed are considered as known and fixed (not random). Our knowledge about the parameters conditional on the data observed is expressed by a probability distribution that quantifies our beliefs about the values of these parameters after (or posterior to) the data.

In contrast, the frequentist framework considers parameters as unknown, fixed quantities and the data as a random sample from a sampling distribution. Inferences are made by referencing the sampling distribution of pivotal quantities that are functions of the model parameters. In the example, the difference in the probability of infection of the two treatment groups is an unknown constant and the data sampled are observed outcomes from a binomial sampling distribution. One consequence of these definitions is that frequentist inference is made conditional on the design from which the sampling distribution derives. Inferences involve probabilities of data that were possible given the study design, including both data observed and data unobserved. Bayesian inferences on the other hand condition only on the data observed and the model that leads to the likelihood. Designs with the same likelihood give the same inferences about model parameters (Berger and Wolpert, 1984; Carlin and Louis, 2009).

A Bayesian model has two components: the likelihood and the prior. The likelihood, introduced in Chapter 5, is the probability of the observed data under a specific model. Because the data are considered fixed at their observed values, the likelihood is a function of the parameters for a given model. As seen in earlier chapters, the likelihood plays a major role in frequentist methods, many of which estimate parameters by the value that maximizes their observed likelihood. The likelihood function is also central to the Bayesian approach because it summarizes all the information that the data can tell us about the parameters. The form of the likelihood follows from the model describing the outcome as a function of predictors and model parameters. A Bayesian must also choose a prior reflecting information about the parameters external to the data. Because the prior and the likelihood are both part of the Bayesian model, it is important to justify and make transparent how and why they were chosen, and perform sensitivity analyses that show how the choice affects inferences. This becomes especially important when the amount of information in the data is small and the likelihood is relatively flat. The prior can then have a large influence on the posterior.

Combining the prior distribution with information from the data through the likelihood, the updated knowledge about the parameters is represented using Bayes' rule by the *posterior* probability distribution

$$p(\Theta \mid \mathbf{y}) = \frac{p(\mathbf{y} \mid \Theta)p(\Theta)}{p(\mathbf{y})} = \frac{p(\mathbf{y} \mid \Theta)p(\Theta)}{\int p(\mathbf{y} \mid \Theta)p(\Theta)d\Theta} \propto p(\mathbf{y} \mid \Theta)p(\Theta) \tag{6.1}$$

The posterior distribution is proportional to the product of the likelihood and the prior because the denominator $p(\mathbf{y})$ involves an integral whose result does not depend on Θ, and is therefore just a constant that ensures that $p(\Theta|\mathbf{y})$ integrates to 1. Because the integral is generally hard to compute in closed form, Bayesians have developed numerical tools such as Markov chain Monte Carlo (MCMC) discussed in Section 6.6 that avoids having to explicitly evaluate the integral.

Often, the model parameters may depend on parameters of their own, called hyperparameters, leading to a multilevel model. The random-effects model introduced in Chapter 4 is an example. The individual study treatment effect parameters $\boldsymbol{\theta} = \{\theta_1, \theta_2, \ldots, \theta_I\}$ are modeled from a common distribution with its own mean μ and variance τ^2. In meta-analysis, we are usually most interested in μ and τ^2 (or τ), but the θ_i are often of interest as well. Note that μ here is the same parameter as θ in Chapter 4. In this chapter, we reserve $\boldsymbol{\theta}$ for the vector of study effects.

We can compute the marginal posterior distribution for μ by integrating over $\boldsymbol{\theta}$ and τ^2 as

$$p(\mu \mid \mathbf{y}) = \frac{\int p(\mathbf{y} \mid \boldsymbol{\theta})p(\boldsymbol{\theta} \mid \mu, \tau^2)p(\mu, \tau^2)d\boldsymbol{\theta}d\tau^2}{\iint p(\mathbf{y} \mid \boldsymbol{\theta})p(\boldsymbol{\theta} \mid \mu, \tau^2)p(\mu, \tau^2)d\boldsymbol{\theta}d\tau^2 d\mu}. \tag{6.2}$$

The posterior for $\boldsymbol{\theta}$ is computed in a similar manner integrating out μ instead. Often the priors for (μ, τ^2) are taken to be independent of each other.

The between-study variance τ^2 is important when making predictive inferences, but is more of a nuisance parameter if the interest is only in μ. In this case, we might condition on an estimate $\hat{\tau}^2$ and draw inferences for $\boldsymbol{\theta}$ and μ using their joint posterior distribution conditional on $\hat{\tau}^2$, $p(\boldsymbol{\theta}, \mu \mid \mathbf{y}, \hat{\tau}^2)$. This approach is referred to as *empirical* Bayes analysis; see Carlin and Louis (2009, chapter 5) for a brief review of empirical Bayes methodology and applications. Many of the approaches for handling between-study variance outlined in Chapter 4 are of this type. One-stage models, described in Chapter 5, have a further set of parameters, the study intercepts, that must be considered.

6.2.1 Choosing a Prior Distribution

Posterior inference depends on both the prior and the likelihood, and a fully specified Bayesian model must specify both. In fact, if two Bayesian analysts collect the same data and use the same likelihood, then their inferences can only differ if they use different priors. Choice of prior is therefore important and must follow the same scientific principles as when choosing a likelihood, relying on evidence and transparency for proper justification. Priors may be strong, or informative, with a tight distribution fully reflecting external information; or they may be weak, or non-informative, with a wide distribution reflecting substantial uncertainty so that information from the likelihood dominates. In general, prior distributions may be constructed from several perspectives (Welton et al., 2012; Carlin and Louis, 2009).

- *Historical evidence or previous studies* Results from a previous study or meta-analysis of previous studies often provide an objective prior distribution for a new study. In fact, cumulative meta-analysis, in which studies are ordered sequentially and the meta-analysis is updated after each new study (Lau et al., 1992), essentially provides a running history of the posterior distribution (Lau et al., 1995). Historical controls, observational studies, and lower quality trials may also provide evidence, although they may need to be adjusted for bias (Turner et al., 2009) and downweighted (Efthimiou et al., 2017).
- *Expert opinion* Prior information can be formally and informally elicited from experts, especially when no suitable data are available to construct an objective prior or when it is important to transparently include expert opinion in a decision-making process. Methodology and examples in which the prior distributions are based upon formally elicited subjective beliefs are discussed by Chaloner et al. (1993) and O'Hagan et al. (2006). This method may be less useful in meta-analysis since the experts' opinions may be based partly on the results of studies in the meta-analysis.
- *Calibrated enthusiasm* This can be considered a type of sensitivity analysis in which two or more parties with a stake in a decision may represent their differing points of view through their prior distributions which are then calibrated through the likelihood. If their posteriors converge, the parties can reach agreement; if they do not converge, then they may better understand their disagreement. Product regulation offers a good example (Spiegelhalter et al., 1994). A government regulatory agency will have a much more skeptical view than a drug manufacturer and will therefore choose a prior that puts a much smaller probability on the possibility of a large effect. One might also choose to place a high prior probability on the point null hypothesis (Berger and Sellke, 1987). The contrasting views of an environmental activist and a coal manufacturer offer an example from ecology

(Schmid and Mengersen, 2013). One can then ask whether the data will persuade a skeptic that the treatment is effective or an enthusiast that it is not effective. Such priors can also be useful when designing a study to ensure that the data are sufficient to be able to overwhelm such priors.

- *Vague/non-informative priors* When little is known about the parameters of interest outside of the data or if one wishes to make inferences relying primarily on the likelihood, then one can choose a prior that is constant over plausible ranges of the parameters and provides little information about them. Models using non-informative priors are dominated by the likelihood and so the inferences made from them resemble frequentist inferences. Many statisticians use non-informative priors as a convenience to retain the advantages of using Bayesian simulation to compute posterior distributions for complex models whose likelihoods are not easily optimized. Often, it is useful to tailor the flat prior to the problem at hand by bounding it to exclude values deemed implausible by experts or by experience from the previous literature.
- *Computational convenience* Certain priors are chosen with specific likelihoods because they simplify computing the posterior, avoiding the need to evaluate the integral in the denominators of (6.1) and (6.2). These *conjugate* priors have the property that the posterior is in the same family as the prior distribution. Many common distributions that we meet in meta-analysis generate likelihoods to which conjugate priors exist (e.g., a normal distribution is conjugate for a normal likelihood, a beta distribution is conjugate to a binomial likelihood, and a gamma distribution is conjugate to a Poisson likelihood). In some problems, conjugacy only applies to a subset of the parameters, but this can still simplify numerical computations by breaking the problem up into pieces in which some parameters can be optimized conditional on others. An example is the normal random-effects model where a normal density conditional on the between-study variance τ^2 can be combined with a conjugate normal prior distribution for $\boldsymbol{\theta}$ and μ to compute the conditional posterior distribution $p(\theta, \mu \mid \mathbf{y}, \tau^2)$ analytically.

Whatever the choice of prior distribution, it should have the property that as the amount of information provided by the data increases in relation to the information provided by the prior, the likelihood will overwhelm the prior and will drive posterior inference. This will hold as long as the prior does not preclude values that are supported by the data. It is also important that sensitivity analyses examine the impact of different reasonable priors on final inferences, just as analysts explore different models and the effect of outlying data points on the likelihood. Strong conclusions should not be drawn from models sensitive to plausible alternative specifications of the prior. Analysts should be transparent about and should justify assumptions made in the model determining the likelihood and the prior.

6.3 Summarizing the Posterior Distribution

A frequentist (or classical) approach produces a point estimate and a standard error of the parameter from which any level of confidence interval can be constructed. A Bayesian approach produces a posterior probability distribution. Sometimes the functional form of the posterior distribution may be known, but more often the posterior is computed via MCMC simulation (see Section 6.6). MCMC returns a set of simulated draws from the posterior that can be used to characterize the entire posterior distribution. Although the set of frequentist confidence intervals can be used to characterize the relative probabilities of different sample statistics, they rely on the probability distribution used to construct the confidence intervals. Bayesian confidence intervals (called credible intervals), on the other hand, derive from simulation and are therefore more flexible.

6.3.1 Point Estimation

The posterior mean, median, or mode can be used as a point estimate for a scalar parameter θ. Under a relatively flat prior, the posterior mode will be close to the maximum likelihood estimate. If the posterior is normal, the three measures are the same, but for multimodal or otherwise non-normal posteriors, such as for a variance parameter, the mode will often be the poorest choice of centrality measure. The posterior mean will sometimes be overly influenced by heavy tails (just as the sample mean is often not robust against outlying observations). As a result, the posterior median will often be the best and safest point estimate and is relatively easy to compute using MCMC.

6.3.2 Interval Estimation

The posterior distribution allows us to make direct statements about not just its median, but any quantile or interval. For instance, a $100\times(1-\alpha)\%$ credible interval for θ is an interval (q_L, q_U) such that $P(q_L < \theta < q_U \mid y) = 1-\alpha$. Such an interval is easy to compute from MCMC simulations and has a direct interpretation as an interval within which θ lies with probability $(1-\alpha)$. Most commonly, the interval is chosen to exclude $\alpha/2$ of the probability on each side. By contrast, a classical confidence interval refers to the probability that the procedure producing the interval will correctly include the true parameter if the procedure is repeated a large number of times. A credible interval and a confidence interval will be numerically similar, though, when little prior information is available about any of the model parameters. Note that for non-symmetric posterior distributions such as odds ratios, a credible interval need not be symmetric about the center of the posterior distribution even though the amount of probability outside the interval is the same on each side. For highly skewed distributions such as those for variances, one-tailed intervals may actually be preferred. In fact, many credible intervals that encompass a given posterior probability exist. For multimodal posterior distributions, the shortest credible interval, meaning the smallest one that includes $100\times(1-\alpha)\%$ of the posterior distribution, may even consist of the union of disjoint intervals.

In general, MCMC simulation will return estimates of Bayesian confidence intervals that are true probabilistic estimates of uncertainty and do not rely on asymptotic normal approximations. Marginal and conditional distributions can be evaluated and probabilities may be computed for different hypotheses. For instance, one can compute probabilities of hypotheses such as the probability is 0.95 that the mean difference between two treatment groups is at least 2.0, or the probability is 0.6 that the mean difference lies between 5.0 and 10.0.

6.4 Two-Stage Common-Effect Model

Recall the two-stage common-effect model for summary effects (e.g., log odds ratios, log hazard ratios, mean differences) following a normal distribution that was introduced in

Chapter 4. In this model, the mean effects $\theta = (\theta_1,\ldots,\theta_I)$ in studies $i = 1,\ldots,I$ are assumed to be the same, equal to a common μ, and the observed $y = (y_1,\ldots,y_I)$ vary only because of within-study variability (i.e., sampling error), so that $y_i \sim N(\mu, s_i^2)$. As noted in Chapter 4, s_i^2 is often assumed known, fixed at its observed value, and we shall assume that here, for now. The likelihood may then be written

$$L(\mu;y) = \prod_{i=1}^{I} \frac{1}{s_i} e^{-\frac{(y_i-\mu)^2}{2s_i^2}}.$$

As a prior distribution for μ, we often take $\mu \sim N(\mu_0, \sigma_\mu^2)$ because it is a *conjugate* prior, meaning that the resulting posterior distribution is in the same family (normal) as the prior distribution. Conjugate priors are computationally convenient because, here, their use avoids the need to numerically evaluate the integral in the denominator of (6.1) and (6.2).

The parameters of the prior distribution form the set of hyperparameters $\lambda = \{\mu_0, \sigma_\mu^2\}$. Combining this with the likelihood gives a posterior distribution

$$p(\mu \mid y, \lambda) \propto \frac{1}{\sigma_\mu} e^{-(\mu-\mu_0)^2/2\sigma_\mu^2} \prod_{i=1}^{I} \frac{1}{s_i} e^{-(y_i-\mu)^2/2s_i^2}$$

which can be seen after some algebra to be a normal distribution with mean

$$\mu_1 = \frac{c_1\bar{y}_w + c_0\mu_0}{c_1 + c_0} \tag{6.3}$$

and precision (inverse of variance)

$$1/\sigma_1^2 = c_0 + c_1 \tag{6.4}$$

with

$$\bar{y}_w = \frac{\sum y_i/s_i^2}{\sum 1/s_i^2}, \quad c_0 = \frac{1}{\sigma_\mu^2} \text{ and } c_1 = \sum \frac{1}{s_i^2}.$$

These expressions are easier to interpret by noting that the weights c_0 and c_1 are precisions (inverse variances). Then $\bar{y}_w$ is just a weighted average of the study effects with weights equal to the study precisions $1/s_i^2$. Treating the prior distribution as a single observation μ_0 with precision c_0, the posterior mean is then a weighted average of this prior "observation" and the data, while the posterior precision is the sum of the prior precision and the data precision. Therefore, for this normal model, the posterior mean will always lie between the prior mean and the data mean, while the posterior precision will always be greater than either the prior precision or the data precision (and thus, the posterior variance will always be less than the prior variance and data variance). If a non-informative prior distribution $N(0,A)$ for large A, say 10^6, is chosen for μ, then $c_0 \approx 0$ and the posterior mean and variance reduce to the likelihood-based common-effect estimates, $\bar{y}_w$ and its squared standard error $1/\sum 1/s_i^2$, respectively, from Chapter 4.

6.4.1 Common-Effect Example

As an example, consider the meta-analysis introduced in Table 4.2 of Chapter 4 summarizing the difference between Alzheimer's patients and controls for mean plasma iron levels. Recall that the inverse variance weighted common effect for the unstandardized difference in means was −6.78 with a standard error of 2.85 and a 95% confidence interval of (−12.37, −1.19). As noted in Chapter 4, the inverse variance estimate is equivalent to the maximum likelihood estimate for normal distributions.

The Bayesian model requires a prior distribution on the mean difference. For now, assume that we know little about this difference and so assume a mean of 0 and a standard deviation of 20. This implies that we are about 99% sure that the average mean difference is between −50 and 50. In practice, it will not matter much how large one chooses the standard deviation of the normal prior distribution as long as the distribution has support that covers all plausible values of the parameter. Using equations (6.3) and (6.4) above, the posterior mean (also the median since the posterior is normal) is −6.62 with standard deviation 2.83 (first line of Table 6.1). This changes little if we increase the prior standard deviation to 100 (second line of Table 6.1). The posterior looks very much like the likelihood, reflecting the weak prior evidence. Inference is very similar whether using a Bayesian or frequentist (likelihood) approach, but the Bayesian approach permits direct probabilistic statements about μ. For example, the probability that the mean iron level in Alzheimer's patients is less than that in controls is $P(\mu < 0 \mid y, \lambda) = \Phi((0 - -6.65)/2.82) = 0.99$ where Φ is the normal distribution function. We could also calculate the probability that the Alzheimer's patients have at least 5 g/dL less levels of iron as $\Phi((-5 - -6.65)/2.82) = 0.72$.

If on the other hand, we had substantial external evidence that differed from the likelihood, the posterior would look more like the prior. For instance, what if we knew of many negative unpublished studies or wanted to investigate the sensitivity of our conclusions to such publication bias (discussed more extensively in Chapter 13). Then we might assume a prior distribution centered at a mean difference of 0 but with a much smaller standard deviation, say 1, implying that it was much less likely that the true difference was large. This prior would make highly unlikely any differences of more than 4 g/dL. Under this prior, the posterior mean and variance are quite different with a posterior mean close to zero and a much tighter credible interval (third line of Table 6.1).

Finally, if auxiliary information provided evidence that iron levels were lower in Alzheimer's patients, we might use a prior mean different from 0. For instance, assume that a prior study indicated that iron levels were 5 g/dL lower and that the standard error of this difference was 2. Line 4 of Table 6.1 shows that the posterior is then weighted more heavily toward lower levels for Alzheimer's patients. In fact, the posterior probability that the mean levels are lower in Alzheimer's patients is now 0.9997.

TABLE 6.1

Posterior Mean (Median), Standard Error, and 95% Credible Intervals for Difference in Iron Levels (μg/dL) in Alzheimer's Patients Compared with Controls

Prior mean	Prior SD	Posterior mean	Posterior SD	95% CrI	$P(\mu < 0 \mid y)$
0	20	−6.65	2.82	−12.18, −1.11	0.99
0	100	−6.78	2.85	−12.37, −1.19	0.99
0	1	−0.74	0.94	−2.59, 1.11	0.78
−5	2	−5.59	1.64	−8.80, −2.38	0.9997

Clearly, choice of prior distribution can make a large difference in the posterior estimates. The prior must be chosen carefully based on the most accurate information one has about the model parameters.

6.5 Two-Stage Random-Effects Models

As was noted in Chapter 4, the common-effect model is not too realistic for most meta-analyses because of heterogeneity among the study effects. Instead, a random-effects model is likely preferable. In a random-effects model, each study has its own treatment effect θ_i. We write the model in two stages. The first describes how each observed effect y_i is generated from a random effect θ_i and within-study error s_i^2 as

$$y_i \sim N(\theta_i, s_i^2). \tag{6.5}$$

The second step describes the generation of the random effects from their common distribution. If this distribution is also normal, then

$$\theta_i \sim N(\mu, \tau^2). \tag{6.6}$$

In such a model, the θ_i are said to be exchangeable because their joint distribution does not depend on their order. In this hierarchical structure, we are interested in both the study level random effects θ_i and the population parameters μ and τ^2. We can think of the second stage model $\theta_i \sim N(\mu, \tau^2)$ as the prior distribution for each study effect, θ_i, with the likelihood supplied by $y_i \sim N(\theta_i, s_i^2)$. The relationships generated by the exchangeability among the parameters in the hierarchical structure and their common random-effects distribution expressed through the hyperparameters allows us to estimate the individual random effects as well as their mean and variance. The posterior distribution of each study treatment effect is thus informed by its own results as well as the results from the other studies. The posterior study effects are weighted averages of the observed study effects and the overall mean and are said to be *shrunk* toward the mean.

Conditioning on the assumed known within-study variances, s_i^2, we can write the posterior as

$$p(\theta_1,\ldots,\theta_I,\mu,\tau^2 \mid y, s_1^2,\ldots,s_I^2) \propto p(y \mid \theta_1,\ldots,\theta_I, s_1^2,\ldots,s_I^2)p(\theta_1,\ldots,\theta_I \mid \mu,\tau^2)p(\mu)p(\tau^2), \tag{6.7}$$

where we have assumed that the prior distributions of μ and τ^2 are independent. To compute the posterior of μ, note that

$$p(\mu \mid y) = \int p(\mu \mid y, \tau^2)p(\tau^2 \mid y)d\tau^2 \tag{6.8}$$

where $\mu \mid y, \tau^2 \sim N(\mu_1, \sigma_1^2)$ takes the same form as in the common-effect model (6.3) except that $\tau^2 + s_i^2$ replaces s_i^2 in the expressions for $\bar{y}_w$ and c_1 because $y_i \sim N(\mu, s_i^2 + \tau^2)$ (Carlin, 1992). Well-estimated studies with small within-study variances s_i^2 get the most weight. As the random effect (between-study) variance τ^2 grows large, however, all studies will tend to be weighted fairly evenly. Under a non-informative prior for μ, the posterior mean and variance for $\mu \mid y, \tau^2$ again reduce to the likelihood-based estimates.

In practice, τ^2 is unknown and is poorly estimated when the number of studies in the meta-analysis is small. Chapter 4 discussed a variety of methods for estimating τ^2 and attempting to incorporate this uncertainty into the analysis. The Bayesian model does this automatically as (6.8) shows that the marginal posterior $\mu|y$ is the conditional posterior $\mu|y,\tau^2$ averaged over $p(\tau^2|y)$. By substituting an estimate $\hat{\tau}^2$ for τ^2, instead, we recover an empirical Bayes estimate of μ. This is theoretically inferior to the full Bayes estimator as it ignores the uncertainty introduced by estimating τ^2. The ability to properly account for the uncertainty in the estimation of between-study heterogeneity is a key feature of Bayesian models for meta-analysis. Because the between-study variance, τ^2, is a model parameter, the posterior distributions of the remaining parameters including the treatment effects of most interest are averaged over the full posterior and neither assume no between-study heterogeneity (common-effect models) nor condition on an estimate of τ^2 as do most of the non-Bayesian methods discussed previously. Both of these last two approaches ignore large values that the data might support for τ^2 which might substantially change the weighting of the different studies. Incorporating the uncertainty about τ^2 leads to wider credible intervals and more uncertainty about the treatment effects than in an empirical Bayes approach, but this is proper given the usually small number of studies and therefore small amount of information about study-to-study variation.

Because their analytic formulas are complex, computing the full marginal posteriors $\mu|y$ and $\tau^2|y$ requires numerical methods, such as MCMC. For instance, one could sample from the joint posterior of $(\theta_1,\ldots,\theta_I,\mu,\tau^2)$ by first sampling τ^2 from $p(\tau^2|y)$ numerically, then sampling μ from $p(\mu|\tau^2,y)$, and finally sampling θ_i from $p(\theta_i|\mu,\tau^2,y)$ (Gelman et al., 2013, chapter 5). Doing so, of course, requires specifying a prior for τ^2. This requires some care. It is worth noting that since τ^2 is a variance, any prior is defined only for $\tau^2>0$ and therefore by definition rules out the common-effect model.

Choosing a non-informative prior distribution for the between-study variance τ^2 or precision $h=1/\tau^2$ poses difficulties because standard choices can actually turn out to be informative. The gamma distribution is a conjugate prior for h and the obvious choice is $h \sim Ga(\alpha,\beta)$ for $\alpha=\beta=\varepsilon$ where ε is some small positive constant, say 0.001. This prior has mean $\alpha/\beta=1$ but variance $\alpha/\beta^2=1/\varepsilon$, making it progressively more diffuse as $\varepsilon \to 0$. It is also a "minimally informative" prior in the sense that choosing a very small ε will have minimal impact on the parameters of the gamma posterior distribution, forcing the data and μ to provide virtually all the information. However, the gamma prior becomes improper as $\varepsilon \to 0$, and its shape also becomes more and more spiked, with an infinite peak at 0 and a very heavy right tail. It will thus tend to favor large variances if these are supported by the data and small variances otherwise. Gelman (2006) suggests instead placing a uniform $U(0,A)$ prior on τ. Often, the problem will suggest an upper bound A. For instance, when μ is a (log) odds ratio, very large values are unlikely, thus suggesting an appropriate bound on the between-study variance.

Using an informative prior removes these difficulties, but the choice must be defensible. One might, for instance, use past meta-analyses to construct empirical distributions of between-study variances to serve as data-based priors. Turner et al. (2012, 2015) derived specific lognormal priors for binary outcomes of various types (e.g., mortality, subjective) comparing different types of interventions (e.g., pharmacologic vs. placebo). Rhodes et al. (2015) derived Student-t priors for continuous outcomes. They found that such informative priors substantially reduced the uncertainty in the estimates of both μ and τ^2.

To compute the conditional posterior of the $\theta_i|y,\mu,\tau^2,s_i^2$ required for MCMC, note that since they have independent priors and contribute separate components to the likelihood, their posterior factors into I components. It can be shown that the posterior distribution of random effects is

$$\theta_i \mid \mu, \tau^2, s_i^2, y \sim N\left(\hat{\theta}_i, 1/(1/s_i^2 + 1/\tau^2)\right) \quad (6.9)$$

where

$$\hat{\theta}_i = (1 - B_i) y_i + B_i \mu \quad (6.10)$$

and

$$B_i = \frac{s_i^2}{s_i^2 + \tau^2} = \frac{1/\tau^2}{1/\tau^2 + 1/s_i^2}. \quad (6.11)$$

These formulas provide insight into the sources of information about each of the parameters. Each study's conditional mean $\hat{\theta}_i$ is a weighted average of the observed study mean y_i and the population mean μ, where the weights B_i reflect the relative weights of the within-study precision $1/s_i^2$ and the between-study precision $1/\tau^2$. Larger weights B_i shrink $\hat{\theta}_i$ back to the population mean μ. Studies with larger within-study precision (smaller within-study variance) are shrunk less than those with smaller within-study precision (larger within-study variance) because they provide more information about θ_i. As τ^2 increases, one is more likely to believe that studies are heterogeneous and therefore each study's mean reflects its own data more closely. Gelman et al. (2013) call this a partial pooling model because the study effects y_i are partially pooled with the population mean μ. They contrast this with a no pooling model in which $\hat{\theta}_i = y_i$, and a complete pooling model in which $\hat{\theta}_i = \bar{y}$. The no pooling model results from assuming $\tau^2 = \infty$ and is the separate-effects model discussed in Chapters 2 and 4. The complete pooling model results from assuming $\tau^2 = 0$ and is the common-effect model. Partial pooling is a type of parameter *shrinkage* toward the average, which has a long history in statistics (James and Stein, 1961; Efron and Morris, 1973). Draper et al. (1993) discuss how one can better estimate each study using information from the others when all the studies are exchangeable.

In addition, the posterior variance of θ_i, $s_i^2(1 - B_i)$ is reduced relative to the observed study variance, s_i^2, by an amount depending on the ratio of s_i^2 to τ^2. When s_i^2 is small relative to τ^2, the reduction is small because θ_i is already estimated precisely and the pooled effect adds little; when s_i^2 is large relative to τ^2, the pooled effect pulls an imprecise estimate of θ_i toward the mean.

One final point about the common and random-effects models concerns the assumption of known within-study variances s_i^2. This assumption is often justified by invoking large study samples. However, in practice, some studies are often too small for the s_i^2 to be well-estimated, let alone known. The known variance assumption will tend to put too much certainty on the observed y_i and will weight smaller studies too heavily. When some studies are very large, this may have little effect on inferences about μ but may have a greater effect on θ_i. In models where y_i is correlated with or even a mathematical function of its empirical standard error, as when y_i is an empirical log odds ratio, the assumption of a known within-study variance is completely unjustified theoretically (see Chapter 5).

As a result, some authors have proposed to model the s_i^2 by assuming them known up to a constant so that $s_i^2 = \phi t_i^2$ for known t_i. For instance, one could take ϕ to be the pooled within-study variance and t_i^2 to be the inverse of the study sample size. This model was proposed by Cochran (1954) in his original paper on meta-analysis in the context of combining small studies. This model circumvents the need to estimate each s_i^2 and instead requires modeling only ϕ. This could be done in either a likelihood-based approach as in Chapter 5 or in a Bayesian model. The Bayesian model requires only placing a prior on ϕ (DuMouchel, 1990; Nam, Mengersen, and Garthwaite, 2003). Zucker et al. (2006) used this model with equal sample sizes (i.e., assuming $t_i^2 = 1$) to combine N-of-1 trials. One can also think of this model as a multiplicative error model (Stanley and Decouliagos, 2015) which in a random-effects context introduces an additive and multiplicative error (Jackson and Baker, 2013: Schmid, 2017).

6.5.1 Random-Effects Example

Returning to the Alzheimer's example from the previous section, we now carry out a random-effects analysis. This requires specifying prior distributions for the hyperparameters μ and τ^2. Here, we use two of the priors for μ that were used for the common-effect model. We also look at two different non-informative priors for the between-study heterogeneity, namely $\tau \sim U(0,100)$ and $1/\tau^2 \sim Ga(.001,.001)$.

Table 6.2 shows results from using a non-informative $N(0,10000)$ prior for μ as well as the two non-informative priors on the between-study variance. The rows give posterior summaries for μ as well as for the study effect estimates θ_i. Overall, there is an 82% chance that iron levels are on average lower in the Alzheimer's group using the uniform prior on τ and a 94% chance with the inverse gamma prior. This higher probability arises from the larger difference and tighter credible interval associated with the inverse gamma prior. The study specific posteriors also differ somewhat between the two analyses with tighter credible intervals leading to higher posterior probabilities of a difference with the inverse gamma prior. In contrast, the frequentist analysis using the restricted maximum likelihood (REML) estimator of variance returns a mean treatment effect of –5.5 with a 95% confidence interval of (–14.3, 3.2). REML, therefore, returns an estimate similar to that using the uniform prior, but returns a confidence interval where the width is closer to the

TABLE 6.2

Posterior Median, 95% Credible Interval, and Posterior Probability That Effect Is Less Than 0 for Overall and Study Differences and Between-Study Standard Deviation in Iron Levels (μg/dL) in Alzheimer's Patients Compared with Controls Using Non-Informative Prior for Mean

	$\mu \sim N(0,10000)$			$\mu \sim N(0,10000)$		
	$\tau \sim U(0,100)$			$1/\tau^2 \sim$ Gamma(.001,.001)		
	Median	**95% CrI**	**Prob < 0**	**Median**	**95% CrI**	$P(\mu < 0 \mid y)$
μ	–5.6	–20.0, 11.1	0.82	–6.5	–13.7, 2.7	0.94
θ_1	–8.9	–26.3, 4.4	0.92	–7.1	–19.5, 1.2	0.96
θ_2	–3.3	–16.0, 13.8	0.67	–6.0	–13.6, 7.6	0.88
θ_3	–6.7	–14.3, 0.8	0.96	–6.7	–13.1, –0.4	0.98
θ_4	2.8	–10.8, 22.6	0.39	–5.2	–12.1, 15.0	0.79
θ_5	–10.9	–21.9, –1.8	0.99	–7.8	–18.3, –1.4	0.99
τ	9.6	0.5, 40.3		1.2	0.04, 18.6	

TABLE 6.3

Posterior Median, 95% Credible Interval, and Posterior Probability That Effect Is Less Than 0 for Overall and Study Differences and Between-Study Standard Deviation in Iron Levels (μg/dL) in Alzheimer's Patients Compared with Controls Using Informative Prior for Mean

	$\mu \sim N(-5,4)$			$\mu \sim N(-5,4)$		
	$\tau \sim U(0,100)$			$1/\tau^2 \sim$ Gamma(0.001,0.001)		
	Median	**95% CrI**	**Prob<0**	**Median**	**95% CrI**	$P(\mu<0 \mid y)$
μ	−5.2	−8.7, −1.5	0.997	−5.4	−8.8, −2.1	0.999
θ_1	−7.5	−23.6, 3.1	0.93	−5.8	−16.3, −0.2	0.98
θ_2	−3.7	−14.3, 11.5	0.72	−5.3	−10.8, 4.9	0.92
θ_3	−6.3	−13.6, 0.4	0.97	−5.7	−11.2, −1.2	0.99
θ_4	0.6	−9.1, 19.8	0.47	−4.8	−9.1, 11.2	0.84
θ_5	−9.7	−20.8, −2.1	0.99	−6.3	−16.6, −2.2	0.997
τ	7.2	0.4, 27.7		0.8	0.03, 13.7	

credible interval based on the inverse gamma prior. The between-study standard deviation τ is very poorly estimated, not surprisingly because it is based on data from only five studies. Moreover, the posterior for τ is very different for the two choices of priors on it. Both also differ from the REML estimate of 6.9.

The two sets of results show that when the number of studies is small and information about between-study variance is therefore weak, posterior inferences can be quite sensitive to the choice of the prior for τ. In this case, the inverse gamma prior with small shape and rate parameters is nearly improper and within the range of variances supported by the data, puts relatively more weight on small variances, resulting in more shrinkage toward the mean for the study estimates and a smaller estimated between-study variance. In such cases, it is better to use a more informative prior by restricting to reasonable values. The uniform prior is more informative and is a better choice in general (Gelman, 2006). However, the main message is that random-effects models are not well-estimated when meta-analyses are based on so few studies.

Table 6.3 shows results using a more informative $N(-5,4)$ prior for μ, chosen here as a prior that matches the likelihood reasonably well. Except for Study 4, the two different variance priors now give more similar results and in general the credible intervals are smaller because of the increased prior information. Study 4 again appears as an outlier in the analysis using the $U(0,100)$ prior for τ. Recall that this was the study in which iron levels were quite a bit higher in the Alzheimer's group. Thus, we are not as sure that the true levels in that study are actually lower in the Alzheimer's group as they are in the other studies. Note that even Study 2 which had minimally higher levels in the Alzheimer's group has a reasonably large posterior probability that the true levels are lower because of the strong evidence of lower levels overall and the resultant shrinkage toward the mean.

6.6 Computation

Historically, the major difficulties with computing posterior distributions were the need to compute complex multidimensional integrals for $p(y)$ in the Bayes' rule denominator (6.1) and the need to integrate out nuisance parameters to get marginal posterior distributions. MCMC solved this problem by providing an iterative numerical integration algorithm that returns simulated values of the entire joint posterior distribution. This permits numerical calculation of any quantities of interest and enables computing extremely complex models provided they can be broken into simpler components in a special way (Gelfand and Smith, 1990). Not only can this algorithm help solve standard Bayesian problems, it also opens whole new classes of models to Bayesian inference that cannot be fit using frequentist methods. Because it is difficult to optimize the likelihoods of many complex models, MCMC has vastly increased the use of Bayesian inference in statistical modeling.

Essentially, MCMC works by repeatedly simulating the parameters in a carefully chosen sequence such that at each step one or more of them is drawn from a known distribution conditional on the data and the current state of the other parameters. Because the sequence of draws forms a Markov chain, all of the information from the previous history of the chain is contained in the most recently sampled values of the parameters, and so the current state is the only part of the chain's history needed to take the next sample. Crucially, it can be shown that the algorithm will converge to a stationary distribution which is the true joint posterior distribution under mild regularity conditions that are generally satisfied for most statistical models (Roberts and Smith, 1993). Once the Markov chain is deemed to have converged, inference is based on further Monte Carlo samples drawn from the Markov chain. The number of further samples should be chosen to be sufficient to obtain results with acceptable precision.

Each complete pass through the algorithm results in a new draw of each parameter. Once the algorithm has converged, each draw is made from the correct posterior. Repeating this procedure many times, we can draw many samples from the posterior and use these draws to make inferences about any parameter or any function of parameters. The sequence of draws provides a random tour of the (high-dimensional) parameter space, visiting locations in that space with frequencies proportional to the joint posterior density. Convergence may be monitored with diagnostics that check whether simulation variability is consistent with that expected from a probability distribution (Gelman, 1996). The simulations taken before convergence, called the *burn-in*, are discarded and the remainder are used to estimate the posterior. Key parts of the process are choosing efficient updating schemes, determining when the algorithm has converged and is sampling from the true posterior, and then taking a sufficiently large number of samples from the posterior distribution to limit Monte Carlo simulation error and thus ensure reliable inferences.

The output of the MCMC sequence is a full set of draws from the posterior distribution. Characteristics of any parameter, set of parameters, or function of parameters can be evaluated by empirical summaries of the drawn samples using the Monte Carlo method. For example, the mean of the marginal distribution of a parameter θ_i can be estimated by the average of its N sampled draws,

$$\hat{\theta}_i = \sum_{t=k+1}^{k+N} \theta_i^{(t)}$$

where the index of the summation indicates that we use only samples after convergence of the Markov chain at iteration k of the simulation (i.e., the burn-in consists of the first k draws). A posterior credible interval for the sample can be constructed based on taking the appropriate quantiles of these same samples as upper and lower bounds. Because the credible interval is constructed directly from the empirical quantiles returned by the simulation, it need not be symmetric and need not rely on asymptotic normality. The empirical posterior distribution also permits inferences to be made about quantities that clearly do not have normal distributions, such as the correlation or ratio between two parameters. For instance, the posterior correlation between θ_i and θ_j can be estimated as the empirical correlation between the sequence of draws for the two parameters, i.e., as

$$r_{\theta_1,\theta_2} = \frac{s_{\theta_1,\theta_2}}{s_{\theta_1} s_{\theta_2}}$$

where

$$s_{\theta_1,\theta_2} = \sum_{t=k+1}^{k+N} \left(\theta_1^{(t)} - \bar{\theta}_1\right)\left(\theta_2^{(t)} - \bar{\theta}_2\right) / N$$

is the empirical covariance and s_{θ_1} and s_{θ_2} are the empirical standard deviations. The posterior distribution of θ_1/θ_2 is calculated from the distribution of the ratio in each simulation. The precision of our estimation of these quantities increases with the number of samples we use to compute them. This ability to evaluate the posterior of any parameter function is a key advantage of MCMC.

For complex, high-dimensional parameter spaces, computational efficiency is crucial both because many simulations may be needed to effectively sample from the entire support of the posterior distribution and because the high autocorrelations introduced by the Markov sampling process may cause the Markov chain to move very slowly through the parameter space. Although we can never fully cover the space except in small discrete problems, we may hope to visit high posterior density regions often enough to adequately represent the posterior distribution, perhaps after clever reparameterization of the problem. For instance, Gilks and Roberts (1996) discuss how centering covariates in regression removes induced correlation between the model intercept and slope.

A variety of MCMC sampling procedures can now be combined to make an efficient overall algorithm depending on the nature of the problem. For instance, slice sampling is useful for bounded parameters, and Hamiltonian MCMC can be more efficient for problems with large numbers of parameters. Many programs also employ adaptive samplers that attempt to accelerate convergence by using the early output of an MCMC chain to refine and improve the sampling as it progresses. The *Handbook of MCMC* in this series (Brooks et al., 2011) nicely summarizes the state of the art. Here, we shall only briefly outline the two most common algorithms, Gibbs sampling and the Metropolis–Hastings algorithm. Section 6.11 discusses software implementations.

6.6.1 Gibbs Sampler

Although MCMC encompasses a variety of algorithms, the Gibbs sampler is the simplest to understand and presents most of the key features. Consider a set of M model parameters $\boldsymbol{\theta} = \{\theta_1, \ldots, \theta_M\}$. The sampler works by drawing in turn each parameter θ_i from its posterior distribution conditional on the most recent draws of all the other parameters $\boldsymbol{\theta}_{-i}$. At iteration t in a normal likelihood model, we draw $\theta_i^{(t)} \sim N\left(\theta_i \mid \boldsymbol{\theta}_{-i}^{(t-1)}, y\right)$ where $\boldsymbol{\theta}_{-i}^{(t-1)}$ indicates that we condition on the most recent draws from $\boldsymbol{\theta}_{-i}$. Note the slight abuse of notation here because each newly drawn θ_i becomes part of the current set of parameters used to draw θ_{i+1} and so some of the parameters in $\boldsymbol{\theta}_{-i}^{(t-1)}$ will actually be drawn from iteration t. These distributions are known as the *full conditional distributions* and are much easier to sample than the joint distribution, either because they are univariate distributions or because they are multivariate distributions such as of a vector of linear regression coefficients from which it is easy to sample.

6.6.2 Metropolis–Hastings Algorithm

When it is not possible to sample from one or more of the full conditional distributions required by the Gibbs sampler, one can use rejection sampling instead. The basic idea is that we will know or can at least compute the form of the required conditional distribution up to its constant of proportionality and can construct a proposal density $q(\theta^* \mid \theta^{t-1}, y)$ that is reasonably close to the true conditional but from which it is easy to sample. One then computes the ratio $r = p(\theta^* \mid y) / p(\theta^{t-1} \mid y)$ of the likelihood times prior of the proposed state to that of the previous state (where the parameters not being sampled are taken at their current values). If the proposed state, θ^* increases the posterior density relative to the previous state θ^{t-1}, the move is accepted with probability 1; if the proposed state decreases the posterior density relative to the previous state, the move is accepted with probability $r<1$. If the move is accepted, we set $\theta^{(t)} = \theta^*$. Otherwise, if rejected, no move is made and $\theta^{(t)} = \theta^{(t-1)}$. Technically, this describes a Metropolis sampler and requires that q be symmetric in its arguments, i.e., that $q(\theta^* \mid \theta^{(t)}, y) = q(\theta^{(t)} \mid \theta^*, y)$. If this condition does not hold, one uses the Metropolis–Hastings sampler which makes a small modification to the acceptance ratio r. Note that the Gibbs sampler is actually a form of Metropolis–Hastings in which q is the full conditional density and $r=1$.

One commonly used Metropolis proposal density is a multivariate normal centered around the current value of the parameters in the chain with variance chosen so that the acceptance rate is around 25 to 40%. Acceptance rates that are much higher or lower than this tend to lead to highly autocorrelated chains which move too slowly around the parameter space (Gelman et al., 2013, pp. 295–297). One can use a pilot adaptation algorithm during burn-in to tune the variance parameter so that it accepts at a reasonable rate.

6.6.3 Convergence Diagnostics

Determining when the algorithm has converged is another crucial part of MCMC. Several diagnostics are in common use. The simplest plots the time series of simulated values for each of a few parallel sampling chains of each parameter. These *trace plots* show the extent of autocorrelation if the chain is covering the entire support of the parameter space or if it is getting stuck or wandering off course. Stability of the chain signals convergence. Because it can be difficult to visually determine convergence from these trace plots, and because chains can be sensitive to simulation starting values, alternative numerical diagnostic methods have been proposed.

Gelman and Rubin (1992) suggested running multiple chains in parallel, starting with overdispersed starting points so that the variance between the chains using different starting points is greater than in the target distribution, and then checking that each chain has stabilized to the same region. The basic idea is that as the different chains converge to the correct posterior distribution and begin to mix together (i.e., overlap), the between-chain variance of the means will decline toward zero and the total variance (within-chain + between-chain) of the posterior samples will approach the within-chain variance. When the ratio of the total and within-chain variances reaches 1, the chains can be said to have converged. The ratio of the estimated posterior variance of the estimand to its pooled within-chain variance estimates how much the standard deviation of the parameter estimate could be reduced by continuing to run the MCMC algorithm longer. One can compute a ratio for each parameter or estimand of interest as well as an omnibus ratio (Brooks and Gelman, 1998). The chains have converged when the ratio for each parameter is less than some value near one. Values of 1.05 are probably sufficient, although some users choose values as low as 1.01. An even more stringent criterion is to require the upper 97.5 percentile of the ratios to achieve the bound. Brooks and Gelman recommend using only the second half of the chain to monitor the sequence, so that upon convergence the second half of each chain provides valid draws from the posterior distribution.

One caveat for users is that the limit may be reached by chance even before the chains have converged if the within-sample variance is overestimated or the between-sample variance is underestimated (hence the importance of choosing overly dispersed starting values that cover the support of the parameter space). Gelman et al. (2013) also suggest that for estimands such as variances or proportions that are highly non-normal (e.g., long-tailed, positive, bounded), monitoring a transformation that makes them more normally distributed (e.g., log variance or logit proportion) may work better because the convergence diagnostics are based on the use of means and variances that are optimized for normally distributed variables. In practice, it is useful to periodically check the diagnostic at prespecified intervals of the length of the chain to check for convergence.

It is also impractical (and unnecessary) to model all parameters when their number is large. Instead, one can monitor a representative subset of the parameters including those of most interest, those that may be expected to converge slowly (e.g., variances), or those which are summaries of others (e.g., random effects). One should examine the autocorrelation functions of each parameter and the cross-correlation function of each pair of parameters to check whether the algorithm is moving or getting stuck near ridges (Berry et al., 2010, p. 49). One must be careful in problems with poorly identified parameters that will converge slowly and may affect the convergence of the parameters of interest (see an example of this in Carlin and Louis, 2009, p. 166, problem 21).

6.6.4 Starting Values for the Chains

Choosing good starting values $\theta^{(0)}$ is another key part of an efficient algorithm. Poor starting values will start the chain far from the correct posterior distribution and will require additional draws for the chain to move toward the correct region. This problem is exacerbated in high-dimensional problems because finding the correct direction is extremely complicated. For a single chain, a good choice is to use starting values that are approximate solutions to simpler problems. For example, in the random-effects model, one could use one of the non-Bayesian procedures outlined in Chapters 4 or 5 to estimate the population parameters θ and τ^2 and empirically estimate the study specific effects θ_i from their sample within-study estimates. It is especially important to ensure that parameters do not have starting values outside their allowed range, such as nonpositive starting values for variances.

When using multiple chains in order to diagnose convergence, the problem is complicated by the need for overdispersed starting values. Because the true posterior is unknown, it is impossible to ever know that the chains are initially overdispersed, as required for a convincing diagnostic. Choosing extreme quantiles from the prior is problematic if the prior is improper. Instead, one might choose to start at values randomly drawn from approximate posterior distributions constructed from the preliminary estimates and their standard errors. Higher level parameters may be initialized at the starting values of the hyperparameters on which they are conditioned. It is always a good idea to try multiple sets of starting values to increase confidence that convergence has been achieved. For transparency, it is helpful, although this has not been standard practice to date, to report how the starting values were chosen including any automated procedures used in the choice (e.g., using least squares to fit an approximate regression model). Failure of the MCMC algorithm to converge often indicates that a model is not fully identifiable. Use of a simpler model or of further prior information/external evidence about those non-identifiable parameters can help.

6.6.5 Estimating Monte Carlo Error

The Gelman–Rubin convergence diagnostic focuses only on detecting bias in the MCMC estimator but provides no information about the precision of the resulting posterior estimate. Because posterior summaries are obtained by inferences from MCMC simulation, the precision in the posterior point estimates will depend on the size of the simulation sample on which they are based. It is therefore important to estimate the size of the simulation error so that one can determine the number of posterior samples required to achieve the desired posterior precision. For example, a Monte Carlo error of 0.001 implies that the estimand is precise to the third decimal place with respect to simulation error. The number of simulations to draw depends on the complexity of the posterior distribution and the nature of the estimand. For a single parameter, we may be interested in estimating its posterior mean and the bounds of its credible interval. The simulation will contain much more information about the mean than about its extreme quantiles (or bounds). Likewise, the posterior mean of a parameter with a bimodal posterior distribution may be of little interest and finding the two modes may require larger samples.

Most of the work on estimating MC error has focused on estimating the posterior mean accurately. This task is complicated by the dependence of the elements of the Markov chains that form the draws. Because of the autocorrelation in the MCMC sequence, the number of MCMC samples needed to form a sample of N independent draws for constructing posterior estimates may be far greater than N and will increase with the degree of autocorrelation. To see this, first note that if the draws were independent, we could calculate the variance of the posterior mean estimator $\hat{E}(\theta \mid y) = \hat{\theta}_N = \frac{1}{N}\sum_{t=1}^{N} \theta^{(t)}$ as

$$\widehat{\mathrm{Var}}_{iid}(\hat{\theta}_N) = s_\theta^2/N = \frac{1}{N(N-1)}\sum_{t=1}^{N}(\theta^{(t)} - \hat{\theta}_N)^2 \tag{6.12}$$

where s_θ^2 is the sample variance of the simulated sequence $\{\theta^{(t)}\}$. However, because of the positive autocorrelation in the MCMC samples, $\widehat{\mathrm{Var}}_{iid}(\hat{\theta}_N)$ will underestimate the true MC error. If k is the approximate lag at which the autocorrelations in the chain die out, one can form an approximately independent sequence by thinning the sample to keep only every kth sample, but this throws away information and increases the variance of the estimators (MacEachern and Berliner, 1994).

Instead, one might use the estimated autocorrelations to adjust the MC variance estimate. Kass et al. (1998) show that one can estimate the effective sample size (ESS) as $N/\kappa(\theta)$ where $\kappa(\theta) = (1 + 2\sum_{k=1}^{\infty} \rho_k(\theta))$ and $\rho_k(\theta)$ is the autocorrelation at lag k for the parameter of interest θ. This may be estimated by the sample autocorrelations from the MCMC chain. The variance estimate for $\hat{\theta}_N$ is then

$$\widehat{\text{Var}}_{ESS}(\hat{\theta}_N) = s_\theta^2 / ESS(\theta) = \frac{1 + 2\sum_{k=1}^{\infty} \rho_k(\theta)}{N(N-1)} \sum_{t=1}^{N} (\theta^{(t)} - \hat{\theta}_N)^2.$$

Note that unless the $\theta^{(t)}$ are uncorrelated, $ESS(\theta) < N$, so that $\widehat{\text{Var}}_{ESS}(\hat{\theta}_N) > \widehat{\text{Var}}_{iid}(\hat{\theta}_N)$ and the variance is inflated by the autocorrelation.

Practically, use of $\widehat{\text{Var}}_{ESS}(\hat{\theta}_N)$ requires truncating the infinite sum in the numerator at some finite lag T. A simple, but often biased, method sets T equal to the lag at which the sum of estimates of autocorrelations at two successive lags is less than zero (Gelman et al., 2013). More sophisticated methods have been proposed but are harder to implement (Berry et al., 2010).

A different method used by the BUGS software (Lunn et al., 2009) divides the chain into m successive batches of length k and calculates the variance from the variance of the batch means. Letting $B_1, \ldots, B_m$ be the batch means, $\hat{\theta}_N = \bar{B} = \frac{1}{m}\sum_{i=1}^{m} B_i$ and the Monte Carlo variance estimate is

$$\widehat{\text{Var}}_{\text{batch}}(\hat{\theta}_N) = \frac{1}{m(m-1)} \sum_{i=1}^{m} (B_i - \hat{\theta}_N)^2.$$

The MC error of the posterior mean can be interpreted as the standard error of the posterior mean but adjusted for the autocorrelation in the posterior samples on which the posterior mean is based. The ratio of the MC error for the estimate of the posterior mean to the estimate of the posterior standard deviation assesses the relative error of the simulation. Increasing the number of simulations will reduce the MC error, but decreasing the posterior standard deviation requires increasing the amount of data. Because additional simulations may be practical even when additional data collection is not, reducing the Monte Carlo error is advisable at least to a level commensurate with the accuracy of the data. One rule of thumb is to run the simulation until the MC error for each parameter of interest is less than 5% of the posterior standard deviation (Spiegelhalter et al., 2003).

It is worth emphasizing that if one is interested in the accuracy of the estimate of the posterior credible interval, then substantially more samples are required, because the bounds of the credible interval being extreme quantiles of the posterior distribution will be much less accurately estimated than its center.

6.6.6 Posteriors of Functions of Parameters

It is also straightforward to obtain posterior distributions for functions of parameters either by algebraic manipulation of a closed form posterior or by applying the same function to the simulated parameter values and then summarizing the transformed simulated values. For instance, if the model parameter is the log of an odds ratio, it is easy to simulate the distribution of the odds ratio by simply taking the antilogarithm of the simulated values of the log odds ratio. We shall see examples of this later.

6.6.7 Choices for Alzheimer's Example

We computed the posteriors in the random-effects model for the Alzheimer's data using MCMC run with three parallel chains. Each chain was run until each parameter of interest (study treatment effects and average treatment effect as well as the total residual deviance and contributions of each data point to residual deviance) had converged using a convergence criterion of 1.05 for the ratio of between chain to within-chain variance. Models were fit with JAGS software (Plummer, 2003) using the second half of each chain after convergence, retaining at least 200,000 samples for each chain. Chains were augmented by additional runs if the second half of the sequence was less than 200,000. The ratios of Monte Carlo standard errors to the posterior standard deviation were all considerably less than 1%. Starting values were chosen by estimating each parameter via a simple regression using the observed study effects and then drawing three values at random from the empirical posterior distributions formed from these regressions. Further details on choosing starting values may be found in Schmid et al., (2014).

6.7 One-Stage Models for Discrete Outcomes

Outcome data are not, of course, always continuous and thus the normal models we have been working with will not always be appropriate. As noted earlier, a strength of the Bayesian approach is the flexible template it provides to work with arbitrary priors and likelihoods, provided they may be computed. As an example, Table 6.4 shows data from 22 trials testing the use of intravenous magnesium to treat patients with acute myocardial infarction (Li et al., 2007). The raw data are the number of deaths and the number studied in the magnesium and no magnesium groups in each trial. The scientific question is whether patients treated with magnesium die less frequently than those not treated. Note that the sizes of the studies vary widely as do their individual odds ratios and 95% confidence intervals calculated by standard maximum likelihood estimation.

We use the notation r_T and n_T to denote the number of deaths and sample size, respectively, in the magnesium (treated) group and r_C and n_C for the corresponding numbers in the no magnesium (control) group. It is natural to assume binomial distributions for the counts within each group in each study so that

$$r_{Ci} \sim Bin(n_{Ci}, \pi_{Ci}) \tag{6.13}$$

and

$$r_{Ti} \sim Bin(n_{Ti}, \pi_{Ti}) \tag{6.14}$$

where the notation indicates the study i. Here π_{Ti} is the probability of death among those receiving magnesium in study i and π_{Ci} is the corresponding probability in the no magnesium group. We usually work on the logit scale and assume

$$\text{logit}(\pi_{Ci}) = \gamma_i \tag{6.15}$$

and

$$\text{logit}(\pi_{Ti}) = \gamma_i + \theta_i, \tag{6.16}$$

TABLE 6.4

Data and Empirical Study Odds Ratios with 95% CI from 22 Studies Comparing with Intravenous Magnesium to No Magnesium

Study	Year	Magnesium deaths/total	No magnesium deaths/total	OR	95% CI
Morton	1984	1/40	2/36	0.44	0.04, 5.02
Smith	1986	2/92	7/93	0.27	0.06, 1.35
Rasmussen	1986	4/56	14/74	0.33	0.10, 1.06
Abraham	1987	1/48	1/46	0.96	0.06, 15.77
Ceremuzynski	1989	1/25	3/23	0.28	0.03, 2.88
Singh	1990	6/81	11/81	0.51	0.18, 1.45
Shechter	1990	1/50	9/53	0.10	0.01, 0.82
Shechter	1991	2/21	4/25	0.55	0.09, 3.37
Feldstedt	1991	10/150	8/148	1.25	0.48, 3.26
Woods	1992	90/1159	118/1157	0.74	0.56, 0.99
Wu	1992	5/125	12/102	0.31	0.11, 0.92
ISIS-4	1995	2216/29,011	2103/29,039	1.06	1.00, 1.13
Shechter	1995	4/96	17/98	0.21	0.07, 0.64
Thorgersen	1995	4/130	8/122	0.45	0.13, 1.54
Bhargava	1995	3/40	3/38	0.95	0.18, 5.00
Urek	1996	1/31	0/30	3.00	0.12, 76.58
Raghu	1999	6/169	18/181	0.33	0.13, 0.86
Gyamiani	2000	2/50	10/50	0.17	0.03, 0.81
Santoro	2000	0/75	1/75	0.33	0.01, 8.20
MAGIC	2000	475/3113	472/3100	1.00	0.87, 1.15
Zhu	2002	101/1691	134/1488	0.64	0.49, 0.84
Nakashima	2004	1/89	3/91	0.33	0.03, 3.27
Pooled (CE)				0.99	0.94, 1.05
Pooled (RE:REML)				0.58	0.43, 0.78

The bottom of the table gives summary common-effect and random-effect estimates. The common effect is estimated by an inverse variance weighted average and the random effect estimate by REML, both applied to empirical logits.
CE = common effect; RE = random effect.

so that θ_i is the study log odds ratio and γ_i is the logit of the probability of events in the control group. The parameters of most interest are the θ_i. The γ_i are often considered nuisance parameters (but see Chapter 14 where we become interested in the relationship between θ_i and γ_i). We can use either a common-effect or random-effects model for the θ_i.

We can then construct a likelihood from these binomial distributions and proceed with Bayesian inference. We could also have proceeded as in Chapter 4 using a normal likelihood based on the empirical study log odds ratios, but this requires knowing the within-study variances which in this case would be functions of the treatment effects since these are proportions. Because many of these trials are small, standard asymptotic variance estimates may not hold. However, using an exact binomial likelihood as in Chapter 5 avoids this difficulty as no within-study variances are required. Furthermore, it eliminates the need to rely on asymptotic normality assumptions for the within-study model.

It is important to emphasize, however, that one must include the transformed control proportions, γ_i, in the binomial model, whereas we can ignore them and work only with the θ_i in the approximate normal model. The binomial model may be written as a generalized-linear mixed model as in Chapter 5 where the γ_i can be considered as intercepts. In practice, two approaches have been taken to modeling the γ_i. The first treats them as fixed, unrelated, independent parameters and puts separate prior distributions directly on them. The second approach instead models them as exchangeable, random effects from a common distribution of study effects, often assumed to be normal, so that $\gamma_i \sim N(\gamma, \sigma_\gamma^2)$ (Jackson et al., 2018). More generally, as noted in Chapter 5, one can also assume a bivariate distribution for the study intercepts and treatment effects allowing for their correlation.

The likelihood for the model with fixed γ_i is then

$$L_1 = \prod_{i=1}^{I} g(\gamma_i)^{r_{C_i}} (1 - g(\gamma_i))^{n_{C_i} - r_{C_i}} g(\gamma_i + \theta_i)^{r_{T_i}} (1 - g(\gamma_i + \theta_i))^{n_{T_i} - r_{T_i}} p(\theta_i \mid \mu, \tau^2)$$

where g is the inverse logit function and we have used equations (6.15) and (6.16) to re-express π_{C_i} and π_{T_i} in terms of γ_i and θ_i. Priors are needed on the γ_i, μ and τ^2.

The likelihood for the model with random γ_i is

$$L_2 = L_1 * \prod_{i=1}^{I} p(\gamma_i \mid \gamma, \sigma_\gamma^2)$$

with priors needed for μ, τ^2, γ, and σ_γ^2.

The model with fixed γ_i is more common in the Bayesian meta-analysis literature, particularly when combining randomized controlled trials whose designs ensure unbiased estimates of study treatment comparisons but may have substantially different control risks that reflect different study characteristics. The model with fixed γ_i minimizes any impact estimation of the γ_i may have on estimation of μ and the θ_i. This assumption reflects the belief that treatment effects are more stable across studies than are control risks (Dias and Ades, 2016; Dias et al., 2018).

If events are rare (with many zero counts), however, the fixed intercept model may not be identifiable, and it may be necessary to make further assumptions, such as an exchangeable random-effects model for the γ_i. Predicting the group event rates in a new study would also require a model on the γ_i. Treating the γ_i as random and modeling them does shrink their estimates toward their mean γ and this may in turn affect the estimation of the θ_i and of μ through the passing of information between studies (see also Section 2.3.3 for more discussion on this point).

Another issue arising with fixed intercepts as noted in Chapter 5 is that the number of parameters to be estimated increases with the number of studies. Because this violates the regularity conditions for maximum likelihood estimators, it may also affect Bayesian estimates when the likelihood dominates the prior, especially with non-informative priors and large numbers of studies. Further work is needed to investigate this possibility.

Returning to the magnesium example, many of the studies in Table 6.4 are small and relatively uninformative with wide confidence intervals. Four studies are quite large and informative but have very different results. The Woods and Zhu trials suggest magnesium is beneficial; the MAGIC trial suggests it has no effect; and the largest trial, ISIS-4, suggests it may be harmful. Figure 6.1 displays a forest plot of the results for each study along with summary estimates using the common-effect model (CE) and the random-effects model (RE) as in Chapter 4 (also displayed at the bottom of Table 6.4). The Bayes estimates in the

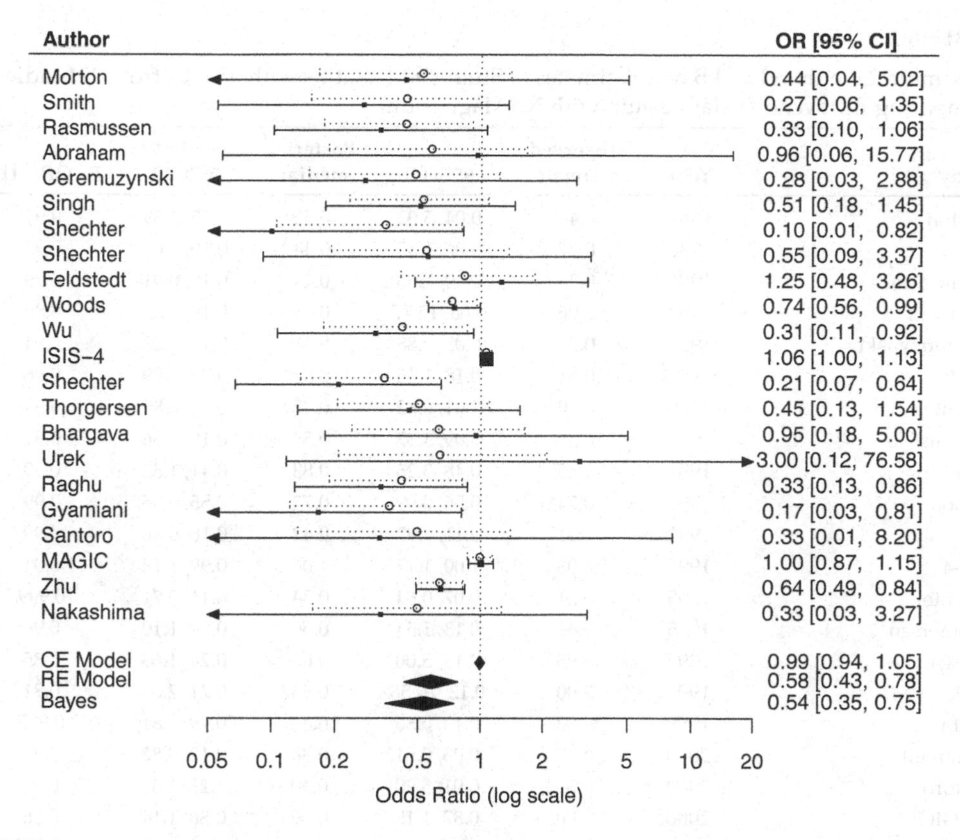

FIGURE 6.1
Forest plot of magnesium studies. Observed study estimates and 95% confidence limits are denoted by the filled circles and solid arrows. Bayes posterior study medians and 95% central credible intervals are given by the open circles and dotted arrows. Summary estimates are given at the bottom for the common effect using an inverse variance weighted average, the random effect using REML, and the Bayes posterior median together with 95% confidence and central credible intervals.

figure will be discussed presently. The random-effects model (using the REML estimate of τ^2) indicates a large reduction in the odds of death after treatment (odds ratio 0.58 with 95% CI: 0.43, 0.78), whereas a common-effect analysis leads to a non-significant summary odds ratio of 0.99 (95% CI: 0.94, 1.05). The two analyses differ because the CE model weights heavily by sample size (thus primarily reflecting ISIS-4), while the RE model weights each study more equally because of substantial between-study variation ($\tau^2 = 0.18$) with $I^2 = 0.82$. A Bayesian analysis that explicitly accounts for the uncertainty in estimating the between-study variation should therefore be useful.

For the Bayesian analysis, we treat the log odds of mortality in each control arm, γ_i, as fixed parameters with non-informative $N(0,10000)$ prior distributions and use a random-effects model for the study treatment effects with non-informative priors $\mu \sim N(0,10000)$ and $\tau \sim U(0,5)$. Table 6.5 displays results after carrying out MCMC (see online materials for code). The posterior study median together with its 95% credible interval and the posterior probability of a positive effect of magnesium under the random-effects model are shown along with the individual study estimates, including those based on the observed data from each study alone repeated from Table 6.4. The estimate for the population median e^{μ} from two different frequentist analyses as well as from the Bayesian analysis are given at

TABLE 6.5

Maximum Likelihood and Bayes Estimates of Study Odds Ratios with 95% CI from 22 Studies Comparing Intravenous Magnesium with No Magnesium

Study	Year	Observed estimate	95% CI	Posterior median	95% CI	$Pr(OR<1\|y)$
Morton	1984	0.44	0.04, 5.02	0.52	0.17, 1.39	0.91
Smith	1986	0.27	0.06, 1.35	0.44	0.15, 1.00	0.98
Rasmussen	1986	0.33	0.10, 1.06	0.44	0.18, 0.91	0.99
Abraham	1987	0.96	0.06, 15.77	0.58	0.19, 1.66	0.86
Ceremuzynski	1989	0.28	0.03, 2.88	0.48	0.15, 1.23	0.94
Singh	1990	0.51	0.18, 1.45	0.53	0.23, 1.09	0.96
Shechter	1990	0.10	0.01, 0.82	0.35	0.11, 0.80	0.99
Shechter	1991	0.55	0.09, 3.37	0.54	0.19, 1.36	0.91
Feldstedt	1991	1.25	0.48, 3.26	0.83	0.41, 1.82	0.69
Woods	1992	0.74	0.56, 0.99	0.72	0.55, 0.95	0.99
Wu	1992	0.31	0.11, 0.92	0.42	0.18, 0.86	0.99
ISIS-4	1995	1.06	1.00, 1.13	1.06	0.99, 1.12	0.04
Shechter	1995	0.21	0.07, 0.64	0.34	0.14, 0.71	0.999
Thorgersen	1995	0.45	0.13, 1.54	0.50	0.20, 1.10	0.96
Bhargava	1995	0.95	0.18, 5.00	0.63	0.24, 1.63	0.85
Urek	1996	3.00	0.12, 76.58	0.63	0.21, 2.09	0.81
Raghu	1999	0.33	0.13, 0.86	0.42	0.19, 0.81	0.995
Gyamiani	2000	0.17	0.03, 0.81	0.36	0.12, 0.82	0.99
Santoro	2000	0.33	0.01, 8.20	0.49	0.13, 1.39	0.92
MAGIC	2000	1.00	0.87, 1.15	0.99	0.86, 1.14	0.56
Zhu	2002	0.64	0.49, 0.84	0.63	0.49, 0.82	0.9997
Nakashima	2004	0.33	0.03, 3.27	0.50	0.16, 1.26	0.93
Pooled (two-stage: REML)		0.58	0.43, 0.78			
Pooled (one-stage)		0.57	0.39, 0.77			
Pooled (Bayes)				0.54	0.35, 0.75	0.9998

The bottom of the table gives summary estimates using a two-stage model fit by REML, one-stage generalized-linear mixed model with fixed study effects (intercepts), and Bayes model with fixed study effects.

the bottom of the table. All estimates are converted to the odds ratios scale by exponentiating the log odds ratio estimates. For these calculations, we ran three Markov chains simultaneously and, after convergence (maximum value of the Gelman–Rubin statistic < 1.01), saved 200,000 simulations from each chain as draws from the correct posterior. The posterior estimate is therefore based on 600,000 samples for each parameter. The simulation error is small with all ratios of MC standard error to the posterior standard deviation for treatment effect parameters less than 0.5%.

The Bayesian random-effects estimate of the population median, 0.54, is close to the frequentist random-effects estimates. The Bayes estimates are slightly closer to those of the one-stage model that uses the binomial likelihood rather than to the two-stage model that uses an approximate normal distribution. The population median e^{μ} is estimated with a bit more uncertainty by the Bayesian analysis as it properly incorporates all of the uncertainty in τ^2. The posterior probability of a beneficial treatment effect for magnesium is 0.9998 in spite of the negative result of the massive ISIS-4 study. The heterogeneity exhibited across the studies leads to downweighting of ISIS-4 in the final average effect.

The study effects θ_i are shrunk considerably toward their average μ, particularly for the smaller studies, and their posterior 95% credible intervals are also much tighter than their individual 95% confidence intervals. Figure 6.2 depicts this graphically. Note the difference in scale compared with Figure 6.1 where the odds ratio scale is much bigger. By incorporating information from other studies as reflected in their common mean μ, we gain more certainty about the location of each study's true effect. Despite the gain in information, the large amount of heterogeneity apparent in the observed study odds ratios does not disappear in the posterior analysis. Only half of the studies have posterior probabilities of benefit greater than 0.95. The two large positive studies, Woods and Zhu, show posterior probabilities of benefit of 0.99; the negative ISIS study has only a probability of 0.04 of benefit; the inconclusive MAGIC study has a probability of 0.56. Thus, the posterior probabilities match with the observed results for the large studies. The smaller studies that exhibit more shrinkage have posterior probabilities of benefit that are high, but not

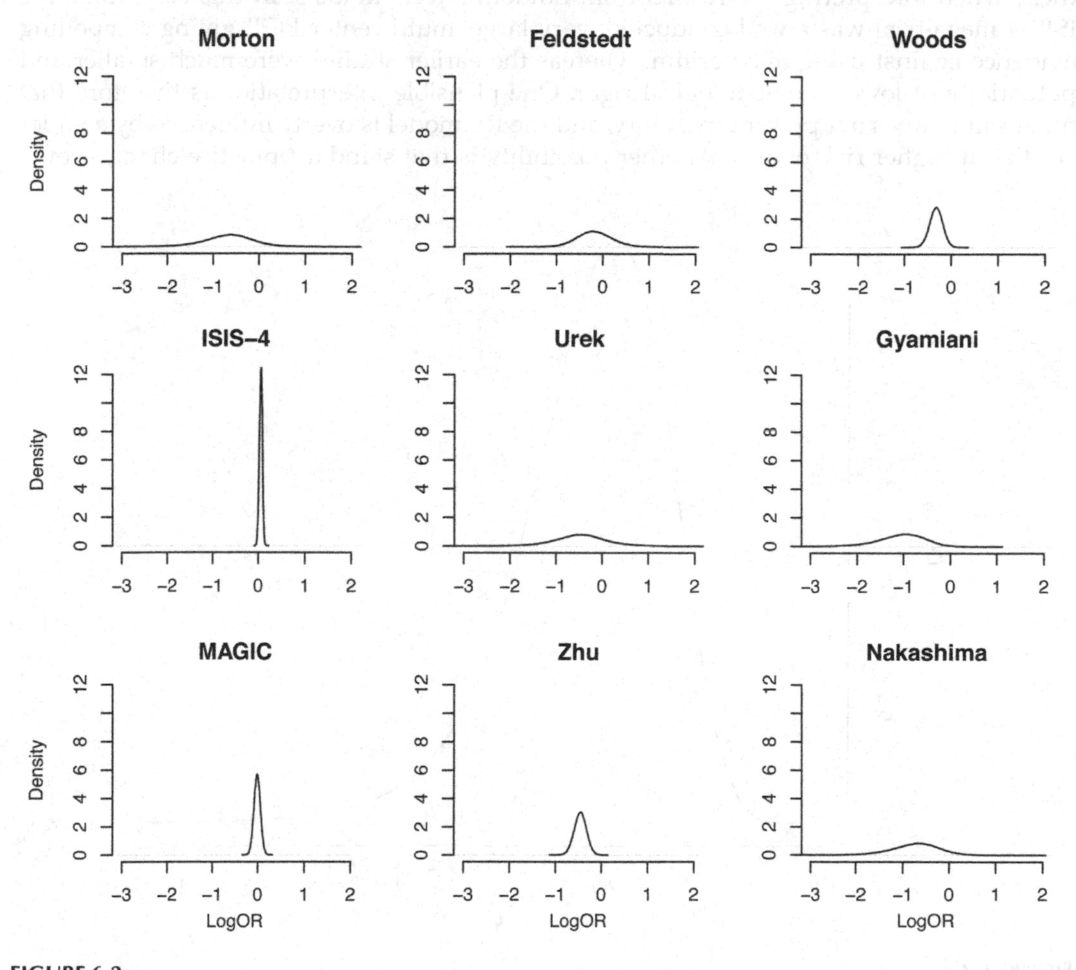

FIGURE 6.2
Kernel density plots of posterior distribution of selected study effects using fixed intercept: magnesium data model.

extremely high, reflecting their lack of information. MAGIC and ISIS-4 are so large that their posterior estimates are the same as their study estimates. They stand out as clear outliers among these studies.

One advantage of the Bayesian analysis is that it does not force the posterior distributions to take a normal form, although most of the posterior study distributions in Figure 6.2 are reasonably symmetric. The large size of the ISIS-4 study makes its posterior estimate very precise; the small size of the Morton study makes its posterior estimate imprecise. The Zhu and Woods studies have most of their probability in the region of treatment benefit (log OR < 0), whereas MAGIC has equal probability on both sides of zero and ISIS-4 is concentrated above zero. The smaller studies are more spread out but have the majority of their probability concentrated below zero, indicating effective treatment.

Figure 6.3 graphs the posterior distribution of the between-study standard deviation τ. This distribution which is centered away from small values, combined with the differences exhibited by the posterior study effects in Figure 6.2, would seem to suggest that τ is likely not equal to zero and therefore rules out the common-effect model.

Heterogeneity may arise for a variety of different reasons, and it is important to explore these when interpreting the results from random-effects models. In this case, the large ISIS-4 mega-trial was a well conducted, very large, multi-center RCT, giving compelling evidence against using magnesium, whereas the earlier studies were much smaller and potentially of lower methodological rigor. One plausible interpretation is therefore that magnesium does not prevent mortality, and the RE model is overly influenced by smaller studies at higher risk of bias. Another possibility is that standard practice changed over

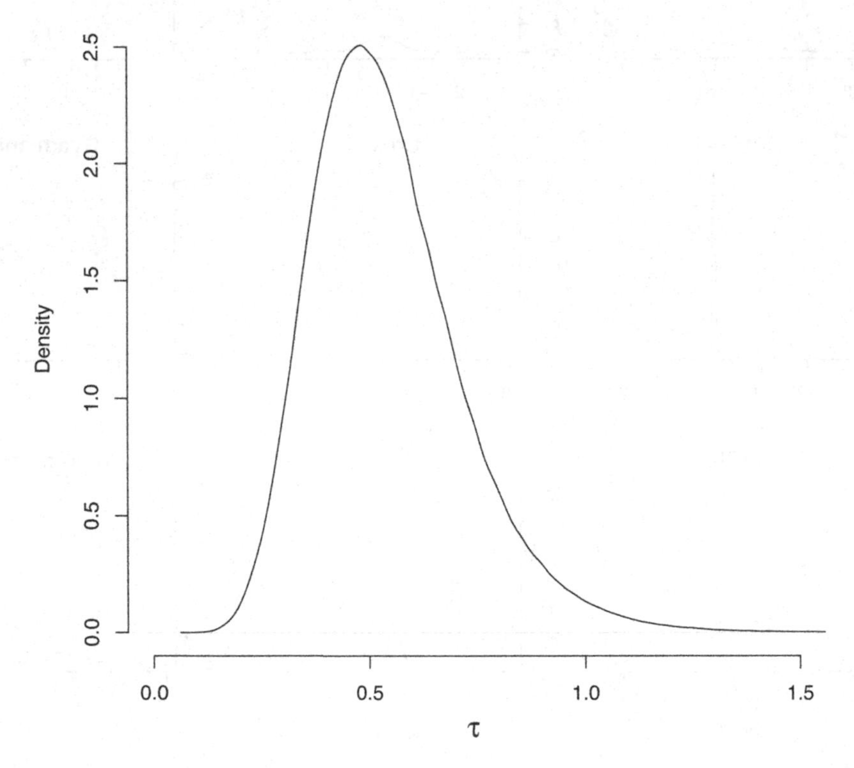

FIGURE 6.3
Kernel density plot of posterior distribution of between-study standard deviation τ using fixed intercept: magnesium data model.

time and that giving magnesium provided less benefit when added to more effective standard therapy. Practically, current clinical practice has moved away from the use of magnesium for treating heart attacks, reflecting the view that other treatments are sufficient. Higgins and Spiegelhalter (2002) present a detailed discussion of the problem from the perspective of the accruing information and investigate various skeptical prior distributions that might be used to temper findings.

6.8 Prediction

Prediction is central to problems requiring decisions. Many regulatory agencies now require a systematic review of the evidence from companies seeking to market a new drug or device. Businesses also use meta-analysis for the planning of new product lines. Such users see meta-analysis as helping them to predict the results of a future study. The Bayesian approach is well suited to problems of prediction through the use of the predictive distribution. For example, when monitoring a single trial, one can use the posterior predictive distribution to determine the probability of reaching a successful result. It is also useful in decision modeling when one needs to predict the results of a new study population that is similar to those in the studies in a meta-analysis (Ades, Lu and Higgins, 2005).

Suppose we want to make a prediction for the result, y^{new}, of a new study that is conditionally independent of past data y given the model parameters $\mathbf{\Theta}$. The posterior prediction for y^{new} given observed data y is found by averaging the conditional predictive distribution $p(y^{new} | \mathbf{\Theta})$ with respect to the posterior $p(\mathbf{\Theta} | y)$:

$$p(y^{new} | y) = \int p(y^{new} | \Theta) p(\Theta | y) d\Theta \tag{6.17}$$

In simple cases, it may be possible to carry out the integration analytically (Berry et al., 2010, pp. 34–37), but in most cases it will be easier to estimate the posterior predictive distribution using the MCMC simulations by drawing from the distribution $p(y^{new} | \mathbf{\Theta})$ for each simulated value of $\mathbf{\Theta}$ and forming a density estimate based on these draws of y^{new}.

By substituting the prior $p(\mathbf{\Theta})$ rather than the posterior $p(\mathbf{\Theta} | y)$ in (6.17), one can make *prior* predictions using:

$$p(y^{new}) = \int p(y^{new} | \Theta) p(\Theta) d\Theta$$

which is the marginal distribution of y^{new}. Prior prediction is useful when designing trials, where it can be used to predict the probability that a trial will demonstrate a difference (Box, 1980).

One might also want to predict the treatment effect θ^{new} in a new study. Then

$$p(\theta^{new} | y) = \int p(\theta^{new} | \Theta) p(\Theta | y) d\Theta.$$

This can be computed using the MCMC output used to find $p(\mathbf{\Theta} | y)$. For example, if $\theta_i \sim N(\mu, \tau^2)$, one can draw θ^{new} by first drawing $\mu^{(k)}$ and $\tau^{(k)}$ and then drawing $\theta^{new} \sim N(\mu^{(k)}, \tau^{(k)^2})$.

In the magnesium example, the ISIS-4 result seemed to be an outlier. However, if we had actually predicted the chance of its result based on the meta-analysis of the other studies, we would have found that a null result was not beyond the bounds of possibility. Assuming the other studies to be exchangeable with ISIS-4, a predictive distribution for ISIS-4 would include values near 1.7 and below 0.2, indicating that the value of 1.06 observed is not at all unexpected.

6.9 Assessing Model Fit and Model Assumptions

After fitting a model, we need to check whether it fits the data well. Chapter 11 discusses model checking in the context of non-Bayesian models. Here, we examine some tools for checking Bayesian models. One key difference is that Bayesian parameters are random variables and have posterior distributions, so predictions are stochastic. As a result, we look at how likely it is that the model could generate the observed outcomes and construct posterior predictive checks to assess ways in which models may fail to reproduce the data. But we still want to calculate measures of goodness of fit and compare different models. Bayesian hierarchical modelers often use the deviance information criterion (DIC) described below. We also want to perform sensitivity analyses to assess model assumptions and their impact on inferences.

6.9.1 Posterior Predictive Checks

External validation in which a model constructed from one dataset is applied to the data from a new dataset with the new predictions compared with the new outcomes is the gold standard for model validation. New datasets are hard to find, though, so we often must make do with internal validation using the data available. Traditionally, analysts have either held out some data from model development as a test set or used a resampling technique such as cross-validation in order to avoid the overfitting that results from having to evaluate model fit using the same data on which the model was developed.

With any internal validation method, a good model should be able to replicate the observed data. For a Bayesian, this entails simulating data from the posterior predictive distribution and checking how well it compares with the observed data. The observed data should look plausible under the posterior predictive distribution. To carry out a posterior predictive check, we first simulate outcomes y^{rep} from the posterior predictive distribution by drawing $\boldsymbol{\Theta}$ from its posterior $p(\boldsymbol{\Theta} \mid y)$, and then drawing $y^{rep} \mid \boldsymbol{\Theta}$. We then evaluate the test statistic $T(y \mid \boldsymbol{\Theta})$ at both y^{rep} and y and compare them. A posterior predictive p-value, $P(T(y^{rep}, y) \geq T(y, \Theta) \mid y)$, is calculated as the proportion of the number of samples $l = 1, 2, \ldots, L$ such that $T(y^{rep}, \Theta^{\lambda}) \geq T(y, \Theta^{\lambda})$. Note that this expression conditions on both y and $\boldsymbol{\Theta}$, so test quantities can also be functions of the unknown parameters $\boldsymbol{\Theta}$ (Gelman et al., 2013).

One set of useful test statistics in meta-analysis is the observed study effects, the y_i. But because each y_i is used to derive the posterior distribution $p(\boldsymbol{\Theta} \mid y)$, the potential lack of fit of any particular study i will be underestimated. Instead, one can use cross-validation to estimate the posterior distribution $p(\Theta \mid y_{(-i)})$ without study i and then use draws $\boldsymbol{\Theta}_{(-i)}$ from this posterior to simulate draws from $y_i^{new} \sim N(\Theta^{new}, s_i^2)$ using the predicted treatment effect $\Theta_{-i}^{new} \sim N(\Theta_{(-i)}, \tau^2)$. The posterior predictive p-value is then calculated by the location of the observed y_i in the simulated posterior for y_i^{new}. To save on having to compute a different posterior model for each dataset that omits one study, one could replace the different $\Theta_{(-i)}^{new}$ with a single Θ^{new} based on the model fit to all the data, although this partially defeats the purpose of avoiding having the observed data contaminate the estimate of the posterior distribution. It is, however, a good first approximation and can suggest for which studies it would be worthwhile performing cross-validation (Marshall and Spiegelhalter, 2003).

6.9.2 Deviance Information Criterion

The deviance of a model is defined as

$$D_{\text{model}} = -2\log(L_{\text{model}})$$

where L_{model} is the likelihood of the model. Smaller values of the deviance indicate higher observed likelihood, hence better model fit. In Bayesian models, each simulated draw from the posterior distribution also returns a draw from the posterior deviance. The residual deviance is defined as $D_{\text{res}} = D_{\text{model}} - D_{\text{sat}}$ where D_{sat} is the deviance for the saturated model which fits the observed data exactly.

The residual deviance therefore provides a standardized measure of how well the model fits the data. If the data follow a normal distribution, the residual deviance is the sum of squared deviations of each data value from its predicted mean normalized by its standard deviation, $\sum_{i=1}^{N}\left[(y_i - \mathbb{E}(y_i \mid \Theta))^2 / \sqrt{V(y_i \mid \Theta)}\right]$. Each of these components is the square of a standard normal distribution, so the sum follows a chi-square distribution with N degrees of freedom if the predicted means are close to the true means. This result also holds asymptotically for non-normal distributions. Since the mean of a chi-square with N degrees of freedom is N, we would expect the posterior mean of the residual deviance $\bar{D}_{\text{res}}$ to be close to the number of unconstrained data points in the data if the model fits well. Contributions to the sum that are much larger than one indicate observations for which the model fits poorly.

To compare two models, we can take the difference in the posterior means of their residual deviances, which by cancellation of the common D_{sat} term is simply the difference in the posterior means of their model deviances. Because a more complex model should have smaller model deviance, we need to include some kind of penalty for model complexity. The penalty in the well-known Akaike information criterion (AIC) adds two times the number of parameters, k, to the deviance evaluated at the maximum likelihood estimate of the parameters, $\text{AIC} = D(\hat{\Theta}) + 2k$. Models with smaller AIC fit better. AIC is a useful model comparison criterion as long as we know the number of parameters in the model. For instance, in the common-effect model with arm-level study effects as in equation (6.15), the number of parameters is clearly one plus the number of study intercepts. In a separate-effects model, the number of parameters is the number of different study effects plus the number of study intercepts. For instance, for the magnesium data with 22 studies, the common-effect model has 23 parameters (and so 21 degrees of freedom), whereas the separate-effects model uses 44 parameters and has no degrees of freedom.

In hierarchical model structures such as random-effects models, though, the number of parameters depends on the degree of between-study heterogeneity. The common-effect and separate-effects models are special cases of a random-effects model with $\tau^2=0$ and $\tau^2=\infty$, respectively. But if $0<\tau^2<\infty$, the number of parameters used for study effects is not explicitly defined. Instead, it varies between 1 and I in a meta-analysis with I studies. The actual value depends on the amount of heterogeneity, i.e., on the size of τ^2. The amount is quantified by the effective number of parameters

$$p_D = \bar{D}_{\text{model}} - D(\bar{\Theta})$$

which is the difference between the posterior mean deviance and the deviance calculated at the posterior mean of the model parameters θ. One can also use the posterior mean of the model predictions $D(\hat{\Theta})$ in place of $D(\bar{\Theta})$. We can then define DIC as the posterior mean deviance plus the effective number of parameters (Spiegelhalter et al., 2002)

$$\text{DIC} = \bar{D}_{\text{model}} + p_D.$$

Models with lower DIC are preferred because they represent a balance between a good fit without too much additional complexity. For two models with equivalent goodness of fit (posterior mean deviance), differences in DIC can be interpreted as the number of effective parameters saved. A general rule of thumb suggested by Spiegelhalter et al. (2002) is that differences in DIC less than five indicate little difference in model quality, although some authors use a cutoff of three (Welton et al., 2012).

The justification for DIC relies on appeals to asymptotic normality, but it seems to work well in practice for a wide array of models. Nevertheless, because it is most useful when comparing hierarchical models, care must be taken in how one defines the likelihood and model parameters as DIC is not invariant to reparameterizations. As a simple example, Berry et al. (2010) note that the likelihood for a hierarchical model having data $\boldsymbol{y}$ with parameters $\boldsymbol{\Theta}$ and hyperparameters $\boldsymbol{\lambda}$ could be written either conditionally as $p(y|\Theta)$ or marginally as $p(y|\lambda)=\int p(y|\Theta)p(\Theta|\lambda)d\Theta$. These two expressions also imply different numbers of model parameters and therefore force the modeler to be careful about which model is more pertinent to the question at hand.

6.9.3 Magnesium Example

We will compare four different models for the magnesium data: 1) random treatment effects, fixed intercepts; 2) random treatment effects, random intercepts; 3) common treatment effect, common intercepts; and 4) common treatment effect, random intercepts. In the second model, we assume independent priors for the treatment effects and intercepts, although it is certainly possible to model these as correlated in Chapter 5.

Table 6.6 shows that random-effects models have much lower DIC than common-effect models and that random intercept models have slightly lower DIC than fixed intercept models. It is difficult, however, to choose between the fixed and random intercepts, because each has $\bar{D}_{\text{model}}$ close to the 44 data points created by the 22 two-arm trials. The slightly better fit of the fixed intercept models is balanced by their greater complexity. The number of parameters used by the random-effects models is midway between the number for the common-effect models and the 44 for the separate-effects model.

One posterior predictive check that focuses on whether the model can detect outliers in the data looks at the number of observed zero counts among the 44 study arms.

TABLE 6.6

Model Fit for Common- and Random-Effects Models Using Fixed and Random Intercepts for the Magnesium Data

Treatment effects	Intercepts	DIC	$\bar{D}_{\text{model}}$	p_D
Random	Fixed	74.4	41.8	32.6
Random	Random	71.1	45.9	25.2
Common	Fixed	116.1	92.4	23.7
Common	Random	112.1	95.1	17.0

The 22 studies produce two zero counts. How many would simulations from a model give and how would this distribution compare with the observed count of zero? We can easily simulate the predictive distribution using equations (6.15) and (6.16) along with equations (6.13) and (6.14).

Figure 6.4 provides a histogram of the predictive distribution of zeroes showing the location of the observed zero count using the random treatment effects, random intercepts model. Among 600,000 simulations, 81,545 returned zero or one zero; 115,949 returned

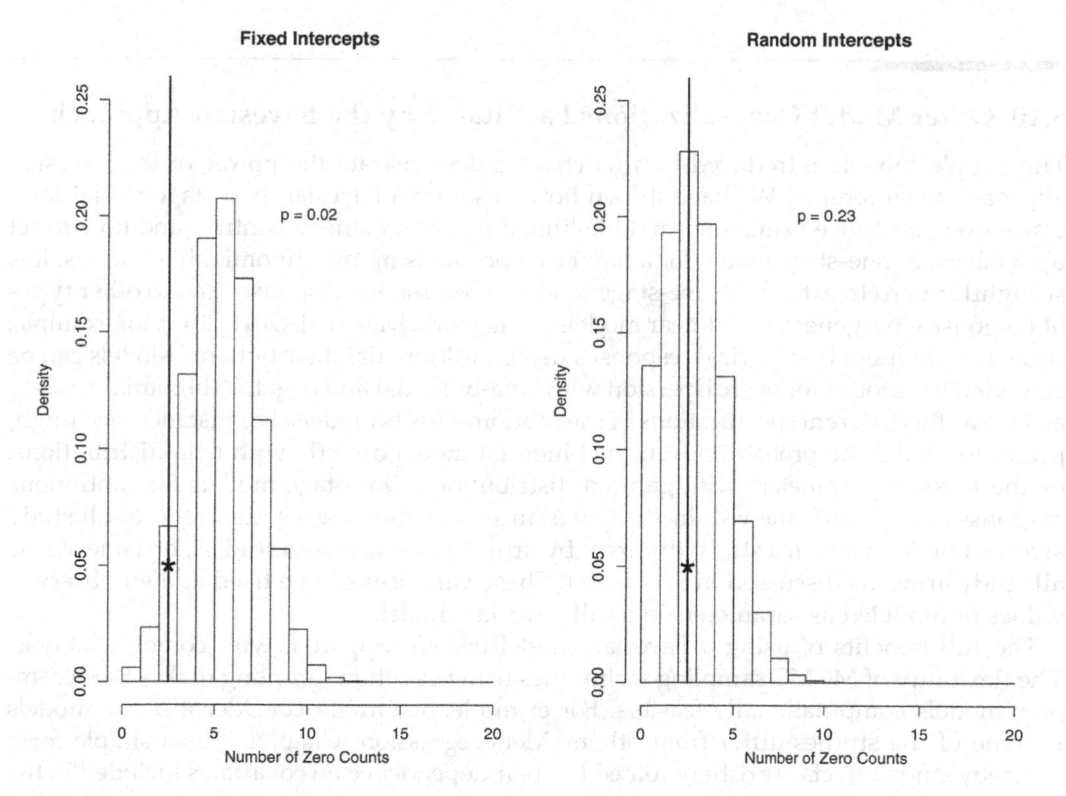

FIGURE 6.4
Location of the observed number of zero counts among 44 study arms from the magnesium data within the distribution simulated from their predictive distribution conditional on the observed data using the random treatment effects, random intercepts model. The posterior predictive p-value for this model is 0.23, indicating acceptable fit. By contrast, the posterior predictive p-value for the random treatment effects, fixed intercepts model is 0.02, indicating poor fit.

two zeroes; and 402,506 returned three or more zeroes. Calculating a posterior predictive p-value as the number of simulations of zero or one zeroes plus half the number of simulations of two zeroes gives a posterior predictive p-value of 0.23, which indicates that the model passes this check. Conversely, the posterior predictive p-value from fitting the random treatment effects, fixed intercepts model is 0.02, as the model tends to simulate more zeroes than were observed.

Thus, both DIC and the posterior predictive check suggest that the random treatment effects model with random intercepts gives a marginally better balance between fit and complexity than the random treatment effects model with fixed intercepts. Because both models give very similar study level and overall treatment effect estimates, however, they are robust to different assumptions about the intercepts. In fact, one might prefer the fixed intercepts model because it fits better, and one does not really mind the extra parameters since the focus is on estimating the treatment effect well. In this problem, the most important findings are the large heterogeneity among the treatment effects that makes an assumption of random treatment effects much more plausible than one of common treatment effects and the sharp dichotomy between the null result of the massive ISIS-4 study and the generally positive results of the smaller studies, leading to the exploration of the reasons for this heterogeneity as discussed in Section 6.7.

6.10 Other Model Generalizations Facilitated by the Bayesian Approach

The simple models introduced by this chapter demonstrate the power of the Bayesian approach for inference. We have shown how to set up a Bayesian two-stage model for a continuous response using a normal likelihood for the treatment contrast and how to set up a Bayesian one-stage model for a binary response using two binomial likelihoods. It is straightforward to extend the one-stage model to continuous responses and to other types of responses in a generalized-linear models framework. Dias et al. (2011, 2013), for example, show how to model categorical responses using multinomial distributions. Models can be extended to account for overdispersion with beta-binomial and negative-binomial models, as well as for different specifications of the random-effects models. For instance, one might prefer to model the probabilities in the binomial model directly with beta distributions, or the Poisson parameters with gamma distributions. One-stage models for continuous responses can handle the within-study variances in various ways (different for all study arms, same for arms in a study, different by arms but same across studies, or same across all study arms) as discussed in Section 5.1. These variances can be fixed at their observed values or modeled as parameters in a full Bayesian model.

The full benefits of using a Bayesian model become apparent with complex models. The flexibility of MCMC sampling techniques using rejection sampling makes these complex models computationally feasible. For example, one might consider mixture models if some of the studies differ from others. Meta-regression (Chapter 7) is a simple form whereby study effects are differentiated by their dependence on covariates included in the model. When some of the studies may have very large or very small effect sizes, the normal assumption for the random-effects distribution as in equation (6.6) may no longer be appropriate. Instead, one might prefer a Student-t distribution, which may be also thought of as a mixture of normal distributions with different variances that themselves follow an inverse gamma distribution. The Student-t is often recommended when one is worried about heavy-tailed distributions, i.e., ones with more outliers than expected. Later chapters will make extensive use of Bayesian models exploring model extensions to topics such as publication bias (Chapter 13), network meta-analysis (Chapter 10), control risk regression (Chapter 14), diagnostic test studies (Chapter 19) and within-study bias (Chapter 12).

6.11 Software

In addition to user-written code, a variety of software packages in Stata, R, and SAS can run MCMC algorithms. Currently, the most widely used Bayesian software is WinBUGS (Windows version of Bayesian inference Using Gibbs Sampling) (www.mrc-bsu.cam.ac.uk /bugs) (Lunn et al., 2000). An open source version called OpenBUGS (http://www.openbugs.info) (Lunn et al. 2009) and a multi-platform (Windows, Mac, Unix) version called JAGS (http://mcmc-jags.sourceforge.net/) (Plummer, 2003) are also available. Generically, we shall call these programs BUGS. Given a user-specified model in the form of likelihood and priors, data, and parameter starting values, BUGS constructs the necessary full conditional distributions, carries out MCMC sampling, provides statistical and graphical summaries of the posterior distributions, and reports some convergence diagnostics. BUGS can also export the simulations to a text file. Welton et al. (2012) and the NICE DSU technical documents (Dias et al., 2011) give many worked examples of Bayesian meta-analyses. BUGS can also be called from within the R and Stata statistical packages, eliminating the need to export and import saved simulation chains, and extending the possibilities for data exploration and presentation of results.

STAN (http://mc-stan.org) based on Hamiltonian MCMC (Brooks et al., 2011) is a more recent entry into Bayesian computational software. It has been gaining users but has not yet been used much for meta-analysis. Since Hamiltonian MCMC is more efficient in complex problems, it probably does not gain much in simple meta-analysis but may be useful for analyzing individual participant data (Chapter 8) and network meta-analysis (Chapter 10) and for problems with sparse or very large datasets.

6.12 Summary

Bayesian models provide a flexible framework for constructing and computing complex models that incorporate information both from the data at hand and from relevant external evidence, thus facilitating principled and transparent inference that aids decision making. Computation of the entire posterior distribution of model parameters through analytic methods or MCMC simulation also provides direct expressions of the probability of hypotheses about unknown parameters conditional on observed data, in contrast to frequentist procedures that describe the probability of the observed data given a specific hypothesis about the unknown parameters. By computing the entire joint probability distribution of all model parameters, Bayesian inference provides full information about them rather than just calculating means and standard errors and then relying on asymptotic normality to produce confidence intervals. The simulation output from MCMC also simplifies construction of functions of parameters, predictive distributions, and the probabilities of non-standard hypotheses. When data are sparse, the likelihood has relatively little information about model parameters and so inferences about them will depend more on external information incorporated through a prior probability distribution that describes knowledge that is auxiliary to the data. In general, the Bayesian paradigm mimics the scientific process of learning through combining new findings (or data) with prior knowledge and assumptions, allowing for the construction of new models that reflect new insights. The use of informative prior input also provides a formal mechanism to incorporate clinical judgements into the analysis, to coherently test the sensitivity of conclusions to those beliefs, and to explore how those beliefs can lead different individuals to interpret the same data differently (Higgins and Spiegelhalter, 2002).

Another important feature of Bayesian models for meta-analysis is that they properly account for uncertainty arising from the estimation of between-study heterogeneity and average over this uncertainty in estimating study effects. Although meta-analysis is often focused on estimating the common study mean, evaluation of heterogeneity is also a crucial part of a meta-analysis (Lau et al., 1998). Thus, the provision of posterior estimates of the true effects in individual studies is another important byproduct of Bayesian models. Obtaining posterior estimates of study effects can aid in determining whether studies really are heterogeneous or whether perceived heterogeneity is an artifact of small sample sizes. Further exploration through meta-regression or individual patient regression may help uncover important treatment effect modifiers and causes of between-study heterogeneity.

Finally, we caution that such powerful tools must be used with care. Bayesian inference is model-based inference; if the model is wrong, the results can be badly misleading. Thus, it is important that model assumptions, including specification of both the likelihood and the prior distribution, be made clear, explicit, and transparent, so that they can be subject to critique. Analysts should always investigate the sensitivity of inferences to model choices and be careful not to overinterpret conclusions. Happily, the MCMC Bayesian paradigm facilitates checking models and making them more robust and realistic.

Bayesian models are becoming more popular in meta-analysis, particularly for more complex problems such as network meta-analysis (Chapter 10). However, the parameters do require specification of prior distributions, which can be tricky for variance parameters. Computation also requires access to specialized software. In simpler problems with a sufficient number of studies, perhaps ten or more, frequentist and Bayesian solutions should be similar most of the time.

References

Ades AE, Lu G and Higgins JPT, 2005. The interpretation of random effects meta-analysis in decision models. *Medical Decision Making* **25**(6): 646–654.

Berger JO and Sellke T, 1987. Testing a point null hypothesis: the irreconcilability of p-values and evidence (with discussion). *Journal of the American Statistical Association* **82**(397): 112–122.

Berger JO and Wolpert R, 1984. *The Likelihood Principle*. Hayward, CA: Institute of Mathematical Statistics Monograph Series.

Berry SM, Carlin BP, Lee JJ and Muller P, 2010. *Bayesian Adaptive Methods for Clinical Trials*. London: Chapman and Hall/CRC Press.

Box G, 1980. Sampling and Bayes' inference in scientific modelling and robustness. *Journal of the Royal Statistical Society, Series A* **143**(4): 383–430.

Brooks SP and Gelman A, 1998. Alternative methods for monitoring convergence of iterative simulations. *Journal of Computational and Graphical Statistics* **7**(4): 434–455.

Brooks S, Gelman A, Jones GL and Meng XL (editors), 2011. *Handbook of Markov Chain Monte Carlo.* Chapman and Hall/CRC Press.

Carlin BP and Louis TA, 2009. *Bayesian Methods for Data Analysis.* 3rd Edition. Boca Raton, FL: Chapman and Hall/CRC Press.

Carlin JB, 1992. Meta-analysis for 2×2 tables: A Bayesian approach. *Statistics in Medicine* **11**(2): 141–158.

Chaloner K, Church T, Louis TA and Matts JP, 1993. Graphical elicitation of a prior distribution for a clinical trial. *Journal of the Royal Statistical Society, Series D (The Statistician)* **42**(4): 341–353.

Cochran WG, 1954. The combination of estimates from different experiments. *Biometrics* **10**(1): 101–129.

Dias S and Ades AE, 2016. Absolute or relative effects? Arm-based synthesis of trial data. *Research Synthesis Methods* **7**(1): 23–28.

Dias S, Ades AE, Sutton AJ and Welton NJ, 2013. Evidence synthesis for decision making 2: A generalized linear modeling framework for pairwise and network meta-analysis of randomized controlled trials. *Medical Decision Making* **33**(5): 607–617.

Dias S, Ades AE, Welton NJ, Jansen JP and Sutton AJ, 2018. *Network Meta-Analysis for Comparative Effectiveness Research.* Hoboken NJ: Wiley.

Dias S, Welton NJ, Sutton AJ and Ades AE, 2011. A generalised linear modelling framework for pairwise and network meta-analysis of randomised controlled trials. *NICE Decision Support Unit Evidence Synthesis Technical Support Document* 2. Accessed at https://scharr.dept.shef.ac.uk/nicedsu/technical-support-documents/evidence-synthesis-tsd-series/.

Draper D, Hodges JS, Mallows CL and Pregibon D, 1993. Exchangeability and data analysis. *Journal of the Royal Statistical Society, Series A* **156**(1): 9–37.

DuMouchel W, 1990. In Berry DA (Ed). *Bayesian Metaanalysis in Statistical Methodology in the Pharmaceutical Sciences.* New York: Marcel Dekker, 509–529.

Efron B and Morris CN, 1973. Stein's estimation rule and its competitors—An empirical Bayes approach. *Journal of the American Statistical Association* **68**(341): 34–38.

Efthimiou O, Mavridis D, Debray T, Samara M, Belger M, Siontis G, Leucht S and Salanti G on behalf of GetReal Work Package 4, 2017. Combining randomized and non-randomized evidence in network meta-analysis. *Statistics in Medicine* **36**(8): 1210–1226.

Gelfand AE and Smith AFM, 1990. Sampling-based approaches to calculating marginal densities. *Journal of the American Statistical Association* **85**(410): 398–409.

Gelman A, 1996. In Gilks WR, Richardson S and Spiegelhalter DJ (Eds). *Inference and Monitoring Convergence.* In *Markov Chain Monte Carlo in Practice.* London: Chapman and Hall.

Gelman A, 2006. Prior distributions for variance parameters in hierarchical models. *Bayesian Analysis* **1**(3): 515–534.

Gelman A, Carlin JB, Stern HS, Dunson DB, Vehtari A and Rubin DB, 2013. *Bayesian Data Analysis.* 3rd Edition. Chapman and Hall/CRC Press.

Gelman A and Rubin DB, 1992. Inferences from iterative simulation using multiple sequences. *Statistical Science* **7**(4): 457–472.

Gilks WR and Roberts GO, 1996. Strategies for improving MCMC. In Gilks WR, Richardson S and Spiegelhalter DJ (Eds). *Markov Chain Monte Carlo in Practice.* London: Chapman and Hall/CRC.

Higgins JPT and Spiegelhalter DJ, 2002. Being sceptical about meta-analyses: A Bayesian perspective on magnesium trials in myocardial infarction. *International Journal of Epidemiology* **31**(1): 96–104.

Jackson D and Baker R, 2013. Meta-analysis inside and outside particle physics: Convergence using the path of least resistance? *Research Synthesis Methods* **4**(2): 125–126.

Jackson D, Law M, Stijnen T, Viechtbauer W and White IR, 2018. A comparison of seven random-effects models for meta-analyses that estimate the summary odds ratio. *Statistics in Medicine* **37**(7): 1059–1085.

James W and Stein C, 1961. Estimation with quadratic loss. *Proceedings of the Fourth Berkeley Symposium on Mathematical Statistics and Probability* **1**: 311–319.

Kass RE, Carlin BP, Gelman A and Neal R, 1998. Markov chain Monte Carlo in practice: A roundtable discussion. *The American Statistician* **52**(2): 93–100.
Lau J, Antman EM, Jiminez-Silva J, Kupelnick B, Mosteller F and Chalmers TC, 1992. Cumulative meta-analysis of therapeutic trials for myocardial infarction. *The New England Journal of Medicine* **327**(4): 248–254.
Lau J, Ioannidis JPA and Schmid CH, 1998. Summing up evidence: One answer is not always enough. *The Lancet* **351**(9096): 123–127.
Lau J, Schmid CH and Chalmers TC, 1995. Cumuslative meta-analysis of clinical trials builds evidence for exemplary medical care. *Journal of Clinical Epidemiology* **48**(1): 45–57.
Li J, Zhang Q, Zhang M and Egger M, 2007. Intravenous magnesium for acute myocardial infarction (review). *The Cochrane Database of Systematic Reviews* Issue 2. Art. No.: CD002755.
Lunn D, Spiegelhalter D, Thomas A and Best N, 2009. The BUGS project: Evolution, critique and future directions. *Statistics in Medicine* **28**(25): 3049–3067.
Lunn DJ, Thomas A, Best N and Spiegelhalter D, 2000. WinBUGS – A Bayesian modelling framework: Concepts, structure, and extensibility. *Statistics and Computing* **10**(4): 325–337.
MacEachern SN and Berliner LM, 1994. Subsampling the Gibbs sampler. *The American Statistician* **48**(3): 188–190.
Marshall EC and Spiegelhalter DJ, 2003. Approximate cross-validatory predictive checks in disease mapping models. *Statistics in Medicine* **22**(10): 1649–1660.
Nam IS, Mengersen K and Garthwaite P, 2003. Multivariate meta-analysis. *Statistics in Medicine* **22**(14): 2309–2333.
O'Hagan A, Buck CE, Daneshkhah A, Eiser JR, Garthwaite PH, Jenkinson DJ, Oalkey JE and Rakow T, 2006. *Uncertain Judgements: Eliciting Experts' Probabilities*. Chichester, UK: John Wiley and Sons.
Plummer M, 2003. JAGS: A program for analysis of Bayesian graphical models using Gibbs sampling. Proceedings of the 3rd International Workshop on Distributed Statistical Computing (DSC 2003), March 20–22, Vienna, Austria.
Rhodes KM, Turner RM and Higgins JPT, 2015. Predictive distributions were developed for the extent of heterogeneity in meta-analyses of continuous outcome data. *Journal of Clinical Epidemiology* **68**(1): 52–60.
Roberts GO and Smith AFM, 1993. Simple conditions for the convergence of the Gibbs sampler and Metropolis-Hastings algorithms. *Stochastic Processes and Their Applications* **49**(2): 207–216.
Schmid CH, 2001. Using Bayesian inference to perform meta-analysis. *Evaluation and the Health Professions* **24**(2): 165–189.
Schmid CH, 2017. Heterogeneity: Multiplicative, additive or both? *Research Synthesis Methods* **8**(1): 119–120.
Schmid CH and Mengersen K, 2013. Bayesian meta-analysis. In Koricheva J, Gurevitch J and Mengersen K (Eds). *The Handbook of Meta-Analysis in Ecology and Evolution*. Princeton, NJ: Princeton University Press, 145–173.
Schmid CH, Trikalinos TA and Olkin I, 2014. Bayesian network meta-analysis for unordered categorical outcomes with incomplete data. *Research Synthesis Methods* **5**(2): 162–185.
Spiegelhalter DJ, Best N, Carlin BP and van der Linde A, 2002. Bayesian measures of model complexity and fit (with discussion). *Journal of the Royal Statistical Society, Series B* **64**(4): 583–639.
Spiegelhalter DJ, Freedman LS and Parmar MKB, 1994. Bayesian approaches to randomised trials (with discussion). *Journal of the Royal Statistical Society, Series A* **157**(3): 357–416.
Spiegelhalter DJ, Thomas A, Best N and Lunn D, 2003. WinBUGS version-1.4 user manual. Cambridge, UK: MRC Biostatistics Unit. Accessed at http://www.mrc-bsu.cam.ac.uk/bugs/.
Stanley TD and Doucouliagos H, 2015. Neither fixed nor random: weighted least squares meta-analysis. *Statistics in Medicine* **34**(13): 2116–2127.
Turner RM, Davey J, Clarke MJ, Thompson SG and Higgins JPT, 2012. Predicting the extent of heterogeneity in meta-analysis, using empirical data from the Cochrane Database of Systematic Reviews. *International Journal of Epidemiology* **41**(3): 818–827.
Turner RM, Jackson D, Wei Y, Thompson SG and Higgins JPT, 2015. Predictive distributions for between-study heterogeneity and simple methods for their application in Bayesian meta-analysis. *Statistics in Medicine* **34**(6): 984–998.

Turner RM, Spiegelhalter DJ, Smith GCS and Thompson SG, 2009. Bias modelling in evidence synthesis. *Journal of the Royal Statistical Society, Series A* **172**(1): 21–47.
Welton NJ, Sutton AJ, Cooper NJ, Abrams KR and Ades AE, 2012. *Evidence Synthesis in Decision Making in Healthcare*. London: John Wiley and Sons.
Zucker DR, Ruthazer R, Schmid CH, Feuer JM, Fischer PA, Kieval RI, Mogavero N, Rapoport RJ, Selker HP, Stotsky SA, Winston E and Goldenberg DL, 2006. Lessons learned combining N-of-1 trials to assess fibromyalgia therapies. *The Journal of Rheumatology* **33**(10): 2069–2077.

7

Meta-Regression

Julian P.T. Higgins, José A. López-López, and Ariel M. Aloe

CONTENTS

7.1 The Role of Meta-Regression

The meta-analysis models and methods described so far in this book focus on the determination of average effect sizes across studies and on the extent of their variation across studies. Since it is inevitable that independent studies of the same problem will differ in many ways, a natural question is to ask whether statistical variation across studies can be explained by specific characteristics of the different studies. Such investigations of heterogeneity are often undertaken using the techniques of meta-regression.

Meta-regression is conceptually similar to traditional linear regression analysis, but with studies, rather than individuals, being the unit of analysis. The numbers of individuals within the studies, however, ought to be somehow reflected, leading to the use of different weights for different studies, i.e., weighted regression. In common with meta-analysis, the weights are usually selected to reflect standard errors of the estimates rather than sample sizes. The dependent (or outcome) variable in the regression is the effect size estimate from each study, and the independent (or explanatory) variables are study characteristics. Sometimes there are multiple effect sizes from each study, with each accompanied by a different covariate value, for example, when there are multiple comparisons of different dose groups against a reference group; in this case, the dependencies between the effect size estimates should be accounted for in the analysis. Regression coefficients describe how the study characteristics are associated with the sizes of effects across the studies. In some fields, the study characteristics are known as moderator variables, although we will generally refer to them as (study-level) covariates in the present chapter.

Study-level covariates might include methodological features of the studies (such as study design or measurement instruments used), descriptors of the study context (such as date or location), descriptors of the participants (such as age group or type of diagnosis), characteristics of interventions or exposures being evaluated in the studies (such as intensity or duration), or aspects of the outcome being measured (such as timing of measurement). Covariates may be numerical or categorical and are coded in the same way as in standard linear regression. Care is needed when the study-level covariates are aggregated summaries of participants within the studies (e.g., average age) due to the possibility of aggregation (also called ecological) bias (see Section 7.4.1).

Candidates for effect size estimates (the dependent variable) are the same as those for a simple meta-analysis and depend on the types of studies included in the analysis and the nature of the data being collected. Clinical trials might each contribute an odds ratio, or a hazard ratio or a difference in mean response, for example. As is the case for simple meta-analyses, ratio measures of effect are usually analyzed on the (natural) log scale. A simple dataset for a meta-regression might then comprise a vector of log odds ratios (one from each study), a vector of their standard errors, and at least one vector of study-level covariates.

7.2 Statistical Models

7.2.1 Common-Effect Model for a Single Covariate

In many areas of research, the number of studies in a meta-analysis is typically small. For this reason, single covariate models are frequently encountered. We will start with such a model, and also with the simplest version, that of a common-effect model based on precomputed effect size estimates with variances assumed known. "Common effect" here is to be interpreted as a single effect size at each value of the covariate, rather than as a single underlying effect across all studies as is the case in Chapter 4. We describe first the statistical models and provide some examples of their implementation, before proceeding to discussion of methods of inference and various extensions that are available.

A meta-regression model can be written as a hierarchical model or a linear model. A hierarchical representation of the model extends the models in Chapter 4 in a natural way. A standard assumption is that the effect size estimates, y_i, have sampling error of the form

$$y_i \sim \mathrm{N}\left(\theta_i, s_i^2\right),$$

with s_i^2 the within-study estimation error variances. At this point we assume the variances, s_i^2, are known without error.

A common-effect meta-analysis model assumes that all θ_i are identical, as discussed in Chapter 4. A common-effect meta-regression model with a single covariate relates the underlying effect sizes, θ_i to observed values, x_i, of a study-level covariate (the independent variable):

$$\theta_i = \beta_0 + \beta_1 x_i.$$

The intercept, β_0, and the coefficient (or slope), β_1, will be familiar from standard linear regression. In the absence of any relationship between the covariate and the effect sizes, i.e., when $\beta_1 = 0$, the model reduces to the standard common-effect meta-analysis model (4.2) of Chapter 4.

A linear model representation relates the observed effect sizes (the dependent variable in the regression context) directly to the covariate:

$$y_i = \beta_0 + \beta_1 x_i + \varepsilon_i \text{ with } \varepsilon_i \sim \mathrm{N}\left(0, s_i^2\right).$$

Note that, unlike standard linear regression where the residuals are assumed independent and identically distributed, in a meta-regression, the residuals reflect the within-study estimation error variances, s_i^2, of the observed effect sizes.

Similar extensions of the linear, logistic, and Poisson modeling approaches described in Chapter 5 follow naturally. For example, in the linear regression model (5.2) and the logistic regression model (5.7), the treatment effect term (β_1 in the notation of Chapter 5) would be replaced by a meta-regression term such as $\beta_1' + \beta_1'' z_i$ (we use z here to denote the covariate, since in Chapter 5 x denotes the treatment group).

7.2.2 Random-Effects Model for a Single Covariate

Recall that a random-effects meta-analysis model can be written as

$$y_i \sim \mathrm{N}\left(\theta_i, s_i^2\right),$$

$$\theta_i \sim \mathrm{N}\left(\theta, \sigma^2\right),$$

where σ^2 reflects the among-studies variability in (true) effect sizes across studies. A random-effects meta-regression model with a single covariate, x_i, can be written in hierarchical form as

$$\theta_i \sim \mathrm{N}\left(\beta_0 + \beta_1 x_i, \sigma_{\mathrm{res}}^2\right).$$

The variance component, σ_{res}^2, now represents heterogeneity among the underlying effect sizes across studies beyond that explained by the covariate, so may be considered to represent residual heterogeneity. This residual heterogeneity variance is estimated as part of the analysis, hence its depiction using a Greek letter in contrast to the within-study variances, s_i^2, which are conventionally assumed known. We discuss inference based on this meta-regression model in Section 7.5.

Written as a linear model (in "marginal form"), the random-effects meta-regression model is

$$y_i = \beta_0 + \beta_1 x_i + \xi_i \text{ with } \xi_i \sim \mathrm{N}\left(0, \sigma_{\mathrm{res}}^2 + s_i^2\right). \tag{7.1}$$

This is a slightly more general model than the hierarchical form, since now σ_{res}^2 may be allowed to be negative.

The random-effects models of Chapter 5 can readily be extended to meta-regression. In the regression models with fixed intercept terms (e.g., 5.4 and 5.10), the treatment effect term (β_1 in the notation of Chapter 5) would be replaced by a meta-regression term in the same way as in the common-effect model in the previous section. In the models with a random intercept term (e.g., 5.5 and 5.11), the covariate and its interaction with treatment would be added to the model equation. The coefficient of the main term for the covariate then describes the association between the covariate and the control group risk, while the interaction term describes the association between the covariate and the treatment effect.

7.2.2.1 Measuring Variance Explained

A convenient summary of the explanatory ability of a covariate in a random-effects meta-regression is a statistic analogous to the R^2 index computed from sums of squared residuals in a standard linear regression. However, the assumption of variance homogeneity across analysis units does not hold for meta-analysis (Aloe et al. 2010) so that a different approach is needed in the meta-analytic context. Instead, an R^2 index can be derived for meta-analysis as the relative decrease in heterogeneity variance from σ^2 to σ_{res}^2; that is after adding one or more covariates to the basic random-effects meta-analysis model. Note that this is also the rationale used in hierarchical modeling (Raudenbush and Bryk 2002). In practice, this ad hoc R^2 statistic can produce (implausible) negative values when covariates are not strong predictors of effect size, and values larger than 1. These problems occur particularly when the number of studies is small: a simulation study has found that approximately 40 studies are required before R^2 becomes stable (López-López et al. 2014).

This R^2 statistic can be related to the I^2 statistic of Chapter 4. The usual I^2 statistic describes approximately the proportion of total variance in effect size estimates that is attributable to heterogeneity rather than within-study estimation error. Of the part that is due to heterogeneity, R^2 describes the proportion that can be explained by the covariate(s) in the model rather than that which remains unexplained. This is illustrated diagrammatically in Figure 7.1. The usual I^2 statistic describes the relationship (A+B)/(A+B+C), and the

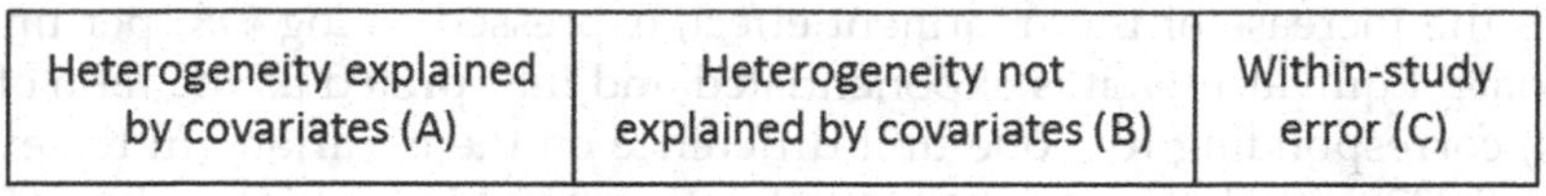

Heterogeneity explained by covariates (A)	Heterogeneity not explained by covariates (B)	Within-study error (C)

FIGURE 7.1
Diagrammatic representation of variance components in a random-effects meta-regression analysis.

R^2 statistic describes the relationship A/(A+B). For more discussion of this relationship, see Chapter 20 of Borenstein et al. (2009).

In fact, a further variant of the I^2 statistic can be obtained from the results of the meta-regression analysis, which very approximately describes the relationship B/(B+C), that is the residual heterogeneity from the regression expressed as a proportion of the variability in effect estimates that is not explained by the covariates. In practice, this meta-regression I^2 statistic, which we might call I^2_{res}, is computed directly and not from results for (the usual) I^2 and R^2, so is only roughly equal to the quantity $I^2\left(1-R^2\right)/\left(1-I^2R^2\right)$ as would be implied by an exact relationship.

7.2.3 Example (Single Covariate)

A widely discussed example of meta-regression in the healthcare field is an investigation of the association between effectiveness of BCG vaccination in preventing tuberculosis and the location of the study (assessed as absolute latitude, i.e., distance from the equator). In Table 7.1, we present data from 13 clinical trials of the BCG vaccine, as presented in a landmark paper on meta-regression methods (Berkey et al., 1995).

A random-effects meta-regression model in which ln(OR) is regressed on absolute latitude as a single covariate gives a coefficient for absolute latitude as −0.032 (95% CI −0.044 to −0.019), using a frequentist approach with restricted maximum likelihood (REML) estimation of σ^2_{res} (see Chapter 4 and Section 7.5.1). The interpretation of the regression

TABLE 7.1
Results of 13 Clinical Trials of the BCG Vaccine

		Vaccinated		Unvaccinated				
Study (i)	Absolute latitude (x_i)	TB	No TB	TB	No TB	OR	ln(OR) (y_i)	SE(ln(OR)) (s_i)
1	44	4	119	11	128	0.39	−0.94	0.60
2	55	6	300	29	274	0.19	−1.67	0.46
3	42	3	228	11	209	0.25	−1.39	0.66
4	52	62	13,536	248	12,619	0.23	−1.46	0.14
5	13	33	5036	47	5761	0.80	−0.22	0.23
6	44	180	1361	372	1079	0.38	−0.96	0.1
7	19	8	2537	10	619	0.20	−1.63	0.48
8	13	505	87,886	499	87,892	1.01	0.01	0.06
9	27	29	7470	45	7232	0.62	−0.47	0.24
10	42	17	1699	65	1600	0.25	−1.40	0.27
11	18	186	50,448	141	27,197	0.71	−0.34	0.11
12	33	5	2493	3	2338	1.56	0.45	0.73
13	33	27	16,886	29	17,825	0.98	−0.02	0.27

coefficient is the increase of the treatment effect, expressed as log OR, per unit increase of the covariate. Equivalently, it is exponentiated and interpreted as the ratio of two odds ratios (ROR) corresponding to a one unit difference on the covariate. In the example the ROR is exp(−0.032) = 0.97, meaning that the treatment OR is multiplied by 0.97 for every degree further away from the equator. There is therefore some evidence of a negative association, suggesting that studies conducted further from the equator find a more beneficial effect of the BCG vaccine. The estimate of heterogeneity variance drops from $\sigma^2=0.338$ to $\sigma^2_{\text{res}}=0.050$ after adding absolute latitude as a covariate, so the R^2 index is 85.1%. This indicates that approximately 85% of the heterogeneity across ln(OR)s is explained by the covariate, although with only 13 data points, this proportion of explained variation is subject to a large amount of uncertainty.

7.2.4 Multiple Covariates

The models above extend naturally to multiple covariates, including the use of multiple dummy covariates to represent levels of a single categorical independent variable. The common-effect meta-regression can be written as follows:

$$y_i \sim \text{N}\left(\theta_i, s_i^2\right),$$

$$\theta_i = \beta_0 + \sum_j \beta_j x_{ji}.$$

Similarly, a random-effects meta-regression model with multiple covariates can be written as follows.

$$\theta_i \sim \text{N}\left(\beta_0 + \sum_j \beta_j x_{ji}, \sigma^2_{\text{res}}\right).$$

As a linear model, the common-effect meta-regression model is

$$y_i = \beta_0 + \sum_j \beta_j x_{ji} + \varepsilon_i \text{ with } \varepsilon_i \sim \text{N}\left(0, s_i^2\right),$$

and the random-effects meta-regression model is

$$y_i = \beta_0 + \sum_j \beta_j x_{ji} + \xi_i \text{ with } \xi_i \sim \text{N}\left(0, \sigma^2_{\text{res}} + s_i^2\right). \tag{7.2}$$

7.2.5 Example (Multiple Covariates)

Table 7.2 presents results from 48 studies assessing the effectiveness of school-based writing-to-learn programs on attainment (Bangert-Drowns et al. 2004). Sample sizes and standardized mean differences (SMDs) between intervention and control groups are presented, along with the standard errors of the SMDs. Positive values of SMDs indicate higher attainment in the intervention group. Three covariates are also included for each study: program length (in weeks), whether writing contained imaginative components (1 = yes; 0 = no), and grade level (1 = elementary; 2 = middle; 3 = high school; 4 = college).

TABLE 7.2

Results of 48 Clinical Trials of Writing-to-Learn Programs

Study (i)	Program length (x_{1i})	Imaginative components (x_{2i})	Grade level	Sample size	Standardized mean difference (y_i)	SE(d) (s_i)
1	15	0	4	60	0.65	0.26
2	10	1	2	34	−0.75	0.35
3	2	0	2	95	−0.21	0.20
4	9	0	4	209	−0.04	0.14
5	14	0	1	182	0.23	0.15
6	1	0	4	462	0.03	0.09
7	4	0	3	38	0.26	0.33
8	15	0	1	542	0.06	0.08
9	4	0	4	99	0.06	0.20
10	9	0	1	77	0.12	0.23
11	15	0	4	40	0.77	0.33
12	15	0	4	190	0.00	0.14
13	8	0	1	113	0.52	0.19
14	4	0	4	50	0.54	0.29
15	14	0	4	47	0.20	0.29
16	15	0	4	44	0.20	0.30
17	4	0	4	24	−0.16	0.41
18	10	0	4	78	0.42	0.23
19	10	0	4	46	0.60	0.30
20	3	0	4	64	0.51	0.25
21	24	1	1	57	0.58	0.27
22	19	0	3	68	0.54	0.25
23	4	0	1	40	0.09	0.32
24	12	0	3	68	0.37	0.24
25	1	0	2	48	−0.01	0.29
26	1	0	3	107	−0.13	0.19
27	1	0	3	58	0.18	0.26
28	1	0	3	225	0.27	0.13
29	14	0	1	446	−0.02	0.09
30	20	0	1	77	0.33	0.23
31	10	1	1	243	0.59	0.13
32	7	0	1	39	0.84	0.33
33	11	0	4	67	−0.32	0.24
34	NA	0	3	91	0.12	0.21
35	NA	0	4	36	1.12	0.36
36	1	0	3	177	−0.12	0.15
37	6	0	2	20	−0.44	0.45
38	15	0	4	120	−0.07	0.18
39	15	0	4	16	0.70	0.51
40	2	0	4	105	0.49	0.20
41	4	0	2	195	0.20	0.14
42	24	0	3	62	0.58	0.26
43	11	0	4	289	0.15	0.12
44	15	0	4	25	0.63	0.41
45	8	0	1	250	0.04	0.13
46	15	0	3	51	1.46	0.31
47	15	0	2	46	0.04	0.29
48	15	0	4	56	0.25	0.27

NA = not available.

A random-effects meta-regression including the first two covariates, again using a frequentist approach with REML estimation of σ^2_{res} (see Section 5.1), gives coefficients of 0.015 (95% CI 0.000 to 0.029) for program length and 0.059 (95% CI −0.313 to 0.431) for the impact of imaginary components. These results suggest that program length is associated with the effect sizes; the positive sign of the coefficient indicates that longer programs are associated with greater effectiveness. Two studies do not provide information on program length and are omitted from this analysis. The among-studies variance in the remaining 46 studies decreases from $\sigma^2 = 0.0465$ in a model with neither covariate to $\sigma^2_{\text{res}} = 0.0446$ after adding both covariates, and the R^2 index of 4% indicates that very little of the among-studies variance is explained by a combination of the two covariates, despite the finding that one covariate is borderline statistically significant.

7.3 Categorical Covariates and Subgroup Analysis

A meta-regression with a single categorical covariate is equivalent to a subgroup analysis, whereby the effect estimates are synthesized within discrete subsets. The meta-regression model for a categorical covariate with C levels can be parameterized either using $C-1$ dummy covariates in addition to the intercept or using C dummy variables and no intercept. In the latter case, the regression coefficient for each category provides the subgroup effect for the studies within that category. Adding the third covariate to the analysis in the example of Section 7.2.5, it would be natural to code grade level using three dummy covariates, choosing one category as the reference category.

The standard random-effects meta-regression model of Section 7.2.4 has a single residual variance term to reflect unexplained heterogeneity. Application of this model to subgroup analyses therefore involves an assumption that there is an equal amount of heterogeneity within each subgroup. Specifically, the true effect sizes for studies i within subgroup c are assumed to be distributed as

$$\theta_{ci} \sim \text{N}\left(\mu_c, \sigma^2_{\text{res}}\right), \tag{7.3}$$

where μ_c is the mean effect size within subgroup c.

An alternative model is to allow the heterogeneity variance to differ by subgroup, and in some fields of application this is the usual approach to undertaking subgroup analyses, equivalent to undertaking each subgroup analysis independently. The model underlying this analysis is

$$\theta_{ci} \sim \text{N}\left(\mu_c, \sigma^2_{\text{res},c}\right), \tag{7.4}$$

where $\sigma^2_{\text{res},c}$ is the among-studies variance specific to subgroup c. This model can be fitted simply by performing separate meta-analyses for each subgroup, or in a regression framework by relaxing the assumption of an equal residual variance term, which requires software that is sufficiently flexible to do this.

An equal among-studies variance, as in (7.3), tends to be more precisely estimated than the subgroup-specific variances in (7.4), since there are more effect size estimates (i.e., more degrees of freedom) available with which to estimate it (Rubio-Aparicio et al., 2017). This approach may be particularly valuable when some of the subgroups have very few studies. This was the case in the example in Section 7.2.5, where only three studies used interventions with imaginative components. However, under the equal among-studies variance model, it is likely that for some of the subgroups the heterogeneity variance will be over-estimated or underestimated. Conversely, estimating subgroup-specific among-studies variances, as in (7.4), allows identification of subsets of studies that are particularly variable, but at the potential cost of estimating the variances poorly. Of course, looking at the estimated among-studies variances brings the danger of over-interpretation, since they will typically be associated with considerable uncertainty. A likelihood ratio test for the null hypothesis of equal residual variances can be done straightforwardly in the regression framework.

7.4 Selection of Appropriate Covariates

Typically, there are many potential covariates that could be included in a meta-regression model. Decisions regarding which variables to be used as potential covariates can be challenging. Covariate selection must be done with input of content experts and the scientific value of investigating each such variable should be justified. To achieve the ultimate goal of explaining all among-studies variation in a meta-analysis, it can be tempting to explore as many covariates as possible. Such a strategy not only may involve a large amount of data collection and coding of variables during the preparation for the meta-analysis, but it risks reaching spurious findings, as explained in Section 7.6. The following sections provide some considerations in the selection of covariates for examination using meta-regression.

7.4.1 Individual-Level versus Study-Level Characteristics

Covariates that represent study characteristics are preferred over covariates that represent individual characteristics aggregated at the study level. Examples of the latter include average age and percentage of the participants who are female. Meta-regression analyses that investigate how effect sizes vary according to aggregated summaries of participant-level characteristics may suffer from aggregation bias, otherwise known as the ecological fallacy. Relationships observed at the study level (in meta-regression) may not reflect relationships that would be observed at the individual level within studies and can easily be misinterpreted. This problem is discussed in more detail in Section 7.6.2.

7.4.2 More Than One Effect Size per Study

Sometimes study-level covariates of interest can also be used to distinguish among multiple effect sizes available from a single study. A common example is dose of an exposure, often encountered in epidemiology (see also Chapter 18). Average dose of the exposure is likely to vary across studies and may also vary across exposure groups within a study. A complication here is that effect size estimates for each dose group against a common reference group are correlated, and so the standard methods above should not be applied. A possible solution is to use the multivariate methods described in Chapter 9. There are several other situations in which multiple effect sizes may be included from the same study, and we discuss some of these along with approaches for accounting for dependencies in meta-regression in more detail in Section 7.5.5.

7.4.3 How Many Covariates?

In common with regression modeling within primary studies, meta-analytical regression models should not attempt to include more covariates than are reasonably permitted by the number of data points. The number of degrees of freedom of the mean squared error in a meta-regression model is $I - p - 1$, where I is the number of studies and p is the number of covariates. Thus, the number of covariates should be selected to be fewer than the number of studies. Rules of thumb for the number of covariates that can be safely modeled follow those recommended for regression modeling in general. A common recommendation is to include at most one covariate for each ten studies in the analysis (Deeks et al., 2019). Unfortunately, meta-analyses in many fields of application typically include very few studies. For example, meta-analyses in Cochrane reviews on the effects of health interventions have a median sample size no greater than three studies (Davey et al., 2011). Therefore, it is often the case that covariates are investigated one at a time. There is often very little ability to investigate interactions between covariates.

Consideration might be given to statistical power and precision. Power analyses can inform how many studies are needed to reject the null hypothesis of no linear association between the covariate and the effect sizes; formulae for the power of a meta-regression with continuous and categorical covariates are available (Hedges and Pigott 2004). Similarly, it is possible to determine how many studies are needed to achieve a desired level of precision in estimation of a meta-regression model coefficient.

7.4.4 Non-Linear Relationships

The majority of meta-regression analyses in practice probably implement simple linear relationships, largely due to the small numbers of studies typically included in a meta-analysis and the subsequent challenges of determining evidence against an assumption of linearity. When sufficient studies are available, non-linear models may be entertained. A starting point for selecting a non-linear model usually stems either from background knowledge or from two-way scatter plots of effect sizes estimates against covariate values. Fractional polynomial and spline models, commonly referred to as "flexible" meta-regression, have been discussed and implemented (Bagnardi et al., 2004; Rota et al., 2010; Vlaanderen et al., 2010). An example of the benefits of specifying fractional polynomial models is presented by Takagi and Umemoto (2013). They illustrate how flexible meta-regression fitted the data better than a simple linear model in an investigation of the dose-response relationship of blood pressure reduction for individuals at risk of cardiovascular events (Takagi and Umemoto, 2013). Careful consideration should be given to including non-linear terms, since the risk of spurious findings may be increased by adding polynomial terms into the regression equation.

7.4.5 The Special Case of Baseline Risk

In meta-analyses of randomized trials with a binary endpoint, a potentially convenient covariate of interest is the observed proportion of the control group who experienced the endpoint during the study. This measure of baseline risk conveniently summarizes both measured and unmeasured prognostic factors at the start of the trial and is a covariate that is always available when faced with a 2×2 table from each study. Furthermore, providing that randomization was appropriately implemented, this baseline risk should summarize the prognostic factors for the full population of the trial.

Meta-regression analysis with baseline risk as a covariate has attracted considerable attention, but it is associated with several problems. First, baseline risk measured as the observed proportion of endpoints in the control group reflects factors other than baseline prognosis. Perhaps the most obvious factor is the length of follow-up. Second, the relationship between effect size and baseline risk depends importantly on the metric used to measure the effect size. Most meta-analyses of randomized trials with binary endpoints opt to use relative effects such as risk ratios or odds ratios, and this is done because of a widespread belief that relative effects are reasonably robust to variation in baseline risk. In other words, absolute measures of effect (such as risk differences) are expected to depend on baseline risk.

Third, a critical statistical problem arises when using baseline risk as a covariate because the effect size estimate and the baseline risk are derived from the same data, namely the same 2×2 table. This introduces an artefactual correlation between the effect size estimate and the covariate (Sharp et al., 1996). For example, an unusually high baseline risk (by chance alone) will simultaneously impact on both the covariate value and the effect size estimate. The meta-regression methods described in this chapter should therefore not be used to associate baseline risk with effect size estimates when both are computed from the same 2×2 tables. Methods to overcome the artefactual correlation are available (McIntosh, 1996; Thompson et al., 1997; Schmid et al., 1998; Arends et al., 2000; Sharp and Thompson 2000), and these are discussed in detail in Chapter 14. These methods also adequately address a different problem: that of measurement error arising because the observed risk in the control group is only an estimate of the true baseline risk.

7.4.6 The Special Case of Study Precision

A further potential covariate that is almost always available in a meta-analysis context is the precision of the effect size estimate. This may be expressed in various forms, for example, as the standard error of the estimate, the variance of the estimate, or the weight (typically inverse variance) awarded to the study in the meta-analysis. An association between effect size and precision reflects asymmetry in a funnel plot (see Chapter 13). Such an association may be due to different magnitudes of bias in smaller studies versus larger studies, different true effects in smaller versus larger studies, or failure to identify results from smaller studies that had smaller (or larger) effects (for example, due to reporting biases). Statistical tests for asymmetry in a funnel plot may be viewed as meta-regression analyses of effect size estimates on measures of precision.

Careful thought is required when performing and interpreting meta-regression analyses that use a measure of precision as a covariate. The validity of the approach depends on whether the estimates of precision are artefactually correlated with the estimates of effect size. For example, such an artefactual correlation exists when unadjusted sample odds ratios are used as the effect metric, since the same data feed into the sample log odds ratio $\ln(ad/bc)$ and its estimated standard error $\sqrt{1/a+1/b+1/c+1/d}$, where a, b, c, and d are the counts in the four cells of a 2×2 table from a single study. In contrast, the standard error of a difference in mean response between two groups is based on independent summary statistics from those contributing to the difference in means. Use of the standard error as a covariate in a meta-regression in this case is therefore valid. The same is not the case for the commonly used computations for the standardized difference between means, because the standardized difference itself is usually used in computation of its standard error, leading to a spurious correlation between the two variables. Further discussion of tests for funnel plot asymmetry appears in Chapter 13.

7.4.7 Missing Data on Covariates

An implicit assumption has been made thus far in the chapter that covariate data are available from all studies. Often this will not be the case, i.e., there will be studies for which information on covariates is missing. If attempts to retrieve missing information from the authors of the primary studies fail to resolve the problem, then decisions will be needed as to how to proceed with the analysis. The simplest approach is to conduct the meta-regression with only the studies that report information on the covariate of interest. When a very small number of studies do not report covariate information, this approach is reasonable providing that the missingness can be regarded as unrelated to the covariate value (a "missing completely at random" or "missing at random" assumption in the parlance of missing data methodology).

Alternatively, statistical methods might be employed to address the missing covariates (Pigott, 2012). One option is to replace the missing information with the mean of the covariate values from the studies that report them. This allows the analysis to proceed with the full set of studies but reduces variability in the covariate and may lead to bias toward lack of association, depending on the weights assigned to the different studies. Preferable is to employ multiple imputation, assuming that the missing covariates are missing at random (or missing completely at random) (Ellington et al., 2015). Hemming et al. take a Bayesian approach to the case of multiple covariates, modeling the joint density of all covariates as a factorization of a meta-regression model and another conditional factorization of the density of covariates (Hemming et al. 2010). When multiple covariates are present, this approach allows for the inter-correlation between covariates to be modeled. The authors provide results of simulation studies demonstrating that their multiple imputation is less biased than removal of the studies with missing information on the covariate. Regardless of the approach used to handle missing data, sensitivity analyses are to be recommended to explore the impact of assumptions made.

7.5 Inference Methods

7.5.1 Frequentist Methods

A common-effect meta-regression—that is, a model in which all among-study variation is assumed to be explained by the covariate(s)—is a weighted linear regression. The effect estimates are the dependent variables, study-level characteristics are the independent variables (covariates), and the weights are assumed known and equal to $1/s_i^2$, where s_i^2 is the within-study estimation error variance in study *i*. If performed using weighted linear regression without specifying the root mean squared error as 1, then the standard errors of the estimated coefficients need to be adjusted by dividing them by the mean squared error from the regression analysis.

A standard random-effects meta-regression is a linear mixed-effects model, in which fixed effects are assumed for the intercept and regression coefficients, and a single random effect across studies allows for heterogeneity not explained by the covariates. Commonly used frequentist implementations are Stata's `metareg` package and R's `metafor` package (Harbord and Higgins, 2008; Viechtbauer, 2010).

These standard frequentist methods are subject to the same criticism of standard meta-analysis methods, in that uncertainty in both variance components is usually ignored. To address sampling errors within studies, direct modeling of the outcome data is desirable (e.g., using random-effects logistic regression for binary outcome data: see Chapter 5); the addition of study-level covariates to such models is usually straightforward. Addressing the uncertainty in the among-studies variance term is less straightforward and is discussed below.

The residual heterogeneity, σ^2_{res}, is estimated first, and several methods proposed in the context of random-effects models can be extended to the meta-regression context (López-López et al., 2014; Veroniki et al., 2016). Most estimators can yield (implausible) negative values, which are typically then set to zero. Some popular options for estimating the residual heterogeneity variance include REML and Paule-Mandel estimators. Alternatives based on method of moments (DerSimonian and Laird, 1986) and maximum likelihood are often implemented, although these options may yield negatively biased estimates (Langan et al., 2017). Conversely, Hunter-Schmidt and Sidik-Jonkman estimators may provide positively biased estimates, with the latter always providing positive values (López-López et al., 2014).

After σ^2_{res} has been estimated, the regression coefficients, β_j, can be obtained using a weighted least squares estimator with weights $1/\left(s_i^2 + \hat{\sigma}^2_{\text{res}}\right)$. Alternatively, β_j and σ^2_{res} can be estimated jointly using a maximum likelihood approach. Traditionally, inference on each coefficient, β_j, has been performed assuming that $\hat{\beta}_j / SE\left(\hat{\beta}_j\right)$ has a standard normal distribution, in line with the assumption that the variance components are estimated without error. Multiparameter Wald tests may be used for collections of covariates (or for the whole model) (Viechtbauer, 2010).

To account for uncertainty in estimation of the among-studies variance, a popular choice is a modification proposed by Knapp and Hartung as discussed in Chapter 4 (Knapp and Hartung, 2003). They introduce an adjustment factor for the covariance matrix of the estimated model coefficients and recommend using a t-distribution (instead of a standard normal distribution) for the test statistic. This approach has been found to perform well in simulation studies, keeping significance levels at the nominal level in most scenarios, even with a small number of studies (Knapp and Hartung, 2003; Sidik and Jonkman, 2005). Knapp and Hartung suggest that the adjustment factor be truncated to 1 if it would otherwise yield an adjusted variance smaller than the unadjusted variance. However, this has been found to lead to an overconservative performance for this method (Knapp and Hartung, 2003; Viechtbauer et al., 2015). Methods to derive confidence and credible intervals for among-studies variance in a random-effects meta-analysis have been extended to random-effects meta-regression (Jackson et al., 2014).

7.5.2 Bayesian Methods

Bayesian approaches make it straightforward to allow for uncertainty in the among-studies variance, and offer the usual advantages of a Bayesian approach such as flexibility to tailor the model directly to the data at hand, and the ability to make direct probability statements about the unknown quantities (see Chapter 6). Bayesian methods require prior distributions to be placed on the parameters β_j and σ^2_{res}. Vague prior distributions are used most commonly, akin to those used for a standard meta-analysis with no study-level covariates. Results can then be expected to be similar to results of frequentist analyses, providing the number of studies is reasonably large. When there are few studies, the acknowledgement of uncertainty in the estimation of among-studies variance tends to cause credible intervals from a Bayesian analysis to be wider than confidence intervals from a frequentist analysis. More detailed discussions of Bayesian meta-analysis are available (Sutton and Abrams, 2001; Spiegelhalter et al., 2004; Higgins et al., 2009). Meta-analytical Bayesian methods can be estimated using WinBUGS, JAGS, STAN, or in R packages such as `bayesmeta` or `bmeta`.

7.5.3 Permutation Test

In practice, the frequentist hypothesis tests mentioned in Section 7.5.1 rely on an asymptotic approximation to a standard normal distribution. For random-effects meta-regression, the test approximation is asymptotic on the number of studies, which is small to moderate in most applications, so that such approximation will often be poor in applied meta-analyses. Consequently, permutation tests have been suggested to improve the performance of hypothesis tests in random-effects meta-regression (Follmann and Proschan, 1999; Higgins and Thompson, 2004).

To carry out the test for a particular model covariate, x_j, we first obtain the test statistic (such as a Wald statistic, $\hat{\beta}_j / SE(\hat{\beta}_j)$). Then, for each of the $k!$ possible permutations of the x_{ij} values (from studies $i=1,\ldots,k$), we refit the model and re-compute the value of the test statistic. By permuting the covariate values, any association found with the effect sizes is expected to be purely a result of chance. Then, the two-sided p-value for the permutation test is twice the proportion of cases in which the test statistic under the permuted data is as extreme or more extreme than under the observed data. In practice, a randomly selected sample of the permutations is typically used. The permutation test has been found to provide an adequate control of the type I error rate in a wide range of scenarios for random-effects meta-regression (Higgins and Thompson, 2004; Viechtbauer et al., 2015). The idea can be extended to sets of covariates (Higgins and Thompson, 2004).

7.5.4 Predictions

Once the parameters of a meta-regression model have been estimated, they can be used to obtain predictions for effect sizes based on specific values (or combinations of values) of the covariate(s) in the model. These predictions will be based only on the fixed effects of the model. It is also possible to obtain best linear unbiased predictions, which combine the fitted values based on the fixed effects and the estimated contribution of the random effects (Viechtbauer 2010). In line with standard practice in regression modeling, predictions should generally be restricted to the range of values of the covariate(s) included in the meta-regression.

A useful process is to examine model residuals after obtaining predicted values, either graphically or by calculating descriptive statistics. Different residuals can be obtained, including raw, standardized, or standardized shrunken residuals. Examination of model residuals enables some checking of the model assumptions and identification of outliers and influential studies (see also Chapter 11), although such explorations may not be particularly helpful when the number of studies is small.

7.5.5 Accounting for Dependencies

Most standard meta-analysis methods assume independence of the effect size estimates, essentially assuming that each study contributes a single data point. Meta-analysts frequently encounter situations in which multiple effect sizes may usefully contribute from each study. This is particularly the case in the context of meta-regression, since covariate values of interest may vary across effect sizes within studies as well as among studies. One example is the presence of multiple exposure groups (or intervention groups) in the primary studies. This may happen in observational studies, for example, when exposure is categorized into several dose groups (see also Chapters 18 and 21), and in experimental studies (including clinical trials) when they have three or more intervention groups (see also Chapter 10). Another common example is multiple outcome measurements made on the same participants. For instance, a meta-analyst may be interested in studying the effects of coaching on the results of the Scholastic Aptitude Test (SAT), for which primary studies often report on multiple scores such as verbal and quantitative subtests.

Multiple comparison groups and multiple outcome measures typically lead to statistical dependence between the effect size estimates, because they are computed from some (or all) of the same individuals. Such statistical dependencies may also arise when different researchers use similar or identical data sources, particularly when a research issue can be studied via large public access datasets. Other, more subtle, types of dependencies can be found in meta-analytic datasets, for example, when the same authors undertake studies that are similar in design, measures used, participants examined, and treatments that are implemented. The essential concern in all of these examples is that subsets of the effect sizes may be related to each other in some way.

There are several ways to account for dependencies in meta-regression, and some of these have been compared in a simulation study (López-López et al., 2017). When information about the covariance between effect sizes is available, multivariate meta-analytical approaches can be used (see Chapter 9). Multivariate meta-regression requires values for the covariances of the effect size estimates arising from the same study, which in practice may not be available. When information about covariances between effect size estimates is not available, or if the meta-analyst decides not to pursue a multivariate approach, two further possibilities are to use robust standard errors and to take a multilevel meta-analysis approach.

Robust approaches are based on the work of Huber (1967) and White (1980), and were proposed for use in meta-analysis by Sidik and Jonkman in order to address potential misspecification of the exact variances of the estimated effect sizes (Sidik and Jonkman, 2006). Hedges, Tipton, and Johnson proposed that these robust standard errors be used for modeling the dependence among effect sizes when the covariances are not available (Hedges et al. 2010). An alternative way to model dependencies is to use a multilevel model, nesting related effect sizes within a higher level (Konstantopoulos, 2011). Three-level meta-analysis has been found to perform similarly to multivariate meta-analysis in terms of parameter and standard error estimates in meta-regression (Van den Noortgate et al., 2015).

7.6 Caveats

7.6.1 Observational Associations

Meta-regression analyses examine observational associations between effect size estimates and study-level covariates. Because the values of covariates are not randomized across studies, and typically do not follow processes comparable with randomization, causal inferences about the influence of study-level covariates on the effect sizes usually cannot be made with confidence. Associations observed may truly be due to the actual covariate included in the analysis, but frequently will be caused by confounding covariates that are correlated with both the covariate and the effect size. Nonetheless, the associations observed can sometimes be used to generate new hypotheses that can be tested in further primary studies (Baker et al., 2009).

7.6.2 Aggregation Bias

When a covariate of interest is an individual-level characteristic, meta-regression entails the use of aggregated summaries of the characteristic at the study level, such as average age or percentage of women in the study sample. Meta-regression results then need to be interpreted cautiously because they may be prone to ecological (or aggregation) problems (Simmonds and Higgins, 2007). Aggregation bias, also called the ecological fallacy, refers to the situation in which relationships between variables that are true at the individual level are different at the study level.

A common situation in which ecological problems occur is when effect sizes are reasonably homogeneous across studies. In such scenarios, meta-regression using study-level covariates will likely fail to detect associations that might be present at the individual level (Schmid et al., 2004). Another situation in which ecological problems will occur is when the aggregated value of the covariate is similar across studies. For example, if average age is similar across studies, then no association is likely to be seen between effect size and average age, even if age is an important effect modifier. Berlin and colleagues illustrated the ecological fallacy in an example involving renal transplant patients (Berlin et al., 2002). At the patient level, they found that treatment was significantly more effective in patients with elevated panel reactive antibodies. However, meta-regression at the study level failed to detect such association.

Ecological problems might also work in the opposite direction. Associations may be found at the study level even if they are not true at the individual level. This is likely to be due to confounding of the study-level covariate value with a true predictor of effect size. The risk of such spurious findings will also increase in scenarios in which participants within studies show substantial variation in the covariate of interest (e.g., a wide age range), and therefore aggregate data (e.g., average age) might not accurately represent each study sample (Baker et al., 2009). Riley and colleagues present a meta-analysis of ten studies on hypertension treatment (Riley et al., 2008). In a study-level meta-regression they found the treatment to be clinically more effective in studies with higher proportions of women. Meta-regression at the patient level did not yield such clinically relevant differences according to sex.

Meta-regression with study-level summaries of individual-level covariates does not allow exploration of variation across individuals. If the goal is to explore relationships between individual characteristics and the effect sizes, ecological biases can be avoided by collecting and analyzing individual participant data (Riley et al., 2010), as discussed in Chapter 8.

7.6.3 False-Positive Rates (Type I Error)

False-positive findings are an important problem in meta-regression, for both practical and theoretical reasons. Many find it tempting to investigate multiple study-level characteristics in an attempt to explain observed heterogeneity across effect sizes. Because information from the units of interest is readily available (in the form of reports from the included studies), it is often possible to continue conceiving and coding new study-level covariates until one is found that appears to correlate well with the effect size estimates. To limit type I error rates, standard advice is that covariates to be included in meta-regression analyses should be prespecified and based on sound rationale. This is particularly important when the number of studies included in the meta-regression model is small to moderate (Raudenbush, 1994). Even with a large number of studies, inclusion of moderators without a strong rationale may lead to spurious findings as a result of chance (Schmidt and Hunter, 2015).

Choices of analysis models and methods can also lead to inflated type I errors in meta-regression. A simulation study demonstrated that implementing common-effect meta-regression analyses in the presence of heterogeneity gives unacceptably high false-positive rates (Higgins and Thompson, 2004). When a moderate amount of inconsistency among studies was simulated (specified as $I^2 = 50\%$), the observed false-positive rate was around 20%, and when a large amount of inconsistency was simulated ($I^2 = 83\%$), false-positive rates exceeded 50%. Random-effects meta-regression analyses assuming normally distributed errors in the estimated mean effect also lead to inflated error rates. Using the Knapp-Hartung correction or a permutation test are appropriate strategies to control this problem in a random-effects meta-regression.

Type I errors can be avoided by avoiding the use of statistical significance tests, and by concentrating instead on the estimated magnitude of the regression coefficient and on the practical importance of values entertained by the confidence interval (Sterne and Davey Smith, 2001). Strategies of limiting the number of covariates, of pre-specifying them, and ensuring a sound rationale for each are nevertheless good practice.

7.6.4 False-Negative Rates (Power)

Ensuring there is a sound rationale for pre-specifying each covariate is important not only to discard irrelevant covariates, but also to ensure that those examined have a realistic chance of having an impact on the effect size under investigation. Failure to identify genuine associations is a common problem in meta-regression. This is typically due to an insufficient number of studies in the meta-analysis, although it may also be due to limited variability in the covariate values observed in the studies to hand.

Meta-regressions will often have limited ability to detect associations, because the number of available studies is small to moderate in most applications. In their survey of meta-analyses from Cochrane and elsewhere in the medical literature, Schmid and colleagues found meta-regression to be most effective for detecting associations between study-level covariates and effect sizes when there are at least ten studies (Schmid et al., 2004). Simulation studies have found the inference methods presented in this chapter generally have power lower than 80% when there were fewer than 20 studies (Huizenga et al., 2011; Viechtbauer et al., 2015). In contrast, the numbers of studies eligible for inclusion in meta-analyses in Cochrane reviews have been shown to be substantially smaller than this (Davey et al., 2011).

7.7 Concluding Remarks

Meta-regression offers a means to address some of the questions of most substantive importance in a meta-analysis, namely whether heterogeneity across studies can be usefully explained by characteristics of the studies. This allows novel research questions to be addressed within a meta-analytic study, with greater understanding of how effect sizes vary across contexts and implementations. Meta-regression analyses are commonly used to try to explain heterogeneity observed across studies brought together with the intention of undertaking a standard meta-analysis. Increasingly, meta-regression analyses are the principal reason for collating studies. For example, when heterogeneity is explicitly anticipated (e.g., in a systematic review of prevalence estimates of a particular disease), then the primary analysis method may be an investigation of potential sources of the heterogeneity.

Study characteristics used as covariates in meta-regression may be categorical or quantitative, and multiple covariates can be included in a single analysis. For categorical characteristics, meta-regression is a generalization of subgroup analyses, where the subgrouping is by studies rather than by participants.

Meta-regression should ideally be conducted only as part of a thorough systematic review to ensure that the studies involved are reliably identified and appraised. Unfortunately, meta-regression is highly problematic in many practical applications. Associations across studies are observational and prone to confounding. These problems are exacerbated in many areas of application (particularly in the health sciences), where fewer studies tend to be available. A high proportion of apparent relationships will then be spurious, confounders difficult to identify, power will be low, and simultaneous effects of multiple study characteristics cannot reliably be evaluated.

References

Aloe AM, Becker BJ and Pigott TD, 2010. An alternative to R^2 for assessing linear models of effect size. *Research Synthesis Methods* **1**(3–4): 272–283.

Arends LR, Hoes AW, Lubsen J, Grobbee DE and Stijnen T, 2000. Baseline risk as predictor of treatment benefit: Three clinical meta-re-analyses. *Statistics in Medicine* **19**(24): 3497–3518.

Bagnardi V, Zambon A, Quatto P and Corrao G, 2004. Flexible meta-regression functions for modeling aggregate dose-response data, with an application to alcohol and mortality. *American Journal of Epidemiology* **159**(11): 1077–1086.

Baker WL, White CM, Cappelleri JC, Kluger J, Coleman CI, P. Health Outcomes and G. Economics Collaborative, 2009. Understanding heterogeneity in meta-analysis: The role of meta-regression. *International Journal of Clinical Practice* **63**(10): 1426–1434.

Bangert-Drowns RL, Hurley MM and Wilkonson A, 2004. The effects of school-based writing-to-learn interventions on academic achievement: A meta-analysis. *Review of Educational Research* **74**(1): 29–58.

Berkey CS, Hoaglin DC, Mosteller F and Colditz GA, 1995. A random-effects regression model for meta-analysis. *Statistics in Medicine* **14**(4): 395–411.

Berlin JA, Santanna J, Schmid CH, Szczech LA and Feldman KA, 2002. Individual patient- versus group-level data meta-regressions for the investigation of treatment effect modifiers: Ecological bias rears its ugly head. *Statistics in Medicine* **21**(3): 371–387.

Borenstein M, Hedges LV, Higgins JPT and Rothstein H, 2009. *Introduction to Meta-Analysis*. Chichester: John Wiley & Sons.

Davey J, Turner RM, Clarke MJ and Higgins JPT, 2011. Characteristics of meta-analyses and their component studies in the Cochrane Database of Systematic Reviews: A cross-sectional, descriptive analysis. *BMC Medical Research Methodology* **11**: 160.

DerSimonian R and Laird N, 1986. Meta-analysis in clinical trials. *Controlled Clinical Trials* **7**(3): 177–188.

Follmann DA and Proschan MA, 1999. Valid inference in random effects meta-analysis. *Biometrics* **55**(3): 732–737.

Ellington EH, Bastille-Rousseau G, Austin C, Landolt KN, Pond BA, Rees EE, Robar N and Murray DL, 2015. Using multiple imputation to estimate missing data in meta-regression. *Methods in Ecology & Evolution* **6**(2): 153–163.

Harbord RM and Higgins JPT, 2008. Meta-regression in Stata. *Stata Journal* **8**(4): 493–519.

Hedges LV and Pigott TD, 2004. The power of statistical tests for moderators in meta-analysis. *Psychological Methods* **9**(4): 426–445.

Hedges LV, Tipton E and Johnson MC, 2010. Robust variance estimation in meta-regression with dependent effect size estimates. *Research Synthesis Methods* **1**(1): 39–65.

Hemming K, Hutton JL, Maguire MG and Marson AG, 2010. Meta-regression with partial information on summary trial or patient characteristics. *Statistics in Medicine* **29**(12): 1312–1324.

Deeks JJ, Higgins JPT and Altman DG on behalf of the Cochrane Statistical Methods Group, 2019. Analysing data and undertaking meta-analyses. In Higgins JPT, Thomas J, Chandler J, Cumpston M, Li T, Page MJ and Welch VA (Eds). *Cochrane Handbook for Systematic Reviews of Interventions*. 2nd Edition. Chichester, UK: John Wiley & Sons. Also online version 6.0 (updated July 2019) available from www.training.cochrane.org/handbook.

Higgins JPT and Thompson SG, 2004. Controlling the risk of spurious findings from meta-regression. *Statistics in Medicine* **23**(11): 1663–1682.

Higgins JPT, Thompson SG and Spiegelhalter DJ, 2009. A re-evaluation of random-effects meta-analysis. *Journal of the Royal Statistical Society: Series A* **172**(1): 137–159.

Huber PJ, 1967. The behavior of maximum likelihood estimates under nonstandard conditions. Proceedings of the Fifth Berkeley Symposium on Mathematical Statistics and Probability, Volume 1. Berkeley, CA: University of California Press, 221–233.

Huizenga HM, Visser I and Dolan CV, 2011. Testing overall and moderator effects in random effects meta-regression. *The British Journal of Mathematical & Statistical Psychology* **64**(Pt 1): 1–19.

Jackson D, Turner R, Rhodes K and Viechtbauer W, 2014. Methods for calculating confidence and credible intervals for the residual between-study variance in random effects meta-regression models. *BMC Medical Research Methodology* **14**: 103.

Knapp G and Hartung J, 2003. Improved tests for a random effects meta-regression with a single covariate. *Statistics in Medicine* **22**(17): 2693–2710.

Konstantopoulos S, 2011. Fixed effects and variance components estimation in three-level meta-analysis. *Research Synthesis Methods* **2**(1): 61–76.

Langan D, Higgins JPT and Simmonds M, 2017. Comparative performance of heterogeneity variance estimators in meta-analysis: A review of simulation studies. *Research Synthesis Methods* **8**(2): 181–198.

López-López JA, Marín-Martínez F, Sánchez-Meca J, Van den Noortgate W and Viechtbauer W, 2014. Estimation of the predictive power of the model in mixed-effects meta-regression: A simulation study. *The British Journal of Mathematical & Statistical Psychology* **67**(1): 30–48.

López-López JA, Van den Noortgate W, Tanner-Smith EE, Wilson SJ and Lipsey MW, 2017. Assessing meta-regression methods for examining moderator relationships with dependent effect sizes: A Monte Carlo simulation. *Research Synthesis Methods* **8**(4): 435–450.

McIntosh MW, 1996. The population risk as an explanatory variable in research synthesis of clinical trials. *Statistics in Medicine* **15**(16): 1713–1728.

Pigott TD, 2012. Missing data in meta-analysis: Strategies and approaches. *Advances in Meta-Analysis*. New York: Springer, 79–107.

Raudenbush SW, 1994. Random effects models. In Cooper H and Hedges LV (Eds). *The Handbook of Research Synthesis*. New York: Russell Sage Foundation, 301–322.

Raudenbush SW and Bryk AS, 2002. *Hierarchical Linear Models: Applications and Data Analysis Methods*. Thousand Oaks, CA: Sage Publications, Inc.

Riley RD, Lambert PC and Abo-Zaid G, 2010. Meta-analysis of individual participant data: Rationale, conduct, and reporting. *BMJ* **340**: c221.

Riley RD, Lambert PC, Staessen JA, Wang J, Gueyffier F, Thijs L and Boutitie F, 2008. Meta-analysis of continuous outcomes combining individual patient data and aggregate data. *Statistics in Medicine* **27**(11): 1870–1893.

Rota M, Bellocco R, Scotti L, Tramacere I, Jenab M, Corrao G, La Vecchia C, Boffetta P and Bagnardi V, 2010. Random-effects meta-regression models for studying nonlinear dose-response relationship, with an application to alcohol and esophageal squamous cell carcinoma. *Statistics in Medicine* **29**(26): 2679–2687.

Rubio-Aparicio M, Sánchez-Meca J, López-López JA, Botella J and Marín-Martínez F, 2017. Analysis of categorical moderators in mixed-effects meta-analysis: Consequences of using pooled versus separate estimates of the residual between-studies variances. *The British Journal of Mathematical & Statistical Psychology* **70**(3): 439–456.

Schmid CH, Lau J, McIntosh MW and Cappelleri JC, 1998. An empirical study of the effect of the control rate as a predictor of treatment efficacy in meta-analysis of clinical trials. *Statistics in Medicine* **17**(17): 1923–1942.

Schmid CH, Stark PC, Berlin JA, Landais P and Lau J, 2004. Meta-regression detected associations between heterogeneous treatment effects and study-level, but not patient-level, factors. *Journal of Clinical Epidemiology* **57**(7): 683–697.

Schmidt FL and Hunter JE, 2015. *Methods of Meta-Analysis: Correcting Error and Bias in Research Findings*. Thousand Oaks, CA: SAGE Publication, Inc.

Sharp SJ and Thompson SG, 2000. Analysing the relationship between treatment benefit and underlying risk in meta-analysis: Comparison and development of approaches. *Statistics in Medicine* **19**(23): 3251–3274.

Sharp SJ, Thompson SG and Altman DG, 1996. The relation between treatment benefit and underlying risk in metaanalysis. *BMJ* **313**(7059): 735–738.

Sidik K and Jonkman JN, 2005. A note on variance estimation in random effects meta-regression. *Journal of Biopharmaceutical Statistics* **15**(5): 823–838.

Sidik K and Jonkman JN, 2006. Robust variance estimation for random effects meta-analysis. *Computational Statistics & Data Analysis* **50**(12): 3681–3701.

Simmonds MC and Higgins JPT, 2007. Covariate heterogeneity in meta-analysis: Criteria for deciding between meta-regression and individual patient data. *Statistics in Medicine* **26**(15): 2982–2999.

Spiegelhalter, DJ, Abrams KR and Myles JP, 2004. *Bayesian Approaches to Clinical Trials and Health-Care Evaluation*. Chichester: John Wiley & Sons.

Sterne JAC and Davey Smith G, 2001. Sifting the evidence: What's wrong with significance tests? *BMJ* **322**(7280): 226–231.

Sutton AJ and Abrams KR, 2001. Bayesian methods in meta-analysis and evidence synthesis. *Statistical Methods in Medical Research* **10**(4): 277–303.

Takagi H and Umemoto T, 2013. The lower, the better?: Fractional polynomials meta-regression of blood pressure reduction on stroke risk. *High Blood Pressure & Cardiovascular Prevention* **20**(3): 135–138.

Thompson SG, Smith TC and Sharp SJ, 1997. Investigating underlying risk as a source of heterogeneity in meta-analysis. *Statistics in Medicine* **16**(23): 2741–2758.

Van den Noortgate W, López-López JA, Marin-Martinez F and Sanchez-Meca J, 2015. Meta-analysis of multiple outcomes: A multilevel approach. *Behavior Research Methods* **47**(4): 1274–1294.

Veroniki AA, Jackson D, Viechtbauer W, Bender R, Bowden J, Knapp G, Kuss O, Higgins JPT, Langan D and Salanti G, 2016. Methods to estimate the between-study variance and its uncertainty in meta-analysis. *Research Synthesis Methods* **7**(1): 55–79.

Viechtbauer W, 2010. Conducting meta-analyses in R with the metafor package. *Journal of Statistical Software* **36**(3): 1–48.

Viechtbauer W, López-López JA, Sánchez-Meca J and Marín-Martínez F, 2015. A comparison of procedures to test for moderators in mixed-effects meta-regression models. *Psychological Methods* **20**(3): 360–374.

Vlaanderen J, Portengen L, Rothman N, Lan Q, Kromhout H and Vermeulen R, 2010. Flexible meta-regression to assess the shape of the benzene-leukemia exposure-response curve. *Environmental Health Perspectives* **118**(4): 526–532.

White H, 1980. A heteroskedasticity-consistent covariance matrix estimator and a direct test for heteroskedasticity. *Econometrica* **48**(4): 817–830.

8

Individual Participant Data Meta-Analysis

Lesley Stewart and Mark Simmonds

CONTENTS

8.1 Introduction

Systematic reviews with individual participant data (IPD) meta-analysis collect, validate, and re-analyze individual-level data recorded for each participant recruited in each included study—rather than using aggregate summary data, commonly obtained from journal articles or trial reports, as in most conventional systematic reviews. Underpinning principles and many supporting processes are the same. However, IPD meta-analyses can incorporate unreported data, allow standardization across studies, and offer considerable potential to carry out more nuanced and sophisticated analyses than are possible with aggregate data.

With an established history of evaluating therapeutic interventions in cancer (Advanced Ovarian Cancer Trialists Group, 1991; Non-small Cell Lung Cancer Collaborative Group, 1995), and in cardiovascular disease (Selker et al., 1997), IPD meta-analysis is being applied increasingly across a broadening range of healthcare areas, although they remain a minority of the systematic reviews undertaken (Riley et al., 2010; Simmonds et al., 2005, 2015). Synthesis of IPD is also being increasingly used in observational, diagnostic, and prognostic research (Fibrinogen Studies Collaboration, 2006; Mant et al., 2009), particularly in risk prediction modeling (see Chapters 19, 21, and 22).

8.2 Benefits of the IPD Approach

A considerable benefit of the IPD approach is that data not reported in publications can be collected and analyzed. This may include data from unpublished studies, outcomes not reported in publications, data on participants that were not included in the reported analyses, and additional long-term follow-up information. This may also be done for reviews that use aggregate data by requesting additional information from study investigators. However, this is seldom done in practice, particularly if it is more than a simple request for, say, one additional outcome. If the request is, for example, for all outcomes with excluded participants reinstated or for extended follow-up, it may in fact be more straightforward for study investigators to provide IPD rather than creating multiple new data tables. Obtaining IPD may also enable inclusion and analysis of data from studies where publications did not present results in a suitable format for meta-analysis. These additional data can help to reduce the impact of publication and reporting bias and mitigate bias associated with missing information.

Collecting individual level data enables inclusion and exclusion criteria to be applied at the individual level, such that relevant subpopulations from more broadly defined studies may be identified and included in the meta-analysis. For example, data from participants aged 18 years or younger could be selected from a trial that recruited participants of all ages and included in a meta-analysis restricted to a pediatric population. Applying inclusion criteria at this level may therefore enable greater use of data from existing studies, increase meta-analysis sample size, and reduce heterogeneity in the populations included in meta-analyses.

Provision of IPD enables data to be recoded and standardized across studies; for example, where an outcome is defined using varying cutpoints or thresholds between studies, obtaining the underlying measurements allows application of a common threshold and outcome definition. This may permit the combination of data that could not otherwise be pooled, and consistent analyses can be performed across studies. IPD enables re-analysis to include updated data, may allow intention-to-treat analysis to be done even if the reported analyses took a different approach, and makes it possible to correct or re-run original study analyses if they are flawed. Data can be checked carefully, and in controversial areas, this independent scrutiny can be a powerful rationale for obtaining IPD.

Individual-level data enables more detailed and flexible analysis, particularly for time-to-event outcomes and of potential effect modifiers—enabling exploration of whether there are particular types of individual who benefit more or less from the intervention being investigated. These important motivations explain why the approach has gained prominence in cancer and cardiovascular disease.

8.3 Conducting a Systematic Review Using IPD

The IPD collected for each study will generally contain demographic information for each participant, including data such as age, sex, and medical condition, details of treatments assigned/received, and details of outcomes measured or events recorded at a series of points in time, or provided with linked dates for each event or measurement. Data may be obtained in a variety of formats—most straightforwardly as flat files with one line of data per participant. They may also be provided in other formats such as a series of separate files created for the original study analyses of differing outcomes that requires those undertaking the meta-analysis to link participants across files in order to explore relationships between outcomes and with participant characteristics. Data may also be provided as electronic (or paper) study forms for each participant, from which required data need to be extracted and analyses files assembled.

Many of the practical aspects are identical or very similar to a standard systematic review and meta-analysis of aggregate data, with the main differences being in the considerable effort that is required to obtain and check data prior to analysis. Although in the future it may be possible to use trial registries and repositories to piece together an IPD dataset, this is unlikely to be feasible in the near future (most IPD-MAs will seek to include older trials and academic trials, neither of which are well represented in existing repositories). IPD-MA therefore usually involves partnership with the investigators who have undertaken the included studies. Establishing and managing these collaborative groups is key to successfully obtaining data and to the success of the project.

It is important to understand that the trickiest aspects of undertaking an IPD-MA are not necessarily the statistical analysis. Meticulous data management is vital and can be time consuming; those managing the data need to have a good understanding of how the underpinning studies are designed and managed with skills that are similar to clinical trial management. It is also important not to underestimate the time and skills required to manage international collaborations on which many IPD-MA build—without building successful relationships and collaboration there will be no data to analyze. Those embarking on an IPD meta-analysis for the first time are encouraged to seek advice from those with experience of managing these projects. The Cochrane IPD Meta-analysis Methods Group website is a good source of information about who has previously carried out IPD-MAs in particular areas (most commonly the IPD-MAs themselves are done outside of Cochrane) (http://methods.cochrane.org/ipdma/welcome-ipd-meta-analysis-methods-group).

IPD-MAs should be reported fully, transparently, and in accordance with the PRISMA-IPD reporting guidelines (Stewart et al., 2015). As IPD-MAs often address key policy questions, it is important that findings are communicated in a way that helps non-technical readers understand and implement findings.

8.4 Principles of IPD Meta-Analysis

There are two key approaches to IPD meta-analysis: the "two-stage" and the "one-stage" approaches (Simmonds et al., 2005). In the two-stage approach, each study is first analyzed separately to obtain effect estimates of interest, such as a relative risk of a treatment compared with a placebo, along with standard errors for these effect estimates. These are then combined in a meta-analysis across studies. In a "two-stage" analysis, any method for published data meta-analysis (as discussed in Chapter 4) may be used, including producing forest plots and assessments of heterogeneity.

As all data from all studies are available, the alternative is to analyze all data from all studies simultaneously in a single regression model. This is the one-stage approach, so-called because only one analysis is applied to all data. The general methodology of the one-stage approach is described in Chapter 5. The models are largely the same when using IPD. The one-stage approach allows for considerable flexibility in the choice of analysis, but is different from conventional meta-analyses based on published data, and so needs careful explanation and presentation when reporting its results.

The general methods for these approaches are the same as described in Chapters 4 and 5; however, having IPD allows a wider range of modeling options than can be used with published data alone.

8.4.1 Example: The PARIS Meta-Analysis

To illustrate how IPD meta-analysis is performed we will consider the data from the PARIS systematic review (Askie et al., 2007). This was an international collaborative systematic review of randomized clinical trials of antiplatelet therapy (aspirin and warfarin) to prevent pre-eclampsia and other morbidities in pregnancy. The review collected the IPD from 24 trials representing over 30,000 pregnant women.

8.4.2 The Two-Stage Approach

We first consider the impact of antiplatelet therapy on pre-eclampsia, which is a dichotomous, or binary, outcome. The original meta-analysis of these data used a two-stage approach: the numbers of women and of pre-eclampsia cases in the antiplatelet and placebo arms were calculated for each trial (Askie et al., 2007). This creates a 2 × 2 contingency table for each trial, as might be reported in publications. From these contingency tables, log odds ratios or relative risks and their standard errors can be calculated (as in Chapter 3).

As an alternative to this contingency table approach, a generalized linear model can be fitted to the data, where the outcome (pre-eclampsia) is regressed against the treatment used in each trial. The model for each study (broadly following the notation in Chapter 5) has the general form:

$$g\left(E\left[Y_{ik}\right]\right) = \beta_{0i} + \beta_{1i}X_{ik} \tag{8.1}$$

where X_{ik} denotes the treatment group for participant k in study i (typically: 0 = control, 1 = treated). The β_{1i} parameters are the treatment effects, and β_{0i} are the intercepts representing the level of the outcome in the control arm. The function g here is any suitable link function: for a dichotomous outcome, this is usually the logit link, in which case β_{1i} is the log odds ratio for the treatment effect.

Once treatment effect estimates are found for each study, they may be combined in a meta-analysis using any of the methods described earlier in this book. In the original analysis of the PARIS data, a common-effect meta-analysis was used to pool relative risks. The Mantel–Haenszel or Peto methods might also be used. In later re-analyses, DerSimonian–Laird random-effects meta-analyses of the log odds ratios and their variances were performed, as described in Chapter 4 (DerSimonian and Laird, 1986; Mantel and Haenszel, 1959). Heterogeneity was assessed using both Cochran's Q test and I^2 (Higgins and Thompson, 2002). Figure 8.1 shows the forest plot for this analysis.

8.4.3 The One-Stage Approach

Instead of applying a generalized linear model to each trial separately, we may expand the regression to incorporate all the data from all studies in a "one-stage" model. This is like analyzing a single large study, except that the model must account for the different studies by including a parameter for each study representing level of the outcome in the control arm (e.g., the risk of pre-eclampsia in the control arm). The models are as in Chapter 5, but are re-presented with the focus on modeling at the individual participant level, and with a

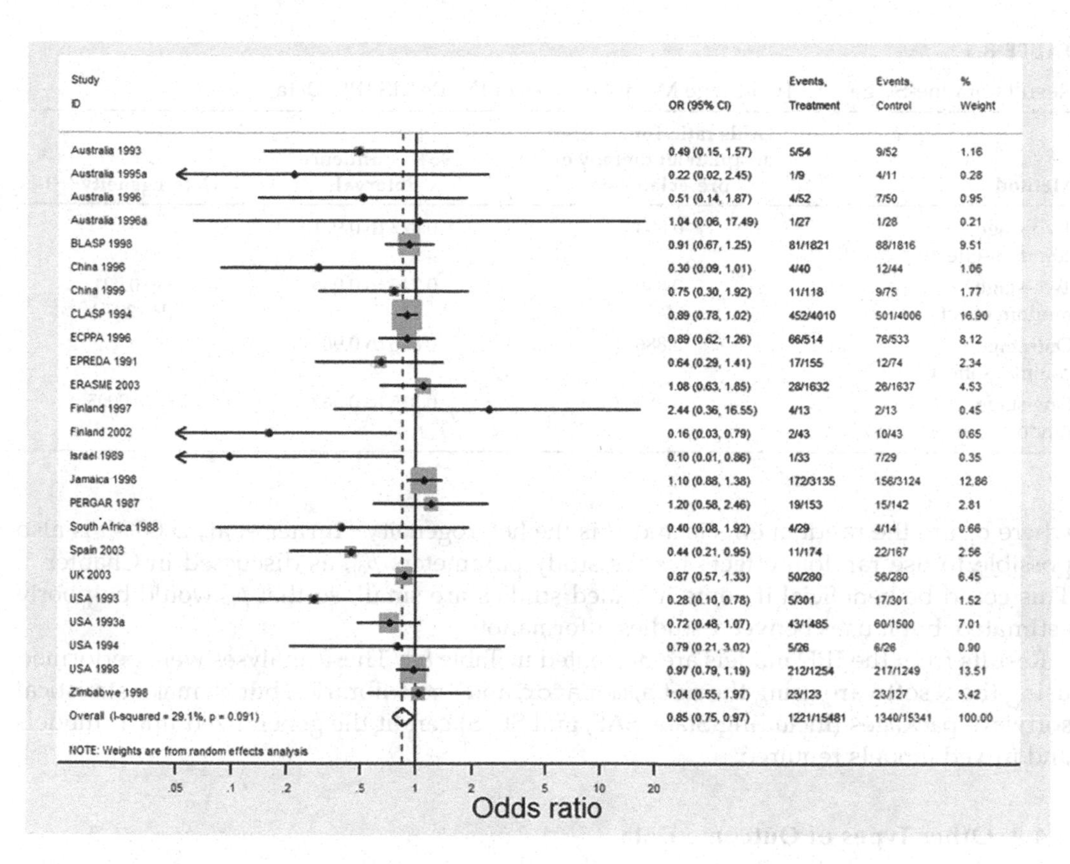

FIGURE 8.1
Forest plot of the two-stage meta-analysis of the impact of antiplatelet therapy on pre-eclampsia in the PARIS IPD meta-analysis.

general link function. A one-stage common-effect model, for example, may have the form (Turner et al., 2000; Higgins et al., 2001):

$$g\left(E\left[Y_{ik}\right]\right) = \beta_{0i} + \beta_1 X_{ik} \tag{8.2}$$

This is identical to the model within each study (8.1), except that now there is a common treatment effect β_1 applied to all studies. This one-stage common-effect model is analogous to the common-effect two-stage meta-analysis in Chapter 4. Note also that there is a separate intercept term for each study, so there are fixed effects for each study. These fixed effect terms ensure that comparisons between individuals are made within each study, rather than across all studies. This is particularly useful in meta-analyses of randomized controlled trials, as it preserves the randomized nature of the comparison in each trial.

As with standard meta-analysis, the common-effect model can be modified to allow for heterogeneity as a random-effects model. When random effects are included, the model becomes a generalized linear mixed model, as discussed in Chapter 5. Model (8.2) is modified to:

$$g\left(E[Y_{ik}]\right) = \beta_{0i} + \left(\beta_1 + b_{1i}\right) X_{ik} \text{ with } b_{1i} \sim N\left(0, \tau^2\right) \tag{8.3}$$

TABLE 8.1

Results of One-Stage and Two-Stage Meta-Analyses of the PARIS IPD Data

Method	Odds ratio for effect of antiplatelet therapy on pre-eclampsia	95% confidence interval	Heterogeneity (τ^2)
Two-stage common-effect	0.893	0.823 to 0.971	–
Two-stage random effects	0.849	0.745 to 0.966	$\tau^2 = 0.021$ ($I^2 = 29\%$)
One-stage common-effect	0.886	0.816 to 0.963	–
One-stage random effects	0.885	0.815 to 0.962	$\tau^2 = 0.0005$

where b_{1i} are the random effects and τ^2 is the heterogeneity (Turner et al., 2000). It is also possible to use random effects for the study parameters β_{0i}, as discussed in Chapter 5. This could be beneficial if some included studies are small, so that β_{0i} would be poorly estimated, but it uses between-studies information.

Results from the IPD models are presented in Table 8.1. These analyses were performed using the R software using the `meta`, `metafor`, and `lme4` libraries, but all major statistical software packages (including Stata, SAS, and SPSS) can fit the generalized linear models and mixed models required.

8.4.4 Other Types of Outcome Data

As with the methods in Chapters 4 and 5, IPD meta-analysis can be used for other types of data. As for dichotomous outcomes, a two-stage approach can be used by selecting a suitable link function in model (8.1) and using the resulting regression model to estimate the treatment effect. For example, for continuous outcomes the identity link $g(y) = y$ may be used, resulting in a linear regression within each study. As before, the estimated mean differences can now be pooled using any available meta-analysis method.

For a one-stage approach, a generalized linear (mixed) model can be fitted to all data simultaneously with either a fixed or a random-effects assumption for the treatment effects, by changing the link function in equations (8.2) or (8.3) to one suited to that data type. For example, continuous data using linear mixed models (Higgins et al., 2001), count data using Poisson regression, or ordinal outcomes (i.e., outcomes in ordered categories) using ordinal regression (Whitehead et al., 2001).

8.4.5 Time-to-Event Outcomes

IPD are often used for the analysis of time-to-event outcomes, because hazard ratios, Kaplan–Meier curves, and other summaries of time-to-event outcomes are often inconsistently reported across studies (Williamson et al., 2002). Perhaps the most straightforward approach to analyzing time-to-event data is to fit a Cox proportional hazards model within each study (Cox, 1972; Simmonds et al., 2011). The resulting hazard ratio estimates for the treatment effect can then be pooled across studies in a meta-analysis. For a one-stage analysis, a common- or random-effects Cox model can be fitted to all the data. Stratification across studies is maintained by having a distinct baseline hazard for each study. For example, if $h_{ik}(t)$ is the hazard at time t for person k in study i:

$$\log\left(h_{ik}(t)\right) = \log\left(h_{i0}(t)\right) + (\beta_1 + b_{1i})X_{ik} \text{ with } b_{1i} \sim N\left(0, \tau^2\right) \tag{8.4}$$

is a one-stage random-effects proportional hazards model, with $h_{i0}(t)$ the stratified baseline hazard for each study and β_1 the summary log hazard ratio.

Parametric survival models such as Weibull models, with either common or random effects, may also be fitted using IPD. If assuming proportional hazards is not appropriate, then other models, such as accelerated failure time models, could, in principle, be fitted using IPD, but these have not been widely studied. Random-effects parametric models are available in Stata (`mestreg`), and random-effects Cox models in R (`coxme`).

8.4.6 Adding Additional Covariates

So far, we have considered estimating only a single covariate, typically the treatment effect in a controlled trial. A major advantage of using IPD over conventional meta-analyses of published results is the ability to consider multiple covariates at once, whether to consider the impact of multiple predictors on the outcome of interest or to adjust for potential confounding factors. Having IPD permits the extension of the models seen in Chapter 5 to include further model parameters. Within each study, the generalized linear (mixed) model (linear, logistic, or otherwise) can easily be extended to include multiple predictor variables, or to adjust for confounders, by adding parameters to the model.

For example, if we wish to adjust for some covariate Z we may add it to the generalized linear mixed model (8.1):

$$g\left(E[Y_{ik}]\right) = \beta_{0i} + \beta_{1i}X_{ik} + \beta_{2i}Z_{ik}. \tag{8.5}$$

This approach is of particular use in meta-analyses of observational data, where the lack of randomization means that accounting for potential confounders is important. Publications may not present results adjusted for confounders, or may be inconsistent in the factors accounted for, and so using IPD is the only way to achieve a consistent analysis across studies. This is discussed in more detail in Chapter 21.

Upon fitting such an extended model in each study separately we obtain estimates of the parameters of interest for each study (each adjusted for the confounding variables), which we can then pool using any desired meta-analysis method.

Alternatively, we can use a one-stage approach by applying the model with multiple parameters to the data from all studies simultaneously. For example, a model with random treatment effects is:

$$g\left(E\left[Y_{ik}\right]\right) = \beta_{0i} + (\beta_1 + b_{1i})X_{ik} + \beta_{2i}Z_{ik} \text{ with } b_{1i} \sim N\left(0, \tau^2\right) \tag{8.6}$$

In this example, there are separate adjustment parameters β_{2i} for each study. However, using a one-stage approach gives considerable flexibility in how each parameter is fitted. For each parameter we may apply either separate (fixed) effects for each study, a common effect across all studies, or random effects to account for heterogeneity across studies. For confounder variables, separate, independent parameters might be preferred for each trial, as this makes the fewest assumptions. However, models may not converge when there are large numbers of studies and parameters, so random or common effects may have to be used.

8.4.7 Interactions

In meta-analyses of clinical trials, there is often interest in whether a covariate can modify the effectiveness of a treatment. For example, the PARIS analysis examined whether the efficacy of antiplatelet therapy improved (or worsened) according to the age of the woman and whether she had diabetes or hypertension.

Parameters for interactions between treatment and covariates can be added to regression models in the same way as adding parameters for confounders. Further extending model (8.1) gives:

$$g\left(E[Y_{ik}]\right) = \beta_{0i} + \beta_{1i}X_{ik} + \beta_{2i}Z_{ik} + \beta_{3i}X_{ik}Z_{ik} \tag{8.7}$$

where the parameter β_{3i} is the interaction between the treatment given X and the covariate of interest Z (Turner et al., 2000).

For a two-stage analysis, model (8.7) can be fitted separately in each study and resulting interaction parameters β_{3i} are then pooled across studies using any available meta-analysis method (Simmonds and Higgins, 2007). It should be noted that model (8.7) will only produce an estimate of β_{3i} if there is variation in Z_{ik} within the study; if instead the covariate is constant for all people in a study then β_{3i} is inestimable and that study will not contribute to the meta-analysis of these interaction terms.

For a one-stage analysis the interaction term can be added to the one-stage regression model with random treatment effects (8.6):

$$g\left(E[Y_{ik}]\right) = \beta_{0i} + \left(\beta_1 + b_{1i}\right)X_{ik} + \beta_{2i}Z_{ik} + \beta_3 X_{ik}Z_{ik} \text{ with } b_{1i} \sim N\left(0, \tau^2\right) \tag{8.8}$$

This model has independent fixed effects for each study for the β_2 parameter. As in model (8.7), common or random effects could be used instead. It is also necessary to decide whether to have a common interaction effect β_3, as in this model, or random effects. Hence, model (8.8) is only one possible model that could be used. In practice, models with independent fixed effects or multiple random effects are less likely to converge, particularly for smaller datasets, and models with fewer parameters may be necessary to achieve convergence by, for example, assuming common effects for β_2 and β_3. The next section discusses a problem that arises in one-stage models.

Some example treatment-covariate interaction models, using model (8.8), from the PARIS data are given in Table 8.2. Separate models were fitted to test the impact of diabetes, chronic hypertension, and maternal and gestational age at randomization. For continuously distributed covariates, such as maternal age, it is conventional to "center" these covariates around the overall average so the main treatment effect β_1 is the estimate at the average maternal age, rather than the (nonsensical) age of zero. The estimates of interaction β_3 are close to unity for hypertension, gestational age, and maternal age, with confidence intervals spanning one. This suggests that the covariates have no impact on the effectiveness of antiplatelet therapy. For diabetes, the interaction odds ratio is less than one (although not conventionally statistically significant), suggesting a possibility that antiplatelet therapy is more effective in people with diabetes.

TABLE 8.2

Results of One-Stage Models (Model [8.8] with Separate Models for Each Covariate) Including Treatment-Covariate Interactions Using the PARIS Data

	Odds ratio (95% confidence interval)			
Maternal covariate (Z)	**Treatment (β_1)**	**Covariate (β_2)**	**Treatment-covariate interaction (β_3)**	**Heterogeneity for treatment effect (τ^2)**
Diabetes	0.904 (0.827 to 0.998)	1.296 (1.005 to 1.671)	0.820 (0.561 to 1.199)	0
Chronic hypertension	0.867 (0.783 to 0.959)	1.542 (1.322 to 1.798)	1.116 (0.901 to 1.381)	0.0003
Gestational age at randomization (weeks)	0.886 (0.815 to 0.963)	0.987 (0.976 to 0.999)	1.005 (0.989 to 1.021)	0.0006
Maternal age (years)	0.885 (0.810 to 0.966)	1.024 (1.1014 to 1.034)	1.002 (0.988 to 1.015)	0.0004

8.4.8 Aggregation Bias

When adding covariate data to IPD models, we must consider the possibility that the association between covariate and outcome within any study, or the nature of the interaction between treatment and covariate, may be different to that observed when applying a model across all studies.

For example, if we are interested in the impact of gender on the effectiveness of treatment there might be no difference, on average, between treatment effects in men and women when they are compared within each study. This is the "within-study" association. By contrast, a regression of treatment effect against proportion of men in each study (a meta-regression as in Chapter 7) might find a significant association, perhaps because of differential effects in all-male or all-female studies. This is the "across-study" association (Fisher et al., 2017). Generally, it is the within-study association that is most relevant and of primary interest. Using information both within and across studies mixes the two associations, and so could give a biased estimate if we assume it to be equal to the desired within-study association. This is known as aggregation bias.

The two-stage approach, fitting model (8.7) in each study and meta-analyzing the resulting β_{3i} estimates, avoids aggregation bias, because each estimate of interaction is made entirely within a study, and so only the within-study association is estimated. However, not using the across-study information means that the two-stage approach does lose power to detect a genuine interaction, particularly if there is little within-study variation in the covariate Z (Simmonds and Higgins, 2007).

Model (8.8) may incorporate aggregation bias because β_3 is estimated using data on Z both within and across studies. The one-stage model (8.8) can be adapted to separate within-study information on the interaction from across-study information (Hua et al., 2017):

$$g\left(E[Y_{ik}]\right) = \beta_{0i} + \left(\beta_1 + b_i\right) X_{ik} + \beta_{2i} Z_{ik} + \beta_A X_{ik} \bar{Z}_i + \beta_W X_{ik}\left(Z_{ik} - \bar{Z}_i\right) \tag{8.9}$$

where $\bar{Z}_i$ is the within-study average of covariate Z. In this model β_W is the treatment-covariate interaction occurring within each study and β_A is the interaction observed across studies (as in a standard meta-regression, see Chapter 7). If these two interaction estimates differ, that is an indication that aggregation bias is present. If there is evidence of aggregation bias then the β_W may be the best summary of the interaction, as it is not affected by aggregation bias. It should be noted that estimating the interaction based only on β_W leads to a loss of statistical power to detect the interaction (Simmonds and Higgins, 2007). Choosing between models (8.8) and (8.9) is therefore a balance between reducing the risk of bias and maximizing statistical power. (Simmonds and Higgins, 2007).

8.5 Missing Data

As in any meta-analysis, data may not be available for all studies, and excluding such studies could lead to bias. In IPD meta-analyses, the IPD for some studies may be unavailable because study authors could not or would not provide the data, or because data have been lost. If these data are missing because they differ from the rest of the IPD, then an analysis of IPD alone will be subject to availability bias (see Chapter 13).

The simplest approach to combining IPD with aggregate data is a two-stage analysis, using treatment effect estimates extracted from publications where IPD are unavailable and pooling those with equivalent estimates obtained for each study where IPD are available. Sensitivity analyses comparing results in studies with and without IPD are generally advisable (Riley et al., 2007). One-stage methods for including studies without IPD exist (see examples in Chapter 5), and involve modeling the IPD and published data separately, while assuming studies with and without IPD have the same treatment effects and parameter values (Riley et al., 2008).

If the results of the analyses incorporating aggregate data from unavailable trials are in line with the IPD, then this adds confidence in the findings of the IPD meta-analyses. If results are substantially different then it raises an important issue that requires further exploration to understand why the results of unavailable trials are different and which analysis is most likely to be the more credible.

Data may also be missing at the individual level. Having IPD makes it possible to account for missing outcome variables through complete case analyses, best/worst case analysis, last observation carried forward, or multiple imputation (see Chapter 12). Multiple imputation could also be used when covariate data are missing in the IPD, potentially including imputation of multiple covariates at once (see Chapter 21).

8.6 Other Fields for IPD Meta-Analysis

This chapter has focused on how IPD is used in analysis of clinical trials, but IPD can be used in any meta-analysis. The use of IPD with observational studies is covered in more detail in Chapter 21. IPD can always be used with any type of study or data by taking a two-stage approach: using the IPD to generate the required effect estimates for each study, and combining these as if they had been extracted from publications.

IPD meta-analyses of clinical trials can be extended to consider comparisons of multiple treatments in a network meta-analysis, and both one-stage and two-stage methods for IPD network meta-analysis have been developed (see Chapter 10). In meta-analysis, outcomes are usually analyzed separately, ignoring any correlation between them. Multivariate meta-analysis, where outcomes are jointly analyzed is gaining in popularity (see Chapter 9). IPD is particularly useful in this area as having full data allows for estimation of the correlations between outcomes, which are rarely reported in publications.

IPD may be particularly useful in the meta-analysis of prognostic models (see Chapter 22). Prognostic models presented in publications generally differ as to which prognostic factors are included in the model and which factors have results reported in the publication, making meta-analysis based on published results difficult. Having the IPD allows for the consistent analysis of prognostic factors across studies. IPD meta-analyses of diagnostic tests have received less attention, but IPD could be useful when comparing different diagnostic tests, where multiple tests are used in sequence, and to assess the impact of participant factors on diagnostic accuracy (see Chapter 19).

References

Advanced Ovarian Cancer Trialists Group, 1991. Chemotherapy in advanced ovarian cancer: An overview of randomised clinical trials. *BMJ (Clinical Research Ed)* **303**(6807): 884–893.

Askie LM, Duley L, Henderson-Smart DJ, Stewart LA and PARIS Collaborative Group, 2007. Antiplatelet agents for prevention of pre-eclampsia: A meta-analysis of individual patient data. *Lancet* **369**(9575): 1791–1798.

Cox DR, 1972. Regression models and life tables. *JRSS B* **34**(2): 187–220.

DerSimonian R and Laird N, 1986. Meta-analysis in clinical trials. *Controlled Clinical Trials* **7**(3): 177–188.

Fibrinogen Studies Collaboration, 2006. Regression dilution methods for meta-analysis: Assessing long-term variability in plasma fibrinogen among 27 247 adults in 15 prospective studies. *International Journal of Epidemiology* **35**(6): 1570–1578.

Fisher DJ, Carpenter JR, Morris TP, Freeman SC and Tierney JF, 2017. Meta-analytical methods to identify who benefits most from treatments: Daft, deluded, or deft approach? *BMJ* **356**: 573.

Higgins JPT and Thompson SG, 2002. Quantifying heterogeneity in a meta-analysis. *Statistics in Medicine* **21**(11): 1539–1558.

Higgins JPT, Whitehead A, Turner RM, Omar RZ and Thompson SG, 2001. Meta-analysis of continuous outcome data from individual patients. *Statistics in Medicine* **20**(15): 2219–2241.

Hua H, Burke DL, Crowther MJ, Ensor J, Tudur Smith C and Riley RD, 2017. One-stage individual participant data meta-analysis models: Estimation of treatment-covariate interactions must avoid ecological bias by separating out within-trial and across-trial information. *Statistics in Medicine* **36**(5): 772–789.

Mant J, Doust J, Roalfe A, Barton P, Cowie MR, Glasziou P, Mant D, McManus RJ, Holder R, Deeks J, Fletcher K, Qume M, Sohanpal S, Sanders S and Hobbs FDR, 2009. Systematic review and individual patient data meta-analysis of diagnosis of heart failure, with modelling of implications of different diagnostic strategies in primary care. *Health Technology Assessment* **13**(32): 1–207, iii.

Mantel N and Haenszel W, 1959. Statistical aspects of the analysis of data from retrospective studies of disease. *Journal of the National Cancer Institute* **22**(4):719–748.

Non-small Cell Lung Cancer Collaborative Group, 1995. Chemotherapy in non-small cell lung cancer: A meta-analysis using updated data on individual patients from 52 randomised clinical trials. *BMJ (Clinical Research Ed)* **311**(7010): 899–909.

Riley RD, Lambert PC and Abo-Zaid G, 2010. Meta-analysis of individual participant data: Rationale, conduct, and reporting. *BMJ* **340**(Feb05 1): c221.

Riley RD, Lambert PC, Staessen JA, Wang J, Gueyffier F, Thijs L and Boutitie F, 2008. Meta-analysis of continuous outcomes combining individual patient data and aggregate data. *Statistics in Medicine* **27**(11): 1870–1893.

Riley RD, Simmonds MC, Look MP, 2007. Evidence synthesis combining individual patient data and aggregate data: A systematic review identified current practice and possible methods. *Journal of Clinical Epidemiology* **60**(5): 431.e1–431.e12.

Selker HP, Griffith JL, Beshansky JR, Schmid CH, Califf RM, D'Agostino RB, Laks MM, Lee KL, Maynard C, Selvester RH, Wagner GS and Weaver WD, 1997. Patient-specific predictions of outcomes in myocardial infarction for real-time emergency use: A thrombolytic predictive instrument. *Annals of Internal Medicine* **127**(7):538–556.

Simmonds M, Stewart G and Stewart L, 2015. A decade of individual participant data meta-analyses: A review of current practice. *Contemporary Clinical Trials* **45**(Pt A): 76–83.

Simmonds MC and Higgins JPT, 2007. Covariate heterogeneity in meta-analysis: Criteria for deciding between meta-regression and individual patient data. *Statistics in Medicine* **26**(15): 2982–2999.

Simmonds MC, Higgins JP, Stewart LA, Tierney JF, Clarke MJ and Thompson SG, 2005. Meta-analysis of individual patient data from randomized trials: A review of methods used in practice. *Clinical Trials* **2**(3): 209–217.

Simmonds MC, Tierney J, Bowden J and Higgins JP, 2011. Meta-analysis of time-to-event data: A comparison of two-stage methods. *Research Synthesis Methods* **2**(3): 139–49.

Stewart LA, Clarke M, Rovers M, Riley RD, Simmonds M, Stewart G, Tierney JF and PRISMA-IPD Development Group, 2015. Preferred reporting items for systematic review and meta-analyses of individual participant data: The PRISMA-IPD statement. *JAMA* **313**(16): 1657–1665.

Turner RM, Omar RZ, Yang M, Goldstein H and Thompson SG, 2000. A multilevel model framework for meta-analysis of clinical trials with binary outcomes. *Statistics in Medicine* **19**(24): 3417–3432.

Whitehead A, Omar RZ, Higgins JPT, Savaluny E, Turner RM and Thompson SG, 2001. Meta-analysis of ordinal outcomes using individual patient data. *Statistics in Medicine* **20**(15): 2243–2260.

Williamson PR, Smith CT, Hutton JL and Marson AG, 2002. Aggregate data meta-analysis with time-to-event outcomes. *Statistics in Medicine* **21**(22): 3337–3351.

9

Multivariate Meta-Analysis

Dan Jackson, Ian R. White, and Richard D. Riley

CONTENTS

9.1 Rationale

Systematic reviews often contain several, or sometimes very many, univariate meta-analyses. For example, a review of 75 systematic reviews published by the Cochrane Pregnancy and Childbirth Group (Riley et al., 2011) found that the median number of forest plots per review was 52 (range 5 to 409). The reason for performing multiple univariate meta-analyses is typically to examine different, but correlated, effects or outcomes of interest. For example, a treatment effect may be of interest for both overall and disease-free survival, or at each of multiple time points. Correlation among the multiple effects of interest can arise because of within-study or between-study correlation, or most usually a combination of both. Within-study correlation results from effect estimates that are calculated using correlated data from the same set of patients. Between-study correlation results from relationships between the true underlying study specific effects; for example, a study with an impressive true underlying effect for one outcome may also have impressive true underlying effects for other outcomes.

Multiple univariate meta-analyses provide valid inferences for the parameters of interest under the assumptions of the models used. Some of these assumptions, such as the absence of reporting bias, may be implicitly made. However, if there are multiple outcomes of interest, there are three main limitations of performing multiple univariate meta-analyses (one for each outcome). First, this is potentially inefficient because separate univariate analyses do not allow the correlated, and sometimes very closely related, outcomes to inform the inferences for other outcomes. This limitation can be especially important when not all studies provide data on all outcomes of interest. Second, because separate univariate meta-analyses do not describe the correlations between the pooled estimates, they do not readily facilitate making joint inferences about multiple outcomes included in the systematic review. These types of joint inferences are often important for decision making. For example, we require *both* the sensitivity and specificity of a diagnostic test to be adequate for clinical use. Finally, the multivariate approach can help to reduce the impact that reporting bias may have on univariate analyses (Kirkham et al., 2012).

Multivariate meta-analysis methods have therefore been proposed (Raudenbush et al., 1988; Gleser and Olkin, 2009; van Houwelingen et al., 2002) to jointly synthesize multiple effects while accounting for their correlation. This chapter explains the fundamental principles of multivariate meta-analysis, while drawing attention to a range of applications. We also highlight the difficulties associated with the multivariate approach, in particular that multivariate models usually require more data and assumptions. We discuss some one-stage methods, but we mainly assume that a "two-stage" approach is adopted, as in Chapter 4. In the first stage, each study must give a vector of study level summaries and their within-study covariance matrices. The study level summaries and their within-study variances are estimated using standard methods (Boronstein et al., 2009); we describe methods for estimating the within-study correlations later in the chapter. In the second stage, we pool the study level summaries.

The development in this chapter initially follows that in Chapter 4: we define the multivariate meta-analysis model and discuss various methods for its estimation. We then look at issues specific to the multivariate model: describing the "borrowing of strength" between outcomes, estimating within-study correlations, and what to do if within-study correlations are unavailable. We discuss, with examples, the advantages and disadvantages of multivariate meta-analysis compared with separate univariate meta-analyses. We examine longitudinal and multinomial data and statistical software and conclude with a short discussion.

9.2 The Random-Effects Model for Multivariate Meta-Analysis

We now describe the standard random-effects model for multivariate meta-analysis (Jackson et al., 2010, 2011); we later describe the common-effect model as a special case. We will denote the vector containing multiple study level summaries for the ith study as the $P \times 1$ column vector $\mathbf{y}_i$, for $i = 1, 2, \ldots I$. For convenience, we assume that these are treatment effects. In the univariate case we have $P = 1$ and we recover the standard univariate random-effects model (Higgins et al., 2009) described in Chapter 4. The multivariate random-effects model follows the corresponding typical univariate model: we make distributional assumptions within and between studies, as explained immediately below.

9.2.1 The Within-Study Model

It is most common to assume multivariate normal within-study distributions. This can be justified as an approximation in large studies by invoking the central limit theorem. This means that within each study we assume

$$\mathbf{y}_i \mid \theta_i \sim N(\theta_i, \mathbf{S}_i) \tag{9.1}$$

where N denotes a *multivariate* normal distribution and $\boldsymbol{\theta}_i$ is the vector of true underlying effects for the ith study. The matrices $\mathbf{S}_i$ are referred to as the within-study covariance matrices; these are $P \times P$ matrices and their entries are estimated in practice but (as in the univariate setting) are typically regarded as fixed and known when pooling the results to make inferences. The main diagonal entries of $\mathbf{S}_i$ are the within-study variances of the $\mathbf{y}_i$ and are typically obtained using standard methods and formulae as for the univariate case. We will return later to the issue of estimating the within-study correlations, so that the off-diagonal entries of $\mathbf{S}_i$ (the within-study covariances) can be obtained as the products of these within-study correlations and the within-study standard errors.

Alternative within-study models that avoid using normal approximations for the within-study distributions can also be used. These are multivariate extensions of the one-stage approaches described in Chapter 5, where arm level summaries or individual patient data (IPD) are modeled. For example, independent logistic models are typically used to model the likelihood with respect to binary sensitivity and specificity outcome data at the within-study level in the multivariate meta-analysis of studies concerned with diagnostic test accuracy (Hamza et al., 2008; Harbord and Whiting, 2009). Multivariate normality is then typically assumed between studies. With the exception of our second example below, we will use the normal approximations in model (9.1) throughout this chapter.

9.2.2 The Between-Study Model

In most applications, the common-effect assumption that all $\boldsymbol{\theta}_i$ are equal, so that all studies estimate the same true underlying quantity, is very strong. The multivariate random-effects model allows the $\boldsymbol{\theta}_i$ to vary from one study to the next and assumes that

$$\boldsymbol{\theta}_i \sim N(\boldsymbol{\theta}, \mathcal{T}) \tag{9.2}$$

where $\boldsymbol{\theta}$ is the (overall) summary treatment effect vector and $\mathcal{T}$ is the between-study covariance matrix, which in an unstructured form contains P between-study variances and the corresponding covariances. The average treatment effects $\boldsymbol{\theta}$ are the parameters of primary inferential interest. We will regard $\mathcal{T}$ as being unstructured in the examples that follow but simplifications are possible: for example, all between-study correlations, or all between-study variances, could be assumed to be the same, as is standard practice in network meta-analysis (Chapter 10). If we assume that $\boldsymbol{\theta}_i = \boldsymbol{\theta}$ for all i, which is equivalent to assuming that $\mathcal{T} = \mathbf{0}$, then we have the common-effect model.

9.2.3 The Marginal Model

Marginally, models (9.1) and (9.2) provide the conventional multivariate random-effects meta-analysis model

$$\mathbf{y}_i \sim N(\boldsymbol{\theta}, \mathbf{S}_i + \mathcal{T}). \tag{9.3}$$

where the $\mathbf{y}_i$ are further assumed to be independent across studies.

This multivariate approach can easily handle studies with missing estimates for some treatment effects, under a missing at random assumption. The model for any studies that provide only some treatment effects is taken as the implied marginal model from (9.3). We will explain below how missing estimates can be conveniently handled computationally.

9.3 Estimation and Making Inferences

As in the univariate case in Chapter 4, one of the main statistical difficulties lies in estimating the between-study covariance matrix, $\mathcal{T}$. Once this has been estimated, the standard procedure for making inferences about the average treatment effect $\boldsymbol{\theta}$ approximates $\mathcal{T}$ with $\hat{\mathcal{T}}$ (Jackson et al., 2011). Writing $\hat{\mathbf{W}}_i^{-1} = \mathbf{S}_i + \hat{\mathcal{T}}$, the maximum likelihood estimate of $\boldsymbol{\theta}$ is

$$\hat{\theta} = \left(\sum_{i=1}^{I} \hat{\mathbf{W}}_i \right)^{-1} \sum_{i=1}^{I} \hat{\mathbf{W}}_i \mathbf{y}_i \tag{9.4}$$

from which we typically calculate confidence intervals and perform hypothesis tests by Wald's method (for each component of $\boldsymbol{\theta}$; joint confidence regions can also be obtained) using the approximation

$$\hat{\theta} \sim N\left(\theta, \left(\sum_{i=1}^{I} \hat{\mathbf{W}}_i \right)^{-1} \right). \tag{9.5}$$

As in Chapter 4, this approximation can be poor: we discuss improvements below. Studies that do not provide all P treatment effect estimates can be incorporated into (9.4) and (9.5) by allocating the missing treatment effect estimates a nominal value (zero, say) with very large within-study variances and zero within-study correlations. There may also be interest in making joint predictive inferences about the effects in a new setting similar to one included in the meta-analysis. Snell et al. (2016) suggest a way to extend the univariate prediction interval (Higgins et al., 2009) for this purpose.

9.3.1 Estimating the Between-Study Covariance Matrix

A variety of multivariate estimators of $\mathcal{T}$, all of which can be used in equations (9.4) and (9.5) in application, have been developed. These correspond to some, but not all, of the methods in Chapter 4. In the univariate setting, the moment estimator proposed by DerSimonian and Laird (1986) and the restricted maximum likelihood (REML) estimator, which was also suggested by DerSimonian and Laird (1986), are now standard methods. Two multivariate extensions of DerSimonian and Laird's univariate estimator have been proposed. The first of these methods (Jackson et al., 2010) estimates each entry of $\mathcal{T}$ using a separate linear equation. The second of these methods was proposed by Chen et al. (2012) who suggest an alternative matrix-based generalization of DerSimonian and Laird's estimator. Jackson et al. (2013) subsequently generalized this second estimator further in order to incorporate missing outcome data and study level covariates. The matrix-based estimator may be considered to be an improvement on the previous estimator because all outcome data directly contribute to all entries of $\hat{\mathcal{T}}$ when using the matrix-based approach. However, the matrix-based estimator is computationally more demanding and so its predecessor may be considered advantageous in very high dimensions. Furthermore, these two DerSimonian and Laird estimators perform similarly in practice (Jackson et al., 2013). Alternative, and also relatively computationally undemanding, estimators of $\mathcal{T}$ have also been proposed (Wouhib, 2014; Makambi and Seung, 2015; Ma and Mazumdar, 2011).

9.3.2 The REML Estimator

Maximum likelihood methods are based on sufficient statistics, are asymptotically efficient, and the asymptotic theory of maximum likelihood provides a way to make approximate inferences about all parameters included in the model. However, likelihood-based methods are fully parametric and are computationally more expensive than the methods described above. Furthermore, maximum likelihood estimates of variance parameters are generally biased downwards. The restricted likelihood is a function of the variance components only (i.e., not $\boldsymbol{\theta}$) and REML helps to correct for the downward bias of maximum likelihood estimates of variance components. This estimation is performed by maximizing the restricted log likelihood (Jackson et al., 2010, 2011) subject to the constraint that the between-study covariance matrix is positive semi-definite:

$$\ell_{\text{REML}} = -\frac{1}{2}\sum_{i=1}^{I}\log\left|\mathbf{S}_i + \mathcal{T}\right| - \frac{1}{2}\log\left|\sum_{i=1}^{I}(\mathbf{S}_i + \mathcal{T})^{-1}\right| - \frac{1}{2}\sum_{i=1}^{I}\mathbf{r}_i'(\mathbf{S}_i + \mathcal{T})^{-1}\mathbf{r}_i,$$

where the $\mathbf{r}_i$ denote the residuals under model (9.3). We suggest that REML should be the preferred estimator of $\mathcal{T}$ in situations where it is computationally feasible but alternative estimators should be considered when between-study normality assumptions are to be explicitly avoided.

9.3.3 Taking into Account the Uncertainty in the Between-Study Variance Components

Regardless of the estimator of $\mathcal{T}$ used, a limitation of using (9.5) to make statistical inferences about θ is that this standard approach makes no allowance for the uncertainty in the estimated between-study variance components. This is exactly the same issue as in Chapter 4. Usually a pressing concern is that there is considerable uncertainty in $\hat{\mathcal{T}}$ that is not reflected in (9.5). A related difficulty is that it is not obvious how to quantify the uncertainty in $\hat{\mathcal{T}}$ when using the DerSimonian and Laird type estimators described above.

When adopting an entirely likelihood-based approach, one straightforward way to reflect, but not fully acknowledge, the uncertainty in the between-study variance components is to invert the entire observed Fisher information matrix (including the components of $\mathcal{T}$) in order to obtain standard errors (White, 2011). An alternative idea, which can be applied regardless of the estimation method used, is the recently proposed multivariate extension of the univariate Hartung–Knapp–Sidik–Jonkman modification (Jackson and Riley, 2014; Hartung and Knapp, 2001). As in the univariate case, this method can result in shorter confidence intervals than the conventional methods for some examples (Wiksten et al., 2016), so caution is required when applying this modification in both the univariate and multivariate settings. A further alternative to reflect the uncertainty in $\hat{\mathcal{T}}$ is to use the profile likelihood to make inferences (The Fibrinogen Studies Collaboration, Jackson et al., 2009); a difficulty here is that the necessary maximizations and iterative methods make this computationally expensive and so unfeasible in high dimensions.

Bayesian multivariate meta-analysis (Nam et al., 2003) allows for all parameter uncertainty but comes at the cost of having to specify prior distributions and use computationally intensive methods such as Markov chain Monte Carlo (MCMC), as implemented in WinBUGS (Lunn et al., 2000). An important advantage of adopting a Bayesian approach is that further inferences, such as prediction intervals (Higgins et al., 2009), can be conveniently obtained from standard MCMC output. For those who would prefer to use an entirely Bayesian approach, WinBUGS code is now available (Mavridis and Salanti, 2013; Riley et al., 2015b).

9.4 Heterogeneity Statistics, Study Weights, and Borrowing of Strength

The methods described above summarize how standard univariate methods for estimating the between-study variance and making inferences about the average effect have been generalized to the multivariate case. However, further results are often presented when communicating the results of univariate meta-analyses and multivariate extensions of some of these have recently been proposed. In particular, descriptive statistics, such as heterogeneity statistics and study weights, have become commonplace in univariate meta-analysis. For example, I^2 statistics (Higgins and Thompson, 2002) are often used to quantify the impact of between-study heterogeneity and study weights are often displayed on forest plots in order to show which studies contribute most to the estimated effect. These descriptive statistics have been extended for use in multivariate meta-analysis and we describe these methods next.

9.4.1 Heterogeneity Statistics

Higgins and Thompson (2002) originally defined three univariate measures of the impact of heterogeneity, R^2, H^2, and I^2. The univariate R^2 statistic generalizes naturally to the multivariate setting (Jackson et al., 2012). We denote the *volumes* of the confidence regions (using normal approximations for estimated effects but any fixed coverage probability) for all P estimated effects in $\boldsymbol{\theta}$ that arise from the random-effects and common-effect models as V_R and V_C. In one dimension V_R and V_C are the lengths of the random-effects and common-effect confidence intervals, and in two and three dimensions they are areas and volumes respectively. In four or more dimensions, V_R and V_F are generalized notions of volumes. In one dimension, Higgins and Thompson's definition of R^2 is equivalent to defining

$$R = \left(\frac{V_R}{V_C}\right)^{1/P}$$

and so we take this to be our multivariate R statistic. By taking the Pth root of V_R/V_C as our measure, we obtain an indication of the "stretching" of confidence regions that results from fitting the multivariate random-effects, rather than the multivariate common-effect, model (Jackson et al., 2012). The R statistic can be conveniently calculated as

$$R = \left(\left|\text{Var}_R(\hat{\theta})\,\text{Var}_C(\hat{\theta})^{-1}\right|\right)^{\frac{1}{2P}} \tag{9.6}$$

where $|\mathbf{A}|$ is the determinant of $\mathbf{A}$ and $\text{Var}_R(\hat{\theta})$ and $\text{Var}_C(\hat{\theta})$ are the covariance matrices of the estimated average effect (see equation [9.5]) under the random-effects and common-effect models respectively. We can then define I^2 statistics (Jackson et al., 2012) that are based on the R statistic as

$$I_R^2 = \frac{R^2 - 1}{R^2}$$

An advantage of these R and I_R^2 statistics is that they can be evaluated for subsets of the pooled effects, by using covariance matrices of reduced dimension in (9.6) and replacing P with this reduced dimension. Furthermore, R and I_R^2 statistics can be calculated for linear combinations of effects by using appropriate covariance matrices of the linear combinations in (9.6) and replacing P with the dimension of these covariance matrices. A disadvantage is that the I_R^2 statistics do not reduce to the conventional univariate I^2 statistic when $P=1$. Alternative multivariate I^2 statistics have also been proposed (White, 2011; Jackson et al., 2012; Gasparrini et al., 2012), including one based on a scalar Q-statistic that results in a single heterogeneity statistic for the multivariate meta-analysis. This alternative has the advantage of reducing to the usual I^2 statistic when $P=1$ but cannot be evaluated for subsets, or linear combinations of, the pooled effects.

9.4.2 Study Weights and Borrowing of Strength Statistics

Study weights are routinely shown on forest plots in univariate meta-analyses and so the question of what constitutes a suitable definition of study weights for a multivariate meta-analysis is inevitably raised. König et al. (2013) suggest using entries of the "hat" matrix to define weights in the context of network meta-analysis. More recently Jackson et al. (2017) present a unified framework for calculating study weights and borrowing of strength (BoS) statistics, where BoS statistics describe the (study specific or total) part of the weights that are due to indirect evidence (for example, due to data for the second effect when making inferences for the first effect in a bivariate meta-analysis). This framework for study weights and BoS statistics is based upon a decomposition of the multivariate score statistic and, unlike the methods proposed by König et al., is invariant to scaling the outcome data (Jackson et al., 2017).

Jackson et al. (2017) show that this framework for BoS can be interpreted as comparing the precision of multivariate and univariate pooled estimates. Hence, the BoS statistics are more intuitively appealing than is immediately apparent. The multivariate R and I_R^2 statistics are conceptualized as comparing the precision of estimates resulting from the random and common-effect multivariate models, so that all descriptive statistics presented in this section share the common theme of comparing the precision of the pooled estimates under various meta-analysis models. Copas et al. (2018) investigate the properties of BoS statistics analytically to clarify the conditions under which borrowing of strength occurs.

9.5 Within-Study Correlations

A major stumbling block for multivariate meta-analysis is that the conventional models require, for every study, the within-study correlations between each pair of observed effect estimates of interest. This adds to the aggregate data that is necessary to perform multiple standard univariate meta-analyses. In some scenarios, the within-study correlations can justifiably be assumed zero or approximately zero (van Houwelingen et al., 2002, Reitsma et al., 2005; Daniels and Hughes, 1997; Korn et al., 2005; Thompson et al., 2005). However, in other settings, such as the meta-analysis of effects at multiple time points (Ishak et al., 2007) or for multiple outcomes (Riley et al., 2007a), it is not at all plausible to assume that the within-study correlations are zero. Unfortunately, estimated within-study correlations are rarely reported in primary study publications. Furthermore, their derivation is at best non-trivial and at worst impossible from other reported information. In this section, we outline some main approaches to obtaining within-study correlations and discuss strategies for performing multivariate meta-analysis when they are unavailable.

We begin by examining a claim made by Ishak et al. (Ishak et al., 2008). They say that, if interest lies in the pooled estimates themselves, one can perform multivariate meta-analysis and assume within-study correlations are zero "without any significant risk of bias or loss of precision in estimates". Our counter to this argument is that all the Ishak et al. simulations contain small within-study variation relative to the between-study variation. In such situations, the total variance is dominated by the between-study variance, and the total correlation is dominated by the between-study correlation; thus, one would not expect the within-study correlations to have much impact. Sohn (2000) has also shown that the within-study correlations have little influence in such situations. More generally, however, the size of within-study correlation is important. For example, Riley (2009) gives an example where the within-study variation is relatively large and the bivariate meta-analysis is sensitive to the value assumed for the within-study correlations. Our position is that, in general, suitable approaches for accounting for non-zero within-study correlations should be used.

9.5.1 Estimating Within-Study Correlations

In some situations, it may be possible to directly estimate the within-study correlations. We examine this situation first, because if all entries of the within-study covariance matrices can be estimated, then the methods described above are immediately applicable.

9.5.2 Derivation from Individual Patient Data

The availability of IPD for studies included in the meta-analysis allows the within-study correlations to be directly estimated, as outlined by Riley et al. and explained below (Riley et al., 2015b). When IPD are available for some studies it is common for other studies to exist where IPD are unavailable (Riley et al., 2007c). In this situation, one strategy is to use the within-study correlations derived from the "IPD studies" to inform the likely value of the within-study correlation in the aggregate data studies. For example, using a Bayesian framework, McDaid et al. (2007) use the observed within-study correlation in their available IPD studies to produce an informative prior distribution for the missing within-study correlations in other studies. Similarly, Bujkiewicz et al. (2013) apply a double bootstrap to the IPD from a single study (the first to estimate the within-study correlation, the second to estimate its uncertainty), to derive a prior distribution for the missing within-study correlations in other studies.

When using IPD, the general strategy for obtaining effect estimates for multivariate meta-analysis is to analyze each study separately to obtain study specific estimates and within-study covariance matrices, where the within-study correlations may be estimated using one of two main approaches: joint modeling or bootstrapping.

9.5.2.1 Joint Modeling

For the joint modeling approach, consider the example of IPD from randomized trials where a treatment effect is of interest for two continuous outcomes ($k=1,2$), such as systolic (SBP) and diastolic blood pressure (DBP). We assume linear models throughout. At baseline (i.e., before randomization), the jth patient in the ith trial provides their SBP and DBP values, which we denote by y_{Bijk}, where $k=1$ for SBP and $k=2$ for DBP. Also, each patient provides their final SBP and DBP values after treatment, which we denote by y_{Fijk}. Let x_{ij} be an indicator for the treatment group. A joint linear regression model can be fitted for the two outcomes, so that the treatment effects for SBP and DBP are estimated simultaneously. For example, a joint model that adjusts for baseline values (analysis of covariance models) could be specified as

$$y_{Fijk} = \phi_{ik} + \beta_{ik} y_{Bijk} + \theta_{ik} x_{ij} + \varepsilon_{ijk}$$

$$\varepsilon_{ijk} \sim N(0, \sigma_{ik}^2)$$

$$\text{Cov}(\varepsilon_{ij1}, \varepsilon_{ij2}) = \sigma_{i12}$$

In this model, ϕ_{ik} is the fixed-trial effect for outcome k, β_{ik} denotes the mean change in y_{Fijk} for a one unit increase in y_{Bijk}, $\boldsymbol{\theta}_{ik}$ is the underlying treatment effect (adjusted for baseline values) for outcome k in trial i, and σ_{ik}^2 is the residual variance of outcome k in trial i after accounting for treatment group and baseline values. The correlation in the responses from the same patient is accounted for by the covariance term, σ_{i12}. For estimation purposes, it is helpful to re-write the model with dummy variables (Riley et al., 2008a). This regression model can be fitted in standard statistical software that allows for correlated errors. From this the treatment effect estimates $\hat{\theta}_{ik}$, which become the entries of the $\mathbf{y}_i$ vectors in the multivariate meta-analysis, and their within-study variances, s_{ik}^2, are obtained. Crucially, the model naturally induces a within-study correlation between the estimated treatment effects, and the within-study covariance matrices $\mathbf{S}_i$ needed for multivariate meta-analysis can be estimated from the inverse of Fisher's information matrix. If baseline values are unavailable, then the model can be fitted excluding the baseline value as a covariate; this will still give unbiased estimates of effect when patients are randomized to the treatment groups, though the resulting estimation will be less efficient (Riley et al., 2013).

9.5.2.2 Bootstrapping

A more general approach for estimating within-study correlations is via non-parametric bootstrapping (The Fibrinogen Studies Collaboration, Jackson et al., 2009; Riley et al., 2015b). Briefly, for each study, participants are sampled with replacement in order to produce study specific bootstrap samples. In each of these bootstrap samples, separate models are fitted to obtain each effect estimate of interest which are the bootstrap replications. The estimated correlations between the bootstrap replications estimate the within-study correlation for each study. The greatest advantage of the bootstrapping approach over the joint modeling approach is that the within-study correlation can easily be obtained for effects relating to different types of outcomes and models (for example, a continuous outcome from a linear regression and a binary outcome from a logistic regression). Illustrative Stata code is provided elsewhere (Riley et al., 2015b). It may be desirable to sample individuals within structural strata such as treatment groups when drawing the bootstrap samples, in order to ensure the same number of participants in these strata within the bootstrap samples and the original data.

9.5.3 Derivation of Within-Study Correlations from Other Information

There are some situations where explicit formulae exist for calculating the within-study correlations from alternative study information that may be more commonly reported. For example, when interested in multiple binary outcomes, some outcomes may be either nested (where one outcome is a subset of the other) or mutually exclusive. In these situations, the number of events should be recorded for each outcome. Within-study correlations can then be estimated by the formulae provided by Trikalinos and Olkin (2008) for mutually exclusive outcomes and Wei and Higgins (2013b) for nested outcomes.

For related binary outcomes that are neither nested nor mutually exclusive, Wei and Higgins (2013b) provide additional formulae for calculating within-study correlations. However, these formulae require, among other information, an estimate of the patient-level correlations of their outcomes. Such correlation may often be reported by primary study authors or elicited from subject experts. Wei and Higgins also provide formulae for how patient-level correlations can be used to calculate within-study correlations between pairs of treatment effect estimates relating to other types of outcomes. Bagos (2012) provides formulae for calculating the covariance of two correlated log odds ratios based on large sample approximations, which is applicable to many situations.

9.6 Options When Within-Study Correlations Are Not Estimable

Despite the strategies described above, very commonly the within-study correlations are not estimable. The results from multivariate meta-analyses then become much more speculative and we suggest three possible strategies in such instances.

9.6.1 Narrow the Range of Possible Values

It may sometimes be possible to narrow the range of the possible values for the unknown within-study correlations, so that suitable values can be adopted in analysis. For example, Raudenbush et al. (1988) use external information for this purpose. Riley (2009) used biological reasoning to impute positive within-study correlations in a multivariate meta-analysis of tumor marker studies. For the special situation where multiple relative risks are to be synthesized, Berrington and Cox (2003) narrow the range of possible values for the within-study correlation by calculating lower and upper bounds from the 2×2 tables available from each study.

9.6.2 Perform a Sensitivity Analysis across the Entire Correlation Range

In situations where it is not possible to narrow the range of values, a further option is to perform sensitivity analyses by imputing correlations over the entire range of values (i.e., –1 to 1), to assess if and how conclusions depend upon the correlations imputed. In a Bayesian framework, Nam et al. (2003) take a related approach by placing a Uniform(–1,1) prior distribution on the within-study correlation, and then assess whether conclusions are robust to changes in the specification of this prior.

9.6.3 Use an Alternative Multivariate Model

An alternative model for multivariate meta-analysis has been proposed which does not require the within-study correlations (Riley et al., 2008b). This model includes only one overall correlation parameter for each pair of outcomes. These correlations are hybrids of the within-study and between-study correlations that can be estimated using only the vectors of estimated effects and the within-study variances, i.e., the same data required to fit separate univariate meta-analyses. Hence this alternative model is very widely applicable.

For example, in the bivariate ($P=2$) case the alternative (to the random-effects model in 3) multivariate model can be specified as $\mathbf{y}_i \sim N(\theta, \Psi_i)$ where

$$\Psi_i = \begin{bmatrix} S_{i11} + T_{11} & \rho\sqrt{(S_{i11} + T_{11})(S_{i22} + T_{22})} \\ \rho\sqrt{(S_{i11} + T_{11})(S_{i22} + T_{22})} & S_{i22} + T_{22} \end{bmatrix}$$

Where interest only lies in the pooled estimates, or some function of them, then unless the estimated total correlation ρ is very close to the edge of the parameter space, the alternative model has been shown to produce appropriate pooled estimates with little bias that are very similar to those from fitting model (9.3); furthermore, these pooled estimates have better statistical properties than those from separate univariate analyses (Riley et al., 2008b). As the full hierarchical structure is lost, we do not recommend using this approach to make predictive inferences, however.

Finally, Chen and colleagues (Chen et al., 2015, 2016) propose methods for accounting for correlation between treatment effects when estimating functions of pooled estimates from separate univariate models. This approach gives correct variances. However, because it derives point estimates from separate univariate models, it does not allow borrowing of strength and assumes that data are missing completely at random.

9.7 The Advantages of Multivariate Meta-Analysis and Three Illustrative Examples

There are numerous advantages of multivariate meta-analysis over univariate meta-analysis that we summarize here. We will illustrate six advantages using three contrasting examples.

9.7.1 Advantage 1: We Can Borrow Strength to Improve Precision and Reduce Bias

Multivariate meta-analysis uses more information than separate univariate analyses, via the within and between-study correlations, which can improve statistical properties of meta-analysis results in terms of greater precision and smaller mean-square error (Jackson et al., 2011). This is especially useful in situations with missing data for some outcomes. For example, Kirkham et al. (2012) conclude that multivariate meta-analysis can be useful for reducing the impact of outcome reporting bias, where some outcomes are selectively missing.

9.7.2 Example 1: Partially and Fully Adjusted Results

The Fibrinogen Studies Collaboration (The Fibrinogen Studies Collaboration, Jackson et al., 2009) use a bivariate meta-analysis to estimate the fully adjusted association when some studies only reported a partially adjusted association (see also Chapter 21). Specifically, they examine whether fibrinogen is a risk factor for cardiovascular disease (CVD) using IPD from 31 studies. Hazard ratios were derived from Cox regressions that estimate the effect associated with a one unit increase in fibrinogen on the rate of CVD. All studies allowed a partially adjusted hazard ratio to be estimated, where the hazard ratio for fibrinogen was adjusted for the same core set of potential confounders, including age, smoking, BMI, and blood pressure. However, a fully adjusted hazard ratio was only estimable in 14 studies, which all additionally recorded the further potential confounders cholesterol, alcohol consumption, triglycerides, and diabetes.

Multivariate and univariate random-effects meta-analyses using log hazard ratios as outcome data gave almost identical estimates for the average log hazard ratios and the between-study standard deviations, for both partially and fully adjusted effects. However, by utilizing the large within-study and between-study correlations, the multivariate meta-analysis substantially increased the precision of the average fully adjusted average hazard ratio: the standard error of the pooled log hazard ratio is about 30% smaller in the multivariate compared with univariate analysis. This is due to the multivariate analysis borrowing considerable strength from the partially adjusted estimates. This is reflected by a large BoS statistic of 53% for the fully adjusted effect, indicating that 53% of the information toward the fully adjusted pooled result is due to the partially adjusted estimates. In contrast, BoS is just 1.9% for the partially adjusted summary result: the partially adjusted estimates are available in all 31 studies so that the corresponding pooled estimate gains little from the fully adjusted results. Figure 9.1 shows the forest plot for the fully adjusted results, and the weighting of each study in the univariate and multivariate analyses. The 17 studies without fully adjusted results contribute heavily in the multivariate analysis; indeed, they contribute around half of the total information toward the fully adjusted summary result.

9.7.3 Advantage 2: We Can Calculate Joint Confidence and Prediction Regions, and Joint Probabilistic Inferences

The multivariate approach allows joint inferences across the multiple effects of interest, which is important when clinical decisions need to be based on two or more measures of interest. We illustrate this using an example in diagnosis.

9.7.4 Example 2: Diagnostic Test Accuracy

Craig et al. (2002) systematically reviewed thermometry studies comparing temperatures taken at the ear and rectum in children, and of clinical interest is the accuracy of infrared ear thermometry for diagnosing fever. Eleven studies (2323 children) evaluated the accuracy of a "FirstTemp" branded ear thermometer in relation to an electronic rectal thermometer. Rectal temperature was the reference measure, as it is a well-established method of measuring temperature in children. However, measuring temperature at the ear is less invasive than measuring temperature at the rectum, and so ear measurements would be preferable if their diagnostic accuracy is adequate. All studies defined patients with an ear temperature of more than 38°C to be test positive, and the gold standard definition of fever was a rectal temperature of more than 38°C, consistent with NHS guidelines for diagnosing fever in children at the time of these studies. The studies included children already in hospital or attending accident and emergency, and so the observed prevalence of fever was high, around 50%.

Multivariate meta-analysis is required to summarize test accuracy across these studies, to appropriately quantify any between-study heterogeneity in test accuracy, and to identify

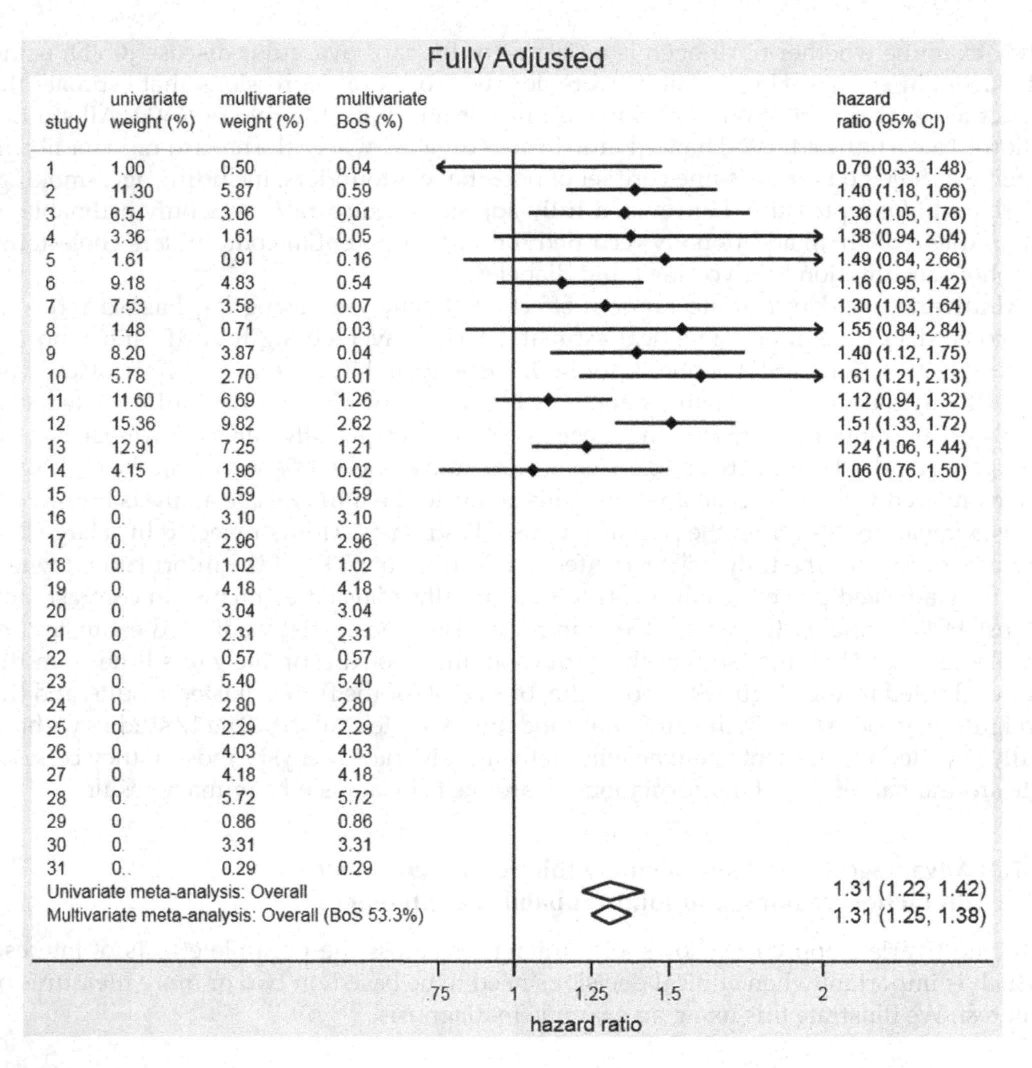

FIGURE 9.1
Example 1. Forest plot for the fully adjusted effects of fibrinogen estimated using the bivariate random-effects meta-analysis.

the potential performance of the test in new populations. The methodology used to produce the fitted model (Harbord and Whiting, 2009) (see also Chapter 5) shown in Figure 9.2 uses binomial within-study distributions, so that for this example the within-study model (9.1) for the logit of the sensitivity and specificity is replaced by two independent intercept only logistic regression models. The joint confidence and prediction intervals in Figure 9.2 allow for an association between the sensitivity and specificity through the between-study correlation. Riley et al. (2015a) applied a Bayesian version of this bivariate random-effects meta-analysis model to these data to make further inferences, but the frequentist model could also be used. The joint probability that in a new population both the sensitivity and specificity will be >80% was found to be 0.18 (Figure 9.2). To provide valid joint inferences, such as those shown in Figure 9.2, we must take into account the association between the outcomes of interest and so use a multivariate approach.

Diagnostic test accuracy is further explored in Chapter 19.

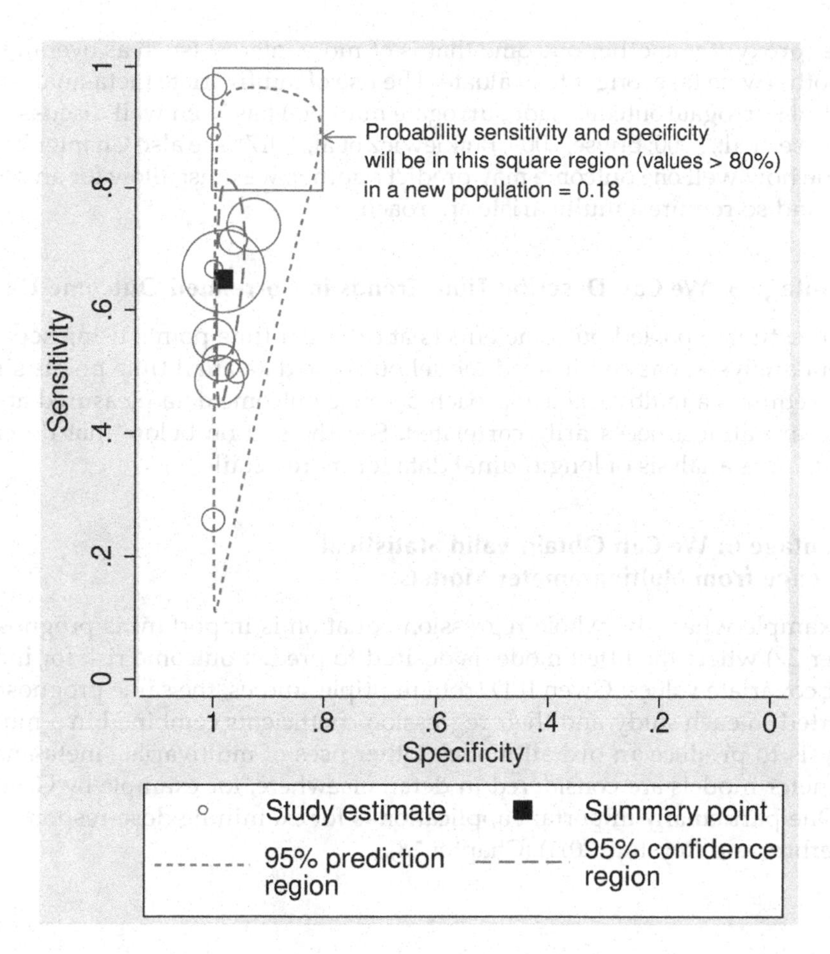

FIGURE 9.2
Example 2. Plot of sensitivity against specificity for the multivariate model fitted to the diagnostic test accuracy data.

9.7.5 Advantage 3: We Can Estimate Appropriate Functions of the Pooled Estimates

As well as deriving joint inferences, one may also be interested in functions of the multiple pooled results. In order to obtain valid inference, the standard error of this function's value must account for the correlations between the pooled results. If one performs separate univariate analyses then, unless one of the approaches suggested by Chen and colleagues (2015, 2016) is adopted, which do not provide borrowing of strength, such covariances are either not taken into account or implicitly assumed to be zero. An important example is where functions of pooled estimates are of interest is network meta-analysis, which can expressed in a multivariate meta-analysis framework (White et al., 2012).

9.7.6 Advantage 4: We Can Predict an Outcome Effect Conditional on a Correlated Outcome Effect

Sometimes the correlation between outcomes may itself be of interest, so that we can determine how well one outcome of interest may predict another. An example of this is the evaluation of surrogate outcomes where one outcome (such as disease-free survival) is used as a surrogate (proxy) for another outcome that is of more interest (such as overall survival), but would otherwise take longer to evaluate. The use of multivariate meta-analysis to evaluate potential surrogate outcomes (or surrogate markers) has been well discussed (Gail et al., 2000; Buyse et al., 2000; Buyse, 2009; Bujkiewicz et al., 2017); see also Chapter 20. In order to determine how well one outcome may predict another, we must allow for an association in the data and so require a multivariate approach.

9.7.7 Advantage 5: We Can Describe Time Trends in Correlated Outcome Data

Rather than estimate pooled outcome effects at different time points using separate univariate meta-analyses, one can instead model outcome data at all time points simultaneously. This requires a multivariate approach because outcome data measured at different time points are almost necessarily correlated. See the section below that describes the multivariate meta-analysis of longitudinal data for more details.

9.7.8 Advantage 6: We Can Obtain Valid Statistical Inference from Multiparameter Models

Another example where the whole regression equation is important is prognostic models (Chapter 22) where the fitted model is desired to predict outcome risk for individuals using their covariate values. Given IPD from multiple studies, the same prognostic model could be fitted in each study and their regression coefficients combined in a multivariate meta-analysis to produce an overall model. Other uses of multivariate meta-analysis for multiparameter models are considered in detail elsewhere, for example by Gasparrini et al. (2012). One particularly important application is for examining dose-response relationships (Sauerbrei and Royston, 2011) (Chapter 18).

9.8 Potential Disadvantages of Multivariate Meta-Analysis

Clearly, there are numerous potential advantages of the multivariate method over the univariate approach. However, possible disadvantages are also prevalent, as follows.

9.8.1 Potential Disadvantage 1: Separate Univariate Meta-Analyses Are Simpler and Often Give Similar Numerical Results for the Individual Pooled Effects

Trikalinos et al. (2014) undertook an empirical evaluation of the impact of multivariate meta-analysis on binary outcomes reported in reviews within the Cochrane library. They conclude that there is generally very little numerical difference between multivariate and univariate pooled results of intervention effects. Very little difference in the pooled estimates and their precision will occur when the borrowing of strength is small, as is usually the case when there is complete data. The simulation studies of Sohn et al. (2000) and Riley et al. (2007b) confirm this.

9.8.2 Potential Disadvantage 2: We Can Encounter Estimation Problems

Sophisticated modeling is extremely difficult in meta-analysis without IPD. All we usually have are a handful of estimates and, if we are lucky, their standard errors. Also, multivariate meta-analysis often requires within-study correlation estimates, but these are rarely available as discussed above. Even when the within-study correlations are available, one particular concern is poor estimation of the multivariate random-effects model. For example, it has been shown that the estimated between-study correlation is often at the edge of the parameter space (Riley et al., 2007b). For this reason, simplified variance-covariance structures may need to be considered or indeed the multivariate approach may not be considered practical.

9.8.3 Potential Disadvantage 3: We Require Additional Modeling Assumptions

In a univariate meta-analysis, the assumption that the random effects are normally distributed is hard to verify. In a multivariate meta-analysis, the multivariate normality assumption is even stronger and harder to verify. Multivariate meta-analysis approaches that use copula representations of the between-study distribution is a current research interest (Kuss et al., 2014; Nikoloulopoulos, 2015), in order to allow for possible non-linear associations between the multiple effects of interest. This new development may be especially important when making joint predictive inferences and/or when the borrowing of strength is large. However, it is clearly going to be difficult to estimate (or identify genuine) non-linear relationships with the few studies that meta-analyses usually have available.

9.8.4 Potential Disadvantage 4: We Might Borrow Bias Instead of Strength

The multivariate model assumes studies are exchangeable and that any missing outcomes are missing at random. This may not be true, and outcomes may be missing not at random if, for example, they are selectively missing due to non-significance. The simulation study of Kirkham et al. (2012) indicates that the multivariate approach may still reduce bias in this situation compared with univariate meta-analysis results, though the bias is not removed completely. However, there is still a concern that poorly reported and highly selectively missing secondary outcomes may result in bias for the primary outcomes if both primary and secondary outcomes are included in the same multivariate meta-analysis.

9.9 Multivariate Meta-Analysis of Longitudinal Data

Multivariate meta-analysis can be used to describe longitudinal (or repeated measures) data, where correlated summary effects at two or more follow-up time points are of interest. This is because studies often report results at multiple times. Therefore, a multivariate approach to meta-analysis is warranted to borrow strength across time points, and improve efficiency compared with separate univariate meta-analyses at each time point. Wrongly assuming time points are independent can result in biased and less precise summary results (Liang and Zeger, 1986; Peters and Mengersen, 2008; Burke et al., 2018).

Jones et al. (2009) describe how to undertake a two-stage multivariate meta-analysis of randomized trials with a longitudinal continuous response (such as blood pressure or pain score) in situations where IPD are available. The approach taken depends on whether or not the effect of time is to be modeled as continuous. If time is to be modeled using a continuous covariate then a repeated measures regression model is fitted to the IPD in each trial separately. Here we usually assume a linear trend in mean responses over time but with a separate intercept and slope for the control and treatment groups. Correlations between multiple responses from the same individual are accounted for in this modeling. This leads to $\mathbf{y}_i$ that contain two estimates per trial: the estimated differences between the intercepts and slopes of the mean regression lines for the treatment and control groups. We also obtain the corresponding $\mathbf{S}_i$ when fitting these regression models. The second step is then conducted using a bivariate meta-analysis model. This leads to an estimated summary mean difference between intercepts $\hat{\theta}_1$ and an estimated summary mean difference between slopes $\hat{\theta}_2$. These estimates can then be used to estimate the treatment effect at any time point of interest (included those not actually reported by any of the original trials). For example, assuming that time was measured in months, the estimated treatment effect at 6 months is $\hat{\theta}_1 + 6\hat{\theta}_2$.

When the effect of time is instead modeled using a categorical covariate then the approach is analogous to a two-stage multivariate meta-analysis of multiple outcomes; data at each of the discrete time points (e.g., 1 month, 3 months, 6 months, and 12 months) provide the "outcomes" in the multivariate meta-analysis. In the first step, a repeated measures linear regression model is fitted to the IPD in each trial separately, including time in the model as a factor, and estimating a separate treatment effect at each time point. This accounts for the correlation between multiple responses over time from the same individual (and could allow a separate residual variance at each time point). This results in a vector of treatment effect estimates in each trial $\mathbf{y}_i$ spanning the time points of interest, and the corresponding estimated within-study covariance matrix, $\mathbf{S}_i$. The second step fits a multivariate meta-analysis model, to estimate the vector of summary (mean) treatment effect $\boldsymbol{\theta}$, assuming either common or random treatment effects across studies. Ideally an unstructured between-study covariance matrix is assumed. However, when data are sparse, it may be necessary to make simplifying assumptions such as a common between-study variance at all time points or an auto-regressive correlation structure, such that between-correlations are largest between pairs of neighboring time points and gradually reduce as time points become further apart. A further difficulty is that trials often report different follow-up times, and therefore "similar" time points may need to be grouped to facilitate meta-analysis. For example, 24 weeks in trial 1, 28 weeks in trial 2, and 30 weeks in trial 3 might need to be classed as "6 months", in order to perform the synthesis across trials.

A variety of other methods are also available. With IPD an alternative one-stage approach is also feasible, where the longitudinal participant-level responses from all studies are analyzed in a single model, accounting for the correlation of multiple responses from the same patient and the clustering of patients within trials (e.g., via stratified intercepts). Trikalinos and Olkin (2012) consider the multivariate meta-analysis of treatment effect estimates for binary outcomes as measured by four common metrics (log odds ratio, log risk ratio, risk difference, and arcsine difference at multiple time points). They provide formulae for estimating within-study variances and within-study correlations between effects at pairs of time points, which depend on the total numbers of participants and the proportion with the outcome of interest in each group at each time point. Such information may be available even without IPD. Similarly, Jones et al. (2009) provide an equation for deriving within-study correlations between treatment effect estimates at a pair of time points within a randomized trial with a continuous outcome. This equation depends on the residual variances and correlation in patient responses at the two time points, as well as the number of individuals contributing to each time point and the number contributing to both times. Again, such information may be available without IPD, for example, from trial reports or authors. Other novel extensions and related work for meta-analysis of longitudinal data include allowing for non-linear trends via fractional polynomials (Jansen et al., 2015), network meta-analysis applications (Jansen et al., 2015; Dakin et al., 2011), and meta-analysis of survival proportions over time (Dear, 1994; Jackson et al., 2014; Arends et al., 2008).

9.9.1 Example 3: Longitudinal Data

Jones et al. (2009) consider an IPD meta-analysis of five trials investigating the effects of selegiline (10 mg/day) versus placebo for the treatment of Alzheimer's disease, with respect to the Mini-Mental State Examination (MMSE) score. This is a measure of cognitive function ranging from 0 to 30, with larger values being regarded as good. When time was modeled using a categorical covariate, the available time points were grouped to create common times across studies: 1 month = weeks 4 and 5; 2 months = weeks 8 and 9; 4 months = week 17; 6 months = weeks 24, 25, and 30; 9 months = week 35 and 43; 12 months = weeks 56 and 65. There was substantial missing data, with not all time points available in all studies.

The resulting two-stage multivariate common-effect meta-analysis provided no strong evidence of differences between selegiline and placebo at any of the time points (Table 9.1). However, the results from a series of separate univariate meta-analyses at each time point are notably different (Jones et al. (2009) present a multivariate common-effect analysis that assumes all within-study covariances are zero; this is equivalent to a series of common-effect univariate analyses). For example, a univariate common-effect analysis estimates a difference at nine months of 0.69 with standard error of 0.63, whereas the corresponding multivariate estimate is 0.34 with standard error of 0.52. Standard errors are consistently smaller when using the multivariate approach, due to the borrowing of strength.

9.10 Multivariate Meta-Analysis for Multinomial Data

Schmid et al. (2014) describe a Bayesian model for combining studies that report data from some or all levels of an outcome with unordered mutually exclusive categories, such as different types of cardiovascular outcomes or numbers of graduates in different academic disciplines. They fit a multinomial model with a probability attached to each outcome type. Probabilities are summed across types reporting incomplete information. For example, if a study reports the total number of deaths but not the specific causes then the identifiable probability is the sum of the cause-specific outcomes. The multinomial assumption identifies the within-study correlations among the outcomes as known functions of the

TABLE 9.1

Summary of Results (Difference in Average MMSE Score for Selegiline vs. Placebo) from a Two-Stage Common-Effect Multivariate Meta-Analysis

	Multivariate analysis	Univariate analyses
1 Month	0.30 (0.47)	0.43 (0.54)
2 Months	−0.47 (0.59)	−0.84 (0.97)
4 Months	0.33 (0.47)	0.75 (0.57)
6 Months	0.19 (0.48)	0.31 (0.50)
9 Months	0.34 (0.52)	0.69 (0.63)
12 Months	−0.03 (0.55)	0.29 (0.66)

The results from a series of six separate univariate meta-analysis are shown for comparison.

category probabilities and therefore incorporates the competing risks nature of the data that a higher likelihood of one outcome makes others less likely.

The probabilities themselves are modeled using baseline category logits $\theta_{jm}^{(k)} = \log\left(\pi_{jm}^{(k)} / \pi_{j0}^{(k)}\right)$ where $\pi_{jm}^{(k)}$ is the probability of an event in category m for treatment j in study k. The baseline category logits are then modeled with random study and treatment effects (with potential covariates in meta-regression). The paper generalizes the model to network meta-analysis (see Chapter 10) and discusses various covariance structures to describe between-study heterogeneity.

9.11 Statistical Software

Until fairly recently, analysts often required bespoke code for performing multivariate meta-analyses. However multivariate meta-analysis models can now be fitted in standard statistics packages such as SAS, R, Stata, and WinBugs. In SAS, the PROC MIXED module allows users enormous flexibility in their model specification and is especially useful for non-standard models. An introduction to the basic use of PROC MIXED for multivariate meta-analysis is given by Van Houwelingen et al. (2002), and more advanced code is also available for newer methods (Riley et al., 2014). The parms statement is crucial for specifying (and holding fixed) the within-study variances and correlations. A potential disadvantage of the PROC MIXED module is that it only allows ML or REML estimation. As it is a general package for mixed models, it does not provide the latest developments such as the multivariate I^2 or BoS statistics.

In contrast, the mvmeta modules in Stata (White, 2011) and R (Gasparrini et al., 2012) are tailored to the multivariate meta-analysis field, and thus have a wider range of estimation options. The Stata module also implements the alternative model of Riley et al. (2008b) described above. However, models with non-standard forms of structured between-study covariance matrices may be more easily specified in PROC MIXED or via user-written likelihoods. The mvmeta module in Stata can also be used to perform network meta-analysis in a frequentist setting, via the module network (White, 2015) (see Chapter 10). Further multivariate procedures also exist, such as those in the R metafor package (Viechtbauer, 2010).

Bayesian multivariate meta-analysis models can be fitted using MCMC, for example, using WinBugs, JAGS, or Stan. Careful specification of prior distributions for the between-study covariance matrix is required, for which a Wishart distribution is a convenient choice but the separation of the components of $\mathcal{T}$ is often thought preferable in order to make the prior less informative (Burke et al., 2018). Wei and Higgins (2013a) discuss a variety of options that allow independent prior distributions for the between-study variances and correlations.

9.12 Discussion

In this chapter, we have described the methods for, and the challenges faced by, multivariate meta-analysis. In situations where all studies provide all estimates of effect and interest lies only in the pooled effects themselves (and not some function of these or joint inferences), then usually multivariate meta-analysis will not add much value to separate univariate analyses. However, in cases where there is substantial missing data, we feel there is a very strong case for using multivariate meta-analysis, due to the potential to borrow strength. If, however, joint inferences, or inferences for functions of the average effects, are required, then the correlations between the estimated average effects must always be taken into account, which requires some form of multivariate analysis regardless of the completeness of the data. Hence, it is clear that the multivariate meta-analysis has the potential to have a considerable impact in applied work. More complicated multivariate modeling is possible, where further possibilities include multivariate meta-regression models, which include study level covariate effects, and models that avoid the use of normal approximations within-studies or use copula approaches to deal with between-study correlation.

The main difficulty for those who would advocate the widespread use of multivariate meta-analysis is that obtaining appropriate within-study correlations is often very difficult. Hence, we have directed much of our attention to dealing with this problem. However, better reporting of the quantities that can be used to either directly estimate or otherwise ascertain suitable values for these correlations would ease this problem. One hope for the future is that study reports will become better at describing the correlations between quantities of interest, and not just their estimates and standard errors, or their means and standard deviations, so that multivariate meta-analysis may become more routinely used.

References

Arends LR, Hunink MG and Stijnen T, 2008. Meta-analysis of summary survival curve data. *Statistics in Medicine* **27**(22): 4381–4396.

Bagos PG, 2012. On the covariance of two correlated log-odds ratios. *Statistics in Medicine* **31**(14): 1418–1431.

Berrington A and Cox DR, 2003. Generalized least squares for the synthesis of correlated information. *Biostatistics* **4**(3): 423–431.

Boronstein M, Hedges LV, Higgins JPT and Rothstein HR, 2009. *Introduction to Meta-Analysis*. Wiley.

Bujkiewicz S, Thompson JR, Sutton AJ, Cooper NJ, Harrison MJ, Symmons DP and Abrams KR, 2013. Multivariate meta-analysis of mixed outcomes: A Bayesian approach. *Statistics in Medicine* **32**(22): 3926–3943.

Bujkiewicz S, Thompson JR, Spata E and Abrams KR, 2017. Uncertainty in the Bayesian meta-analysis of normally distributed surrogate endpoints. *Statistical Methods in Medical Research* **26**(5): 2287–2318.

Burke D, Bujkiewicz S and Riley RD, 2018. Bayesian bivariate meta-analysis of correlated effects: Impact of the prior distributions on the between-study correlation, borrowing of strength, and joint inferences. *Statistical Methods in Medical Research* **27**(2): 428–450.

Buyse M, Molenberghs G, Burzykowski T, Renard D and Geys H, 2000. The validation of surrogate endpoints in meta-analyses of randomized experiments. *Biostatistics* **1**(1): 49–67.

Buyse M, 2009. Use of meta-analysis for the validation of surrogate endpoints and biomarkers in cancer trials. *Cancer Journal* **15**(5): 421–425.

Chen H, Manning AK and Dupuis J, 2012. A method of moments estimator for random effect multivariate meta-analysis. *Biometrics* **68**(4): 1278–1284.

Chen Y, Hong C and Riley RD, 2015. An alternative pseudolikelihood method for multivariate random-effects meta-analysis. *Statistics in Medicine* **34**(3): 361–380.

Chen Y, Cai Y, Hong C and Jackson D, 2016. Inference for correlated effect sizes using multiple univariate meta-analyses. *Statistics in Medicine* **35**: 1405–1422.

Copas JB, Jackson D, White IR and Riley RD, 2018. The role of secondary outcomes in multivariate meta-analysis. *Journal of the Royal Statistical Society: Series C (Applied Statistics)* **27**: 1177–1205.

Craig JV, Lancaster GA, Taylor S, Williamson PR and Smyth RL, 2002. Infrared ear thermometry compared with rectal thermometry in children: A systematic review. *Lancet* **360**(9333): 603–609.

Dakin HA, Welton NJ, Ades AE, Collins S, Orme M and Kelly S, 2011. Mixed treatment comparison of repeated measurements of a continuous endpoint: An example using topical treatments for primary open-angle glaucoma and ocular hypertension. *Statistics in Medicine* **30**(20): 2511–2535.

Daniels MJ and Hughes MD, 1997. Meta-analysis for the evaluation of potential surrogate markers. *Statistics in Medicine* **16**(17): 1965–1982.

Dear KB, 1994. Iterative generalized least squares for meta-analysis of survival data at multiple times. *Biometrics* **50**(4): 989–1002.

Dersimonian R and Laird N, 1986. Meta-analysis in clinical trials. *Controlled Clinical Trials* **7**(3): 177–188.

Gail MH, Pfeiffer R, Van Houwelingen HC and Carroll RJ, 2000. On meta-analytic assessment of surrogate outcomes. *Biostatistics* **1**(3): 231–246.

Gasparrini A, Armstrong B and Kenward MG, 2012. Multivariate meta-analysis for non-linear and other multi-parameter associations. *Statistics in Medicine* **31**: 3821–3839.

Gleser LJ and Olkin I, 2009. Stochastically dependent effect sizes. In Cooper H and Olkin I (Eds). *The Handbook of Research Synthesis*. Russell Sage Foundation.

Hamza TH, van Houwelingen HC and Stijnen T, 2008. The binomial distribution of meta-analysis was preferred to model within-study variability. *Journal of Clinical Epidemiology* **61**(1): 41–51.

Harbord RM and Whiting P, 2009. metandi: Meta-analysis of diagnostic accuracy using hierarchical logistic regression. *The STATA Journal* **9**(2): 211–229.

Hartung J and Knapp G, 2001. A refined method for the meta-analysis of controlled clinical trials with binary outcome. *Statistics in Medicine* **20**: 3875–3889.

Higgins JPT and Thompson SG, 2002. Quantifying heterogeneity in a meta-analysis. *Statistics in Medicine* **21**(11): 1539–1558.

Higgins JPT, Thompson SG and Spiegelhalter DJ, 2009. A re-evaluation of random-effects meta-analysis. *Journal of the Royal Statistical Society: Series A*, **172**: 137–159.

Ishak KJ, Platt RW, Joseph L, Hanley JA and Caro JJ, 2007. Meta-analysis of longitudinal studies. *Clinical Trials* **4**(5): 525–539.

Ishak KJ, Platt RW, Joseph L and Hanley JA, 2008. Impact of approximating or ignoring within-study covariances in multivariate meta-analyses. *Statistics in Medicine* **27**(5): 670–686.

Jackson D and Riley RD, 2014. A refined method for multivariate meta-analysis and meta-regression. *Statistics in Medicine* **33**(4): 541–554.

Jackson D, White IR and Thompson SG, 2010. Extending DerSimonian and Laird's methodology to perform multivariate random effects meta-analyses. *Statistics in Medicine* **29**(12): 1282–1297.

Jackson D, Riley R and White IR, 2011. Multivariate meta-analysis: Potential and promise (with discussion). *Statistics in Medicine* **30**(20): 2481–2510.

Jackson D, White IR and Riley RD, 2012. Quantifying the impact of between-study heterogeneity in multivariate meta-analyses. *Statistics in Medicine* **31**: 3805–3820.

Jackson D, White IR and Riley RD, 2013. A matrix based method of moments for fitting the multivariate random effects model for meta-analysis and meta-regression. *The Biometrical Journal* **55**(2): 231–245.

Jackson D, Rollins K and Coughlin P, 2014. A multivariate model for the meta-analysis of study level survival data at multiple times. *Research Synthesis Methods* **5**(3): 264–272.

Jackson D, White IR, Price M, Copas J and Riley R, 2017. Borrowing of strength and study weights in multivariate and network meta-analysis. *Statistical Methods in Medical Research* **26**: 2853–2868.

Jansen JP, Vieira MC and Cope S, 2015. Network meta-analysis of longitudinal data using fractional polynomials. *Statistics in Medicine* **34**(15): 2294–2311.

Jones AP, Riley RD, Williamson PR and Whitehead A, 2009. Meta-analysis of individual patient data versus aggregate data from longitudinal clinical trials. *Clinical Trials* **6**(1): 16–27.

Kirkham JJ, Riley RD and Williamson PR, 2012. A multivariate meta-analysis approach for reducing the impact of outcome reporting bias in systematic reviews. *Statistics in Medicine* **31**(20): 2179–2195.

König J, Krahn U and Binder H, 2013. Visualizing the flow of evidence in network meta-analysis and characterizing mixed treatment comparisons. *Statistics in Medicine* **32**: 5414–5429.

Korn EL, Albert PS and McShane LM, 2005. Assessing surrogates as trial endpoints using mixed models. *Statistics in Medicine* **24**(2): 163–182.

Kuss O, Hoyer A and Solms A, 2014. Meta-analysis for diagnostic accuracy studies: A new statistical model using beta-binomial distributions and bivariate copulas. *Statistics in Medicine* **33**(1): 17–30.

Liang KY and Zeger SL, 1986. Longitudinal data analysis using generalized linear models. *Biometrika* **73**(1): 13–22.

Lunn DJ, Thomas A, Best N and Spiegelhalter D, 2000. WinBUGS – A Bayesian modelling framework: Concepts, structure, and extensibility. *Statistics in Computing* **10**(4): 325–337.

Ma Y and Mazumdar M, 2011. Multivariate meta-analysis: A robust approach based on the theory of U-statistic. *Statistics in Medicine* **30**(24): 2911–2929.

Makambi KH and Seung H, 2015. A non-iterative extension of the multivariate random effects meta-analysis. *Journal of Biopharmaceutical Statistics* **25**(1): 109–123.

Mavridis D and Salanti G, 2013. A practical introduction to multivariate meta-analysis. *Statistical Methods in Medical Research* **22**(2): 133–158.

McDaid C, Griffin S, Weatherley H, Duree K, van der Burgt M, van Hout S, et al., 2007. Sleep apnoea continuous positive airways pressure (CPAP) ACD: Assessment report. NICE Report.

Nam IS, Mengerson K and Garthwaite P, 2003. Multivariate meta-analysis. *Statistics in Medicine* **22**: 2309–2333.

Nikoloulopoulos AK, 2015. A mixed effect model for bivariate meta-analysis of diagnostic test accuracy studies using a copula representation of the random effects distribution. *Statistics in Medicine* **34:** 3842–3865.

Peters JL and Mengersen KL, 2008. Meta-analysis of repeated measures study designs. *Journal of Evaluation in Clinical Practice* **14**(5): 941–950.

Raudenbush SW, Becker BJ and Kalaian H, 1988. Modeling multivariate effect sizes. *Psychological Bulletin* **103**(1): 111–120.

Reitsma JB, Glas AS, Rutjes AW, Scholten RJ, Bossuyt PM and Zwinderman AH, 2005. Bivariate analysis of sensitivity and specificity produces informative summary measures in diagnostic reviews. *Journal of Clinical Epidemiology* **58**(10): 982–990.

Riley RD, Abrams KR, Lambert PC, Sutton AJ and Thompson JR, 2007a. An evaluation of bivariate random-effects meta-analysis for the joint synthesis of two correlated outcomes. *Statistics in Medicine* **26**(1): 78–97.

Riley RD, Abrams KR, Sutton AJ, Lambert PC and Thompson JR, 2007b. Bivariate random effects meta-analysis and the estimation of between-study correlation. *BMC Medical Research Methodology* **7**: 3.

Riley RD, Simmonds MC and Look MP, 2007c. Evidence synthesis combining individual patient data and aggregate data: A systematic review identified current practice and possible methods. *Journal of Clinical Epidemiology* **60**(5): 431–439.

Riley RD, Lambert PC, Staessen JA, Wang J, Gueyffier F, Thijs L and Boutitie F, 2008a. Meta-analysis of continuous outcomes combining individual patient data and aggregate data. *Statistics in Medicine* **27**(11): 1870–1893.

Riley RD, Thompson JR and Abrams KR, 2008b. An alternative model for bivariate random-effects meta-analysis when the within-study correlations are unknown. *Biostatistics* **9**(1): 172–186.

Riley RD, Gates S, Neilson J and Alfirevic Z, 2011. Statistical methods can be improved within cochrane pregnancy and childbirth reviews. *Journal of Clinical Epidemiology* **64**(6): 608–618.

Riley RD, Kauser I, Bland M, Thijs L, Staessen JA, Wang J, Gueyffier F and Deeks JJ, 2013. Meta-analysis of continuous outcomes according to baseline imbalance and availability of individual participant data. *Statistics in Medicine* **32**(16): 2747–2766.

Riley RD, Takwoingi Y, Trikalinos T, Guha A, Biswas A, Ensor J, Morris RK and Deeks J, 2014. Meta-analysis of test accuracy studies with multiple and missing thresholds: A multivariate-normal model. *Journal of Biometrics and Biostatistics* **5**: 3.

Riley RD, Ahmed I, Debray TPA, Willis BH, Noordzij JP, Higgins JPT and Deeks JJ, 2015a. Summarising and validating the accuracy of a diagnostic or prognostic test across multiple studies: A new meta-analysis framework. *Statistics in Medicine* **34**(13): 2081–2103.

Riley RD, Price MJ, Jackson D, Wardle M, Gueyffier F, Wang J, Staessen JA and White IR, 2015b. Multivariate meta-analysis using individual participant data. *Research Synthesis Methods* **6**(2): 157–174.

Riley RD, 2009. Multivariate meta-analysis: The effect of ignoring within-study correlation. *Journal of the Royal Statistical Society: Series A* **172**(4): 789–811.

Sauerbrei W and Royston P, 2011. A new strategy for meta-analysis of continuous covariates in observational studies. *Statistics in Medicine* **30**(28): 3341–3660.

Schmid CH, Trikalinos TA and Olkin I, 2014. Bayesian network meta-analysis for unordered categorical outcomes with incomplete data. *Research Synthesis Methods* **5**(2): 162–185.

Snell KIE, Hua H, Debray TPA, Ensor J, Look MP, Moons KGM and Riley RD, 2016. Multivariate meta-analysis of individual participant data helped externally validate the performance and implementation of a prediction model. *Journal of Clinical Epidemiology* **69**: 40–50.

Sohn SY, 2000. Multivariate meta-analysis with potentially correlated marketing study results. *Naval Research Logistics* **47**(6): 500–510.

The Fibrinogen Studies Collaboration, Jackson D, White I, Kostis JB, Wilson AC, Folsom AR et al., 2009. Systematically missing confounders in individual participant data meta-analysis of observational studies. *Statistics in Medicine* **28**(8): 1218–1237.

Thompson JR, Minelli C, Abrams KR, Tobin MD and Riley RD, 2005. Meta-analysis of genetic studies using Mendelian randomization – A multivariate approach. *Statistics in Medicine* **24**(14): 2241–2254.

Trikalinos TA and Olkin I, 2008. A method for the meta-analysis of mutually exclusive binary outcomes. *Statistics in Medicine* **27**(21): 4279–4300.

Trikalinos TA and Olkin I, 2012. Meta-analysis of effect sizes reported at multiple time points: A multivariate approach. *Clinical Trials* **9**(5): 610–620.

Trikalinos TA, Hoaglin DC and Schmid CH, 2014. An empirical comparison of univariate and multivariate meta-analyses for categorical outcomes. *Statistics in Medicine* **33**: 1441–1459.

van Houwelingen HC, Arends LR and Stijnen T, 2002. Advanced methods in meta-analysis: Multivariate approach and meta-regression. *Statistics in Medicine* **21**: 589–624.

Viechtbauer W, 2010. Conducting meta-analyses in R with the metafor package. *Journal of Statistical Software* **36**: 1–48.

Wei Y and Higgins JP, 2013a. Bayesian multivariate meta-analysis with multiple outcomes. *Statistics in Medicine* **32**(17): 2911–2934.

Wei Y and Higgins JP, 2013b. Estimating within-study covariances in multivariate meta-analysis with multiple outcomes. *Statistics in Medicine* **32**(7): 1191–1205.

White IR, Barrett JK, Jackson D and Higgins JPT, 2012. Consistency and inconsistency in network meta-analysis: Model estimation using multivariate meta-regression. *Research Synthesis Methods* **3**(2): 111–125.

White IR, 2011. Multivariate random-effect meta-regression: Updates to mvmeta. *The STATA Journal* **11**: 255–270.

White IR, 2015. Network meta-analysis. *The STATA Journal* **15**(4): 951–985.

Wiksten A, Rucker G and Schwarzer G, 2016. Hartung Knapp method is not always conservative compared with fixed-effect meta-analysis. *Statistics in Medicine* **35**: 2503–2515.

Wouhib A, 2014. Estimation of a matrix of heterogeneity parameters in multivariate meta-analysis of random-effects models. *Journal of Statistical Theory and Applications* **13**(1): 46–64.

10

Network Meta-Analysis

Adriani Nikolakopoulou, Ian R. White, and Georgia Salanti

CONTENTS

10.1 Introduction

"What is the best treatment for this condition?" is a key question asked by patients and clinicians (Del Fiol et al., 2014). In the era of multiple treatment options in nearly all clinical fields, pairwise meta-analyses may not be sufficient to convey the broad picture of the possible treatments or interventions and answer this question (Salanti, 2012; Caldwell et al., 2005;

Naci and Fleurence, 2011). Network meta-analysis (NMA) aspires to synthesize evidence about multiple treatments using valid statistical methods and can provide a broader picture of the available treatment options for a condition, identify research gaps, produce a treatment hierarchy, and inform healthcare decision making (Salanti, 2012; Mavridis et al., 2015; Cipriani et al., 2013; Higgins and Welton, 2015). Similar questions are asked in other scientific areas, for example, which is the best education or wildlife protection program. The methods described in this chapter may be applied to all such problems, although we will draw our examples from healthcare where these methods were originally developed.

The earliest example of NMA, in its contemporary form, dates back to 1996. Higgins and Whitehead synthesized 26 studies that examined the effectiveness of three treatments for reducing bleeding in cirrhosis (Higgins and Whitehead, 1996). The method aimed to more precisely estimate relative treatment effects and heterogeneity. Higgins and Whitehead suggested that when meta-analyses include data from three or more treatments "there would be little reason not to combine all treatments into a single analysis". Several publications reported NMA during the following years and an exponential increase occurred after 2008 (Petropoulou et al., 2016; Zarin et al., 2017; Nikolakopoulou et al., 2014a).

Alongside the proliferation of NMA applications, new analytical approaches and novel synthesis tools are evolving. The continuous enrichment of NMA methodology has enabled researchers to answer complex questions in health sciences and has placed NMA at the front line of research in biostatistics (Jackson et al., 2014; Lu et al., 2011; Higgins et al., 2012; White et al., 2012; König et al., 2013; Rücker and Schwarzer, 2014; Simmonds and Higgins, 2016). Tutorials and guidance papers have contributed to the accessibility and applicability of NMA to clinical practice (Salanti, 2012; Caldwell et al., 2005; Mavridis et al., 2015; Cipriani et al., 2013, Chaimani et al., 2013; Dias et al., 2013a, 2013d; Senn et al., 2013).

NMA has been characterized as the new norm in comparative effectiveness research (Higgins and Welton, 2015) and its role in reimbursement decisions has been well recognized (Laws et al., 2014) with various organizations such as the UK's National Institute for Health and Care Excellence adopting its use (Petropoulou et al., 2016; Kanters et al., 2016). Naci and O'Connor suggested that NMA could also be used to estimate the efficacy and safety of new drugs relative to existing alternatives at the time of market entry and consequently inform approval decisions (Naci and O'Connor, 2013); this would facilitate the comparison of new drugs with their alternatives in the regulatory setting and maximize the healthcare value of clinical trials.

Two main factors constrain the usefulness of NMA. First, the technicalities underlying the method are resource-demanding and require statistical expertise. Second, the credibility of the results depends on the plausibility of the underlying assumptions of transitivity and consistency. In this chapter, we describe the technical and conceptual aspects of NMA methodology and illustrate the methods in an application. Section 10.2 describes a running example. Section 10.3 introduces the concepts of direct, indirect, and network evidence and discusses heterogeneity in NMA. Section 10.4 describes the assumptions underlying NMA, including transitivity and consistency. Section 10.5 describes the statistical models used in NMA. Section 10.6 describes extensions of these statistical models used to explore and test for inconsistency. We conclude with Section 10.7 on presentation and interpretation of NMA results, Section 10.8 on special topics, and Section 10.9 on software.

10.2 Example

We illustrate the methodology using a published NMA of topical antibiotics without steroids for treatment of ears that chronically discharge fluid with underlying eardrum perforations (Table 10.1) (Macfadyen et al., 2005). A total of 1539 patients are included in 13 randomized controlled trials (RCTs) comparing four topical interventions: no treatment (A); quinolone antibiotic (B); non-quinolone antibiotic (C); and antiseptic (D). The evidence is visualized using a network plot, where nodes represent treatments and lines between nodes represent studies directly comparing those treatments (Figure 10.1). Evidence comparing two treatments is direct when the treatments are compared in the same study and indirect otherwise, for example, when the treatments are compared with a common comparator in different studies. Figure 10.1 shows that the network contains loops: cycles of

TABLE 10.1

Data from the Network of Topical Antibiotics

Design	Study	Treatment	Events	Sample size
Design 1: AB	1. Kasemsuwan 1997	A	14	16
		B	3	19
Design 2: BC	4. Kaygusuz 2002	B	10	20
		C	10	20
	5. Lorente 1995	B	8	159
		C	9	149
	6. Tutkun 1995	B	3	24
		C	14	20
	7. VH 1998 daily	B	9	32
		C	12	36
	8. VH 1998 weekly	B	16	39
		C	13	31
Design 3: BD	12. Jaya 2003	B	6	21
		D	6	19
	13. Macfadyen 2005	B	66	196
		D	108	198
Design 4: CD	10. Clayton 1990	C	19	60
		D	14	42
	11. Browning 1983a	C	15	18
		D	13	20
Design 5: ABD	2. van Hasselt 2002	A	66	83
		B	32	79
		D	77	91
Design 6: BCD	3. van Hasselt 1997	B	7	14
		C	24	40
		D	34	39
	9. Fradis 1997	B	10	19
		C	8	18
		D	13	17

A = no treatment, B = quinolone antibiotic, C = non-quinolone antibiotic, D = antiseptic.

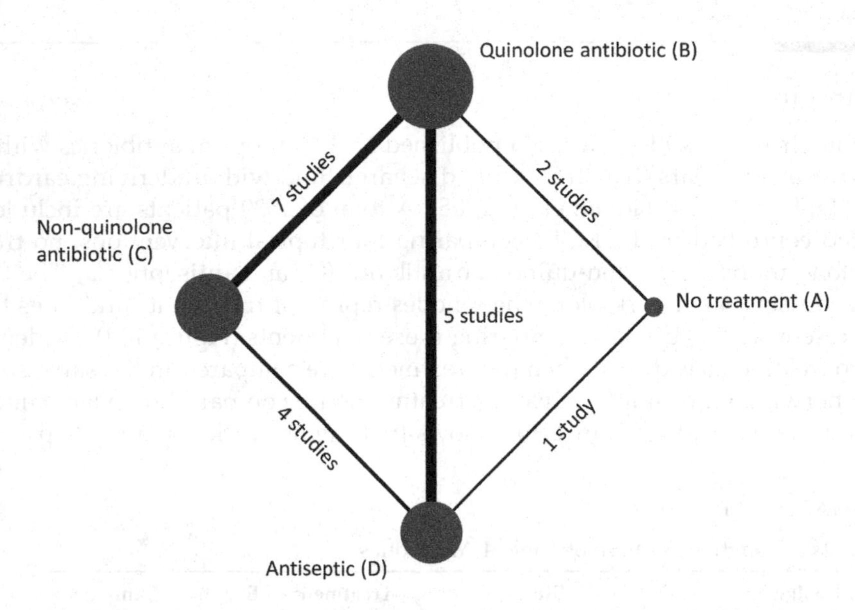

FIGURE 10.1
Network plot of topical antibiotics. Nodes are weighted according to the number of studies involved in each treatment. Edges are weighted according to the number of patients in each comparison and labeled with the number of studies. The graph was produced in Stata using the `networkplot` command.

direct evidence such as B-C-D or B-D-A. The outcome is treatment failure, defined as the presence of persistent discharge at one week after treatment. Relative treatment effects are measured using odds ratios (OR). Pairwise meta-analysis results suggest that quinolone antibiotic is superior to both no treatment and antiseptic; other direct comparisons do not show significant differences between the interventions (Table 10.2, upper triangle). Boxes 10.1–10.4 throughout the chapter illustrate the application of methodology to the example. Figure 10.2 shows a reduced network excluding three-arm trials and excluding treatment A, which we use to explain the simpler methods.

TABLE 10.2

Summary Odds Ratios (OR) and 95% Confidence Intervals Comparing Topical Antibiotics: Pairwise Meta-Analyses (Upper Triangle) and Network Meta-Analysis as a Multivariate Model in Stata (Lower Triangle)

No treatment (A)	**0.09 (0.01, 0.51)**	–	1.42 (0.65, 3.09)
0.15 (0.04, 0.54)	Quinolone antibiotic (B)	1.46 (0.80, 2.67)	**3.47 (1.71, 7.07)**
0.26 (0.07, 1.04)	1.73 (0.91, 3.30)	Non-quinolone antibiotic (C)	1.69 (0.59, 4.83)
0.48 (0.13, 1.79)	**3.14 (1.56, 6.34)**	1.82 (0.85, 3.88)	Antiseptic (D)

A random-effects model is used for all comparisons with two studies or more. The heterogeneity standard deviation estimated using the methods of moments is $\hat{\tau}_{AB} = 0.10$, $\hat{\tau}_{BC} = 0.56$, $\hat{\tau}_{BD} = 0.62$, $\hat{\tau}_{CD} = 0.87$ for the pairwise meta-analyses and $\hat{\tau} = 0.74$ for network meta-analysis estimated using the restricted maximum likelihood estimator. For estimates above the diagonal, ORs greater than 1 favor the treatment specified in the row. For estimates below the diagonal, ORs greater than 1 favor the treatment specified in the column. Bold results indicate statistical significance.

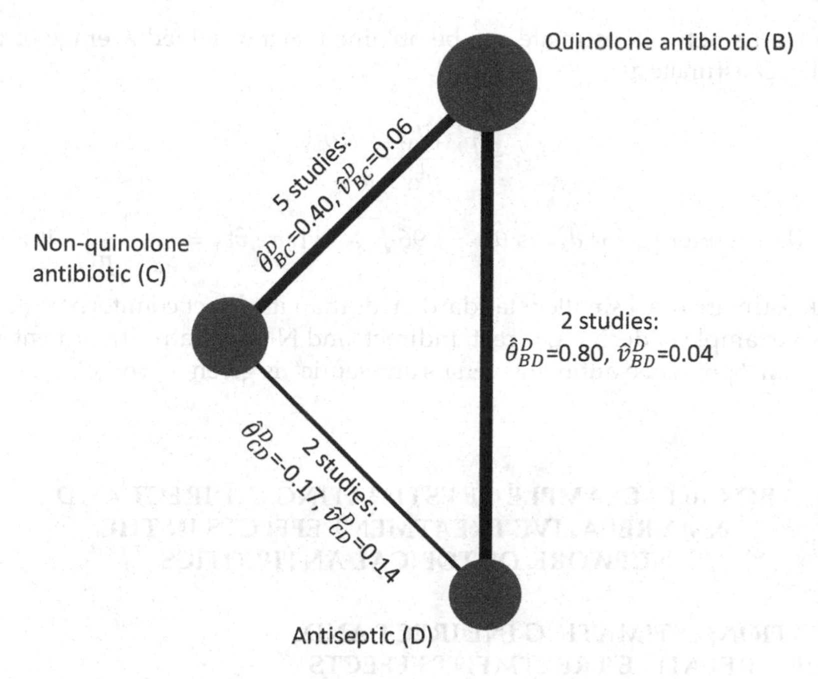

FIGURE 10.2
Reduced network plot of topical antibiotics, omitting treatment A and three-arm studies, and showing direct estimates of treatment effects ($\hat{\theta}^D_{BC}$ etc.) and their variances ($\hat{v}^D_{BC}$ etc.) under common-effect models.

10.3 The Concept of Indirect and Network Comparison

10.3.1 Estimating Indirect and Network Relative Treatment Effects

We let the letter θ denote a generic relative treatment effect: in the example, this would be a log OR. We use X, Y, Z to denote generic treatments in an NMA. Synthesis of studies directly comparing treatments X and Y produces a "direct estimate" $\hat{\theta}^D_{XY}$ (defined as the log OR or other measure on Y minus on X) with variance $\hat{v}^D_{XY}$ and precision $\hat{p}^D_{XY} = 1/\hat{v}^D_{XY}$; the superscript *D* refers to the use of direct evidence and it holds that $\hat{\theta}^D_{XY} = -\hat{\theta}^D_{YX}$. We defer until Section 10.5 details of the statistical methods used to estimate this and other quantities. Indirect evidence refers to the derivation of a relative treatment effect between X and Y through one or more intermediate comparator treatments that form an indirect path between X and Y. An "indirect estimate" for "X versus Y" is denoted as $\hat{\theta}^I_{XY}$ (where superscript *I* refers to the use of indirect evidence) and is derived as the sum of the direct estimates along the indirect path between X and Y. The variance of $\hat{\theta}^I_{XY}$ is estimated as the sum of the variances of the direct estimates involved in the indirect path and its precision is $\hat{p}^I_{XY} = 1/\hat{v}^I_{XY}$.

Synthesis of direct and indirect evidence using NMA instead of multiple pairwise meta-analyses provides "NMA relative treatment effects" $\hat{\theta}^N_{XY}$ (where the superscript *N* refers to NMA). Under certain simplifications, noted below, direct and indirect estimates are independent, so a network estimate can be obtained as a weighted average of the direct and the indirect estimate as

$$\hat{\theta}_{XY}^{N} = \frac{\hat{p}_{XY}^{I}\hat{\theta}_{XY}^{I} + \hat{p}_{XY}^{D}\hat{\theta}_{XY}^{D}}{\hat{p}_{XY}^{I} + \hat{p}_{XY}^{D}}.$$

A 95% confidence interval for $\hat{\theta}_{XY}^{N}$ is $\hat{\theta}_{XY}^{N} \pm 1.96\sqrt{\hat{v}_{XY}^{N}}$ where $\hat{v}_{XY}^{N} = \frac{1}{\hat{p}_{XY}^{I} + \hat{p}_{XY}^{D}}$. It follows that the network estimate has a smaller standard error than its direct counterpart (Caldwell et al., 2005). An example of deriving direct, indirect, and NMA relative treatment effects for the comparison "quinolone antibiotic versus antiseptic" is given in Box 10.1.

BOX 10.1 EXAMPLE OF ESTIMATING INDIRECT AND NMA RELATIVE TREATMENT EFFECTS IN THE NETWORK OF TOPICAL ANTIBIOTICS

APPLICATION: ESTIMATING INDIRECT AND NETWORK RELATIVE TREATMENT EFFECTS

Consider estimating the comparison, as a log odds ratio, of quinolone antibiotic (B) with antiseptic (D) using the two-arm trials shown in Figure 10.2 under a common-effect model. Figure 10.2 shows that the direct evidence estimates the comparison as $\hat{\theta}_{BD}^{D} = 0.80$ with variance $\hat{v}_{BD}^{D} = 0.04$. There is also indirect evidence through non-quinolone antibiotic (C). The indirect estimate for this path is $\hat{\theta}_{BD}^{I} = \hat{\theta}_{BC}^{D} + \hat{\theta}_{CD}^{D} = 0.40 + -0.17 = 0.23$. The variance of $\hat{\theta}_{BD}^{I}$ is estimated as $\hat{v}_{BD}^{I} = \hat{v}_{BC}^{D} + \hat{v}_{CD}^{D} = 0.06 + 0.14 = 0.20$. The NMA combines the direct and indirect evidence to give a network estimate $\hat{\theta}_{BD}^{N} = \frac{0.80/0.04 + 0.23/0.20}{1/0.04 + 1/0.20} = 0.71$ with variance $\hat{v}_{BD}^{N} = \frac{1}{1/0.04 + 1/0.20} = 0.03$. Note that the direct evidence receives about $0.20/0.04 = 5$ times more weight than the indirect evidence.

NMA relative treatment effects are always weighted averages of direct and indirect evidence, but the simplifications required to make the above weights exactly correct are (1) there is only one indirect path between X and Y; (2) each study compares only two treatments (two-arm studies); and (3) heterogeneity in the relative treatment effects has fixed and known variance. These are usually unrealistic. Networks are typically complex structures and we need a general process that can estimate all pairwise comparisons in the network taking into account all possible routes of indirect evidence. Moreover, multi-arm studies (studies with three or more arms) yield at least two relative treatment effects whose dependency needs to be accounted for when estimating the direct treatment effects. Finally, the model needs to estimate heterogeneity. As a result, the equation above is inadequate for synthesis of direct and indirect evidence, and in general the alternative methods introduced in Section 10.5 are needed.

10.3.2 Allowing for Heterogeneity in Estimating Indirect and Network Relative Treatment Effects

To allow for heterogeneity, estimation of direct relative treatment effects (e.g., $\hat{\theta}^D_{XY}$) is undertaken in a random-effects model, under which the parameters θ_{XY} are interpreted as the mean of a distribution of relative treatment effects of "X versus Y" across studies comparing X with Y. The variance of this heterogeneity distribution is written τ^2_{XY}. Across a network with many treatments, we could assume either different comparison-specific heterogeneity parameters (e.g., $\tau^2_{XY}, \tau^2_{XZ}, \tau^2_{YZ}$) or a single common parameter τ^2. The latter assumption implies that the variability in the relative treatment effects is independent of the comparison being made. Heterogeneity can be estimated using a variety of frequentist methods (Jackson et al., 2012) or with informative or vague priors in a Bayesian setting (Lu and Ades, 2004).

These different assumptions and estimation methods lead to different estimates of heterogeneity: the question then arises as to which assumption and method to use. Assuming different heterogeneity parameters requires that each treatment comparison has been studied in a sufficiently large number of studies to enable estimation of heterogeneity. Thus, in practice, a common heterogeneity is frequently assumed, even though it has been empirically shown that heterogeneity depends on the subject matter of the treatment comparison (Turner et al., 2012; Lu and Ades, 2009). The various estimation methods are discussed in Section 10.5.

10.4 Assumptions Underlying Indirect Comparisons and Network Meta-Analysis

We express the assumptions in a different way from much existing literature. The assumption underlying the computation of any indirect estimate $\hat{\theta}^I_{XY}$ is the transitivity assumption: that there is an underlying true relative treatment effect θ_{XY} which applies to all studies regardless of the treatments compared. A more general case, which we do not consider further, arises if any systematic differences between the θ_{XY} can be explained by observed treatment effects. The unique set of treatments compared in a study is termed here the "design" and is denoted $d = 1,\ldots,D$ (although "design" is a much broader concept). Let θ^d_{XY} be the underlying relative treatment effect comparing X with Y in design d, defined even if design d did not include these treatments. Heterogeneity is variation in relative treatment effects between studies of the same design; in models with heterogeneity, θ^d_{XY} is the mean relative treatment effect across studies of design d. The transitivity assumption is

$$\theta^d_{XY} = \theta^{d'}_{XY} \quad \forall d, d', X, Y \tag{10.1}$$

which means that there is an underlying true relative treatment effect $\theta_{XY} := \theta^d_{XY} = \theta^{d'}_{XY}$ that is the same in any pairwise comparison of any design, irrespective of whether it is observed or not. The equation $\theta^d_{XY} = \theta^d_{XZ} - \theta^d_{YZ}$ always holds by the definition of relative treatment effects, whether X, Y, and Z are observed or hypothetical treatments in design d. Equation 10.1 implies that parameters θ_{XY}, θ_{XZ} and θ_{YZ} have a *transitive relationship*, meaning that

$$\theta_{XY} = \theta_{XZ} - \theta_{YZ}. \tag{10.2}$$

Equations 10.1 and 10.2 can be violated due to variation across studies of interventions, populations, and outcomes. Thus, evaluation of the plausibility of transitivity relies on clinical and epidemiological judgments. For instance, non-transitivity may occur when the nature of a treatment varies according to the alternative it is compared against; for example, when X differs systematically between "X versus Y" and "X versus Z" studies (Salanti, 2012; Jansen et al., 2011; Hoaglin et al., 2011). Important insights about transitivity can be gained from examining the distribution of effect modifiers across study populations. For instance, consider a healthcare condition where weight is an effect modifier. If "X versus Z" studies include mainly normal-weight participants and "X versus Y" studies include mainly obese participants, then the indirect comparison "Y versus Z" may not be valid.

If equations 10.1 and 10.2 hold, the *consistency equation*

$$\theta_{XY}^{XY} = \theta_{XZ}^{XZ} - \theta_{YZ}^{YZ} \tag{10.3}$$

is implied. Equation 10.3 can be statistically evaluated as the parameters θ_{XY}^{XY} and so on are estimable. Statistical tests to detect inconsistency are discussed in Section 10.6. Note however, that consistency does not ensure transitivity. In this chapter, this basic assumption is called transitivity, where much of the literature would call it consistency.

10.5 Statistical Models for Network Meta-Analysis

10.5.1 Notation

Let the entire evidence base consist of I studies comparing a set $\Omega = \{X, Y, Z, \ldots\}$ of J treatments. Each design $d = 1, \ldots D$ corresponds to a set of J_d treatments $\Omega_d \subseteq \Omega$ and the number of studies of design d is I_d. Assuming transitivity is essential for performing NMA as it allows the meta-analysis model to borrow strength across comparisons by reducing the number of parameters. In a network with J treatments there are $\binom{J}{2}$ relative treatment effects, but under transitivity, all are linear combinations of a set of $J-1$ relative treatment effects which we collect in $\boldsymbol{\theta}$, termed basic parameters. $\boldsymbol{\theta}$ is the target of inference and is often chosen to include the effects of all treatments relative to a reference (arbitrarily set as treatment X here). For example, the vector of basic parameters in a network of three interventions X, Y, and Z could be $\boldsymbol{\theta} = (\theta_{XY}, \theta_{XZ})$; other choices are valid as long as there are $J-1$ independent basic parameters (Lu et al., 2011; Lu and Ades, 2006; van Valkenhoef et al., 2016).

Data may come in two forms. Contrast-level data are estimated treatment contrasts arising in a two-stage analysis (Chapter 4). Let $y_{i,d,XY}$ be the observed relative treatment effect for "X versus Y" in study i of design d, with associated standard error $s_{i,d,XY}$. Arm-level summary data are sufficient statistics per treatment group, as in Chapter 5. The arm-level statistics are denoted as $y_{i,d,j}$ and can be arm-level mean, number of events, and so on, with associated statistics such as sample size, where the arms included in study i are denoted by $j \in \Omega_d$. The suitable likelihoods for fitting NMA using contrast-level and arm-level type of data are described in Section 10.5.5.

10.5.2 Design-Specific Meta-Analysis

Before building a full NMA, we first consider the special case where all studies to be synthesized involve the same set of treatments or, in other words, have the same design d. Under the common-effect assumption, the observed relative treatment effect in any given study differs from the overall mean by the sampling error ($\varepsilon_{i,d,XY}$). Under the random-effects assumption, there is a second source of variation between underlying true relative treatment effects ($\delta_{i,d,XY}$).

Using contrast-level data, we model the observed relative treatment effects under a random-effects model by

$$y_{i,d,XY} = \theta_{XY} + \delta_{i,d,XY} + \varepsilon_{i,d,XY}. \quad (10.4)$$

When study i is a multi-arm study ($J_i > 2$), the correlation between relative treatment effects within the study needs to be taken into account. A study with J_i treatments produces $\binom{J_i}{2}$ relative treatment effects, but only $J_i - 1$ relative treatment effects need to be estimated: their linear combinations give the remaining $\binom{J_i}{2} - J_i + 1$ relative treatment effects. The $J_i - 1$ observed $y_{i,d,XY}$ can be collected in a vector $\boldsymbol{y}_{i,d}$. The vectors of within-study errors $\boldsymbol{\varepsilon}_{i,d}$ and random effects $\boldsymbol{\delta}_{i,d}$ are assumed to follow multivariate normal distributions $\boldsymbol{\varepsilon}_{i,d} \sim N(0, \boldsymbol{S}_{i,d})$ and $\boldsymbol{\delta}_{i,d} \sim N(0, \boldsymbol{T}_d)$ with $\boldsymbol{S}_{i,d}$ and $\boldsymbol{T}_d$ being $(J_i - 1) \times (J_i - 1)$ within-study and between-study variance-covariance matrices.

For instance, suppose design d includes arms X, Y, and Z. A study i with this design contributes two independent relative treatment effects $y_{i,d,XY}$ and $y_{i,d,XZ}$ which we collect in a vector $\boldsymbol{y}_{i,d} = \left(y_{i,d,XY}, y_{i,d,XZ}\right)'$ of length $J_i - 1 = 2$. The corresponding variances and covariances between relative treatment effects are given in a 2×2 matrix $\boldsymbol{S}_{i,d} = \begin{pmatrix} s^2_{i,d,XY} & s_{i,d,XY,XZ} \\ s_{i,d,XY,XZ} & s^2_{i,d,XZ} \end{pmatrix}$ and are assumed known, while the variance-covariance matrix of the random effects $\boldsymbol{\delta}_{i,d}$ is $\boldsymbol{T}_d = \begin{pmatrix} \tau^2_{XY} & \tau_{XY,XZ} \\ \tau_{XY,XZ} & \tau^2_{XZ} \end{pmatrix}$ and must be estimated. The within-study covariances $s_{i,d,XY,XZ}$ are computed from the data within each study. If, as in Section 10.3.2, we assume that the between-study variances are common across comparisons, then $\tau^2_{XY} = \tau^2_{XZ} = \tau^2$ (Lumley, 2002; Salanti et al., 2008). Because the between-study variance of "Y versus Z" comparisons is $\tau^2_{XY} - 2\tau_{XY,XZ} + \tau^2_{XZ}$, this further implies that $\tau_{XY,XZ} = \tau^2/2$ and consequently $T_d = \begin{pmatrix} \tau^2 & \frac{\tau^2}{2} \\ \frac{\tau^2}{2} & \tau^2 \end{pmatrix}$. The assumption of common heterogeneity allows better estimation of heterogeneity for comparisons with few studies and will be made for all the NMA models for the rest of the chapter.

10.5.3 Overview of Statistical Models for Network Meta-Analysis under Transitivity

Several models have been developed for conducting NMA and have been applied in practice. First, Lumley derived NMA relative treatment effects from a modified meta-regression model (Lumley, 2002), where the study-level covariates are the treatments being compared (Chapter 7). Second, NMA can be seen as a multivariate meta-analysis model for contrast-level data, where different treatment comparisons in the network can be thought of as different "outcomes" (White et al., 2012). Third, NMA can be fitted as a hierarchical model considering two levels of estimation for the random-effects cases where the basic parameters are estimated under the constraint of the transitivity relationships (Higgins and Whitehead, 1996; Lu and Ades, 2004). Because of its flexibility, the hierarchical model is a very popular choice when conducting NMA (Petropoulou et al., 2016; Nikolakopoulou et al., 2014a; Song et al., 2009; Sobieraj et al., 2013; Coleman et al., 2012).

Decisions regarding the choice of NMA model to be employed are guided by software availability and the network structure. Nikolakopoulou et al. found that nearly one-third of star networks (networks without loops, where treatments are only compared with a common reference but not between themselves) synthesized evidence using simple indirect comparisons (as in Section 10.3), whereas only one in ten networks that contained at least one loop chose this method (Nikolakopoulou et al., 2014a). The hierarchical model is most commonly implemented in a Bayesian framework using arm-level summary data, whereas meta-regression is conveniently fitted using frequentist software using contrast-level summary data (and therefore making a normal approximation to the within-study likelihood, as in Chapter 4). Fitting NMA using standard meta-regression software routines is only valid in the absence of multi-arm studies (Lu et al., 2011; Rücker and Schwarzer, 2014; Simmonds and Higgins, 2016; Lu and Ades, 2004; Lumley, 2002), which frequently appear in the medical literature: Chan and Altman (Chan and Altman, 2005) found that approximately one-quarter of randomized trials have multiple arms and Nikolakopoulou et al. found that networks include a median of two multi-arm studies (Nikolakopoulou et al., 2014a). For this reason, we do not discuss the Lumley model further. The other two models are defined in the next two subsections.

10.5.4 Network Meta-Analysis as a Multivariate Meta-Analysis Model

Relative treatment effects against a common comparator can be modeled by multivariate meta-analysis where different "outcomes" are different basic parameters (White et al., 2012; White, 2011; Mavridis and Salanti, 2013). For example, when all basic parameters are the relative treatment effects of any treatment against a reference treatment X, an "X versus Y versus Z" study reports on two "outcomes", "X versus Y" and "X versus Z". Data can be "augmented" if needed so that each study includes the reference treatment as long as the augmenting data (in studies that do not include the common comparator) carry very little information. The validity of the augmentation process relies on the transitivity assumption (Salanti, 2012; Lu and Ades, 2009). This does not require that the choice of arms in a study is made in a completely random way, but that the selection of treatments to be included is independent of their effectiveness compared with those included (Jansen et al., 2012).

Within each design, relative treatment effects against the common comparator can be modeled by multivariate meta-analysis as in Section 10.5.2 (White et al., 2012; White, 2011; Mavridis and Salanti, 2013). We can then assemble the design-specific models into a multivariate meta-analysis model with $J-1$ parameters (here treatment comparisons). All the observed effects $y_{i,d,XY}$ are collected in a column vector $\boldsymbol{y}$ of length $L=\sum_{i=1}^{I}\left(J_i^*-1\right)$ where J_i^* refers to the arms in study i after potential augmentation. The NMA model is written as

$$\boldsymbol{y} = \boldsymbol{X\theta} + \boldsymbol{\delta} + \boldsymbol{\varepsilon} \tag{10.5}$$

where vectors $\boldsymbol{\varepsilon} \sim N(0,\boldsymbol{S})$ and $\boldsymbol{\delta} \sim N(0,\boldsymbol{T})$ include the random errors and the random effects and are of the same dimensions as $\boldsymbol{y}$. Matrices $\boldsymbol{S}$ and $\boldsymbol{T}$ are block-diagonal variance-covariance matrices of dimensions $L \times L$. The design matrix $\boldsymbol{X}$ is a $L \times (J-1)$ matrix, in which each study contributes $\left(J_i^* - 1\right)$ rows and each column refers to a basic parameter. The design matrix has 1 where the observed effect in $\boldsymbol{y}$ estimates the basic parameter in $\boldsymbol{\theta}$, and 0 elsewhere; this simplicity is an advantage of the augmentation approach. In Box 10.2, we exemplify the model of equation 10.5 for the topical antibiotics example.

BOX 10.2 EXAMPLE OF NETWORK META-ANALYSIS AS A MULTIVARIATE MODEL IN THE NETWORK OF TOPICAL ANTIBIOTICS

APPLICATION: NETWORK META-ANALYSIS AS A MULTIVARIATE META-ANALYSIS MODEL

Let designs $d = 1,\ldots,6$ represent AB, BC, BD, CD, ABD, and BCD, respectively. Let A be the reference treatment. Note that design BC contributes two estimates (AB and AC) after augmentation. We focus on studies 1 and 2 (as numbered in Table 10.1), of design $d=1$ (AB) and $d=5$ (ABD), respectively. Equation 10.5 takes the form

$$\begin{pmatrix} y_{1,1,AB} \\ y_{2,5,AB} \\ y_{2,5,AD} \end{pmatrix} = \begin{pmatrix} 1 & 0 \\ 1 & 0 \\ 0 & 1 \end{pmatrix} \begin{pmatrix} \theta_{AB} \\ \theta_{AD} \end{pmatrix} + \begin{pmatrix} \varepsilon_{1,1,AB} \\ \varepsilon_{2,5,AB} \\ \varepsilon_{2,5,AD} \end{pmatrix} + \begin{pmatrix} \delta_{1,1,AB} \\ \delta_{2,5,AB} \\ \delta_{2,5,AD} \end{pmatrix}$$

assuming transitivity, with

$$\begin{pmatrix} \varepsilon_{1,1,AB} \\ \varepsilon_{2,5,AB} \\ \varepsilon_{2,5,AD} \end{pmatrix} \sim N\left\{0, \begin{pmatrix} s_{1,1,AB}^2 & 0 & 0 \\ 0 & s_{2,5,AB}^2 & s_{2,5,AB,AD} \\ 0 & s_{2,5,AB,AD} & s_{2,5,AD}^2 \end{pmatrix}\right\}$$

and assuming a common between-study variance

$$\begin{pmatrix} \delta_{1,1,AB} \\ \delta_{2,5,AB} \\ \delta_{2,5,AD} \end{pmatrix} \sim N\left\{0, \begin{pmatrix} \tau^2 & 0 & 0 \\ 0 & \tau^2 & \frac{\tau^2}{2} \\ 0 & \frac{\tau^2}{2} & \tau^2 \end{pmatrix}\right\}.$$

Relative treatment effects from fitting the model to all studies using the command `network meta consistency` in Stata are given in the lower triangle of Table 10.2.

Methods to estimate the variance-covariance matrix of the random effects $\boldsymbol{T}$ include likelihood based methods and extensions of the DerSimonian and Laird method (see Chapter 9) (Lu et al., 2011; van Houwelingen et al., 2002), although estimation may be difficult when the data are sparse. For the special case of a common-effect model, $\boldsymbol{T}=0$.

Using standard multivariate meta-regression (extending ideas in Chapter 9), we first estimate $\boldsymbol{T}$, and then estimate the vector of basic parameters as

$$\hat{\theta} = \left(X' \left(S + \hat{T} \right)^{-1} X \right)^{-1} X' \left(S + \hat{T} \right)^{-1} y$$

and all NMA relative treatment effects $\hat{\theta}^N$ are derived as linear combinations of $\hat{\theta}$ imposing the transitivity relationships. The variance-covariance matrix ignoring uncertainty in estimation of $\boldsymbol{T}$ is

$$\mathbf{var}\left(\hat{\theta}\right) = \left(X' \left(S + \hat{T} \right)^{-1} X \right)^{-1}.$$

10.5.5 Network Meta-Analysis as a Hierarchical Model

If we instead have arm-level data, we can use the exact likelihood of the data instead of the approximate within-study normal likelihood, as in Chapter 5. The exact likelihood is defined by the probability distribution of the arm-level data $y_{i,d,j}$ given its mean $\mu_{i,d,j}$ and a link function g relating the mean to network parameters. For a study i comparing reference group X with other treatments Y, we have:

$$g\left(\mu_{i,d,X}\right) = u_i$$

$$g\left(\mu_{i,d,Y}\right) = u_i + \theta_{XY} + \delta_{i,d,XY}$$

where u_i are study-specific effects. The basic parameter θ_{XY} and the random effects $\delta_{i,d,XY}$ are as in previous sections.

Several link functions, such as the logit and the identity, can be used depending on the nature of the data, as described in Chapter 5 (Dias et al., 2013a). For example, if the outcome is dichotomous, we assume a binomial likelihood for the number of events:

$$y_{i,d,j} \sim \text{Bin}\left(\mu_{i,d,j}, n_{i,d,j}\right)$$

where $y_{i,d,j}$ is the number of events, $\mu_{i,d,j}$ is the probability of an event and $n_{i,d,j}$ is the total number of participants in arm j of study i with design d. Then, we use the model above with $g(\cdot) = \text{logit}(\cdot)$.

Observations in different arms of the same study are independent and hence no adjustment is required in the likelihood. However, accounting for correlations between random effects is still required. The model is usually implemented in a Bayesian framework; software options to fit this model are discussed in Section 10.9. Details on viewing NMA as a (one-stage or two-stage) meta-regression model are available (Lu et al., 2011; Salanti et al., 2008).

10.5.6 Network Meta-Regression Models

NMA models can be extended to include covariates that may act as potential effect modifiers (Chaimani and Salanti, 2012; Salanti et al., 2010; Nixon et al., 2007). For example, equation 10.4 is extended to

$$y_{i,d,XY} = x_i \beta_{XY} + \theta_{XY} + \delta_{i,d,XY} + \varepsilon_{i,d,XY}$$

where x_i is a study-level covariate, β_{XY} is the change in the comparison "X versus Y" per one-unit change in x_i, and θ_{XY} now represents the "X versus Y" comparison in studies with $x_i=0$. Covariates can be dichotomous, categorical, or continuous and can include characteristics such as study quality, baseline severity, and risk of bias (Chapter 12).

Different assumptions can be employed regarding the regression coefficients β_{XY}. The weakest assumption is to fit fixed and independent regression coefficients for all comparisons. An assumption often considered is that of transitive comparison-specific coefficients (Chaimani, 2014; Cooper et al., 2009). Under this assumption, regression coefficients are forced to satisfy transitivity relationships, $\beta_{YZ}=\beta_{XZ}-\beta_{XY}$ (Chaimani, 2014; Cooper et al., 2009). One may further assume a common regression coefficient $\beta_{XZ}=\beta$ for all Y if it is considered plausible that the examined characteristic acts in the same way in all treatment contrasts with the reference treatment. However, this assigns a special role to the reference treatment X, since it implies $\beta_{YZ}=0$ for $Y,Z\neq X$. Relaxing this approach, one may assume exchangeable regression coefficients $\beta_{XY} \sim N(\xi,\eta^2)$ for all Y, where β_{XY} share a common distribution with mean ξ and between comparison variance η^2. This also assigns a special role to the reference treatment X, since it implies $\beta_{YZ} \sim N(0,2\eta^2)$ for $Y,Z\neq X$. According to the context, such an assumption might or might not be plausible. If, for example, placebo controlled trials are prone to publication bias, as has been argued for antidepressants (Turner et al., 2008), assigning a special role to the reference treatment when modeling bias might be a sensible decision. A model handling the treatments symmetrically, not previously described, is $\beta_{XY} \sim N(0,\eta^2)$ for all Y, with $\text{corr}(\beta_{XY},\beta_{XZ})=1/2$ where X is the reference treatment.

Comparing the magnitude of the estimated regression coefficients and the between-study variance between the unadjusted and the adjusted models can give insight into the potential impact of the examined study characteristics (Salanti et al., 2010; Nixon et al., 2007; Cooper et al., 2009; Salanti et al., 2009; Dias et al., 2013b). Donegan et al. propose models to assess the transitivity assumptions by testing the consistency assumption of the regression coefficients along with testing the consistency assumption of equation 10.3 (Donegan et al., 2018). They demonstrate how to assess agreement between direct and indirect evidence depending on the covariate value at which it is assessed (Donegan et al., 2017).

Like meta-regression in pairwise meta-analysis (see Chapter 7), network meta-regression can suffer from inflated false positive rates when multiple covariates are examined, and aggregation bias when patient-level covariates are summarized at study level. Thus, it is recommended that potential effect modifiers are specified *a priori* with a clear scientific rationale for exploring their impact on NMA relative treatment effects, and that individual patient data are sought if possible, to examine the role of patient-level characteristics.

10.6 Statistical Models and Tests for Inconsistency

As discussed in Section 10.4, the assumption of transitivity is central to NMA. While transitivity is not fully testable, the implied property of consistency can be evaluated statistically unless the data form a star network. In particular, equation 10.2 refers to the relationship between the true values of the parameters; this relationship underlies NMA models but cannot be tested. Thus, tests for inconsistency focus on testing equation 10.3 (consistency assumption). Below we distinguish two categories of statistical tests for inconsistency, local and global, and describe a measure of inconsistency that quantifies the amount of inconsistency in the data (Jackson et al., 2014). The tests and models described below are applied to the topical antibiotics example in Box 10.3.

BOX 10.3 EXAMPLE OF STATISTICAL MODELS AND TESTS FOR INCONSISTENCY IN THE NETWORK OF TOPICAL ANTIBIOTICS

APPLICATION: STATISTICAL MODELS AND TESTS FOR INCONSISTENCY

Local Tests and Models for Inconsistency

We illustrate the loop-specific approach for the comparison between B and D using the simplified data in Figure 10.2 and a common-effect analysis. As in Box 10.1, the direct and indirect evidence give log OR estimates $\hat{\theta}_{BD}^{D} = 0.80$ and $\hat{\theta}_{BD}^{I} = 0.23$ with variances $\hat{v}_{BD}^{D} = 0.04$ and $\hat{v}_{BD}^{I} = 0.20$. The inconsistency factor is $\widehat{IF}_{BCD} = 0.80 - 0.23 = 0.57$ with variance $0.04 + 0.20 = 0.24$. This means that the OR for BD estimated directly is on average $\exp(0.57) = 1.77$ times larger than the OR estimated indirectly. The calculated 95% confidence interval for this estimate (0.67 to 4.67) indicates no evidence of difference between direct and indirect evidence in the BCD loop. The same result can be obtained from the "separating indirect and direct evidence" approach, fitted using the `network sidesplit` command in Stata.

We now return to the full network in Figure 10.1 and a random-effects analysis. Direct calculations of indirect evidence are not now possible, so we use the "separating indirect and direct evidence" approach (Table 10.3). The comparisons "A versus B" and "A versus D" are associated with statistically significant inconsistency. However, the "A versus D" comparison is only informed by ABD studies; thus, not

TABLE 10.3

Evaluation of Inconsistency Using the "Separating Indirect and Direct Evidence" Method for the Network of Topical Antibiotics Obtained Using Stata

	Odds ratio		
Comparisons	Direct evidence	Indirect evidence	Ratio of odds ratios, direct/indirect
No treatment (A) versus quinolone antibiotic (B)	0.11 (0.03–0.36)	2.15 (0.14–31.95)	**0.05 (0.00–0.94)**
No treatment (A) versus antiseptic (D)	1.42 (0.48–4.21)	0.08 (0.02–0.41)	**17.2 (2.46–120)**
Quinolone antibiotic (B) versus non-quinolone antibiotic (C)	1.46 (0.73–2.95)	4.02 (0.82–19.65)	0.36 (0.06–2.06)
Quinolone antibiotic (B) versus antiseptic (D)	3.42 (1.48–7.89)	2.38 (0.52–10.84)	1.44 (0.25–8.09)
Non-Quinolone antibiotic (C) versus antiseptic (D)	1.69 (0.63–4.59)	2.02 (0.53–7.69)	0.84 (0.16–4.42)

Bold results indicate statistically significant evidence of inconsistency. Numbers are odds ratios or their ratios and 95% confidence intervals.

accounting for multi-arm studies poses a barrier to the interpretation of the loop-specific and the separating indirect and direct evidence approaches.

Global Tests and Models for Inconsistency

We fitted the DBT and the LA inconsistency models for the worked example in Stata as presented in Section 10.6.2.1 using the command `network meta inconsistency`. Designs are AB, BC, BD, CD, ABD, and BCD (Table 10.1) and it can be shown that the DBT model has five *IFs*. The global test results in $W = 13.15$ with five degrees of freedom corresponding to $p = 0.02$ indicating statistically significant inconsistency in the network. The LA model has two inconsistency factors, one for each loop, and the test for inconsistency is 7.20 with two degrees of freedom ($p = 0.03$). Figure 10.3 displays, for each comparison, the study-specific estimates, results under the consistency model ("all studies", light gray diamonds) and results under the DBT inconsistency model ("all B C" etc., dark gray diamonds). It helps us to understand the heterogeneity and inconsistency. From the "C vs. B" part of the figure, we see that one outlying study of B versus C (Tutkun, 1995) causes heterogeneity. From the "D vs. C" part, we see design inconsistency between BCD and CD studies. From the "D vs. A" part, we see possible loop inconsistency, since the direct evidence about A versus D (dark gray diamond) is markedly different from the overall network evidence (light gray diamond).

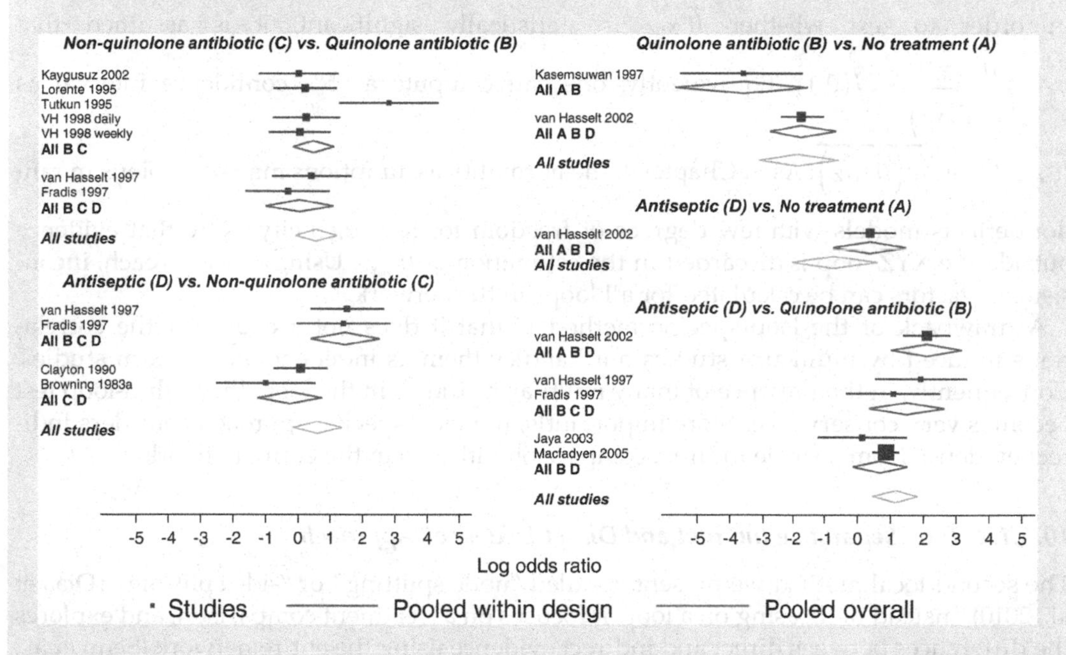

FIGURE 10.3
Design-specific and network meta-analysis ORs for all comparisons in the network of topical antibiotics. The method used to obtain the results is a multivariate meta-analysis and the graph is produced in Stata using the `network forest` command.

10.6.1 Local Tests and Models for Inconsistency

Local tests for inconsistency measure and evaluate the statistical agreement between different sources of evidence in specific parts of the network. We present below two frequently applied approaches, one not model-based and one model-based. Other methods to present inconsistency in the network have been described (Krahn et al., 2013, 2014; Dias et al., 2010).

10.6.1.1 The Loop-Specific Approach

The simplest approach is the loop-specific approach (Bucher et al., 1997). The idea originates from inconsistency being a property of a loop of evidence where direct and indirect estimates can be contrasted (Higgins et al., 2012; Bucher et al., 1997). In an XYZ loop, the inconsistency factor IF_{XYZ} is defined as the disagreement between estimates of θ_{XY} using indirect versus direct evidence. In particular,

$$\widehat{IF}_{XYZ} = \hat{\theta}^{I}_{XY} - \hat{\theta}^{D}_{XY}$$

with variance

$$\hat{v}\left(\widehat{IF}_{XYZ}\right) = \hat{v}^{I}_{XY} + \hat{v}^{D}_{XY}.$$

In order to test whether $\widehat{IF}_{XYZ}$ is statistically significant, it is assumed that $z = \frac{\widehat{IF}_{XYZ}}{\sqrt{\hat{v}\left(\widehat{IF}_{XYZ}\right)}} \sim N(0,1)$. Equivalently, one can compute a 95% confidence interval as $\widehat{IF}_{XYZ} \pm 1.96\sqrt{\hat{v}\left(\widehat{IF}_{XYZ}\right)}$. As in Chapter 4, the normality assumptions may be dubious in random-effects models with few degrees of freedom for heterogeneity. Note that evidence outside the XYZ loop is discarded in the estimation of IF_{XYZ}. Using this approach, inconsistency factors can be calculated for all loops in the network.

A drawback of the loop-specific method is that it does not account for the correlations induced by multi-arm studies but handles them as independent two-arm studies. Consequently, in the presence of many multi-arm studies in the loop, the within-loop test becomes very conservative. More importantly, the loop-specific approach considers indirect evidence from a single indirect comparison rather than the entire network.

10.6.1.2 The "Separating Indirect and Direct Evidence" Approach

The second local method we present is called "node splitting" or "side splitting" (Dias et al., 2010). Instead of focusing on a loop, it focuses on a treatment comparison, and explores the differences between direct and indirect evidence using the entire network (Senn et al., 2013; Dias et al., 2010). Consider, for example, a comparison "X versus Y". An inconsistency model is constructed including different parameters for this comparison in studies containing both X and Y (the direct comparison θ^{D}_{XY}) and in other studies (the indirect comparison θ^{I}_{XY}). The inconsistency factor $\widehat{IF}_{XY}$ is then estimated as the difference $\hat{\theta}^{D}_{XY} - \hat{\theta}^{I}_{XY}$; as its sign is often not of interest some authors report its absolute value. The estimated $\widehat{IF}_{XY}$

and its standard error give a z-test to evaluate the assumption that there is agreement between direct and indirect estimates of the "X versus Y" relative treatment effect. The main difference between this test and the loop-specific approach is that here we derive indirect estimates using the entire network and not only one loop. With multi-arm studies, the method is valid but can be implemented in more than one way (White, 2015).

10.6.2 Global Tests and Models for Inconsistency

Global approaches to inconsistency aim to assess the plausibility of the consistency assumption in the entire network. They use "inconsistency models" that relax the consistency assumption either by removing all transitivity relationships or by allowing extra variability in the NMA model of Section 10.5.4 to account for differences between the various sources of evidence. The consistency and inconsistency models can be compared in a frequentist framework using goodness of fit and trade-off between model fit and parsimony, or in a Bayesian framework using the deviance information criterion (Dias et al., 2013d; Spiegelhalter et al., 2002).

A simple inconsistency model is constructed by removing the transitivity relationships: this is called the *unrelated mean effects model*. It is equivalent to a series of pairwise meta-analyses with common heterogeneity (Dias et al., 2013a). A better fit or reduction of heterogeneity for the unrelated mean effects model compared with a network model assuming consistency suggests that the consistency assumption may not be plausible (Dias et al., 2013a). The presence of multi-arm studies complicates the unrelated mean effects model, since including all contrasts in a multi-arm study uses some data twice, but omitting contrasts is arbitrary. In the following sections, we discuss two inconsistency models that relax the consistency assumption.

10.6.2.1 Global Models for Inconsistency

The consistency assumption in equation 10.3 refers to the notion of "loop consistency" in a loop of evidence; that is a path from X to Y of at least two comparisons for which direct evidence also exists. A different notion of consistency, which arises when a network includes multi-arm studies, is design consistency: similarity of a treatment effect estimated across studies that each include the effect but in different designs (Higgins et al., 2012). Design inconsistency refers to the differences in the underlying relative treatment effects in different designs; for example, it occurs when the relative treatment effect of "X versus Y" is different in "X versus Y" and "X versus Y versus Z" studies. Loop and design inconsistency cannot be formally distinguished; their distinction is intuitive and is primarily made for practical reasons (Higgins et al., 2012).

The *design by treatment (DBT) interaction model* accounts for both loop and design inconsistency (Jackson et al., 2014; Higgins et al., 2012; White et al., 2012; Jackson et al., 2016). The general form for the observed treatment effect for the comparison "X versus Y" using the DBT model is

$$y_{i,d,XY} = \theta_{XY} + IF_{d,XY} + \delta_{i,d,XY} + \varepsilon_{i,d,XY} \tag{10.6}$$

where $IF_{d,XY}$ represents disagreements in the estimation of the relative treatment effect of "X versus Y" obtained in a loop (loop inconsistency) or in different designs (design inconsistency). A model needs to include only a subset of all the potential inconsistency factors $IF_{d,XY}$. For example, a triangular network with four designs XY, XZ, YZ, XYZ has the potential of four conflicts (each two-arm study design with the three-arm study design plus loop inconsistency) but only three inconsistency factors $IF_{d,XY}$ can be included in the model. The total number of degrees of freedom for inconsistency is denoted as b and is derived as the difference between the number of identified fixed parameters in the DBT interaction model and number of identified fixed parameters in the model of Section 10.5.4 (transitivity model), $b = \sum_d (J_d - 1) - (J - 1)$.

The b inconsistency factors $IF_{d,XY}$ can be handled as fixed effects or as random effects. Under the fixed-effects assumption, b linearly independent *IFs* are included in the model as separate parameters; different choices lead to different numerical values of parameters, but the same overall model fit. Under the random effects (or exchangeability) assumption, all possible inconsistency factors are included in the model as drawn from a common distribution

$$IF_{d,XY} \sim N\left(0, \zeta^2\right)$$

where ζ^2 is the inconsistency variance. The choice between fixed and random inconsistency terms has been debated (Jackson et al., 2014; Higgins et al., 2012; White et al., 2012; Lumley, 2002; Lu and Ades, 2006). Higgins et al. argued for the fixed-effects approach as the interpretation as well as the *a priori* likelihood of being zero may differ across inconsistency factors (Higgins et al., 2012). Jackson et al. adopted the random-effects approach to facilitate conceptual understanding of inconsistency as additional variation similar to the notion of heterogeneity (Jackson et al., 2014; Jackson et al., 2016). As the inconsistency factors can be placed in different ways in a network of interventions, the DBT model is useful to summarize the evidence against consistency and (with random inconsistency terms) to describe its overall magnitude, rather than describing in which comparisons it arises.

The *Lu and Ades (LA) model* is a special case of the DBT model that assesses only loop inconsistency (Lu and Ades, 2006). The number of independent loops in a network is identified: for example, in Figure 10.1, loops ABD and BCD are independent and can have separate evidence for inconsistency, but inconsistency around loop ABCD is constrained to equal the sum of inconsistencies around ABD and BCD and is not independent. Each independent loop in the network is used to add a single inconsistency factor to the model. In Figure 10.1, this might assign one inconsistency parameter to the BD studies (differing from AB and AD in the ABD loop) and one inconsistency parameter to the CD studies (differing from BC and BD in the BCD loop). In the absence of multi-arm studies, the LA model is identical to the DBT model. With multi-arm studies, however, the model, and hence its results, depends on the choice of reference treatment in the network and the order in which treatments are considered (Higgins et al., 2012).

10.6.2.2 Global Tests for Inconsistency

We collect all $IF_{d,XY}$ based on the DBT model with fixed IFs in a column vector $\boldsymbol{G}$. In order to assess the presence of inconsistency in the entire network, we assess the null hypothesis H_0: $\boldsymbol{G} = 0$ using a χ^2-test with b degrees of freedom: $W = \boldsymbol{G}'\boldsymbol{Z}^{-1}\boldsymbol{G}$ where $\boldsymbol{Z}$ is the $b \times b$ variance-covariance matrix of $\boldsymbol{G}$. Different parameterizations for the DBT model may result in different $IF_{d,XY}$ values, but the overall W test is the same for all parameterizations (Higgins et al., 2012; White et al., 2012). The respective χ^2-test with f degrees of freedom for evaluating loop inconsistency globally using the LA model is obtained if G contains the f loop-inconsistency factors from the LA model and Z is its $f \times f$ variance-covariance matrix.

The LA model is intuitive and has a straightforward interpretation of the inconsistency factors. However, the problems with multi-arm studies described above mean that different implementations can lead to different conclusions about inconsistency (Higgins et al., 2012). Moreover, the LA model cannot take into account design inconsistency and therefore may not represent all sources of potential disagreement in a network of interventions. Thus, when applying the LA model, variability due to design inconsistency might result in a larger estimate for the between-study variance (Veroniki, 2014). Accounting for design inconsistency, the DBT model has more degrees of freedom and may have lower power compared with the LA model (Jackson et al., 2014). The loss in power might be an undesirable property of the DBT model in particular when inconsistency is genuinely due to loop inconsistency (Higgins et al., 2012).

Several measures have been proposed to distinguish between different sources of variation in NMA. The Q-statistic, used for heterogeneity in meta-analysis, has been extended to test for heterogeneity and inconsistency in NMA (Lu et al., 2011; Krahn et al., 2013).

10.6.2.3 I^2 Measure for Heterogeneity and Inconsistency

A particular challenge in NMA is to capture the trade-off between heterogeneity and inconsistency, namely the variation in underlying true relative treatment effects between studies within a design (heterogeneity) and across designs (inconsistency). These two notions are interwoven, and both the magnitude and estimation of heterogeneity affect inferences for inconsistency.

The I^2 measure of the impact of heterogeneity on the total variance (Chapter 4) has been extended to measure the impact of heterogeneity, inconsistency, and both in the total variability of NMA relative treatment effects (Jackson et al., 2014, 2016). Let $\left|\mathbf{var}(\hat{\theta}_{CT})\right|$ be the determinant of the variance-covariance matrix of $\hat{\theta}$ under the common-effect transitivity model (described by equation 10.4 assuming $\tau^2=0$) and $\left|\mathbf{var}(\hat{\theta}_{RT})\right|$ the respective determinant under the random-effects transitivity model (described by equation 10.4 with $\tau^2 \neq 0$). Then

$$I^2_{\text{het}} = \frac{R^2_{\text{het}} - 1}{R^2_{\text{het}}}$$

where $R^2_{\text{het}} = \left(\frac{\left|\mathbf{var}\left(\hat{\theta}_{RT}\right)\right|}{\left|\mathbf{var}\left(\hat{\theta}_{CT}\right)\right|} \right)^{\frac{1}{2(J-1)}}$ (Jackson et al., 2014). I^2_{het} measures the variability due to heterogeneity rather than sampling error. In the same spirit, the determinant of the variance-covariance matrix under the random-effects inconsistency model (described by equation 10.6) is denoted as $\left|\mathbf{var}(\hat{\theta}_{RI})\right|$ and the ratio

$$I^2_{\text{inc}} = \frac{R^2_{\text{inc}} - 1}{R^2_{\text{inc}}}$$

with $R^2_{\text{inc}} = \left(\frac{\left| \textbf{var}\left(\hat{\theta}_{RI}\right) \right|}{\left| \textbf{var}\left(\hat{\theta}_{RT}\right) \right|} \right)^{\frac{1}{2(J-1)}}$ measures the variability due to inconsistency rather than heterogeneity or sampling error (Jackson et al., 2014). A measure of both heterogeneity and inconsistency can be expressed by the ratio

$$I^2_{\text{inc+het}} = \frac{R^2_{\text{inc+het}} - 1}{R^2_{\text{inc+het}}}$$

where $R^2_{\text{inc+het}} = \left(\frac{\left| \textbf{var}\left(\hat{\theta}_{RI}\right) \right|}{\left| \textbf{var}\left(\hat{\theta}_{CT}\right) \right|} \right)^{\frac{1}{2(J-1)}}$ and expresses the variability due to inconsistency and heterogeneity on top of random error.

10.6.3 Empirical Evidence on Inconsistency, Simulations, and Interpretation of Test Results

Empirical studies published up to 2014 highlighted that in most applications, researchers had not evaluated the assumptions underlying NMA using adequate methods (Nikolakopoulou et al., 2014a; Song et al., 2009; Coleman et al., 2012; Donegan et al., 2010; Bafeta et al., 2013). However, reporting and testing of NMA assumptions has considerably improved over time: a study of the years 2005–2015 found that 86% of NMAs published in 2015 discussed transitivity and/or consistency (compared with < 50% in 2005–2009) and 74% used appropriate methods to test for inconsistency (compared with <30% in 2005–2009) (Petropoulou et al., 2016).

Empirical studies have re-analyzed published networks of interventions to evaluate the presence of inconsistency (Song et al., 2009, Song et al., 2003, 2011; Veroniki et al., 2013). Song et al. examined loop inconsistency in 112 loops of evidence from previously published NMAs, using the OR as effect measure, and found that 14% show evidence of inconsistency at the 5% level (Song et al., 2011; Veroniki et al., 2013). In a similar study of 303 loops, Veroniki et al. found evidence of inconsistency in 2% to 9% of loops (under different estimators and assumptions employed about heterogeneity) (Song et al., 2011; Veroniki et al., 2013). Veroniki et al. also evaluated the consistency assumption in 40 published networks using the W test from the DBT model (see Section 10.6.2.2) and found that approximately one in eight networks show evidence of inconsistency (Veroniki et al., 2013). A simulation study found that the power of the loop-specific approach to detect inconsistency in the "typical" loop in published networks (with eight studies) is 14–21% (under different estimators) (Veroniki et al., 2014). Simulations also showed that large heterogeneity reduces the power to detect inconsistency (Veroniki et al., 2014; Song et al., 2012).

Based on these empirical studies and simulation results, non-statistically significant results should be interpreted with caution and conclusions regarding inconsistency should be based on the magnitude and the confidence interval of the inconsistency factors, or comparisons of direct and indirect evidence, as well as the statistical significance of the tests. Lack of power limits the usefulness of statistical approaches, especially if the network is sparse (containing few studies compared with the number of treatments so that the network is "thin" and poorly connected) and if heterogeneity is high. Non-statistical considerations, such as comparison of the distribution of effect modifiers, should complement the tests for inconsistency.

Finding evidence of inconsistency suggests several strategies for addressing it. If clinically relevant covariates and sufficient data are available, attempts to explain inconsistency through network meta-regression is recommended (Salanti et al., 2009; Dias et al., 2013b; Donegan et al., 2012). Alternatives include splitting the network into smaller, more consistent networks, removing inconsistent parts of the network (although this must be justified on the basis of pre-specified study characteristics to avoid bias), or undertaking only pairwise meta-analyses. Presentation of relative treatment effects from a random inconsistency model is only suggested in large networks with small amounts of inconsistency (Jackson et al., 2014; Higgins et al., 2012).

10.7 Presentation and Interpretation of NMA Results

A series of tables and graphical tools has been proposed for presenting the data, the underlying assumptions, and the results of an NMA (Chaimani et al., 2013; Chaimani and Salanti, 2015). Drawing a network plot (Figure 10.1) assists visualization of the evidence base when nodes and edges are conveniently weighted according to characteristics of treatments and comparisons, such as sample size (Chaimani et al., 2013). Relative treatment effects along with confidence and prediction intervals can be displayed in a "league table" or in a forest plot (e.g., as in Figure 10.4

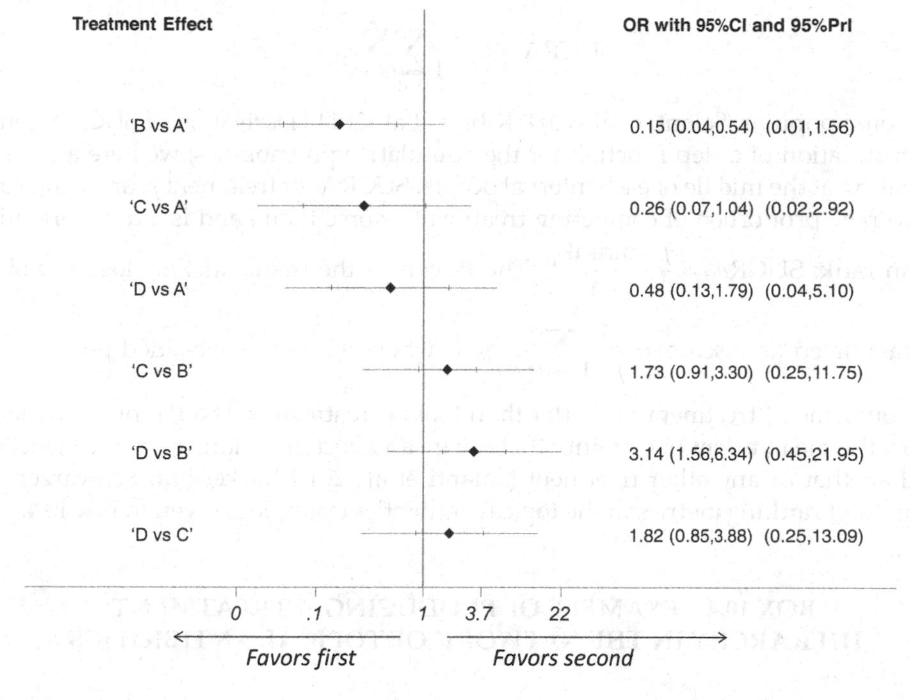

FIGURE 10.4
Network meta-analysis effect estimates with confidence intervals and prediction intervals for all comparisons in the network of topical antibiotics. The method used to obtain the results is the meta-analysis as a multivariate meta-analysis and the graph is produced in Stata using the `intervalplot` command. OR = odds ratio; CI = confidence interval; PrI = prediction interval.

and Table 10.2). The league table may include the relative treatment effects from two different outcomes or from pairwise and network meta-analysis.

Alongside relative treatment effects, ranking probabilities aids the interpretation of NMA results. These are the probabilities p_{jk} of each treatment j being at a rank k in the list of relative treatment effects for $k=1,\ldots J$. Ranking probabilities can be calculated using simulation techniques; a large number of samples are drawn from either the posterior distributions for the treatment effects in a Bayesian framework or from the assumed multivariate normal distributions with means and standard deviations from the estimated NMA results in a frequentist framework (White et al., 2012; White, 2015) and can be plotted in rankograms (Salanti et al., 2011).

It is best to describe the full rank distribution. Considering only the probability of being at the first rank over-emphasizes the probability of the treatment producing favorable relative effects and ignores the probabilities of large unfavorable effects. Consequently, a treatment hierarchy using the probability of being at the first rank will favor treatments whose effects are estimated with large uncertainty. The cumulative rank (the probability that a treatment is ranked at least as high as a specific rank, e.g., at least as high as second) is better and if applied for all possible cumulative ranks uses all the ranking data.

Other measures such as the mean rank $\left(\text{mrank}_j = \sum_{k=1}^{J} p_{jk} \times k\right)$ or median rank (Jansen et al., 2014) that summarize across the full ranking distribution also use the full distribution. Cumulative ranking probability curves and the surface area under the cumulative ranking curve (SUCRA), calculated as

$$\text{SUCRA}_j = \frac{1}{J-1} \sum_{k=1}^{J-1} \sum_{x=1}^{k} p_{jx}$$

have become popular (Salanti et al., 2011; Kibret et al., 2014). Derivation of SUCRA_j relies on the determination of a step function for the cumulative probabilities; we here assume that extrapolating at the middle of each interval occurs. SUCRA for treatment j can be interpreted as the average proportion of competing treatments worse than j and is a transformation of the mean rank: $\text{SUCRA}_j = \frac{J - \text{mrank}_j}{J-1}$. The P-score is the frequentist analogue of SUCRA and is calculated as $\text{P-score}_j = \frac{1}{J-1} \sum_{r, r \neq j}^{J} p_{j>r}$, where $p_{j>r}$ is the one-sided p-value testing that the outcome of treatment j is better than that of treatment r. The P-score of treatment j expresses the mean extent of certainty that the mean effect of j is larger (for a beneficial outcome) than that of any other treatment (Salanti et al., 2011; Rücker and Schwarzer, 2015). Application of ranking metrics in the topical antibiotics example is given in Box 10.4.

BOX 10.4 EXAMPLE OF PRODUCING A TREATMENT HIERARCHY IN THE NETWORK OF TOPICAL ANTIBIOTICS

APPLICATION: RANKING TREATMENTS

Figure 10.5 shows that in the worked example, treatment B has the highest probability to be associated with the most treatment success (probability of first rank 95.4%),

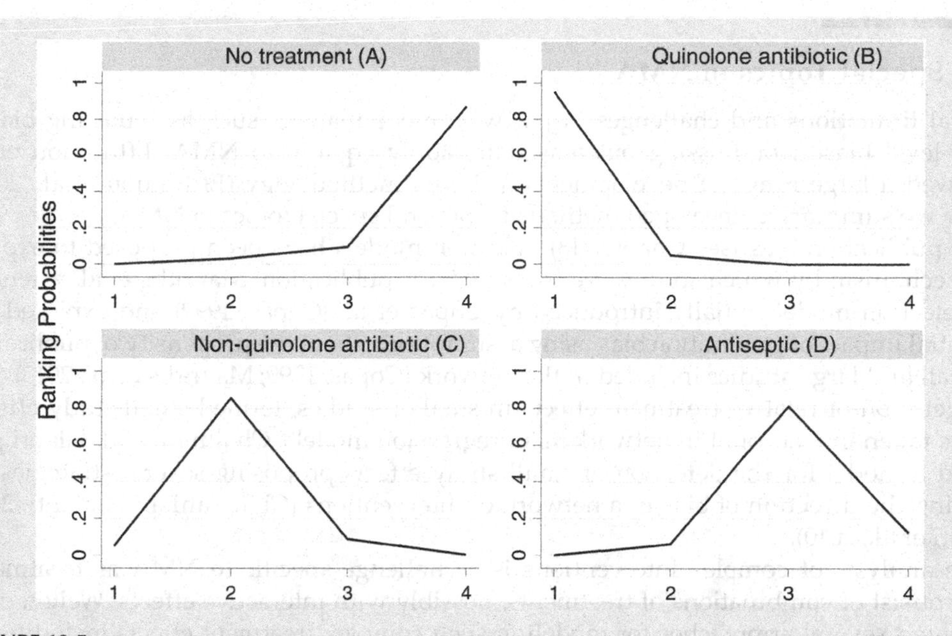

FIGURE 10.5
Rankograms for the treatments in the network of topical antibiotics. The graph is produced in Stata using the `sucra` command.

has 4.6% probability of being at the second place, and is less than 0.1% likely to rank third or fourth. Cumulative probabilities for each treatment are given in Figure 10.6. The area under these curves represent the SUCRA values, which are calculated to be 0.06, 0.99, 0.65, and 0.31 for treatments A, B, C, and D, respectively (using the `sucra` command in Stata).

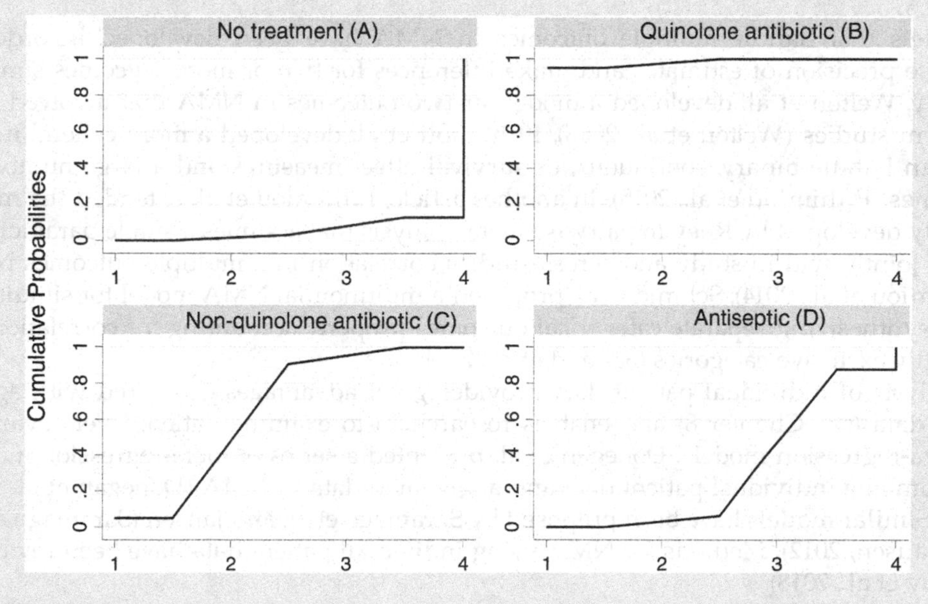

FIGURE 10.6
Cumulative ranking probabilities for the treatments in the network of topical antibiotics. The graph is produced in Stata using the `sucra` command.

10.8 Special Topics in NMA

Several limitations and challenges in pairwise meta-analysis, such as reporting biases, study-level biases, or missing outcome data, apply equally to NMA. Efthimiou et al. reviewed a large range of developments in NMA methodology (Efthimiou et al., 2016). Below we summarize developed methods for several special topics in NMA.

For publication bias (see Chapter 13), selection models have been proposed to explore the mechanism by which studies are "selected" for publication. Mavridis et al. extended the selection model initially introduced by Copas et al. (Copas, 1999) and explored the potential impact of publication bias using assumptions about the propensity of publication of small and large studies included in the network (Copas, 1999; Mavridis et al., 2013). The exaggeration of relative treatment effects in smaller studies, termed small study effects, can be taken into account in network meta-regression models. Chaimani and Salanti presented a model for the detection of small study effects proposing several strategies for defining the direction of bias in a network of interventions (Chaimani and Salanti, 2012; Salanti et al., 2010).

The analysis of complex interventions is a challenge specific to NMA as treatments may consist of combinations of treatments, possibly with interactive effects. Welton et al. presented several approaches for modeling such complex treatment effects including an additive model, a two-way interaction model, and a full interaction model (Welton et al., 2009). A similar challenge is deciding whether to regard treatments, which differ slightly across studies, as the same or different. In particular, the definition of "control" (e.g., usual care) across studies in an NMA might substantially differ over time. Several alternative approaches have been proposed by Del Giovane et al. and Warren et al. to account for variability in treatment definition (Del Giovane et al., 2013; Warren et al., 2014). The inclusion of concomitant treatments in the definition of treatments constitutes a further challenge in NMA.

Models that handle multiple outcomes in NMA have been developed in order to increase precision of estimates and make inferences for two or more outcomes simultaneously. Welton et al. developed a model for two outcomes in NMA that involved only two-arm studies (Welton et al., 2008). Efthimiou et al. developed a more general model that can handle binary, continuous, or survival effect measures and a large number of outcomes (Efthimiou et al., 2015). In another article, Efthimiou et al. extended the model initially developed by Riley in pairwise meta-analysis that assumes a single parameter to model jointly within-study and across-studies correlation in a multiple outcomes NMA (Efthimiou et al., 2014). Schmid et al. proposed a multinomial NMA model for simultaneously synthesizing separate categorical outcomes properly accounting for correlations in mutually exclusive categories (Schmid et al., 2014).

Analysis of individual patient data provides great advantages compared with aggregated data (see Chapter 8) and enables researchers to examine patient-level covariates in meta-regression models. Donegan et al. presented a series of meta-regression models that combine individual patient data and aggregated data in NMA (Donegan et al. 2012, 2013). Similar models have been proposed by Saramago et al. and Jansen (Saramago et al., 2012; Jansen, 2012). Methods for NMA using individual patient data have been reviewed (Debray et al., 2018).

NMA models have been proposed to combine information from observational studies and RCTs (Dias et al., 2013c; Schmitz et al., 2013; Efthimiou et al., 2017). Schmitz et al. proposed three alternative methods for combining data from various study designs in an NMA model: naïve pooling; incorporation of evidence from observational studies as prior information; and a three–level hierarchical model that properly accounts for between-study heterogeneity while allowing bias adjustment (Schmitz et al., 2013). In the same spirit, Efthimiou et al. presented and compared alternative methods for including non-randomized evidence in NMA and discussed advantages and disadvantages of the proposed approaches (Efthimiou et al., 2017).

NMA also has a role in the design of a new study, which should take into account all available existing knowledge and evidence (Clarke, 2004; Clarke et al., 2007; Jones et al., 2013). The derivation of conditional power of a new study given previous studies has been proposed for meta-analysis by Roloff et al. and Sutton et al., and has been extended for NMA by Nikolakopoulou et al. (Roloff et al., 2013; Sutton et al., 2007; Nikolakopoulou et al., 2014b). Naci and O'Connor suggested the prospective design of NMA to compare new drugs with existing alternatives at the time of market entry (Naci and O'Connor, 2013), and an extension of sequential methods for NMA has also been suggested (Nikolakopoulou et al., 2016).

The rapid evolution of NMA has raised concerns regarding the conduct and reporting of NMA (Hutton et al., 2014). Several empirical research articles have described the characteristics of published networks of interventions and attempted to evaluate the quality of reporting (Petropoulou et al., 2016; Zarin et al., 2017; Nikolakopoulou et al., 2014a; Bafeta et al., 2013; Hutton et al., 2014; Bafeta et al., 2014). Tan et al. underlined the need for standardization in reporting methods employed in NMA (Tan et al., 2013). Hutton et al. reviewed guideline articles and empirical studies addressing methodological issues in NMA and concluded that guidance in reporting NMA is required (Hutton et al., 2014). In response to this, a PRISMA extension for NMA has been published (Hutton et al., 2015).

The GRADE (Grading of Recommendations, Assessment, Development and Evaluation) system is widely used for assessing the confidence to be placed in the results from pairwise meta-analysis (Guyatt et al., 2008, 2011). Based on this system, two frameworks have been developed for evaluating the confidence in the findings from NMA (Puhan et al., 2014; Nikolakopoulou et al., 2020). A further alternative approach to assess the reliability of NMA has been suggested though the use of threshold analysis; Caldwell et al. proposed the exploration of the robustness of treatment recommendations according to potential degrees of bias in the evidence (Caldwell et al., 2016).

10.9 Software

Statistical synthesis of data about multiple interventions using NMA is possible in R, Stata, WinBUGS/OpenBUGS, Stan, JAGS, and SAS (Efthimiou et al., 2016). The `mvmeta` command in Stata (called by the `network` package) performs NMA as a multivariate meta-analysis model (White, 2011, 2015); completely "structured" (common between-study variance) and "unstructured" (separate between-study variances) forms of $\boldsymbol{T}$ are available with the option `bscov()`. Inconsistency models (DBT and LA) as well as the "separating indirect and direct evidence" approach can also be fitted using the `network` package. A series of Stata routines for presenting the evidence base, evaluating its assumptions, and presenting the results has been developed (Chaimani et al., 2013; Chaimani and Salanti, 2015).

The `netmeta` package in R (Rücker et al., 2018) performs NMA using a graph-theoretical approach (Rücker, 2012), equivalent to a two-stage meta-regression model (Lu et al., 2011;

Rücker and Schwarzer, 2014). Functionality of `netmeta` includes, but is not limited to, presenting the evidence base, fitting the NMA model, evaluating inconsistency, ranking treatments, presenting results, and implementing the component NMA model for complex interventions (Rücker et al., 2018). The `metafor` package also fits NMA models via multivariate meta-analysis commands (Viechtbauer, 2010). Automated packages for performing NMA in R have been reviewed (Neupane et al., 2014).

Performing NMA in a Bayesian framework is a popular approach (Petropoulou et al., 2016; Nikolakopoulou et al., 2014a); relevant code is available for the consistency model (Dias et al., 2013a, 2013d), the Lu–Ades inconsistency model (Senn et al., 2013), and the DBT inconsistency model (Jackson et al., 2014). GeMTC is an online platform for performing NMA with a substantial amount of functionality that covers a wide range of statistical methods (van Valkenhoef et al., 2012). The `gemtc` R package has also been developed to perform NMA in a Bayesian framework using JAGS (van Valkenhoef and Kuiper, 2016). NetMetaXL is a freely available Microsoft-based tool for performing and critically appraising NMA (Brown et al., 2014). A framework for evaluating NMA results has been proposed (Nikolakopoulou et al., 2020) and is implemented in a web application, CINeMA (Confidence in Network Meta-Analysis), which aims to simplify the process through semi-automation of the framework's steps (Papakonstantinou et al., 2020).

References

Bafeta A, Trinquart L, Seror R and Ravaud P, 2013. Analysis of the systematic reviews process in reports of network meta-analyses: Methodological systematic review. *BMJ* **347**: f3675.

Bafeta A, Trinquart L, Seror R and Ravaud P, 2014. Reporting of results from network meta-analyses: Methodological systematic review. *BMJ* **348**: g1741.

Brown S, Hutton B, Clifford T, Coyle D, Grima D, Wells G and Cameron C, 2014. A Microsoft-Excel-based tool for running and critically appraising network meta-analyses—An overview and application of NetMetaXL. *Systematic Reviews* **3**(1): 110.

Bucher HC, Guyatt GH, Griffith LE and Walter SD, 1997. The results of direct and indirect treatment comparisons in meta-analysis of randomized controlled trials. *Journal of Clinical Epidemiology* **50**(6): 683–691.

Caldwell DM, Ades AE and Higgins JPT, 2005. Simultaneous comparison of multiple treatments: Combining direct and indirect evidence. *BMJ* **331**(7521): 897–900.

Caldwell DM, Ades AE, Dias S, Watkins S, Li T, Taske N, Naidoo B and Welton NJ, 2016. A threshold analysis assessed the credibility of conclusions from network meta-analysis. *Journal of Clinical Epidemiology* **80**: 68–76.

Chaimani A and Salanti G, 2012. Using network meta-analysis to evaluate the existence of small-study effects in a network of interventions. *Research Synthesis Methods* **3**(2): 161–176.

Chaimani A and Salanti G, 2015. Visualizing assumptions and results in network meta-analysis: The network graphs package. *Stata Journal* **15**(4): 905–950.

Chaimani A, Higgins JPT, Mavridis D, Spyridonos P and Salanti G, 2013. Graphical tools for network meta-analysis in STATA. *PLOS ONE* **8**(10): e76654.

Chaimani A, 2014 *Investigating Bias in Network Meta-Analysis*. University of Ioannina.

Chan A-W and Altman DG, 2005. Epidemiology and reporting of randomised trials published in PubMed journals. *The Lancet* **365**(9465): 1159–1162.

Cipriani A, Higgins JPT, Geddes JR and Salanti G, 2013. Conceptual and technical challenges in network meta-analysis. *Annals of Internal Medicine* **159**(2): 130–137.

Clarke M, Hopewell S and Chalmers I, 2007. Reports of clinical trials should begin and end with up-to-date systematic reviews of other relevant evidence: A status report. *Journal of the Royal Society of Medicine* **100**(4): 187–190.

Clarke M, 2004. Doing new research? Don't forget the old. *PLOS Medicine* **1**(2): e35.

Coleman CI, Phung OJ, Cappelleri JC, Baker WL, Kluger J, White CM and Sobieraj DM, 2012. *Use of Mixed Treatment Comparisons in Systematic Reviews* [Internet]. Rockville (MD): Agency for Healthcare Research and Quality (US) [cited 2015 Apr 2]. (AHRQ Methods for Effective Health Care). Available from: http://www.ncbi.nlm.nih.gov/books/NBK107330/.

Cooper NJ, Sutton AJ, Morris D, Ades AE and Welton NJ, 2009. Addressing between-study heterogeneity and inconsistency in mixed treatment comparisons: Application to stroke prevention treatments in individuals with non-rheumatic atrial fibrillation. *Statistics in Medicine* **28**(14): 1861–1881.

Copas J, 1999. What works?: Selectivity models and meta-analysis. *Journal of the Royal Statistical Society* **162**(1): 95–109.

Debray TP, Schuit E, Efthimiou O, Reitsma JB, Ioannidis JP, Salanti G and Moons KG, 2018. An overview of methods for network meta-analysis using individual participant data: When do benefits arise? *Statistical Methods in Medical Research* **27**(5): 1351–1364.

Del Fiol G, Workman TE and Gorman PN, 2014. Clinical questions raised by clinicians at the point of care: A systematic review. *JAMA Internal Medicine* **174**(5): 710–718.

Del Giovane C, Vacchi L, Mavridis D, Filippini G and Salanti G, 2013. Network meta-analysis models to account for variability in treatment definitions: Application to dose effects. *Statistics in Medicine* **32**(1): 25–39.

Dias S, Welton NJ, Caldwell DM and Ades AE, 2010. Checking consistency in mixed treatment comparison meta-analysis. *Statistics in Medicine* **29**(7–8): 932–944.

Dias S, Sutton AJ, Ades AE and Welton NJ, 2013a. Evidence synthesis for decision making 2: A generalized linear modeling framework for pairwise and network meta-analysis of randomized controlled trials. *Medical Decision Making* **33**(5): 607–617.

Dias S, Sutton AJ, Welton NJ and Ades AE, 2013b. Evidence synthesis for decision making 3: Heterogeneity – Subgroups, meta-regression, bias, and bias-adjustment. *Medical Decision Making* **33**(5): 618–640.

Dias S, Welton NJ, Sutton AJ and Ades AE, 2013c. Evidence synthesis for decision making 5: The baseline natural history model. *Medical Decision Making* **33**(5): 657–670.

Dias S, Welton NJ, Sutton AJ, Caldwell DM, Lu G and Ades AE, 2013d. Evidence synthesis for decision making 4: Inconsistency in networks of evidence based on randomized controlled trials. *Medical Decision Making* **33**(5): 641–656.

Donegan S, Williamson P, Gamble C and Tudur-Smith C, 2010. Indirect comparisons: A review of reporting and methodological quality. *PLOS ONE* **5**(11): e11054.

Donegan S, Williamson P, D'Alessandro U and Tudur Smith C, 2012. Assessing the consistency assumption by exploring treatment by covariate interactions in mixed treatment comparison meta-analysis: Individual patient-level covariates versus aggregate trial-level covariates. *Statistics in Medicine* **31**(29): 3840–3857.

Donegan S, Williamson P, D'Alessandro U, Garner P and Smith CT, 2013. Combining individual patient data and aggregate data in mixed treatment comparison meta-analysis: Individual patient data may be beneficial if only for a subset of trials. *Statistics in Medicine* **32**(6): 914–930.

Donegan S, Welton NJ, Tudur Smith C, D'Alessandro U and Dias S, 2017. Network meta-analysis including treatment by covariate interactions: Consistency can vary across covariate values. *Research Synthesis Methods* **8**(4): 485–495.

Donegan S, Dias S and Welton NJ, 2018. Assessing the consistency assumptions underlying network meta-regression using aggregate data. *Research Synthesis Methods* **10**(2): 207–224.

Efthimiou O, Mavridis D, Cipriani A, Leucht S, Bagos P and Salanti G, 2014. An approach for modelling multiple correlated outcomes in a network of interventions using odds ratios. *Statistics in Medicine* **33**(13): 2275–2287.

Efthimiou O, Mavridis D, Riley RD, Cipriani A and Salanti G, 2015. Joint synthesis of multiple correlated outcomes in networks of interventions. *Biostatistics* **16**(1): 84–97.

Efthimiou O, Debray TPA, van Valkenhoef G, Trelle S, Panayidou K, Moons KGM, Reitsma JB, Shang A and Salanti G, 2016. GetReal in network meta-analysis: A review of the methodology. *Research Synthesis Methods* **7**(3): 236–263.

Efthimiou O, Mavridis D, Debray TPA, Samara M, Belger M, Siontis GCM, Leucht S and Salanti G, 2017. Combining randomized and non-randomized evidence in network meta-analysis. *Statistics in Medicine* **36**(8): 1210–1226.

Guyatt GH, Oxman AD, Vist GE, Kunz R, Falck-Ytter Y, Alonso-Coello P and Schünemann HJ, 2008. GRADE: An emerging consensus on rating quality of evidence and strength of recommendations. *BMJ* **336**(7650): 924–926.

Guyatt G, Oxman AD, Akl EA, Kunz R, Vist G, Brozek J, Norris S, Falck-Ytter Y, Glasziou P, DeBeer H, Jaeschke R, Rind D, Meerpohl J, Dahm P and Schünemann HJ, 2011. GRADE guidelines: 1. Introduction-GRADE evidence profiles and summary of findings tables. *Journal of Clinical Epidemiology* **64**(4): 383–394.

Higgins JPT and Welton NJ, 2015. Network meta-analysis: A norm for comparative effectiveness? *The Lancet* **386**(9994): 628–630.

Higgins JP and Whitehead A, 1996. Borrowing strength from external trials in a meta-analysis. *Statistics in Medicine* **15**(24): 2733–2749.

Higgins JPT, Jackson D, Barrett JK, Lu G, Ades AE and White IR, 2012. Consistency and insconsistency in network meta-analysis: Concepts and models for multi-arm studies. *Research Synthesis Methods* **3**(2): 98–110.

Hoaglin DC, Hawkins N, Jansen JP, Scott DA, Itzler R, Cappelleri JC, Boersma C, Thompson D, Larholt KM, Diaz M and Barrett A, 2011. Conducting indirect-treatment-comparison and network-meta-analysis studies: Report of the ISPOR Task Force on Indirect Treatment Comparisons Good Research Practices: Part 2. *Value in Health* **14**(4): 429–437.

Hutton B, Salanti G, Chaimani A, Caldwell DM, Schmid C, Thorlund K, Mills E, Catalá-López F, Turner L, Altman DG and Moher D, 2014. The quality of reporting methods and results in network meta-analyses: An overview of reviews and suggestions for improvement. *PLOS ONE* **9**(3): e92508.

Hutton B, Salanti G, Caldwell DM, Chaimani A, Schmid CH, Cameron C, Ioannidis JPA, Straus S, Thorlund K, Jansen JP, Mulrow C, Catalá-López F, Gøtzsche PC, Dickersin K, Boutron I, Altman DG and Moher D, 2015. The PRISMA extension statement for reporting of systematic reviews incorporating network meta-analyses of health care interventions: Checklist and explanations. *Annals of Internal Medicine* **162**(11): 777–784.

Jackson D, White IR and Riley RD, 2012. Quantifying the impact of between-study heterogeneity in multivariate meta-analyses. *Statistics in Medicine* **31**(29): 3805–3820.

Jackson D, Barrett JK, Rice S, White IR and Higgins JPT, 2014. A design-by-treatment interaction model for network meta-analysis with random inconsistency effects. *Statistics in Medicine* **33**(21): 3639–3654.

Jackson D, Law M, Barrett JK, Turner R, Higgins JPT, Salanti G and White IR, 2016. Extending DerSimonian and Laird's methodology to perform network meta-analyses with random inconsistency effects. *Statistics in Medicine* **35**(6): 819–839.

Jansen JP, Fleurence R, Devine B, Itzler R, Barrett A, Hawkins N, Lee K, Boersma C, Annemans L and Cappelleri JC, 2011. Interpreting indirect treatment comparisons and network meta-analysis for health-care decision making: Report of the ISPOR Task Force on Indirect Treatment Comparisons Good Research Practices: Part 1. *Value in Health* **14**(4): 417–428.

Jansen JP, Schmid CH and Salanti G, 2012. Directed acyclic graphs can help understand bias in indirect and mixed treatment comparisons. *Journal of Clinical Epidemiology* **65**(7): 798–807.

Jansen JP, Trikalinos T, Cappelleri JC, Daw J, Andes S, Eldessouki R and Salanti G, 2014. Indirect treatment comparison/network meta-analysis study questionnaire to assess relevance and credibility to inform health care decision making: An ISPOR-AMCP-NPC Good Practice Task Force report. *Value in Health* **17**(2): 157–173.

Jansen JP, 2012. Network meta-analysis of individual and aggregate level data. *Research Synthesis Methods* **3**(2): 177–190.

Jones AP, Conroy E, Williamson PR, Clarke M and Gamble C, 2013. The use of systematic reviews in the planning, design and conduct of randomised trials: A retrospective cohort of NIHR HTA funded trials. *BMC Medical Research Methodology* **13**(1): 50.

Kanters S, Ford N, Druyts E, Thorlund K, Mills EJ and Bansback N, 2016. Use of network meta-analysis in clinical guidelines. *Bulletin of the World Health Organization* **94**(10): 782–784.

Kibret T, Richer D and Beyene J, 2014. Bias in identification of the best treatment in a Bayesian network meta-analysis for binary outcome: A simulation study. *Clinical Epidemiology* **6**: 451–460.

König J, Krahn U and Binder H, 2013. Visualizing the flow of evidence in network meta-analysis and characterizing mixed treatment comparisons. *Statistics in Medicine* **32**(30): 5414–5429.

Krahn U, Binder H and König J, 2013. A graphical tool for locating inconsistency in network meta-analyses. *BMC Medical Research Methodology* **13**: 35.

Krahn U, Binder H and König J, 2014. Visualizing inconsistency in network meta-analysis by independent path decomposition. *BMC Medical Research Methodology* **14**: 131.

Laws A, Kendall R and Hawkins N, 2014. A comparison of national guidelines for network meta-analysis. *Value in Health* **17**(5): 642–654.

Lu G and Ades AE, 2004. Combination of direct and indirect evidence in mixed treatment comparisons. *Statistics in Medicine* **23**(20): 3105–3124.

Lu G and Ades AE, 2006. Assessing evidence inconsistency in mixed treatment comparisons. *Journal of the American Statistical Association* **101**(474): 447–459.

Lu G and Ades A, 2009. Modeling between-trial variance structure in mixed treatment comparisons. *Biostatistics* **10**(4): 792–805.

Lu G, Welton NJ, Higgins JPT, White IR and Ades AE, 2011. Linear inference for mixed treatment comparison meta-analysis: A two-stage approach. *Research Synthesis Methods* **2**(1): 43–60.

Lumley T, 2002. Network meta-analysis for indirect treatment comparisons. *Statistics in Medicine* **21**(16): 2313–2324.

Macfadyen CA, Acuin JM and Gamble C, 2005. Topical antibiotics without steroids for chronically discharging ears with underlying eardrum perforations. *Cochrane Database of Systematic Reviews* **4**: CD004618.

Mavridis D and Salanti G, 2013. A practical introduction to multivariate meta-analysis. *Statistical Methods in Medical Research* **22**(2): 133–158.

Mavridis D, Sutton A, Cipriani A and Salanti G, 2013. A fully Bayesian application of the Copas selection model for publication bias extended to network meta-analysis. *Statistics in Medicine* **32**(1): 51–66.

Mavridis D, Giannatsi M, Cipriani A and Salanti G, 2015. A primer on network meta-analysis with emphasis on mental health. *Evidence-Based Mental Health* **18**(2): 40–46.

Naci H and Fleurence R, 2011. Using indirect evidence to determine the comparative effectiveness of prescription drugs: Do benefits outweigh risks? *Health Outcomes Research in Medicine* **2**(4): e241–e249.

Naci H and O'Connor AB, 2013. Assessing comparative effectiveness of new drugs before approval using prospective network meta-analyses. *Journal of Clinical Epidemiology* **66**(8): 812–816.

Neupane B, Richer D, Bonner AJ, Kibret T and Beyene J, 2014. Network meta-analysis using R: A review of currently available automated packages. *PLOS ONE* **9**(12): e115065.

Nikolakopoulou A, Chaimani A, Veroniki AA, Vasiliadis HS, Schmid CH and Salanti G, 2014a. Characteristics of networks of interventions: A description of a database of 186 published networks. *PLOS ONE* **9**(1): e86754.

Nikolakopoulou A, Mavridis D and Salanti G, 2014b. Using conditional power of network meta-analysis (NMA) to inform the design of future clinical trials. *Biometrical Journal* **56**(6): 973–990.

Nikolakopoulou A, Mavridis D, Egger M and Salanti G, 2016. Continuously updated network meta-analysis and statistical monitoring for timely decision-making. *Statistical Methods in Medical Research* **27**(5): 1312–1330.

Nikolakopoulou A, Higgins JPT, Papakonstantinou T, Chaimani A, Del Giovane C, Egger M, Salanti G, 2020. CINeMA: An approach for assessing confidence in the results of a network meta-analysis. *PLoS Med* **17**(4): e1003082.

Nixon RM, Bansback N and Brennan A, 2007. Using mixed treatment comparisons and meta-regression to perform indirect comparisons to estimate the efficacy of biologic treatments in rheumatoid arthritis. *Statistics in Medicine* **26**(6): 1237–1254.

Papakonstantinou T, Nikolakopoulou A, Higgins JPT, Egger M, Salanti G, 2020. CINeMA: Software for semiautomated assessment of the confidence in the results of network meta-analysis. *Campbell Systematic Reviews* **16**(1): e1080.

Petropoulou M, Nikolakopoulou A, Veroniki A-A, Rios P, Vafaei A, Zarin W, Giannatsi M, Sullivan S, Tricco AC, Chaimani A, Egger M and Salanti G, 2016. Bibliographic study showed improving statistical methodology of network meta-analyses published between 1999 and 2015. *Journal of Clinical Epidemiology* **82**: 20–28.

Puhan MA, Schünemann HJ, Murad MH, Li T, Brignardello-Petersen R, Singh JA, Kessels AG and Guyatt GH, 2014. A GRADE Working Group approach for rating the quality of treatment effect estimates from network meta-analysis. *BMJ* **349**: g5630.

Roloff V, Higgins JPT and Sutton AJ, 2013. Planning future studies based on the conditional power of a meta-analysis. *Statistics in Medicine* **32**(1): 11–24.

Rücker G and Schwarzer G, 2014. Reduce dimension or reduce weights? Comparing two approaches to multi-arm studies in network meta-analysis. *Statistics in Medicine* **33**(25): 4353–4369.

Rücker G and Schwarzer G, 2015. Ranking treatments in frequentist network meta-analysis works without resampling methods. *BMC Medical Research Methodology* **15**: 58.

Rücker G, Schwarzer G, Krahn U and König J, 2018. *netmeta: Network Meta-Analysis Using Frequentist Methods* [Internet]. Available from https://CRAN.R-project.org/package=netmeta.

Rücker G, 2012. Network meta-analysis, electrical networks and graph theory. *Research Synthesis Methods* **3**(4): 312–324.

Salanti G, Higgins JPT, Ades AE and Ioannidis JPA, 2008. Evaluation of networks of randomized trials. *Statistical Methods in Medical Research* **17**(3): 279–301.

Salanti G, Marinho V and Higgins JPT, 2009. A case study of multiple-treatments meta-analysis demonstrates that covariates should be considered. *Journal of Clinical Epidemiology* **62**(8): 857–864.

Salanti G, Dias S, Welton NJ, Ades AE, Golfinopoulos V, Kyrgiou M, Mauri D and Ioannidis JPA, 2010. Evaluating novel agent effects in multiple-treatments meta-regression. *Statistics in Medicine* **29**(23): 2369–2383.

Salanti G, Ades AE and Ioannidis JP, 2011. Graphical methods and numerical summaries for presenting results from multiple-treatment meta-analysis: An overview and tutorial. *Journal of Clinical Epidemiology* **64**(1878-5921 (Electronic)): 163–171.

Salanti G, 2012. Indirect and mixed-treatment comparison, network, or multiple-treatments meta-analysis: Many names, many benefits, many concerns for the next generation evidence synthesis tool. *Research Synthesis Methods* **3**(2): 80–97.

Saramago P, Sutton AJ, Cooper NJ and Manca A, 2012. Mixed treatment comparisons using aggregate and individual participant level data. *Statistics in Medicine* **31**(28): 3516–3536.

Schmid CH, Trikalinos TA and Olkin I, 2014. Bayesian network meta-analysis for unordered categorical outcomes with incomplete data. *Research Synthesis Methods* **5**(2): 162–185.

Schmitz S, Adams R and Walsh C, 2013. Incorporating data from various trial designs into a mixed treatment comparison model. *Statistics in Medicine* **32**(17): 2935–2949.

Senn S, Gavini F, Magrez D and Scheen A, 2013. Issues in performing a network meta-analysis. *Statistical Methods in Medical Research* **22**(2): 169–189.

Simmonds MC and Higgins JP, 2016. A general framework for the use of logistic regression models in meta-analysis. *Statistical Methods in Medical Research* **25**(6): 2858–2877.

Sobieraj DM, Cappelleri JC, Baker WL, Phung OJ, White CM and Coleman CI, 2013. Methods used to conduct and report Bayesian mixed treatment comparisons published in the medical literature: A systematic review. *BMJ Open* **3**(7): e003111.

Song F, Altman DG, Glenny A-M and Deeks JJ, 2003. Validity of indirect comparison for estimating efficacy of competing interventions: Empirical evidence from published meta-analyses. *BMJ* **326**(7387): 472.

Song F, Loke YK, Walsh T, Glenny A-M, Eastwood AJ and Altman DG, 2009. Methodological problems in the use of indirect comparisons for evaluating healthcare interventions: Survey of published systematic reviews. *BMJ* **338**: b1147.

Song F, Xiong T, Parekh-Bhurke S, Loke YK, Sutton AJ, Eastwood AJ, Holland R, Chen Y-F, Glenny A-M, Deeks JJ and Altman DG, 2011. Inconsistency between direct and indirect comparisons of competing interventions: Meta-epidemiological study. *BMJ* **343**: d4909.

Song F, Clark A, Bachmann MO and Maas J, 2012. Simulation evaluation of statistical properties of methods for indirect and mixed treatment comparisons. *BMC Medical Research Methodology* **12**: 138.

Spiegelhalter DJ, Best NG, Carlin BP and Van Der Linde A, 2002. Bayesian measures of model complexity and fit. *Journal of the Royal Statistical Society: Series B* **64**(4): 583–639.

Sutton AJ, Cooper NJ, Jones DR, Lambert PC, Thompson JR and Abrams KR, 2007. Evidence-based sample size calculations based upon updated meta-analysis. *Statistics in Medicine* **26**(12): 2479–2500.

Tan SH, Bujkiewicz S, Sutton A, Dequen P and Cooper N, 2013. Presentational approaches used in the UK for reporting evidence synthesis using indirect and mixed treatment comparisons. *Journal of Health Services Research & Policy* **18**(4): 224–232.

Turner EH, Matthews AM, Linardatos E, Tell RA and Rosenthal R, 2008. Selective publication of antidepressant trials and its influence on apparent efficacy. *The New England Journal of Medicine* **358**(3): 252–260.

Turner RM, Davey J, Clarke MJ, Thompson SG and Higgins JP, 2012. Predicting the extent of heterogeneity in meta-analysis, using empirical data from the Cochrane Database of Systematic Reviews. *International Journal of Epidemiology* **41**(3): 818–827.

van Houwelingen HC, Arends LR and Stijnen T, 2002. Advanced methods in meta-analysis: Multivariate approach and meta-regression. *Statistics in Medicine* **21**(4): 589–624.

van Valkenhoef G and Kuiper J, 2016. *gemtc: Network Meta-Analysis Using Bayesian Methods* [Internet]. Available from https://CRAN.R-project.org/package=gemtc.

van Valkenhoef G, Lu G, de Brock B, Hillege H, Ades AE and Welton NJ, 2012. Automating network meta-analysis. *Research Synthesis Methods* **3**(4): 285–299.

van Valkenhoef G, Dias S, Ades AE and Welton NJ, 2016. Automated generation of node-splitting models for assessment of inconsistency in network meta-analysis. *Research Synthesis Methods* **7**(1): 80–93.

Veroniki AA, Vasiliadis HS, Higgins JPT and Salanti G, 2013. Evaluation of inconsistency in networks of interventions. *International Journal of Epidemiology* **42**(1): 332–345.

Veroniki AA, Mavridis D, Higgins JPT and Salanti G, 2014. Characteristics of a loop of evidence that affect detection and estimation of inconsistency: A simulation study. *BMC Medical Research Methodology* **14**: 106.

Veroniki AA, 2014. *Study of the Heterogeneity and Inconsistency in Networks of Multiple Interventions*. University of Ioannina.

Viechtbauer W, 2010. Conducting meta-analyses in R with the metafor Package. *Journal of Statistical Software* **36**(1): 1–48.

Warren FC, Abrams KR and Sutton AJ, 2014. Hierarchical network meta-analysis models to address sparsity of events and differing treatment classifications with regard to adverse outcomes. *Statistics in Medicine* **33**(14): 2449–2466.

Welton NJ, Cooper NJ, Ades AE, Lu G, Sutton AJ, 2008. Mixed treatment comparison with multiple outcomes reported inconsistently across trials: Evaluation of antivirals for treatment of influenza A and B. *Statistics in Medicine* **27**(27): 5620–5639.

Welton NJ, Caldwell DM, Adamopoulos E and Vedhara K, 2009. Mixed treatment comparison meta-analysis of complex interventions: Psychological interventions in coronary heart disease. *American Journal of Epidemiology* **169**(9): 1158–1165.

White IR, Barrett JK, Jackson D and Higgins JPT, 2012. Consistency and inconsistency in network meta-analysis: Model estimation using multivariate meta-regression. *Research Synthesis Methods* **3**(2): 111–125.

White IR, 2011. Multivariate random-effects meta-regression: Updates to mvmeta. *Stata Journal* **11**(2): 255–270.

White IR, 2015. Network meta-analysis. *Stata Journal* **15**(4): 951–985.

Zarin W, Veroniki AA, Nincic V, Vafaei A, Reynen E, Motiwala SS, Antony J, Sullivan SM, Rios P, Daly C, Ewusie J, Petropoulou M, Nikolakopoulou A, Chaimani A, Salanti G, Straus SE and Tricco AC, 2017. Characteristics and knowledge synthesis approach for 456 network meta-analyses: A scoping review. *BMC Medicine* **15**(1): 3.

11

Model Checking in Meta-Analysis

Wolfgang Viechtbauer

CONTENTS

11.1 Introduction

In previous chapters, methods have been described to compute and model various outcome or effect size measures, such as risk differences, (log transformed) risk/odds ratios, raw or standardized mean differences, and correlation coefficients. The observed values of such measures may reflect the size of a treatment effect, the degree to which a risk factor is related to the chances of being afflicted by (or the severity of) a particular disease, or more generally the size of group differences. Some measures, such as the correlation coefficient, simply reflect the degree to which two variables of interest are (linearly) related to each other.

The process of analyzing such data involves fitting one or more models to the observed outcomes,* based on which we can draw conclusions about the effectiveness of a treatment, the relevance of a risk factor, the degree to which groups differ, the strength of the association between two variables, and so on. In addition, it is typically of interest to examine whether the phenomenon being studied (e.g., the treatment effect) is relatively homogeneous across studies or varies, possibly as a function of one or more variables that can account for this heterogeneity. However, the models used in such analyses make various assumptions. In practice, assumptions may be violated, which in turn may affect the statistical properties of the inferential methods used to draw conclusions from the data at hand. It is therefore important to carefully consider to what extent the various assumptions may be violated and what impact this may have on the results and conclusions.

Similarly, when we fit a model to our data, we make the implicit assumption that the model represents an adequate approximation to some underlying data generating process. Naturally, reality is more complex than any model we can envision, but gross mismatch between the data and model should warn us that we are far from providing an adequate description of how the data may have arisen. Therefore, assessing model fit should be an essential step in any analysis. For some aspects of the models, this can be done by means of a statistical test. An examination of the residuals and standardized versions thereof can also provide clues about the presence of an incongruity between the data and the assumed model and/or may indicate that the model is not appropriate for certain data points.

Finally, it is important to examine whether one or more studies exert a disproportionally large influence on the conclusions of a meta-analysis. If we find that certain findings hinge on only one or two studies being present in our dataset, then this may call into question the robustness of the findings, in which case the corresponding conclusions should be framed more cautiously. An important tool in this context is to examine how the removal of studies from the dataset would alter the results. We will therefore consider a variety of diagnostic measures on the basis of this idea.

11.2 Models Assuming Normal Sampling Distributions

In this section, some of the meta-analytic models introduced in previous chapters will be reviewed, with emphasis on the common-, random-, and (mixed-effects) meta-regression models. The models considered here are all based on the assumption that the observed outcomes represent draws from normal sampling distributions (models that assume other types of sampling distributions will be briefly discussed in Section 11.7). Moreover, the models are not tied to any particular outcome or effect size measure. Therefore, in general, let y_i denote the observed value of the chosen outcome measure (e.g., log odds ratio) in the ith study and let θ_i denote the value of the corresponding (unknown) true outcome (e.g., the true log odds ratio).

* As described in Chapter 5, it is also possible to model the raw data directly where available (or where it can be reconstructed based on the available information) using appropriate generalized linear mixed-effects models. We will return to a discussion of such models at the end of this chapter.

11.2.1 Common-Effect Model

The first model we will consider is the common-effect model, which is given by

$$y_i = \theta + e_i, \tag{11.1}$$

where θ denotes the underlying true value of the outcome measure and $e_i \sim N(0, s_i^2)$. Therefore, as the name implies, the model assumes that the true outcomes are equal (homogeneous) across studies (i.e., $\theta_i = \theta$ for $i = 1, \ldots, k$). Moreover, the sampling error in the ith study (i.e., e_i) is assumed to be normally distributed with sampling variance equal to s_i^2, which in turn we consider to be a known and fixed quantity. Finally, assuming independence between studies implies $\text{Cov}[e_i, e_{i'}] = 0$ for $i \neq i'$.

11.2.2 Random-Effects Model

The common-effect model can be considered a special case of the more general random-effects model, where potential heterogeneity in the true outcomes is accounted for by adding a random effect for each study to the model. In particular, the random-effects model is given by

$$y_i = \mu + u_i + e_i, \tag{11.2}$$

where μ denotes the average true outcome and $u_i \sim N(0, \tau^2)$ is a normally distributed random effect by which the true outcome in the ith study differs from the average true outcome (so that $\theta_i = \mu + u_i$). Therefore, τ^2 denotes the amount of variance (or "heterogeneity") in the true outcomes (hence, if $\tau^2 = 0$, then the random-effects model simplifies to the common-effect model, so that $\mu \equiv \theta$). Assumptions about e_i are as described previously, with the addition that we assume independence between different u_i values (and hence, $\text{Cov}[u_i, u_{i'}] = 0$ for $i \neq i'$) and between u_i and e_i (which implies $\text{Cov}[u_i, e_i] = 0$).

11.2.3 Meta-Regression Model

An alternative approach to account for heterogeneity in the true outcomes is to explicitly model such differences by means of one or more predictor (or "moderator") variables. This leads to the meta-regression model, which is given by

$$y_i = \beta_0 + \beta_1 x_{i1} + \cdots + \beta_q x_{iq} + e_i, \tag{11.3}$$

where β_1 through β_q are model coefficients that denote how the true outcome changes for a one-unit increase in the corresponding moderator variables x_{i1} through x_{iq} and β_0 denotes the model intercept, which corresponds to the true outcome when all moderator variables take on the value 0.

11.2.4 Mixed-Effects Model

Finally, analogous to the random-effects model, the meta-regression model can also be extended by the addition of a random effect for each study. Doing so yields the mixed-effects model (also called random-effects meta-regression model), which is given by

$$y_i = \beta_0 + \beta_1 x_{i1} + \cdots + \beta_q x_{iq} + u_i + e_i, \tag{11.4}$$

with $u_i \sim N(0,\tau^2)$ as before, except that τ^2 should now be interpreted as the amount of variance in the true outcomes that is not accounted for by the moderator variables included in the model (hence, τ^2 is often described as the amount of "residual heterogeneity").

11.2.5 Model Fitting

Methods for fitting the various models described above were discussed in earlier chapters but will be briefly restated here for completeness sake and to fix notation. We start with the mixed-effects model and then treat the remaining models as special cases thereof. Matrix notation is used, as it provides a compact way of writing out the equations. When it is helpful for understanding, algebraic expressions are also provided.

The mixed-effects model implies that $\boldsymbol{y} \sim N(\boldsymbol{X\beta}, \boldsymbol{V} + \tau^2\boldsymbol{I})$, where $\boldsymbol{y}$ is a $k \times 1$ column vector with the observed outcomes y_1 through y_k, $\boldsymbol{X}$ is a $k \times (q+1)$ matrix containing the values of the moderator variables (with the constant 1 in the first column for the model intercept), $\boldsymbol{\beta}$ is a $(q+1) \times 1$ column vector with the model coefficients $\beta_0, \beta_1, \ldots, \beta_q$, $\boldsymbol{V}$ is a diagonal matrix with the s_i^2 values along the diagonal, and $\boldsymbol{I}$ is a $k \times k$ identity matrix.

To fit the mixed-effects model, we must first estimate τ^2 using one of the various estimators that have been described in the literature for this purpose (e.g., Thompson and Sharp, 1999; Viechtbauer et al., 2015) and that are reviewed in Chapter 4. For the purposes of this chapter, we will just consider a relatively simple method-of-moments estimator, which is given by

$$\hat{\tau}^2 = \frac{Q_E - (k-q-1)}{\text{trace}[\boldsymbol{P}]}, \tag{11.5}$$

where $Q_E = \boldsymbol{y}'\boldsymbol{P}\boldsymbol{y}$ and $\boldsymbol{P} = \boldsymbol{V}^{-1} - \boldsymbol{V}^{-1}\boldsymbol{X}(\boldsymbol{X}'\boldsymbol{V}^{-1}\boldsymbol{X})^{-1}\boldsymbol{X}'\boldsymbol{V}^{-1}$. In case $\hat{\tau}^2$ is negative, the estimate is set to 0.

Once $\hat{\tau}^2$ has been calculated, estimates of the model coefficients in $\boldsymbol{\beta}$ can be obtained using the weighted least squares estimator

$$\hat{\boldsymbol{\beta}} = (\boldsymbol{X}'\boldsymbol{W}\boldsymbol{X})^{-1}\boldsymbol{X}'\boldsymbol{W}\boldsymbol{y}, \tag{11.6}$$

where $\boldsymbol{W} = (\boldsymbol{V} + \hat{\tau}^2\boldsymbol{I})^{-1}$. The variance-covariance matrix of $\hat{\boldsymbol{\beta}}$ can be estimated with

$$\text{Var}[\hat{\boldsymbol{\beta}}] = (\boldsymbol{X}'\boldsymbol{W}\boldsymbol{X})^{-1}. \tag{11.7}$$

Taking the square root of the diagonal elements of $\text{Var}[\hat{\boldsymbol{\beta}}]$ then yields standard errors of the model coefficients (i.e., $SE[\hat{\beta}_0], \ldots, SE[\hat{\beta}_q]$). Note that (11.7) ignores the uncertainty in the estimate of τ^2, which is discussed in more detail in Chapter 4.

By comparing the test statistic $z_j = \hat{\beta}_j / SE[\hat{\beta}_j]$ against appropriate percentiles of a standard normal distribution (e.g., ±1.96 for a two-sided test at $\alpha = 0.05$), we can test H_0: $\beta_j = 0$, that is, whether there is a significant relationship between a moderator variable and the outcomes. Analogously, an approximate 95% confidence interval (CI) for β_j can be constructed with $\hat{\beta}_j \pm 1.96 SE[\hat{\beta}_j]$. Simultaneous tests of multiple coefficients can also be conducted by computing

$$Q_M = \hat{\boldsymbol{\beta}}_{[2]}'(\text{Var}[\hat{\boldsymbol{\beta}}]_{[2]})^{-1}\hat{\boldsymbol{\beta}}_{[2]}, \tag{11.8}$$

where $\hat{\beta}_{[2]}$ is a column vector containing the m coefficients to be tested and $\text{Var}[\hat{\beta}]_{[2]}$ denotes an $m \times m$ matrix with the corresponding rows and columns from (11.7). Under the null hypothesis that the true values of the coefficients tested are all equal to zero, Q_M follows (approximately) a chi-square distribution with m degrees of freedom. A common application of (11.8) is to test all coefficients excluding the intercept (i.e., H_0: $\beta_1 = \cdots = \beta_q = 0$), which (analogous to the omnibus F-test in multiple regression) can be used to examine whether at least one of the moderator variables included in the model is related to the outcomes, or put differently, whether the set of moderators included in the model actually accounts for any heterogeneity in the true outcomes.

The fitted values based on the model can be computed with $\hat{y} = X\hat{\beta}$ or equivalently with $\hat{y} = Hy$, where

$$H = X(X'WX)^{-1}X'W \tag{11.9}$$

is the hat matrix, whose relevance will be discussed in more detail further below. Similarly, for any row vector x_i (with x_i not necessarily a row from X), we can compute the corresponding predicted value with $\hat{\mu}_i = x_i\hat{\beta}$, with variance equal to $\text{Var}[\hat{\mu}_i] = x_i\text{Var}[\hat{\beta}]x_i'$ and standard error $SE[\hat{\mu}_i] = \sqrt{\text{Var}[\hat{\mu}_i]}$. Therefore, an approximate 95% CI for the expected true outcome given vector x_i can be computed with $\hat{\mu}_i \pm 1.96SE[\hat{\mu}_i]$. Note that in the mixed-effects model, $\hat{\mu}_i$ denotes the predicted *average* outcome for a particular combination of moderator values. The true outcome for a *particular* study can still differ from $\hat{\mu}_i$ due to residual heterogeneity.

The other three models can be considered to be special cases of the mixed-effects model. First, fixing $\tau^2 = 0$ yields the meta-regression model. The model coefficients and corresponding variance-covariance matrix are then estimated with (11.6) and (11.7) with $W = V^{-1}$. All other equations work accordingly, although notationally, it is now more appropriate to denote a predicted value with $\hat{\theta}_i$ (i.e., in the absence of residual heterogeneity, there is no longer a distinction between the predicted average outcome and the predicted outcome for a single study).

Second, the random-effects model results when X is only a column vector with the constant 1, in which case $\hat{\beta}_0 \equiv \hat{\mu} = \sum w_i y_i / \sum w_i$ with corresponding standard error $SE[\hat{\mu}] = \sqrt{1/\sum w_i}$, where $w_i = 1/(s_i^2 + \hat{\tau}^2)$. Note that (11.5) then simplifies to the well-known DerSimonian–Laird estimator of τ^2 in the random-effects model (DerSimonian and Laird, 1986).

Finally, when fixing $\tau^2 = 0$ and X is only a column vector with the constant 1, then we obtain the common-effect model, where $\hat{\beta}_0 \equiv \hat{\theta} = \sum w_i y_i / \sum w_i$ with corresponding standard error $SE[\hat{\theta}] = \sqrt{1/\sum w_i}$ as before, but now with $w_i = 1/s_i^2$.

An additional statistic that is often reported is the I^2 statistic (Higgins and Thompson, 2002). It is given by

$$I^2 = 100\% \times \left(\frac{\hat{\tau}^2}{\hat{\tau}^2 + \tilde{s}^2} \right), \tag{11.10}$$

where $\hat{\tau}^2$ is the estimate of τ^2 from the random-effects model and

$$\tilde{s}^2 = \frac{k-1}{\sum w_i - \sum w_i^2 / \sum w_i} \tag{11.11}$$

is a way of quantifying the "typical" sampling variance across the k studies, which is computed with $w_i = 1/s_i^2$. The I^2 statistic estimates what percentage of the total variability (which is composed of heterogeneity plus sampling variability) can be attributed to heterogeneity among the true outcomes. Also, for the mixed-effects model, a pseudo R^2-type measure can be computed with

$$R^2 = 100\% \times \left(\frac{\hat{\tau}_{RE}^2 - \hat{\tau}_{ME}^2}{\hat{\tau}_{RE}^2} \right), \tag{11.12}$$

where $\hat{\tau}_{RE}^2$ and $\hat{\tau}_{ME}^2$ are the estimates of τ^2 in the random- and mixed-effects models, respectively (López-López et al., 2014; Raudenbush, 2009). The R^2 statistic estimates the proportional reduction in the amount of heterogeneity when including moderators in a random-effects model, or put differently, what percentage of the total heterogeneity can be accounted for by the moderators included in the mixed-effects meta-regression model.

11.3 Example Dataset

For didactic purposes, a meta-analytic dataset was constructed that will be used to illustrate the application of the aforementioned models and the methods to be discussed in more detail below. The dataset, given in Table 11.1, can be thought of as a set of $k=20$ randomized controlled trials (e.g., patients receiving a medication versus a placebo) where a dichotomous response variable of interest was measured within the individual studies (e.g., remission versus persistence of symptoms). Columns n_i^T and n_i^C denote the total number of patients in the treatment and control group, respectively, while columns x_i^T and x_i^C denote the number of patients within the respective groups that experienced the outcome of interest (e.g., remission). The outcome measure to be used for the meta-analysis will be the log odds ratio.

The observed log odds ratios can be computed with

$$y_i = \ln\left[\frac{(x_i^T + 0.5)/(m_i^T + 0.5)}{(x_i^C + 0.5)/(m_i^C + 0.5)} \right], \tag{11.13}$$

where $m_i^T = n_i^T - x_i^T$ and $m_i^C = n_i^C - x_i^C$, and are given in the corresponding column in the table. A positive value for y_i therefore indicates higher odds of remission in the treatment compared with the control group. The sampling variances of the log odds ratios, given in the adjacent column, were computed with

$$s_i^2 = \frac{1}{x_i^T + 0.5} + \frac{1}{m_i^T + 0.5} + \frac{1}{x_i^C + 0.5} + \frac{1}{m_i^C + 0.5}. \tag{11.14}$$

TABLE 11.1

Illustrative Data for a Meta-Analysis of 20 Trials

Study	n_i^T	n_i^C	x_i^T	x_i^C	y_i	s_i^2	Dose
1	66	59	42	24	0.922	0.133	100
2	59	65	42	34	0.796	0.141	200
3	253	257	96	32	1.447	0.052	250
4	137	144	51	44	0.296	0.063	125
5	327	326	47	39	0.209	0.053	50
6	584	588	38	87	−0.907	0.041	25
7	526	532	390	323	0.617	0.018	125
8	28	30	10	3	1.495	0.471	125
9	191	201	165	126	1.316	0.065	125
10	86	94	58	39	1.059	0.096	150
11	229	221	72	60	0.206	0.043	100
12	153	144	79	56	0.514	0.055	150
13	93	95	48	35	0.597	0.087	200
14	40	40	8	4	0.752	0.398	25
15	85	88	44	21	1.214	0.108	175
16	100	107	10	13	−0.208	0.191	25
17	72	64	11	9	0.088	0.226	25
18	80	74	47	23	1.134	0.113	200
19	191	195	144	116	0.730	0.049	100
20	85	85	48	49	−0.047	0.095	75

The 0.5 term in the equations above serves two purposes. First, it reduces the bias in y_i as an estimate of θ_i, the true log odds ratio in the ith study (Walter, 1985; Walter and Cook, 1991). A second, more practical reason for the addition of the 0.5 term is that it allows the computation of the log odds ratio (and its corresponding sampling variance) even in studies where one of the 2×2 table cells (i.e., x_i^T, m_i^T, x_i^C, m_i^C) is equal to zero (although not applicable here, this issue will become relevant further below).

The last column in Table 11.1 reflects the dosage of the medication provided to patients in the treatment group (e.g., in milligrams per day), which we can envision as a potentially relevant moderator variable in this context. We will examine this moderator further with (mixed-effects) meta-regression models. In addition, common- and random-effects models were fitted to the data. Results for the various models are given in Table 11.2.

Under the assumption that the true outcomes are homogeneous, the common-effect model yields an estimate of the true log odds ratio equal to $\hat{\theta}=0.529$ (95% CI: 0.415 to 0.644). Since the estimate is positive and the CI excludes the value 0, this suggests that the treatment significantly increases the odds of remission. Similarly, the random-effects model yields an estimate of $\hat{\mu}=0.587$ (95% CI: 0.302 to 0.871), but the value should now be interpreted as the estimated *average* true effect of the treatment. In fact, we estimate that $I^2=81.70\%$ of the total variability can be attributed to heterogeneity, so it seems implausible that the true outcomes are homogeneous.

We can try to explain some of the heterogeneity with the dosage moderator. The meta-regression model suggests a significant relationship between dosage and the treatment effect, with the true log odds ratio increasing by $\hat{\beta}_1=0.008$ points for each additional

TABLE 11.2

Results for the Common-Effect, Random-Effects, Meta-Regression, and Mixed-Effects Models When Applied to the Data in Table 11.1 (Standard Error of the Estimates Are Given in Parentheses)

Common-effect model	Random-effects model	Meta-regression model	Mixed-effects model
$\hat{\theta}=0.529$	$\hat{\mu}=0.587$	$\hat{\beta}_0=-0.423$	$\hat{\beta}_0=-0.304$
(0.0586)	(0.1451)	(0.1333)	(0.2061)
$z=9.03$	$z=4.04$	$z_0=-3.18$	$z_0=-1.48$
		$\hat{\beta}_1=0.008$	$\hat{\beta}_1=0.007$
		(0.0010)	(0.0015)
		$z_1=7.96$	$z_1=4.76$
	$\hat{\tau}^2=0.317$		$\hat{\tau}^2=0.090$

milligram ($z_1=7.96$, so we can reject H_0: $\beta_1=0$). The mixed-effects model leads to a similar conclusion, except that $\hat{\beta}_1=0.007$ now reflects how the *average* true effect changes as a function of the treatment dosage (with $z_1=4.76$, the relationship is still significant). Both the random- and mixed-effects models indicate the presence of (residual) heterogeneity, although the dosage moderator appears to account for a substantial amount thereof. The pseudo R^2 statistic indicates that about $100\% \times (0.317-0.090)/0.317=72\%$ of the total amount of heterogeneity is accounted for by the dosage moderator alone.

Results for the models are also illustrated graphically in Figure 11.1. The left-hand side of the figure shows a forest plot of the observed outcomes for the individual studies (with approximate 95% CI bounds for θ_i given by $y_i \pm 1.96 s_i$) and the results from the common-effect (CE) and random-effects (RE) models indicated at the bottom of the figure in terms of polygon shapes (with the center corresponding to the estimate and the ends corresponding to the 95% CI bounds). The right-hand side of Figure 11.1 shows a scatterplot of the dosage moderator on the x-axis versus the observed outcomes on the y-axis with the area of the points drawn proportional to the inverse sampling variances (i.e., $1/s_i^2$).

11.4 Checking Model Assumptions

The models described in Section 11.2 could be considered the main "workhorses" in meta-analytic applications. However, the models make various assumptions, which may be violated in practice. One of the crucial assumptions concerns the form of the sampling distribution of the chosen effect size or outcome measure. In particular, all of the models assume that $y_i \mid \theta_i \sim N(\theta_i, s_i^2)$, that is, conditional on θ_i, each observed outcome y_i is assumed to be drawn from a sampling distribution that is normal with expected value θ_i and variance s_i^2, with s_i^2 assumed to be a known and fixed quantity. We will now examine these assumptions in more detail and consider their plausibility.

11.4.1 Normal Sampling Distributions

Let us first examine the normality assumption in more detail. To begin with, it is important to emphasize that this assumption does not pertain to the collection of observed

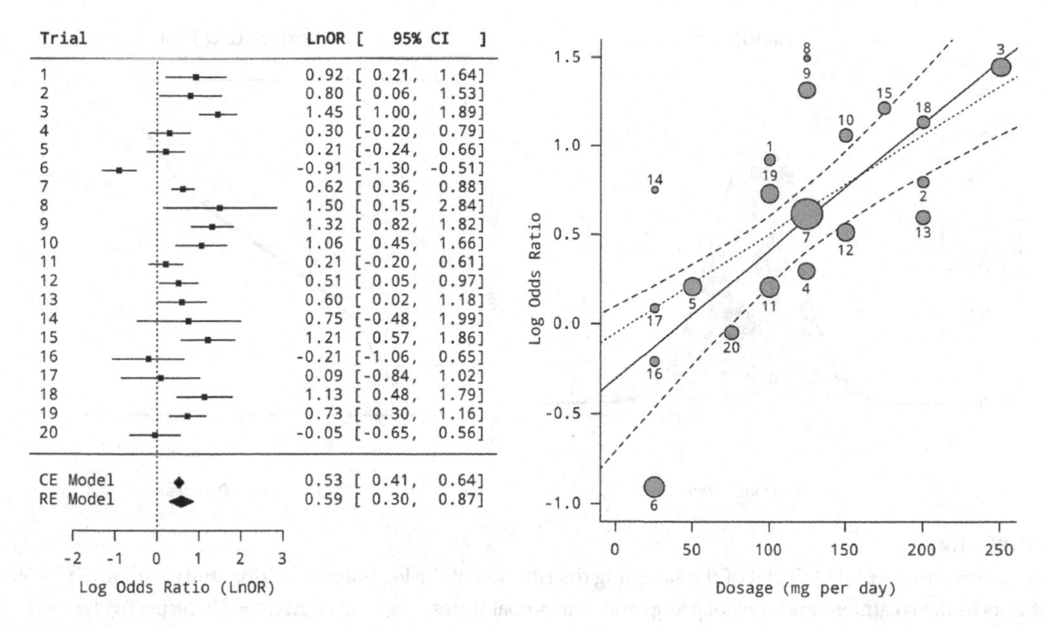

FIGURE 11.1
Forest plot showing the results of the CE and RE models and a scatterplot of the dosage moderator against the log odds ratios for the illustrative data in Table 11.1. The solid line in the scatterplot corresponds to the regression line (with intercept $\hat{\beta}_0$ and slope $\hat{\beta}_1$) for the mixed-effects model, with the dashed lines indicating (pointwise) 95% CIs around the predicted values. For reasons to be discussed further, the dotted line shows the regression line from the mixed-effects model when omitting study 6 from the dataset.

outcomes included in a meta-analysis (i.e., we do not assume that the observed y_i values themselves are normally distributed), but the theoretical distribution that would arise if a certain study were to be repeated a large number of times under identical circumstances. We shall construct this distribution now for one of the studies included in the illustrative dataset, again focusing on the log odds ratio as the chosen outcome measure.

Consider study 14, which included $n_i^T = n_i^C = 40$ patients in each group. Now, in order to construct the sampling distribution of the log odds ratio for this study, we would have to know π_i^T and π_i^C, the true probabilities of remission for patients in the treatment and control group, respectively. Obviously, these values are unknown to us, but we can use the observed proportions from this study (i.e., x_i^T / n_i^T and x_i^C / n_i^C) as an indication of the true probabilities and, for illustration purposes, simply assume that $\pi_i^T = .20$ and $\pi_i^C = .10$. Therefore, the true log odds ratio for this study would be equal to $\theta_i = \ln[(.20/(1-.20))/(.10/(1-.10))] = 0.811$. By taking random draws from the binomial distributions $B(n_i^T = 40, \pi_i^T = .20)$ and $B(n_i^C = 40, \pi_i^C = .10)$, we can then easily generate values for x_i^T and x_i^C, which we can use to calculate the log odds ratio as given by (11.13). By repeating this process a large number of times under identical circumstances (i.e., keeping n_i^T, n_i^C, π_i^T, and π_i^C the same), we can generate the sampling distribution of the log odds ratio under the described scenario. Use of the 0.5 adjustment term as described earlier guarantees that we can compute the log odds ratio in every iteration of such a simulation.

Figure 11.2 shows the shape of the distribution generated in this manner (after 10^7 iterations) in terms of a histogram and a normal quantile-quantile (Q-Q) plot. The solid line

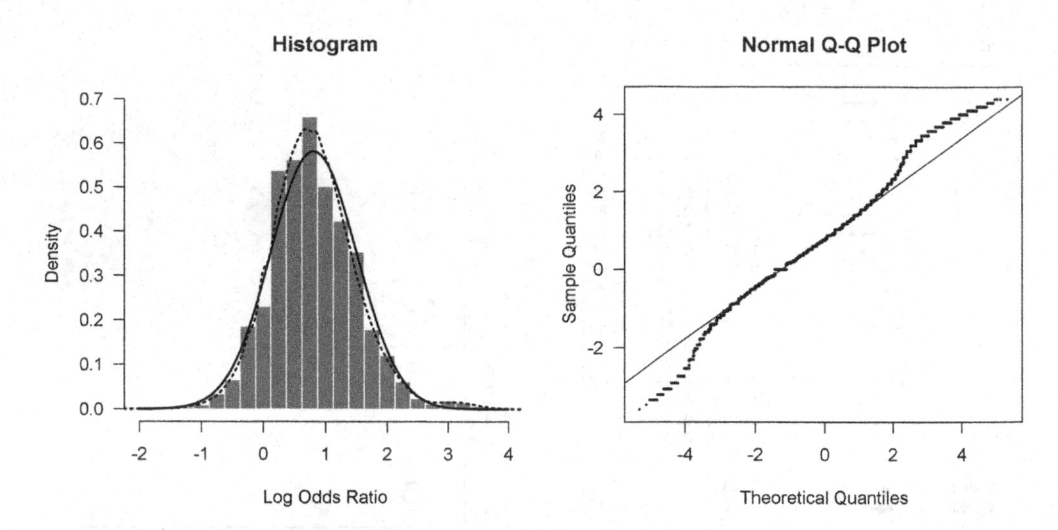

FIGURE 11.2
Histogram and normal Q-Q plot of the sampling distribution of the log odds ratio for a study with $n_i^T = n_i^C = 40$ patients in the treatment and control group and true probabilities of $\pi_i^T = .20$ and $\pi_i^C = .10$, respectively.

superimposed on the histogram corresponds to a normal distribution with mean and standard deviation equal to that of the simulated log odds ratios, while the dotted line represents a kernel density estimate of the underlying distribution. The log odds ratios ranged from −3.570 to 4.394 with a mean of 0.812 (SD=0.687). Therefore, under the simulated conditions, (11.13) provides an essentially unbiased estimate of $\theta_i = 0.811$. However, the shape of the sampling distribution deviates slightly from that of a normal distribution, especially in the tails. Although the departure from normality may be negligible in this case, the example does illustrate that we cannot assume that the normality assumption is automatically fulfilled.

In fact, the normality assumption is only approximately true for most of the commonly used outcome measures that are applied in the meta-analytic context. Certainly, for measures that are bounded (e.g., the raw correlation coefficient or the risk difference), the assumption cannot be true in general, but it may still hold as a rough approximation. At the same time, for unbounded measures (such as the log odds ratio or the standardized mean difference), it may be quite inaccurate under certain circumstances. In particular, when sample sizes within studies are small and/or when the underlying parameters are very large or small (e.g., π_i^T and π_i^C are close to 0 or 1 for measures such as the log odds/risk ratio or the true standardized mean difference is far from 0), the assumption can break down altogether. Also, due to the discrete nature of the counts on which they are based, measures such as the log odds ratio can only generate a finite number of unique values. For example, in the simulation above, only 310 unique values of the log odds ratio were observed across the 10^7 iterations.* Strictly speaking, the sampling distribution cannot be normal then, but we can still consider the values to be discretized observations from an underlying normal distribution.

* Technically, $41^2 = 1681$ combinations of x_i^T and x_i^C are possible, although some of them (e.g., those where $x_i^T = x_i^C$) lead to the same value of y_i. Simple enumeration of all possibilities shows that there are only 1083 unique values of y_i that could be observed, but many of them are so unlikely to happen that they do not occur even once across such a large number of iterations.

11.4.2 Unbiased Estimates

The second assumption implied by the models is that y_i is an unbiased estimate of θ_i. In the simulation above, the log odds ratio computed with (11.13) was found to be essentially unbiased in the given scenario. Moreover, prior work has shown that (11.13) yields an approximately unbiased estimator across a wide range of conditions (e.g., Walter, 1985; Walter and Cook, 1991). Interestingly, this work also indicates that the 0.5 adjustment used in the equation should always be applied, not just in studies where the presence of a zero cell in the 2×2 table would necessitate its use.

Similarly, the bias in estimators for other outcome measures has also been examined, leading to known bias corrections for the standardized mean difference (Hedges, 1981), the raw correlation coefficient (Olkin and Pratt, 1958), and Fisher's r-to-z transformed correlation coefficient (Hotelling, 1953). On the other hand, the risk difference and the raw mean difference are unbiased by construction. Generally, for estimators that are biased, the amount of bias usually diminishes quickly as the within-study sample size increases.

11.4.3 Known Sampling Variances

Next, we will consider the assumption that the sampling variances of the observed outcomes are known and fixed quantities. This assumption is often not exactly true for two reasons. First, for many outcome measures, commonly used equations to compute the sampling variances are based on derivations that describe the asymptotic (i.e., large-sample) properties of the outcome measures. Hence, in finite samples, the equations may not be accurate. Second, the equations often depend on unknown parameters. To compute the sampling variances, these unknown parameters must be replaced by corresponding estimates, which introduces error into the values.

Consider again the log odds ratio. The asymptotic sampling variance of y_i can be shown to be equal to

$$s_i^2 \overset{\infty}{=} \frac{1}{n_i^T \pi_i^T} + \frac{1}{n_i^T (1-\pi_i^T)} + \frac{1}{n_i^C \pi_i^C} + \frac{1}{n_i^C (1-\pi_i^C)}, \tag{11.15}$$

where π_i^T and π_i^C are the true probabilities of the event of interest occurring in the treatment and control group, respectively. Note that this equation is technically only correct when n_i^T and n_i^C are sufficiently large, which raises the question how large the groups need to be for the equation to be accurate. Moreover, since π_i^T and π_i^C are unknown parameters, the observed probabilities $p_i^T = x_i^T / n_i^T$ and $p_i^C = x_i^C / n_i^C$ are typically substituted, leading to equation (11.14) (with the addition of the 0.5 term to make the computation of s_i^2 possible under all circumstances). The resulting s_i^2 values are therefore not truly fixed and known quantities but are estimates themselves.

We can illustrate this again using the simulated data from the previous section. Given the assumed values of $\pi_i^T = 0.20$ and $\pi_i^C = 0.10$, we can compute the (large-sample) variance with (11.15), which is equal to $s_i^2 \overset{\infty}{=} 0.434$ in this case. However, taking the variance of the simulated y_i values yields $\text{Var}[y_i] = 0.472$, which we can consider to be the true sampling variance of the log odds ratio under the given scenario (the simulation error is negligible due to the very large number of values generated). Therefore, the actual sampling variance is about 9% larger than what we obtain with (11.15), so some inaccuracy is introduced by basing our computations on an equation that describes the large-sample properties of the log odds ratio.

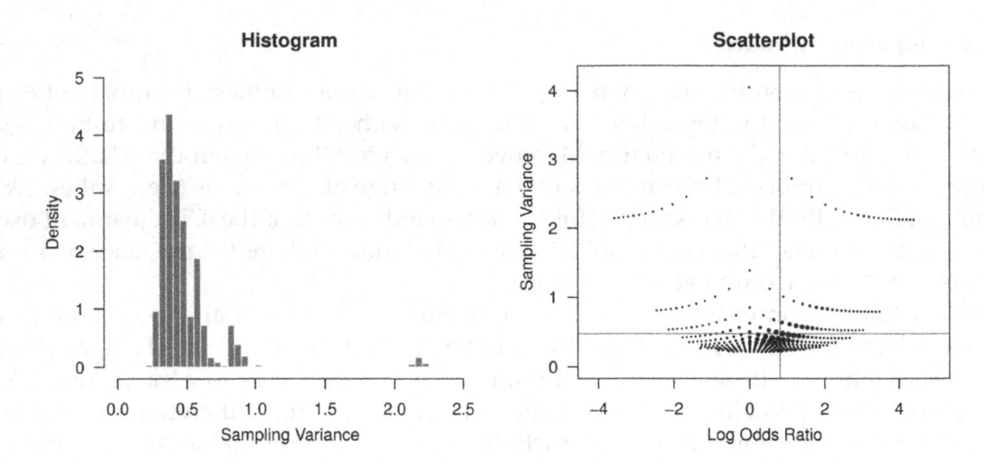

FIGURE 11.3
Distribution of the observed sampling variances of the log odds ratio for a study with $n_i^T = n_i^C = 40$ patients in the treatment and control group and true probabilities of $\pi_i^T = .20$ and $\pi_i^C = .10$, respectively.

However, neither of these two values would be available in practice. Instead, one would compute the sampling variance with (11.14) by plugging in the x_i^T and x_i^C values observed in a given sample. As a result, the calculated s_i^2 value may under- or overestimate the true sampling variance, sometimes to a considerable degree. The simulation above allows us to explore the range of values one could obtain under the stated scenario. Across all iterations, the s_i^2 values ranged from 0.210 to 4.049, although very large values were rare, as can be seen in the histogram of the observed s_i^2 values in Figure 11.3. Values above 2.5 were only seen in 316 out of the 10^7 iterations (and hence, the x-axis in the histogram was restricted to a range of 0 to 2.5) and usually fell below 1 in 98% of the cases. However, even then, the observed s_i^2 value could still provide a rather inaccurate estimate of the true sampling variance (i.e., 0.472 as noted above). At least somewhat reassuring is the finding that the average of the observed values was 0.482, so (11.14) provides an approximately unbiased estimate of the true sampling variance in this scenario.

Interestingly, if we plot the observed log odds ratios against the estimated sampling variances, a peculiar pattern emerges as can be seen in the right-hand side of Figure 11.3 (the vertical and horizontal lines indicate the true log odds ratio and true sampling variance, respectively). The plot illustrates the discrete nature of possible values that can actually arise (and therefore, the size of the points was drawn proportional to the number of observed values for a particular $\langle y_i, s_i^2 \rangle$ combination). Clearly, the two sets of values are related to each other (in this example, the correlation between the y_i and s_i^2 values is 0.57), a fact that has been noted before (Berkey et al., 1995; Rücker et al., 2008). In summary then, it seems inappropriate to assume that the sampling variances of the observed outcomes are really fixed and known quantities, a point that has been emphasized in Chapter 5.

In general, the problem that the sampling variances are inaccurately estimated (and hence, falsely treated as fixed and known quantities) tends to become more severe in smaller samples and when the underlying parameters are very large or small. The same issue applies, to a larger or lesser extent, to many other outcome measures used in meta-analysis, such as the risk difference, the log risk ratio, the raw and the standardized mean difference, and the raw correlation coefficient. An exception to this are measures based on a variance stabilizing transformation, most notably Fisher's r-to-z transformed correlation coefficient (Fisher, 1921) and the arcsine (and square root) transformed risk difference (Rücker et al., 2009), although even here we need to be precise: The variance stabilizing transformation eliminates the unknown parameters from the equation used to compute the sampling variance, but the resulting equation is still in principle a large-sample approximation. However, for both the r-to-z correlation coefficient and the arcsine transformed risk difference, the approximation is surprisingly accurate even for relatively small studies.

11.4.4 Uncorrelated Errors and Random Effects

One additional assumption pertains to the random- and mixed-effects models. Recall that these models contain two random effects, namely u_i, which represents a deviation of the study-specific true outcome from the average true outcome of all studies (or from the average true outcome for those studies that share a particular combination of moderator variables) and e_i, the sampling error that represents a deviation of the observed outcome from the study-specific true outcome. The models assume that u_i and e_i are independent.

Once again, we will examine whether this assumption is appropriate for the log odds ratio by means of a simulation study. Here, we proceed as follows. As before, assume we are dealing with a study with $n_i^T = n_i^C = 40$ patients in each group and that $\pi_i^C = .10$. For a given true log odds ratio, θ_i, we can compute the implied value of the treatment group probability with $\pi_i^T = \pi_i^C \exp(\theta_i) / (1 - \pi_i^C + \pi_i^C \exp(\theta_i))$. For example, for $\theta_i = 0.811$, this yields $\pi_i^T = .20$ as expected. However, instead of fixing θ_i, we now let θ_i follow a normal distribution, as assumed by the random-effects model. Specifically, we draw u_i from $N(0, \tau^2 = .20)$ and then compute $\theta_i = \mu + u_i$ with $\mu = 0.811$. Next, we compute π_i^T and then draw x_i^T and x_i^C values from the respective binomial distributions. Finally, we can compute y_i as described earlier and then $e_i = y_i - \theta_i$ yields the sampling error for one iteration.

Repeating this process 10^6 times yields pairs of $\langle u_i, e_i \rangle$ values, whose bivariate distribution we can now examine, for example, by means of the scatterplot shown on the left-hand side of Figure 11.4 (due to the large number of points drawn, alpha blending* was used to make differ-

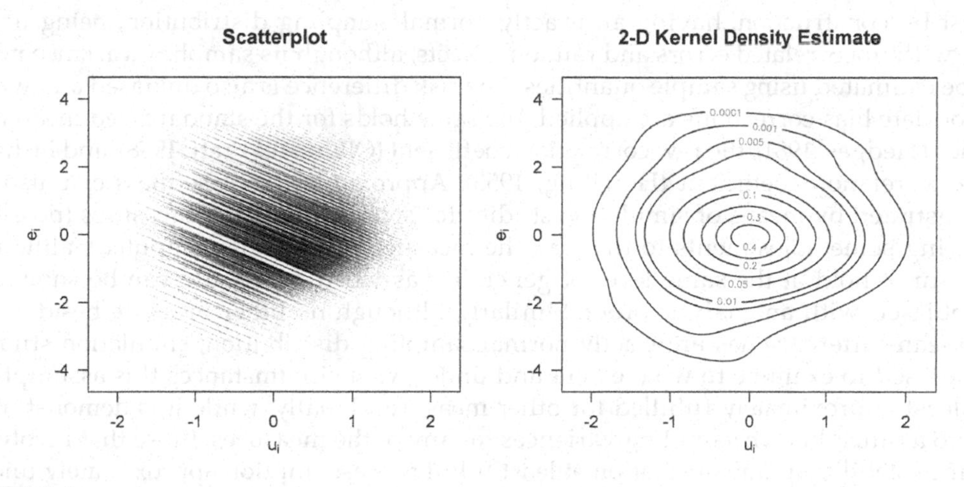

FIGURE 11.4
Scatterplot and 2D kernel density estimate of the bivariate distribution of u_i and e_i.

* In alpha blending, each point is drawn with a certain degree of transparency. Multiple points drawn on top of each other then blend together to create increasing darker shades of the plotting color. As a result, the color intensity indicates the density of points within a particular region of the plot.

ences in the density of the distribution more apparent). The right-hand side shows a 2D kernel density estimate of the bivariate distribution. The patterns in the scatterplot are again a result of the discrete nature of the distribution of y_i values that can arise. However, the kernel density estimate suggests a unimodal, roughly symmetric distribution with no apparent relationship between the u_i and e_i values (the correlation is zero to three decimals). Although this does not demonstrate independence (zero correlation only implies independence if $\langle u_i, e_i \rangle$ would follow a bivariate normal distribution), it does provide some support for the independence assumption in this scenario.

To what extent the independence assumption holds in other scenarios, and more generally, for other outcome measures besides the log odds ratio, has not been examined in detail in the literature. For the raw mean difference, u_i and e_i are independent by construction. On the other hand, for measures where the shape of the sampling distribution depends on the size of the underlying parameter estimated (e.g., the raw correlation coefficient), it is easy to reason that u_i and e_i will not form a bivariate normal distribution with zero correlation, especially in small samples. However, further research is needed to determine if or to what degree u_i and e_i are dependent in such cases.

11.4.5 Implications and Some General Remarks

The log odds ratio was given particular attention in this section, since this was the outcome measure of choice in the example meta-analysis. However, analogous considerations apply to other outcome measures for dichotomous response variables (e.g., the log risk ratio and the risk difference), outcome measures for continuous response variables (e.g., the raw mean difference and the standardized mean difference), and outcome measures used to quantify the relationship between variables (e.g., the raw or Fisher's r-to-z transformed correlation coefficient).

In fact, a careful examination shows that there is not a single outcome measure used in meta-analysis that fulfills all of the assumptions described above exactly. Most of them actually fulfill very few of them! The raw mean difference could be argued to comes closest by construction, having an exactly normal sampling distribution, being unbiased, with uncorrelated errors and random effects, although its sampling variance must still be estimated using sample quantities. The risk difference is also unbiased and when appropriate bias corrections are applied, the same holds for the standardized mean difference (Hedges, 1981), the raw correlation coefficient (Olkin and Pratt, 1958), and Fisher's r-to-z correlation coefficient (Hotelling, 1953). Approximate unbiasedness can also be demonstrated by means of simulation studies for some of the other measures (possibly requiring some adjustments to the way the measure is typically computed), although this cannot hold at the same level of generality as when a measure can be shown to be unbiased with an analytic proof. Similarly, although no other measure besides the raw mean difference has an exactly normal sampling distribution, simulation studies can be used to examine to what extent and under what circumstances this assumption is at least approximately fulfilled for other measures. Finally, while it is demonstrably false to assume known sampling variances for any of the measures, those that involve a variance stabilizing transformation at least fulfill this assumption approximately under most circumstances.

At the same time, it can be shown for essentially all measures that the underlying estimators are consistent, asymptotically unbiased, and that their sampling distribution approaches the shape of a normal distribution as the within-study sample size increases (note that for measures that reflect a contrast between two independent samples—such as the log odds ratio or the standardized mean difference—this requires that both n_i^T and n_i^C increase with similar rates).

The discussion above then raises two important questions. First, under what circumstances do the assumptions break down to such an extent that we should no longer apply these methods and models? And second, what are the consequences when one or more of the assumptions are violated? Unfortunately, there are no simple answers to these questions, because the answers depend on various factors, including the outcome measure under consideration, the size of the studies, and whether underlying parameters may be extreme in some sense and/or close to their natural bounds. However, for all measures, there are circumstances where serious concerns should be raised about their use, whether due to bias, non-normality, or other violations. As a result, point estimates may not be trustworthy and/or tests and CIs may not have their nominal properties (i.e., the actual type I error rate of hypothesis tests and the actual coverage of CIs may deviate from the chosen significance/confidence level).

There is a large amount of literature that has examined the statistical properties of the methods and models described above for a wide variety of outcome measures and conditions (e.g., Berkey et al., 1995; Field, 2001; Friedrich et al., 2008; Hauck et al., 1982; Hedges, 1982a, 1982b, 1982c; Sánchez-Meca and Marín-Martínez, 2001 just to give a small selection). While it would require many more pages to discuss the details, the general conclusion is that for many of the measures, the methods perform adequately, even when the assumptions discussed above are not exactly fulfilled. However, this should not be taken as an *a priori* truth. Moreover, most studies have examined the properties of the methods under circumstances where the assumptions underlying the *construction* of the outcome measures are exactly fulfilled (e.g., for the standardized mean difference, the data within the two groups was simulated from normal distributions). More work is still needed to examine the robustness of the methods when such underlying assumptions are also violated.

11.5 Checking Model Fit

The assumptions discussed in the previous section relate to the statistical properties of the outcome measure chosen for the meta-analysis and should be carefully considered before we even start fitting models of the type described in Section 11.2. Once we are satisfied that these assumptions are (at least approximately) fulfilled, the actual analysis can then proceed by fitting one or more models to the data at hand.

It is important to emphasize that a statistical model in essence represents an assumption about the underlying data generating process. For example, when we fit the common-effect model to the data from the illustrative example, we implicitly assume that the data within the individual studies arose in such a way that the true treatment effect (as measured in terms of the log odds ratio) is constant across the trials, irrespective of any differences in the way the studies were designed, conducted, how the treatment was implemented/administered, how the dependent variable was measured, or any relevant patient characteristics. This is unlikely to be exactly true but may hold as a rough approximation in some applications (e.g., when a series of trials is conducted using identical methods in very similar patient populations).

Similarly, if we reject the assumption that the true outcomes are homogeneous across trials and decide to fit a meta-regression model with medication dose as a moderator, we implicitly assume that any differences in treatment effectiveness across trials can be accounted for with this single explanatory variable. Again, this is likely to be a great over-simplification of a much more complex reality, but it may serve as an adequate approximation. To check whether these assumptions are actually appropriate, we can apply tests for model misspecification, which are covered next.

11.5.1 Testing for Model Misspecification

The homogeneity assumption underlying the common-effect model can be tested by means of the Q-test. Under the assumptions of the model, the test statistic

$$Q = \sum w_i (y_i - \hat{\theta})^2 \tag{11.16}$$

follows a chi-square distribution with $k-1$ degrees of freedom, where $w_i = 1/s_i^2$ and $\hat{\theta}$ is the estimate of θ under the common-effect model. However, when the observed outcomes fluctuate more around $\hat{\theta}$ than expected based on their sampling variances alone (which should be the only source of variance affecting the y_i values under the common-effect model), the Q-statistic increases. Therefore, a large Q-statistic should lead us to question the correctness of the assumed model.

Similarly, the meta-regression model (11.3) assumes that all of the heterogeneity in the true outcomes can be accounted for with the moderator variables included in the model, or analogously, that the model $\theta_i = x_i\beta$ is correctly specified. Again, this assumption can be tested, using a generalization of the Q-test to the meta-regression model. Here, the test statistic is computed with

$$Q_E = \sum w_i (y_i - \hat{\theta}_i)^2, \tag{11.17}$$

where $\hat{\theta}_i = x_i\hat{\beta}$ is the fitted value for the ith study from the meta-regression model (see Section 11.2.5) and $w_i = 1/s_i^2$ as before (note that this is the same Q_E statistic that is also involved in the estimator for τ^2 given by equation 11.5). If the assumed meta-regression model correctly describes the underlying data generating process, then the Q_E statistic follows a chi-square distribution with $k-q-1$ degrees of freedom. Again, as the degree of mismatch between the observed outcomes and the fitted values under the assumed model increases (i.e., more so than would be expected based on sampling variability alone), the Q_E statistic grows larger.

Although not frequently described in this manner, we can therefore consider (11.16) and (11.17) to be tests of model misspecification (Hedges, 1992). Rejection of the null hypothesis (that the model is adequately specified) should then be taken as an indication that the assumed model does not provide an adequate approximation to the underlying data generating process. At the same time, non-rejection must be cautiously interpreted, as the tests may lack power especially when the number of studies is small.

For the example data, we find $Q = 103.81$, a value so extreme that the chances of it (or an even larger value) occurring under a chi-square distribution with 19 degrees of freedom are extremely small (i.e., $p < 0.0001$). Consequently, we would reject the common-effect model as a plausible approximation. The same applies to the meta-regression model, for which we find $Q_E = 40.44$, also a rather unlikely occurrence under a chi-square distribution with 18 degrees of freedom (i.e., $p = 0.002$).

There are no analogous versions of these types of tests for the random- and mixed-effects models. However, we can resort to a different approach by examining the residuals (and standardized versions thereof) computed from the fitted model, which may reveal a mismatch between particular data points and the assumed model.

11.5.2 Residuals and Standardized Versions Thereof

The raw residual for the *i*th study is $y_i - \hat{\mu}_i$ (in the common- and random-effects models, $\hat{\mu}_i$ is simply $\hat{\theta}$ and $\hat{\mu}$, respectively). The raw residuals should scatter randomly around the fitted values, but are not very useful for diagnostic purposes, as they do not account for differences in the sampling variances across studies. Moreover, in the random-effects and mixed-effects models, (residual) heterogeneity represents an additional source of variability, which needs to be taken into consideration. Instead, we can compute Pearson (or semi-standardized) residuals, which are given by

$$r_i = \sqrt{w_i}(y_i - \hat{\mu}_i), \tag{11.18}$$

where $w_i = 1/s_i^2$ in the common-effect and meta-regression models and $w_i = 1/(s_i^2 + \hat{\tau}^2)$ in the random- and mixed-effects models. As can be seen from (11.16) and (11.17), for the common-effect and meta-regression models (where $\hat{\mu}_i = \hat{\theta}$ and $\hat{\mu}_i = \hat{\theta}_i$, respectively), the Q and Q_E statistics are just the sum of the squared Pearson residuals, or put differently, r_i^2 is the contribution of the *i*th study to these statistics.

However, Pearson residuals are not properly standardized (in the sense of having unit variances), as they do not account for the imprecision in the $\hat{\mu}_i$ values. In fact, it follows from the results laid out in Section 11.2.5 that the variance of the raw residual in the *i*th study can be estimated with $(1-h_i)(s_i^2 + \hat{\tau}^2)$, where h_i is the *i*th diagonal element from the hat matrix $\boldsymbol{H}$ (for the common-effect and meta-regression models, $\hat{\tau}^2 = 0$ by definition). Therefore, the (internally) standardized residuals can be computed with

$$\tilde{r}_i = \frac{y_i - \hat{\mu}_i}{\sqrt{(1-h_i)(s_i^2 + \hat{\tau}^2)}}, \tag{11.19}$$

which have approximately unit variances (imprecision is introduced due to s_i^2 and $\hat{\tau}^2$ being estimates themselves).

We will consider one other type of residual, which is related to the deletion diagnostics to be discussed in more detail further below. Here, we compute the residual of a study based on a model that excludes the study during the model fitting process. Therefore, we first delete the *i*th study from the dataset and then fit the model of choice using the remaining $k-1$ studies. When fitting a random- or mixed-effects models, let $\hat{\tau}^2_{(-i)}$ denote the estimate of τ^2 from this model (i.e., the $(-i)$ part in this and other subscripts will be used to indicate that the value was computed from the fitted model that excluded the *i*th study from the model fitting). Next, we compute the predicted value for the study that was deleted, which we denote by $\hat{\mu}_{i(-i)}$. Furthermore, let $\text{Var}[\hat{\mu}_{i(-i)}]$ denote the corresponding variance of the predicted value. Then we define the "deleted residual" as $r_{i(-i)} = y_i - \hat{\mu}_{i(-i)}$. Finally,

$$t_i = \frac{y_i - \hat{\mu}_{i(-i)}}{\sqrt{s_i^2 + \hat{\tau}^2_{(-i)} + \text{Var}[\hat{\mu}_{i(-i)}]}} \tag{11.20}$$

yields the standardized deleted residual (or externally standardized residual), which again has approximately unit variance (Viechtbauer and Cheung, 2010). Note that $\hat{\tau}^2_{(-i)} = 0$ for the common-effect and meta-regression models by definition. Also, the notation can be further simplified depending on the model (i.e., $\hat{\mu}_{i(-i)} = \hat{\theta}_{i(-i)}$ for the meta-regression model, $\hat{\mu}_{i(-i)} = \hat{\mu}_{(-i)}$ for the random-effects model, and $\hat{\mu}_{i(-i)} = \hat{\theta}_{(-i)}$ for the common-effect model).

11.5.3 Checking for Outliers

Standardized (deleted) residuals are useful for detecting outliers, that is, studies that do not fit the assumed model. The advantage of the deleted residuals is that they are more sensitive to detecting outliers. In particular, if a study does not fit the assumed model, then this affects the results in two ways. First, the study will introduce additional heterogeneity into the data (i.e., $\hat{\tau}^2$ tends to increase), which will get subsumed into the standard deviation of the residuals, as shown in the denominator of (11.19). This will shrink the standardized residuals toward 0 to some degree, making it more difficult to detect the outlying study. In addition, the y_i value of an outlying study will pull $\hat{\mu}_i$ toward it, leading to a smaller raw residual, and hence, a smaller standardized residual. By first deleting a potentially outlying study from the dataset, both of these effects are eliminated, making the standardized deleted residual for the study a more sensitive indicator of whether the study fits the model or not.

Assuming that the fitted model is the correct one for the data at hand, the probability that a standardized (deleted) residual is larger than ±1.96 is approximately 5%. Therefore, while we may expect to observe one or maybe two large values in a set of $k=20$ studies, the presence of a large number of such values would indicate that the assumed model does not represent a good approximation to the underlying data generating process. Based on the binomial distribution (with $\pi=0.05$ and $k=20$), the chances of observing one or more, two or more, and three or more standardized (deleted) residuals this large are 64%, 26%, and 8%, respectively.

For the example data, we find four standardized deleted residuals larger than ±1.96 for the common-effect model and three for the meta-regression model, again suggesting a mismatch between the data and these models. How do the random- and mixed-effects models fare? Figure 11.5 shows the standardized deleted residuals for these models. Only one and two values larger than ±1.96 are found in these models, respectively, so we do not have grounds to question the adequacy of these models in general terms. However, study 6 appears to be an outlier in both of these models, which is also recognizable from inspection of the plots in Figure 11.1. In the mixed-effects model, the standardized deleted residual of study 9 is also larger than 1.96, although just barely.

Interestingly, the scatterplot in Figure 11.1 shows that the estimate of study 8 is further away from the regression line than that of study 9 (which may suggest that study 8 is more of an outlier than study 9), but the standardized deleted residual for the latter is larger. This seeming contradiction can be explained by the fact that study 9 included almost seven times as many participants and hence has a much smaller sampling variance than study 8. Therefore, a deviation as large as the one for study 8 could be accounted for based on its sampling variability (plus residual heterogeneity and variance in the predicted value), while the residual of study 9 is more unusual given its smaller sampling variance. Also, Figure 11.1 suggests that study 14 has an unusually large outcome for a study with such

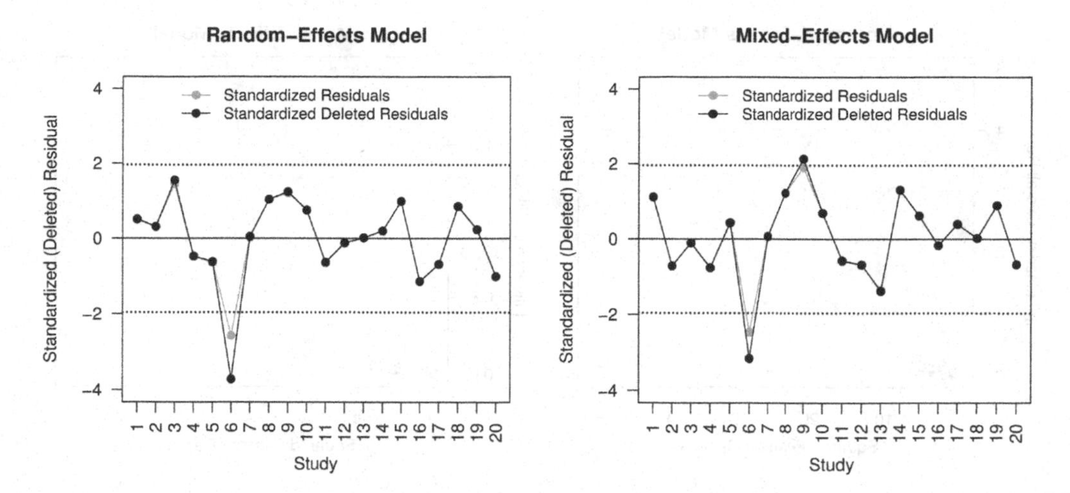

FIGURE 11.5
Standardized (deleted) residuals for the random- and mixed-effects models.

a low dosage, but its standardized deleted residual indicates that this deviation is not anomalous under the model. Hence, while forest and scatterplots can be useful visual aids for detecting outlying studies, they can also be deceptive.

Besides the standardized deleted residuals, Figure 11.5 also shows the regular standardized residuals computed with (11.19). While there is not much of a difference between these two types of residuals for most studies, we do see how the standardized deleted residuals are more sensitive to potential outliers, especially for studies 6 and 9. For example, for study 6, the deleted residual (i.e., the deviation of the outcome of the study from the dotted regression line in the scatterplot in Figure 11.1) is considerably larger than the regular residual (i.e., the deviation from the solid regression line). In addition, $\hat{\tau}^2_{(-6)} = 0.032$ is much smaller than the estimate of residual heterogeneity when all studies are included (i.e., $\hat{\tau}^2 = 0.090$), further leading to a more sensitive test.

11.5.4 Baujat and GOSH Plots

Several other graphical tools have been suggested in the literature for detecting outliers and sources of (residual) heterogeneity in meta-analytic data. We will now consider two of these devices and illustrate their use by applying them to the example dataset.

First, we will consider a type of plot suggested by Baujat and colleagues (Baujat et al., 2002). As originally described, the plot shows the contribution of each study to the Q-statistic on the x-axis versus the influence of each study on the overall estimate from a common-effect model on the y-axis. However, the idea underlying this type of plot can be easily generalized to random-effects models and models including moderators (with the common-effect model again forming a special case). In general, we plot the squared Pearson residual of each study (i.e., r_i^2) on the x-axis against

$$\frac{(\hat{\mu}_i - \hat{\mu}_{i(-i)})^2}{\mathrm{Var}[\hat{\mu}_{i(-i)}]}, \tag{11.21}$$

that is, the standardized squared difference between the predicted/fitted value for a study with and without the study included in the model fitting. Hence, a study whose observed

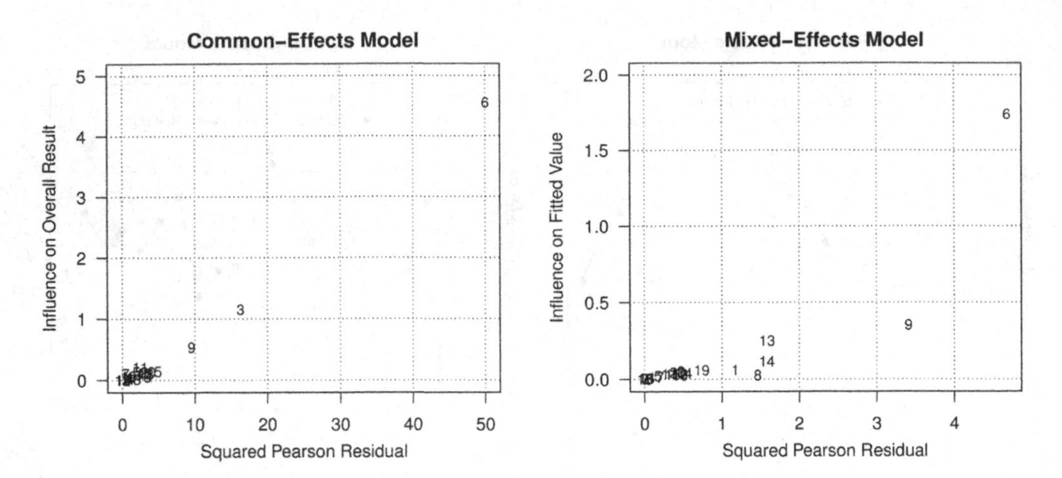

FIGURE 11.6
Baujat plots for the common-effect and mixed-effects models.

outcome deviates strongly from its predicted value based on the model will be located on the right-hand side of the plot. For the common-effect and the meta-regression models, these are the studies that contribute most to the Q (i.e., equation 11.16) and Q_E (i.e., equation 11.17) statistics. Furthermore, a study whose predicted value changes strongly depending on whether it is included or excluded from the dataset will be located on the top-hand side of the plot.

Figure 11.6 shows Baujat plots corresponding to the common-effect and mixed-effects models for the illustrative dataset. As we saw earlier, studies 6 and 9 show up again as apparent outliers in the context of the mixed-effects model. On the other hand, the plot for the common-effect model indicates that study 3 also contributes considerably to the overall amount of heterogeneity.

Another type of plot that is useful for detecting outliers and sources of heterogeneity is the so-called GOSH (graphical display of study heterogeneity) plot (Olkin et al., 2012). As originally described, the plot is constructed as follows. First, we fit the common-effect model to all possible subsets of size $1,\ldots,k$ of the k studies included in a meta-analysis. Therefore, at the one extreme, this will include k models that are each fitted to a single observed outcome (in which case $\hat{\theta} = y_i$), then the $\binom{k}{2}$ models fitted to all pairwise combinations of two observed outcomes, and so on, until we get to the original model using all k outcomes. In total, there are then $\sum_{i=1}^{k} \binom{k}{i} = 2^k - 1$ models that need to be fitted. We can then plot the model estimates obtained this way (e.g., as a histogram and/or using a kernel density estimate) to examine the resulting distribution. In a homogeneous set of studies, the distribution should be roughly symmetric, contiguous, and unimodal. On the other hand, when the distribution is multimodal, then this suggests the presence of heterogeneity, possibly due to the presence of outliers and/or distinct subgroupings of studies. Plotting the estimates against some measure of heterogeneity (e.g., I^2) computed within each subset can also help to reveal subclusters, which are indicative of heterogeneity.

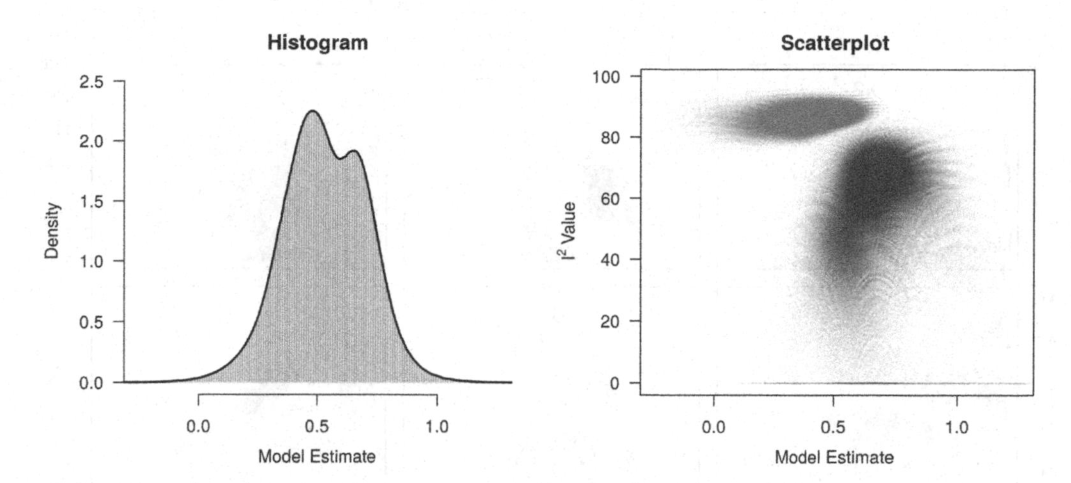

FIGURE 11.7
GOSH plot showing the distribution of estimates from the common-effect model based on all possible subsets and the bivariate distribution of the estimates and the corresponding I^2 values (results from subsets including study 6 are shown in light and dark gray otherwise).

For the illustrative dataset, a total of $2^{10}-1=1{,}048{,}575$ subsets can be constructed. The left-hand side of Figure 11.7 shows a histogram of the model estimates when fitting the common-effect model to each of these subsets (with a kernel density estimate of the distribution superimposed). The bimodal shape of the distribution is a result of study 6, which has considerable impact on the model estimate depending on whether the study is included in a subset or not. The influence of this study becomes even more apparent in the plot on the right-hand side of Figure 11.7, which shows a scatterplot of the model estimates against the corresponding I^2 values (using alpha blending). Results from subsets including study 6 are shown in light and dark gray otherwise. Inclusion of study 6 in a subset not only tends to reduce the size of the model estimate, but also increases the percentage of variability that can be attributed to heterogeneity.

The idea underlying the GOSH plot can be generalized to other types of models (including models with moderator variables) by examining the distribution of all model coefficients across subsets, plotting them against each other, and against some measure of (residual) heterogeneity (e.g., $\hat{\tau}^2$). Note that for a model with q moderator variables and an intercept term, a subset must include at least $q+1$ studies for the model to be estimable. When fitting a mixed-effects model, at least $q+2$ studies must be included in a subset so that τ^2 can also be estimated. Even then, a model may not be estimable in certain subsets. For example, for the (mixed-effects) meta-regression model with the dosage moderator, the subset including studies 6, 14, 16, and 17 does not allow estimation of β_1, since all four studies were conducted at the same dosage level of 25 mg. Therefore, only subsets where all parameters are estimable can be used for creating the plot.

Figure 11.8 shows such a generalized GOSH plot for the mixed-effects model (a total of 1,048,353 models could be fitted). The figure shows the distribution of $\hat{\tau}^2$, $\hat{\beta}_0$, and $\hat{\beta}_1$, and all pairwise scatterplots. Again, subsets including study 6 are shown in light and dark gray otherwise. As expected (cf. the scatterplot in Figure 11.1), for subsets that do include study 6, the estimate of residual heterogeneity tends to be higher, the model intercept tends to be lower, and the slope tends to be steeper. As a result, the distributions are again bimodal.

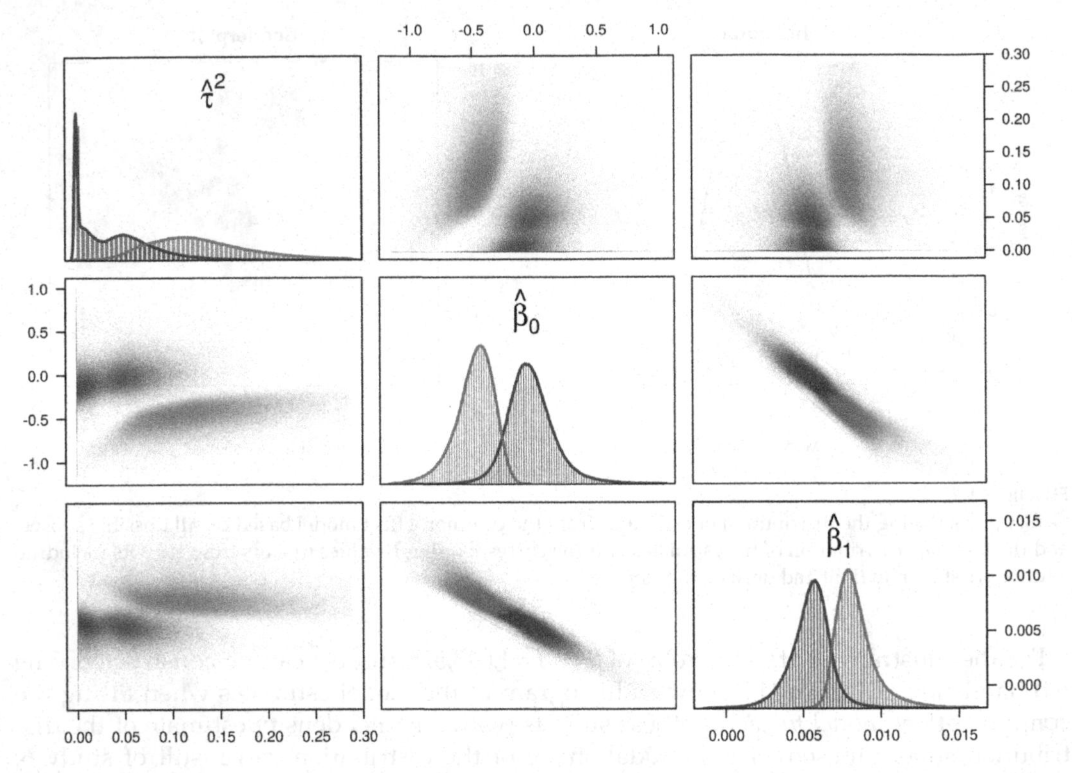

FIGURE 11.8
GOSH plot for the mixed-effects model showing the distribution of $\hat{\tau}^2$, $\hat{\beta}_0$, and $\hat{\beta}_1$ and all pairwise scatterplots based on all possible subsets (results from subsets including study 6 are shown in light and dark gray otherwise).

11.5.5 Testing for Lack of Linearity

For the illustrative dataset, we saw earlier that the meta-regression models suggest an increasing relationship between the treatment dosage and the (average) log odds ratio. However, when fitting meta-regression models that include continuous moderators, an aspect to consider is the linearity of the assumed relationship. Although the scatterplot in Figure 11.1 does not suggest any non-linearity of the relationship, several approaches can be used to examine the linearity assumption more systematically.

As in regular regression modeling (e.g., Kutner et al., 2004), one can examine a plot of each continuous moderator variable against the residuals from the (mixed-effects) meta-regression model to examine whether patterns are visible that may suggest potential non-linearity (e.g., a U- or an upside-down U-shape). For a model with a single continuous moderator variable, such a plot is not fundamentally different than just the scatterplot of the moderator variable against the observed outcomes, although it can be easier to detect patterns when the linear trend has been removed from the data by computing the residuals. Also, since the residuals are heteroscedastic (in part due to the heteroscedastic nature of the sampling variances), it can be useful to place the standardized (deleted) residuals on the y-axis (which should have roughly unit variance). The left-hand side of Figure 11.9 shows such a plot of the standardized residuals from the mixed-effects model. Again, we notice the outlier (study 6) in the lower left-hand corner of the plot, but otherwise no apparent curvature in the point cloud.

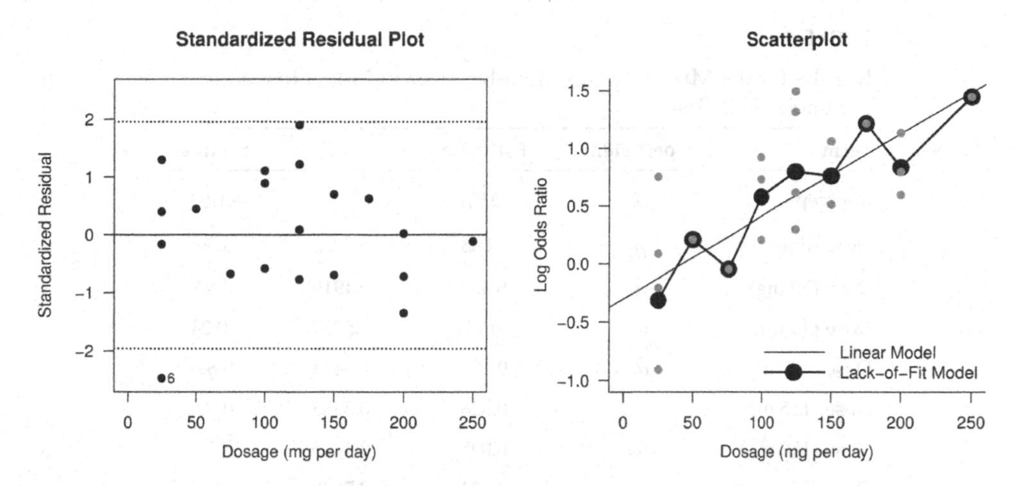

FIGURE 11.9
Plot of treatment dosage against the standardized residuals (from the mixed-effects model) and against the observed log odds ratios.

It is also possible to test more formally whether the relationship between dosage and the outcomes deviates from linearity. As a simple first approximation, one could consider fitting polynomial meta-regression models (e.g., adding the squared dosage as an additional moderator to the mixed-effects model). Doing so in the present case leads to a non-significant coefficient for the squared dosage term of the model ($z_2 = -1.26$) and hence no evidence of non-linearity.

The type of non-linearity that a polynomial model is most sensitive to is constrained by the degree of the polynomial included in the model (i.e., a quadratic polynomial model will be most sensitive to departures from linearity that are roughly quadratic in nature). Therefore, further models involving higher polynomial terms could be examined. However, when the dataset includes multiple observed outcomes at the same level of the moderator (i.e., replicates), we can also conduct a more general lack-of-fit test that is much more flexible in terms of the types of non-linearity it can detect. For this, we need to extend the lack-of-fit test from regular regression (e.g., Fisher, 1922; Kutner et al., 2004) to the mixed-effects meta-regression setting, which can be done as follows.

We start with the linear model (i.e., $y_i = \beta_0 + \beta_1 x_{i1} + u_i + e_i$) and add dummy variables to the model for each level of the continuous moderator. However, this will lead to an overparameterized model. To obtain a model where all parameters are estimable, two of the dummy variables need to be removed, for example, those for the first and the last level (for the purposes of the test, it is completely arbitrary which two levels are removed). Therefore, if there are $j = 1, \ldots, \ell$ levels, then this yields the model

$$y_i = \beta_0 + \beta_1 x_{i1} + \sum_{j=2}^{\ell-1} \alpha_j d_{ij} + u_i + e_i, \tag{11.22}$$

where $d_{ij} = 1$ if the ith outcome is at level j and 0 otherwise. This model can be fitted with the methods described in Section 11.2.5.

For the example data, we therefore include dummy variables corresponding to the 50, 75, 100, 125, 150, 175, and 200 mg levels of the dosage moderator (leaving out the 25 mg and 250 mg levels). The results for this model are shown in Table 11.3. The right-hand side

TABLE 11.3

Results for the Mixed-Effects Meta-Regression Model to Conduct the Lack-of-Fit Test

Term	Coefficient	Estimate	SE	z-value
Intercept	$\hat{\beta}_0$	−0.510	0.3035	−1.68
Dose (linear)	$\hat{\beta}_1$	0.008	0.0022	3.50
Dose (50 mg)	$\hat{\alpha}_1$	0.328	0.4916	0.67
Dose (75 mg)	$\hat{\alpha}_2$	−0.125	0.5257	−0.24
Dose (100 mg)	$\hat{\alpha}_3$	0.305	0.3433	0.89
Dose (125 mg)	$\hat{\alpha}_4$	0.328	0.3333	0.99
Dose (150 mg)	$\hat{\alpha}_5$	0.095	0.4135	0.23
Dose (175 mg)	$\hat{\alpha}_6$	0.354	0.5698	0.62
Dose (200 mg)	$\hat{\alpha}_7$	−0.221	0.4392	−0.50

of Figure 11.9 shows a scatterplot of the dosage moderator against the observed outcomes with the regression line from the linear model and the fitted values from the lack-of-fit model superimposed. As can be seen from the plot, (11.22) yields a "saturated" model that provides estimates of the average true log odds ratio for each level of the dosage moderator without any kind of implied shape. In fact, the lack-of-fit model is just a different parameterization of the model

$$y_i = \sum_{j=1}^{\ell} \alpha_j d_{ij} + u_i + e_i, \tag{11.23}$$

which includes a fixed effect for each level of the moderator variable. However, the advantage of the parameterization in (11.22) is that it allows for a direct test of the linearity assumption. In particular, using (11.8), we can conduct an omnibus test of $H_0: \alpha_2 = \cdots = \alpha_{\ell-1} = 0$. This yields $Q_M = 3.57$, which we compare against a chi-square distribution with seven degrees of freedom. This yields a p-value of 0.83 and hence no evidence that the relationship between dosage and outcomes is non-linear.

11.5.6 Checking the Normality Assumptions

Various assumptions underlying the models were discussed in Section 11.4, including the assumption that the sampling distributions are approximately normal. Let us assume that all of the assumptions discussed in that section are fulfilled. Still, this does not imply that the observed outcomes themselves are normally distributed. Even in the simplest case of the common-effect model, the y_i values are a mixture of normally distributed random variables with the same mean, θ, but different variances, s_i^2, which does not yield a normal marginal distribution. In addition, in models involving moderators, the marginal distribution is a mixture of variables with different means. Accordingly, there is no use in examining the distribution of the y_i values directly.

However, when a particular model indeed represents a rough but adequate approximation to the underlying data generating process, then this implies that the standardized

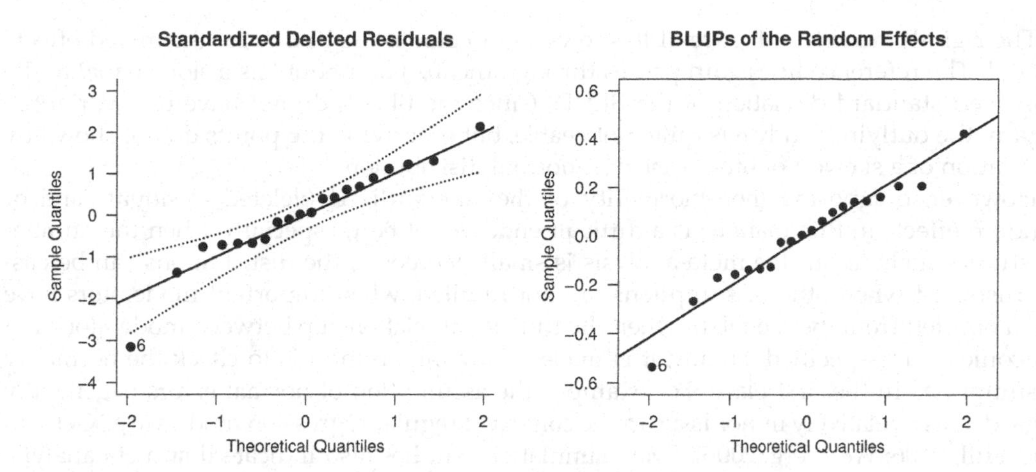

FIGURE 11.10
Q-Q plots of the standardized deleted residuals and the BLUPs of the random effects for the mixed-effects model.

(deleted) residuals should, at least approximately, follow a standard normal distribution. Q-Q plots can be used to examine whether this assumption holds. The left-hand side of Figure 11.10 shows such a plot of the standardized deleted residuals from the mixed-effects model. A diagonal reference line with an intercept of 0 and a slope of 1 was added to the plot. In addition, the dotted lines correspond to an approximate 95% pseudo confidence envelope, which was constructed based on the quantiles of sets of pseudo residuals simulated from the given model (for details, see Cook and Weisberg, 1982). Except for study 6, the points are roughly linear and fall close to the reference line. Therefore, there is no indication of non-normality in the standardized deleted residuals.

One other normality assumption not discussed so far underlies the random- and mixed-effects models. Besides assuming normally distributed sampling errors, the random- and mixed-effects models also make the additional assumption that the random effects, u_i, are normally distributed. In part, this assumption is often made purely because of convenience, that is, it greatly simplifies model fitting. However, this assumption can also be justified if we imagine that the (residual) heterogeneity in the true outcomes is a result of the influence of a large number of moderating factors, where each individual factor only has a small influence on the size of the true outcomes. When taken together, an approximately normal distribution could then emerge due to the central limit theorem. On the other hand, when outliers are present and/or when an important and strongly influential moderator has been omitted from the model, then this could lead to a non-normal random-effects distribution.

One possible approach to check this assumption is to compute the best linear unbiased predictions (BLUPs) of the random effects for a given model (Raudenbush and Bryk, 1985; Robinson, 1991) and then examine their distribution. Following the results in section (2.5), it can be shown that

$$\hat{u}_i = \lambda_i(y_i - \hat{\mu}_i) \tag{11.24}$$

provides predictions of the u_i values which have minimum mean squared error (among the class of linear unbiased estimators), where $\lambda_i = \hat{\tau}^2/(s_i^2 + \hat{\tau}^2)$. If the assumed model is correct, then the BLUPs should follow (approximately) a normal distribution.*

* There are two technical issues here. First, the computed values are really so-called empirical BLUPs (eBLUPs), since the unknown value of τ^2 is replaced by an estimate. Second, the eBLUPs do not have constant variance, so their marginal distribution may not be normal, even if the correct model is fitted and all assumptions hold.

The right-hand side of Figure 11.10 shows a Q-Q plot of the BLUPs for the mixed-effects model. The reference line again passes through the (0,0) point but has a slope equal to the observed standard deviation of the BLUPs (since the BLUPs do not have unit variance). Again, the outlying study 6 is quite noticeable, but otherwise, the points do not show any indication of a skewed or otherwise non-normal distribution.

However, diagnosing (non-)normality of the standardized (deleted) residuals and/or random effects in this manner is a difficult endeavor at best, especially when the number of studies included in the meta-analysis is small. Moreover, the distributions can be easily distorted when other assumptions are not fulfilled, when important moderators have been omitted from the model, or when the functional relationship between moderators and outcomes is misspecified. Finally, it is unclear how important it is to check the normality assumptions in the first place. For example, the assumption of normal errors is generally regarded as a relatively minor issue in the context of regular regression models (e.g., Gelman and Hill, 2006; Weisberg, 2006). Some simulations studies also indicate that meta-analytic models are quite robust to violations of the normality assumption of the random-effects distribution (Kontopantelis and Reeves, 2012; van den Noortgate and Onghena, 2003). However, further research is needed before more general recommendations can be made.

11.6 Checking for Influential Studies

So far, emphasis has been on assessing model fit and detecting outliers. Another issue to consider is the influence of each study on the results from the meta-analysis. Roughly speaking, an influential study is a study that exerts a considerable influence on the parameter estimates, test statistics, and/or conclusions that can be drawn from a given model. In some situations, certain findings (e.g., about the overall effectiveness of a treatment or the relevance of a particular moderator) may actually hinge on only one (or a few) of the studies in the dataset. In that case, it would be prudent to frame the corresponding conclusions more cautiously and to point out their volatility.

While it is often the case that outliers are also at least somewhat influential, it is important to properly distinguish between these concepts. For example, an outlier may not necessarily have much impact on the results if it comes from a very small study. Therefore, the presence of an outlier does not automatically call into question the conclusions drawn from the analyses. At the same time, a study whose observed outcome falls close to the fitted value based on the model (i.e., it is not an outlier) might still be influential, for example, if it is unusual in terms of its moderator values and its removal from the dataset might lead to considerable changes in any observed relationships.

An examination of a forest or scatterplot can already provide some indication whether influential studies may be present in a given dataset. However, as was the case for outliers, such informal approaches can be deceiving. We will therefore consider more rigorous methods for assessing and quantifying the influence of each study on various aspects of the fitted model.

11.6.1 Weights and Hat Values

As described in Section 11.2.5, model fitting is done by means of weighted least squares, with weights equal to $w_i = 1/s_i^2$ for the common-effect and meta-regression models and $w_i = 1/(s_i^2 + \hat{\tau}^2)$ for the random- and mixed-effects models. An examination of the weights (either in their raw form or scaled to a percentage with $100\% \times w_i / \sum w_i$) can reveal which studies have the potential for exerting a strong influence on the results. However, in the context of a (mixed-effects) meta-regression model, the values of the moderator variables also play a prominent role in determining the potential influence of each study on the results. For example, when examining the relationship between the treatment dosage and the size of the outcomes in the example dataset, studies with very low or very high dosages will tend to be more influential than those with intermediate dosage levels.

Therefore, instead of just examining the weights, a more general approach is to compute the hat matrix $\boldsymbol{H}$ with (11.9). The values in the ith row of the hat matrix indicate how the fitted value of a particular study is a function of the observed values of all k studies (i.e., $\hat{\mu}_i = \sum_{j=1}^{k} h_{ij} y_j$, where h_{ij} is the jth value in the ith row of $\boldsymbol{H}$). Often, only the diagonal elements of $\boldsymbol{H}$ are examined (i.e., h_{ii} for $i, \ldots, k$, which is often abbreviated to just h_i; cf. Section 11.5.2), which are called the hat values (or "leverages") and which indicate to what extent the fitted value of a study is influenced by its own observed value. In fact, for the common- and random-effects models, each row of the hat matrix (and hence also the diagonal) is equal to $w_i / \sum w_i$, so for these models it is fully sufficient to examine the hat values and not the entire hat matrix. Moreover, as can be seen, the hat values for these models are simply the scaled weights.

However, for meta-regression models, the values in the hat matrix are not only a function of the weights, but also the moderator variables. Especially studies with unusual values for the moderator variable(s) (in essence, studies that are outliers in terms of the moderator space) will then tend to receive larger hat values. An interesting property of the hat values is that they always add up to $q+1$ (i.e., $\sum_{i=1}^{k} h_{ii} = q+1$). Hence, the reference value $(q+1)/k$ represents the (hypothetical) scenario where each study would have the same leverage on the results.

The left-hand side of Figure 11.11 shows the hat values for the common- and random-effects models for the example dataset, with the reference line drawn at $1/k$, corresponding to the case where each study has the same weight in the analysis. This is clearly not the case for the common-effect model. In particular, study 7 has a considerably larger leverage due to its relatively small sampling variance (and hence larger weight) compared with the rest of the studies. On the other hand, for the random-effects model, we see that the hat values are nearly equalized. This is a consequence of the relatively large estimate of τ^2 compared with the sampling variances of the studies, in which case the weights, and therefore the leverages, are dominated by $\hat{\tau}^2$. As a result, each individual study has approximately the same influence on the results from the random-effects model.

For the mixed-effects model, the right-hand side of Figure 11.11 provides a heatmap constructed on the basis of the entire hat matrix. The hat values are located along the diagonal and are marked with dots to make them easier to locate. Especially studies 3 and 6 have large leverages, which partly reflects their larger weights, but also their position in the moderator space (i.e., at the very low and high ends of the dosage continuum). Studies 14, 16, and 17 are equally extreme in terms of their dosages (all at the very low end), but their larger sampling variances (and hence lower weights) limits their leverages. By examining the corresponding rows of the hat matrix, we see that the fitted values for these studies are actually mostly a function of the observed outcomes of the other studies (especially

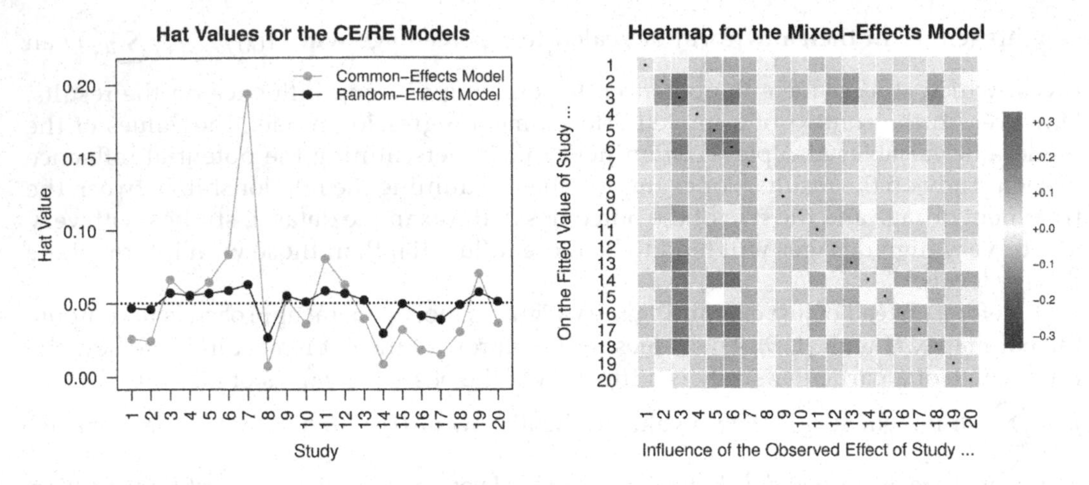

FIGURE 11.11
Plot of the hat values for the common- and random-effects models and a heatmap of the hat matrix for the mixed-effects model.

studies 3, 5, and 6) and not their own. Finally, study 7, which had high leverage in the context of the common-effect model, now only plays a relatively minor role. Therefore, the potential impact of a study must be considered with respect to a specific model.

11.6.2 Cook's Distances

A high leverage study has the *potential* to exert considerable influence on the results but does not necessarily do so. A study actually becomes influential if the estimates obtained from the model change substantially depending on whether the study is included in the dataset or not. To determine whether this is the case for a particular study, we can make use of an approach introduced earlier where we examine the consequences of deleting the study from the dataset.

There are various aspects of a model that can be influenced by a study. Of particular interest are the model coefficients themselves (i.e., $\hat{\beta}$ in (mixed-effects) meta-regression models or $\hat{\theta}$ and $\hat{\mu}$ in the common- and random-effects models). In order for a study with high leverage to become influential on this aspect of a model, its observed outcome must also deviate from the fitted value to a noteworthy degree, that is, it must be an outlier. As we have seen earlier, standardized deleted residuals are especially useful for detecting the latter. We can put these two ideas together and plot the leverages against the standardized deleted residuals. Influential studies will then be located at the top or bottom right-hand corner of the plot.

A measure that combines these two pieces of information into a single influence measure is Cook's distance (Cook and Weisberg, 1982; Viechtbauer and Cheung, 2010). It can be computed with

$$D_i = (\hat{\beta} - \hat{\beta}_{(-i)})'(X'WX)(\hat{\beta} - \hat{\beta}_{(-i)}), \tag{11.25}$$

where $\hat{\beta}_{(-i)}$ denotes the estimate of β computed with (11.6) when excluding the ith study from the model fitting. Written this way, the Cook's distance of a study can be interpreted

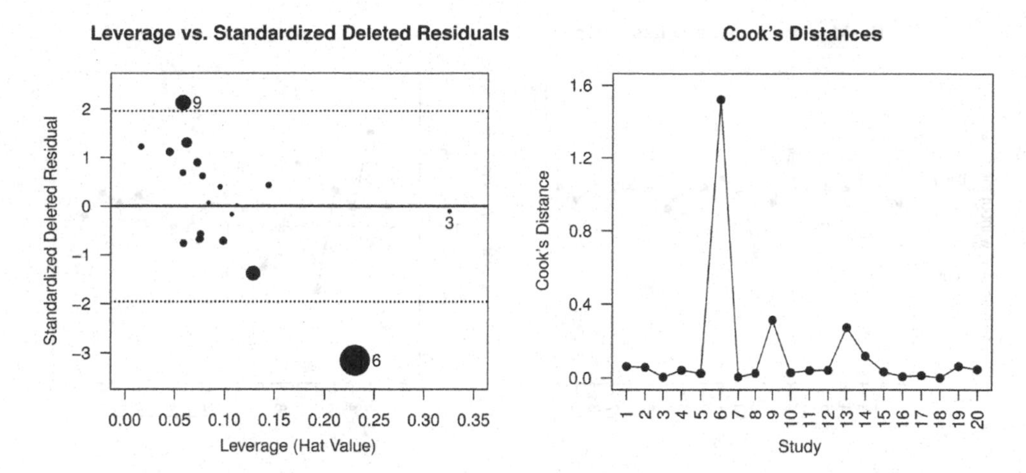

FIGURE 11.12
Scatterplot of the leverages versus the standardized deleted residuals and a plot of Cook's distances for the mixed-effects model.

as the Mahalanobis distance between the estimated model coefficients based on all k studies (i.e., $\hat{\beta}$) and the estimates obtained when the study is excluded from the model fitting (i.e., $\hat{\beta}_{(-i)}$). An equivalent way of expressing Cook's distance is

$$D_i = \sum_{j=1}^{k} \frac{(\hat{\mu}_j - \hat{\mu}_{j(-i)})^2}{s_j^2 + \hat{\tau}^2}, \tag{11.26}$$

which in turn can be interpreted as the Mahalanobis distance between the fitted values computed based on all k studies (i.e., $\hat{\mu}_j$ for $j=1,\ldots,k$) and the fitted values when the ith study is excluded from the model fitting (i.e., $\hat{\mu}_{j(-i)}$ for $j=1,\ldots,k$). Accordingly, a large value of D_i indicates notable changes in the model coefficients and fitted values depending on whether a study is included or removed from the dataset.

The left-hand side of Figure 11.12 shows a scatterplot of the leverages against the standardized deleted residuals for the mixed-effects model, with the points drawn proportional in size to Cook's distances. The right-hand side of the figure shows Cook's distances themselves. As we can see, high leverage points that are not outliers are not influential (study 3). Moreover, studies that are outliers but with relatively low leverage also do not exert much influence on the results (study 9). However, high leverage combined with poor fit results in an influential case (study 6).

11.6.3 Covariance Ratios and Other Deletion Diagnostics

Cook's distance quantifies the influence of each study on the model coefficients (and hence, the fitted values). However, this is not the only aspect of a model that can be affected by a study. In fact, a study may have relatively little influence on this part of a model, yet its removal can have other noteworthy consequences, which we would also want to be aware of.

For example, in random- and mixed-effects models, another important parameter besides the model coefficients is the estimate of (residual) heterogeneity. On the left-hand side of Figure 11.13, we can see a plot of the τ^2 estimates from the mixed-effects model

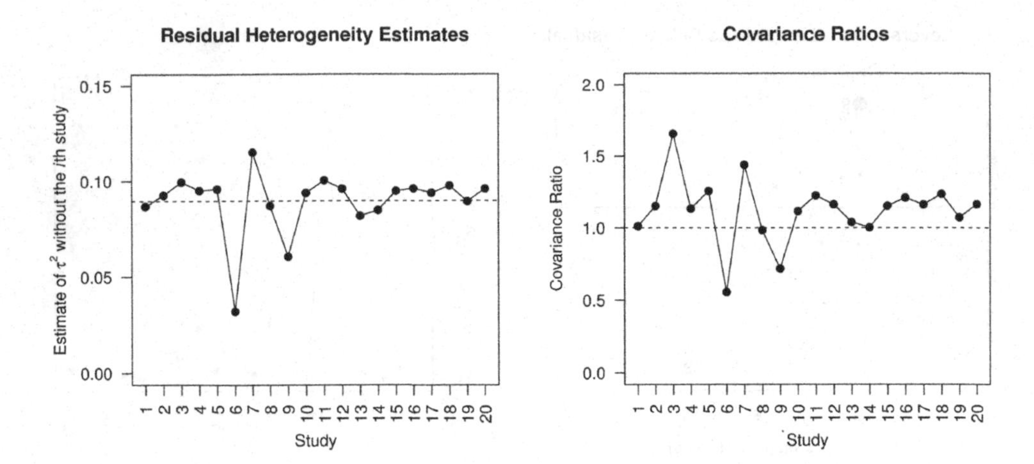

FIGURE 11.13
Plot of the leave-one-out estimates of τ^2 as each study is removed in turn and a plot of the covariance ratios for the mixed-effects model.

when each study is removed in turn. The horizontal dashed line corresponds to the estimate of τ^2 based on all k studies. Clearly, removal of study 6 leads to a substantial decrease in the amount of heterogeneity that is unaccounted for. This also applies to study 9, but to a lesser extent. On the other hand, removing study 7 would lead to a less pronounced but still discernible *increase* in the estimate.

Another aspect of all models worth considering is the precision with which we are able to estimate the model coefficients. At first sight, one would expect more data to lead to better (i.e., more precise) estimates, but this is not always the case. For random- and mixed-effects models, studies that introduce considerable (residual) heterogeneity into the data can actually lead to *decreases* in precision. At the same time, it is also informative to determine if there are studies in the dataset that are most responsible for driving up the precision of the estimates.

A useful measure for quantifying the effect of each study on this part of a model is the covariance ratio (Belsley et al., 1980; Viechtbauer and Cheung, 2010), given by

$$\text{COVRATIO}_i = \frac{\det\left[\text{Var}[\hat{\beta}_{(-i)}]\right]}{\det\left[\text{Var}[\hat{\beta}]\right]}, \tag{11.27}$$

where the numerator and denominator are generalized variances (i.e., the determinants of equation 7) for the reduced dataset (with the ith study excluded) and the full dataset, respectively. Since a smaller (generalized) variance is desirable, covariance ratios below one indicate that removal of a study leads to higher precision, while values above one indicate a decrease in precision.

The covariance ratios for the mixed-effects model are plotted on the right-hand side of Figure 11.13. Not surprisingly, removal of studies 6 and 9, which we identified earlier as sources of residual heterogeneity, would lead to increases in precision. Removing study 7, on the other hand, would have the opposite effect. Interestingly, study 3 also has a considerable covariance ratio. Recall that despite its high leverage, this study had essentially no influence on the model coefficients (i.e., its Cook's distance was very low). Moreover, it had no noteworthy effect on the estimate of τ^2. Yet, this study plays an important role, as its exclusion would result in substantially less precise estimates of the model coefficients.

Therefore, as this example demonstrates, some studies may only have a notable influence on this aspect of a model.

11.7 Other Types of Models

As discussed in Section 11.4, the models considered in this chapter assume that the sampling distributions of the observed outcomes are normal, that the observed outcomes are unbiased estimates, and that their sampling variances are known constants. Under certain circumstances, at least some of these assumptions are likely to break down.

An important case where we should be worried about violations of these assumption is in the context of meta-analyses examining the occurrence of rare events (e.g., Lane, 2013; see also Chapter 5). The sampling distributions of measures such as the risk difference and the log odds/risk ratio will then be poorly approximated by normal distributions. Moreover, estimates of the sampling variances will be very inaccurate, so that it is no longer acceptable to treat the variances as known constants. As a result of these assumption violations, inferential procedures (i.e., tests and confidence intervals) will no longer have nominal properties and the results/conclusions obtained cannot be trusted.

Fortunately, there is a wide variety of alternative models and methods available that can be used in this context, including Peto's method, the Mantel–Haenszel method, generalized linear mixed-effects models (i.e., mixed-effects logistic and Poisson regression), the non-central hypergeometric model, and the beta-binomial model (e.g., Mantel and Haenszel, 1959; Ma et al., 2016; Stijnen et al., 2010; Yusuf et al., 1985; see also Chapter 5). These methods relax certain underlying assumptions and try to model the observed data using more appropriate distributions. For example, for the data in Table 11.1, we could assume that x_i^T and x_i^C follow binomial distributions and then use logistic mixed-effects regression to model the log odds for remission in the treatment and control arms of each study. See Jackson et al., 2018 and Chapter 5 for more details.

Analogous model checking procedures as the ones described in this chapter can (and should) also be applied for such models. For example, akin to the tests for model misspecification described in Section 11.5.1, one can conduct likelihood ratio tests of $H_0\colon \tau^2 = 0$ in random- and mixed-effects logistic regression models. Outliers can again be detected by means of an examination of the residuals, although some additional complications arise in this context. For example, in logistic regression, we model the log odds in each study arm and hence the residuals will reflect deviations from the predicted log odds (or from the predicted event probabilities) for each arm. These residuals therefore do not directly address the question whether the log odds ratio of a particular study should be considered an outlier under a given model.

On the other hand, other methods generalize in a straightforward manner to logistic models. For example, GOSH plots could be generated based on the estimated model coefficients (although the computational burden would be increased considerably, especially when fitting random/mixed-effects logistic regression models), the lack of linearity test could be applied in the same manner, BLUPs of the random effects can be obtained and checked for normality, and influence measures such as Cook's distances and covariance ratios can be easily computed.

As discussed earlier, an assumption that applies specifically to random- and mixed-effects models concerns the nature of the random-effects distribution (i.e., the distribution of the underlying true effects). In practice, we typically assume that the random effects are normally distributed. Although it remains unclear how important it is to assess this assumption, it nevertheless can be worrisome if a diagnostic procedure (such as a Q-Q plot of the BLUPs) suggests considerable non-normality. In case non-normality is detected, one could explore models that allow for other types of distributional assumptions with respect to the random effects (Baker and Jackson, 2008; Beath, 2014; Lee and Thompson, 2008).

Finally, one additional assumption underlies all of the models described in Section 11.2 that warrants attention. As described, the models assume that the observed outcomes are independent. However, the independence assumption may be violated in various ways. For example, multiple outcomes extracted from the same sample of subjects are likely to be correlated. Returning to the example dataset, suppose that remission was defined and measured in two different ways in a particular study, leading to two log odds ratios that can be computed from the study. Then the sampling errors for the two outcomes are probably correlated (if patients who went into remission under the first definition also have higher chances of remission under the second definition). Similarly, if remission was measured at two different time points within a study, then the two corresponding log odds ratios are also probably correlated (if the occurrence of remission at the first time point is correlated with remission at the second time point).

Even if each study only provides a single outcome, it is still possible that the independence assumption is violated. For example, the underlying true outcomes may be correlated when multiple studies were conducted by the same author. Due to similarities in patient populations, methods, and procedures across studies (that are not captured by relevant moderator variables), the underlying true treatments effects may then be more alike for studies conducted by the same author than those conducted by different authors, which in essence again violates the independence assumption.

A common approach to deal with such dependencies is to reduce the dataset to one where dependencies are avoided (e.g., by selecting only one log odds ratio per study and/or author). Alternatively, multilevel and multivariate structures in a given dataset can be accounted for using appropriate models (e.g., Berkey et al., 1998; Jackson et al., 2011; Konstantopoulos, 2011). Model checking (including the detection of outliers and the identification of influential studies) also remains an important issue in the context of such analyses. Many of the methods discussed in this chapter can be generalized to such models, but the details of this are beyond the scope of this chapter.

11.8 Final Comments

Some final issues are worth commenting on. First of all, many meta-analyses involve only a relatively small number of studies. For example, a review of the Cochrane Database of Systematic Reviews indicated that the majority of Cochrane reviews contains only a handful of studies (Davey et al., 2011). That number tends to be somewhat higher for meta-analyses published in other outlets and/or for other disciplines (e.g., Cafri et al., 2010; Page et al., 2016), but meta-analyses with less than a dozen studies are still commonly encountered. Some of the techniques described in this chapter may be less informative or relevant in that context. At the same time, it is then even more important to check the data for outliers and influential studies, as their impact will tend to be larger in smaller datasets. Hence, standardized (deleted) residuals, Cook's distances, and other influence measures remain useful diagnostic tools. Still, the best one can hope to accomplish in such a situation is to identify cases where one particular study yields rather different results than the rest of the studies. If multiple studies yield very disparate results, then this will usually be indistinguishable from a situation where there is a high amount of heterogeneity in the data.

On the other hand, when k is large, it is important to realize that many of the techniques discussed in this chapter are "deletion diagnostics" that remove individual studies from the dataset and then examine the consequences thereof. These methods can also be effective for detecting multiple outliers and influential studies, but the presence of multiple outliers can lead to distortions such that none of the true outliers are detected (a phenomenon known as masking) or that some studies are incorrectly labeled as outliers (a phenomenon known as swamping) (Barnett and Lewis, 1978). Deletion diagnostics involving the simultaneous removal of multiple studies will then be required to sort out such intricacies. The GOSH plot may be especially useful in this context, as it is based on all possible subsets and hence does not require the specification of the number of studies to remove *a priori*.

For the most part, specific decision rules or cutoffs for deciding when a study should be considered an outlier and/or influential have been avoided throughout this chapter. Any such guidelines would ultimately be arbitrary (which also applies to the ±1.96 value to which the standardized deleted residuals were compared earlier). Instead, emphasis has been on visual inspection of the various diagnostic measures. By comparing the relative magnitude of the values across studies, one can often easily identify those studies that stand out with respect to a particular measure. The plot of Cook's distances (i.e., Figure 11.12) is exemplary for this approach.

An important point not discussed so far is the question what one should do when some studies are identified as potential outliers and/or influential. To begin with, one should check that the data are not contaminated in some way, for example, due to errors in reporting or coding. For example, a standard error of the mean that is mistaken for a standard deviation can lead to a substantial overestimate of the true standardized mean difference or the precision of a mean difference. In the context of the illustrative example, a study author may have reported a dosage of "50 mg tid" (for a daily dosage of 150 mg), but this may have accidentally been coded as a daily dosage of 50 mg.

However, in many cases, no such simple explanations will be found. In that case, one approach that one should never take is to simply remove the unusual studies from the analysis. All studies that fit the initial inclusion criteria should be reported and described. However, one may still probe the *robustness* of the analyses by excluding outliers or influential studies from the dataset. For example, is dosage still a significant when studies 3 and/or 6 are removed from the illustrative dataset? If the conclusion about the relevance of this moderator would be overturned, it would indicate that this finding is not particularly robust, and the finding should be described more cautiously. Fortunately, removal of either or both studies still yields the same conclusion in this example, which lends more credibility to the hypothesis that medication dosage is related to the treatment effectiveness.

Moreover, studies yielding unusual results may actually point toward (or rather, raise interesting new hypotheses about) conditions under which the effect or association of interest is particularly large or small (Hedges, 1986; Light and Pillemer, 1984). Such post-hoc explanations should, of course, be treated with caution, but they can be an opportunity to learn something new about the phenomenon being studied. Hence, it is actually in the best interest of the reviewer to apply some of the model checking methods described in the present chapter.

As a final practical point, it is worth noting that all of the methods described in this chapter are implemented in the R package `metafor` (Viechtbauer, 2010) (code corresponding to the analyses conducted is provided on the book website). General purpose software packages could also be used for model fitting and extracting diagnostic measures such as standardized (deleted) residuals, BLUPs, and Cook's distances (e.g., with PROC MIXED from SAS), but certain specialized plots for meta-analysis (e.g., forest, Baujat, and GOSH plots) are then not directly available.

References

Baker R and Jackson D, 2008. A new approach to outliers in meta-analysis. *Health Care Management Science* **11**(2): 121–131.

Barnett V and Lewis T, 1978. *Outliers in Statistical Data*. New York: Wiley.

Baujat B, Mahé C, Pignon J-P and Hill C, 2002. A graphical method for exploring heterogeneity in meta-analyses: Application to a meta-analysis of 65 trials. *Statistics in Medicine* **21**(18): 2641–2652.

Beath KJ, 2014. A finite mixture method for outlier detection and robustness in meta-analysis. *Research Synthesis Methods* **5**(4): 285–293.

Belsley DA, Kuh E and Welsch RE, 1980. *Regression Diagnostics: Identifying Influential Data and Sources of Collinearity*. New York: Wiley.

Berkey CS, Hoaglin DC, Antczak-Bouckoms A, Mosteller F and Colditz GA, 1998. Meta-analysis of multiple outcomes by regression with random effects. *Statistics in Medicine* **17**(22): 2537–2550.

Berkey CS, Hoaglin DC, Mosteller F and Colditz GA, 1995. A random-effects regression model for meta-analysis. *Statistics in Medicine* **14**(4): 395–411.

Cafri G, Kromrey JD and Brannick MT, 2010. A meta-meta-analysis: Empirical review of statistical power, type I error rates, effect sizes, and model selection of meta-analyses published in psychology. *Multivariate Behavioral Research* **45**(2): 239–270.

Cook RD and Weisberg S, 1982. *Residuals and Influence in Regression*. London: Chapman and Hall.

Davey J, Turner RM, Clarke MJ and Higgins JP, 2011. Characteristics of meta-analyses and their component studies in the Cochrane Database of Systematic Reviews: A cross-sectional, descriptive analysis. *BMC Medical Research Methodology* **11**: 160.

DerSimonian R and Laird N, 1986. Meta-analysis in clinical trials. *Controlled Clinical Trials* **7**(3): 177–188.

Field AP, 2001. Meta-analysis of correlation coefficients: A Monte Carlo comparison of fixed- and random-effects methods. *Psychological Methods* **6**: 161–180.

Fisher RA, 1921. On the "probable error" of a coefficient of correlation deduced from a small sample. *Metron* **1**: 1–32.

Fisher RA, 1922. The goodness of fit of regression formulae, and the distribution of regression coefficients. *Journal of the Royal Statistical Society* **85**(4): 597–612.

Friedrich JO, Adhikari NKJ and Beyene J, 2008. The ratio of means method as an alternative to mean differences for analyzing continuous outcome variables in meta-analysis: A simulation study. *BMC Medical Research Methodology* **8**: 32.

Gelman A and Hill J, 2006. *Data Analysis Using Regression and Multilevel/Hierarchical Models*. Cambridge: Cambridge University Press.

Hauck WW, Anderson S and Leahy FJ, 1982. Finite-sample properties of some old and some new estimators of a common odds ratio from multiple 2×2 tables. *Journal of the American Statistical Association* **77**(377): 145–152.

Hedges LV, 1981. Distribution theory for Glass's estimator of effect size and related estimators. *Journal of Educational Statistics* **6**(2): 107–128.

LV Hedges. Estimation of effect size from a series of independent experiments. *Psychological Bulletin* **92**(2): 490–499, 1982a.

Hedges LV, 1982b. Fitting categorical models to effect size from a series of experiments. *Journal of Educational Statistics* **7**(2): 119–137.

Hedges LV, 1982c. Fitting continuous models to effect size data. *Journal of Educational Statistics* **7**(4): 245–270.

Hedges LV, 1986. Issues in meta-analysis. *Review of Research in Education* **13**: 353–398.

Hedges LV, 1992. Meta-analysis. *Journal of Educational Statistics* **17**(4): 279–296.

Higgins JPT and Thompson SG, 2002. Quantifying heterogeneity in a meta-analysis. *Statistics in Medicine,* **21**(11): 1539–1558.

Hotelling H, 1953. New light on the correlation coefficient and its transforms. *Journal of the Royal Statistical Society, Series B* **15**(2): 193–232.

Jackson D, Law M, Stijnen T, Viechtbauer W and White IR, 2018. A comparison of seven random-effects models for meta-analyses that estimate the summary odds ratio. *Statistics in Medicine* **37**(7): 1059–1085.

Jackson D, Riley R and White IR, 2011. Multivariate meta-analysis: Potential and promise. *Statistics in Medicine* **30**(20): 2481–2498.

Konstantopoulos S, 2011. Fixed effects and variance components estimation in three-level meta-analysis. *Research Synthesis Methods* **2**(1): 61–76.

Kontopantelis E and Reeves D, 2012. Performance of statistical methods for meta-analysis when true study effects are non-normally distributed: A simulation study. *Statistical Methods in Medical Research* **21**(4): 409–426.

Kutner MH, Nachtsheim CJ, Neter J and Li W, 2004. *Applied Linear Statistical Models*. 5th Edition. New York: McGraw-Hill.

Lane PW, 2013. Meta-analysis of incidence of rare events. *Statistical Methods in Medical Research* **22**(2): 117–132.

Lee KJ and Thompson SG, 2008. Flexible parametric models for random-effects distributions. *Statistics in Medicine* **27**(3): 418–434.

Light RJ and Pillemer DB, 1984. *Summing Up: The Science of Reviewing Research*. Cambridge: Harvard University Press.

López-López JA, Marín-Martínez F, Sánchez-Meca J, van den NoortgateW and Viechtbauer W, 2014. Estimation of the predictive power of the model in mixed-effects meta-regression: A simulation study. *British Journal of Mathematical and Statistical Psychology* **67**(1): 30–48.

Ma Y, Chu H and Mazumdar M, 2016. Meta-analysis of proportions of rare events: A comparison of exact likelihood methods with robust variance estimation. *Communications in Statistics, Simulation and Computation* **45**(8): 3036–3052.

Mantel N and Haenszel W, 1959. Statistical aspects of the analysis of data from retrospective studies of disease. *Journal of the National Cancer Institute* 22(4): 719–748.

Olkin I, Dahabreh IJ and Trikalinos TA, 2012. Gosh – A graphical display of study heterogeneity. *Research Synthesis Methods* **3**(3): 214–223.

Olkin I and Pratt JW, 1958. Unbiased estimation of certain correlation coefficients. *Annals of Mathematical Statistics* **29**(1): 201–211.

Page MJ, Shamseer L, Altman DG, Tetzlaff J, Sampson M, Tricco AC, Catalá-López F, Li L, Reid EK, Sarkis-Onofre R and Moher D, 2016. Epidemiology and reporting characteristics of systematic reviews of biomedical research: A cross-sectional study. *PLoS Medicine* **13**(5): e1002028.

Raudenbush SW, 2009. Analyzing effect sizes: Random-effects models. In Cooper H, Hedges LV and Valentine JC (Eds). *The Handbook of Research Synthesis and Meta-Analysis*. 2nd Edition. New York: Russell Sage Foundation, 295–315.

Raudenbush SW and Bryk AS, 1985. Empirical Bayes meta-analysis. *Journal of Educational Statistics* **10**(2): 75–98.

Robinson GK, 1991. That BLUP is a good thing: The estimation of random effects. *Statistical Science* **6**(1): 15–32.

Rücker G, Schwarzer G and Carpenter J, 2008. Arcsine test for publication bias in meta-analyses with binary outcomes. *Statistics in Medicine* **27**(5): 746–763.

Rücker G, Schwarzer G, Carpenter J and Olkin I, 2009. Why add anything to nothing? The arcsine difference as a measure of treatment effect in meta-analysis with zero cells. *Statistics in Medicine* **28**(5): 721–738.

Sánchez-Meca J and Marín-Martínez F, 2001. Meta-analysis of 2×2 tables: Estimating a common risk difference. *Educational and Psychological Measurement* **61**(2): 249–276.

Stijnen T, Hamza TH and Ozdemir P, 2010. Random effects meta-analysis of event outcome in the framework of the generalized linear mixed model with applications in sparse data. *Statistics in Medicine* **29**(29): 3046–3067.

Thompson SG and Sharp SJ, 1999. Explaining heterogeneity in meta-analysis: A comparison of methods. *Statistics in Medicine* **18**(20): 2693–2708.

van den Noortgate W and Onghena P, 2003. Multilevel meta-analysis: A comparison with traditional meta-analytic procedures. *Educational and Psychological Measurement* **63**(5): 765–790.

Viechtbauer W, 2010. Conducting meta-analyses in R with the metafor package. *Journal of Statistical Software* **36**(3): 1–48.

Viechtbauer W and Cheung MW-L, 2010. Outlier and influence diagnostics for meta-analysis. *Research Synthesis Methods* **1**(2): 112–125.

Viechtbauer W, López-López JA, Sánchez-Meca J and Marín-Martínez F, 2015. A comparison of procedures to test for moderators in mixed-effects meta-regression models. *Psychological Methods* **20**(3): 360–374.

Walter SD, 1985. Small sample estimation of log odds ratios from logistic regression and fourfold tables. *Statistics in Medicine* **4**(4): 437–444.

Walter SD and Cook RJ, 1991. A comparison of several point estimators of the odds ratio in a single 2×2 contingency table. *Biometrics* **47**(3): 795–811.

Weisberg S, 2006. *Applied Linear Regression*. 3rd Edition. Hoboken, NJ: Wiley.

Yusuf S, Peto R, Lewis J, Collins R and Sleight P, 1985. Beta blockade during and after myocardial infarction: An overview of the randomized trials. *Progress in Cardiovascular Disease* **27**(5): 335–371.

12

Handling Internal and External Biases: Quality and Relevance of Studies

Rebecca M. Turner, Nicky J. Welton, Hayley E. Jones, and Jelena Savović

CONTENTS

12.1 Introduction

Meta-analysis assumes either that all studies estimate a common effect or exchangeability across studies, meaning that underlying treatment effects from included studies are expected to be similar and their magnitudes cannot be differentiated in advance (Higgins et al., 2009). The presence of within-study biases can potentially invalidate the assumption of exchangeability, since underlying treatment effects may differ systematically across studies. There are two forms of within-study biases: biases to internal validity (internal biases) and biases to external validity (external biases). Internal biases are caused by methodological flaws in the studies, such as inadequacy of randomization or lack of blinding, which reduce the study's ability to estimate its intended parameters accurately. External biases are caused by differences between the designs of available studies and the target research question. Internal biases require careful assessment and consideration in all meta-analyses. External biases should be dealt with similarly in meta-analyses informing policy decisions, but may be accepted and regarded as between-study heterogeneity if the aim of the analysis is to summarize the literature rather than to address a particular research question. In this chapter, we will focus primarily on internal biases.

We begin by discussing methods for assessing the presence of internal biases in meta-analyses of randomized clinical trials (RCTs). We then describe meta-epidemiological research studies which have assembled and modeled collections of meta-analyses in order to estimate the magnitudes of internal biases associated with various methodological flaws in RCTs. Next, we present a method for using meta-epidemiological evidence to adjust for internal biases in RCTs, and also present a method for adjusting for both internal and external biases in RCTs and observational studies using elicited expert opinion. We then discuss methods developed for addressing biases caused by missing data within studies. Finally, we describe methods for adjusting for internal biases in the special case of network meta-analysis. Most of the methods presented are inherently Bayesian and involve incorporating external evidence on the extent of particular biases as prior distributions.

12.2 Assessing Risk of Bias in Randomized Clinical Trials

Systematic reviews and meta-analyses of RCTs are considered the best source of evidence of effectiveness for healthcare interventions. But not all trials are conducted with the same methodological rigor and sometimes trials can produce biased estimates of treatment effects. This could be due to flaws in the design, conduct, and analysis of the trials, or biased selection of the results they report. Inclusion of biased trials in meta-analyses can lead to misleading meta-analysis estimates. Assessing the risk of bias of included trials is thus an essential part of a systematic review. We will now discuss the most likely sources of bias in RCTs and how these can be assessed.

12.2.1 Bias Arising from the Randomization Process

The unique advantage of randomization is that, when implemented successfully, it prevents biased allocation of participants into the intervention groups. Two components are needed for successful randomization: an unpredictable random sequence, used to determine the order of treatment allocations of consecutive participants, and concealment of that allocation sequence from both the participant and the enrolling investigator. The former is referred to as *sequence generation* and the latter as *allocation concealment*. If one or both of these components were compromised, the investigators (and in some cases also participants) may be able to predict the forthcoming intervention assignment. This can lead to selective enrolment of participants influenced by factors that predict the outcome (e.g., severity of illness). For example, if investigators were aware of the forthcoming allocation and did not think the patient would do so well on that treatment, then they may retrospectively class that participant as ineligible, whereas they had considered them eligible prior to their knowledge of allocation. This is known as *selection bias* in epidemiology. Alternatively, participants' allocation to the "desired" intervention group may be manipulated if the allocation schedule is predictable, for example, by delaying a participant's entry into the trial until the next entry to the "correct" intervention is available.

The risk of bias arising from the randomization process is judged by assessing the appropriateness of the methods for generating the random sequence and the concealment of its allocation and their perceived success. These days, a random sequence is usually created by a computer algorithm, but random number tables are still in use and some other methods (drawing lots or cards or coin toss) may still be used occasionally. The most reliable modern method for concealment of allocation is the use of remote randomization services, independent from the investigators, usually accessed via telephone or the internet (Higgins and Altman, 2008; Schulz et al., 1995; Schulz, 1995).

12.2.2 Bias Due to Deviations from Intended Interventions

When the provision of care differs systematically between the treatment groups, or if there are differences in exposure to factors other than the intervention of interest (e.g., co-interventions that differ between groups), this may in some cases lead to a biased estimate of treatment effect. This type of bias is sometimes referred to as *performance bias*.

When feasible, blinding of participants and care providers prevents this type of bias because lack of knowledge of allocation prevents other care decisions from being influenced by the group to which a patient was allocated (Hrobjartsson et al., 2012, 2014a; Nuesch et al., 2009; Savović et al., 2012a, 2012b; Wood et al., 2008). When blinding is not feasible other measures can help reduce differential behavior by patients and care providers, for example, treating patients according to a strict protocol.

12.2.3 Bias in Measurement of Outcomes

Systematic differences between treatment groups in how outcomes are measured can lead to bias (often referred to as *detection bias*). This bias is outcome-specific, and some outcomes measured in the study may be affected while others are not. The knowledge of the intervention received can introduce bias in the measurement of outcomes, and prevention of that knowledge (through blinding of the outcome assessor) can avoid it. In trials where blinding of patients and care providers is not feasible, it may still be possible to blind the assessment of certain outcomes (e.g., blinded outcome adjudicating committee).

There is some evidence that objective outcomes, such as all-cause mortality, are less likely to be biased even in the absence of blinding (Hrobjartsson et al., 2012, 2013, 2014b; Savović et al., 2010, 2012b; Wood et al., 2008). If the outcome is deemed to be objective, the risk of bias is likely to be low. For subjective outcomes, the next step is to determine who the assessor was and whether they were blinded. If outcome assessment was blinded, the risk of bias is likely to be low. If the assessment was not blinded it is likely that the risk of bias is high. A final consideration is whether the outcome and measurement procedures were well-defined and measured in the same way for all treatment arms.

12.2.4 Bias Due to Missing Outcome Data

Bias due to missing outcome data can occur when there are systematic differences in occurrence or handling of protocol deviations, withdrawals, or losses to follow-up. Data could be missing from analyses either due to exclusions or attrition. Exclusions occur when participants for whom the outcome data are available are omitted from analyses (e.g., due to ineligibility, protocol deviation, or lack of compliance). Attrition occurs when data are not available (e.g., due to patients being lost to follow-up) (Higgins and Altman, 2008).

The risk of bias due to missing data is likely to be low under any of the following circumstances: there are no missing outcome data; reasons for missing data are unlikely to be related to outcome; missing data are balanced across the groups in terms of numbers and reasons; the proportion of missing data is small enough that it is unlikely to have changed the observed treatment effect estimate had it been available; or missing data have been imputed using appropriate methods, of which the statistical assumptions are likely to be fulfilled. The risk of bias is likely to be high if none of these criteria are met, or if the "as treated" analysis was carried out in a trial where a substantial number of participants switched from their allocated treatment arm, or inappropriate methods of imputation were used (Higgins and Altman, 2008). Methods for dealing with missing data in meta-analysis are covered in Section 12.5.

12.2.5 Bias Due to Selective Reporting

Bias due to selective reporting occurs when outcomes or analyses are preferentially selected for publication because of a significant test result or withheld from publication due to a non-significant test result. This can happen when several outcome measures are collected to measure the same outcome domain (e.g., several depression scales are used to measure the severity of depression), but only those that show the benefit of treatment (e.g., significant p-value) are reported. Similarly, the "modified" intention-to-treat analysis or per-protocol analysis may be reported because it gives a significant p-value compared with the pre-planned intention-to-treat analysis. This type of bias is difficult to assess, and assessors need to rely on the availability of published trial protocols or information on trial registers, which are often not available. In the absence of the study protocol, content expertise may be important to understand which outcomes are expected to be reported in a specific clinical field. The criteria for judging low risk of bias are that reported outcomes match those in the protocol, or if the study protocol is not available, that it is clear that all expected outcomes are reported (Kirkham et al., 2010).

The most extreme example of selective reporting is when the trial is not published at all due to a non-significant finding (publication bias). The risk that certain outcomes are unreported should be assessed across all included studies in each systematic review, in a similar way to investigating publication bias (see Chapter 13).

12.2.6 Tools for Assessing Risk of Bias in Trials Included in Meta-Analyses

With the growth of systematic reviews of RCTs in the 1980s and 1990s, there was an increased need for identification of RCTs with potentially biased results so that bias can be minimized in the meta-analysis estimates and review conclusions. This resulted in a rapid development of quality assessment scales, instruments, and checklists aimed at distinguishing high-quality trials from low-quality trials. By 1995, a large number of scales and checklists of variable complexity were already available (Moher et al., 1995). The scales allocated numerical scores to included items, thus introducing a form of weighting that is difficult to interpret. Jüni and colleagues used 25 different scales to assess quality of 17 trials included in a single meta-analysis and examined the association between treatment effect and summary quality scores (Jüni et al., 1999). They identified large discrepancies between the scales (e.g., the same trial could be labeled as high or low quality depending on the scale used). The authors concluded that the use of such scales is problematic and should be discouraged.

Informed by these findings and empirical evidence of bias from meta-epidemiological studies (described in Section 12.3), Cochrane (www.cochrane.org) developed a component-based approach to assess validity of trials included in Cochrane reviews. The Cochrane risk of bias tool was first released in 2008 (Higgins and Altman, 2008), with a slightly revised version published in 2011 (Higgins et al., 2011). Its use is mandatory for all Cochrane reviews (Chandler et al., 2013) and a recent survey suggests it is also the most commonly used tool for assessing risk of bias in non-Cochrane reviews (Hopewell et al., 2013). Bias assessments consist of a domain-specific judgment of "Low", "High", or "Unclear" risk of bias, accompanied by information in support of the judgment. This is likely to include a description of what was done in the study and how this has informed the judgment, thus adding transparency to what is inevitably a subjective opinion of an assessor (Higgins and Altman, 2008). An example of a completed risk of bias assessment using the Cochrane tool is shown in Table 12.1.

12.3 Meta-Epidemiological Evidence on Sources of Bias in RCTs

Do "high risk of bias" studies really produce biased estimates of an intervention effect and, if so, by how much? Ideally this would be assessed within sets of trials. For example, Hrobjartsson et al. compared results across 21 trials in which the outcome had been assessed by both blinded and non-blinded assessors, and across 12 trials in which some patients were blinded and others were not (Hrobjartsson et al., 2012, 2014a). However, trials are not usually designed to assess the impact of risk of bias indicators and doing so reduces power to detect relative treatment effects.

The association between a potential study flaw (as assessed in the last section) and intervention effect estimates can be assessed using a meta-regression (see Chapter 7). Frequently, "unclear" risk of bias assessments are combined with "high" risk for this, to form a dichotomous covariate. The meta-regression then provides an estimate of the average difference in results between trials in the meta-analysis assessed to be at high or unclear risk of bias relative to low risk, quantified for example as a ratio of odds ratios (ROR) or difference in mean differences (see 7.3.2 and 12.4.1). Unfortunately, since the number of trials within a meta-analysis is usually small, the power to detect true associations is low. At the same time, if assessing the effect of multiple potential flaws using separate meta-regressions, the chance of false positive findings is high. Investigation of potential bias associated with a study characteristic might be more feasible within a network meta-analysis (see 12.6).

TABLE 12.1

Example of a Completed Risk of Bias Table for a Trial of Strength Training versus Aerobic Exercise versus Relaxation for Treatment of Depression (Krogh et al., 2009; Sterne et al., 2002)

Bias domain	Risk of bias judgment	Description of what was done/support for judgment
Random sequence generation	Low`	Quote: "Randomization was carried out by the CTU using computerized restricted randomization with a block size of 6. The block size and thus the allocation sequence were unknown to the trial staff."
Allocation concealment	Low	See quote above. Comment: Allocation was carried out centrally by a clinical trials unit.
Blinding of participants and personnel	High	Comment: Due to the nature of the intervention, it was not possible to blind participants and personnel to treatment assignment. Mean participation was significantly lower in the relaxation group, which may be indicative of performance bias.
Blinding of outcome assessment *Outcome: Depression severity by BDI*[a]	High	Quote: "The patient administered Becks [sic] Depression Inventory (BDI) scale" (protocol) Comment: BDI is a patient self-report depression scale and this is a subjective outcome. Participants were not blind to treatment assignment.
Incomplete outcome data *Outcome: Depression severity by BDI*[a] *Comparison: Active groups vs. relaxation*[b]	High	Comment: Intention-to-treat analysis was carried out. Missing data were imputed with an assumption that data were missing at random. Reasons for attrition are similar for the two active groups, but substantially higher in the relaxation group (refused to participate or lost to follow-up). As the reason for high attrition in relaxation group may be related to both the allocated intervention and the outcome, there is potential for bias in the comparison of either of the active interventions with the relaxation group. It is not clear if imputation could solve this problem.
Incomplete outcome data *Outcome: Depression severity by BDI*[a] *Comparison: Strength vs. aerobic*[b]	Low	Comment: Numbers and reasons for attrition were similar for the two active groups. No bias is expected in comparisons between the strength and the aerobic group.
Selective reporting	Low	Comment: The study protocol is available and all of the study's pre-specified (primary and secondary) outcomes that are of interest have been reported in the pre-specified way.
Other bias	Low	No other issues identified.

BDI = Beck depression inventory (depression symptom scale); CTU, clinical trials unit.

[a] Blinding of outcome assessment and incomplete outcome date is outcome-specific and should be done separately for each outcome (or group of outcomes if appropriate).

[b] This was a three-arm trial comparing strength exercise, aerobic exercise, and relaxation (inactive arm). The implications for attrition bias were different in the comparisons of active versus inactive treatment arms and the comparison of the two active arms, thus, it was necessary to assess the risk of bias separately for these comparisons.

To obtain a more general understanding of how bias might operate "on average", researchers have turned to larger bodies of evidence, in what has become known as "meta-epidemiological" research. This involves quantification of the *average* association between a study characteristic and intervention effect estimates across a large number of meta-analyses. Early meta-epidemiological studies used logistic regression with interaction terms (Schulz et al., 1995) or a two-stage approach in which the ROR from each meta-analysis was pooled using a random-effects model, allowing bias to vary across meta-analyses (Sterne et al., 2002), and models were fitted within a frequentist framework. Some more recent meta-epidemiological studies (Savović et al., 2012a, 2012b, 2018) have used a Bayesian hierarchical model developed by Welton et al (2009). This models the estimated intervention effect (e.g., log odds ratio for binary outcomes) in trial i of meta-analysis m as $\delta_{im} + \beta_{im}x_{im}$, where x_{im} is an indicator variable for a study characteristic (e.g., inadequate or unclear allocation concealment, coded 1 for inadequate or unclear, 0 for adequate). Hence δ_{im} is the intervention effect in studies without the characteristic (e.g., those with adequate allocation concealment). A standard random-effects structure across studies is assumed for these intervention effects: $\delta_{im} \sim N(\mu_m, \sigma_m^2)$. β_{im} is the bias in intervention effect associated with the study characteristic in trial i of meta-analysis m. The following two-level structure allows the bias to vary *within* each meta-analysis (standard deviation $= \kappa$) but also the average bias to vary *across* meta-analyses (standard deviation $= \phi$):

$$\begin{aligned} \beta_{im} &\sim N(b_m, \kappa^2) \\ b_m &\sim N(b_0, \phi^2) \end{aligned} \tag{12.1}$$

The model assumes exchangeability of study-specific biases across trials within each meta-analysis and of average bias across meta-analyses. b_0 represents the average bias across all meta-analyses (such that, for an analysis of odds ratios, $\exp(b_0)$ is the average ROR). It is assumed that between-study heterogeneity among trials in each meta-analysis, m, with the characteristic ($= \sigma_m^2 + \kappa^2$) is at least as large as the heterogeneity among trials without the characteristic ($= \sigma_m^2$). This was motivated by a finding from the landmark meta-epidemiological study of Schulz et al. (1995) that treatment effect estimates were more variable among trials with inadequate relative to adequate allocation concealment. For a variation on the model that does not make this assumption, see Rhodes et al. (2018).

The Schulz et al. (1995) study, based on 250 trials from 33 meta-analyses, estimated a 41% exaggeration of intervention effect estimates in trials with inadequate allocation concealment relative to adequate (ROR = 0.59, 95% CI 0.48–0.73) and a 33% exaggeration in trials in which allocation concealment was "unclear" relative to adequate (ROR = 0.67, 95% CI 0.60–0.75). After excluding trials with inadequate allocation concealment from the analysis, Schulz et al. further found an association between lack of double blinding and intervention effect estimates (ROR = 0.83, 95% CI 0.71–0.96) but little or no evidence for bias associated with inadequate randomization methods or exclusions after randomization (Schulz et al., 1995).

A number of subsequent meta-epidemiological studies, across a range of clinical areas, reported varying associations between bias assessments and intervention effect estimates (Moher et al., 1998; Balk et al., 2002; Egger et al., 2003; Pildal et al., 2007; Kjaergard et al., 2001), some of which were considerably smaller than those found by Schulz (Pildal et al., 2007; Gluud, 2006). The BRANDO study (Bias in Randomized and Observational studies) was a comprehensive re-analysis of data from seven previously reported meta-epidemiological studies (Savović et al., 2012a, 2012b). After removal of overlapping meta-analyses, 1973 trials from 234 unique meta-analyses were analyzed. The summary results indicated an average exaggeration of treatment effect estimates in trials with inadequate or unclear generation of randomization sequence versus adequate (ROR = $\exp(b_0)$ = 0.89, 95% CrI 0.82–0.96), with inadequate or unclear allocation concealment versus adequate (ROR = 0.93, 95% CrI 0.87–0.99) and with lack of or unclear double blinding versus double blind

(ROR = 0.87, 95% CrI 0.79–0.96). Similar results were also found in a more recent study of 2443 trials in 228 meta-analyses published in the Cochrane library (Savović et al., 2018).

Meta-epidemiological studies have provided some evidence that bias tends to be greatest in trials with subjectively measured outcomes (Savović et al., 2012a, 2012b; Wood et al., 2008). This suggests that it could be particularly important to consider risk of bias indicators when working with subjective outcomes, although the more recent study of Cochrane meta-analyses (Savović et al., 2018) did not replicate this finding. The BRANDO study also examined whether bias estimates varied by clinical area or type of intervention (pharmaceutical versus non-pharmaceutical), but found little evidence for this, although credible intervals were wide (Savović et al., 2012a, 2012b).

The evidence base for associations with other study characteristics is, to date, considerably smaller. For example, a comparison of results from single versus multi-center trials in the 54 meta-analyses in the BRANDO database in which this information had been assessed provided an average ROR of 0.91 (95% CrI 0.79–1.04) (Savović et al., 2012b). In a separate analysis of 48 meta-analyses, Dechartres et al. (2011) found an average ROR of 0.73 (95% CI 0.64–0.83). Several meta-epidemiological studies have examined evidence for attrition bias in trials in which intention-to-treat analysis was not used, or in which a relatively large proportion of patients had missing outcome data (Nuesch et al., 2009; Savović et al., 2012b). However, evidence for this domain is currently not consistent (Berkman et al., 2014). Study size has also been used in meta-epidemiological research, with evidence suggesting that smaller studies, on average, produce larger intervention effect estimates (Dechartres et al., 2013), but there are multiple potential reasons for this, for example, publication bias, selective reporting, smaller studies tending to be lower quality, or larger trials only being conducted when the anticipated intervention effect is small. For systematic overviews of meta-epidemiological evidence across a range of domains see Berkman et al. (2014) and Page et al (2016).

Meta-epidemiological studies rely on accurate reporting of trial characteristics such as randomization procedure. Often trial characteristics are unclear from research reports: for example, in the BRANDO study allocation concealment was assessed as "unclear" in 69% of all trials (Savović et al., 2012a). Although the effect of "double blinding" has been assessed, it has not yet been possible to separate the potential effects of blinding patients, care providers, and outcome assessors, due to unclear reporting. There is potential for our understanding of bias to increase with improved reporting of trial characteristics (Schulz et al., 2010). The potential for meta-epidemiological research has also been greatly improved by the introduction of routine and standardized bias assessment such as that described in the last section (Savović et al., 2018): previously, authors of a meta-epidemiological study would need to produce their own bias assessments for all contributing trials, which was necessarily very time-consuming.

These studies are of course intrinsically observational: observed associations may well be confounded by other factors. In particular, study characteristics relating to risk of bias are often correlated with each other (Savović et al., 2012a), so that an observed association between one characteristic and intervention effect estimates could be due to a different, correlated, characteristic. Our knowledge of how multiple study characteristics might interact with each other is also very limited at present. Exploratory stratified or multivariable analyses to account for multiple study characteristics simultaneously, or to adjust for one characteristic when assessing the effect of another, have been performed (Nuesch et al., 2009; Savović et al., 2012a; Dechartres et al., 2013; Siersma et al., 2007). For example, study size was still found to be associated with treatment effect estimates after adjusting for internal bias indicators, providing evidence that such internal biases are not the only driver of such associations (Dechartres et al., 2013). Siersma et al. (2007) found that univariable and multivariable analyses of five distinct bias indicators gave similar results, although they note that these bias indicators were only modestly correlated with each other in their sample. The BRANDO study suggested that the effects of allocation concealment and blinding might be less than additive, although there was considerable uncertainty (Savović et al., 2012a). Bias adjustment (see next section) would be considerably easier if we were able to summarize risk of bias using a single proxy measure, but this does not appear to be the case.

12.4 Adjusting for Bias Using Data-Based Evidence or Elicited Opinion

We now consider quantitative methods for adjusting for within-study biases in a meta-analysis. Although sensitivity analyses give insight into the possible impact of biases, it is often desirable to present numerical results from an analysis which includes all eligible studies, while adjusting for potential biases.

Methods for synthesizing evidence from a set of studies of varying quality have been proposed by a number of authors. The pioneering work in this area was carried out by Eddy et al. (1992), who proposed a range of models allowing for biases affecting internal and external validity, as part of their confidence profile method for meta-analysis. Eddy et al. (1992) constructed detailed models for each potential bias, in which the parameters are informed by data or by subjective opinion. Adjustment for biases in the meta-analysis is carried out through a full likelihood approach, by incorporating the bias models into the likelihood. Wolpert and Mengersen (2004) used a similar approach to adjust for biases due to eligibility violations, and misclassification of exposure or outcome status, in a meta-analysis combining case-control and cohort studies. Greenland (2005) modeled biases due to uncontrolled confounding, non-response, and misclassification of exposure, in a meta-analysis of case-control studies, and used Monte Carlo simulation methods and approximate adjustments to allow for the biases.

Another approach to adjusting for bias is to use hierarchical modeling to allow for variability among different study designs, as suggested by Prevost et al. (2000). For example, this method can be used to combine comparative evidence from randomized trials, prospective cohort studies, and case-control studies. The hierarchical structure of the standard random-effects meta-analysis model is extended by adding an extra level for study type, so that variation between and within-study types is acknowledged. Prior distributions can be incorporated in the model, to express beliefs that certain study types are more biased than others or to change the relative weighting of the study types in estimation of the overall pooled result (Prevost et al., 2000). Choice of such priors could be informed by empirical data or by subjective opinion.

12.4.1 Adjusting for Bias in RCTs Using Data-Based Evidence

Suppose we have a pairwise meta-analysis m^*, where studies are categorized according to their risk of bias on a particular indicator, denoted by x_i, where $x_i = 1$ indicates a study at high risk of bias, and $x_i = 0$ indicates a study at low risk of bias, as in Section 12.3. As in 12.3, the model for the relative treatment effect in trial i is $\delta_{im^*} + \beta_{im^*} x_i$, where the treatment effects in studies at low risk of bias are $\delta_{im^*} \sim N(\mu_{m^*}, \sigma^2_{m^*})$ and the study-specific biases are $\beta_{im^*} \sim N(b_{m^*}, \kappa^2)$.

Although it is possible to estimate the mean bias, b_{m^*}, and between-study variability in bias, κ^2, within the meta-analysis and obtain an adjusted pooled mean treatment effect, μ_{m^*}, there is limited ability to identify the bias parameters, resulting in estimates that are similar to an analysis including low risk of bias studies only. It has therefore been proposed (Welton et al., 2009) that estimated bias parameters from meta-epidemiological studies (Section 12.3) can provide empirically based prior information on the bias parameters, b_m and κ^2, allowing for between meta-analysis variability in bias $b_m \sim N(b_0, \phi^2)$, as in equation 12.1.

A Bayesian analysis using priors for b_0, κ^2, and φ^2 provides prior information on the study-specific bias parameters β_{im^*} and thus allows the meta-analysis to borrow strength from the studies at high risk of bias, while simultaneously adjusting for and down-weighting the evidence from those studies.

As discussed in Section 12.3, Savović et al. (2012a, 2012b) report predictions for b_0, κ^2, and φ^2 in their Table 25, that can be used as priors for a meta-analysis with binary outcome data modeled on a log odds scale, stratified by risk of bias indicator and outcome type.

To illustrate, Table 12.2 shows estimated odds ratios for the outcome all-cause mortality based on a meta-analysis of studies comparing intravenous immunoglobin (IVIG) with control in patients with severe sepsis, where the results are stratified by a risk of bias indicator as to whether blinding in the studies was adequate, or unclear/inadequate (Soares et al., 2012). The estimated odds ratio for the low risk of bias studies is 0.55, but this is very uncertain (95% CrI: 0.12 to 2.06). The high risk of bias studies estimate a more favorable odds ratio of 0.41 and this estimate is more precise (95% CrI: 0.23 to 0.64). For blinding, Savović et al. (2012a) report, for all outcome types combined, a mean ratio of odds ratios ($\exp(b_0)$) for unclear/inadequate versus adequate studies of 0.87 (95% CrI 0.79, 0.96) (as reported in Section 12.3), between-studies standard deviation in bias, κ, of 0.14 (95% CrI 0.02, 0.30), and between meta-analysis standard deviation in mean bias, φ, of 0.12 (95% CrI 0.03, 0.28), which can be used as informative priors for the bias model as described above, where m^* is the IVIG meta-analysis.

We compared results from three different models for bias-adjustment: (i) ignoring bias (all studies taken at face value); (ii) using uninformative priors for b_{m^*} and κ (flat normal prior for b_{m^*}, and Un(0,1) for κ) so that they are estimated from meta-analysis m^* alone; and (iii) using informative priors for b_{m^*} and κ from Savović et al. (2012a) as described above. Table 12.2 shows, for each of the models, the estimated ratio of odds ratios for studies with unclear/inadequate blinding relative to studies with adequate blinding, and the estimated treatment effect odds ratio of IVIG versus control. Using a bias model (rows (ii) and (iii) in Table 12.2) leads to a bias-adjusted estimate of the odds ratio that is closer to 1 (no effect) and with 95% CrIs that contain 1, compared with a model that ignores bias where there is strong evidence of a beneficial effect of IVIG. Using an informative prior based on meta-epidemiological evidence (row (iii)) gives a similar estimate of the odds ratio compared with using uninformative priors for b_{m^*} and κ (so they are estimated from meta-analysis m^* alone), however, with a slightly narrower credible interval, due to prior information from the meta-epidemiological evidence. In this example, there is evidence of bias within the IVIG meta-analysis, and this dominates the prior evidence from meta-epidemiological evidence. Both adjustment methods give very similar estimates of the ratio of odds ratios for studies with inadequate/unclear blinding compared with studies with adequate blinding.

TABLE 12.2

Posterior Median and 95% Credible Intervals for the Odds Ratio of All-Cause Mortality for IVIG versus Control

Model	Odds ratio (IVIG vs. control)	Ratio of odds ratios
Low risk of bias studies only	0.55 (0.12, 2.06)	
High risk of bias studies only	0.41 (0.23, 0.64)	
(i) Ignoring bias (all studies at face value)	0.45 (0.28, 0.67)	1 by assumption
(ii) Uninformative priors for bias parameters (estimation using IVIG meta-analysis only)	0.59 (0.26, 1.13)	0.64 (0.27, 1.75)
(iii) BRANDO priors for bias parameters	0.58 (0.27, 1.11)	0.68 (0.29, 1.64)

Results are shown for low risk of bias studies only (adequate blinding); high risk of bias studies only (inadequate/unclear blinding); and for three bias-adjustment models: (i) ignoring bias (all studies taken at face value); (ii) using uninformative priors for the bias parameters (flat normal prior for b_{m^*}, and Un(0,1) for κ), so b_{m^*} and κ are estimated from meta-analysis m* alone; and (iii) using informative priors for b_{m^*} and κ based on Savović et al. (2012a). For the bias-adjustment models, posterior median and 95% credible intervals are given for the ratio of odds ratios for studies with unclear/inadequate blinding relative to studies with adequate blinding.

The approach assumes that the study-specific biases in the meta-analysis can be considered exchangeable with those in the meta-epidemiological data used to provide the prior distributions used for adjustment. Restricting the meta-epidemiological data used to obtain priors to relevant disease areas and populations can help ensure the validity of the exchangeability assumption, however, this comes at a cost of basing the prior information on a smaller dataset leading to a loss of precision. Sensitivity analysis to assumptions is essential.

12.4.2 Adjusting for Bias in RCTs Using Elicited Opinion

An alternative to using empirical data-based evidence to inform bias adjustment is to use elicited opinion on the magnitude of potential biases. Turner et al. (2009) developed an approach for using expert opinion to inform adjustment for both internal biases caused by methodological flaws and external biases caused by deviation of study designs from the research question of interest. The method of bias adjustment was implemented through the following steps: define the target research question; describe an idealized version of each available study; identify internal and external biases; elicit opinion on the magnitude and uncertainty of the biases; and perform a bias-adjusted meta-analysis. These steps will be discussed in more detail below and in the following section and illustrated through an example.

Rather than using data-based evidence to inform distributions for suspected biases, as in Section 12.4.1, expert opinion is sought on the magnitude and uncertainty of the biases identified. An advantage of this approach over using data-based evidence is that adjustment can be made for external as well as internal biases: since external biases are specific to a particular target setting, data-based evidence on these would rarely be available. Turner et al. (2009) recommended that internal biases are judged by assessors with methodological expertise, while external biases are judged by assessors with specialist knowledge of the clinical area. A disadvantage over using data-based evidence is that the process of obtaining expert opinion is time-consuming, and that it must be carried out separately for each new meta-analysis since the biases are specific to the studies included.

Opinion on biases is obtained during elicitation meetings, in which assessors discuss each study in turn and review bias checklists, while discussing any queries and resolving misunderstandings. Assessors are asked to provide their independent opinions on the magnitude and uncertainty of each bias in each study by marking ranges on bias elicitation scales. Opinions are marked using 67% intervals, such that the assessor believes the true bias was twice as likely to lie inside rather than outside the range. For each bias, assessors also discuss and agree whether the bias could change the direction as well as magnitude of the intervention effect (an additive bias), or whether it could only change the magnitude of the effect (a proportional bias). These two bias types are handled separately in the adjustment models below.

For each assessor, the 67% range limits elicited for individual biases are used to calculate means and variances for the corresponding biases in each study. The means and variances for individual biases are used to obtain means and variances for total internal and total external bias in each study, and for additive and proportional biases separately (Turner et al., 2009). We denote total internal additive bias in study i by $\beta_i^I \sim f\left[\mu_{i\beta}^I, {\sigma_{i\beta}^I}^2\right]$ and total external additive bias by $\beta_i^E \sim f\left[\mu_{i\beta}^E, {\sigma_{i\beta}^E}^2\right]$, where the notation $f[\mu,\sigma^2]$ represents a generic distribution with mean μ and variance σ^2. Similarly, we denote total internal proportional bias in study i by $\gamma_i^I \sim f\left[\mu_{i\gamma}^I, {\sigma_{i\gamma}^I}^2\right]$ and total external proportional bias by $\gamma_i^E \sim f\left[\mu_{i\gamma}^E, {\sigma_{i\gamma}^E}^2\right]$. Writing y_i for the observed intervention effect in study i, with sampling variance s_i^2 assumed known, we assume the model $y_i \sim f\left[\gamma_i^I\theta_i + \beta_i^I, s_i^2\right]$, where θ_i is the underlying intervention effect in study i, to allow adjustment for internal additive bias β_i^I and internal proportional bias γ_i^I. To allow for external additive bias β_i^E and external proportional bias γ_i^E, reflecting differences between the design of study i and the target research question, we model θ_i as $\theta_i = \gamma_i^E\theta + \beta_i^E$, where θ represents the underlying target intervention effect. A model incorporating adjustment for both internal and external biases within the same study is:

$$y_i \big| \beta_i^I, \beta_i^E, \gamma_i^I, \gamma_i^E \sim f\left[\gamma_i^I\left(\gamma_i^E\theta + \beta_i^E\right) + \beta_i^I, s_i^2\right]$$

When using bias distributions elicited from a single assessor for study i, the bias-adjusted estimate of the target intervention effect is calculated as:

$$\hat{\theta} = \left(y_i - \mu_{i\gamma}^I\mu_{i\beta}^E - \mu_{i\beta}^I\right) \Big/ \mu_{i\gamma}^I\mu_{i\gamma}^E$$

The corresponding bias-adjusted standard error is calculated as:

$$SE\left(\hat{\theta}\right) = \left(\frac{1}{\mu_{i\gamma}^I\mu_{i\gamma}^E}\right)^2\left(s_i^2 + \left(\mu_{i\gamma}^{I2} + \sigma_{i\gamma}^{I2}\right)\left(\hat{\theta}^2\sigma_{i\gamma}^{E2} + \sigma_{i\beta}^{E2}\right) + \sigma_{i\gamma}^{I2}\left(\hat{\theta}\mu_{i\gamma}^E + \mu_{i\beta}^E\right)^2 + \sigma_{i\beta}^{I2}\right)$$

For each study, Turner et al. (2009) pooled results across assessors by taking the medians of the assessors' bias-adjusted estimates and standard errors, in order to obtain a bias-adjusted result based on the opinions of a "typical" assessor. More conventional approaches to pooling opinion include the linear opinion pool and the logarithmic opinion pool, in which the pooled opinion from multiple assessors represents weakened or strengthened knowledge respectively compared with a single assessor (O'Hagan et al., 2006). Finally, a conventional random-effects meta-analysis model is used to combine the bias-adjusted results across studies. This approach will be illustrated in the next section.

The approach described above makes no allowance for the correlations between multiple biases. In principle, it would be possible to extend the bias-adjustment model to allow for correlations, but these would be very difficult to elicit.

12.4.3 Adjusting for Bias in RCTs and Observational Studies Using Elicited Opinion

An advantage of the opinion-based approach to adjusting for bias is that it can be applied to meta-analyses including both RCTs and observational studies. Currently, very little data-based evidence is available on biases affecting observational studies. As an illustrative example of bias adjustment using opinion, we consider evidence for the effectiveness of routine antenatal anti-D prophylaxis (RAADP) for preventing sensitization in pregnant Rhesus negative women. The UK National Institute for Health and Clinical Excellence (NICE) has carried out two appraisals on this topic, to inform UK health policy, which identified ten studies evaluating the clinical effectiveness of RAADP compared with control (Chilcott et al., 2003; Pilgrim et al., 2009). However, study quality was generally low: many of the studies had used historical rather than concurrent controls; only one study was randomized; and there were often post-hoc exclusions and high losses to follow-up. In addition to these flaws which lead us to suspect internal biases, the ten studies differed substantially in the doses and timing of administering RAADP, in the characteristics of the women recruited, and in timing of assessing the outcome. There were important differences between the study settings and the target UK setting, which cause external biases.

The first step in the approach proposed by Turner et al. (2009) is to define the target research setting by describing the population to which findings will be applied, the intervention and control policies under comparison, and the outcome of interest. In the RAADP evaluation, the target setting reflects the UK health policy question, defined as follows: (i) population: non-sensitized pregnant Rhesus negative women in the UK; (ii) intervention: dose of 500 IU anti-D immunoglobulin offered intramuscularly at 28 and 34 weeks' gestation, in addition to control antenatal care; (iii) control: anti-D immunoglobulin offered postpartum and after potentially sensitizing events during pregnancy, according to 2002 UK policy; and (iv) outcome: prevention of Rhesus sensitization which would affect a subsequent pregnancy.

The next step is to define an idealized version of each study in the meta-analysis, by describing the intended population, interventions compared, and outcome, as determined from the published study report. The idealized study is a tool which assists identification of internal and external biases and may be viewed as an imagined repeat of the original study, in which all sources of internal bias (i.e., methodological flaws) have been eliminated. As an example, we define the idealized version of one of the RAADP studies by Trolle (1989): (i) population: non-sensitized Rhesus negative women delivered of Rhesus positive babies at Kolding Hospital, Denmark; (ii) intervention: dose of 1500 IU anti-D immunoglobulin given at 28 weeks' gestation, in addition to control antenatal care; (iii) control: dose of at least 1000 IU anti-D immunoglobulin given postpartum and anti-D offered after potentially sensitizing events during pregnancy; (iv) outcome: sensitization at ten months postpartum.

Next, to identify internal biases, each study is compared against its idealized version and an internal bias checklist is completed. This involves answering a series of questions about potential sources of bias and justifying the answers by extracting relevant details from the original study publications. Turner et al. considered five categories of internal bias (Deeks et al., 2003): biases caused by differences between intervention and control groups at baseline; biases related to lack of blinding of participants or caregivers; biases caused by exclusions and drop-outs; biases related to measurement of the outcome; and an extra category of "other bias". To identify external biases, the idealized version of each study in the meta-analysis is compared against the target setting. The four characteristics of the target setting lead us to consider four categories of external bias, each caused by differences between the idealized studies and the target setting in population type, intervention type, control type, and outcome type. We note that identification of internal biases could now be carried out using the Cochrane risk of bias tool for RCTs (see Section 12.2) and the recently developed ACROBAT-NRSI tool for observational studies (Sterne et al., 2014).

Once internal and external biases had been identified and relevant supporting information had been extracted, expert opinion on the magnitude and uncertainty of the biases was sought, as described in Section 12.4.2. Details of the potential internal and external biases identified in the ten studies were reported by Turner et al. (2012). Biases caused by differences between intervention and control groups at baseline was a major concern in seven studies, because women receiving RAADP were compared against historical rather than concurrent controls, who were very likely to have differed from the intervention women in a variety of ways; also, the analyses of the original studies had not adjusted for confounding. None of the studies had used blinding of subjects or caregivers, which means that the handling of potentially sensitizing events during pregnancy was likely to have differed between intervention and control arms in all studies. Additionally, there were losses to follow-up. External biases arose from the substantial differences between the available RAADP studies and the target UK setting with respect to populations, details of interventions, control group care, and the timing of the outcome.

Opinions on the magnitude and uncertainty of the identified internal biases were provided by four assessors with methodological expertise, while opinions on external biases were provided separately by four assessors chosen for their knowledge of anti-D prophylaxis (Turner et al., 2012). Adjustment was made for internal and external biases, using the models presented in Section 12.4.2, and a bias-adjusted meta-analysis was performed.

Figure 12.1 shows the impact of bias adjustment on the results of the ten RAADP studies and on the overall pooled result. The figure presents three versions of the meta-analysis: (i) unadjusted for biases; (ii) adjusted for internal biases; and (iii) adjusted for both internal and external biases. In each analysis, a random-effects model was fitted to the unadjusted or bias-adjusted results, using the DerSimonian and Laird estimate for between-study heterogeneity (see Chapter 4). In the unadjusted meta-analysis, the pooled odds ratio for sensitization in pregnant Rhesus negative women, comparing RAADP with control, was estimated as 0.25 (95% CI 0.18, 0.36), and the between-study variance was estimated as 0.06. This result does not acknowledge the uncertainty caused by suspected biases, so we judge the confidence interval to be inappropriately narrow. After adjusting for internal biases, the pooled odds ratio was estimated as 0.28 (95% CI 0.15, 0.53), and the between-study variance was estimated as 0. When allowing additionally for external biases arising from varying relevance of the studies to the target UK setting, the pooled odds ratio was estimated as 0.31 (95% CI 0.17, 0.56), and the between-study heterogeneity variance was estimated as 0. The confidence interval is substantially wider than the unadjusted result, but still provides strong evidence for the effectiveness of RAADP, after allowing for the expected effects of internal and external biases.

The approach described above assumes independence for the multiple internal biases and multiple external biases considered and does not allow for correlations between these.

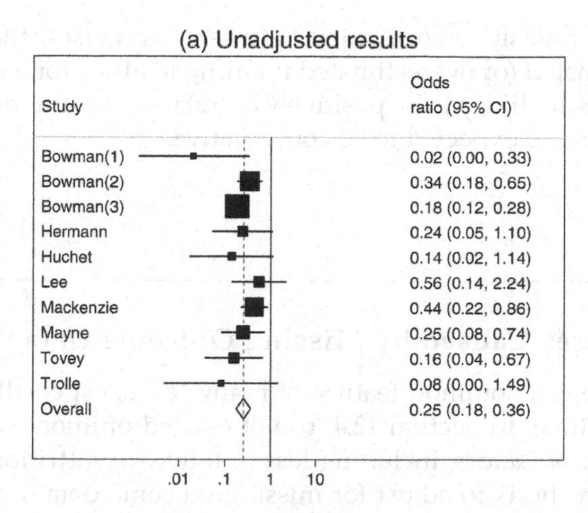

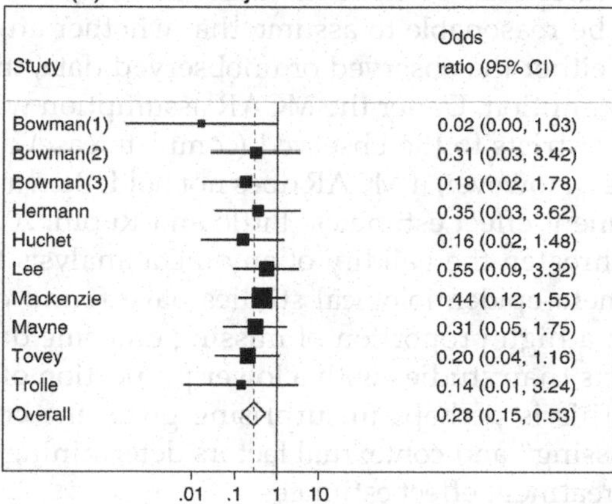

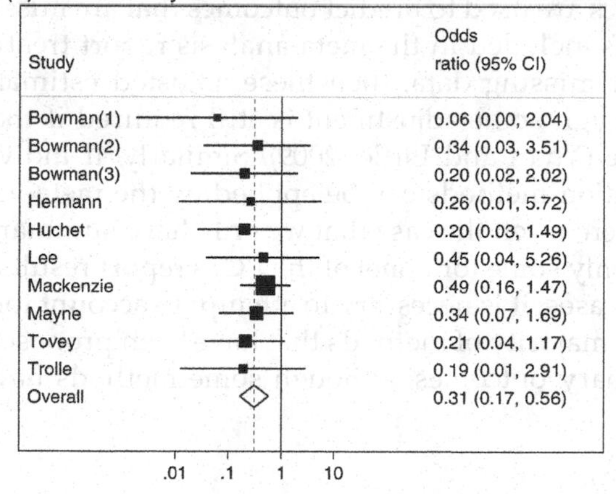

FIGURE 12.1
Impact of adjusting for biases in a meta-analysis of 10 studies comparing routine antenatal anti-D prophylaxis to control.

If failing to allow for a positive correlation between two biases, the total bias-adjusted variance is underestimated (or overestimated if failing to allow for a negative correlation). Since individual biases are likely to be positively correlated, the amount of variance adjustment to the study results is expected to be conservative.

12.5 Handling Biases Caused by Missing Outcome Data within Studies

Missing outcome data is a common feature of many RCTs, especially when participants are followed up over time. In Section 12.4, use of elicited opinion was used to adjust for biases due to a variety of factors, including loss to follow-up/attrition. In this section, we consider alternative methods to adjust for missing outcome data (noting that this is distinct from missing covariates which will be discussed in Chapters 21 and 22).

In some cases, it may be reasonable to assume that whether an outcome is missing or not does not depend on either the observed or unobserved data, the "missing completely at random" (MCAR) assumption. Under the MCAR assumption within each randomized group, an analysis that restricts to the observed (complete case) data provides unbiased treatment effect estimates. However, if MCAR does not hold, then a complete-case analysis can lead to biased treatment effect estimates (Little and Rubin, 2002). Studies with missing outcome data also threaten the validity of any meta-analysis that includes them. As noted in Section 12.3, meta-epidemiological studies have found conflicting evidence as to whether studies with a high proportion of missing outcome data are associated with different treatment effects than studies with a lower proportion of missing outcome data (Berkman et al., 2014). This is perhaps unsurprising given the crudeness of the indicator "high proportion missing" and contextual factors determining the direction in which missing data may bias treatment effect estimates.

Methods to adjust for missing data in RCTs typically center around imputation, where other observed variables are used to predict outcomes that are missing (Little and Rubin, 2002). If all of the RCTs included in the meta-analysis report treatment effect estimates *adequately* adjusted for missing data, then these adjusted estimates can be combined in meta-analysis, although some adjustment is still required if the proportion missing depends on effect size (Yuan and Little, 2009). Similarly, if individual patient data is available, then imputation methods can be applied by the meta-analyst (Burgess et al., 2013). However, it is more often the case that we only have summary level data available from each study, and only some (or none) of the RCTs report results adjusted for missing outcome data. In such cases, it is necessary to attempt to account for missing data within the meta-analysis. The majority of methods that have been proposed in the literature are specific to missing binary outcomes, although some methods have been proposed for continuous outcomes.

12.5.1 Missing Binary Outcome Data

12.5.1.1 Sensitivity Analyses

It is good practice to conduct a number of sensitivity analyses that make different assumptions about the missing outcomes (Deeks et al., 2019; National Institute for Health and Care Excellence, 2013; National Institute for Health and Clinical Excellence, 2012) to see how robust the results of the meta-analysis are. Common assumptions to explore with binary outcomes are:

(i) "All failures analysis" where all missing outcomes are assumed failures
(ii) "All successes analysis" where all missing outcomes are assumed successes
(iii) "Worst-case analysis" where all missing data in the active arms are assumed to be failures, and all missing outcomes in the control arms are assumed to be successes
(iv) "Best-case analysis" where all missing data in the active arms are assumed to be successes, and all missing outcomes in the control arms are assumed to be failures

12.5.1.2 Two-Stage Adjustment

Higgins et al. (2008) proposed the informative missingness odds ratio (IMOR) as a measure to adjust for bias arising from missing data on a binary outcome. The $\text{IMOR}_{i,j}$ for study *i* arm *j* is defined as the odds of an event in the missing individuals divided by the odds of an event in the observed individuals, and can depend on study and treatment arm:

$$\text{IMOR}_{i,j} = \frac{\left(\pi_{i,j}^{\text{miss}} \big/ \left(1-\pi_{i,j}^{\text{miss}}\right)\right)}{\left(\pi_{i,j}^{\text{obs}} \big/ \left(1-\pi_{i,j}^{\text{obs}}\right)\right)}$$

where $\pi_{i,j}^{\text{miss}}$ is the probability of an event in the missing individuals and $\pi_{i,j}^{\text{obs}}$ the probability of an event in the observed individuals. It is not possible to estimate the IMOR from the observed data, and so it must be based on assumptions. An IMOR of 1 means that the odds of an event is the same for the observed and the missing individuals, and so the data is taken at face value. An IMOR assumed not equal to 1 can be used to adjust the estimate for that study and arm.

White et al. (2008a) suggest using prior information on the IMORs to reflect departures from MCAR in a two-stage analysis. In the first stage, study-specific estimates of treatment effect are adjusted using prior beliefs on the IMOR. In the second stage, these adjusted estimates are pooled in a meta-analysis. Robustness of results to the priors used is recommended to be explored in a sensitivity analysis (Higgins et al., 2008; White et al., 2008a).

The advantage of the two-stage approach is that it is transparent, in that the assumed prior beliefs must be clearly stated and can be easily subjected to sensitivity analyses to alternative prior beliefs. A potential limitation is that the prior beliefs on the IMORs are "fixed" and cannot be updated, or learnt about, from the observed data. Furthermore, eliciting beliefs on missingness mechanisms is challenging. One option is to ask experts to weight values on a plausible range (Turner et al., 2009; White et al., 2007), while another approach is to elicit the lowest and highest value an expert would conceivably expect (Mavridis et al., 2013). White et al. (2008a) suggest providing an expert with a plausible estimate for the probability of success given a subject was observed and then ask, given that estimate, their opinion on the probability of success given a subject was missing. There may also be information in the RCT publications as to the reasons for missingness that could be used to inform priors.

12.5.1.3 One-Stage Adjustment

There have been a few attempts to estimate missingness parameters within a meta-analysis. White et al. (2008b) proposed a one-stage hierarchical model for the IMORs in an attempt to "learn" about the IMORs from the observed data, although found there was limited ability to do so in a pairwise meta-analysis. Spineli et al. (2013) used a network meta-analysis model to estimate the IMORs, but again found that the data was barely sufficient to identify the IMOR parameters, even though a network meta-analysis in theory provides more degrees of freedom for estimation. In random treatment effect models, there is almost complete confounding between the random treatment effect and the random IMOR elements. In fixed treatment effect models, particularly when some trials have only small amounts of missing data, then the data is sufficient to identify missingness parameters and "learning" can take place.

Turner et al. (2015) present a general framework for a Bayesian one-stage estimation for any definition of the missingness parameter, such as the IMOR, probability of an event in the missing individuals, and so on. Although the IMOR is a special case, the model differs from that proposed by White et al. (2008b) because it is a "pattern-mixture" model rather than a "selection" model (Little, 1993). The strength of the pattern-mixture approach is that it simultaneously allows both the incorporation of prior beliefs and the estimation of the missingness parameters, so that the prior beliefs may be updated in light of the observed data.

12.5.1.4 In the Absence of Prior Information on Missingness

In the absence of any prior information on the missingness mechanism, it is important to reflect the additional uncertainty in effect estimates as a result of the missing data. Gamble and Hollis (2005) proposed down-weighting studies where "best-case" and "worst-case" analyses give wide limits on treatment effect, to reflect the uncertainty associated with missing data. Turner et al. (2015) instead put a flat prior on the probability of an event in the missing individuals in their one-stage adjustment method to reflect the uncertainty induced by the missing outcomes.

12.5.2 Missing Continuous Outcome Data

12.5.2.1 Sensitivity Analyses

Sensitivity analyses are equally important when missing outcome data is continuous, however, individual patient data is usually required in order to carry this out. The following assumptions can be explored with individual patient data (Molenberghs and Kenward, 2007):

(i) Complete case (CC) analysis, where only individuals without missing data are included in the analysis. A complete case analysis relies on the validity of the MCAR.

(ii) Last observation carried forward (LOCF), where the most recent previous observation for each individual is assumed for subsequent missing observations for that individual. LOCF analyses the missing data as if it was known, and so overestimates precision. It also relies on the assumption that no further changes in outcome occur over time since the last observed measurement.

(iii) Baseline observation carried forward (BOCF), where the baseline (time 0) observation for each individual is assumed for subsequent missing observations for that individual. BOCF analyses the missing data as if it was known, and so overestimates precision. It also relies on the assumption that any benefit (or decrement) obtained since the start of the study is lost in missing observations.

(iv) Multiple imputation (MI), where predictions are made for the missing observations, based on observed values. The data are re-analyzed repeatedly for different sampled predictions, and a pooled estimate and its uncertainty obtained. MI is applicable when the missing observations depend on observed quantities but not on unobserved quantities (the missing at random (MAR) assumption), and so predictions can be obtained from the observed data. The performance of MI therefore depends on the validity of the MAR assumption, and on the validity of the imputation model. MI has the benefit of representing the uncertainty arising from missing data.

Jorgensen et al. (2014) compared these different approaches in the analysis of a placebo-controlled trial of an anti-obesity drug, and found that both the point estimate and precision were sensitive to choice of imputation method.

12.5.2.2 Two-Stage Adjustment

Recent work (Mavridis et al., 2014b) has extended the concept of the IMOR to continuous outcomes, defining the informative missingness difference of means (IMDoM) and the informative missingness ratio of means (IMRoM), and proposed a pattern-mixture model for pairwise and network meta-analysis, using a two-stage estimation procedure.

12.6 Modeling Bias in Network Meta-Analysis

Network meta-analysis (NMA) pools evidence from studies that form a connected network of treatment comparisons (Chapter 10). The consistency (or transitivity) equations (Section 10.4) made in NMA mean that for networks with loops of evidence and sufficient numbers of studies at high and low risk of bias on treatment comparisons that form loops, there are spare degrees of freedom to estimate interactions with risk of bias indicators. This presents the opportunity to estimate bias parameters without requiring the strong "exchangeability" assumptions needed to adjust for bias using meta-epidemiological data. The bias model described in Section 12.4.1 can be adapted to network meta-analysis as follows:

$$\theta_{ik} = \mu_i + (\delta_{ik} + \beta_{ik} x_i) \quad \text{where } \delta_{i1} = \beta_{i1} = 0$$

where θ_{ik} is a measure of underlying outcome in arm k of study i (e.g., log odds or log rate), with risk of bias indicator x_i, study-specific treatment effect δ_{ik}, and study-specific bias β_{ik}, for the treatment t_{ik} in arm k relative to the treatment t_{i1} in arm 1 of trial i. We note that a meta-analysis index m is not required here, since we are using data from only one network meta-analysis. We can assume either a common treatment effect: $\delta_{ik} = d_{t_{i1},t_{ik}}$ or a hierarchical model across studies: $\delta_{ik} \sim N(d_{t_{i1},t_{ik}}, \tau^2)$, and in both cases that the consistency equations hold for any pair of treatments: $d_{k_1 k_2} = d_{1k_2} - d_{1k_1}$. We assume a hierarchical model for the study-specific biases:

$$\beta_{ik} \sim N(b_{t_{i1},t_{ik}}, \kappa^2)$$

with between-study variance in bias, κ^2, and where the mean bias $b_{t_{i1},t_{ik}}$ also follows the consistency equations: $b_{k_1 k_2} = b_{1k_2} - b_{1k_1}$ (Section 10.5.6). In order to be able to identify the bias parameters, we need to make some simplifying assumptions on the mean biases. One possibility is that the mean bias is the same for all active treatments which are compared with a standard or placebo treatment 1 ("active vs. placebo" trials), so that $b_{1,k_2} = b_1$ for $k_2 = 2, \ldots$ It is less clear what to assume about bias in trials which make comparisons between active treatments. One approach might be to assume a mean bias of 0 for active versus active comparisons, based on the assumption that the mean bias against placebo is the same for the two active treatments. An alternative is to estimate a separate mean bias term which is the same for all active versus active comparisons (Dias et al., 2010), but a judgment as to which direction the bias would act is necessary. One way to do this is to assume the average bias is always in favor of the newer treatment ("optimism bias") (Salanti et al., 2010; Song et al., 2008).

12.6.1 Example: Fluoride Therapies for the Prevention of Caries in Children

Dias et al. (2010) present bias-adjustment models in a network meta-analysis of 130 trials of fluoride therapies to prevent the development of caries in children (Figure 12.2).

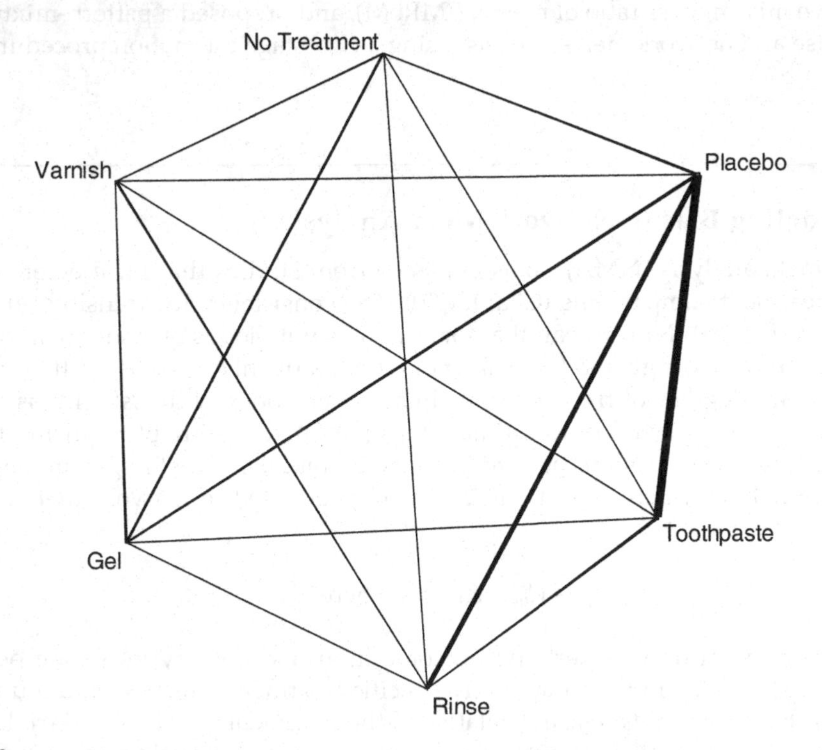

FIGURE 12.2
Fluoride example. Network of treatment comparisons (drawn using R code from Salanti (Salanti, 2011)). The thickness of the lines is proportional to the number of trials making that comparison.

Dias et al. (2010) explored two models for the bias. In the first model, they assume that the mean bias in active versus active comparisons is zero, and in the second that the mean bias is equal for all of the active versus active comparisons. In both models, they assume the mean bias is equal for all active versus inactive (placebo or no treatment) comparisons. Risk of bias indicators are usually categorized as high, low, or unclear, and a judgment has to be made as to whether to group "unclear" with "low" or "high" risk of bias. Dias et al. (2010) presented an alternative which estimates a *probability* that each unclear study is at high risk of bias.

For the allocation concealment risk of bias domain, the model that allowed a probability that unclear studies were at high risk of bias gave the best fit to the data. There was no evidence of bias for the active versus active comparisons (rate ratio 1 with 95% CrI (0.52, 1.77)). The estimated mean bias (rate ratio) for active versus (placebo or no treatment) was 0.83 with 95% CrI (0.70, 0.98), suggesting that trials with high risk of bias due to allocation concealment have a tendency to over-estimate treatment effects relative to placebo or no treatment.

Figure 12.3 shows the estimated log rate ratios from the unadjusted network meta-analysis model (solid lines) and from the bias-adjustment model, where the unclear studies have a probability of being at risk of bias (dashed lines). It can be seen that the main impact of the bias adjustment is to move the treatment effect estimates in toward the null effect, and this is especially the case for varnish. The bias-adjusted results show evidence of a placebo effect, likely to be due to the placebo involving brushing or other treatment of the teeth, albeit without fluoride. Based on the bias-adjusted analysis, we conclude that there is evidence that fluoride is effective in reducing caries, but that there is no evidence that any one fluoride intervention is more effective than another (as observed by their overlapping credible intervals).

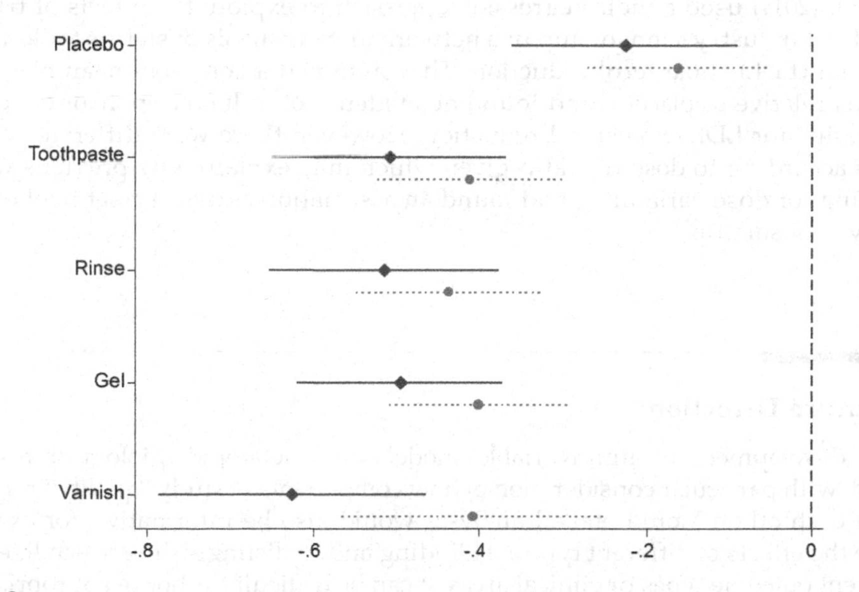

FIGURE 12.3
Estimated posterior means and 95% credible intervals for log rate ratios compared with No Treatment. Results from a network meta-analysis model with no bias adjustment shown with diamonds and solid lines. Circles and dotted lines represent results from bias-adjustment model one with common mean bias term for the Active versus Placebo or No Treatment comparisons, zero mean bias for Active versus Active comparisons, a probability of being at risk of bias in studies rated as unclear. The vertical dotted line represents no effect.

12.6.2 Other Applications

Salanti et al. (2010) used a bias-adjustment model to explore the existence of novel agent effects in three network meta-analyses of treatments for cancer (ovarian, colorectal, and breast cancer). They found some evidence of novel agent effects, with hazard ratios for overall survival exaggerated by 6% (95% CrI 2%, 11%) for newer treatments. Note that overall survival is an objective measure, and we would expect to see bigger effects on subjective outcomes (Savović et al., 2012b). Salanti et al. (2010) assumed that the novel agent effect did not interact with other indicators of risk of bias, and it would be interesting to explore such interactions in future work.

Several authors have used bias-adjustment methods in NMA to explore the existence of "small-study effects", a phenomenon whereby stronger treatment effects are observed in smaller studies. Moreno et al. (2009a, 2009b) show that, in the context of anti-depressants, the bias-adjusted estimate from this approach approximates closely to the results found in a simple meta-analysis based on a register of prospectively reported data. Chaimani et al. (2013) found that in their collection of 32 star-shaped networks, imprecise studies were associated with larger treatment effects, and also that imprecise studies tended to be those studies with inadequate conduct. One explanation for small study effects is "publication bias", where studies with statistically significant findings are more likely to be published (see Chapter 13). Copas and Shi (2001) present a selection model to explore sensitivity of results to publication bias in pairwise meta-analysis, and these models have been extended to network meta-analysis (Mavridis et al., 2013; Mavridis et al., 2014a; Trinquart et al., 2012; Chootrakool et al., 2011), including networks with loops (Mavridis et al., 2014a).

It has been suggested that trials sponsored by industry tend to favor the product of the sponsor (Gartlehner et al., 2010; Gartlehner and Fleg, 2010; Flacco et al., 2015). Naci et al. (2014) used a meta-regression approach to explore the effects of trials with and without industry sponsorship in a network meta-analysis of statins for low-density lipoprotein (LDL) cholesterol reduction. They assumed a common mean bias term for all statins relative to placebo and found no evidence of industry sponsor effects in trials of statins for LDL cholesterol reduction. However, there were differences in effectiveness according to dose of statin given which may explain why previous work, not accounting for dose variability, had found an association between treatment effect and industry sponsorship.

12.7 Future Directions

Further development of multivariable models for meta-epidemiological research is required, with particular consideration of how components of study "quality" might interact with each other. More detailed analyses would also be informative, for example, to separate the effects of different types of blinding and to distinguish between likely biases in different outcome types or clinical areas. It can be difficult to choose appropriate priors for variance parameters in complex hierarchical models and further exploration of this would aid future meta-epidemiological research.

The data-based and opinion-based methods of bias adjustment discussed in Section 12.4 both have certain advantages and disadvantages. For example, adjusting for bias in a new meta-analysis using generic data-based priors is much less time-consuming than using elicited opinion; however, the assumption of exchangeability between biases in the new meta-analysis and those in the meta-epidemiological data must be considered carefully. It would be beneficial to develop a method for bias adjustment which is informed by both generic data-based evidence and opinion based on knowledge of the specific trials in the meta-analysis.

References

Balk EM, Bonis PAL, Moskowitz H, Schmid CH, Ioannidis JPA, Wang C and Lau J, 2002. Correlation of quality measures with estimates of treatment effect in meta-analyses of randomized controlled trials. *Journal American Medical Association* **287**(22): 2973–2982.

Berkman ND, Santaguida PL, Viswanathan M and Morton SC, 2014. *The Empirical Evidence of Bias in Trials Measuring Treatment Differences*. Rockville, MD: Agency for Healthcare Research and Quality.

Burgess S, White IR, Resche-Rigon M and Wood AM, 2013. Combining multiple imputation and meta-analysis with individual participant data. *Statistics in Medicine* **32**(26): 4499–4514.

Chaimani A, Vasiliadis HS, Pandis N, Schmid CH, Welton NJ and Salanti G, 2013. Effects of study precision and risk of bias in networks of interventions: A network meta-epidemiological study. *International Journal of Epidemiology* **42**(4): 1120–1131.

Chandler J, Churchill R, Higgins J, Lasserson T and Tovey D, 2013. *Methodological Standards for the Conduct of New Cochrane Intervention Reviews*, Version 2.3. Cochrane Editorial Unit.

Chilcott J, Lloyd Jones M, Wight J, Forman K, Wray J, Beverley C and Tappenden P, 2003. A review of the clinical effectiveness and cost-effectiveness of routine anti-D prophylaxis for pregnant women who are rhesus-negative. *Health Technology Assessment* **7**(4): 1–62.

Chootrakool H, Shi JQ and Yue R, 2011. Meta-analysis and sensitivity analysis for multi-arm trials with selection bias. *Statistics in Medicine* **30**(11): 1183–1198.

Copas JB and Shi JQ, 2001. A sensitivity analysis for publication bias in systematic reviews. *Statistical Methods in Medical Research*, **10**(4): 251–265.

Dechartres A, Boutron I, Trinquart L, Charles P and Ravaud P, 2011. Single-centre trials show larger treatment effects than multicenter trials: Evidence from a meta-epidemiological study. *Annals of Internal Medicine* **155**(1): 39–51.

Dechartres A, Trinquart L, Boutron I and Ravaud P, 2013. Influence on trial sample size of treatment effect estimates: Meta-epidemiological study. *British Medical Journal* **346**: f2304.

Deeks JJ, Dinnes J, D'Amico R, Sowden AJ, Sakarovitch C, Song F, Petticrew M, Altman DG and International Stroke Trial Collaborative Group; European Carotid Surgery Trial Collaborative Group, 2003. Evaluating non-randomised intervention studies. *Health Technology Assessment* **7**(27): 1–5.

Deeks JJ, Higgins JPT, Altman DG on behalf of the Cochrane Statistical Methods Group, 2019. Analysing data and undertaking meta-analyses. In Higgins JPT, Thomas J, Chandler J, Cumpston M, Li T, Page MJ, Welch VA (Eds). *Cochrane Handbook for Systematic Reviews of Interventions*. 2nd Edition. Chichester, UK: John Wiley & Sons. Also online version 6.0 (updated July 2019) available from www.training.cochrane.org/handbook.

Dias S, Welton NJ, Marinho VCC, Salanti G, Higgins JPT and Ades AE, 2010. Estimation and adjustment of bias in randomized evidence by using mixed treatment comparison meta-analysis. *Journal of the Royal Statistical Society: Series A* **173**(3): 613–629.

Eddy DM, Hasselblad V and Schachter R, 1992. *Meta-Analysis by the Confidence Profile Method: The Statistical Synthesis of Evidence*. San Diego, CA: Academic Press.

Egger M, Jüni P, Bartlett C, Holenstein F and Sterne J, 2003. How important are comprehensive literature searches and the assessment of trial quality in systematic reviews?: Empirical study. *National Coordinating Centre for Health Technology Assessment* **36**: 847–857.

Flacco ME, Manzoli L, Boccia S, Capasso L, Aleksovska K, Rosso A, Scaioli G, De Vito C, Siliquini R, Villari P and Ioannidis JPA, 2015. Head-to-head randomized trials are mostly industry sponsored and almost always favor the industry sponsor. *Journal of Clinical Epidemiology* **68**(7): 811–820.
Gamble C and Hollis S, 2005. Uncertainty method improved on best-worst case analysis in a binary meta-analysis. *Journal of Clinical Epidemiology* **58**(6): 579–588.
Gartlehner G and Fleg A, 2010. Pharmaceutical company-sponsored drug trials: The system is broken. *Journal of Clinical Epidemiology* **63**(2): 128–129.
Gartlehner G, Morgan L, Thieda P and Fleg A, 2010. The effect of sponsorship on a systematically evaluated body of evidence of head-to-head trials was modest: Secondary analysis of a systematic review. *Journal of Clinical Epidemiology* **63**(2): 117–125.
Gluud LL, 2006. Bias in clinical intervention research. *American Journal of Epidemiology* **163**(6): 493–501.
Greenland S, 2005. Multiple-bias modelling for analysis of observational data (with discussion). *Journal of the Royal Statistical Society: Series A* **168**(2): 267–306.
Higgins JPT and Altman DG, 2008. Assessing risk of bias in included studies. In: Higgins JPT, Green S (Eds). *Cochrane Handbook for Systematic Reviews of Interventions*. Chichester: John Wiley & Sons.
Higgins JPT, White IR and Wood A, 2008. Imputation methods for missing outcome data in meta-analysis of clinical trials. *Clinical Trials* **5**(3): 225–239.
Higgins JPT, Thompson SG and Spiegelhalter DJ, 2009. A re-evaluation of random-effects meta-analysis. *Journal of the Royal Statistical Society: Series A* **172**(1): 137–159.
Higgins JPT, Altman DG, Gotzsche PC, Jüni P, Moher D, Oxman AD, Savović J, Schulz KF, Weeks L, Sterne JAC, Cochrane Bias Methods Group and Cochrane Statistical Methods Group, 2011. The Cochrane Collaboration's tool for assessing risk of bias in randomized trials. *BMJ* **343**: d5928.
Hopewell S, Boutron I, Altman DG and Ravaud P, 2013. Incorporation of assessments of risk of bias of primary studies in systematic reviews of randomised trials: A cross-sectional study. *British Medical Journal* **3**(8): e003342.
Hrobjartsson A, Thomsen ASS, Emanuelsson F, Tendal B, Hilden J, Boutron I, Ravaud P and Brorson S, 2012. Observer bias in randomised clinical trials with binary outcomes: Systematic review of trials with both blinded and non-blinded outcome assessors. *British Medical Journal* **344**: e1119.
Hrobjartsson A, Thomsen ASS, Emanuelsson F, Tendal B, Hilden J, Boutron I, Ravaud P and Brorson S, 2013. Observer bias in randomized clinical trials with measurement scale outcomes: A systematic review of trials with both blinded and nonblinded assessors. *Canadian Medical Association Journal* **185**(4): E201–E211.
Hrobjartsson A, Emanuelsson F, Thomsen ASS, Hilden J and Brorson S, 2014. Bias due to lack of patient blinding in clinical trials: A systematic review of trials randomizing patients to blind and non-blind substudies. *International Journal of Epidemiology* **43**(4): 1272–1283.
Hrobjartsson A, Thomsen ASS, Emanuelsson F, Tendal B, Rasmussen JV, Hilden J, Boutron I, Ravaud P and Brorson S, 2014. Observer bias in randomized clinical trials with time-to-event outcomes: Systematic review of trials with both blinded and non-blinded outcome assessors. *International Journal of Epidemiology* **43**(3): 937–948.
Jorgensen AW, Lundstrom LH, Wetterslev J, Astrup A and Gotzsche PC, 2014. Comparison of results from different imputation techniques for missing data from an anti-obesity drug trial. *PLOS ONE* **9**(11): e111964.
Jüni P, Witschi A, Bloch R and Egger M, 1999. The hazards of scoring the quality of clinical trials for meta-analysis. *Journal American Medical Association* **282**(11): 1054–1060.
Kirkham JJ, Dwan KM, Altman DG, Gamble C, Dodd S, Smyth R and Williamson PR, 2010. The impact of outcome reporting bias in randomised controlled trials on a cohort of systematic reviews. *British Medical Journal* **340**: c365.
Kjaergard LL, Villumsen J and Gluud C, 2001. Reported methodologic quality and discrepancies between large and small randomized trials in meta-analyses. *Annals of Internal Medicine* **135**(11): 982–989.
Krogh J, Petersen L, Timmermann M, Saltin M and Nordentoft M, 2007. Design paper: The DEMO trial: A randomized, parallel-group, observer-blinded clinical trial of aerobic versus non-aerobic versus relaxation training for patients with light to moderate depression. *Contemporary Clinical Trials* **28**(1): 79–89.

Krogh J, Saltin B, Gluud C and Nordentoft M, 2009. The DEMO trial: A randomized, parallel-group, observer-blinded clinical trial of strength versus aerobic versus relaxation training for patients with mild to moderate depression. *The Journal of Clinical Psychiatry* **70**(6): 790–800.

Little RJA and Rubin DB, 2002. *Statistical Analysis with Missing Data*. Hoboken, NJ: Wiley.

Little RJA, 1993. Pattern-mixture models for multivariate incomplete data. *Journal of the American Statistical Association* **88**(421): 125–134.

Mavridis D, Sutton A, Cipriani A and Salanti G, 2013. A fully Bayesian application of the Copas selection model for publication bias extended to network meta-analysis. *Statistics in Medicine* **32**(1): 51–66.

Mavridis D, White IR, Higgins JPT, Cipriani A and Salanti G, 2014b. Allowing for uncertainty due to missing continuous outcome data in pairwise and network meta-analysis. *Statistics in Medicine* **34**(5): 721–741.

Mavridis D, Welton NJ, Sutton A and Salanti G, 2014a. A selection model for accounting for publication bias in a full network meta-analysis. *Statistics in Medicine* **33**(30): 5399–5412.

Moher D, Jadad AR, Nichol G, Penman M, Tugwell P and Walsh S, 1995. Assessing the quality of randomized controlled trials: An annotated bibliography of scales and checklists. *Controlled Clinical Trials* **16**(1): 62–73.

Moher D, Pham B, Jones A, Cook DJ, Jadad AR, Moher M, Tugwell P and Klassen TP, 1998. Does quality of reports of randomised trials affect estimates of intervention efficacy reported in meta-analyses? *The Lancet* **352**(9128): 609–613.

Molenberghs G and Kenward M, 2007. *Missing Data in Clinical Studies*. Chichester: Wiley.

Moreno SG, Sutton AJ, Ades AE, Stanley TD, Abrams KR, Peters JL and Cooper NJ, 2009a. Assessment of regression-based methods to adjust for publication bias through a comprehensive simulation study. *BMC Medical Research Methodology* **9**(2).

Moreno SG, Sutton AJ, Turner EH, Abrams KR, Cooper NJ, Palmer TM and Ades AE, 2009b. Novel methods to deal with publication biases: Secondary analysis of antidepressant trials in the FDA trial registry database and related journal publications. *British Medical Journal* **339**: b2981.

Naci H, Dias S and Ades AE, 2014. Industry sponsorship bias in research findings: A network meta-analytic exploration of LDL cholesterol reduction in the randomised trials of statins. *British Medical Journal* **349**: g5741.

National Institute for Health and Care Excellence, 2013. *Guide to the Methods of Technology Appraisal 2013*. London: National Institute for Health and Care Excellence.

National Institute for Health and Clinical Excellence, Nov 2012. *The Guidelines Manual*. London: National Institute for Health and Care Excellence.

Nüesch E, Trelle S, Reichenbach S, Rutjes AWS, Burgi E, Scherer M, Altman DG and Jüni P, 2009. The effects of excluding patients from the analysis in randomised controlled trials: Meta-epidemiological study. *British Medical Journal* **339**: b3244.

O'Hagan A, Buck CE, Daneshkhah A, Eiser JR, Garthwaite PH, Jenkinson DJ, Oakley JE and Rakow T, 2006. *Uncertain Judgements: Eliciting Experts' Probabilities*. Chichester: Wiley.

Page MJ, Higgins JPT, Clayton G, Sterne JAC, Hrobjartsson A and Savović J, 2016. Empirical evidence of study design biases in randomized trials: Systematic review of meta-epidemiological studies. *PLOS ONE* **11**(7): e0159267.

Pildal J, Hrobjartsson A, Jorgensen KJ, Hilden J, Altman DG and Gotzsche PC, 2007. Impact of allocation concealment on conclusions drawn from meta-analyses of randomized trials. *International Journal of Epidemiology* **36**(4): 847–857.

Pilgrim H, Lloyd-Jones M and Rees A, 2009. Routine antenatal anti-D prophylaxis for RhD-negative women: A systematic review and economic evaluation. *Health Technology Assessment* **13**(10).

Prevost TC, Abrams KR and Jones DR, 2000. Hierarchical models in generalized synthesis of evidence: An example based on studies of breast cancer screening. *Statistics in Medicine* **19**(24): 3359–3376.

Rhodes KM, Mawdsley D, Turner RM, Jones HE, Sterne JAC and Savović J, 2018. Label-invariant models for the analysis of meta-epidemiological data. *Statistics in Medicine* **37**(1): 60–70.

Salanti G, Dias S, Welton NJ, Ades AE, Golfinopoulos V, Kyrgiou M, Mauri D and Ioannidis JPA, 2010. Evaluating novel agent effects in multiple treatments meta-regression. *Statistics in Medicine* **29**(23): 2369–2383.

Salanti G, 2011. Multiple treatments meta-analysis of a network of interventions. Available from http://www.mtm.uoi.gr/.

Savović J, Harris RJ, Wood L, Beynon R, Altman D, Als-Nielsen B, Balk EM, Deeks J, Gluud LL, Gluud C, Ioannidis JPA, Jüni P, Moher D, Pildal J, Schulz KF and Sterne JAC, 2010. Development of a combined database for meta-epidemiological research. *Research Synthesis Methods* **1**(3–4): 212–225.

Savović J, Jones HE, Altman D, Harris R, Jüni P, Pildal J, Als-Nielsen B, Balk E, Gluud C, Gluud L, Ioannidis J, Schulz K, Beynon R, Welton N, Wood L, Moher D, Deeks J and Sterne J, 2012a. Influence of reported study design characteristics on intervention effect estimates from randomised controlled trials: Combined analysis of meta-epidemiological studies. *Health Technology Assessment* **16**(35): 1–82.

Savović J, Jones HE, Altman DG, Harris RJ, Jüni P, Pildal J, Als-Nielsen B, Balk E, Gluud C, Gluud L, Ioannidis J, Schulz K, Beynon R, Welton N, Wood L, Moher D, Deeks J and Sterne J, 2012b. Influence of reported study design characteristics on intervention effect estimates from randomised controlled trials: Combined analysis of meta-epidemiological studies. *Annals of Internal Medicine* **157**(6): 429–438.

Savović J, Turner RM, Mawdsley D, Jones HE, Beynon R, Higgins JPT and Sterne JAC, 2018. Association between risk-of-bias assessments and results of randomized trials in cochrane reviews: The ROBES meta-epidemiologic study. *American Journal of Epidemiology* **187**(5): 1113–1122.

Schulz KF, Chalmers I, Hayes RJ and Altman DG, 1995. Empirical evidence of bias: Dimensions of methodological quality associated with estimates of treatment effects in controlled trials. *Journal American Medical Association* **273**(5): 408–412.

Schulz KF, Altman DG, Moher D and CONSORT Group, 2010. CONSORT 2010 statement: Updated guidelines for reporting parallel group randomised trials. *British Medical Journal* **340**: c332.

Schulz KF, 1995. Subverting randomization in controlled trials. *Journal American Medical Association* **274**(18): 1456–1458.

Siersma V, Als-Nielsen B, Chen W, Hilden J, Gluud LL and Gluud C, 2007. Multivariable modelling for meta-epidemiological assessment of the association between trial quality and treatment effects estimated in randomized clinical trials. *Statistics in Medicine* **26**(14): 2745–2758.

Soares M, Welton NJ, Harrison DA, Peura P, Hari MS, Harvey SE, Madan JJ, Ades AE, Palmer SJ and Rowan KM, 2012. An evaluation of the feasibility, cost and value of information of a multicentre randomised controlled trial of intravenous immunoglobulin for sepsis (severe sepsis and septic shock): Incorporating a systematic review, meta-analysis and value of information analysis. *Health Technology Assessment* **16**(7): 1–186.

Song F, Harvey I and Lilford R, 2008. Adjusted indirect comparison may be less biased than direct comparison for evaluating new pharmaceutical interventions. *Journal of Clinical Epidemiology* **61**(5): 455–463.

Spineli LM, Higgins JPT, Cipriani A, Leucht S and Salanti G, 2013. Evaluating the impact of imputations for missing participant outcome data in a network meta-analysis. *Clinical Trials* **10**(3): 378–388.

Sterne JAC, Jüni P, Schulz KF, Altman DG, Bartlett C and Egger M, 2002. Statistical methods for assessing the influence of study characteristics on treatment effects in 'meta-epidemiological' research. *Statistics in Medicine* **21**(11): 1513–1524.

Sterne JAC, Higgins JPT and Reeves BC, 2014. ACROBAT-NRSI obotdgf. A Cochrane Risk of Bias Assessment Tool: For Non-Randomized Studies of Interventions (ACROBAT-NRSI). Available from http://www.riskofbias.info.

Trinquart L, Chatellier G and Ravaud P, 2012. Adjustment for reporting bias in network meta-analysis of antidepressant trials. *BMC Medical Research Methodology* **12**: 150.

Trolle B, 1989. Prenatal Rh-immune prophylaxis with 300 µg immune globulin anti-D in the 28th week of pregnancy. *Acta Obstetricia et Gynecologica Scandinavica* **68**(1): 45–47.

Turner RM, Spiegelhalter DJ, Smith GCS and Thompson SG, 2009. Bias modelling in evidence synthesis. *Journal of the Royal Statistical Society: Series A* **172**(1): 21–47.

Turner RM, Lloyd Jones M, Anumba DOC, Smith GCS, Spiegelhalter DJ, Squires H, Stevens JW, Sweeting MJ, Urbaniak SJ, Webster R and Thompson SG, 2012. Routine antenatal anti-D prophylaxis in women who are Rh(D) negative: Meta-analyses adjusted for differences in study design and quality. *PLOS ONE* **7**(2): e30711.

Turner NL, Dias S, Ades AE and Welton NJ, 2015. A Bayesian framework to account for uncertainty due to missing binary outcome data in pairwise meta-analysis. *Statistics in Medicine* **34**(12): 2062–2080.

Welton NJ, Ades AE, Carlin JB, Altman DG and Sterne JAC, 2009. Models for potentially biased evidence in meta-analysis using empirically based priors. *Journal of the Royal Statistical Society: Series A* **172**(1): 119–136.

White IR, Carpenter J, Evans S and Schroter S, 2007. Eliciting and using expert opinions about dropout bias in randomized controlled trials. *Clinical Trials* **4**(2): 125–139.

White IR, Higgins JPT and Wood A, 2008a. Allowing for uncertainty due to missing data in meta-analysis – Part 1: Two-stage methods. *Statistics in Medicine* **27**(5): 711–727.

White IR, Wood A, Welton NJ, Ades AE and Higgins JPT, 2008b. Allowing for uncertainty due to missing data in meta-analysis – Part 2: Hierarchical models. *Statistics in Medicine* **27**(5): 728–745.

Wolpert RL and Mengersen KL, 2004. Adjusted likelihoods for synthesizing empirical evidence from studies that differ in quality and design: Effects of environmental tobacco smoke. *Statistical Science* **19**(3): 450–471.

Wood L, Egger M, Gluud LL, Schulz KF, Jüni P, Altman DG, Gluud C, Martin RM, Wood AJG and Sterne JAC, 2008. Empirical evidence of bias in treatment effect estimates in controlled trials with different interventions and outcomes: Meta-epidemiological studies. *British Medical Journal* **336**: 601–605.

Yuan Y and Little RJA, 2009. Meta-analysis of studies with missing data. *Biometrics* **65**(2): 487–496.

13

Publication and Outcome Reporting Bias

Arielle Marks-Anglin, Rui Duan, Yong Chen, Orestis Panagiotou, and Christopher H. Schmid

CONTENTS

13.1 Introduction

Systematic reviews and meta-analyses rely on the availability of all relevant information to provide an accurate summary of evidence. When entire studies or some of the results of studies are not published or discoverable, they cannot be incorporated into the evidence synthesis potentially leading to biased conclusions and poor decisions. Failure to publish has several other negative consequences for the scientific process, including the inability to empirically replicate or test previous work and the potential waste of financial and human resources for research that does not contribute to scientific knowledge (Brassington, 2017). It also breaks the implicit contract made with study participants that they undertake risk in order to advance human knowledge (Dickersin and Chalmers, 2011).

In this chapter, we consider statistical methods for identifying and correcting two related but distinct types of non-reporting bias: publication and outcome reporting bias. Publication bias occurs when the reports of studies undertaken remain unpublished. Outcome reporting bias involves selective reporting of results from completed studies. Both types of bias threaten the validity of systematic reviews because they prevent reviewers from being able to assess the entire body of scientific evidence. In essence, they create missing data and require the application of principled missing data methods that must consider the mechanism by which the missing data have arisen. As we will see, some of the methods that have been used define these mechanisms more explicitly in terms of the information available (e.g., reported p-values) and in ways that can be more easily evaluated than others.

The chapter proceeds by first discussing some of the consequences for meta-analysis created by publication and outcome reporting bias. We then discuss some of the empirical evidence for reporting bias and the use of registries to lessen the impact of the potential bias before investigating statistical approaches to evaluate and adjust for bias. Statistical treatment of potential bias falls into three main categories: sensitivity analyses that explore the number of unpublished studies that would be needed to overcome the existing evidence; graphical and statistical testing approaches to detect publication bias; and models to correct estimates for bias under specified missing data mechanisms. The chapter concludes with a short discussion of available software to implement the various methods.

13.2 Consequences of Publication and Outcome Reporting Bias

A study may go unpublished for a variety of reasons. It may never have been finished; it may have been finished but the results may have been disappointing and it was decided not to submit them; the authors might have not finished writing a manuscript; or the authors may have submitted the manuscript for publication but had it rejected and decided not to resubmit elsewhere. Bias ensues in a meta-analysis when the reason for not publishing relates to the results of the study, primarily failure to publish non-significant results. For example, study sponsors may suppress results when they go against their interests, financial or otherwise (Goldacre, 2013). Failure to publish then leads to an observed results distribution that is unrepresentative of the complete distribution from all studies undertaken.

Outcome reporting bias is a closely related problem often overlooked or downplayed in the literature (Kirkham et al., 2010; Schmid, 2017). Outcome reporting bias arises when some studies fail to numerically summarize the results of some or all of their outcomes. The studies are known and reported so there is no publication bias, but the results for outcomes are incomplete and inferences from the observed data may not apply to the entire population of studies. This bias applies not just to the studies that provide some data to analyze, but also to studies omitted entirely by the exclusion criteria of the meta-analysis that eliminate any studies that do not supply usable data. As most meta-analyses routinely analyze only the results reported (i.e., they do a complete case analysis of outcomes), they necessarily assume that the missing outcomes are missing completely at random, i.e., they are unrelated to any of the results observed or unobserved.

Again, the potential for bias depends upon the reason for non-reporting of the results. Sometimes, outcomes of importance for a review were not considered important in a study which might have been investigating a different question. For instance, because hypertension is an important condition to treat in both cardiovascular and kidney disease, many systematic reviews of hypertensive agents for treating kidney disease incorporate studies whose primary purpose was to study cardiovascular disease. Some of these studies focus solely on cardiovascular endpoints and so may fail to report data on kidney endpoints such as changes in serum creatinine levels. This would not bias a summary estimate of the kidney outcome unless the reason for not reporting the kidney outcome was related to the size of its effect. On the other hand, when studies plan to or do collect an outcome but do not report it, bias is often suspected because the non-publication may be related to the size of the effect or its statistical significance. Results from abstracts are particularly susceptible because they usually report only key results that are often large effects. Adverse event outcomes are also often problematic because many studies will either only mention adverse events that differ significantly between groups or may note those that are not different without providing numerical summaries, thus, precluding use of their numerical results in the meta-analysis.

13.3 Empirical Evidence of Publication and Outcome Reporting Bias

Investigators have uncovered many instances of biased publication practices (Dickersin, 2005). An early example found that published trials showed a statistically significant advantage for combination over single-agent chemotherapy for advanced ovarian cancer, but that unpublished studies identified through an international registry did not (Simes, 1986). Sterne et al. (2002) used meta-analyses from the Cochrane Library that included both published and unpublished studies to demonstrate that results from published trials showed more beneficial effects than those from unpublished studies. Vedula et al. (2009) used legal discovery of internal company documents to document systematic suppression of studies involved in off-label uses of gabapentin. Eyding et al. (2010) examined published and unpublished studies comparing reboxetine versus placebo for acute treatment of major depression and found 13 trials, of which five were unpublished including data from 74% of the patients studied. They concluded that "published data overestimated the benefit of reboxetine versus placebo by up to 115% and reboxetine versus SSRIs by up to 23%, and also underestimated harm". The investigators were able to get unpublished data from the manufacturer of reboxetine because they were affiliated with a German government agency, the Institute for Quality and Efficiency in Health Care (IQWiG). Most investigators will not have such legal resources and so will not be able to discover studies "hidden away in a file drawer". Research also suggests that trials that appear to demonstrate beneficial effects of an intervention being tested are more likely to be published (Hopewell et al., 2009), published more quickly (Hopewell et al., 2007), and disclosed in greater detail (Dwan et al., 2008) than those that fail to show such benefit or which identify harm.

Outcome reporting bias is also a potentially large problem in the literature. Hahn et al. (2002) and Chan et al. (2004) demonstrated its existence when comparing publications with protocols submitted to research and ethics boards. Inspection of many published systematic reviews indicates that different outcomes are reported in different proportions. For example, Xie et al. (2016) compared angiotensin-converting enzyme inhibitors, angiotensin II receptor blockers, active controls, and placebo for treating chronic kidney disease in 119 randomized controlled trials. They reported on kidney failure outcomes from 85 studies, cardiovascular outcomes from 94 studies, mortality outcomes from 104 studies, and adverse events from 99 studies, but did not explore why some outcomes were missing or their potential impact on conclusions. In their study of reboxetine, Eyding et al. (2010) also noted outcome reporting bias as publications of three trials of reboxetine only gave results for subpopulations or for selected outcomes.

13.4 Using Registries to Reduce the Impact of Publication and Outcome Reporting Bias

While the careful and thorough literature searches outlined in Section 1.3 of Chapter 1 are the most effective way to uncover reports of studies and their results, other resources are needed to find what is unreported. The recognition that many studies were never published prompted scientific agencies to call for and mandate the creation of registries listing studies that they sponsored and funded. Thus, the protocols of all clinical trials of humans funded and regulated by the United States government through the National Institutes of Health (NIH) and the Food and Drug Administration (FDA) are required by law to be registered with the National Library of Medicine at Clinicaltrials.gov. ClinicalTrials.gov is now the largest database of ongoing and completed clinical trials in the world and includes nearly 300,000 research studies from 204 countries. Since 2009, many studies are also required to submit information about the progress of participants through the study, baseline characteristics, outcome measures, statistical analysis, and adverse events within one year of the study's completion date (Clinicaltrials.gov, 2017). The European Medicines Agency (EMA) of the European Union (EU) has similar regulations and has a registry of its own called the EU Clinical Trials Register which started in 2004. The World Health Organization (WHO) has established the International Clinical Trials Registry Platform (ICTRP) to facilitate a network of registries across the world that meet specific criteria for content, quality and validity, accessibility, unique identification, technical capacity, and administration. Registration is also supported by the requirement of journals that are members of the International Committee of Medical Journal Editors (ICMJE) that studies be prospectively registered in order to be eligible for publication (De Angelis et al., 2004). Other study registries have been established by private companies (e.g., GlaxoSmithKline at www.gsk-clinicalstudyregister.com).

Registries reduce the risk of publication and outcome reporting bias in several ways. First, they allow comparing the set of trials started with those published, thus offering a chance to find studies that have been completed but remain unpublished. Some of these unpublished studies may have reported their results to the registry, in which case such studies can be included in the review. Second, they provide another source of results which might differ from published results and might provide additional information, including unreported outcomes. Discrepancies may need to be resolved by contacting investigators. Third, the availability of full protocols can reduce the risk of outcome reporting bias by determining which outcomes reported were prespecified, which were not prespecified but added later, and which were modified from their original specification (Ioannidis et al., 2017). Identifying unreported outcomes from protocols can help assess their planned importance; examining unpublished results can help determine whether they were unreported because of their lack of statistical significance. Finally, by providing a list of the studies that were started, registries provide a means to determine the number of unpublished studies and unreported outcomes and thereby judge the extent of potential bias. Their presence alone therefore may encourage better research practice.

Even when protocols cannot be located, investigators cannot be reached, or study data are no longer available, analysis of the available data may be able to uncover potential bias. For example, one could check if studies reporting certain outcomes differ systematically from those that do not on baseline or study design features that are generally reported. These comparisons should include studies identified as eligible but excluded from the meta-analysis because they did not report usable outcomes. Careful modeling with this available data could also potentially construct an imputation scheme assuming the data are missing at random, although the likely dependence of the missing outcome on its statistical significance suggests that the data are probably missing not at random.

Having properly conducted this extensive search with existing tools and resources, one might still expect some studies to be undiscovered, particularly older ones started before the registries were established. Registries are not panaceas. Unfortunately, the majority of trials go unregistered (Dal-Re et al., 2015; De Oliveira et al., 2015); most journals do not require trial registration (Hooft et al., 2014); and few journals require protocols to be submitted. Work has also shown that many study results are often not deposited (Anderson et al., 2015) or not deposited promptly (Prayle et al., 2012). Surveys of the literature have demonstrated frequent changes between protocol and publication in primary outcomes, target sample sizes, and study interventions (De Oliveira et al., 2015), suggesting that investigators may determine what to publish based on study results. Finally, registries are primarily useful for trials and few observational studies are registered anywhere even though the need for registration of observational studies has been a long-standing debate in the literature regarding the need for registration of observational studies (Dal-Re et al., 2014:, Editors of Epidemiology, 2010; Williams et al., 2010; Loder et al., 2010) and trials of non-regulated interventions (Dal-Re et al., 2015).

13.5 Statistical Approaches for Estimating the Number of Unpublished or Unreported Studies

The oldest and simplest approach for assessing the potential impact of publication bias is to assess how much negative information must be missing in order to overturn current evidence. The earliest and probably still most popular method is called the fail-safe N method (or the file-drawer method) (Rosenthal, 1979). It aims to calculate the number N of unpublished negative studies that would be needed to transform a meta-analysis with a statistically significant effect into one without significance. It uses a test statistic based on study p-values developed by Stouffer and Hovland (1949) to compute the overall significance. Assume a meta-analysis of I studies indexed by $i = 1,2,\ldots,I$ where each study i examines the null hypothesis $H_0\colon \theta_i = 0$ and θ_i is the treatment effect (log odds ratio, log risk ratio, mean difference, etc.) in study i. The statistical test for the directional null hypothesis $H_0\colon \theta_i \le 0$ has one-tailed p-value p_i. The Stouffer statistic is

$$Z_S = \frac{\sum_{i=1}^{I} z_i}{\sqrt{i}}$$

where z_i is the normal deviate associated with the one-tailed p-value p_i in study i, i.e., $z_i = \Phi^{-1}(p_i)$. Under the null hypothesis $H_0\colon \theta_1 = \cdots = \theta_I = 0$, Z_S has a standard normal distribution and the hypothesis can be tested by comparing Z_S with the α-level upper-tail critical value of the normal distribution, Z_α. The fail-safe N method essentially asks the following question: if an observed value of Z_S is already larger than Z_α, how many studies with z_i values averaging zero would need to be added to reduce the value of Z_S to Z_α (Sutton et al., 2000)? This can be answered by solving the following equation for N:

$$\frac{\sum_{i=1}^{I} z_i}{\sqrt{I+N}} < Z_\alpha$$

which has the solution

$$N > \left(\frac{\sum z_i}{Z_\alpha}\right)^2 - I = I\left(\frac{Z_S}{Z_\alpha}\right)^2 - I.$$

Stouffer's test statistic increases when individual studies produce large z_i values, which can happen either because studies have larger sample sizes and thus small standard errors or have large treatment effects. A fail-safe N much larger than the number of studies found implies a small chance that publication bias could overturn the significance of the results. Rosenthal did not provide specific statistical criteria for the threshold of the fail-safe N, but suggested that N should be larger than $5I + 10$. Although adding N studies with an average of $z_i = 0$ is arithmetically equivalent to adding N studies each of which has $z_i = 0$, those two sets of studies are not identical in reality and in practice it is unlikely that we find a set of studies with identical null results (Becker, 2005).

Instead of the Stouffer statistic, one could also use the Fisher statistic $\sum -2\ln(p_i)$ which under the same composite null hypothesis follows a chi-square distribution with $2I$ degrees of freedom (Fisher, 1932 and Chapter 3). One then adds in N studies with one-sided $p = 0.5$ corresponding to a null result. Each addition increases the Fisher statistic by $-2\ln(0.5) = 1.386$ and adds one degree of freedom. This can be programmed to find the value of N that has a Fisher test statistic with p-value at the α critical level and the appropriate degrees of freedom.

The fail-safe N has several serious drawbacks that limit its utility. First, it relies on test statistics derived only from study p-values, ignoring the size of effects and ignoring a study's size except as it might affect its p-value. In other words, two studies with the same p-value will make the same contribution regardless of whether the study is small with a large effect or large with a small effect. It also assumes that the missing studies have no effect and ignores between-study heterogeneity. Finally, it is not based on any statistical model of study effect distributions and so can only provide heuristic guides to the importance of the size of N found (Becker, 2005).

A variety of other methods have been proposed to estimate the number of unpublished studies. Orwin (1983) based his statistic on the number of effects N with a specified average $\bar{d}$ that would be needed to reduce an observed mean effect $\bar{d}_0$ based on I studies to a particular level d_C. Because

$$d_C = \frac{I(\bar{d}_0) + N(\bar{d})}{N_0 + N},$$

then

$$N = \frac{I(\bar{d}_0 - d_C)}{d_C - \bar{d}}.$$

This method uses the effect sizes but ignores the uncertainty associated with the estimate of $\bar{d}_0$ as well as the distribution of effects and their variances in each of the studies.

Gleser and Olkin (1996) derived two estimators for N, one assuming that one only has the p-values from the k studies with the smallest p-values and the other in which one has the m smallest p-values plus a random sample of $k-m$ of the remaining $N+k-m$ p-values. Both estimators assume the null hypothesis of no effect to be true. They show that an unbiased estimate is

$$\hat{N} = \frac{I(1-p_{(I)})-1}{p_{(I)}}$$

where $p_{(I)}$ is the largest observed p-value from a set of I studies. Intuitively, if all p-values are small then unpublished studies likely exist if the null hypothesis is true. $\hat{N}$ will then be large when p_I is small and will increase as the number of studies observed with small p-values increases. But this crucially relies on the assumption of the true effect being zero.

Eberly and Casella (1999) proposed a Bayesian model computed via a Gibbs sampler to calculate the posterior distribution of N given the number of significant and non-significant published studies and assuming a beta prior distribution for the probability of publication, that all significant studies are published, and that we have an estimate of the probability of publishing a non-significant result. This is a simple form of a selection model (see Section 13.10) in which the probability of publishing non-significant results is constant and does not depend on the size of the estimated study effect.

13.6 Funnel Plots

The funnel plot graphs the treatment effects of individual studies in the meta-analysis along the horizontal axis with the corresponding measures of their precision (e.g., inverse variance, inverse standard deviation, or sample size) on the vertical axis. In the common-effect model, all studies are estimating the same underlying effect and variability is completely due to random error. In this case, the most precisely estimated effect at the top of the plot will provide a good estimate of the true mean. Smaller, less precise studies spread out at the bottom of the plot so that the overall appearance of the plot is symmetric around the average effect and looks like an inverted funnel (Figure 13.1).

If publication bias were to preferentially hide small non-significant studies, the plot would exhibit asymmetry as one side of the bottom of the funnel would lose data points (Light and Pillemer, 1984). Figure 13.2 shows the data from Figure 13.1 with the smallest effects removed. This is an example of a contour-enhanced funnel plot in which the gray and white areas describe 90% and 95% confidence regions about the random-effects estimate (indicated by a vertical line), respectively.

One can generalize a bit further to say that publication bias can lead to asymmetry in a funnel plot assuming that the observed effects share a common mean. This holds even

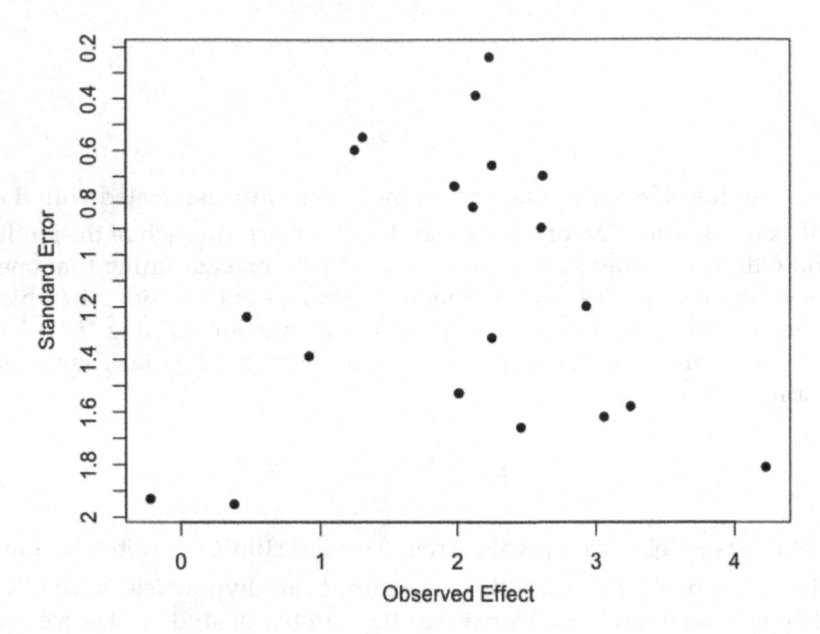

FIGURE 13.1
Symmetric funnel plot based on artificial data.

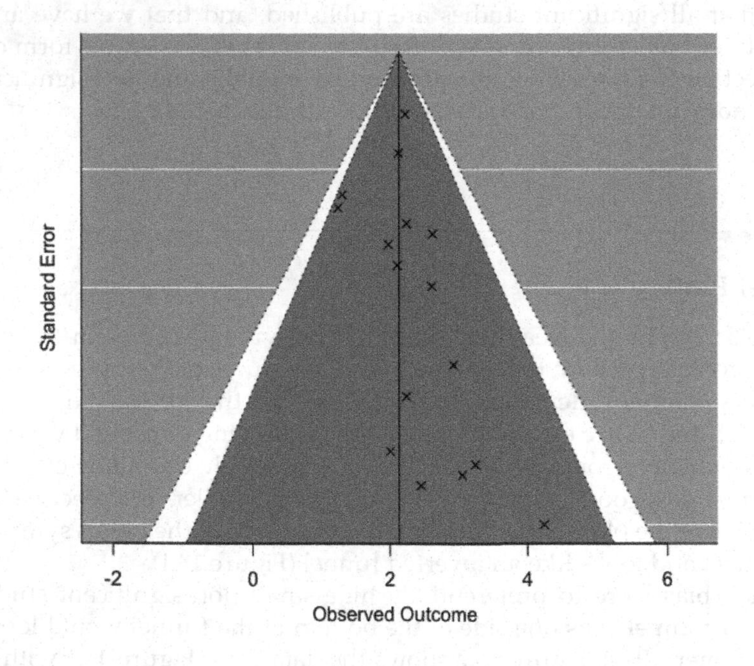

FIGURE 13.2
Asymmetric contour-enhanced funnel plot based on the artificial data generating Figure 13.1 with all studies with effects less than one removed. The vertical line indicates the random-effects estimate of the mean treatment effect. The gray area describes a 90% confidence region about the random effect (indicated by a vertical line). The white area describes a 95% confidence region.

if the true study effects differ so long as they come from a common random-effects distribution with a single mean. In the random-effects model, the marginal variance of each observed effect is the sum of its study variance and the common between-study variance. As a consequence, the funnel plot under a random-effects model must account for this between-study variance in the calculation of each study's precision. Because the between-study variance is likely to be poorly estimated unless the number of studies is large, funnel plots may be misleading when the number of studies is small and heterogeneity may be present.

Unfortunately, funnel plots are often misused and misinterpreted for detecting publication bias. Most users fail to recognize the importance of the assumption of a common mean in the interpretation of asymmetry. In particular, a significant meta-regression or subgroup effect invalidates the assumption of a common mean around which the funnel is centered. In such cases, multiple funnels may develop around the different means and the overall plot may take on an asymmetric aspect with respect to the most precise estimate which is now no longer the single common mean.

As an example, consider Figure 13.3, which clearly shows asymmetry in a funnel plot of study effects versus the standard error. Under the publication bias assumption, it appears that small studies with positive effects are missing. In fact, the asymmetry is completely an artifact of the mixture of studies coming from two different distributions. In one set, 20 effects y_i were simulated from $N(\theta_{1i}, \sigma_{1i}^2)$ with $\theta_{1i} \sim N(2, 1/100)$ and $\sigma_{1i} \sim U(0.5, 2)$. The other set of ten studies had $y_{i2} \sim N(\theta_{i2}, \sigma_{i2}^2)$ with $\theta_{i2} \sim N(-5, 1)$ and $\sigma_{i2} \sim U(1, 10)$. The funnel plot was then generated by plotting (y_{i1}, y_{i2}) versus $(\sigma_{1i}, \sigma_{2i})$. Thus, it is incorrect to conclude that publication bias is indicated when a funnel plot is asymmetric. Publication bias is only one of many potential causes of asymmetry in a funnel plot (Sterne et al., 2011; Lau et al., 2006). Failing to incorporate the between-study variance in study variance estimates leads to data plotted in the wrong locations according to within-study and not total variance.

It is also rarely noted that funnel plots assume that the plotted study effects are uncorrelated with the study sample sizes (Terrin et al., 2003; Lau et al., 2006; Simonsohn, 2017). In fact, though, prospective sample size calculations usually reflect anticipated sizes of

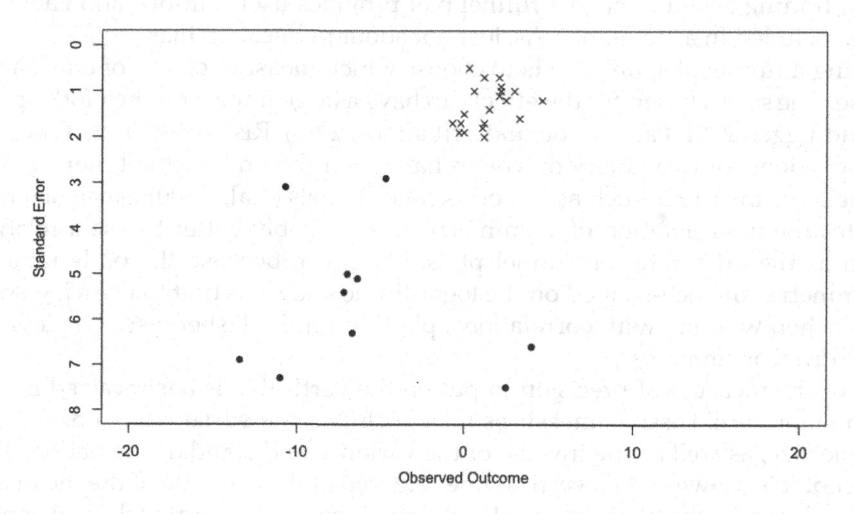

FIGURE 13.3
Asymmetry arising from a two-class mixture with data generated according to the procedure described in the text.

treatment effects that can generate sufficient power. Bigger expected effects will lead to smaller studies, whereas smaller expected effects will require larger studies. For example, early in the course of development of a treatment when clinical practice has yet to determine optimal treatment conditions, small studies may target high-risk populations to determine whether large gains are possible. If the treatment is shown to be effective, larger studies may be undertaken to test the effect in more diverse populations. If this treatment program is correctly powered, then one would expect asymmetry since the small studies will have big effects and the large studies will have smaller effects. A funnel plot of the data will then be asymmetric, not because of publication bias but because of successful experimental management.

Because the variance of many common-effect measures such as the odds ratio and correlation involve the effect measures themselves (see Chapter 4), estimates of the effect and its variance are necessarily correlated. In such cases, asymmetry may arise naturally. Sterne et al. (2000) investigated this for odds ratios by simulation and found that asymmetry was common even in the absence of any relationship between-study sample and effects when there were large average effects, small prevalences, or similar sample sizes.

Bias introduced as a consequence of poor study procedures and choice of effect measure may also introduce asymmetry. It has been shown that more rigorously performed studies often show smaller effects (see Chapter 12). If larger studies are more carefully designed, perhaps because the groups carrying them out are more experienced, then the larger, more precise studies may show smaller effects and a plot of the results may show a gap in a funnel plot where smaller studies with smaller effects are missing. Inadequate analysis and fraud are other potential causes of asymmetry (Sterne et al., 2011).

As a consequence of these different potential etiologies of asymmetry, it is not a good idea to use a funnel plot as a diagnostic for publication or outcome reporting bias. Moreover, unsurprisingly, visual interpretation of a given funnel plot by different observers injects considerable uncertainty and subjectivity. In fact, Terrin et al. (2005) found that clinical and methodological researchers were unable to distinguish publication bias from naturally occurring asymmetry using a set of constructed funnel plots. All of these etiologies are consistent with the finding that the distribution of sizes of effects in studies with small precision (large studies) differs from those in studies with large precision (small studies). Therefore, finding asymmetry in a funnel plot provides useful information about the set of studies included in a meta-analysis, just not about publication bias.

In making a funnel plot, one needs to choose which measure of size of effect and precision to use. The scale chosen for the effect can have a large impact on the plot's appearance (Sterne and Egger, 2001; Panagiotou and Trikalinos, 2015). Risk differences for comparing event proportions for two binary outcomes have been shown to exhibit more heterogeneity than relative measures such as the odds ratio (Engels et al., 2000). Since heterogeneity complicates the interpretation of asymmetry, it is probably better to use a relative measure such as the odds ratio on funnel plots. Moreover, because the odds ratio is more likely symmetric and bell-shaped on the logarithmic scale, it is best to use a log odds ratio. Likewise, when working with correlations, plotting on the Fisher z scale is less likely to lead to artificial asymmetry.

Choice of the measure of precision to put on the vertical axis has received more attention in the literature. Possible measures have included the variance, standard deviation, and sample size, as well as the inverse of the variance and standard deviation. Typically, the funnel plot is drawn with a vertical line centered at the estimate of the mean and with boundaries drawn so that they cover a $(1-\alpha)\%$ confidence interval at each level of precision. When precision is measured as the standard deviation, these boundaries will be lines with slope 1/1.96. Using the variance returns curved boundaries which are a bit harder to envision as a funnel than straight boundaries. Inverses of the variance or standard deviation also return curves and also fail to spread points well in the vertical direction. Boundaries cannot be drawn using the sample size as the measure of precision because the standard error is unknown. Because standard errors depend on the number of events for binary outcomes and on the standard deviation for continuous outcomes, they may also not be ordered exactly as sample sizes so use of the sample size may not reflect actual precision.

Consequently, the standard error has tended to be the recommended choice (Sterne and Egger, 2001). In a funnel plot where the standard error is plotted as the measure of precision, this is done on a reversed scale so that larger (i.e., most powerful studies) are placed on the top of the plot. When standard errors are used, 95% of the studies will lie within the contours of a triangle that is centered on the fixed-effect summary estimate and extend 1.96 standard errors on either side. If a random-effects estimate is used, it will be necessary to plot with standard errors that incorporate the between-study variance.

13.7 Tests and Regression Methods for Publication Bias

Because chance plays a large role in the appearance of funnel plot asymmetry, a variety of statistical tests have been suggested to evaluate whether reasons other than chance may explain an observed asymmetry. Simply put, a statistical test for funnel plot asymmetry (small study effects) formally examines whether the association between estimated treatment effects and some measure of precision is greater than what might be expected by chance alone (Higgins et al., 2019). In the following, we discuss several of the most commonly used asymmetry tests.

The most common test can be formulated as a weighted regression of the treatment effect y_i on the standard error s_i where each point is weighted by the inverse of its standard error $w_i = 1/s_i$. The slope measures the association between the effect and the standard error. This is equivalent to the original formulation by Egger et al. (1997) in which the test statistic $z_i = y_i/s_i$ is regressed on the inverse of the standard error so that $E[z_k] = \beta_0 + \beta_1(1/s_k)$. If there is no association between y_i and $1/s_i$, then Egger's model will have a slope equal to the common mean of μ and an intercept of zero since the studies with less precision will tend to have a test statistic z_i near zero. A test of H_0: $\beta_0 = 0$ is then a test of asymmetry and is equivalent to a test of zero slope in the weighted regression using y_i as the outcome. If studies with less precision and negative effects are missing, then the plot will become asymmetric and the intercept will have a positive value. If the missing studies are those with positive effects, then the intercept will be negative. The greater the association between treatment effects and their standard errors, the more the line shifts vertically away from the origin, corresponding to asymmetry in a related funnel plot. Figure 13.4 shows the regression line resulting from the asymmetric plot resulting from the mixture distribution in Figure 13.3. The intercept is estimated as –1.1 with a standard error of 0.4.

As an asymmetry test for the funnel plot, Egger's regression test suffers from the same limitations as funnel plots in general and should not be assumed to be a test for publication bias unless other potential causes of asymmetry can be ruled out. In particular, heterogeneity between studies is again problematic unless the regression model can incorporate a between-study component of variance. One modification of the weighted form of the Egger regression that allows for a multiplicative between-study component

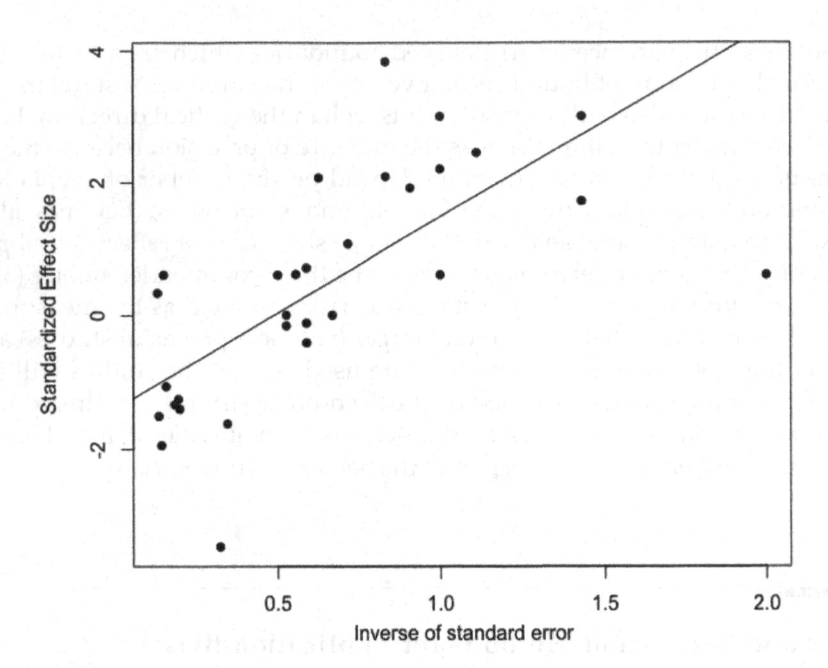

FIGURE 13.4
Egger regression for asymmetry. Data are the same as for the two-class mixture distribution from Figure 13.3.

assumes $y_i \sim N(\beta_0 s_i + \beta_1, v_i \phi)$ where ϕ is a multiplicative overdispersion parameter (Thompson and Sharp, 1999). A drawback to this formulation is that ϕ may be estimated to be less than one which implies negative between-study heterogeneity. A different formulation uses a random-effects regression model with an additive between-study variance component τ^2 replacing the study variance $v_i \phi$ by $v_i + \tau^2$. When working with outcomes for which the effect and its standard error share common parameters (such as log odds ratios), the regression also suffers from inherent correlation between the dependent and independent variables (Higgins and Spiegelhalter, 2002).

Begg and Mazumdar (1994) proposed a non-parametric test using the rank correlation between the statistic $s(y_i) = (y_i - \hat{\theta}_{CE}) / \sqrt{v_i^*}$ and the variance v_i of the study statistic y_i where $\hat{\theta}_{CE}$ is the average treatment effect under a common-effect model, and $v_i^* = v_i - \text{var}(\hat{\theta}_{CE})$ is the variance of the numerator of $s(y_i)$. Although the original test utilizes Kendall's τ as the measure of correlation, other rank correlation coefficients (e.g., Spearman's ρ) can also be used. The basis of this test is comparing the ranks of $s(y_i)$ and v_i, which are ordered values where the largest value of $s(y_i)$ is ranked as 1, the next is ranked as 2, and so on. For I studies, there are $I(I-1)/2$ possible pairs of studies for which their $s(y_i)$ and s_i^* can be ranked. For studies i and j, the pairs $\{s(y_i), v_i\}$ and $\{s(y_j), v_i\}$ are concordant if their ranks are in the same direction, i.e., $\{s(y_i) > s(y_j) \text{ and } v_i > v_j\}$ or $\{s(y_i) < s(y_j) \text{ and } v_i < v_j\}$. The ranks are discordant if the two variables rank in opposite directions. If x and y are the numbers of concordant and discordant ranks respectively, then the test statistic is $z = (x-y)/\sqrt{I(I-1)(2I+5)/18}$. The test statistic z follows a standard normal distribution under the null hypothesis of no small-study effects. Because it is non-parametric, the Begg and Mazumdar test makes fewer assumptions and is less sensitive to some biases, but also has less power than the regression test if the latter's assumptions hold (Sterne et al., 2000).

In general, neither test has much power for a small number of studies, certainly for fewer than ten studies. Unfortunately, many health-related meta-analyses are this small, but researchers persist in using these tests anyway. Other tests based on the funnel plot have been proposed to examine asymmetry, but all fail as tests of publication bias for the reasons outlined for the Egger test.

13.8 Trim-and-Fill

The trim-and-fill method is a simple non-parametric approach that tries to both identify and correct for funnel plot asymmetry (Duval and Tweedie, 2000a, 2000b). It first estimates the number of unpublished studies using an iterative algorithm based on one of several rank-based procedures and then imputes their missing values in order to provide an estimate of the treatment effect adjusted for publication bias.

Trim-and-Fill Algorithm

1. Order the observed treatment effects by size and potentially recode them so that the missing studies have negative effects corresponding to asymmetry on the left-hand side of the funnel plot
2. Estimate the mean treatment effect, μ, using a common or random-effects model
3. Estimate the number of unpublished studies, k
4. Trim off or remove the k most extreme studies causing asymmetry
5. Re-estimate μ and k using the trimmed data, repeating the trimming process until k stabilizes
6. Replace the trimmed studies and impute their missing counterparts symmetrically about μ
7. Calculate a final adjusted estimate of μ using both observed and imputed studies

The imputed studies are counterparts symmetric in size but opposite in direction to the k studies chosen to be omitted in order to make the original plot symmetric. Different formulas can be used in step 3 of the algorithm to estimate k. These formulas involve signs and ranks of the absolute values of the study effects after these have been centered about μ. The simplest estimates k as one less than the number of positive effects that are larger than the absolute value of the most negative effect. Unlike the fail-safe N method, which assumes that the suppressed studies have an average null effect, trim-and-fill assumes suppression in the direction of asymmetry about a true common effect.

The method makes several strong assumptions, however, that reduce its utility. First and foremost, it assumes that funnel plot asymmetry (small-study effects) occurs only because of publication bias and thus ignores all other mechanisms that may give rise to asymmetry. Second, it assumes that there is always a "target" symmetric funnel plot and that the size of the missing study effects are the same as those of the largest positive effects.

Figure 13.5 shows an example of how trim-and-fill can be fooled using the two-class mixture data from Figure 13.2. Note that the procedure estimates seven missing studies with large positive outcomes and large standard errors. The algorithm imputes these

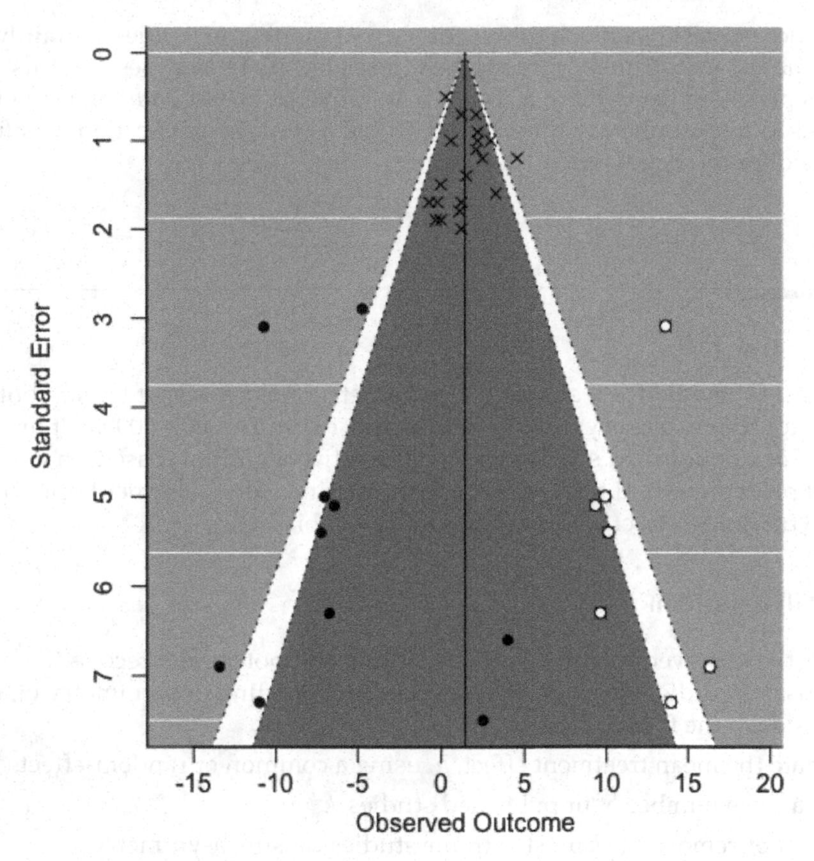

FIGURE 13.5
Trim-and-fill imputing studies (open circles) to the two-class mixture data.

because it is fooled into thinking that the studies with negative outcomes and large standard errors come from the same distribution as those with positive outcomes and small standard errors.

13.9 Example: Magnesium Data

As an application of the use of funnel plots, regression tests and trim-and-fill to real data, let us reconsider the studies evaluating the efficacy of intravenous magnesium for treating patients with myocardial infarction considered in Chapter 6. Recall that the outcome was mortality and that data were analyzed on the log odds ratio scale. For the purposes of checking for publication bias, we shall compute the empirical log odds ratios and their asymptotic standard errors and use these as the effects and standard errors for the funnel plot diagnostics.

Figure 13.6 is a funnel plot of the magnesium results. The largest studies with smallest standard errors at the top of the plot are concentrated between effect sizes of 0 and −0.5. The plot suggests asymmetry in the sense that the triangular shape of points to the left of an effect size of zero is not reflected to the right of zero. In fact, one could claim that

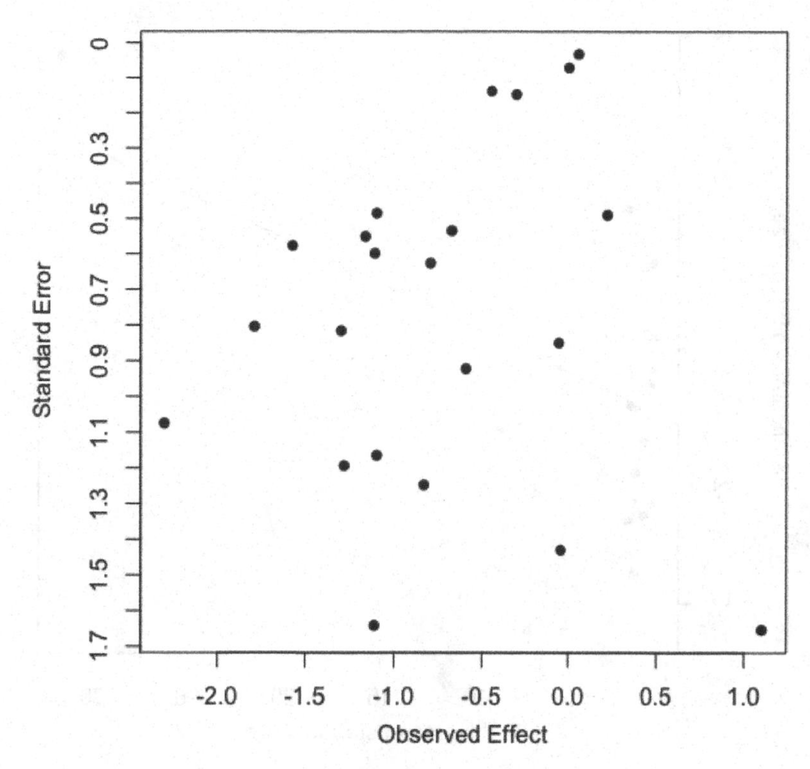

FIGURE 13.6
Funnel plot for magnesium data. Data are on the log odds scale. A 0.5 correction was applied to zero cells.

almost all of the studies showing magnesium to be harmful (effect > 0) are missing. One could interpret this as publication bias, since the largest studies indicate little effect and yet many of the smaller studies show large effects.

Fitting the Egger regression of the test statistics on the inverse of the standard errors returns a slope of 0.079 with a standard error of 0.035 together with an intercept of –1.38 with a standard error of 0.28 (Figure 13.7). The slope indicates a small effect size, but the highly significant negative intercept indicates considerable asymmetry.

Applying trim-and-fill produces Figure 13.8. All of the imputed studies have positive effect sizes corresponding to studies for which magnesium is harmful. The estimated mean odds ratio after applying trim-and-fill is 0.73 with a 95% confidence interval of (0.54, 0.97), quite different from the random-effects estimate of 0.54 with upper bound of 95% interval reaching only to 0.75 as computed by the Bayesian model in Chapter 6.

13.10 Selection Models

An alternative and arguably more sophisticated approach to the missing studies problem uses a class of models called selection models, which assume that studies are selected into the publication pool with a probability that is a function of the statistical significance of their results (Vevea and Hedges, 1995; Copas, 1999; Citkowicz and Vevea, 2017). For instance, one might

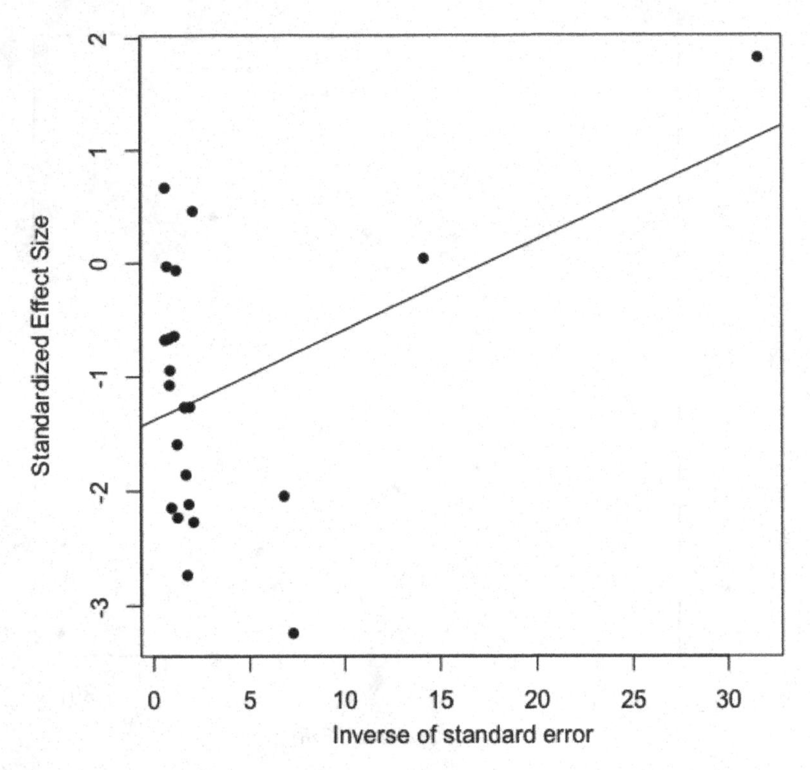

FIGURE 13.7
Egger regression for asymmetry for magnesium data.

assume that all studies with a p-value less than 0.05 are published and other studies are published with probabilities less than or equal to one. The complete model typically consists of two parts: an outcome model seeking to describe the distribution of effect estimates in all studies ever conducted (corresponding to a standard meta-analysis), and a separate selection model for the publication process. The publication probability might depend on the level of significance as well as the size of the study so that larger trials are more likely to be published irrespective of their results. This approach therefore seeks to model the underlying missing-not-at-random mechanism and is, of course, strongly dependent on assumptions about the likelihood of publication and the form of the publication probabilities attached to non-significant studies.

13.10.1 Selection Models for Publication Bias

The first class of selection models based study publication only on the p-value as a measure of study significance, under the premise that the test statistic in a study is monotonically related to the magnitude of the effect estimate. Initially, only extreme forms of bias were considered, with strict censoring of all non-significant study results at level α (Lane and Dunlap, 1978; Hedges, 1984), and assigning a probability of 1 for observing significant results.

Recognizing the need for a less simplistic framework in practice, Iyengar and Greenhouse (1988) proposed two parametric functions describing the probability of observing a t-test statistic x, where x follows a central t distribution with q degrees of freedom:

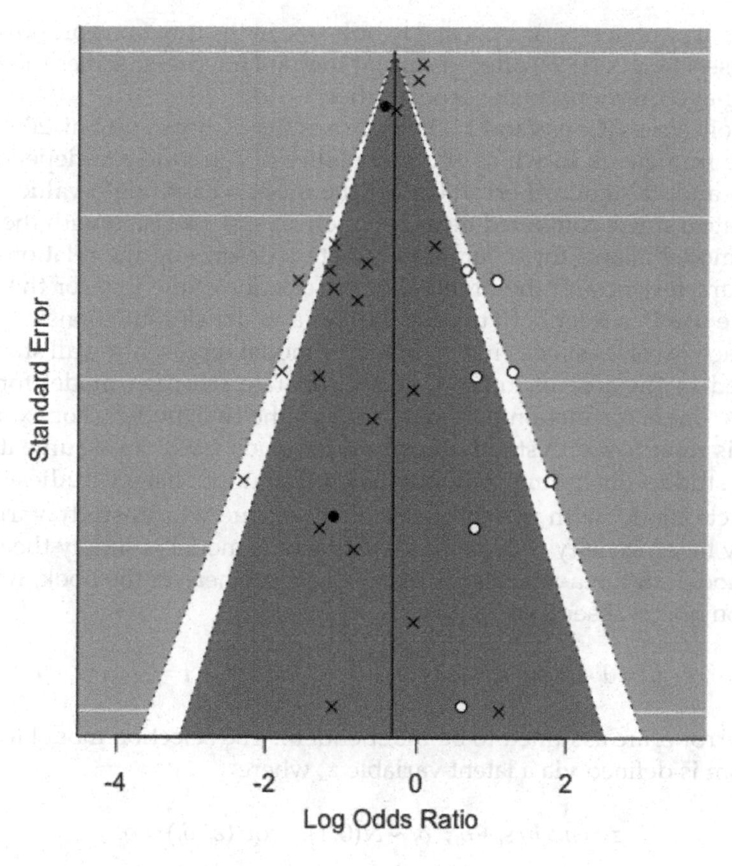

FIGURE 13.8
Trim-and-fill imputing studies (open circles) to the magnesium data.

$$w_1(x;\beta,q,\alpha)=\begin{cases}\dfrac{|x|^{\beta}}{t(q,\alpha)^{\beta}}, & \text{if } |x|\le t(q,\alpha)\\ 1, & \text{otherwise}\end{cases}$$

or

$$w_2(x;\beta,q,\alpha)=\begin{cases}e^{-\gamma}, & \text{if } |x|\le t(q,\alpha)\\ 1, & \text{otherwise}\end{cases}$$

where $t(q,\alpha)$ is the critical value for a size α two-sided t-test with q degrees of freedom. If x does not meet the critical value, the probability of non-observation under w_1 monotonically increases as $|x|$ moves farther away from t (i.e., p-value increases), and under w_2 remains a constant positive value. Both models assume that all studies with statistically significant results are observed.

Later approaches sought to impose fewer constraints *a priori* on the form of the probability functions, allowing the data to dictate the shape of the functions as much as possible. Dear and Begg (1992) suggested a left-continuous step-function for the probabilities, with discontinuities determined by the p-values. In this case, both the probabilities and the positioning/discontinuities need to be estimated from the data, requiring optimization of a large number of parameters, as opposed to only one in the functions proposed by Iyengar and Greenhouse. Hedges (1992) offers a similar method but pre-specifies the regions of the p-value scale that correspond to the probabilities.

Copas and colleagues (Copas and Li, 1997; Copas 1999; Copas and Shi, 2000, 2001) developed a flexible framework in which the probability of selection can depend on both the effect estimate and its standard error. Unlike the models based on p-values or test statistics, which relate a single combined measure of effect and precision with the weight functions, Copas' model allows for separate parameters describing the relationship between each study characteristic and the probability of selection. Note that for the remainder of this section, we use P to refer both to probabilities and density functions.

Their approach involves specifying an outcome model representing all studies that have been conducted in the area of interest, and a separate selection model for the publication process. A single correlation parameter relates the two models. For example, if treatment effect y_i is reported with standard error s_i for study i, we can assume it came from a larger pool of studies (including both published and unpublished studies) which follow a random-effects model with overall mean effect θ, true within-study variance σ_i^2, and between-study heterogeneity τ^2. Note that the outcome model is simply the standard random-effects model for meta-analysis as discussed elsewhere in the book, which assumes that publication bias is absent, i.e.,

$$y_i = \theta_i + \sigma_i \varepsilon_i, \;\; \varepsilon_i \sim N(0,1), \;\; \theta_i \sim N(\theta, \tau^2), \quad i = 1, \ldots, m$$

where θ_i and error ε_i are assumed to be independent. The selection model for the publication mechanism is defined via a latent variable z_i, where

$$z_i = a + b/s_i + \delta_i, \;\; \delta_i \sim N(0,1), \;\; \text{corr}(\varepsilon_i, \delta_i) = \rho.$$

This postulates that study i is published if the latent variable $z_i > 0$, thus the above model characterizes the probability of selection of study i, although z_i itself is not bounded. If $\rho = 0$, then the outcome and selection models are unrelated and the magnitude and/or direction of y_i have no bearing on whether a study is reported. However, $\rho > 0$ (or $\rho < 0$) indicates that among studies of similar precision, those with more positive (or negative) estimated treatment effects are more likely to be published. Finally, parameter b characterizes how the propensity for publication depends on study precision/sample size (s_i). If $b > 0$, the framework assumes that very large studies will be published with a probability close to 1.

Together, these two models imply that the observed effect estimates follow a conditional distribution given that $z_i > 0$. Thus the authors differentiate between the true σ_i and a reported s_i, defined as an estimate of the *conditional* standard deviation, given that the study is observed (Copas and Shi, 2000). This concept is not without controversy, since in practice investigators do not condition their estimates on study publication. An alternative formulation of Copas' model allows the observed within-study standard deviation to approximate the true standard deviation, replacing σ_i with s_i (Ning et al., 2017).

Selection models go beyond the capabilities of graph-based methods by explicitly defining the missing-not-at-random mechanism underlying selection bias, which enables bias correction using the assumed selection process as opposed to relying on asymmetry alone. The ease of interpretation and flexibility of Copas' model in particular makes it an attractive approach for handling publication bias. In theory, one would maximize the following observed data log-likelihood to estimate the true, unbiased treatment effect, θ:

$$
\begin{aligned}
L(\theta,\tau,\rho,a,b) &= \sum_{i=1}^{n} \log P(y_i \mid z_i > 0, s_i) \\
&= \sum_{i=1}^{n} \log \frac{P(y_i)P(z_i > 0 \mid y_i, s_i)}{P(z_i > 0 \mid s_i)} \\
&= \sum_{i=1}^{n} -\frac{1}{2}\log(\sigma_i^2 + \tau^2) - \frac{(y_i - \theta)^2}{2(\tau^2 + \sigma_i^2)} \\
&\quad + \log \Phi(v_i) - \log \Phi(w_i),
\end{aligned}
\tag{13.1}
$$

where

$$
v_i = \frac{a + \frac{b}{s_i} + \rho\sigma_i(\tau^2 + \sigma_i^2)(y_i - \theta)}{\sqrt{(1 - \rho^2\sigma_i^2/(\tau^2 + \sigma_i^2))}} \quad \text{and} \quad w_i = a + b/s_i. \tag{13.2}
$$

However, the lack of information from missing studies and limited sample size for the observed studies often lead to the likelihood being flat over a wide range of selection parameter values, resulting in identifiability issues. Thus, instead of attempting to directly maximize the likelihood in equation (13.1), Copas and Shi (2000) propose a sensitivity analysis, basing inference for θ on the profile likelihood over a range of specified values for (a,b),

$$
L_{a,b}(\theta) = \max_{\rho,\tau|\theta,a,b} L(\theta,\rho,\tau,a,b). \tag{13.3}
$$

The authors offer only a graph-based testing approach for selecting the best adjusted treatment effect in the sensitivity analysis, which they refer to as a goodness of fit test for the funnel plot (Copas and Shi, 2000).

Data augmentation methods have been proposed as alternatives to sensitivity analysis of selection models, simulating the number and outcomes of unpublished studies to create a "complete" dataset for inference. Such imputation is dependent on the publication process proposed by selection models such as Copas' model. However, one should note that the strong assumptions made in Copas' model may allow for interchangeable formulation between a selection model and a pattern-mixture model, which specifies the distribution of missing outcomes conditional on their non-observation.

Givens et al. (1997) and Silliman (1997) proposed Bayesian hierarchical data augmentation approaches for some of the earlier selection models. By allowing prior knowledge to supplement the information available in the data, estimation of the bias-corrected effect is made feasible and uncertainty about the selection mechanism is propagated through the standard error for the adjusted pooled treatment effect estimate. Both authors assume that for each observed study there are m unpublished studies with latent outcomes U_i, $i = 1,\ldots,m$. Givens' approach uses the random-effects model and frequentist selection mechanism from Hedges (1992), in which the weights ($\boldsymbol{\omega}$) for selection follow a step-function according to regions of the p-values, and the latent data (m,$\boldsymbol{U}$) are treated as nuisance variables to be marginalized out of the final inference on θ. Prior distributions are placed on θ, τ^2, σ^2, $\boldsymbol{\omega}$, and $\boldsymbol{U}$. The joint posterior distribution implied by the priors and complete data log-likelihood is then derived through numerical sampling techniques and marginalized over the nuisance parameters, such that final inference is made on the observed data posterior.

More recently, Mavridis et al. (2013) developed a Bayesian approach to Copas' model, albeit for the expanded framework of network meta-analysis, which is discussed in Section 13.11. In the context of Copas' model, they put priors on θ, τ, and ρ, but not on a and b. Rather, they identify P_{low} and P_{large}, which are the propensities for publication corresponding to a study with the largest and smallest standard errors (often equating to the smallest and largest sample sizes), respectively. They argue it is easier to specify prior knowledge for such quantities as opposed to the parameters themselves. Then a range of (a,b) can be calculated through the inequality,

$$P_{\text{low}} \leq P(z_i > 0 \mid s_i) \leq P_{\text{large}} \quad \forall i.$$

Informative priors for ρ can be used when information on the direction of publication bias is available. All of the above approaches suggest using a non-informative prior for θ, due to the observed data providing sufficient information on the location of θ to eclipse the influence of any weak or moderately informative priors (Carlin, 1992).

Ning et al. (2017) introduced a frequentist data-augmentation approach to a Copas-like model, using an expectation-maximization (EM) algorithm to obtain a maximum likelihood estimate of the bias-corrected θ. This development makes use of both the parameterization of selection models and the symmetry of funnel plots, as the EM algorithm imputes or "fills in" the missing studies according to the symmetry principle to overcome the identifiability issue. Although similar to trim-and-fill in assuming that imputed studies have treatment effect estimates of equal magnitude and opposite direction to their observed counterparts (recall the limitations of using asymmetry as a proxy for publication bias in Section 13.6), the number of imputed studies and the distribution of treatment effects are based on a parametric model for the probability of publication. Such parameterization also means it is less sensitive to outliers, which is a known limitation of trim-and-fill that often leads to inflated standard errors and conservative adjustment (Schwarzer et al., 2010).

Like the approaches by Givens and Silliman, Ning and colleagues assume that for each observed y_i, there are a corresponding m_i unobserved studies, although they allow for each study i to have a different m_i. They further surmise that the missing outcomes are equal to $-(y_i - 2\theta)$, according to the symmetry principle. This is similar to the imputation step in trim-and-fill, although only a single study is imputed per trimmed study in trim-and-fill, whereas m_i is assumed to have a geometric distribution parameterized with the probability of publication from Copas' model, $(1 - P(z_i > 0 \mid y_i, s_i))^{m_i} P(z_i > 0 \mid y_i, s_i) = (1 - \Phi(v_i))^{m_i} \Phi(v_i)$, $m_i = 0, 1, \ldots,$ where v_i is defined in equation (13.2). The *complete-data* log-likelihood for parameters in Copas' model $\psi = (\theta, \tau, \rho, a, b)$ can then be written as

$$\log L(\psi) = \sum_{i=1}^{n} \log P(z_i > 0, y_i \mid s_i) + m_i \log P(z_i < 0, -(y_i - 2\theta) \mid s_i), \tag{13.4}$$

where $P(z > 0, y)$ means the joint probability of the event $z > 0$ and the value $Y = y$. The EM algorithm iterates between taking the expectation of equation (13.4) given the observed data and current parameter estimates for the iteration, ψ^*, which equates to calculating $E[m_i \mid y_i, s_i, \psi^*,\ i = 1, \ldots, n]$, and then maximizing the conditional expected complete-data log-likelihood, until convergence of the estimates is achieved. According to the geometric distribution of m_i, we have for the E-step

$$E[m_i \mid y_i, s_i, \psi^*, i = 1,\ldots,n] = \frac{1-\Phi(v_i \mid \psi^*)}{\Phi(v_i \mid \psi^*)}.$$

Then the M-step maximizes the following expected complete-data log-likelihood:

$$\log L(\psi^*) = \sum_{i=1}^{n} \log P(z_i > 0, y_i \mid s_i) + \frac{1-\Phi(v_i \mid \psi^*)}{\Phi(v_i \mid \psi^*)} \log P(z_i < 0, -(y_i - 2\theta) \mid s_i).$$

This approach was shown to achieve substantial bias reduction and improved convergence rates relative to directly maximizing equation (13.1), while also being more computationally efficient than sensitivity analysis using the profile likelihood in equation (13.3) and a grid of selection model parameter values. A limitation of the algorithm is that it does incorporate the assumption of symmetry to improve numerical stability, which may be an issue if the cause of asymmetry is not publication bias, however, the simulation studies evaluating its performance generated publication bias according to Copas' selection model, not asymmetry. Furthermore, the EM algorithm can be sensitive to initial values, and a range of starting points should be considered, and final estimates chosen which minimize the negative expected complete-data log-likelihood.

It should be noted that while the validity of selection models may depend on the correctness of model specification, it was observed by Copas and Shi (2001) that a standard random-effects analysis involves the much stronger assumption of no publication bias. Should a sensitivity analysis under a weakening of that assumption yield different results, at minimum it would cast doubt on the naïve effect estimate. Thus, one could argue that it is less imperative that the model be completely and correctly specified than that some attempt is made to characterize the selection mechanism and thus detect and correct for possible publication bias.

13.10.2 Example: Magnesium Data

In this section, we illustrate the application of Copas' selection model for publication bias to the dataset used in Section 13.9, evaluating the effect of intravenous magnesium on mortality among patients with myocardial infarction. Recall that the observed estimated treatment effects are distributed largely to the left of 0 on the log odds ratio scale (see Figure 13.6), with smaller studies having larger, more negative treatment effects than studies with greater precision.

Using the sensitivity analysis approach to Copas' model, we estimate the adjusted treatment effect for a range of values for a and b (reflecting varying probabilities of publication) by maximizing the profile log-likelihood using the `copas` function in R. A sample of estimates are shown in Table 13.1 with their corresponding publication probabilities (summarized by the probability of publishing the study with the largest standard error), and the estimated number of unpublished studies.

The usefulness of a sensitivity approach is in assessing the robustness of results to relaxed assumptions, in this case, the assumption of no existing publication bias. We see that the adjusted odds ratio attenuates substantially toward the null as the probability of non-publication (and correspondingly the number of unpublished studies) increases.

TABLE 13.1

Estimated Adjusted Treatment Effects Using the Copas Model Given Various Publishing Probabilities

Publishing probability	Adjusted OR (95% CI)	p-value OR	N unpublished
0.00	0.59 (0.42, 0.83)	0.002	0
0.97	0.61 (0.43, 0.86)	0.005	0
0.94	0.64 (0.43, 0.94)	0.022	1
0.90	0.67 (0.47, 0.95)	0.025	2
0.86	0.70 (0.50, 0.99)	0.041	3
0.82	0.74 (0.54, 1.01)	0.056	4
0.78	0.78 (0.58, 1.05)	0.097	5
0.75	0.82 (0.65, 1.03)	0.085	6
0.71	0.86 (0.71, 1.04)	0.126	8
0.71	0.87 (0.72, 1.04)	0.127	8
0.61	0.86 (0.68, 1.10)	0.226	12
0.49	0.90 (0.72, 1.13)	0.378	20

OR = odds ratio; N unpublished = estimated total number of unpublished studies, based on publishing probability.

The trim-and-fill result of 0.73 from Section 13.9 would correspond to a publishing probability between 86% and 82% under Copas' model, with fewer estimated studies missing (three to four, compared with the seven imputed by trim-and-fill). Apart from offering a stress test for the unadjusted result, little more can be gleaned without additional knowledge of the publication process from experts in the relevant field of study.

Using the EM algorithm method proposed by Ning et al. (2017) as an alternative sensitivity analysis, we arrive at an estimated adjusted odds ratio of 0.77 with a 95% confidence interval of (0.61, 0.97). Due to the sensitivity of EM to initial values, a range of initial values was considered, and the final estimates were chosen that minimized the expected complete-data log-likelihood. This places the adjusted effect closer to the null than trim-and-fill, with a slightly narrower confidence interval.

13.10.3 Selection Models for Outcome Reporting Bias

Although greater attention has been given to publication bias in meta-analyses, some model-based methods have been proposed for the case of outcome reporting bias (ORB). Copas et al. (2014) initially presented a likelihood-based sensitivity analysis on the basis of the ORB in Trials (ORBIT) study classification system for missing or incomplete outcome reporting in randomized trials that describes the extent of measurement and reporting for an outcome. Details on study classifications can be found at http://outcome-reporting-bias.org/.

Unlike the simpler scheme for publication bias, Copas' ORB model classifies each known study based on the following criteria.

Copas' ORB Model

1. The outcome of interest is measured (M) or not measured (m)
2. If measured, the outcome of interest is significant (S) or not significant (s), and reported (R) or not reported (r)

3. An unreported outcome (r) is classified as having high risk for bias (High) or low risk for bias (Low).

According to the above framework, only an outcome that is measured (M) can be assessed for significance (S/s) by the trialists and/or reported (R/r). When an outcome is measured, it is assumed to be reported if significant (i.e., $P(R \mid M \cap S) = 1$). They also parameterize the following probabilities:

$$P(M) = \alpha, \quad P(R \mid M \cap s) = \beta,$$

where α is the probability of an outcome being measured and β refers to the probability that a result is reported if measured but non-significant.

Although the trialists may know M/m for all outcomes and S/s for all measured outcomes, the reader only observes M and S/s for the reported outcomes (R). For the subgroup defined by r, a risk of bias assessment is proposed, whereby an outcome is believed to be unreported either because (i) it was not measured in the first place or (ii) it was measured and failed to be significant. Copas' ORB model assesses scenario (i) to have a low risk of bias, since the decision to report cannot depend on the significance of the result if the outcome was not measured. Accordingly, scenario (ii) is perceived to have a high risk of bias, since the significance of the measured result can factor into the decision to report. The uncertainty of such risk assessments is modeled by two reliability parameters,

$$P(\text{High} \mid M \cap s \cap r) = \rho_1, \quad P(\text{Low} \mid m \cap r) = \rho_2,$$

where $\rho_1 = \rho_2 = 1$ means perfect risk assessment.

The four aforementioned probabilities ($\alpha, \beta, \rho_1, \rho_2$) are then used to construct the log-likelihood for four groupings of studies in a meta-analysis: (a) those reported and significant; (b) those reported and non-significant; (c) those unreported and classified as high risk; and (d) those unreported and classified as low risk. Under a random-effects model for outcome y_i of study i (whether or not it is measured or reported), centered on true effect θ,

$$y_i \sim N(\theta, \sigma_i^2 + \tau^2),$$

the complete data log-likelihood can be shown to have the following form:

$$\begin{aligned} L_{\rho_1,\rho_2}(\theta,\tau,\alpha,\beta) &= \sum_{S \cap R} P(y_i \cap M \cap S \cap R) + \sum_{s \cap R} P(y_i \cap M \cap s \cap R) \\ &\quad + \sum_{r \cap \text{low}} P(r \cap \text{low}) + \sum_{r \cap \text{high}} P(r \cap \text{high}) \\ &= \sum_{S \cap R} \log\{\alpha f_i(y_i)\} + \sum_{s \cap R} \log\{\alpha\beta f_i(y_i)\} \\ &\quad + \sum_{r \cap \text{high}} \log\{\rho_1 \alpha Q_i (1-\beta) + (1-\rho_2)(1-\alpha)\} \\ &\quad + \sum_{r \cap \text{low}} \log\{(1-\rho_1)\alpha Q_i (1-\beta) + \rho_2 (1-\alpha)\}, \end{aligned}$$

where

$$Q_i = P(|y_i| < 1.96\sigma_i) = \Phi\left\{\frac{1.96\sigma_i - \theta}{\sqrt{\sigma_i^2 + \tau^2}}\right\} - \Phi\left\{\frac{-1.96\sigma_i - \theta}{\sqrt{\sigma_i^2 + \tau^2}}\right\}.$$

θ is then estimated by maximizing the profile log-likelihood for a given pair (ρ_1,ρ_2). Hence, sensitivity analysis can be performed over a range of combinations for ρ_1 and ρ_2. The authors show that the accuracies of these risk assessments do not unduly affect the overall bias, although it does depend on the number of studies identified as high risk.

Though an intuitive approach, in 2017, Copas and colleagues (Copas et al., 2017) further distinguished an outcome as a benefit or harm outcome and provided a simpler model for sensitivity analysis. For benefit outcomes, the risk assessment is the same as in the original ORB model presented above, where an unreported outcome is classified as being high risk if it is measured but not significant. However, they assume perfect classification of high and low risk studies ($\rho_1 = \rho_2 = 1$). In doing so, they offer a simplification of the original ORB model, since under correct identification of high risk studies (and assuming the direction of bias is away from the null), bias adjustment can be performed by adding to the observed log-likelihood the probabilities of high-risk unreported outcomes being non-significant. Thus, the log-likelihood function for this new model includes only studies in subgroups R and $r \cap \text{High}$, shown below:

$$L_B(\theta) = -\frac{1}{2}\sum_{R} \log\left\{(\sigma^2 + \tau^2) + \frac{(y_i - \theta)^2}{\sigma_i^2 + \tau^2}\right\}$$

$$+\sum_{\text{high}} \log\left[\Phi\left\{\frac{z_\alpha\sigma_i - \theta}{\sqrt{\sigma_i^2 + \tau^2}}\right\} - \Phi\left\{\frac{-z_\alpha\sigma_i - \theta}{\sqrt{\sigma_i^2 + \tau^2}}\right\}\right].$$

Conversely to benefit outcomes, harm outcomes are more likely to be suppressed if the effect is positive (whether or not it is significant), as this indicates a greater incidence of side effects. Therefore, the risk for an unreported harm outcome is high if it is measured and positive. Accordingly, the likelihood function is modified as

$$L_H(\theta) = -\frac{1}{2}\sum_{R} \log\left\{(\sigma^2 + \tau^2) + \frac{(y_i - \theta)^2}{\sigma_i^2 + \tau^2}\right\} + \sum_{\text{high}} \log\left[\Phi\left\{\frac{\theta}{\sqrt{\sigma_i^2 + \tau^2}}\right\}\right].$$

By conditioning on the number of studies that have been identified, these models only adjust for outcome reporting bias, and not publication bias. The other approaches discussed in this section account for publication bias, but not outcome reporting bias. To our knowledge, no one has yet developed a model that incorporates both publication and outcome reporting bias. Such a model would need to separately account for studies discovered with the outcome reported, discovered with the outcome unreported, and undiscovered.

The Copas ORB model relies on being able to obtain information on whether the unreported outcome is measured, significant, or positive. Such information is becoming more readily available for clinical trials through the high-standard data registries mentioned in Section 13.4. However, it is not guaranteed that each trial will have detailed information on which outcomes were measured and analyzed, and on the reason for not reporting a certain outcome. If such information is not available for a relatively large number of studies, then the Copas ORB model will be under-powered.

13.11 Publication Bias in Network Meta-Analysis

Network meta-analysis (NMA) extends the traditional MA framework to compare more than two treatments simultaneously, consolidating direct evidence from head-to-head trials as well as indirect evidence through a common comparator, under the assumption of evidence consistency (see Chapter 10). There is an added layer of complexity to handling publication bias in the context of network meta-analysis, as more than one treatment comparison is being evaluated, sometimes within a single multi-arm study. Thus, publication bias cannot be thought of in relation to one pairwise comparison, but rather a collective network of studies and contrasts.

Graphical methods to detect publication bias in NMAs are limited, as asymmetry cannot be judged according to a single reference line. In practice, funnel plots are assessed separately for each comparison in the network, perhaps with some distinctive labeling for two-arm and multi-arm studies (Mavridis et al., 2014; Trinquart et al., 2012). In an attempt to consolidate the information into a single graph, Chaimani et al. (2013) proposed the "comparison-adjusted funnel plot", in which the difference between study-specific effects and their comparison-specific summary estimates are plotted against the standard errors, with symmetry expected around the zero bias line in the absence of publication bias. However, the use of this method requires a meaningful ordering of the treatments of interest and assumptions regarding the direction of small-study effects. Otherwise, it offers little additional information over plotting the contrasts separately.

In light of the difficulties of graphically evaluating asymmetry in the network setting, far more attention has been devoted to modeling approaches for dealing with publication bias, particularly selection models. As in the univariate MA setting, selection models can be used in sensitivity analyses for detecting and correcting for publication bias. The unique statistical challenges encountered when extending models to NMAs include accounting for multiple study designs (which differ by the types and number of treatments compared), thereby increasing the number of estimable parameters, and accounting for the within-study correlation structure in multi-arm studies.

Trinquart et al. (2012) extended a logistic selection model to a Bayesian NMA of antidepressants, assuming the propensity of selection to be a decreasing function of the standard error of the effect estimate, and upweighting trials with a lower selection probability to correct for bias. The effect-size model and selection model were comparison-specific, such that each treatment comparison had its own set of model parameters. The Copas model described in Section 13.10 has also been extended to NMAs, dividing the studies into design groups based on the sets of treatments compared, and specifying multiple selection models with design-specific parameters (Chootrakool et al., 2011; Mavridis et al., 2014). A Bayesian approach to estimation is often used, as the large number of selection model parameters leads to identifiability issues with frequentist methods.

13.12 Software for Publication Bias Analyses

Several computer programs can help to carry out the methods described in this chapter. The Review Manager (RevMan) program used by the Cochrane Collaboration and compatible with the Cochrane Library can generate funnel plots but does not do any of the other methods discussed. Comprehensive Meta-Analysis (CMA), a program specifically for meta-analysis, provides high quality forest and funnel plots, regression testing, fail-safe N, and trim-and-fill. These methods as well as more complex methods such as selection models can be implemented using macros (or coding packages) that have been written by experts in the field for use within general purpose statistical programs like Stata and R. Although less user-friendly than the dedicated software, they offer greater flexibility and user-control. Some examples of more common commands include `metafunnel`, `metabias`, and `metatrim` in Stata, which create funnel plots, run regression tests, and perform trim-and-fill respectively. The `metafor` package in R can create funnel plots and do regressions and trim-and-fill. For selection modeling, the `metasens` package is available in R to evaluate the sensitivity of meta-analysis results to potential publication bias. In this package the `copas` function, developed by Carpenter et al. (2009), implements a grid-search algorithm for sensitivity analysis and bias correction using the model by Copas and Shi (2001), maximizing the likelihood over a grid of (a,b) values to reflect varying selection strength. Since programs like R are commonly used by statistical researchers for methods development and testing, macros for new and cutting-edge methods are regularly produced and released to the platform for public use.

13.13 Conclusion

In this chapter, we have reviewed methods that have been used to help identify and correct for publication and outcome reporting bias. Although graph-based methods such as the funnel plot, Egger's test, and trim-and-fill are common approaches for diagnosing such bias, their use is insufficient for drawing conclusions about the presence or absence of such bias. Fundamentally, these methods focus on identifying and correcting for asymmetry in a funnel plot, but asymmetry can be generated by many causes other than publication or outcome reporting bias. In general, funnel plots can identify an association between the sizes of treatment effects and their associated uncertainty which may suggest that larger studies are returning results different from those of smaller studies, i.e., that there is a small-study effect. This in itself is useful to know, but it is not a marker for publication or outcome reporting bias.

Selection models are more principled approaches, but choice of a good selection model requires a trade-off between flexibility in capturing the publication or reporting mechanism and parsimony of the model. More complex models are less analytically and computationally tractable. Because meta-analyses often have a small number of studies, computational algorithms may be unstable. When the number of studies is sufficient (say greater than 10), selection models can help to quantify the sensitivity of conclusions to different amounts of potential bias. Further study and development of these models are needed.

Because the *post hoc* measures described in this chapter depend on assumptions that are often unmet, it is best to take preemptory measures to reduce the scope of bias by carefully searching databases, registries, gray literature, professional contacts, and regulatory reports for all studies undertaken. The knowledge thereby gained about the research landscape may either reduce concern about information loss or confirm that such loss has occurred. In either case, this will reduce the need for publication bias diagnostics and may help to bound the range of plausible inferences that need to be considered.

References

Anderson ML, Chiswell K, Peterson ED, Tasneem A, Topping J and Califf RM, 2015. Compliance with results reporting at ClinicalTrials.gov. *New England Journal of Medicine* **372**(11): 1031–1039.

Becker, BJ, 2005. Failsafe N or file-drawer number. In Rothstein HR, Sutton AJ and Borenstein M (Eds), 2006. *Publication Bias in Meta-Analysis: Prevention, Assessment and Adjustments.* John Wiley & Sons, 111–125.

Begg CB and Mazumdar M, 1994. Operating characteristics of a rank correlation test for publication bias. *Biometrics* **50**(4): 1088–1101.

Brassington I, 2017. The ethics of reporting all the results of clinical trials. *British Medical Bulletin* **121**(1): 19–29.

Carlin JB, 1992. Meta-analysis for 2×2 tables: A Bayesian approach. *Statistics in Medicine* **11**(2): 141–158.

Carpenter JR, Rücker G and Schwarzer G, 2009. copas: An R package for fitting the Copas selection model. *The R Journal* **2**(2): 31–36.

Chaimani A, Higgins JP, Mavridis D, Spyridonos P and Salanti G, 2013. Graphical tools for network meta-analysis in STATA. *PLOS ONE* **8**(10): e76654.

Chan AW, Krleza-Jeric K, Schmid I and Altman DG, 2004. Outcome reporting bias in randomized trials funded by the Canadian Institutes of Health Research. *Canadian Medical Association Journal* **171**(7): 735–740.

Chootrakool H, Shi JQ and Yue R, 2011. Meta-analysis and sensitivity analysis for multi-arm trials with selection bias. *Statistics in Medicine* **30**(11): 1183–1198.

Citkowicz M and Vevea JL, 2017. A parsimonious weight function for modeling publication bias. *Psychological Methods* **22**(1): 28–41.

ClinicalTrials.gov, 2017. About the Results Database. Available from https://clinicaltrials.gov/ct2/about-site/results (accessed 12 April 2017).

Copas J and Li HG, 1997. Inference for non-random samples. *Journal of the Royal Statistical Society: Series B* **59**(1): 55–95.

Copas J, 1999. What works?: Selectivity models and meta-analysis. *Journal of the Royal Statistical Society: Series A* **162**(1): 95–109.

Copas J and Shi JQ, 2000. Meta-analysis, funnel plots and sensitivity analysis. *Biostatistics* **1**(3): 247–262.

Copas J and Shi JQ, 2001. A sensitivity analysis for publication bias in systematic reviews. *Statistical Methods in Medical Research* **10**(4): 251–265.

Copas J, Dwan K, Kirkham J and Williamson P, 2014. A model-based correction for outcome reporting bias in meta-analysis. *Biostatistics* **15**(2): 370–383.

Copas J, Marson A, Williamson P and Kirkham J, 2017. Model-based sensitivity analysis for outcome reporting bias in the meta analysis of benefit and harm outcomes. *Statistical Methods in Medical Research* **28**(3): 889–903.

Dal-Ré R, Ioannidis JP, Bracken MB, Buffler PA, Chan AW, Franco EL, La Vecchia C and Weiderpass E, 2014. Making prospective registration of observational research a reality. *Science Translational Medicine* **6**: 224cm1.

Dal-Ré R, Bracken MB and Ioannidis JP, 2015. Call to improve transparency of trials of non-regulated interventions. *BMJ* **356**: h1323.

De Angelis C, Drazen JM, Frizelle FA, Haug C, Hoey J, Horton R, Kotzin S, Laine C, Marusic A, Overbeke AJPM, Schroeder TV, Sox HC and Van Der Weyden MB, 2004. Clinical trial registration: A statement from the International Committee of Medical Journal Editors. *Annals of Internal Medicine* **141**(6): 477–478.

De Oliveira GS Jr., Jung MJ and McCarthy RJ, 2015. Discrepancies between randomized controlled trial registry entries and content of corresponding manuscripts reported in anesthesiology journals. *Anesthesia and Analgesia* **121**(4): 1030–1033.

Dear KB and Begg CB, 1992. An approach for assessing publication bias prior to performing a meta-analysis. *Statistical Science* **7**(2): 237–245.

Dickersin K, 2005. Publication bias: Recognizing the problem, understanding its origins and scope, and preventing harm. In Rothstein HR, Sutton AJ and Borenstein M (Eds), 2006. *Publication Bias in Meta-Analysis: Prevention, Assessment and Adjustments*. John Wiley & Sons.

Dickersin K and Chalmers I, 2011. Recognizing, investigating and dealing with incomplete and biased reporting of clinical research: From Francis Bacon to the WHO. *Journal of the Royal Society of Medicine* **104**(12): 532–538.

Duval S and Tweedie R, 2000a. A nonparametric "trim and fill" method of accounting for publication bias in meta-analysis. *Journal of the American Statistical Association* **95**(449): 89–98.

Duval S and Tweedie R, 2000b. Trim and fill: A simple funnel-plot-based method of testing and adjusting for publication bias in meta-analysis. *Biometrics* **56**(2): 455–463.

Dwan K, Altman DG, Arnaiz JA, Bloom J, Chan AW, Cronin E, Decullier E, Easterbrook PJ, Von Elm E, Gamble C, Ghersi D, Ioannidis JPA, Simes J and Williamson PR, 2008. Systematic review of the empirical evidence of study publication bias and outcome reporting bias. *PLOS ONE* **3**(8): e3081.

Eberly LJ and Casella G, 1999. Bayesian estimation of the number of unseen studies in a meta-analysis. *Journal of Official Statistics* **15**(4): 477–494.

Editors, 2010. The registration of observational studies–when metaphors go bad. *Epidemiology* **21**(5): 607–609.

Egger M, Davey Smith G, Schneider M and Minder C, 1997. Bias in meta-analysis detected by a simple, graphical test. *British Medical Journal* **315**(7109): 629–634.

Engels EA, Schmid CH, Terrin N, Olkin I and Lau J, 2000. Heterogeneity and statistical significance in meta-analysis: An empirical study of 125 meta-analyses. *Statistics in Medicine* **19**(13): 1707–1728.

Eyding D, Lelgemann M, Grouven U, Härter M, Kromp M, Kaiser T, Kerekes MF, Gerken M and Wieseler B, 2010. Reboxetine for acute treatment of major depression: Systematic review and meta-analysis of published and unpublished placebo and selective serotonin reuptake inhibitor controlled trials. *BMJ* **341**: c4737.

Fisher RA, 1932. *Statistical Methods for Research Workers*. 4th Edition. London: Oliver and Boyd.

Givens GH, Smith DD and Tweedie RL, 1997. Publication bias in meta-analysis: A Bayesian data-augmentation approach to account for issues exemplified in the passive smoking debate. *Statistical Science* **12**(4): 221–240.

Gleser LJ and Olkin I, 1996. Models for estimating the number of unpublished studies. *Statistics in Medicine* **15**(23): 2493–2507.

Goldacre B, 2013. Are clinical trial data shared sufficiently today? No. *BMJ* **347**: f1880.

Hahn S, Williamson PR and Hutton JL, 2002. Investigation of within-study selective reporting in clinical research: Follow-up of applications submitted to a local research ethics committee. *Journal of Evaluation in Clinical Practice* **8**(3): 353–359.

Hedges LV, 1984. Estimation of effect size under nonrandom sampling: The effects of censoring studies yielding statistically insignificant mean differences. *Journal of Educational Statistics* **9**(1): 61–85.

Hedges LV, 1992. Modeling publication selection effects in meta-analysis. *Statistical Science* **7**(2): 246–255.

Higgins JPT, Thomas J, Chandler J, Cumpston M, Li T, Page MJ and Welch VA (editors), 2019. *Cochrane Handbook for Systematic Reviews of Interventions*. 2nd Edition. Chichester, UK: John Wiley & Sons. Also online version 6.0 (updated July 2019) available from www.training.cochrane.org/handbook.

Higgins JPT and Spiegelhalter DJ, 2002. Being sceptical about meta-analyses: A Bayesian perspective on magnesium trials in myocardial infarction. *International Journal of Epidemiology* **31**(1): 96–104.

Hooft L, Korevaar DA, Molenaar N, Bossuyt PM and Scholten RJ, 2014. Endorsement of ICMJE's clinical trial registration policy: A survey among journal editors. *The Netherlands Journal of Medicine* **356**(7): 349–355.

Hopewell S, Clarke M, Stewart L and Tierney J, 2007. Time to publication for results of clinical trials. *The Cochrane Database of Systematic Reviews* **2**(2): MR000011.

Hopewell S, Loudon K, Clarke MJ, Oxman AD and Dickersin K, 2009. Publication bias in clinical trials due to statistical significance or direction of trial results. *The Cochrane Database of Systematic Reviews* **1**(1): MR000006.

Ioannidis JP, Caplan AL and Dal-Ré R, 2017. Outcome reporting bias in clinical trials: Why monitoring matters. *BMJ* **356**: j408.

Iyengar S and Greenhouse JB, 1988. Selection models and the file drawer problem. *Statistical Science* **3**(1): 109–117.

Kirkham JJ, Dwan KM, Altman DG, Gamble C, Dodd S, Smyth R and Williamson PR, 2010. The impact of outcome reporting bias in randomised controlled trials on a cohort of systematic reviews. *BMJ* **340**: c365.

Lane DM and Dunlap WP, 1978. Estimating effect size: Bias resulting from the significance criterion in editorial decisions. *British Journal of Mathematical and Statistical Psychology* **31**(2): 107–112.

Lau J, Ioannidis JPA, Terrin N, Schmid CH and Olkin I, 2006. The case of the misleading funnel plot. *British Medical Journal* **333**(7568): 597–600.

Light RJ and Pillemer DB, 1984. *Summing Up: The Science of Reviewing Research*. Cambridge, MA: Harvard University Press.

Loder E, Groves T and Macauley D, 2010. Registration of observational studies. *BMJ* **340**: c950.

Mavridis D, Sutton A, Cipriani A and Salanti G, 2013. A fully Bayesian application of the Copas selection model for publication bias extended to network meta-analysis. *Statistics in Medicine* **32**(1): 51–66.

Mavridis D, Welton NJ, Sutton A and Salanti G, 2014. A selection model for accounting for publication bias in a full network meta-analysis. *Statistics in Medicine* **33**(30): 5399–5412.

Ning J, Chen Y and Piao J, 2017. Maximum likelihood estimation and EM algorithm of Copas-like selection model for publication bias correction. *Biostatistics* **18**(3): 495–504.

Orwin RG, 1983. A fail-safe N for effect size in meta-analysis. *Journal of Educational Statistics* **8**(2): 157–159.

Panagiotou OA and Trikalinos TA, 2015. Commentary: On effect measures, heterogeneity, and the laws of nature. *Epidemiology* **26**(5): 710–713.

Prayle A, Hurley MN and Smyth AR, 2012. Compliance with mandatory reporting of clinical trial results on ClinicalTrials.gov. *BMJ* **344**: d7373.

Rosenthal R, 1979. The file drawer problem and tolerance for null results. *Psychological Bulletin* **86**(3): 638–641.

Schmid CH, 2017. Outcome reporting bias: A pervasive problem in published meta-analyses. *American Journal of Kidney Diseases* **69**(2): 172–174.

Schwarzer G, Carpenter J and Rücker G, 2010. Empirical evaluation suggests Copas selection model preferable to trim-and-fill method for selection bias in meta-analysis. *Journal of Clinical Epidemiology* **63**(3): 282–288.

Silliman NP, 1997. Hierarchical selection models with applications in meta-analysis. *Journal of the American Statistical Association* **92**(439): 926–936.

Simes RJ, 1986. Publication bias: The case for an international registry of clinical trials. *Journal of Clinical Oncology* **4**(10): 1529–1541.

Simonsohn U, 2017. The Funnel Plot is Invalid because of This Crazy Assumption: r(n,d)=0. Available from http://datacolada.org/58 (posted 21 March 2017).

Sterne JA and Egger M, 2001. Funnel plots for detecting bias in meta-analysis: Guidelines on choice of axis. *Journal of Clinical Epidemiology* **54**(10): 1046–1055.

Sterne JAC, Gavaghan D and Egger M, 2000. Publication and related bias in meta-analysis: Power of statistical tests and prevalence in the literature. *Journal of Clinical Epidemiology* **53**(11): 1119–1129.

Sterne JAC, Jüni P, Schulz KF, Altman DG, Bartlett C and Egger M, 2002. Statistical methods for assessing the influence of study characteristics on treatment effects in 'meta-epidemiological' research. *Statistics in Medicine* **21**(11): 1513–1524.

Sterne JAC, Sutton AJ, Ioannidis JPA, Terrin N, Jones DR, Lau J, Carpenter J, Rücker G, Harbord RM, Schmid CH, Tetzlaff J, Deeks JJ, Peters J, Macaskill P, Schwarzer G, Duval S, Altman DG, Moher D and Higgins JPT, 2011. Recommendations for examining and interpreting funnel plot asymmetry in meta-analyses of randomised controlled trials. *British Medical Journal* **343**: d4002.

Stouffer SA and Hovland CI, Social Science Research Council (US), United States and Army Service Forces, & Information and Education Division, 1949. *Studies in social psychology in World War II*. Princeton, NJ: Princeton University Press.

Sutton AJ, Song F, Gilbody SM and Abrams KR, 2000. Modelling publication bias in meta-analysis: A review. *Statistical Methods in Medical Research* **9**(5): 421–445.
Terrin N, Schmid CH, Lau J and Olkin I, 2003. Adjusting for publication bias in the presence of heterogeneity. *Statistics in Medicine* **22**(13): 2113–2126.
Terrin N, Schmid CH and Lau J, 2005. In an empirical evaluation of the funnel plot, researchers could not visually identify publication bias. *Journal of Clinical Epidemiology* **58**(9): 894–901.
Thompson SG and Sharp SJ, 1999. Explaining heterogeneity in meta-analysis: A comparison of methods. *Statistics in Medicine* **18**(20): 2693–2708.
Trinquart L, Chatellier G and Ravaud P, 2012. Adjustment for reporting bias in network meta-analysis of antidepressant trials. *BMC Medical Research Methodology* **12**: 150.
Vedula SS, Bero L, Scherer RW and Dickersin K, 2009. Outcome reporting in industry-sponsored trials of gabapentin for off-label use. *The New England Journal of Medicine* **361**(20): 1963–1971.
Vevea JL and Hedges LV, 1995. A general linear model for estimating effect size in the presence of publication bias. *Psychometrika* **60**(3): 419–435.
Williams RJ, Tse T, Harlan WR and Zarin DA, 2010. Registration of observational studies: Is it time? *CMAJ* **182**(15): 1638–1642.
Xie X, Liu Y, Perkovic V, Li X, Ninomiya T, Hou W, Zhao N, Liu L, Lv J, Zhang H and Wang H, 2016. Renin-angiotensin system inhibitors and kidney and cardiovascular outcomes in patients with CKD: A Bayesian network meta-analysis of randomized clinical trials. *American Journal of Kidney Diseases* **67**(5): 728–741.

14

Control Risk Regression

Annamaria Guolo, Christopher H. Schmid, and Theo Stijnen

CONTENTS

Heterogeneity of treatment effects across studies is a consequence of differences among studies arising from a variety of sources including study designs, patient characteristics, and types of interventions. As noted in Chapter 7, meta-regression can help quantify the contributions of these patient, study, and design factors to the overall heterogeneity. Unfortunately, not all sources of heterogeneity are always measured and available, but some may be related to differences in how severely ill patients in each study are. Severity of illness can often be approximated by the underlying risk with which outcomes occur in a given study. In such cases, the proportion or rate of events in the control or reference group may serve as a useful surrogate for the underlying, latent risk (Brand and Kragt, 1992; Schmid et al., 1998, 2004; van Houwelingen et al., 2002; Chaimani, 2015). The inclusion of the control risk in meta-analysis gives rise to what is called *control risk regression*.

14.1 Model and Notation

Consider a meta-analysis of I independent studies about the effectiveness of a treatment and let η_i and ξ_i denote the true unobserved measure of risk in the treatment group and the true unobserved measure of risk in the control group, respectively, for each study $i = 1,\ldots,I$. The risk measure can be any measure of risk, such as the proportion of individuals with the event, the event rate (i.e., number of events per time unit), the odds, or the log odds. The control risk is often also called the underlying risk, and we will use these two terms interchangeably. The relationship between η_i and ξ_i is typically defined through the linear regression model (e.g., Arends et al., 2000; van Houwelingen et al., 2002)

$$\eta_i = \beta_0 + \beta_1 \xi_i + \varepsilon_i, \varepsilon_i \sim N(0,\sigma^2), \quad i = 1,\ldots,I, \tag{14.1}$$

where the parameter σ^2 is the residual variance that describes the variation in the treatment risk unexplained by the control risk. Usually, the inferential interest is in the parameter β_1 associated with the underlying risk. If $\beta_1 = 1$, then if the control risk increases by a certain amount, the treatment group risk increases by the same amount. Thus, interesting cases are usually those where β_1 deviates from 1. Model (14.1) predicts the true risk in the treatment group based on the true risk in the control group. However, as the true treatment effect is a function of the true risks in treatment and control group, model (14.1) can also be reformulated with the true treatment effect as the dependent variable in order to predict the true treatment effect from the true risk in the control group. In addition, the independent variable is often centered by subtracting the mean control risk μ_ξ, in order to let the intercept correspond to the average treatment effect (e.g., McIntosh, 1996; Schmid et al., 1998). This model is then written as

$$\eta_i^* = \beta_0^* + \beta_1^*(\xi_i - \mu_\xi) + \varepsilon_i^*, \ \varepsilon_i^* \sim N(0,\tau^2), \quad i = 1,\ldots,I, \tag{14.2}$$

where η_i^* is the treatment effect and the intercept β_0^* is the average treatment effect when the control risk is at the average level μ_ξ.

Note that in (14.2) the residual variance is denoted by τ^2, in order to stick to the convention in this book to denote the between-studies variance of the treatment effects by τ^2. If the treatment effect is chosen to be $\eta_i^* = \eta_i - \xi_i$, then (14.2) is a reparametrization of model (14.1): $\beta_0 = \beta_0^* - \beta_1^* \mu_\xi$, $\beta_1 = \beta_1^* + 1$, $\varepsilon_i = \varepsilon_i^*$, and $\sigma^2 = \tau^2$. Centering ξ_i can also improve computational efficiency. In model (14.2), β_1^* quantifies how much the treatment effect increases for every unit increase of the control risk, and $\left(\beta_0^*, \beta_1^*\right) = (0,0)$ is a (strong) claim of no treatment effect (Ghidey et al., 2011). That is, the treatment effect $\eta_i^* = 0$ on average, and there is no relation between η_i^* and ξ_i. Note that if the overall treatment effect β_0^* is zero, there still might be a relation between control risk and treatment effect, i.e., $\beta_1^* \neq 0$. This can be thought of as a qualitative interaction between treatment and control risk such that positive effects and negative effects cancel out overall.

In the literature, some authors use model (14.1) and others model (14.2). Model (14.1) has the advantage that the estimates of the true treatment risk and control risk are independent, as they are based on different groups of subjects. In this chapter, we will work with both (14.1) and (14.2), choosing the one which is most convenient in a given context.

These linear models are computationally convenient and are a reasonable option since treatment risks or treatment effects are often monotonically related to the underlying risk. When the number of studies in the meta-analysis is small, the data may be insufficient to fit anything but a linear trend. When data are sufficient, non-linear relationships such as U-shaped curves that describe treatment effects that are similar for high and low risk individuals but different for moderate risk individuals or relationships that have a threshold below or above which the relationship is constant may be useful (Wang et al., 2009).

14.1.1 L'Abbé Plot

In this chapter, the data consist of either the sample size and number of participants with an event for two groups per study, or the number of events and the total follow-up time in the two groups. The L'Abbé plot (L'Abbé et al., 1987) is a bivariate representation of the data, plotting an observed measure of the treatment risk $\hat{\eta}_i$ on the y-axis against an observed measure of the control group risk $\hat{\xi}_i$ on the x-axis. Mostly, for the purpose of graphical illustration, the plotted risk measures are just the event proportions or event rates, but these may not necessarily be the same as the $\hat{\eta}_i$ and $\hat{\xi}_i$ used in the regression calculations. Typically, the observations in the L'Abbé plot are represented by a circle with size proportional to the amount of information carried by the study (e.g., inverse standard error of the treatment effect or the sample size), so that larger circles correspond to more precise estimates or larger studies. If $\eta_i - \xi_i$ is chosen as the treatment effect measure, homogeneity of the treatment effect is indicated when the points are close to a $y = c + x$ line. If the ratio η_i/ξ_i is chosen as the treatment effect measure instead, then homogeneity would follow if the points lie close to $y = cx$. Deviations from these lines suggest heterogeneity. Note that homogeneity on one scale implies heterogeneity on another unless $y = x$. When working on a different scale in the regression modeling, as with log odds ratios, it might make sense to make the plot on the transformed scale as well, for example, plotting the logits of the two proportions.

Figure 14.1 shows a L'Abbé plot for the data in Table 14.1 that refer to a meta-analysis of clinical trials evaluating a drug treatment for preventing cardiovascular mortality in middle-aged patients with mild to moderate hypertension (Pagliaro et al., 1992). The data have been used in Sharp et al. (1996) to illustrate the relationship between treatment risk

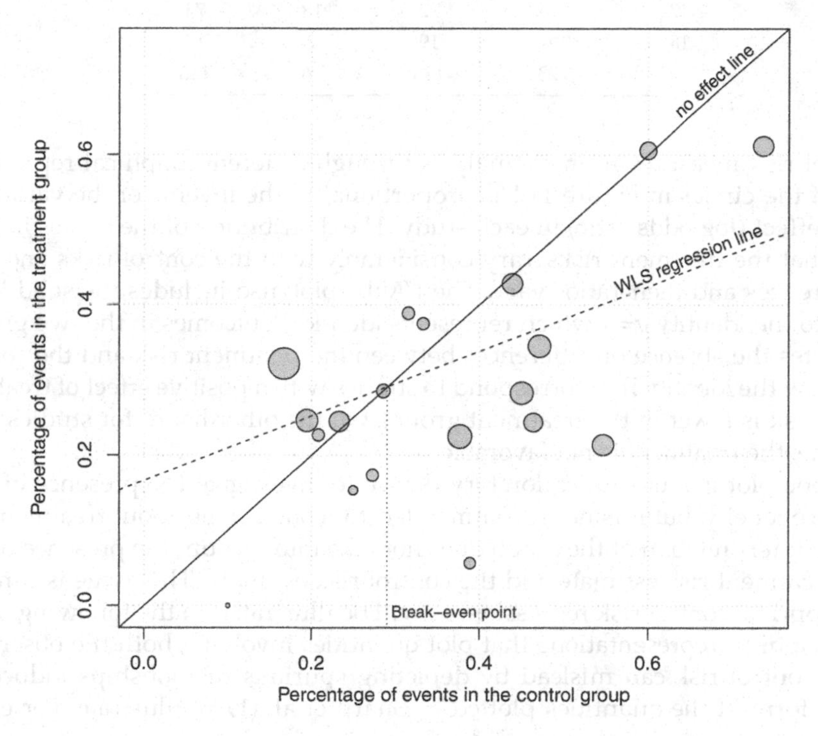

FIGURE 14.1
L'Abbé plot for sclerotherapy data. Circle size indicates the inverse of the variance of the treatment effect (log odds ratio) in each study. The solid line corresponds to the identity line. The dashed line is the weighted least squares regression line (with treatment proportion as dependent variable, control proportion as independent variable, and weights equal to the inverse standard error of the treatment proportion), whose drawbacks are discussed in Section 14.1.2.

TABLE 14.1

Sclerotherapy Data (Pagliaro et al., 1992)

	Controls		Treated	
Study	**Deaths**	**Total**	**Deaths**	**Total**
1	14	36	2	35
2	2	20	0	21
3	29	53	12	56
4	19	41	10	41
5	27	60	15	53
6	26	69	16	71
7	4	16	2	13
8	6	22	4	23
9	34	46	30	49
10	24	51	19	55
11	18	41	18	42
12	8	28	6	21
13	21	35	20	33
14	14	60	13	53
15	5	24	5	22
16	6	18	6	16
17	14	72	18	73
18	6	19	7	18
19	23	138	46	143

and control risk measures in meta-analysis through different graphical representations. The size of the circles in Figure 14.1 is proportional to the inverse of the variance of the treatment effect (log odds ratio) in each study. The distribution of the circles in the graph indicates that the treatment risks vary considerably with the control risks and that both their differences and their ratios vary. The L'Abbé plot also includes the solid line corresponding to the identity $y=x$ which represents identical outcomes in the two groups, that is, it indicates the absence of differences between the treatment risk and the control risk. Circles below the identity line correspond to studies with a positive effect of treatment (i.e., where the risk is lower in the treatment group). On the other hand, for studies above the identity line, the treatment is not favorable.

The L'Abbé plot is a useful exploratory device to investigate the presence of between-study heterogeneity, but it is not recommended for conclusions about treatment benefits. In fact, any interpretation of the graph does not take into account the presence of errors in both the treatment risk estimate and the control risk estimate. This issue is fundamental to an appropriate control risk regression, as will be illustrated in the following sections.

Other graphical representations that plot quantities involving both the observed treatment and control risk can mislead by depicting spurious relationships induced by the functional form of the quantities plotted as Sharp et al. (1996) illustrate. For example, a plot of the observed treatment effect $\hat{\eta}_i - \hat{\xi}_i$ against $\hat{\xi}_i$ may indicate a spurious relationship between the two simply because the observed control risk is a component of both. A preferable solution then is to plot the treatment effect against the average of the treatment risk and control risk, as in the Bland and Altman plot (1986, 1999) or in the SROC method of Littenberg and Moses (Chapter 19) to meta-analyze pairs of sensitivity and specificity.

14.1.2 Weighted Least Squares Regression and Its Limitations

A simple approach for analyzing the relationship between the treatment effect and the underlying risk measure is weighted least squares (WLS) regression, with weights given by the inverse of the variance of the treatment effect. Instead of using the estimated treatment effect, one could take the estimated treatment group risk as dependent variable, with the inverse variance of the treatment risk as weight. The dashed line in Figure 14.1 represents the line estimated in this way. The line has a slope smaller than one and intersects the line of no-treatment effect. The underlying risk corresponding to the intersection point is the so-called *break-even point*. It defines a cut-off value of the control risk, above which the treatment is beneficial and below which it is not. Note that in this example we have chosen the event proportion as risk measure. Alternatively, we could have chosen the log odds or another transformation of the event proportion.

Deriving conclusions about the treatment effect on the basis of a WLS regression in this way has been largely criticized, because of two major drawbacks related to the approach.

(1) If the treatment effect is chosen as the dependent variable, the WLS regression carried out in this way is inappropriate as the treatment effect measure naturally includes control risk information. For instance, following the approach originally suggested by Brand and Kragt (1992) (e.g., Senn, 1994; Sharp et al., 1996), suppose that the log odds is taken as the risk measure and the log odds ratio as the treatment effect measure. In this case, in fact, there is a negative correlation between the estimation error in $\hat{\eta}_i - \hat{\xi}_i$ and the estimation error in the log odds in the control group $\hat{\xi}_i$, which leads to a negative bias in the estimated slope of the relationship. Since the estimation errors of $\hat{\eta}_i$ and $\hat{\xi}_i$ are uncorrelated, it would be more appropriate to calculate the WLS regression of $\hat{\eta}_i$ on $\hat{\xi}_i$ (with weights given by the inverse variance of $\hat{\eta}_i$) and rewrite the result to the regression line of $\hat{\eta}_i - \hat{\xi}_i$ on $\hat{\xi}_i$.

Alternatively, one can formulate the problem in terms of uncorrelated quantities such as regression of the risk difference on the average of the treatment risk and the control risk, and then rewrite the estimated regression equation as an equation relating the treatment effect to the control risk.

(2) Even if the WLS is done using $\hat{\eta}_i$ as dependent and $\hat{\xi}_i$ as independent variable, the analysis still does not take into account that the control risks are estimated rather than being true values, as they are based on a sample. If this measurement error is not properly incorporated in the analysis, it leads to bias in the regression coefficient of the control risk. Accordingly, the observed control risk measure $\hat{\xi}_i$ cannot be considered a fixed covariate. The next section discusses this point more extensively.

In this section, we have shown what the fundamental difficulty is in control risk regression and what makes it different from ordinary meta-regression. We are interested in the relation between the true η_i (or η_i–ξ_i) with the true ξ_i, while these quantities are both not directly observed but measured with error. Therefore, we need methods that properly account for the measurement errors, and the remainder of this chapter is devoted to methods which do that in a variety of ways.

14.2 Correlated Measurement Errors

Measurement error is a widespread problem in many areas of research, as inaccurate measures result from different sources, for example, laboratory instruments, self-reported data, and summary information. A huge literature discusses the impact of measurement errors on inferences, as Carroll et al. (2006) and Buonaccorsi (2010) illustrate in detail. The most common effect of ignoring the presence of measurement errors affecting covariates in linear models is the *attenuation* phenomenon, sometimes called *regression dilution bias*. The measurement error biases the slope of the weighted least squares regression line toward zero. In control risk regression, this reduces the estimated effect of the control risk and implies more homogeneous treatment effects than really exist.

A hierarchical modeling approach combining model (14.1) or (14.2) with a model for measurement error (e.g., McIntosh, 1996; Schmid et al., 1998; Sharp and Thompson, 2000; van Houwelingen et al., 2002) can properly account for the presence of measurement errors. For model (14.2), these approaches also account for the correlation between the treatment effect and the control risk. The literature distinguishes between the choice of the exact measurement error structure, leading to one-stage meta-analysis, and the choice of an approximate measurement error structure, leading to two-stage meta-analysis.

Exact measurement error model. The specification of the exact measurement error structure is strictly related to the available information from each study included in the meta-analysis, see Arends et al. (2000) for an illustration. Consider, for example, when the information from study i comprises the observed number of events y_i and x_i in the treatment and control groups, respectively, and the number of treated and controls, n_{i1} and n_{i0}, respectively. Let η_i and ξ_i denote the true unobserved log odds in the treatment group and in the control groups, respectively. Thus, $\hat{\eta}_i = \log \frac{y_i}{n_{i1} - y_i}$ and $\hat{\xi}_i = \log \frac{x_i}{n_{i0} - x_i}$. The exact measurement error model relating $(\hat{\eta}_i, \hat{\xi}_i)$ to (η_i, ξ_i) is implicitly given by considering the binomial distribution for the variables Y_i and X_i (Arends et al., 2000; Schmid et al., 2004)

$$Y_i \sim \text{Binomial}\left(n_{i1}, \frac{e^{\eta_i}}{1 + e^{\eta_i}} \right), \tag{14.3}$$

$$X_i \sim \text{Binomial}\left(n_{i0}, \frac{e^{\xi_i}}{1 + e^{\xi_i}} \right). \tag{14.4}$$

Often, the measurement error model specification is completed by assuming a normal distribution for the underlying risks ξ_i (in addition, of course, to the regression relationship between η_i and ξ_i which incorporates the measurement error in η_i). The resulting model, though with another parametrization, is also discussed in other chapters, for example, Chapter 5 (Section 14.3) and Chapter 19 (Section 14.2.3).

As a second example, assume that, together with the observed number of events y_i and x_i, each study i provides information about the number of person-years in the treatment and control groups, T_{i1} and T_{i0}, respectively. Let η_i and ξ_i denote the true unobserved log event rates in the treatment and control groups, respectively. Thus, $\hat{\eta}_i = \log \frac{y_i}{T_{i1}}$ and

$\hat{\xi}_i = \log\frac{x_i}{T_{i0}}$. Then, the exact measurement error model relating $(\hat{\eta}_i, \hat{\xi}_i)$ to (η_i, ξ_i) is implicitly given by using the Poisson distribution for the variables Y_i and X_i (Arends et al., 2000; Chapter 5 (Section 14.4))

$$Y_i \sim \text{Poisson}\left(T_{i1}e^{\eta_i}\right), \tag{14.5}$$

$$X_i \sim \text{Poisson}\left(T_{i0}e^{\xi_i}\right). \tag{14.6}$$

Approximate measurement error model. A common measurement error structure specifies the relationship between the observed $\hat{\eta}_i$ and $\hat{\xi}_i$ and the corresponding error-free variables η_i and ξ_i through a bivariate normal distribution,

$$\begin{pmatrix} \hat{\eta}_i \\ \hat{\xi}_i \end{pmatrix} \sim N\left(\begin{pmatrix} \eta_i \\ \xi_i \end{pmatrix}, \Gamma_i\right), \text{ where } \Gamma_i = \begin{pmatrix} s^2_{\hat{\eta}_i} & 0 \\ 0 & s^2_{\hat{\xi}_i} \end{pmatrix}. \tag{14.7}$$

The relationship is an approximation of the exact model, with the variance/covariance matrix Γ_i based on the asymptotic distributions of the risk estimates. For example, if the log odds is chosen as risk measure, the approximation of the exact binomial measurement error model specified in (14.3) and (14.4) has variance/covariance matrix

$$\Gamma_i = \begin{pmatrix} y_i^{-1} + (n_{i1} - y_i)^{-1} & 0 \\ 0 & x_i^{-1} + (n_{0i} - x_i)^{-1} \end{pmatrix}.$$

The approximation of the exact Poisson measurement error model specified in (14.5) and (14.6) for the event rates has variance/covariance matrix

$$\Gamma_i = \begin{pmatrix} y_i^{-1} & 0 \\ 0 & x_i^{-1} \end{pmatrix}.$$

Note that using the treatment risk as the dependent variable leads to zero covariances in Γ_i; using the treatment effect $\hat{\eta}_i^*$ as the dependent variable leads to correlation in the sample quantities.

In the approximate measurement error models, the variance/covariance matrices Γ_i are assumed to be fixed and known. Thus, in the terminology of Chapters 4 and 5, the methods using an approximate measurement error model lead to two-stage methods and the methods using an exact measurement error model result in one-stage methods. Chapter 5 discussed extensively the disadvantages of two-stage methods and these apply here as well. In particular, one disadvantage is that the variance/covariance matrix Γ_i may be correlated with the risk measure estimates $\hat{\eta}_i$ and $\hat{\xi}_i$, which may result in bias.

In this section, we do not extensively cover the continuous outcome case. In that case $\hat{\eta}_i$ and $\hat{\xi}_i$ are the sample means of the treatment and control groups. An exact measurement model would use pseudo IPD as presented in Chapter 5. The approximate measurement error model uses (14.7) with the two squared standard errors of the mean on the main diagonal of Γ_i. In this particular case there is no correlation between Γ_i and $(\hat{\eta}_i, \hat{\xi}_i)$.

McIntosh (1996) provides an interesting evaluation of the measurement error effects in the model relating the treatment risk to the control risk as well as the model relating the treatment effect to the control risk. In particular, the estimate of the slope obtained ignoring the presence of measurement error is biased, with bias being positive or negative. If the measurement errors are not correlated, the bias is always toward zero. The magnitude of the bias depends on the relative size of the within-study variance $s^2_{\hat{\xi}_i}$ of the control risk and the residual variance σ^2. A small bias is obtained with small $s^2_{\hat{\xi}_i}$ relative to σ^2, while the opposite is obtained in case of small σ^2 as the control risks do not provide sufficient variation to fit a regression line. See also Schmid et al. (1998).

Example. To illustrate the effect of measurement error on the weighted least squares regression, consider the following simulation example based on the sclerotherapy data in Section 14.1.1. Data are simulated for 19 studies, with sample size equal to that of the sclerotherapy data. The mortality risk for controls is simulated from a uniform distribution on the interval [0.05,0.3]. Let η_i and ξ_i denote the true log odds for treated and controls, respectively. The relationship between η_i and ξ_i is $\eta_i = 0.17 + 0.43\,\xi_i + \varepsilon_i$, with regression coefficients chosen equal to the values provided by the weighted least squares regression on the estimated mortality risk for the sclerotherapy data. Errors ε_i are assumed to be observations from a normal variable with mean zero and variance $\sigma^2 = 0.02$. The number of events for the treatment and the control groups are simulated from a binomial distribution as in (14.3) and (14.4) and used to compute the summary measures of η_i and ξ_i at

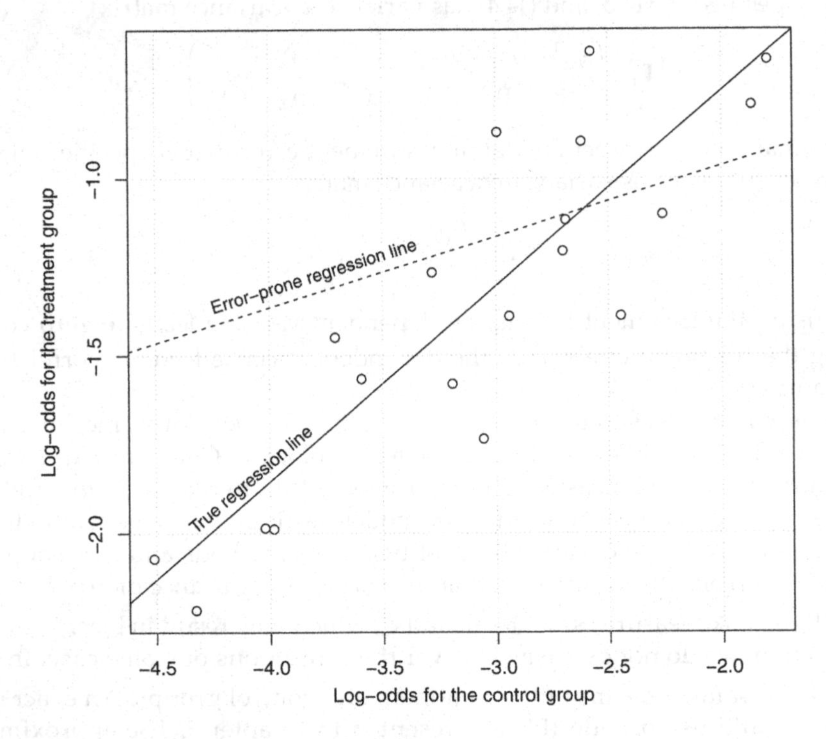

FIGURE 14.2
Simulated example. Scatter plot of the simulated values (η_i, ξ_i), $i = 1, \ldots, 19$, on the log odds scale. The true weighted least squares regression line is superimposed (solid line), together with the weighted least squares regression line (dashed line) based on the error-prone summary measure $(\hat{\eta}_i, \hat{\xi}_i)$, obtained from each study included in the meta-analysis.

the study-level, namely, $\hat{\eta}_i$ and $\hat{\xi}_i$. Figure 14.2 plots the simulated data, together with the weighted least squares regression line based on the true data (solid line). The weighted least squares regression line based on the observed log odds $\left(\hat{\xi}_i, \hat{\eta}_i\right)$ obtained from the single studies is superimposed (dashed line). The measurement error arising from using $\hat{\xi}_i$ as a proxy for ξ_i gives rise to the attenuation phenomenon, with a slope for the regression line using the error-prone measures smaller than the true one.

Similar effects of not properly accounting for measurement error on the estimation of control risk regression are highlighted in empirical studies in Schmid et al. (1998).

14.3 Measurement Error Correction Methods

Given the impact of measurement errors on inferential conclusions, a substantial literature has focused on correction techniques with book-length reviews by Gustafson (2003), Carroll et al. (2006), and Buonaccorsi (2010). The measurement error literature distinguishes between the structural approach and the functional approach to error correction. In the structural case, the unobserved true quantities are interpreted as random variables and the specification of a distribution function for them is a substantial component of the inferential procedures. Conversely, in the functional case, no assumptions are made on the unobserved quantities, which can be thought of as nuisance parameters, which do not need a distributional specification. Following the same convention, the subsequent sections review inferential approaches proposed in the literature to address the error problem affecting the measurements from the treatment and control groups by distinguishing structural and functional methods.

14.3.1 Structural Approaches

Structural methods for measurement error correction entail the specification of a model for both the observed and unobserved variables in the model.

Let $p\left(\eta_i \mid \xi_i; \beta_0, \beta_1, \sigma^2\right)$ denote the density function for model (14.1) relating the treatment risk and the control risk and let $p\left(\hat{\eta}_i, \hat{\xi}_i \mid \eta_i, \xi_i\right)$ denote the density function for the measurement error model relating the estimated effects to the true values. Finally, the structural approach requires the specification of a distribution for the unobserved underlying risk ξ_i. Let $p(\xi_i; \boldsymbol{\delta})$ denote the density function for the model for ξ_i, with parameter vector $\boldsymbol{\delta}$. The likelihood function for the whole parameter vector $\theta = \left(\beta_0, \beta_1, \sigma^2, \boldsymbol{\delta}\right)$ based on the observed $\left(\hat{\eta}_i, \hat{\xi}_i\right)$ is obtained by integrating out the joint model density $p\left(\eta_i, \xi_i, \hat{\eta}_i, \hat{\xi}_i; \theta\right)$ with respect to (η_i, ξ_i),

$$\begin{aligned} L(\theta) &= \prod_{i=1}^{I} \iint p\left(\eta_i, \xi_i, \hat{\eta}_i, \hat{\xi}_i; \theta\right) d\eta_i d\xi_i \\ &= \prod_{i=1}^{I} \iint p\left(\hat{\eta}_i, \hat{\xi}_i \mid \eta_i, \xi_i\right) p\left(\eta_i \mid \xi_i; \beta_0, \beta_1, \sigma^2\right) p\left(\xi_i; \boldsymbol{\delta}\right) d\eta_i d\xi_i. \end{aligned} \tag{14.8}$$

A mathematically and computationally convenient choice is the specification of the approximate normal measurement error model (14.7) and a normal distribution for the underlying risk, $\xi_i \sim N(\mu_\xi, \sigma_\xi^2)$, so that $p(\xi_i;\delta) = p(\xi_i;\mu_\xi,\sigma_\xi^2)$ with $\delta = (\mu_\xi, \sigma_\xi^2)$ (e.g., McIntosh, 1996; van Houwelingen et al., 2002). With such a choice, the resulting likelihood function (14.8) can be re-written in a closed-form as the likelihood function obtained from the bivariate normal marginal density function of $(\hat{\eta}_i, \hat{\xi}_i)$, $p(\hat{\eta}_i, \hat{\xi}_i;\theta)$,

$$L(\theta) = \prod_{i=1}^{n} p(\hat{\eta}_i, \hat{\xi}_i;\theta),$$

with

$$\begin{pmatrix} \hat{\eta}_i \\ \hat{\xi}_i \end{pmatrix} \sim N\left(\begin{pmatrix} \beta_0 + \beta_1\mu_\xi \\ \mu_\xi \end{pmatrix}, \Gamma_i + \begin{pmatrix} \sigma^2 + \beta_1^2\sigma_\xi^2 & \beta_1\sigma_\xi^2 \\ \beta_1\sigma_\xi^2 & \sigma_\xi^2 \end{pmatrix} \right). \qquad (14.9)$$

McIntosh (1996) used the treatment effect as dependent variable according to (14.2). His model is essentially the same, but with another parametrization. When introducing the above model, McIntosh (1996) called the pair of equations given by (14.1) and the underlying risk distribution the *structural model*, in line with the terminology in the measurement error literature, to distinguish it from the (approximate) measurement error model (14.7), thus emphasizing the bivariate two-level hierarchical structure of the model.

The closed-form expression of the likelihood function is lost when the exact measurement error structure is adopted (Arends et al., 2000) or when particular non-normal underlying risk distributions are chosen, but the model can still be fitted through the likelihood method, using numerical integration techniques. Alternatively, Bayesian methods could be applied.

The structural approach based on the normal specification of the distribution for ξ_i balances computational advantages with strong limitations. Since the number of studies in a meta-analysis is typically small, it is often difficult in practice to check normality assumptions. More importantly, the normality assumption for ξ_i cannot account for common forms of non-normality arising in applications, as, for example, bi-modality or skewness. Finally, deviations from a normal distribution at the population level can arise as a consequence of a case-control sampling scheme (Guolo, 2008).

The *risk of misspecification* of the true underlying risk distribution is thus a serious drawback of the structural approach. While misspecification may not have a strong impact on the estimation of the regression coefficients, the estimation of the variance components can be affected and, consequently, inferential results on all the model parameters can be inaccurate (Ghidey et al., 2007; Guolo, 2013). The issue of misspecification of the latent variable distribution in random-effects models is deeply investigated in Verbeke and Lesaffre (1997), Zhang and Davidian (2001), and Lee and Thompson (2008). Several proposals in the literature address the misspecification problem by replacing the normal assumption with more flexible choices.

Arends et al. (2000) suggested using a *mixture of two normal variables* with the aim of describing a wide range of possible distributions for ξ_i, including unimodal, bimodal, symmetric, and skewed distributions. Bayesian inference was then carried out with non-informative prior distributions. Both the approximate measurement error model (14.7) and the exact measurement error model were considered within different scenarios, with no substantial differences in a number of meta-analysis examples.

The semi-parametric model of Ghidey et al. (2007) assumes that the density function of ξ_i is approximated by a *mixture of J normal variables* with unknown mixture weights c_j, $j = 1,\ldots,J$, namely,

$$\xi_i \sim \sum_{j=1}^{J} c_j \mathrm{N}\left(\mu_j, \sigma_j^2\right), \quad c_j = \frac{e^{a_j}}{\sum_{k=1}^{J} e^{a_k}},$$

for some unknown parameter vector $a = (a_1,\ldots,a_K)$. The resulting likelihood function interestingly maintains a closed-form expression. The number of normal densities J is a tuning parameter chosen by imposing a penalty on the resulting log-likelihood function $\ell(\theta)$, namely,

$$\ell(\theta) - \frac{\lambda}{2} \sum_{j=d+1}^{J} \left(\Delta^d a_j\right)^2,$$

with λ a tuning parameter and Δ^d a difference operator of order d, for example, $\Delta^1 = a_j - a_{j-1}$. The order d of the difference operator is suggested to be set to three, see Ghidey et al. (2004). The maximum likelihood estimates of the parameters are obtained by an Expectation–Maximization algorithm for each λ. The optimal λ is determined on the basis of the Akaike information criterion. The penalized log-likelihood can alternatively be interpreted in a Bayesian framework with a suitable penalty.

The choice of a mixture of normal distributions for the unobserved ξ_i adds some flexibility to the structural approach but gives rise to some drawbacks. As pointed out by Lee and Thompson (2008), the choice of the number of mixture components can complicate the analysis. To avoid the mixture complications, they specify the underlying risk distribution via a *skew extension* of the normal and the t distributions, according to the formulation by Fernandez and Steel (1998). Inference is then performed using a Bayesian Markov chain Monte Carlo approach with vague prior distributions. Despite the flexibility of the methodology, some disadvantages are related to characteristics of the Fernandez and Steel family of distributions. In fact, the model is parameterized in terms of the mode, thus making the mean a complex function of the parameters with a difficult interpretation. Moreover, the discontinuity of the even derivatives prevents using standard likelihood inference.

Recently, Guolo (2013) addressed the non-normality of the underlying risk distribution due to skewness by specifying the distribution of ξ_i via the Skew-normal family (Azzalini, 1985). This flexible choice includes the normal distribution as a special case and does not limit the application of the standard likelihood theory. The maximum likelihood estimators of the parameters of interest are obtained via numerical procedures as the likelihood function cannot be expressed in closed-form. The standard errors of the maximum likelihood estimators are provided via the sandwich formula, in order to account for possible misspecification of the true underlying risk distribution.

14.3.2 Functional Approaches

Functional methods for measurement error correction differ from structural approaches by not putting any assumption on the distribution of the unobserved underlying risk and, consequently, $\xi_1,\ldots,\xi_I$ are interpreted as nuisance parameters. Such an approach guarantees a gain in robustness of the inferential conclusions as there is no risk of misspecification of the control risk distribution. A loss of efficiency compared with fully parametric solutions is the price for the added flexibility. Moreover, dealing with an increased number of nuisance components leads to inferential difficulties of its own. Maximizing the likelihood function for the complete vector of parameters including the nuisance parameters, $\left(\beta_0, \beta_1, \sigma^2, \xi_1, \ldots, \xi_I\right)$, leads to *inconsistent estimators* or unreliable inferential results, see, for example, Ghidey et al. (2011). In the spirit of the functional approach, Thompson et al. (1997) take a Bayesian approach by putting non-informative priors on the nuisance parameters. With respect to the nuisance components ξ_i, they assume that they are independently drawn from a normal distribution with fixed mean and large value of the variance. However, this does not help in solving the inconsistency problem (van Houwelingen and Senn, 1999). See also Arends et al. (2000). In fact, this is directly related to the choice between fixed-intercepts modeling and random-intercepts modeling discussed at several places in Chapter 5.

Walter (1997) considers likelihood inference on the vector of parameters of interest (β_0,β_1) under a linear relationship between η_i and ξ_i as in (14.1) and a normal measurement error model for the observed η_i and ξ_i, treating ξ_i as unknown parameters to be estimated. The approach has some shortcomings, as Bernsen et al. (1999) and Sharp and Thompson (2000) highlight. The most relevant is the focus on a common-effect model, with no variability in the treatment effect other than the variation explained by the underlying risk.

Cook and Walter (1997) start from a logistic model to describe the relationship between the treatment effect and the underlying risk across a series of 2×2 tables. In order to account for the sampling variability of the estimated risks, Cook and Walter (1997) develop an unconditional product-binomial likelihood for the main parameters of interest and the control risks as well. Accordingly, the risk of inconsistent estimators remains when the number of studies involved in the meta-analysis increases. Similar to Walter (1997), a main limitation of the approach is the focus on a common-effect modeling strategy.

In order to avoid the inconsistency problem, Ghidey et al. (2011) use the theory of unbiased estimating equations for inference on the parameter vector $(\beta_0,\beta_1,\sigma^2)$, adopting two well-known correction techniques developed in the measurement error literature, namely, the *corrected score approach* and the *conditional score approach* (chapter 7 of Carroll et al., 2006). The approach considers the model relating the treatment and control risks as in (14.1) and the approximate measurement error model as in (14.7). The corrected score approach modifies the estimating equations one would obtain in case of no measurement error in order to guarantee their unbiasedness in the presence of measurement error. The conditional score estimating equations, instead, are obtained by conditioning on the sufficient statistic for the parameters representing the underlying risk. In both cases, the standard error of the estimator is based on the sandwich methodology. The estimators from both approaches are practically identical in cases of small measurement error. Simulation experiments show that for large measurement error and a small number of studies the conditional score approach is more efficient. However, the measurement error literature highlights that the conditional approach can suffer from the presence of multiple roots, not all guaranteed to be consistent (Stefanski and Carroll, 1987; Tsiatis and Davidian, 2001).

Guolo (2014) investigates the use of SIMEX, a simulation-based technique to correct for measurement errors in the control risk estimates, starting from model (14.1) and the approximate measurement error model (14.7). SIMEX was originally developed (Cook and Stefanski, 1994; Stefanski and Cook, 1995) to deal with additive errors, and so it is well-suited to handle the measurement error structure in (14.7). The method uses a two-step procedure: (1) a simulation step involving a resampling-like strategy obtains (naive) estimates of the parameters conditional on increasing measurement errors added to control risk estimate; and (2) the relationship between the estimated parameters and the amount of the extra errors is established and used to extrapolate the corrected estimate back to the no measurement error case. Details about the choice of the amount of error to add, the number of simulations in step (1), and the relationship needed in step (2) have been extensively discussed in the literature (Carroll et al., 2006, chapter 5), but do not seem to seriously affect the inferential conclusions. Simulation experiments show a satisfactory behavior of SIMEX in terms of robustness under different specifications of the true baseline risk distribution. Improvements compared with structural solutions based on the normality assumption for ξ_i are more evident in the estimation of the between-study variance component σ^2. However, the measurement error literature warns against the possibility of a non-positive definite SIMEX estimated variance/covariance matrix (Carroll et al., 2006, section B.4.1), although this is a very infrequent case.

14.4 Extension to Network Meta-Analysis

A recent portion of the literature on meta-analysis is concerned with the impact of covariates on the treatment effects across studies included in a network meta-analysis (see Chapter 10). Imbalance in the distribution of the covariates across treatments, in fact, is a violation of the transitivity assumption that underlies the method. Detailed illustrations of the risks of violating network meta-analysis assumptions in the presence of prognostic covariates can be found in Song et al. (2003), Jansen et al. (2012), and Jansen and Naci (2013). Taking advantage of the results in standard pairwise meta-analysis with underlying risk information, regression-based techniques have been proposed in network meta-analysis to account for (1) differences in underlying risk *across* comparisons as a source of inconsistency and (2) differences *within* comparisons as a source of between-study heterogeneity. Examples include the findings in Jansen (2006), Cooper et al. (2009), Salanti et al. (2009), and Salanti et al. (2010). A recent illustration in Jansen et al. (2012) using directed acyclic graphs shows that adjusting for the underlying risk as a good surrogate for effect modifiers across comparisons can remove bias in estimates.

The proposal by Achana et al. (2013) constitutes an extension of previous meta-regression models to adjust for baseline imbalances in the underlying risk across comparisons. The model is designed to take account of missing data, that is, the inclusion in the network meta-analysis of studies without a placebo or non-active treatment control and thus without underlying risk information. The missing information is treated as an unknown parameter, to be estimated within a Bayesian framework. The model is flexible enough to allow different distributional assumptions for the unknown underlying risk measure, thus addressing a much-debated issue in control risk regression, as described in Section 14.3.

14.5 Software

Despite the development of innovative methods for control risk regression, their diffusion in practice is still limited in favor of simplified solutions. Regarding the frequentist approaches, the only models that can be easily fitted in practice are the structural models with the distribution of the underlying control risk assumed to be normal. If the approximate measurement error model (14.9) is assumed, the model is just the bivariate meta-analysis model introduced by Van Houwelingen et al. (1993). In fact, the model is a linear mixed model which can be fitted in linear mixed model programs such as SAS PROC MIXED or Stata `gllamm`, or with R-packages such as `metafor`. If the exact measurement error model (14.3) and (14.4) is adopted, the model becomes a generalized linear mixed model (GLMM), which can be fitted in the GLMM programs of many statistical packages such as SAS, Stata, SPSS, and R. As far as we know there are no user-friendly generally available programs to fit the other methods. The availability of user-friendly implementations would help to promote the use of these alternative methodologies. Regarding the Bayesian methods, all of them can be implemented in, for instance, WinBUGS, JAGS, or STAN, with the help of code or descriptions given in the original publications, though considerable experience and skills in Bayesian methods are required. At the time of writing, only some meta-analysis approaches are accompanied by code provided by the authors as supporting information or available upon request. Examples include the methods developed in Sharp and Thompson (2000), Arends et al. (2000), Achana et al. (2013), Ghidey et al. (2011), and Guolo (2008, 2013).

14.6 Summary

The inclusion of the control risk as a measure of the underlying risk in meta-analysis represents an effective approach to account for between-study heterogeneity. Correcting for measurement error affecting the treatment risk and the control risk is a well-recognized necessary step for providing reliable inferences. In practice, control risk regression typically makes simple assumptions for the distribution of the model components, which are not guaranteed to hold: a linear relationship between the treatment risk and the control risk measure, a normally distributed measurement error, and a normally distributed underlying risk. The recent literature provides interesting solutions for increasing the robustness against deviations from the assumption of normal underlying risk, either from a frequentist or a Bayesian point of view. Results by Arends et al. (2000) based on real data analyses suggest that deviations from the normality of the measurement error model do not impact inferences, although a deeper investigation under different simulation scenarios might shed more light on the risk of misspecification. The linear specification of the control risk regression has been generally assumed, probably because of the typical small number of studies in a meta-analysis. The possibility of alternative relationships has been suggested, for example, Boissel et al. (2008) and Wang et al. (2009), but the performance of the methods in non-linear cases has not been investigated in detail.

References

Achana FA, Cooper NJ, Dias S, Lu G, Rice SJC, Kendrick D and Sutton AJ, 2013. Extending methods for investigating the relationship between treatment effect and baseline risk from pairwise meta-analysis to network meta-analysis. *Statistics in Medicine* **32**(5): 752–771.

Arends LR, Hoes AW, Lubsen J, Grobbee DE and Stijnen T, 2000. Baseline risk as predictor of treatment benefit: Three clinical meta-re-analyses. *Statistics in Medicine* **19**(24): 3497–3518.

Azzalini A, 1985. A class of distributions which includes the normal ones. *Scandinavian Journal of Statistics* **12**(2): 171–178.

Bernsen RMD, Tasche MJA and Nagelkerke NJD, 1999. Variation in baseline risk as an explanation of heterogeneity in meta-analysis (letter). *Statistics in Medicine* **18**(2): 233–238.

Bland JM and Altman DG, 1986. Statistical method for assessing agreement between two methods of clinical measurement. *The Lancet* **i**(8476): 307–310.

Bland JM and Altman DG, 1999. Measuring agreement in method comparison studies. *Statistical Methods in Medical Research* **8**(2): 135–160.

Boissel J-P, Cucherat M, Nony P, Chabaud S, Gueyffier F, Wright JM, Lièvre M and Leizorovicz A, 2008. New insights on the relation between untreated and treated outcomes for a given therapy effect model is not necessarily linear. *Journal of Clinical Epidemiology* **61**(3): 301–307.

Brand R and Kragt H, 1992. Importance of trends in the interpretation of an overall odds ratio in the meta-analysis of clinical trials. *Statistics in Medicine* **11**(16): 2077–2082.

Buonaccorsi JP, 2010. *Measurement Error: Models, Methods and Applications*. Boca Raton, FL: Chapman & Hall, CRC Press.

Carroll RJ, Ruppert D, Stefanski LA and Crainiceanu C, 2006. *Measurement Error in Nonlinear Models: A Modern Perspective*. Boca Raton, FL: Chapman & Hall, CRC Press.

Chaimani A, 2015. Accounting for baseline differences in meta-analysis. *Evidence-Based Mental Health* **18**(1): 23–26.

Cook JR and Stefanski LA, 1994. Simulation extrapolation estimation in parametric measurement error models. *Journal of the American Statistical Association* **89**(428): 1314–1328.

Cook RJ and Walter SD, 1997. A logistic model for trend in 2×2×K tables with applications to meta-analysis. *Biometrics* **53**(1): 352–357.

Cooper NJ, Sutton AJ, Morris D, Ades AE and Welton NJ, 2009. Addressing between-study heterogeneity and inconsistency in mixed treatment comparisons: Application to stroke prevention treatments in individuals with non-rheumatic atrial fibrillation. *Statistics in Medicine* **28**(14): 1861–1881.

Fernandez C and Steel MFJ, 1998. On Bayesian modelling of fat tails and skewness. *Journal of American Statistical Association* **93**(441): 359–371.

Ghidey W, Lesaffre E and Eilers P, 2004. Smooth random effects distribution in a linear mixed model. *Biometrics* **60**(4): 945–953.

Ghidey W, Lesaffre E and Stijnen T, 2007. Semi-parametric modelling of the distribution of the baseline risk in meta-analysis. *Statistics in Medicine* **26**(30): 5434–5444.

Ghidey W, Stijnen T and van Houwelingen HC, 2011. Modelling the effect of baseline risk in meta-analysis: A review from the perspective of errors-in-variables regression. *Statistical Methods in Medical Research* **22**(3): 307–323.

Guolo A, 2008. A flexible approach to measurement error correction in case-control studies. *Biometrics* **64**(4): 1207–1214.

Guolo A, 2013. Flexibly modeling the baseline risk in meta-analysis. *Statistics in Medicine* **32**(1): 40–50.

Guolo A, 2014. The SIMEX approach to measurement error correction in meta-analysis with baseline risk as covariate. *Statistics in Medicine* **33**(12): 2062–2076.

Gustafson P, 2003. *Measurement Error and Misclassification in Statistics and Epidemiology: Impacts and Bayesian Adjustments*. Chapman & Hall, CRC Press.

Jansen JP, 2006. Self-monitoring of glucose in type 2 diabetes mellitus: A Bayesian meta-analysis of direct and indirect comparisons. *Current Medical Research and Opinion* **22**(4): 671–681.

Jansen JP and Naci H, 2013. Is network meta-analysis as valid as standard pairwise meta-analysis? It all depends on the distribution of effect modifiers. *BMC Medicine* **11**(159).

Jansen JP, Schmid CH and Salanti G, 2012. Directed acyclic graphs can help understand bias in indirect and mixed treatment comparisons. *Journal of Clinical Epidemiology* **65**(7): 798–807.

L'Abbé KA, Detsky, AS and O'Rourke K, 1987. Meta-analysis in clinical research. *Annals of Internal Medicine* **107**(2): 224–233.

Lee JL and Thompson SG, 2008. Flexible parametric models for random-effects distributions. *Statistics in Medicine* **27**(3): 418–434.

McIntosh MW, 1996. The population risk as an explanatory variable in research synthesis of clinical trials. *Statistics in Medicine* **15**(16): 1713–1728.

Pagliaro L, D'Amico G, Sorensen T, Lebrec D, Burroughs AK, Morabito A, Tiné F, Politi F and Traina M, 1992. Prevention of first bleeding in cirrhosis – A meta-analysis of randomized trials of non-surgical treatment. *Annals of Internal Medicine* **117**(1): 59–70.

Salanti G, Dias S, Welton NJ, Ades AE, Golfinopoulos V, Kyrgiou M, Mauri D and Ioannidis JPA, 2010. Evaluating novel agent effects in multiple-treatment meta-regression. *Statistics in Medicine* **29**(23): 2369–2383.

Salanti G, Marinho V and Higgins JPT, 2009. A case study of multiple-treatments meta-analysis demonstrates that covariates should be considered. *Journal of Clinical Epidemiology* **62**(8): 857–864.

Schmid CH, Lau J, McIntosh MW and Cappelleri JC, 1998. An empirical study of the effect of the control rate as a predictor of treatment efficacy in meta-analysis of clinical trials. *Statistics in Medicine* **17**(17): 1923–1942.

Schmid CH, Stark PC, Berlin JA, Landais P and Lau J, 2004. Meta-regression detected associations between heterogeneous treatment effects and study-level, but not patient-level, factors. *Journal of Clinical Epidemiology* **57**(7): 683–697.

Senn S, 1994. Importance of trends in the interpretation of an overall odds ratio in the meta-analysis of clinical trials. Letters to the editor (with author's reply). *Statistics in Medicine* **13**(3): 293–296.

Sharp SJ and Thompson SG, 2000. Analysing the relationship between treatment effect and underlying risk in meta-analysis: Comparison and development of approaches. *Statistics in Medicine* **19**(23): 3251–3274.

Sharp SJ, Thompson SG and Altman DG, 1996. The relation between treatment benefit and underlying risk in meta-analysis. *British Medical Journal* **313**(7059): 735–738.

Song F, Altman DG, Glenny AM and Deeks JJ, 2003. Validity of indirect comparison for estimating efficacy of competing interventions: Empirical evidence from published meta-analyses. *British Medical Journal* **326**(7387): 472.

Stefanski LA and Carroll RJ, 1987. Conditional scores and optimal scores for generalized linear measurement-error models. *Biometrika* **74**(4): 703–716.

Stefanski LA and Cook JR, 1995. Simulation-extrapolation: The measurement error jackknife. *Journal of the American Statistical Association* **90**(432): 1247–1256.

Thompson SG, Smith TC and Sharp SJ, 1997. Investigating underlying risk as a source of heterogeneity in meta-analysis. *Statistics in Medicine* **16**(23): 2741–2758.

Tsiatis AA and Davidian M, 2001. A semiparametric estimator for the proportional hazards model with longitudinal covariates measured with error. *Biometrika* **88**(2): 447–458.

van Houwelingen, HC, Arends, LR and Stijnen, T, 2002. Advanced methods in meta-analysis: Multivariate approach and meta-regression. *Statistics in Medicine* **21**(4): 589–624.

van Houwelingen HC and Senn S, 1999. Investigating underlying risk as a source of heterogeneity in meta-analysis. *Statistics in Medicine* (Letter) **18**(1): 107–113.

Van Houwelingen HC, Zwinderman KH, Stijnen T, 1993. A bivariate approach to meta-analysis. *Statistics in Medicine* **12**(24): 2273–2284.

Verbeke G and Lesaffre E, 1997. The effect of misspecifying the random effects distribution in linear models for longitudinal data. *Computational Statistics and Data Analysis* **23**(4): 541–556.

Walter SD, 1997. Variation in baseline risk as an explanation of heterogeneity in meta-analysis. *Statistics in Medicine* **16**(24): 2883–2900.

Wang H, Boissel J-P and Nony P, 2009. Revisiting the relationship between baseline risk and risk under treatment. *Emerging Themes in Epidemiology* **6**, 1.

Zhang D and Davidian M, 2001. Linear mixed models with flexible distributions of random effects for longitudinal data. *Biometrics* **57**(3), 795–802.

15

Multivariate Meta-Analysis of Survival Proportions

Marta Fiocco

CONTENTS

15.1 Introduction

In this chapter, meta-analysis methods that use aggregated survival data extracted from publications are discussed. Chapters 21 and 22 consider meta-analysis of survival data using individual patient data (IPD). Mostly in practice, meta-analysis of survival data focuses on a single summary, for instance, the hazard ratio or the between group survival difference or ratio at one particular time, for example, five years. The data are then analyzed by standard methods, using either a univariate common-effect or, as preferred by most statisticians, a univariate random-effects model (Chapter 4). These meta-analyses are often hampered by the hazard ratio not being given in all publications, or by heterogeneity in the times for which survival proportions are provided. If more than one endpoint is considered, than these are mostly meta-analyzed separately, for example, the three-year survival differences using publications that provide information on three-year survival proportions. Although simple, this approach is suboptimal. If there are multiple outcomes, such as survival proportions at multiple times, multivariate meta-analysis of all endpoints together is preferred above separate univariate meta-analyses (Chapter 9). One advantage is that all information is used in one analysis, while separate univariate meta-analyses are based on only a part of the data. Thus, a multivariate meta-analysis is more efficient.

Another advantage is that the correlation between the different endpoints is taken into account.

In this chapter, we consider meta-analysis methods that simultaneously try to use all relevant information on survival proportions, including the information from survival curve plots given in a publication. The methods fall into two types. The first type of methods uses, in a multivariate meta-analysis, all survival proportions reported in the publications for which a standard error is also provided. The reported survival proportions are correlated over time, implying that the data have a multivariate nature. In fact, there are two kinds of correlation, within-study correlation and between-studies correlation. The within-study correlations are the correlations between the estimated survival probabilities given the true underlying survival probabilities of a study. The between-studies correlations are the correlations between the true survival probabilities of the studies. In multivariate random-effects models, both the within-study correlations and the between-studies correlations are incorporated. Some of the correlations might be assumed to be zero. Correlations between survival proportions of different groups within a study are zero, since they are calculated using separate sets of participants. However, survival proportions observed within the same group of participants have non-zero correlations and these should be accounted for in the meta-analysis. In multivariate meta-analysis, the within-study correlations are usually assumed to be known, but in practice they may be difficult to obtain, especially from published information. Even if IPD are available, calculation of the within-study correlation may still be non-trivial, perhaps requiring bootstrap methods. A number of articles and also Chapter 9 consider the problem of unavailable within-study correlations. For instance, Berkey et al. (1996) assess how results change for a range of different within-study correlation values, and Nam et al. (2003) perform sensitivity analysis using a range of different prior distributions for the unknown correlations. Based on the standard errors, Dear (1994) presented a method for estimating the within-group correlations between the reported survival probabilities.

Dear (1994) proposed to analyze all reported survival proportions of all groups and all studies together in a common-effect multivariate meta-analysis model. The parameters are estimated by weighted least squares. Arends et al. (2008) proposed a multivariate random-effects model that can be seen as a generalization of Dear's model. The model is applicable on data with an arbitrary number of survival estimates at arbitrary time points, possibly different between studies, and for one or more groups. The method fits in the framework of the linear mixed model.

The second type of methods is based on the reconstruction of a life table per group and study. The life tables are made for a chosen partition of the time axis uniformly across groups and studies. Parmar et al. (1998) and Earle et al. (2000) described several methods for calculating a log hazard ratio and its variance from the reconstructed life tables of two groups in a study. Next, a univariate random-effects meta-analysis on the log hazard ratios along the lines of Chapter 4 is performed. Williamson et al. (2002) extended this approach to incorporate information from the number at risk for each treatment group. Assuming a constant hazard per life table interval, Fiocco et al. (2009b) proposed a multivariate method based on the number of events and the number of person-years per interval. This method employs a Poisson correlated gamma frailty process and is linked to the well-known gamma frailty models for survival data. It can model non-proportional hazard ratios. In fact, it leads to estimation of the entire overall survival curves together with the between-studies heterogeneity in the individual study specific curves.

This chapter is organized as follows. In Section 15.2 univariate meta-analysis of one survival proportion per study is reviewed. In Section 15.3 we discuss multivariate meta-analysis that is directly based on all reported survival proportions and their standard errors. In Section 15.4 we discuss univariate and multivariate meta-analysis methods based on reconstructed life tables. Section 15.5 gives an example and the chapter ends with a conclusion in Section 15.6.

15.2 Univariate Meta-Analysis

In this section, univariate meta-analysis for survival probabilities is reviewed. The data consist of one or more follow-up times per study along with an estimate of the survival probability and its standard error. The general structure of the data is as follows: (1) there are I studies $(i=1,\ldots,I)$ contributing information; and (2) each study i provides a set of survival probabilities $\hat{S}_i(t_k)$ and corresponding standard errors s_{ik} at time point t_k $(k=1,\ldots,K)$. Some or many of the survival probabilities may be missing. The information provided for the meta-analysis can be summarized as in Table 15.1.

The univariate approach combines, for each time point t_k separately, the estimates $\hat{S}_i(t_k)$ and standard errors s_{ik} with the methods described in Chapter 4. Under the common-effect model, where the unknown survival probabilities are the same for all studies, i.e., $S_1(t_k)=S_2(t_k)=\cdots=S_I(t_k)=S(t_k)$, the overall estimated survival probability $\hat{S}(t_k)$ and its estimated standard error are then given as

$$\hat{S}(t_k)=\frac{\sum_{i=1}^{I}\hat{S}_i(t_k)/s_{ik}^2}{\sum_{i=1}^{I}1/s_{ik}^2}$$

$$SE(\hat{S}(t_k))=\frac{1}{\sqrt{\left(\sum_{i=1}^{I}1/s_{ik}^2\right)}}.$$

To allow for heterogeneity, the random-effects model is used, which considers $S_i(t_k)$ $(i=1,\ldots,I)$ to be an independent random sample from a normal population

TABLE 15.1
Data Structure: Estimated Survival Probabilities for I Studies at Times t_k $(k=1,\ldots,K)$

Study	Observed survival proportion (standard error)			
1	$\hat{S}_1(t_1)(s_{11})$	$\hat{S}_1(t_2)(s_{12})$	...	$\hat{S}_1(t_K)(s_{1K})$
2	$\hat{S}_2(t_1)(s_{21})$	$\hat{S}_2(t_2)(s_{22})$	...	$\hat{S}_2(t_K)(s_{2K})$
.	...	...	...	...
i	$\hat{S}_i(t_1)(s_{i1})$	$\hat{S}_i(t_2)(s_{i2})$	...	$\hat{S}_i(t_K)(s_{iK})$
.	...	...	...	...
I	$\hat{S}_I(t_1)(s_{I1})$	$\hat{S}_I(t_2)(s_{I2})$	...	$\hat{S}_I(t_K)(s_{IK})$

$S_i(t_k) \sim N(S(t_k), \tau_k^2)$. The overall estimated survival probability and the corresponding standard error at time point t_k are given by

$$\hat{S}(t_k) = \frac{\sum_{i=1}^{I} \hat{S}_i(t_k)/(\hat{\tau}_k^2 + s_{ik}^2))}{\sum_{i=1}^{I} 1/(\hat{\tau}_k^2 + s_{ik}^2)}$$

$$SE(\hat{S}(t_k)) = \frac{1}{\sqrt{\left(\sum_{i=1}^{I} 1/(\hat{\tau}_k^2 + s_{ik}^2)\right)}}.$$

Chapter 4 discusses several methods to estimate the between-studies variance τ_k^2. If survival proportions are reported for two groups per study, differences between survival proportions, or ratios between proportions, can be analyzed analogously. In practice these types of meta-analysis are often performed on transformed survival probabilities, for instance, using a logit or log(-log) transformation. With this univariate approach, the potential correlations between outcomes are not considered and the monotonicity of the estimated overall survival estimates is not guaranteed. In practice, the follow-up times and the number of follow-up times can be different among studies, leading to many times t_k and many missing values per t_k. An approach that uses all data simultaneously is therefore desirable to improve power and precision. This is the topic of the remainder of this chapter.

15.3 Multivariate Meta-Analysis of Reported Survival Proportions with Standard Errors

In this section, two multivariate models are discussed that are directly based on the observed survival proportions and their standard errors. This is in contrast to the methods of the next section, where first life tables are constructed for a chosen partition of the time axis.

Dear (1994) proposed a general linear model with the survival estimate as dependent variable, and follow-up time, treatment, and study as categorical covariates. The parameters are estimated by weighted least squares, using the inverse squared standard errors as weights. Arends et al. (2008) proposed a flexible random-effects model which is a generalization of Dear's method. Both methods can be applied on data with an arbitrary number of survival estimates for one or more groups at specific time points, possibly different between studies. The methods fit in the framework of the linear model (Dear) or the linear mixed model (Arends). Details of the two methods are given in the following two subsections.

15.3.1 Iterative Generalized Least Squares

Dear (1994) proposed a generalized linear regression model to relate the estimated (non-transformed) survival proportions $\hat{S}_{ijk}$ to covariates through a design matrix $\mathbf{X}_i$, where i, j and k denote study, group and time point respectively. The standard errors s_{ijk} of the $\hat{S}_{ijk}$

are assumed to be given. The covariates setting up the columns in the design matrix typically include the main effects of study, group, and time. The model is as follows:

$$\hat{\mathbf{S}}_i = \mathbf{X}_i \boldsymbol{\beta} + \boldsymbol{\epsilon}_i, \tag{15.1}$$

where $\hat{\mathbf{S}}_i$ is the column vector of estimated survival probabilities $\hat{S}_{ijk}$ for study i; $\boldsymbol{\beta}$ is a vector of regression coefficients to be estimated; $\boldsymbol{\epsilon}_i$ is a vector containing residual terms ϵ_{ijk}, which are assumed to be normally distributed with expectation zero and with covariance matrix $\mathbf{V}_i$ and are assumed to be independent between-studies and treatment groups. The matrix $\mathbf{V}_i$ is block diagonal with blocks corresponding to the treatment groups j ($j = 0,1$),

$$\mathbf{V}_i = \begin{pmatrix} \mathbf{V}_{i1} & \mathbf{0} \\ \mathbf{0} & \mathbf{V}_{i2} \end{pmatrix}. \tag{15.2}$$

The main diagonal contains the squares of the published standard errors s_{ijk} of the survival proportions for each study i, group j, and time point k. The correlations are functions of the true survival probabilities S_{ijk} and need to be estimated. Dear derived the following formula for the correlation between the survival proportions at time points t_k and $t_{k'}$ within a treatment group:

$$\text{corr}(\hat{S}_{ijk}, \hat{S}_{ijk'}) \approx \sqrt{\frac{S_{ijk}(1 - S_{ijk'})}{(1 - S_{ijk})S_{ijk'}}}. \tag{15.3}$$

This formula was derived under the assumption of no censoring. However, on the basis of a little simulation study, Dear argues that the effect on the correlation of a moderate amount of censoring will not be large. An alternative formula, that also holds in the presence of censoring and uses the fact that the increments of the estimated cumulative hazard over different intervals are uncorrelated, is given by

$$\text{corr}(\hat{S}_{ijk}, \hat{S}_{ijk'}) \approx \frac{s_{ijk} S_{ijk'}}{s_{ijk'} S_{ijk}}. \tag{15.4}$$

However, when the standard errors are reported in low precision, formula (15.4) is inaccurate and (15.3) is preferred. A generalized least squares (GLS) model requires the correlation matrix of the response variables to be known. The correlations can be estimated by plugging the reported survival proportions into (15.3) or (15.4). Once the model is fitted, new estimates of the S_{ijk} have become available and these can be used to update the correlations matrix. This leads to an iterative procedure that is continued until the change is smaller than a specified tolerance. It is advantageous to estimate the correlations in this way. First, more precise estimates can be obtained using all the available data. Second, correlation estimates obtained directly from the survival proportions reported in the individual studies will be exactly 1 whenever a study reports equal survival probabilities at consecutive times, which is troublesome since the covariance matrix is not positive definite. Let $\mathbf{s}$ be the column vector of all the reported standard errors s_{ijk} for the survival probabilities, and $\mathbf{C}^{(m)}$ the correlation matrix of all S_{ijk} at iteration m, respectively. The covariance matrix $\mathbf{V}^{(m)}$ is then obtained as

$$\mathbf{V}^{(m)} = \mathbf{s}'\mathbf{C}^{(m)}\mathbf{s}.$$

The estimated regression coefficients $\hat{\beta}^{(m)}$ at iteration m are then calculated by

$$(\mathbf{X}'\mathbf{V}^{(m)-1}\mathbf{X})^{-1}\mathbf{X}'\mathbf{V}^{(m)-1}\mathbf{S}^{(m)} \quad (15.5)$$

with variance $(X'\mathbf{V}^{(m)-1}\mathbf{X})^{-1}$, and these provide estimated survival probabilities $\hat{\mathbf{S}}^{(m)} = X\hat{\beta}^{(m)}$. Here X denotes the total design matrix combining all $\mathbf{X}_i$. An updated correlation matrix $\mathbf{C}^{(m+1)} = \mathbf{C}(\hat{\mathbf{S}}^{(m)})$ to be used in the linear model in the next iteration is calculated as a function of these probabilities. See Dear (1994) for more technical details.

Dear's method is very flexible. For instance, it is possible to combine studies which provide data on treatment $j=0$ for two years and on treatment $j=1$ for three years, implying that missing values do not require special handling. Though Dear handled the factor time as categorical, it can also be modeled as continuous, for instance, using splines. Study level or group level covariates can be incorporated in the model by extending the design matrix with extra columns. A shortcoming is that it is a common-effects model. To account for heterogeneity between studies, a dummy variable for each study is introduced. This implies that there may be many parameters to be estimated compared with the number of data points.

15.3.2 Multivariate Random-Effects Model

The approach of Arends et al. (2008) is a straightforward extension of Dear's method. The model relates transformed survival estimates to both common and random covariates by assuming

$$\ln(-\ln(\hat{S}_i)) = X_i\beta + Z_ib_i + \epsilon_i, \quad (15.6)$$

where $\hat{S}_i$, X_i, β and ϵ are defined as in Dear's model (equation [15.1]). The term Z_ib_i is used to model the heterogeneity in the treatment effect(s) and possibly the heterogeneity between studies. The term $b_i \sim N(0, T_i)$ is a vector of random coefficients with expectation zero and between-studies covariance matrix T_i, and is independent of ϵ_i. Usually, the assumed model for T_i is unstructured, meaning that all variances and covariances are freely estimated. The term $\mathbf{Z}_i$ stands for the design matrix of the random effects. It is used to model the between-studies heterogeneity in treatment effect (or effects if the treatment effect is allowed to be time dependent). It can also be used to model between-studies heterogeneity in baseline survival, as an alternative to doing that through fixed parameters. The covariance matrix V_i of ϵ_i is block diagonal with blocks V_{ij} corresponding to groups and is again given by (15.2). Element (k,k') of the within-group covariance matrix V_{ij} is given by

$$\frac{s_{ijk}}{\hat{S}_{ijk}\ln(\hat{S}_{ijk})}\sqrt{\frac{S_{ijk}(1-S_{ijk'})}{(1-S_{ijk})S_{ijk'}}}\frac{s_{ijk'}}{\hat{S}_{ijk'}\ln(\hat{S}_{ijk'})}.$$

Note that the inner term is the correlation between S_{ijk} and $S_{ijk'}$ as before (equation [15.3]). The outer terms are the standard errors of S_{ijk} and $S_{ijk'}$ on the log(-log) scale. The inner term might also be replaced by (15.4).

Model (15.6) is an extension of (15.1) where the random part Z_ib_i has been included. Also note that the observed survival proportions are transformed, which has two advantages. First the estimated survival probabilities are forced to lie between 0 and 1, which is not guaranteed in Dear's approach. Second, the within-study distributions of the transformed proportions are better approximated by a normal distribution, in particular, for survival proportions close to zero or one. Of course, any other transformation than the log(-log) might be used, for instance the logit. However, the log(-log) transformation can lead to a nice interpretation of the treatment effect parameter as discussed at the end of this section.

Analogous to Dear's approach, the correlations between the log(-log) transformed survival proportions are estimated iteratively. The only difference with Dear's fitting algorithm is that at each iteration a linear mixed model program is used instead of an ordinary weighted linear regression program.

As for Dear's method, a disadvantage of the method of Arends et al. (2008) is that the estimated survival proportions are not necessarily non-increasing in time. Another disadvantage that only survival proportions are used for which a standard error is also given in the publication. Usually there are more survival proportions given, for example, through Kaplan–Meier plots, but that information is not used. The methods in the next section do use that information, but the price to pay for that are extra assumptions on the censoring distribution. An advantage is that the model can be fitted using the linear mixed model procedures available in standard statistical packages. Another advantage is that the model is very flexible, since it inherits all properties of the linear mixed model. There are many choices possible for the design matrices $\boldsymbol{X}_i$ and $\boldsymbol{Z}_i$. For instance, the model can easily be changed into a meta-regression model by adding group or study level covariates. The factor time can be modeled as categorical, which is a straightforward choice when there are not too many different time points in all studies together. However, time can also be modeled continuously, using linear and quadratic time terms, or splines, or fractional polynomials. As an example, Arends et al. (2008) fit the following model on meta-analysis data with two groups per study:

$$\log(-\log(\hat{S}_{ijk})) = \beta_0 + \beta_1 g_{ijk} + \beta_2 \log(t_{ijk}) + b_0 + b_1 g_{ijk} + b_2 \log(t_{ijk}) + \epsilon_{ijk}.$$

Here g is a dummy variable for the group ($j=0,1$) and t is the time. There are three random effects: b_0 models the heterogeneity between studies, b_1 models the between-studies heterogeneity in the treatment effect, and b_2 models the between-studies heterogeneity in the slopes of logarithmic time. Note that this model assumes a linear relation between the log(-log) survival function and logarithmic time, which is a defining property for a Weibull distribution. Thus, this model is a parametric survival model assuming Weibull distributions. This implies that the overall treatment effect β_2 has a nice log hazard ratio interpretation. See Arends et al. (2008) for more details.

15.4 Methods Based on Reconstructed Life Tables

In the previous section, methods were discussed which only use the explicitly reported survival proportions for which a standard error was also given. In this section, we discuss methods that are based on reconstructed life tables per study and group, according to a chosen partition of the time axis uniformly across all groups and studies. All relevant information on survival given in the publication is used, including figures of survival curves. The first subsection describes the method to reconstruct the life tables.

15.4.1 Reconstructing Numbers at Risk, Number of Censored, and Numbers of Events

In this section, we follow the line of Fiocco et al. (2009a). Essentially the same method was presented earlier by Parmar et al. (1998). The time axis is split into K disjoint intervals $I_1 = [0,t_1),\ldots,I_K = [t_{K-1},t_K)$ such that the event probability within each interval is relatively small. We assume that survival estimates are available for all t_k, for instance, from Kaplan–Meier plots. We want to calculate estimates for the number of events and the number of participants at risk (or the number of person-years) for all intervals. Often, the numbers at risk are given below the Kaplan–Meier graph for a number of time points. If this is the case, the method presented below can and should be adapted to take advantage of that.

We first allow for administrative censoring, i.e., termination of study data collection at a specific date, assuming that participants enter the study over a period. Let $\max_{\text{FUP}}$ be the maximum follow-up, i.e., the time from the date that the first participant entered the study until the date of termination of the study, and let $\min_{\text{FUP}}$ denote the minimal follow-up defined as the time from the date of inclusion of the last participant to the date of termination of the study. If this information is not present, these quantities may be estimated by looking at dates of accrual (if given) and at the date of submission, or perhaps publication of the manuscript. To compute the number of participants alive and at risk for each time point, a model for the administrative censoring distribution based on the minimum and the maximum follow-up is assumed. The complement of the cumulative censoring distribution function is assumed to be

$$C(t) = \begin{cases} 1 & \text{if } t \le \min_{\text{FUP}} \\ 1 - \dfrac{t - \min_{\text{FUP}}}{\max_{\text{FUP}} - \min_{\text{FUP}}} & \text{if } \min_{\text{FUP}} < t < \max_{\text{FUP}} \\ 0 & \text{if } t \ge \max_{\text{FUP}}. \end{cases} \tag{15.7}$$

The so-called completeness function $C(t)$ (van Houwelingen et al., 2005) expresses the proportion of participants at time t with at least t time units of (potential) follow-up. We define $C_k = C(t_k)$, $S_k = S(t_k)$, and $\tilde{r}_k$ as the completeness, estimated survival, and the number of participants alive and at risk at time t_k, respectively. Let n denote the sample size and assume that the censoring is non-informative. Then the number at risk at time t_k, the number of events in time interval I_k, and the number of censorings are estimated, respectively, as

$$\tilde{r}_k = nS_kC_k, \tag{15.8}$$

$$d_k = n(S_{k-1} - S_k)\frac{C_{k-1} + C_k}{2}, \tag{15.9}$$

and

$$c_k = n(C_{k-1} - C_k)\frac{S_{k-1} + S_k}{2}. \tag{15.10}$$

This method does not yet take censoring due to drop-out into account, but in order to do so the method can be refined. Using the information on the drop-outs given in the publication, an estimate of the complement of the cumulative distribution function $C_{do}(t)$ of censoring due to drop-out has to be made. For instance, usually the number of drop-outs n_{do} is given, and the total number of person-years T is given or can be estimated. Assuming a constant hazard for dropping-out, $C_{do}(t)$ is estimated by $\exp(-n_{do}t/T)$. Then $C(t)$ should be replaced by $C(t)C_{do}(t)$ in the above formulae.

To estimate the number at risk and deaths in each interval, Parmar et al. (1998) distinguish between the number of participants at risk at the start and during a time interval. Especially when using a coarse set of time intervals their method is to be preferred. Obviously, in cases where the numbers of participants at risk and/or the number of events are given in the manuscript, it is not necessary to estimate them.

Throughout the remainder of this section, we assume a piecewise constant hazard model, i.e., we assume a constant hazard per interval I_k. The hazard on I_k can then be estimated as d_k/r_k, with r_k the number of person-years over I_k estimated as $r_k = \Delta_k(\tilde{r}_k - c_k/2)$, with $\Delta_k = t_k - t_{k-1}$ the length of I_k. The squared standard error of the log estimated hazard is $1/d_k$. Different kinds of meta-analysis models, univariate or multivariate, with common effect or random effects, can then work with the estimated log hazard and its standard error in two-stage modeling, or with number of deaths and number of person-years in one-stage modeling.

15.4.2 Pooling Survival Curves under Homogeneity

Consider a meta-analysis of I trials reporting survival data for one or more groups j. Suppose that it was possible by the methodology of the previous subsection to calculate for each group j and each study i the number of person-years r_{ijk} and the number of deaths d_{ijk} in interval I_k. In this subsection we assume that for each group the survival curves are equal across studies.

Then estimation of an overall survival function per group is very straightforward. For each interval I_k and group j, compute the aggregate number of person-years and number of deaths as

$$r_{jk} = \sum_{i=1}^{I} r_{ijk}, \quad d_{jk} = \sum_{i=1}^{I} d_{ijk}.$$

The estimated hazard on I_k and the cumulative hazard function up to and including I_k are given respectively by

$$\hat{\lambda}_{jk} = \frac{d_{jk}}{r_{jk}}, \quad \hat{\Lambda}_{jk} = \sum_{l \le k} \hat{\lambda}_{jl}\Delta_l.$$

The overall survival function easily follows from $\hat{S}(t_{jk}) = -\ln(\hat{\Lambda}_{jk})$. To estimate the standard error of the overall survival at each time point Greenwood's formula can be applied (chapter 4 of Klein and Moeschberger, 2005). For times t within intervals, say $t_{j,k-1} < t < t_{jk}$, the survival function is estimated by $\hat{S}(t_{j,k-1})\exp(-\hat{\lambda}_{jk}(t - t_{k-1}))$.

15.4.3 Univariate Meta-Analysis of Hazard Ratios

To compare the survival of two groups (j=0,1), Parmar et al. (1998) proposed to calculate first per study a hazard ratio from the reconstructed life tables assuming proportional hazards, and next to combine these in a univariate meta-analysis using the methods of Chapter 4. The log hazard at I_k for group j of study i is estimated as $\log(d_{ijk}/r_{ijk})$ with standard error $1/d_{ijk}$. Thus, the log hazard ratio for time interval I_k and study i and its variance are estimated respectively as

$$\log(HR_{ik}) = \log\left(\frac{d_{i0k}/r_{i0k}}{d_{i1k}/r_{i1k}}\right),$$

$$\mathrm{var}(\log(HR_{ik})) = \frac{1}{d_{i1k}} + \frac{1}{d_{i0k}}.$$

Since the hazard ratios HR_{ik} for $k=1,\ldots,K$ are uncorrelated, they can be combined over intervals under the assumption of a common hazard ratio across intervals as in a univariate common-effect meta-analysis (Section 4.3.1). Then the log (HR_i) for study i along with its variance can be obtained respectively as

$$\log(HR_i) = \frac{\sum_{k=1}^{K} \frac{\log(HR_i)}{\mathrm{var}(\log(HR_{ik}))}}{\sum_{k=1}^{K} \frac{1}{\mathrm{var}(\log(HR_{ik}))}}, \tag{15.11}$$

$$\mathrm{var}(\log(HR_i)) = \left[\sum_{k=1}^{K} \frac{1}{\mathrm{var}(\log(HR_{ik}))}\right]^{-1}. \tag{15.12}$$

The study specific log hazard ratios can be pooled using one of the common-effect or random-effects meta-analysis methods of Chapter 4.

15.4.4 Multivariate Meta-Analysis Based on Poisson Correlated Gamma Frailty Processes

We write the study, group, and interval specific hazards as $\lambda_{ijk} = Z_{ijk} \cdot \lambda_{jk}$, where λ_{ijk} is the overall hazard and Z_{ijk} a random frailty variable for group j at time t_k. The random frailty aims to describe the heterogeneity between studies. To model the heterogeneity and correlations between studies, Fiocco et al. (2009a, 2009b) proposed a bivariate gamma frailty process $\mathbf{Z}_i = (\mathbf{Z}_{i0}, \mathbf{Z}_{i1}) = (Z_{i01},\ldots,Z_{i0K}; Z_{i11},\ldots,Z_{i1K})$ with marginal distributions $\Gamma(\theta_j,\theta_j)$ with mean 1 and variance $\xi_j = \theta_j^{-1}$. The model for the correlations between the Z_{ijk} contains two parameters, ρ_{time} and ρ_{group}. More specifically it is assumed that

$$\mathrm{corr}(Z_{ijk}, Z_{ij'k'}) = \rho_{\text{group}}^{|j-j'|}\, \rho_{\text{time}}^{|k-k'|}. \tag{15.13}$$

It is seen that within a group a first-order autoregressive correlation structure with parameter ρ_{time} is assumed, and that ρ_{group} is the correlation between the frailties of the two different groups at the same time k. For technical details concerning the construction of the frailty process see Fiocco et al. (2009b).

Assuming a constant hazard per interval, we can assume that conditional on the unobserved frailties the event counts are independent Poisson random variables:

$$D_{ijk} \mid Z_{ijk} \sim Po(\lambda_{jk} Z_{ijk} r_{ijk}). \tag{15.14}$$

In fact, the distribution of D_{ijk} is not really Poisson, but its likelihood is a Poisson likelihood, as pointed out in Section 5.4.1 of Chapter 5. If the overall average hazards λ_{jk} are not subject to any constraints other than $\lambda_{jk} > 0$, the model allows a time-varying hazard ratio

$$\frac{\lambda_{1k}}{\lambda_{0k}} = HR_k, \quad k = 1,\ldots,K, \tag{15.15}$$

that varies freely with k. By constraining λ_{jk} alternative meaningful models can be specified. The most important of these models is obtained by specifying a constant hazard ratio at each time point

$$\frac{\lambda_{1k}}{\lambda_{0k}} = HR. \tag{15.16}$$

The resulting proportional hazards model is similar to other proportional hazards models that have been proposed in the context of IPD meta-analysis for survival curves.

To estimate the meta-analytic population parameters in the Poisson correlated gamma frailty model, a two-stage estimation procedure based on the composite likelihood approach can be used (Lindsay, 1988). First, the vector of hazards λ_{jk} and the variances ξ_j are simultaneously estimated from the marginal distributions of the event counts obtained by integrating out the frailties. In the second stage, these estimated values are plugged into the composite likelihood based on pairs of observations at time k and l within the same treatment groups and into the pairwise composite likelihood based on the joint distribution at time k between groups, for estimating respectively the correlation in time (ρ_{time}) and between groups (ρ_{group}). For details we refer to Fiocco et al. (2009b). The approach based on the correlated gamma frailty allows for heterogeneity in baseline risk, which in the literature has only been proposed for individual patient data (Tudur Smith et al., 2005) but not for aggregate data. R code to fit the Poisson correlated gamma frailty model and to perform parametric bootstrap to estimate the standard errors was developed by the author of this chapter and is available on request.

Under the assumption of proportional hazards, the Poisson correlated gamma frailty model will likely result in an estimated overall hazard ratio very similar to that of the method of Parmar presented in Section 15.4.3. However, there is an obvious difference in the random structure. Parmar's model assumes that the deviation of the study specific log hazard ratio from the overall log hazard ratio is constant over time and follows a normal distribution. In the Poisson correlated gamma frailty model, the random effects may vary over time. The log hazard ratio of study i at time k including random effects equals

$$\log\left(\frac{\lambda_{1k}Z_{i1k}}{\lambda_{0k}Z_{i0k}}\right) = \log(HR) + \log\left(\frac{Z_{i1k}}{Z_{i0k}}\right). \tag{15.17}$$

Compared with Parmar's model, the normal random error term is replaced by the log of a ratio of two correlated mean 1 gamma variables with possibly different variances. The variance of the error terms in (15.17) is constant over time and increases with increasing variances of $\log Z_{ijk}$ and with decreasing covariance of $\log Z_{ijk}$ ($j=0,1$). For more details about the random structure, see Fiocco et al. (2009b).

The methods discussed in this section, compared with those of the previous section, are elegant, but come with extra assumptions on the censoring distribution needed in reconstructing the life tables. The methods of this section are based on the hazards instead of the survival proportions, which avoids estimation of the within-study correlation. Moreover, this guarantees that the estimated survival proportions or curves are non-increasing in time.

15.5 Data Illustration

The Poisson correlated gamma frailty model described in Section 15.4.4 is applied to a meta-analysis of published survival curves on high grade osteosarcoma patients. The aim of the studies was to assess the efficacy of different chemotherapy regimens: two versus three drugs (Anninga et al., 2011). For each study, the estimates of the survival probabilities at each of a predetermined set of time points are read off from Kaplan–Meier curves. The information on length of follow-up and number of patients are extracted from each manuscript. Allowing heterogeneity between studies, our aim is to obtain an overall survival curve for each chemotherapy regimen. The survival curves for each study are illustrated in Figure 15.1. The set of time points (0.5, 1, 1.5, 2, 2.5, 3, 3.5, 4, 4.5, 5 years) were suggested by clinicians on the basis of survival curves and the follow-up. Table 15.2 shows the number

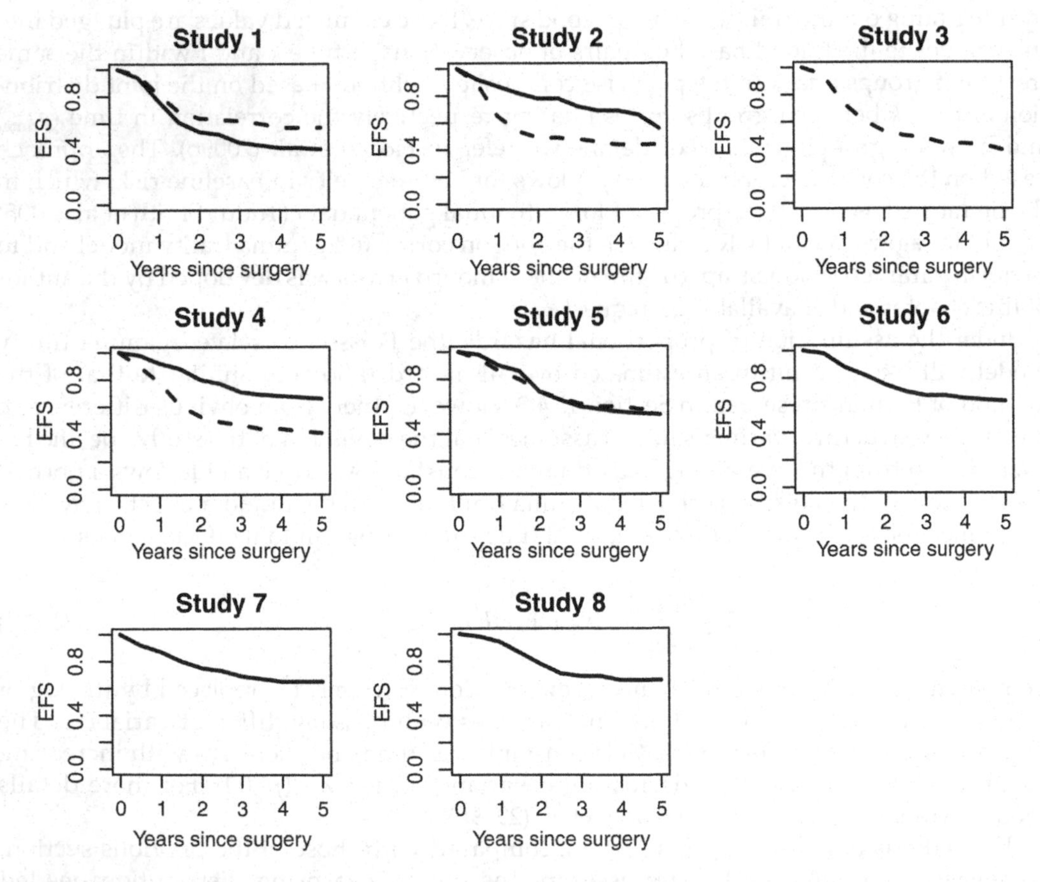

FIGURE 15.1
Survival at a predetermined set of time points for all studies and both groups. Comparison made: two drugs regimens (dotted line) versus three drugs (solid line).

TABLE 15.2

Accrual Information (Sample Size n, $\min_{FUP}$ and $\max_{FUP}$ in Months) and Survival Proportions for Each Group at Time t: 0.5, 1, 1.5, 2, 2.5, 3, 3.5, 4, 4.5, and 5 Years

Study	n	$\min_{FUP}$	$\max_{FUP}$	$\hat{S}_1$	$\hat{S}_2$	$\hat{S}_3$	$\hat{S}_4$	$\hat{S}_5$	$\hat{S}_6$	$\hat{S}_7$	$\hat{S}_8$	$\hat{S}_9$	$\hat{S}_{10}$
1	89	36	55	0.96	0.82	0.69	0.64	0.63	0.62	0.60	0.57	0.57	0.57
2	199	53	79	0.89	0.65	0.54	0.49	0.49	0.48	0.46	0.44	0.44	0.44
3	250	9	103	0.95	0.74	0.61	0.53	0.50	0.48	0.44	0.44	0.43	0.39
4	254	9	103	0.90	0.78	0.61	0.53	0.49	0.47	0.44	0.44	0.42	0.41
5	116	36	120	0.95	0.86	0.80	0.71	0.66	0.62	0.60	0.60	0.58	0.58
1	90	36	55	0.94	0.77	0.61	0.54	0.52	0.46	0.44	0.42	0.42	0.41
2	48	24	54	0.92	0.87	0.82	0.78	0.78	0.72	0.69	0.69	0.66	0.56
3	66	18	52	1.00	0.89	0.75	0.68	0.68	0.68	0.68	0.68	0.64	0.55
4	164	120	156	0.98	0.92	0.88	0.80	0.77	0.71	0.69	0.68	0.67	0.66
5	95	72	106	0.98	0.90	0.86	0.71	0.65	0.60	0.58	0.56	0.55	0.54
6	113	42	124	0.98	0.88	0.82	0.75	0.70	0.67	0.67	0.65	0.64	0.63
7	172	26	86	0.92	0.87	0.80	0.75	0.73	0.69	0.67	0.65	0.65	0.65
8	118	36	120	0.98	0.94	0.86	0.78	0.71	0.69	0.69	0.66	0.66	0.66

Upper panel: two drugs regimens; lower panel: three drugs regimens.

of patients at the beginning of each study, the length of follow-up and the estimated survival probabilities at the specific time points for each group and all trials involved in the meta-analysis. By using the techniques described in Section 15.4.1 for each group in each trial and each time interval I_k, $k=1,\ldots,K$, the effective number of patients at risk, the number of deaths, and the number of censored patients during the time interval for the two and three drugs group were reconstructed. Tables 15.3 and 15.4 show the numbers of deaths, the number of patients at risk, and the number of censored for each of I studies and $K=10$ time intervals for the two groups. The number of person-years has then been computed from Tables 15.3 and 15.4.

The Poisson correlated gamma frailty model described in Section 15.4.4 is applied without putting constraints on the overall hazard parameters λ_{jk}. Note that three studies present only one group but are nevertheless included in the meta-analysis. Table 15.5 shows the estimates of the mean hazards (λ_{jk}) for group j ($j=0,1$) at each interval k. The estimated marginal frailty variances (se) $\hat{\xi}_0$ and $\hat{\xi}_1$ are equal to 0.128(0.030) and 0.107(0.046) respectively. They give an indication about the heterogeneity between trials which is in this case not very high. The correlation in time is equal to 0.419(0.232) which implies that the hazard in the earlier intervals is moderately correlated with the hazard in later intervals. Correlation between groups is equal to 0.947(0.034) indicating a strong frailty correlation between groups. To estimate the standard errors parametric, bootstrap techniques are employed. For details see Fiocco et al. (2009b).

The estimation of the overall event free survival function for each group j is obtained by using the estimated $\hat{\lambda}_{jk}$'s as parameters of the piecewise exponential distribution. The overall estimated survival functions $\hat{S}_j(t)$ for each group j are plotted in Figure 15.2 along with their 95% pointwise confidence intervals.

TABLE 15.3

Effective Number of Patients at Risk, Deaths, and Censored, Estimated for Each of the Ten Intervals for Two Drug Regimen

	Time interval									
	0–0.5	**0.5–1**	**1–1.5**	**1.5–2**	**2–2.5**	**2.5–3**	**3–3.5**	**3.5–4**	**4–4.5**	**4.5–5**
Study	N. of patients at risk									
1	84.14	75.65	60.11	46.48	38.94	33.73	27.95	20.81	11.30	2.59
2	191.44	162.59	113.42	89.60	76.86	72.16	65.75	57.91	50.03	44.02
3	242.72	223.10	168.05	133.64	111.70	101.05	92.67	80.77	76.36	70.06
4	246.60	214.36	179.66	135.54	113.29	100.44	92.02	81.91	77.45	69.40
5	113.10	104.48	91.95	83.02	71.38	64.14	58.10	54.06	51.81	47.81
	N. of deaths									
1	3.56	11.03	9.53	3.37	0.61	0.54	0.90	1.04	0.00	0.00
2	21.89	43.84	19.20	8.30	0.00	1.47	2.74	2.52	0.00	0.00
3	12.50	49.32	29.52	17.53	6.32	4.04	7.72	0.00	1.74	0.52
4	25.40	28.58	39.16	17.78	8.55	4.10	5.87	0.00	3.52	1.65
5	5.80	9.90	6.42	9.34	5.03	3.89	1.87	0.00	1.73	0.00
	N. of censored									
1	4.85	4.93	4.51	4.10	4.17	4.60	5.24	6.24	8.47	3.89
2	7.56	6.97	5.32	4.63	4.43	4.71	4.93	5.11	5.36	4.00
3	7.28	7.12	5.73	4.89	4.41	4.33	4.34	4.18	4.41	3.57
4	7.40	6.84	6.12	4.96	4.47	4.30	4.31	4.24	4.47	3.53
5	2.90	2.82	2.63	2.52	2.30	2.21	2.15	2.16	2.25	1.28

15.6 Discussion

In this chapter, the need for complex models that can be used to perform meta-analysis of survival proportions reported at multiple times in published studies was discussed. A simple approach to the analysis of such data considers each outcome measure separately. An approach that uses all data simultaneously may improve power and precision and helps in providing unified conclusions. These multivariate methods may also recover more information from the available data.

We discussed two types of methods. The first type only uses reported survival proportions for which a standard error is given. Two generalized linear models for this type of data were discussed, one with fixed effects only, and the other one with additional random effects. The second type of method uses all information on survival proportions that is provided in the publications, including Kaplan–Meier graphs, to reconstruct life tables for a chosen partition of the time. Two hazard-based methods which use these reconstructed life tables were discussed. The first method assumes proportional hazards, and, in a two-stage fashion, calculates for each study a hazard ratio and standard error, which are then analyzed with one of the methods discussed in Chapter 4. The second method is a one-stage multivariate method, which does not assume proportional hazards and is based on a Poisson correlated gamma frailty process.

All methods considered in this chapter can be extended to a meta-regression analysis where study level or group level covariates are incorporated in the analysis. This aspect has been investigated in Fiocco et al. (2012). The Poisson correlated gamma frailty model

TABLE 15.4

Effective Number of Patients at Risk, Deaths, and Censored Estimated for Each of the Ten Intervals for Three Drug Regimen

	Time interval (years)									
	0–0.5	0.5–1	1–1.5	1.5–2	2–2.5	2.5–3	3–3.5	3.5–4	4–4.5	4.5–5
Study					N. of patients at risk					
1	85.09	74.81	57.01	41.50	31.95	29.20	20.95	15.41	8.41	6.81
2	45.33	38.90	34.16	29.51	25.27	22.11	17.01	12.22	6.11	2.59
3	62.19	58.14	47.86	36.77	29.77	25.71	20.89	14.62	3.66	8.60
4	160.85	154.41	141.94	132.82	118.00	110.87	99.67	94.32	90.37	86.42
5	92.31	87.70	77.97	71.96	57.24	50.33	44.47	40.97	37.51	34.72
6	110.27	105.26	91.99	83.29	73.89	66.77	61.73	59.47	55.42	52.23
7	166.00	146.53	132.95	116.86	104.25	96.04	85.33	77.21	68.99	62.52
8	115.05	109.72	102.32	90.86	79.83	70.25	65.83	63.30	58.02	55.38
					N. of deaths					
1	5.40	13.53	11.85	6.12	5.23	4.33	0.91	0.70	0.00	0.40
2	3.84	2.11	1.96	1.44	0.00	1.70	0.71	0.00	0.27	0
3	0.00	6.39	7.53	3.43	0.00	0.00	0.00	0.00	0.22	1.21
4	3.28	9.45	6.17	12.07	4.42	8.64	2.81	1.37	1.33	1.29
5	1.90	7.16	3.47	12.55	4.84	3.87	1.48	1.41	0.67	0.63
6	2.26	10.74	6.27	7.11	4.93	2.86	0.00	1.78	0.85	0.82
7	13.76	7.96	10.70	7.30	2.78	5.26	2.47	2.30	0.00	0.00
8	2.36	4.48	8.71	8.45	7.16	1.98	0.00	2.75	0.00	0.00
					N. of censored					
1	4.91	4.88	4.28	3.66	3.42	3.98	3.93	4.62	6.31	5.22
2	2.67	2.59	2.63	2.68	2.81	3.16	3.40	4.07	6.11	1.00
3	3.81	4.06	3.88	3.56	3.57	4.06	4.82	6.27	10.97	5.16
4	3.15	3.15	3.02	2.95	2.74	2.70	2.56	2.55	2.58	2.62
5	2.69	2.71	2.57	2.54	2.17	2.07	1.99	2.02	2.05	2.13
6	2.73	2.75	2.53	2.43	2.29	2.20	2.18	2.26	2.28	2.34
7	6.00	5.71	5.62	5.39	5.30	5.44	5.45	5.65	5.91	6.47
8	2.95	2.97	2.92	2.75	2.58	2.42	2.44	2.53	2.52	2.64

TABLE 15.5

Estimated Hazards and Standard Errors (Within Brackets) for Two ($\hat{\lambda}_{0k}$) and Three ($\hat{\lambda}_{1k}$) Drugs Regimens

Time interval	
$\hat{\lambda}_{0k}$	$\hat{\lambda}_{1k}$
0.165(0.029)	0.158(0.028)
0.266(0.034)	0.254(0.041)
0.277(0.042)	0.264(0.042)
0.224(0.034)	0.213(0.033)
0.130(0.026)	0.123(0.024)
0.090(0.021)	0.086(0.018)
0.091(0.022)	0.087(0.019)
0.065(0.017)	0.061(0.015)
0.071(0.018)	0.067(0.017)
0.051(0.014)	0.048(0.015)

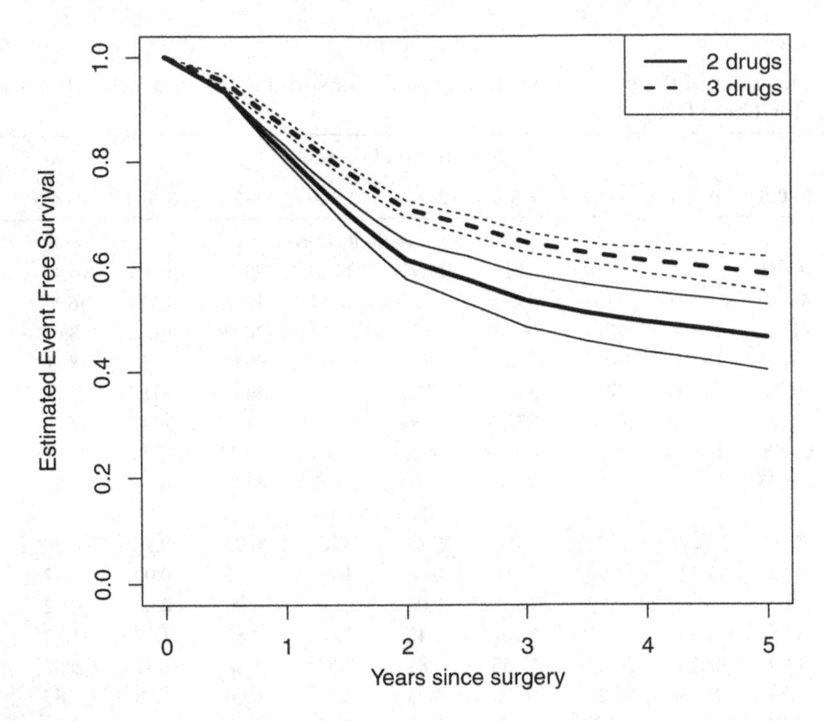

FIGURE 15.2
Overall event free survival.

has also been used in the context of meta-analysis of diagnostic tests with multiple thresholds (see Putter et al., 2010).

References

Anninga JK, Gelderblom H, Fiocco M, Kroep JR, Taminiau AH, Hogendoorn PC and Egeler RM, 2011. Chemotherapeutic adjuvant treatment for osteosarcoma: Where do we stand? *European Journal of Cancer* **47**: 2431–2445.

Arends LR, Hunink MG and Stijnen T, 2008. Meta-analysis of summary survival curve data. *Statistics in Medicine* **27**: 4381–4396.

Berkey CS, Anderson JJ and Hoaglin DC, 1996. Multiple-outcome meta-analysis of clinical trials. *Statistics in Medicine* **15**: 537–557.

Dear KBG, 1994. Alternative generalized least squares for meta-analysis of survival data at multiple times. *Biometrics* **50**: 989–1002.

Earle CC, Pham B and Wells GA, 2000. An assessment of methods to combine published survival curves. *Medical Decision Making* **20**: 104–111.

Fiocco M, Putter H and van Houwelingen JC, 2009a. A new serially correlated gamma frailty process for longitudinal count data. *Biostatistics* **10**: 245–247.

Fiocco M, Putter H and van Houwelingen JC, 2009b. Meta-analysis of pairs of survival curves under heterogeneity: A Poisson correlated gamma-frailty approach. *Statistics in Medicine* **28**: 3782–3797.

Fiocco M, Stijnen T and Putter H, 2012. Meta-analysis of time-to-event outcomes using a hazard-based approach: Comparison with other models, robustness and meta-regression. *Computational Statistics and Data Analysis* **56**: 1028–1037.

Klein PK and Moeschberger ML, 2005. *Survival Analysis: Techniques for Censored and Truncated Data*. Springer.

Lindsay BG, 1988. Composite likelihood methods. *Contemporary Mathematics* **80**: 221–239.

Nam I, Mengersen K and Garthwaite P, 2003. Multivariate meta-analysis. *Statistics in Medicine* **22**: 2309–2333.

Parmar MKB, Torri V and Stewart L, 1998. Extracting summary statistics to perform meta-analysis of the published literature for survival endpoints. *Statistics in Medicine* **17**: 2815–2834.

Putter H, Fiocco M and Stijnen T, 2010. Meta-analysis of diagnostic test accuracy studies with multiple thresholds using survival methods. *Biometrical Journal* **52**: 95–110.

Tudur Smith C, Williamson PR and Marson AG, 2005. Investigating heterogeneity in an individual patient data meta-analysis of time to event outcomes. *Statistics in Medicine* **24**: 1307–1319.

van Houwelingen HC, van de Velde CJH and Stijnen T, 2005. Interim analysis on survival data: Its potential bias and how to repair it. *Statistics in Medicine* **24**: 2747–2910

Williamson PR, Tudur Smith C, Hutton JL and Marson AG, 2002. Aggregate data meta-analysis with time-to-event outcomes. *Statistics in Medicine* **21**: 3337–3351.

16

Meta-Analysis of Correlations, Correlation Matrices, and Their Functions

Betsy Jane Becker, Ariel M. Aloe, and Mike W.-L. Cheung

CONTENTS

16.1 Introduction

Since the term *meta-analysis* was coined by Glass in 1976, the idea of synthesizing effect measures from related studies has evolved from a technique applied to relatively simple studies that examine basic research questions, to a set of multivariate approaches used to bring together series of sophisticated studies representing complex chains of events such as the prediction of behaviors based on *sets* of precursor variables (e.g., Brown et al., 2016); the latter are the focus of this chapter.

A key part of these multivariate meta-analyses in social sciences is often a pooled correlation matrix, whereas biomedical research tends to focus on regression coefficients. Becker (1992) first presented a generalized least squares (GLS) approach to obtaining the mean correlation matrix from a series of studies, as well as path models based on that mean and its variance-covariance matrix. Others have examined and offered modifications or alternatives to that original GLS method (e.g., Becker and Fahrbach, 1994; Cheung and Chan, 2005; Furlow and Beretvas, 2005; Prevost et al., 2007). Yet others have taken the approach of exploring and summarizing measures from individual studies that represent association via complex models—these partial effect sizes (Aloe and Becker, 2012; Aloe and Thompson, 2013) share some properties with bivariate correlations. Thus, we also touch on them in this chapter.

We review asymptotic distribution theory for the elements of the correlation matrix and functions thereof. We present methods for combining correlation matrices and using them to estimate structural equation models and show some newer techniques that enable reviewers to extract and synthesize partial effect measures directly from regression analyses in primary studies.

16.2 Motivating Examples

Suppose a reviewer wanted to draw on the accumulated evidence to understand the predictors of metabolic control and blood sugar levels in patients with diabetes, as in work by Brown and colleagues (Brown and Hedges, 1994; Brown et al., 2015, 2016). Let predictors Y_1 through Y_3 represent health beliefs, knowledge of diabetes, and compliance, and let the outcome Y_4 represent metabolic control. The reviewer might want to use meta-analysis to estimate a correlation matrix among Y_4 and the three predictors across studies, or possibly to estimate models such as those shown in Figures 16.1 and 16.2. The latter models can be estimated using the approach of model-based meta-analysis (Becker and Schram, 1994;

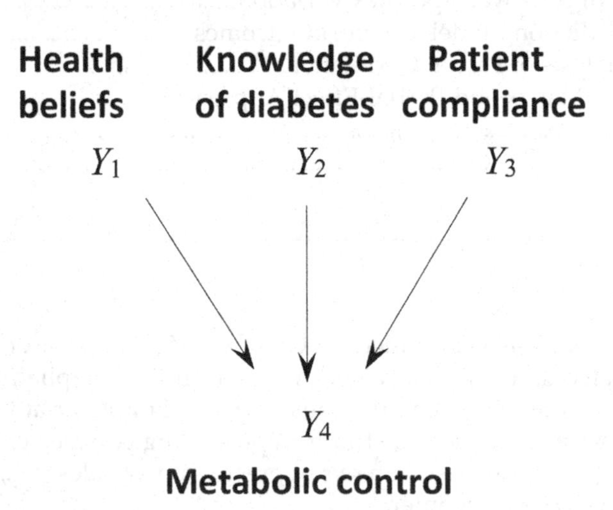

FIGURE 16.1
Prediction of metabolic control from Y_1, Y_2, and Y_3.

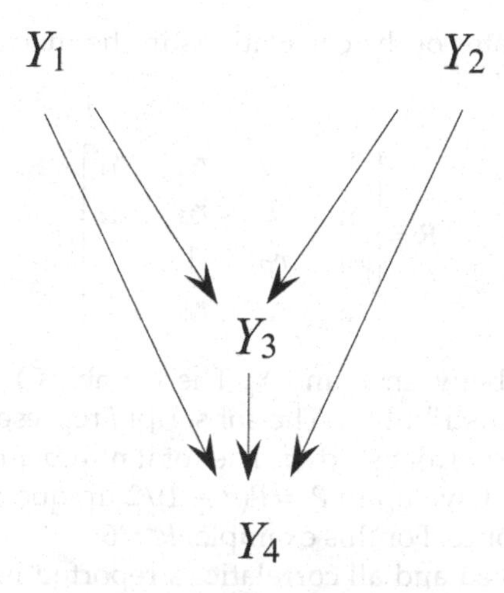

FIGURE 16.2
Y_3 plays a mediating role in metabolic control.

Becker, 2009). If instead regression coefficients are of interest then the model in Figure 16.1 can also be estimated using multivariate meta-analysis (Becker and Wu, 2007; Chapter 9).

On other occasions, the meta-analyst may be interested in a function of the correlation matrix other than Pearson's r, such as a partial or multiple correlation. For example, in the model in Figure 16.1, one may want to examine only the partial correlation of compliance with metabolic control, adjusting for health behaviors and knowledge about diabetes. Partial effect sizes can often be directly extracted or computed from the data in the primary studies, as we describe later on.

Because analyses like multiple regression and structural equation models statistically adjust for the presence of other covariates, their results also produce partial effect sizes (Aloe and Thompson, 2013; Keef and Roberts, 2004). Examples of such indices (also called effect measures or metrics, Chapter 3) are the partial correlation, semi-partial correlation (Aloe and Becker, 2012), and the partial d which is computed based on the analysis of covariance model (Keef and Roberts, 2004). These indices partial out or "adjust for" other variables in addition to the two of primary interest. The main advantage of directly synthesizing partial effect sizes is that one can rely on traditional univariate techniques such as weighted means and meta-regression analyses to synthesize these relatively simple indices. Analyses would parallel those for other effect indices (see, e.g., Chapter 4). Moreover, under this approach the variance of each partial-r effect size can still be partitioned into within-study and between-studies variance components.

16.3 Notation

We next introduce notation for the univariate and multivariate approaches to the synthesis of correlations and correlation matrices. The first step in a meta-analysis of correlation matrices is to gather estimates of the correlations in the matrix **R**, shown here for a case with $P=4$ variables. Here

$$\mathbf{R}=\begin{bmatrix} 1 & r_{12} & r_{13} & r_{14} \\ r_{21} & 1 & r_{23} & r_{24} \\ r_{31} & r_{32} & 1 & r_{34} \\ r_{41} & r_{42} & r_{43} & 1 \end{bmatrix},$$

where r_{st} is the correlation between Y_s and Y_t. The variables $Y_1, \ldots, Y_P$ are assumed to follow a multivariate normal distribution. The subscript i represents the ith study, so $\mathbf{R}_i$ represents the correlation matrix from study i. The total number of variables in the matrix is P (here $P=4$). A study may have up to $P^*=P(P-1)/2$ unique correlations assuming each variable is measured only once. For this example, $P^*=6$.

If all variables are observed and all correlations reported in study i, the matrix $\mathbf{R}_i$ will contain $m_i=P^*$ unique off-diagonal elements. If one or more variables are not examined in study i or if some correlations are not reported, $\mathbf{R}_i$ will contain $m_i<P^*$ elements. The total number of observed correlations across studies is thus $m=\Sigma m_i \leq IP^*$, where I is the number of independent studies or samples. A typical pattern when data are missing is that the study may report only the correlations of predictors with an outcome (but not the correlations among the predictors). From the collection of $\mathbf{R}_i$ matrices, whether all are complete or not, the reviewer can estimate an average correlation matrix and its associated variance-covariance matrix.

On some occasions, the entire matrix is not of interest. One common index that may be of interest is the partial correlation (e.g., Pearson, 1915), shown here with one variable, Y_u, being partialled. Here

$$r_{st.u}=\frac{r_{st}-r_{su}r_{tu}}{\sqrt{\left(1-r_{su}^2\right)\left(1-r_{tu}^2\right)}}, \tag{16.1}$$

which is the correlation between the residuals of regressions of Y_s on Y_u and Y_t on Y_u (rather than the correlation between Y_s and Y_t). This index "removes" the effects of the third variable (here Y_u). The partial r can also be considered as the correlation of Y_s with Y_t within levels of Y_u, or as phrased by Pearson, "the intensity of association between two variates when other correlated variates are considered as constant" (p. 492). For simplicity, the formula above is for a partial r with only one control variable, but in practice more than one additional variable may be adjusted or "partialled out", as in our example below.

On some occasions, the target partial r values may be reported in the primary studies of the meta-analysis, and in others the meta-analyst may need to compute the partial index as part of the data-extraction process. Regardless of the number of predictors in the model, the partial correlation can easily be estimated from regression results as $r_{st.uvw\ldots}=t/\sqrt{t^2+(n-c-2)}$ where t is the t-test statistic associated with the slope of Y_t as a predictor of Y_s in a regression model, n is the sample size, and c is the total number of variables being partialled out (i.e., one less than the number of predictors; see Gustafson, 1961 or Aloe and Thompson, 2013).

The semi-partial correlation, $r_{s(t.u)}$, also known as the part correlation (Griffin, 1932), is another index that can be quite useful. In this index, variance shared between the predictors (here variance shared with the third variable Y_u) is removed from the variable Y_t, but not from Y_s. The formula for $r_{s(t.u)}$ for three variables is very similar to (16.1), except for its denominator:

$$r_{s(t.u)} = \frac{r_{st} - r_{su}r_{tu}}{\sqrt{\left(1 - r_{tu}^2\right)}}. \tag{16.2}$$

The semi-partial correlation can also easily be computed from information typically reported in multiple regression analyses (Aloe and Becker, 2012). Specifically, $r_{s(t.u)} = t\sqrt{\left(1-R_s^2\right)/\left(n-c-2\right)}$, where R_s^2 is the squared multiple correlation for predicting Y_s from Y_t and the partialled variable(s), and all other terms were previously defined.

16.4 Analyses

16.4.1 Univariate Approach

Consider the goal of estimating a mean correlation—summarizing either r_{st}, $r_{st.u}$, or $r_{s(t.u)}$—from a series of studies which all estimate the same index, for example, all estimate r_{st}. The simplest approach is to use standard meta-analytic techniques to estimate the means of each relationship individually, ignoring the other correlations; this is referred to as the univariate approach. We consider both common-effect and random-effects analyses.

A variety of estimators of the mean correlation have been examined; see Field (2001) for an excellent study and review of this literature. Possibilities include the sample-size weighted mean across studies (with elements $\bar{r}_{st} = \sum_{i=1}^{I} n_i r_{ist} / \sum_{i=1}^{I} n_i$), the inverse-variance weighted mean shown below, and the mean obtained using Fisher's (1932) variance stabilizing transformation z(r). These univariate estimators ignore the dependence between pairs of correlations in the correlation matrix, because each relationship is treated separately.

The large-sample sampling distribution of a single correlation provides a basis for the univariate approach for summarizing Pearson's r values and also partial and semi-partial correlations. Assume that variables Y_s and Y_t are normally distributed, and that r_{ist} estimates the population value ρ_{ist}. Then in large samples the model for a single correlation $r_{ist} = \rho_{ist} + \varepsilon_{ist}$ implies that r_{ist} has expected value ρ_{ist}, and it also has a variance that depends on ρ_{ist}. Olkin and Siotani (1976) showed that in large samples of size n_i, the correlation r_{ist} is approximately normally distributed with mean ρ_{ist} and variance

$$s_{ist}^2 = \mathrm{Var}\left(r_{ist}\right) = \left(1 - \rho_{ist}^2\right)^2 / n_i. \tag{16.3}$$

Sometimes the denominator in (16.3) uses $(n_i - 1)$ rather than n_i.

The variance in (16.3) has been estimated in two ways—either by substituting the study specific correlation into (16.3), or by using a mean correlation such as $\bar{r}_{st}$. The latter provides a somewhat more stable estimator (Becker and Fahrbach, 1994; Furlow and Beretvas, 2005), which can be useful given that the variance is treated as known in further analyses; it is more appropriate in common-effect models where ρ_{ist} is the same for all studies than in random-effect models. (Because we treat the variance estimate as known, we use s_{ist}^2 rather than σ_{ist}^2 to denote the variance.) Treating s_{ist}^2 as known is only safe for larger samples, as is the assumption of normality of the correlation, especially for large correlation values. However, concerns about normality and treating s_{ist}^2 as known are reduced when the mean is used to compute (16.3). This is because the distribution of the mean r better approximates the normal distribution regardless of the shape of the distribution of r_i. In addition, using the mean to compute the within-study variance removes the dependence between each study's correlation and its variance; similar issues are discussed in Section 4.3.1.

The distribution of the partial r for normally distributed data is approximately normal with mean $\rho_{ist.u}$ and variance $\text{Var}(r_{ist.u}) = (1 - \rho_{ist.u}{}^2)^2/(n_i - c - 2)$, where c is the number of additional controlled (partialled) variables (Olkin and Siotani, 1976). As for r_{ist}, the mean of the set of partial correlations can be used to compute the variance of $r_{ist.u}$ if desired.

As noted above, both of these commonly used variances have a property that is less than optimal— their dependence on the parameter to be estimated. One common approach that avoids this issue is to transform the r values using the Fisher z transformation (Chapter 3), which is both normalizing and variance-stabilizing. This approach has been extensively studied for univariate approaches to combining Pearson's r (e.g., Field, 2001, 2005), and performs well with little bias. Fisher's transformation is not often used in the synthesis of correlation matrices, for reasons discussed below.

For the semi-partial r, the asymptotic mean of $r_{s(t.u)}$ is $\rho_{s(t.u)}$ and the variance is more complex than those above. Aloe and Becker (2012) showed that for large samples

$$\text{Var}\left(r_{s(t.u)}\right) = \frac{1}{(n-c-2)\left(R_s^2 - R_{s(t)}^2\right)}$$

$$\left[R_s^2\left(1-R_s^2\right)^2 + R_{s(t)}^2\left(1-R_{s(t)}^2\right)^2\right.$$

$$\left.-2R_sR_{s(t)}\left[0.5\left(\left(2\frac{R_{s(t)}^2}{R_s^2}\right)^{\frac{1}{2}} - R_sR_{s(t)}\right)\left[1-R_s^2-R_{s(t)}^2-\frac{R_{s(t)}^2}{R_s^2}\right]+\left(\frac{R_{s(t)}^2}{R_s^2}\right)^{3/2}\right]\right],$$

where R_s^2 is the multiple correlation for predicting Y_s from Y_t and any partialled variables, and $R_{s(t)}^2$ is the multiple correlation for predicting Y_s from the partialled variables but not Y_t. Aloe and Becker (2012) showed how the quantities involved can be computed from reports of multiple regression analyses.

16.4.1.1 The Univariate Mean under the Common-Effect Model

Consider now a series of I independent correlations r_{ist}, or a series of partial or semi-partial correlations as defined above. The inverse-variance weighted mean of the correlations between Y_s and Y_t under the univariate common-effect (CE) model $r_{ist} = \rho_{st} + \varepsilon_{ist}$ is

$$\bar{r}_{\text{CE}st} = \sum_{i=1}^{I} w_{ist} r_{ist} \bigg/ \sum_{i=1}^{I} w_{ist}\,, \tag{16.4}$$

where the weight $w_{ist} = 1/s_{ist}^2$, and s_{ist}^2 is defined in (16.3).

16.4.1.2 The Univariate Mean under Random Effects

In current practice, it is unusual for meta-analysts to adopt the common-effect model, where all effects arise from a single population and share the parameter ρ_{st}. The univariate random-effects (RE) model for the *i*th correlation, $r_{ist} = \rho_{ist} + \varepsilon_{ist}$, is more likely to be appropriate and also allows for broader generalizations than the common-effect model (Hedges and Vevea, 1998). This model can also be written as $r_{ist} = \rho_{st} + \varepsilon_{ist} + u_{ist}$, which makes explicit the two sources of error—within (ε_{ist}) and between studies (u_{ist}).

As described elsewhere (Chapter 4; Hedges and Vevea, 1998), an estimate of τ^2_{st}, the between-studies variance for the correlation between Y_s and Y_t (i.e., for the error u_{ist}), must be incorporated into the analysis if this model is adopted. The quantity τ^2_{st} reflects true differences across studies among the values of ρ_{ist}, and is specific to the relationship between Y_s and Y_t (though other models are possible). For a correlation matrix with P^* unique elements, P^* variance components must be estimated to obtain all means under the random-effects model. The univariate random-effects mean correlation between Y_s and Y_t is thus

$$\bar{r}_{\text{RE}\,st} = \sum_{i=1}^{I} w_{\text{RE}ist} r_{ist} \bigg/ \sum_{i=1}^{I} w_{\text{RE}ist}, \tag{16.5}$$

where $w_{REist} = 1/(s^2_{ist} + \tau^2_{st})$ which is estimated as $1/(s^2_{ist} + \hat{\tau}^2_{st})$, with s^2_{ist} given in (16.3). The random-effects estimator in (16.5) simplifies to the common-effect mean in (16.4) when $\hat{\tau}^2_{st}$ is zero.

This mean is a best estimator in the sense that it has an asymptotic variance that is the minimum across all weighted combinations (Hedges and Olkin, 1985).

A similar approach can be used to estimate the mean partial or semi-partial *r* by substituting the analogous estimator in place of r_{ist}, and using the reciprocal of the appropriate variance as the weight. The appropriate within-study variance weight would be added to a between-studies component, as for (16.5). In practice, it would be unusual for a meta-analyst to estimate a full matrix of partial correlations, though in theory it can be done. It is more typical for just one or two partial relationships to be of interest in such syntheses (e.g., Bowman, 2010).

16.4.2 Multivariate Approach

If the meta-analyst aims to compare several of the elements in a correlation matrix, or to conduct a model-based meta-analysis, computing the average for each correlation in the **R** matrix separately does not account for the correlations among the elements of the matrix. Ignoring dependency among pairs of correlations, or among any other dependent effect sizes, is not recommended (Becker, 2000; Riley, 2009; see also Chapter 9). To model dependence, the meta-analysis must incorporate the variance-covariance matrix of the matrix $\mathbf{R}_i$. Typical multivariate approaches can be used (e.g., Raudenbush et al., 1988; Riley, 2009), and more sophisticated Bayesian methods are also available (Burke et al., 2018; Prevost et al., 2007). We present a simple approach to modeling dependence within correlation matrices using GLS methods (Becker, 1992, 1995, 2009).

16.4.2.1 Multivariate Distribution of r_i and Models for r_i

We list the unique elements of the matrix $\mathbf{R}_i$ in a vector, $\mathbf{r}_i = (r_{i12}, r_{i13}, \ldots, r_{i1P}, r_{i23}, \ldots, r_{i(P-1)P})$, so $\mathbf{r}_i$ contains the same information as the square matrix $\mathbf{R}_i$. The vectors $\boldsymbol{\rho}_i = (\rho_{i12}, \rho_{i13}, \ldots, \rho_{i(P-1)P})$

and $\boldsymbol{\rho}=(\rho_{12}, \rho_{13}, \ldots, \rho_{(P-1)P})$ are the study-specific and overall population correlation vectors. If only m_i correlations are measured or reported in study i, the vector $\mathbf{r}_i$ and its covariance matrix are reduced in dimension accordingly. Below we also use $\mathbf{P}_i$ and $\mathbf{P}$ to represent the square forms of $\boldsymbol{\rho}_i$ and $\boldsymbol{\rho}$.

The asymptotic covariance matrix $\mathbf{S}_i$ of the vector of correlations $\mathbf{r}_i$ was given by Olkin and Siotani (1976) for correlations among the normal variables Y_s, Y_t, Y_u, and Y_v. The variances (diagonal elements of $\mathbf{S}_i$) are defined in (16.3) above. The within-study covariance between r_{ist} and r_{iuv} requires all cross-correlations involving the indexes s, t, u, and v. Specifically,

$$\begin{aligned} S_{istuv} &= \mathrm{Cov}\left(r_{ist}, r_{iuv}\right) \\ &= \Big[0.5\rho_{ist}\rho_{iuv}\left(\rho_{isu}^2+\rho_{isv}^2+\rho_{itu}^2+\rho_{itv}^2\right) \\ &\quad + \rho_{isu}\rho_{itv}+\rho_{isv}\rho_{itu} \\ &\quad \Big[-\left(\rho_{ist}\rho_{isu}\rho_{isv}+\rho_{its}\rho_{itu}\rho_{itv}+\rho_{ius}\rho_{iut}\rho_{iuv}+\rho_{ivs}\rho_{ivt}\rho_{ivu}\right)\Big]/n_i, \end{aligned} \tag{16.6}$$

for s, t, u, and $v=1$ to P. Sometimes the denominator is shown as (n_i-1) rather than n_i. The covariances in (16.6) can be estimated using study specific correlations or mean correlations. The inverse-variance weighted univariate means shown in (16.4) and (16.5) essentially treat the covariance terms in (16.6) as if they were all zero.

As mentioned above, the Fisher z transform has been widely recommended for univariate analyses of correlations. However, its use in the context of synthesizing correlation matrices presents challenges if one needs to obtain variances for complex (non-linear) functions of the mean matrix (e.g., for Becker's model-driven meta-analysis or meta-analytic structural equation modeling; Cheung and Chan, 2005). Additionally, the covariances of z-transformed correlations are more complex than those for raw r values. Although in the case of univariate meta-analysis the use of Fisher's variance-stabilizing z transformation makes the effect size (i.e., r) and its variance independent, this cannot be said for multivariate meta-analysis because the within-study covariance for the transformed r's is still a complex function of the population effect sizes. Thus, we prefer to use untransformed correlation coefficients in our analyses, recognizing that standard construction of confidence intervals may lead to values outside the range –1 to 1.

The variance-covariance matrix of the vector $\mathbf{r}_i$ may include only the variances and covariances (σ_{ist}^2 and σ_{istuv}, in the matrix $\boldsymbol{\Sigma}_i$) of the within-study sampling errors under the common-effect model $\mathbf{r}_i=\boldsymbol{\rho}+\boldsymbol{\varepsilon}_i$, or it can include sampling error plus between-studies variation (in matrix form, $\boldsymbol{\Sigma}_i+\mathbf{T}$) under the random-effects model $\mathbf{r}_i=\boldsymbol{\rho}+\boldsymbol{\varepsilon}_i+\mathbf{u}_i$. We assume multivariate normality with $\mathbf{r}_i|\boldsymbol{\rho}_i \sim \mathrm{N}(\boldsymbol{\rho}_i, \mathbf{S}_i)$ and $\boldsymbol{\rho}_i \sim \mathrm{N}(\boldsymbol{\rho}, \mathbf{T})$. Estimation of $\mathbf{T}$ is discussed in Chapter 9 as well as by Becker and Schram (1994), Becker (2009) and many others (Knapp and Hartung, 2003; Viechtbauer, 2005). Easily accessible R routines such as the `rma.mv` function in the `metafor` package provide a variety of estimators. Below we use the restricted maximum likelihood (REML) estimator from `metafor` (Viechtbauer, 2010).

We present next the widely used generalized least squares estimation approach to summarizing matrices across the set of I studies. Specifically, to estimate a common correlation vector $\boldsymbol{\rho}$ of length P^*, Becker (1992, 1995) initially proposed using the model $\mathbf{r}=\mathbf{X}\boldsymbol{\rho}+\boldsymbol{\varepsilon}$, where $\mathbf{r}$ is formed by concatenating the I correlation vectors as $\mathbf{r}=(\mathbf{r}_1, \ldots, \mathbf{r}_I)$ and $\mathbf{X}$ is a $m \times P^*$

matrix created by stacking I identity matrices, each of dimension $P^* \times P^*$, denoted as $\mathbf{I}_{P^*}$. We consider here the more commonly used random-effects model $\mathbf{r} = \mathbf{X}\,\boldsymbol{\rho} + \boldsymbol{\varepsilon} + \mathbf{u}$, as discussed by Becker (2009). Thus, under random effects we have

$$\begin{bmatrix} r_{112} \\ r_{113} \\ \vdots \\ r_{212} \\ r_{213} \\ \vdots \\ r_{I(P-1)P} \end{bmatrix} = \begin{bmatrix} 1 & 0 & \cdots & 0 \\ 0 & 1 & \cdots & 0 \\ \vdots & \vdots & \vdots & \vdots \\ 1 & 0 & \cdots & 0 \\ 0 & 1 & \cdots & 0 \\ \vdots & \vdots & \vdots & \vdots \\ 0 & 0 & \cdots & 1 \end{bmatrix} \begin{bmatrix} \rho_{12} \\ \rho_{13} \\ \rho_{14} \\ \vdots \\ \vdots \\ \rho_{(P-1)P} \end{bmatrix} + \begin{bmatrix} e_{112} \\ e_{113} \\ \vdots \\ e_{212} \\ e_{213} \\ \vdots \\ e_{I(P-1)P} \end{bmatrix} + \begin{bmatrix} u_{112} \\ u_{113} \\ \vdots \\ u_{212} \\ u_{213} \\ \vdots \\ u_{I((P-1)P} \end{bmatrix}.$$

The design matrix $\mathbf{X}$ can also be expanded to include study characteristics as moderator variables (effect modifiers), but this is rarely done.

As above if the original outcomes Y_1 through Y_P follow a normal distribution, then $\mathbf{r}$ has a multivariate normal distribution in large samples. We form a blockwise diagonal variance-covariance matrix using either $\hat{\mathbf{W}} = \text{diag}(\mathbf{W}_1, \ldots, \mathbf{W}_I)$ under random-effects, where $\hat{\mathbf{W}}_i = (\mathbf{S}_i + \hat{\mathbf{T}})^{-1}$, or under common-effect assumptions $\mathbf{S}^{-1} = \text{diag}(\mathbf{S}_1^{-1}, \ldots, \mathbf{S}_I^{-1})$ for studies $i = 1$ to I. Estimation of $\mathbf{T}$ is discussed below. If the data are complete in all studies $\hat{\mathbf{W}}$ and $\mathbf{S}^{-1}$ will be square matrices of dimension IP^*, and otherwise their dimensions are reduced by the number of missing correlations as described below.

16.4.2.2 The Multivariate Mean under Random Effects

Under the random-effects model (i.e., using $\hat{\mathbf{W}}$ as the weight matrix), the mean correlation matrix is estimated as

$$\bar{\mathbf{r}}_{\text{RE}} = (\mathbf{X}'\hat{\mathbf{W}}\mathbf{X})^{-1}\mathbf{X}'\hat{\mathbf{W}}\mathbf{r}, \tag{16.7}$$

where $\bar{\mathbf{r}}_{\text{RE}}$ is the $P^* \times 1$ vector of mean correlations, $\mathbf{r}$ is the $m \times 1$ vector of all correlations, and $\mathbf{X}$ and $\hat{\mathbf{W}}$ are defined above. With some minimal matrix algebra we can show that under normality of the correlations $\bar{\mathbf{r}}_{\text{RE}}$ is also the maximum likelihood estimator.

If a study has reported fewer than P^* correlations, the rows of $\mathbf{X}$ and the rows and columns of $\mathbf{S}_i$ and $\hat{\mathbf{T}}$ corresponding to the missing r values are deleted. For example, if $m_i < P^*$ correlations are reported in study i, only m_i rows are included in $\mathbf{X}$, and $\mathbf{S}_i$ will be $m_i \times m_i$. For such a study the matrix $\hat{\mathbf{T}}$ would also be reduced accordingly. These deletions assume that unreported values are missing at random; in most cases, it is difficult to assess how reasonable this assumption actually is, because researchers do not often tell us why they did not include particular variables in their studies.

To obtain the common-effect estimate, substitute $\mathbf{S}^{-1}$ for the weight matrix $\hat{\mathbf{W}}$ in (16.7). However, if between-studies variances are small or zero, the random-effects weights converge to the common-effect weights, so this substitution is rarely necessary.

The variance covariance matrix of $\bar{\mathbf{r}}_{\text{RE}}$ under random effects is estimated as

$$V(\bar{\mathbf{r}}_{\text{RE}}) = (\mathbf{X}'\hat{\mathbf{W}}\mathbf{X})^{-1},$$

where $\mathbf{X}$ and $\hat{\mathbf{W}}$ are defined above. This matrix (or its common-effect analogue $(\mathbf{X}'\,\mathbf{S}^{-1}\,\mathbf{X})^{-1}$, if the common-effect model is chosen) can then be used to obtain confidence intervals for each element of the mean vector. As an example, the approximate $(1-\alpha)\%$ confidence interval for the first element of $\bar{\mathbf{r}}_{\text{RE}}$ is obtained as $\bar{r}_{12} \pm z_{\alpha/2}\sqrt{\text{V}(\bar{r}_{12})}$, where $\text{V}(\bar{r}_{12})$ is the first element of $\text{V}(\bar{\mathbf{r}}_{\text{RE}})$ and $z_{\alpha/2}$ is the $\alpha/2$ percentile of the normal distribution.

16.4.2.3 Multivariate Tests about Correlation Matrices

It is also possible to evaluate whether the correlation matrices from the I studies arise from the same population. A test of the hypothesis H_0: $\boldsymbol{\rho}_1 = \ldots = \boldsymbol{\rho}_I$ is computed as

$$Q_{CE} = \mathbf{r}'\left[\mathbf{S}^{-1} - \mathbf{S}^{-1}\mathbf{X}(\mathbf{X}'\mathbf{S}^{-1}\mathbf{X})^{-1}\mathbf{X}'\mathbf{S}^{-1}\right]\mathbf{r}, \tag{16.8}$$

where the inverse of the common-effect variance matrix $\mathbf{S}$ is used along with $\mathbf{X}$ defined above (Becker, 1992). This is a multivariate version of the homogeneity statistic computed using Fisher z values described in Becker (2009, formula 20.9). Here Q_{CE} has approximately a chi-square distribution with $(I-1)P^*$ degrees of freedom when all studies observe the full set of correlation values and $m - P^*$ if they are not complete. However, the Q-test in the univariate analysis of correlations has been found to reject at higher than nominal levels, especially for small samples (Viechtbauer, 2007). Also, some argue that Q in general has frequently been misused (e.g., Hoaglin, 2016) so one should not rely solely on this test as an indication of homogeneity or heterogeneity. Rejection of the null hypothesis is evidence in favor of adopting a random-effects model, but the converse should not be assumed to be true. That is, when Q_{CE} fails to reject H_0, average correlations can be estimated under either of the two models—a model that incorporates only within-sample variability (i.e., common-effect model) or a model that also incorporates between-studies variability (i.e., random-effects model). In general, it is sensible to use other indices such as $\hat{\mathbf{T}}$ itself, or indices like I^2 (Higgins et al., 2003) to track the degree of dispersion in the correlations.

Finally, we can obtain a test of the hypothesis that all correlation elements are zero in the population, specifically, a test of whether $\boldsymbol{\rho} = \mathbf{0}$. Under random-effects assumptions, the test, $Q_B = \bar{\mathbf{r}}_{\text{RE}}'(\mathbf{X}'\hat{\mathbf{W}}\mathbf{X})^{-1}\bar{\mathbf{r}}_{\text{RE}}$, follows an asymptotic chi-square distribution with P^* degrees of freedom. Rejecting this null model implies that at least one of the elements of the correlation matrix is non-zero in the population.

16.4.3 Linear and Structural Equation Models Based on Series of Correlations

Becker (1992, 1995) proposed to use the mean correlation matrix to compute linear models among variables of interest in the correlation matrix. These linear models can represent multiple regression models or path analyses, such as the model shown in Figure 16.2. Such models are common in research in education and psychology, and often are used to represent complex sets of interrelated constructs. The path model in Figure 16.2 allows one to examine whether predictors like Y_1 (variables not influenced by other variables in the model, known as exogenous variables in the parlance of structural equation modeling [SEM]) have more than one type of influence on the outcome. Specifically, Y_1 may have a direct influence (shown as the arrow going directly to Y_4) or an indirect influence via the arrow passing through the variable Y_3, or both. The meta-analyst may want to ask whether the variable Y_3 mediates (enhances or reduces) the effects of the predictors Y_1 and Y_2.

16.4.3.1 Estimating the Path Model

Starting with the square form of the mean correlation matrix (from either a common-effect or random-effects model), we order the variables so that correlations involving the potential outcome are in the last row and column. Denoting the square form of either (16.7) or its common-effect analogue as $\bar{\mathbf{R}}$, the matrix $\bar{\mathbf{R}}$ is partitioned so that the mean correlations of the predictors with the outcome of interest are put into a vector $\bar{\mathbf{R}}_{XY}$, and correlations among the predictors appear in $\bar{\mathbf{R}}_{XX}$. Complex path models can be estimated using different submatrices of $\bar{\mathbf{R}}$; these will depend on the model(s) to be estimated.

Consider here an illustration with four variables, where Y_4 is treated as an outcome. Here $\bar{\mathbf{R}}_{XY}$ is the vector of three values $(\bar{r}_{14}\,\bar{r}_{24}\,\bar{r}_{34})'$ and $\bar{\mathbf{R}}_{XX}$ is the upper square matrix of correlations among Y_1, Y_2, and Y_3 which are now viewed as predictors:

$$\bar{\mathbf{R}} = \begin{bmatrix} \bar{\mathbf{R}}_{XX} & \bar{\mathbf{R}}_{XY} \\ \bar{\mathbf{R}}_{XY}' & 1 \end{bmatrix} = \left[\begin{array}{ccc|c} 1 & \bar{r}_{12} & \bar{r}_{13} & \bar{r}_{14} \\ \bar{r}_{21} & 1 & \bar{r}_{23} & \bar{r}_{24} \\ \bar{r}_{31} & \bar{r}_{32} & 1 & \bar{r}_{34} \\ \hline \bar{r}_{41} & \bar{r}_{42} & \bar{r}_{43} & 1 \end{array}\right], \tag{16.9}$$

For this partitioning, the product $\mathbf{b}^* = \bar{\mathbf{R}}_{XX}^{-1}\bar{\mathbf{R}}_{XY}$ gives the standardized regression slopes in the regression model that predicts Y_4 from Y_1 through Y_3. Becker and Schram (1994) provide further details about the estimation of variances and tests associated with such models; the variances were obtained using the delta method and are functions of Var($\bar{\mathbf{R}}$) that depend on the specific mean used as $\bar{\mathbf{R}}$ and on the model being estimated.

Any of the variables in a matrix can serve as an outcome, and it is not necessary to include all variables in the model(s) to be estimated. In the example below, the primary outcome is labeled as Y_1, which is also convenient for the coding of data, as the correlations most likely to be completely reported are those between the outcome (Y_1) and the predictors (Y_2 through Y_4 in our example). However, the labeling is completely arbitrary.

16.4.3.2 Estimation of Structural Equation Models

Cheung and Chan (2005) also realized that series of correlation matrices could be synthesized and incorporated into the SEM framework. Their method, termed the two-stage structural equation modeling (TSSEM) approach, is conceptually similar to that used by Becker, but Cheung and Chan drew on methods developed for SEM and thus provided model fit tests and goodness-of-fit indices on models with both observed and latent variables that had not been provided by Becker (1992, 1995).

Briefly, SEM methods are based on examination of a model for relationships among variables proposed by the data analyst. The proposed model can be, for example, a multiple regression model, path model, confirmatory factor-analytic model, or a structural equation model with latent variables. Regardless of the form, the relations among variables are conceived of as being determined by a set of parameters denoted as $\boldsymbol{\theta}$ (e.g., $\boldsymbol{\theta}$ may represent a set of path coefficients in a structural equation model).

In general applications, SEM models are tested against the data by comparing $\boldsymbol{\Sigma}(\boldsymbol{\theta})$—the covariance matrix implied by the proposed model—to the sample covariance matrix $\mathbf{S}_s$. A likelihood ratio statistic and various goodness-of-fit indices may be used to test the exact and approximate fit of the proposed model. In addition, the difference between $\mathbf{S}_s$ and $\boldsymbol{\Sigma}(\boldsymbol{\theta})$ is used in the process of estimating the $\boldsymbol{\theta}$ elements. In the context of meta-analysis, correlation matrices may be used in place of covariance matrices, because it is unusual to find a set of studies where all variables are on a common scale and thus would have potentially comparable covariance matrices. This is why direct application of (multigroup) SEM methods based on covariances is often not possible in meta-analysis.

Cheung and Chan (2005) drew on the SEM literature in which the true covariance matrix in the ith study is decomposed into a product of a population correlation matrix ($\mathbf{P}_i$) and a diagonal matrix of standard deviations ($\mathbf{D}_i$) as $\boldsymbol{\Sigma}_i(\boldsymbol{\theta}) = \mathbf{D}_i\,\mathbf{P}_i\,\mathbf{D}_i$. As in typical practice $\boldsymbol{\Sigma}_i(\boldsymbol{\theta})$ is the variance covariance matrix implied by the hypothesized relationships among the variables. When the inputs to the SEM analysis are correlation matrices, the diagonal elements of $\mathbf{D}_i$ are close to 1. This model for $\boldsymbol{\Sigma}_i(\boldsymbol{\theta})$ allows researchers to analyze correlation matrices correctly based on statistical theory for the covariance matrix. Constraining all I correlation matrices $\mathbf{P}_i$ to be equal to $\mathbf{P}$, the square form of the common effect vector $\boldsymbol{\rho}$,

$$\mathbf{P} = \begin{bmatrix} 1 & \rho_{12} & \rho_{13} & \rho_{14} \\ \rho_{21} & 1 & \rho_{23} & \rho_{24} \\ \rho_{31} & \rho_{32} & 1 & \rho_{34} \\ \rho_{41} & \rho_{42} & \rho_{43} & 1 \end{bmatrix}$$

enables the meta-analyst to estimate a common mean matrix. This approach is typically implemented by replacing the Olkin and Siotani covariance of correlations used for weighting with an alternative method, which relies on a structural equation modeling approach (see Cheung and Chan, 2004). Regardless of whether a common- or a random-effects model is used, we have an estimate of the (average) correlation vector $\boldsymbol{\rho}$, which is taken from the off-diagonal elements of $\bar{\mathbf{R}}$, and its sampling covariance matrix $\mathrm{Var}(\bar{\mathbf{R}})$ after applying either the GLS or the TSSEM approach. This step is usually known as the stage-1 analysis.

In the stage-2 analysis, researchers may fit structural equation models on $\bar{\mathbf{R}}$ and its sampling covariance matrix $\mathrm{Var}(\bar{\mathbf{R}})$. Suppose we are fitting a population correlation structure model $\rho(\theta)$, meaning that the population correlation vector $\boldsymbol{\rho}$ is a function of the parameters $\boldsymbol{\theta}$. The proposed model can be, for instance, a multiple regression (like that shown above with slope $\mathbf{b}^*$), a path analysis, or a confirmatory factor analytic model. We fit the data with the following GLS fit function (i.e., discrepancy function): $F(\hat{\theta}) = (\bar{r} - \rho(\hat{\theta}))'\,\hat{\Psi}(\bar{r} - \rho(\hat{\theta}))$ where $\bar{\mathbf{r}}$ contains the observed elements (mean correlations) in vectorized form, $\rho(\hat{\theta})$ includes model-implied correlation values, and $\hat{\Psi}$ is a weight matrix, which can take different forms. Cheung and Chan (2005) have suggested using the inverse of the asymptotic covariance matrix of the average correlation matrix obtained in stage 1 as the weight matrix. The parameter estimates $\hat{\theta}$ (and thus $\rho(\hat{\theta})$) are obtained by minimizing the differences between the observed correlation mean vector $\bar{\mathbf{r}}$ and the model-implied correlation vector $\rho(\hat{\theta})$, weighted by the inverse of $\bar{\mathbf{r}}$'s sampling covariance matrix. After fitting the proposed model, test statistics and goodness-of-fit indices can be used to judge whether the proposed model fits the data reasonably well.

16.5 Example

The dataset for our example consists of eight studies drawn from Craft et al. (2003), who examined the relationship between anxiety and sport performance. Sport performance (Y_1) was measured in a variety of ways, including both objective measures of individual skills such as race times, numbers of foul shots made, or targets hit, and subjective ratings by judges. All studies used the Competitive Sport Anxiety Index or CSAI (Martens et al., 1990), a measure that taps cognitive aspects of anxiety (Y_2) and somatic anxiety (or physical symptoms, Y_3), as well as self-confidence (Y_4). We do not explore all of the complexities of the original data in our example; the reader is referred to Craft et al. (2003) for the full analysis and results. The data from the eight studies are shown in Table 16.1 in row form. In addition, we show the partial correlations of self-confidence with performance, partialling out cognitive and somatic anxiety (labeled S-c/Perf.CS and $r_{14.23}$ in the table). These partial correlations are computed from the bivariate correlations. A typical primary study may report bivariate correlations but not the desired partial correlations.

16.5.1 Estimating the Study Covariances

Applying the methods described above, we collect the correlations for all studies and compute their covariance matrices. For instance, for the study with ID 3, the 6×1 vector $\mathbf{r}_3$ and its 6×6 indicator matrix $\mathbf{X}_3$ are

$$\mathbf{r}_3 = \begin{bmatrix} 0.53 \\ -0.12 \\ 0.03 \\ 0.52 \\ -0.48 \\ -0.40 \end{bmatrix}, \text{ and } \mathbf{X}_3 = \begin{bmatrix} 1 & 0 & 0 & 0 & 0 & 0 \\ 0 & 1 & 0 & 0 & 0 & 0 \\ 0 & 0 & 1 & 0 & 0 & 0 \\ 0 & 0 & 0 & 1 & 0 & 0 \\ 0 & 0 & 0 & 0 & 1 & 0 \\ 0 & 0 & 0 & 0 & 0 & 1 \end{bmatrix}.$$

The columns of $\mathbf{X}_3$ reflect that each of the six correlations was studied; the first three elements of $\mathbf{r}_3$ are the correlations of sport performance (Y_1) with Y_2 through Y_4. The intercorrelations among the CSAI scales follow.

Per the discussion in Section 16.4.1, the covariance matrix of $\mathbf{r}_3$ was computed using formulas (16.3) and (16.6), with the vector $\bar{\mathbf{r}}_n$ of sample-size weighted mean correlations (with elements $\bar{r}_{st} = \sum_{i=1}^{I} n_i r_{ist} \Big/ \sum_{i=1}^{I} n_i$) across all studies instead of the r values from the study with ID 3. Specifically $\bar{\mathbf{r}}_n = (-0.102 \;\; -0.177 \;\; 0.362 \;\; 0.519 \;\; -0.416 \;\; -0.414)'$, and the study's sample size was $n_3 = 37$. The covariance matrix for study with ID 3, computed using $\bar{\mathbf{r}}_n$ is

$$\text{Cov}(\mathbf{r}_3) = \begin{bmatrix} 0.026 & 0.013 & -0.009 & -0.003 & 0.008 & 0.004 \\ 0.013 & 0.025 & -0.009 & -0.001 & 0.003 & 0.007 \\ -0.009 & -0.009 & 0.020 & 0.001 & -0.001 & -0.002 \\ -0.003 & -0.001 & 0.001 & 0.014 & -0.005 & -0.005 \\ 0.008 & 0.003 & -0.001 & -0.005 & 0.018 & 0.008 \\ 0.004 & 0.007 & -0.002 & -0.005 & 0.008 & 0.019 \end{bmatrix}.$$

TABLE 16.1

Data from Craft et al. (2003) on Relationships between CSAI Subscales and Sport Behavior

		Variable names and corresponding correlations						
		Performance/ cognitive anxiety	Performance/ somatic anxiety	Performance/ self-confidence	Cognitive anxiety/ somatic anxiety	Cognitive anxiety/ self-confidence	Somatic anxiety/ self-confidence	Performance/ self-confidence partial
ID	n_i	r_{i12}	r_{i13}	r_{i14}	r_{i23}	r_{i24}	r_{i34}	$r_{14.23}$
1	142	−0.55	−0.48	0.66	0.47	−0.38	−0.46	0.54
3	37	0.53	−0.12	0.03	0.52	−0.48	−0.40	0.33
10	14	−0.39	−0.17	0.19	0.21	−0.54	−0.43	−0.07
22	100	0.23	0.08	0.51	0.45	−0.29	−0.44	0.65
26	51	−0.52	−0.43	0.16	0.57	−0.18	−0.26	0.04
28	128	0.14	0.02	0.13	0.56	−0.53	−0.27	0.25
36	70	−0.01	−0.16	0.42	0.62	−0.46	−0.54	0.43
70	30	−0.27	−0.13	0.15	0.63	−0.68	−0.71	−0.02

If a study does not report all correlations, its variance matrix can still be computed when the mean is used in the calculations. (This approach assumes data to be missing completely at random and is most appropriate under the common-effect model.)

16.5.2 Estimating the Mean Correlations and Their Covariance Matrix

For completeness, we first compute the univariate means using inverse-variance weighting. The common-effect mean is $\bar{r}_{CE} = (-0.176\ \ -0.228\ \ 0.465\ \ 0.535\ \ -0.452\ \ -0.454)'$. It is fairly similar to the sample-size weighted univariate mean, but all elements are slightly further from zero. The random-effects univariate mean from (16.5) is $\bar{r}_{RE} = (-0.099\ \ -0.181\ \ 0.314\ \ 0.535\ \ -0.445\ \ -0.448)'$. In practice, we prefer to interpret the estimates obtained from the multivariate approach.

Next, we show both the random-effects and common-effect multivariate means for the set of correlation matrices, and their variance matrices. The mean assuming random effects for all correlations was computed using the multivariate GLS approach shown in (16.7), with between-studies variances estimated via restricted maximum likelihood. The square form of the mean matrix is

$$\bar{\mathbf{R}}_{RE} = \begin{bmatrix} 1 & -0.098 & -0.176 & 0.319 \\ -0.098 & 1 & 0.527 & -0.418 \\ -0.176 & 0.527 & 1 & -0.401 \\ 0.319 & -0.418 & -0.401 & 1 \end{bmatrix}$$

The elements of $\bar{\mathbf{R}}_{RE}$ do not differ greatly from the multivariate common-effect means, which are

$$\bar{\mathbf{R}}_{CE} = \begin{bmatrix} 1 & -0.102 & -0.177 & 0.362 \\ -0.102 & 1 & 0.519 & -0.416 \\ -0.177 & 0.519 & 1 & -0.414 \\ 0.362 & -0.416 & -0.414 & 1 \end{bmatrix}.$$

A bigger difference is seen in the covariance matrices of the means. The covariance of the random-effects means, $\mathrm{Cov}(\bar{\mathbf{R}}_{RE}) = (\mathbf{X}'\hat{\mathbf{W}}\mathbf{X})^{-1}$, is

$$\mathrm{Cov}(\bar{\mathbf{R}}_{RE}) = \begin{bmatrix} 0.0190 & 0.0108 & -0.0048 & 0.0002 & -0.0018 & 0.0006 \\ 0.0108 & 0.0074 & -0.0028 & 0.0001 & -0.0010 & 0.0006 \\ -0.0048 & -0.0028 & 0.0071 & -0.0010 & 0.0012 & -0.0018 \\ 0.0002 & 0.0001 & -0.0010 & 0.0011 & -0.0005 & -0.0000 \\ -0.0018 & -0.0010 & 0.0012 & -0.0005 & 0.0021 & 0.0002 \\ 0.0006 & 0.0006 & -0.0018 & -0.0000 & 0.0002 & 0.0017 \end{bmatrix},$$

and the (inappropriate) covariance of the common-effect means is

$$\text{Cov}\left(\overline{\mathbf{R}}_{CE}\right)=\begin{bmatrix} 0.0017 & 0.0009 & -0.0006 & -0.0002 & 0.0005 & 0.0003 \\ 0.0009 & 0.0016 & -0.0006 & -0.0001 & 0.0002 & 0.0005 \\ -0.0006 & -0.0006 & 0.0013 & 0.0000 & -0.0000 & -0.0001 \\ -0.0002 & -0.0001 & 0.0000 & 0.0009 & -0.0003 & -0.0003 \\ 0.0005 & 0.0002 & -0.0000 & -0.0003 & 0.0012 & 0.0005 \\ 0.0003 & 0.0005 & -0.0001 & -0.0003 & 0.0005 & 0.0012 \end{bmatrix},$$

which has much smaller variances for the first three elements—the correlations of the CSAI scales with performance. Omitting the between-studies variances greatly underestimates the uncertainty in these three relationships. Further evidence against omitting the between-studies variances from our model are the multivariate I^2 statistics (Jackson et al., 2012). Because we adopted an unstructured variance-covariance matrix for our models, we obtained I^2 statistics for each of our six sets of correlations. These I^2 values show that the proportion of observed dispersion varies widely across those relationships (91%, 78%, 81%, 16%, 43%, and 31%), suggesting that ignoring between-studies variance in this model would not be advisable. Though in current practice the test of homogeneity is not often used, we note that it also confirms that the common-effect model is not appropriate ($Q_{CE}=176.7$, df$=42$). For further analyses, we use the random-effects means.

16.5.3 Estimating Between-Studies Variation

Next, we display the estimate of the between-studies variance. The diagonals represent the variances, the elements above the diagonal represent the covariances, and the elements below the diagonal represent the correlations computed from the covariances and variances. Thus, the estimate of the between-studies variance **T** estimated using restricted maximum likelihood is

$$\hat{\text{T}}=\begin{bmatrix} 0.132 & 0.076 & -0.031 & -0.005 & 0.007 & 0.005 \\ 0.999 & 0.044 & -0.042 & -0.005 & -0.001 & 0.008 \\ -0.421 & -0.979 & 0.042 & -0.004 & 0.002 & 0.001 \\ -0.393 & -0.626 & -0.517 & 0.001 & -0.003 & 0.002 \\ 0.232 & -0.587 & 0.097 & -0.938 & 0.006 & -0.002 \\ 0.201 & 0.586 & 0.063 & 0.988 & -0.485 & 0.004 \end{bmatrix}.$$

Here **T** was estimated without constraints, using a REML estimator. The smallest variance is that of the correlation between cognitive and somatic anxiety. The method-of-moments estimate of this variance component (not shown) is just below zero. This illustrates that even when the overall test of heterogeneity is significant, correlations for specific relationships may appear homogeneous. Indeed, all of the variances for the correlations among the CSAI subscales (the last three diagonal elements) are much smaller than the anxiety-outcome variances.

The between-studies correlations involving these last three r values have a few quite extreme values near 1 and –1. These may appear inconsistent with the idea that the final three sets of r values each arise from a single population, with variances of zero.

Technically, if the true values of the between-studies variances are zero, the covariances would also be expected to be zero, and the between-studies correlations would be undefined, as we cannot divide by zero. Prior experience has suggested that, with small numbers of studies, between-studies covariances are challenging to estimate well (see also Riley et al., 2007), particularly if the studies are also small. Such cases may produce values of between-studies correlations below –1 or above 1 (especially with method-of-moments estimators). Becker (2009) and Becker and Aloe (2019) have suggested that in such cases, meta-analysts may want to treat the off-diagonal elements as zero and use only the variances in $\hat{\mathbf{T}}$ in further estimation.

Under the more appropriate random-effects model, $Q_B = 401.67$ with df = 6, still indicating that at least one population correlation differs from zero. Individual tests or confidence intervals can be used to identify elements whose population values are likely to differ from zero. An individual test of whether the first correlation (of cognitive anxiety with performance) is zero is not significant under random-effects ($z = -0.71$), but all other mean correlations differ from zero at $\alpha = 0.05$.

16.5.4 The Prediction Model for Sport Outcomes

Next, we estimate the regression model for predicting sport outcomes from the three components of anxiety, using the random-effects mean matrix. For our data, the partitioned matrix laid out in (16.9) above (under random effects) is

$$\bar{\mathbf{R}}_{RE} = \left[\begin{array}{c|ccc} 1 & -0.098 & -0.176 & 0.319 \\ \hline -0.098 & 1 & 0.527 & -0.418 \\ -0.176 & 0.527 & 1 & -0.401 \\ 0.319 & -0.418 & -0.401 & 1 \end{array}\right]$$

and the regression slopes $\mathbf{b}^* = \bar{\mathbf{R}}_{XX}^{-1}\bar{\mathbf{R}}_{XY}$ for predicting physical or sport performance from cognitive and somatic anxiety and self-confidence are 0.083 (SE = 0.154), –0.093 (SE = 0.064), and 0.316 (SE = 0.102). Only the third slope, for self-confidence, is significantly different from zero.

Figure 16.3 shows the full path model with both direct and indirect effects based on Becker's approach to path estimation. While only one direct effect (for self-confidence) is significant under the random-effects model, indirect effects for the other two components of anxiety appear to flow through self-confidence, at least in these studies. The negative slopes on the inter-anxiety paths show that the two negatively worded subscales (cognitive and somatic anxiety) relate negatively to self-confidence, which then relates positively to sport outcomes.

The model in Figure 16.3 is just identified with zero degrees of freedom. Therefore, the fit of this model is perfect. However, the variance explained in self-confidence and in performance can be used as an indicator of the model quality (i.e., the larger the percent of variance explained, the more these sets of predictors tell us about the outcomes). An estimator of variance explained for each outcome in our path model can be obtained by incorporating the proper elements of $\bar{\mathbf{R}}_{RE}$ into the formula $R^2 = \mathbf{b}^*\bar{\mathbf{R}}_{XY}$. Not surprisingly given the relatively low correlations in $\bar{\mathbf{R}}_{RE}$, the variance explained by both of these component models appears small, with 22% of the variance in self-confidence explained, and 11% of the variance explained for predicting sport performance. These relatively low values are

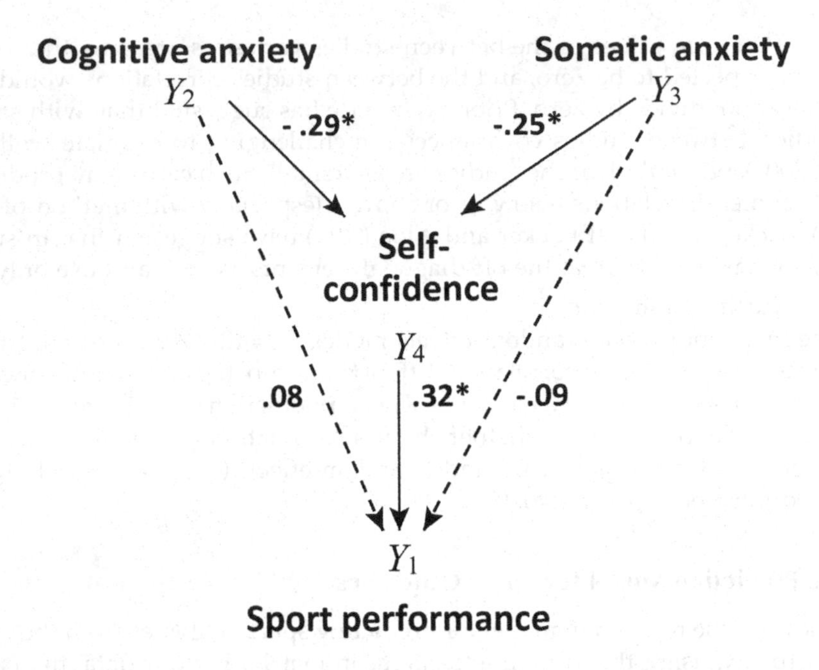

FIGURE 16.3
Random-effects path coefficients for prediction of sport performance. Asterisks mark coefficients significant at the 0.05 level.

not unreasonable, as many other factors could affect sport self-confidence and sports performance (e.g., previous performance, injuries, and stress).

16.5.5 An Analysis of Partial Correlations

Alternately, in cases when a complex model is not of interest, it may make sense to examine a single partial relationship rather than a model based on the full correlation matrix. Partial effects can be computed at the study level, using formulas (16.1) and (16.2) above, then summarized, or "synthetic partials" can be obtained by substituting means from a summary Pearson's correlation matrix into (16.1) and (16.2) in place of single r values.

For example, our dataset includes the partial correlation of self-confidence and performance, adjusting both for cognitive anxiety and somatic anxiety ($r_{14.23}$). The partial correlations for this relationship in each study are shown in the last column of Table 16.1. The values reveal the complexity of the partial-r formula—there is no consistent trend for the partial $r_{14.23}$ to be larger or smaller than the bivariate r_{14}, because the partial r depends on all other relevant relationships (i.e., on r_{13}, r_{23}, etc.), and in our data the signs and magnitudes of the component correlations are not consistent across studies. For the relationship between performance and self-confidence, partialling out cognitive anxiety and somatic anxiety, a univariate analysis under random-effects produces a mean of 0.328 (SE = 0.090). The random-effects mean is relatively close to the mean bivariate r from the random-effects mean correlation matrix (0.319), suggesting that adjusting for cognitive anxiety and somatic anxiety does little on average to impact the relationship of self-confidence to performance. This is also consistent with the finding from our model above of a significant partial slope of 0.316 for self-confidence on performance, and non-significant slopes for the other anxiety components.

We can also use the summary correlation matrix to compute "synthetic" partial or semi-partial correlations. These can be computed using formulas 16.1 and 16.2 (or more complex versions of these if two or more covariates are to be partialled out), but with elements of the mean correlation matrix substituted for individual correlation values. A formula in matrix notation for the full matrix of synthetic partial correlations is

$$-\left(\mathrm{diag}\left(\bar{\mathbf{R}}_{\mathrm{RE}}^{-1}\right)\right)^{-\frac{1}{2}} \bar{\mathbf{R}}_{\mathrm{RE}}^{-1}\left(\mathrm{diag}\left(\bar{\mathbf{R}}_{\mathrm{RE}}^{-1}\right)\right)^{-\frac{1}{2}},$$

where $\bar{\mathbf{R}}_{\mathrm{RE}}$ is the average bivariate correlation matrix under the random-effects model. In this matrix, each partial correlation controls for all additional variables in the mean matrix (for instance, its elements estimate $\rho_{12.34}$, $\rho_{13.24}$, etc.). This way of estimating the synthetic partial correlation matrix can yield negative diagonal elements (i.e., −1), which are typically ignored.

The variance-covariance matrix of the synthetic partial-correlation matrix has been derived using the multivariate delta method by Aloe and Toro Rodriguez (under review). The element of this synthetic partial-correlation matrix corresponding to the correlation between performance and self-confidence, partialling out cognitive anxiety and somatic anxiety is 0.28, with SE = 0.089; these values are similar to those from the univariate analysis of the partials. This coefficient is the partial correlation corresponding to the path from self-confidence to performance (the center arrow) in Figure 16.3.

16.6 Comparison of Matrix and Partial-Effect Approaches

Until recently, the synthesis of correlation matrices (Becker, 1992) and the synthesis of indices such as slopes or semi-partial *r*s from regression results (Aloe and Becker, 2012; Becker and Wu, 2007) were perceived as two separate approaches to the synthesis of complex studies. However, when a meta-analyst wants to understand, say, a relationship between two variables adjusting for the effects of others, these two approaches can yield comparable results for point estimates, standard errors, and heterogeneity measures (Aloe, 2014; Aloe and Toro Rodriguez, under review). Each approach (i.e., computing partial effect estimates from synthesized correlation matrices, and synthesizing partial effect sizes directly) has advantages and disadvantages.

The advantage of synthesizing the full correlation matrix is flexibility. A variety of models and indices can be computed from the summary matrix across studies, allowing for exploration of different theoretical and empirical models (Becker, 2009). Models that have not been examined in any prior primary study can be constructed with this approach. In contrast, when meta-analysts synthesize partial effects such as partial or semi-partial correlations directly, they must commit to a specific focal relationship, or set of relationships, that have been examined in all studies. Each relationship is examined in turn using univariate analyses. Multivariable path or regression models are not easily constructed with this approach, unless similar models are estimated in each study (Becker and Wu, 2007).

Another major difference between these two approaches relates to the treatment of missing data. Many primary studies do not report full sufficient statistics for all variables that may interest the meta-analyst. Gozutok, Alghamdi, and Becker (under revision) examined 163 recent multiple-regression studies in top education and psychology journals and found that only 36% of studies reported full correlation matrices. Also, some primary researchers may fully report summary statistics, but may not have investigated all variables of interest to the meta-analyst. This lack of data will impact analyses in different ways depending on the approach taken.

When synthesizing correlation matrices, one can incorporate all available correlations from each study, but missing data will cause different elements of the average correlation matrix to be based on different numbers of studies, and potentially on different populations. If no studies have examined a particular pair of variables, the correlation matrix will have an empty entry, making it impossible to examine models that include both of the variables in the missing correlation (see, for example, Whiteside and Becker, 2000). On the other hand, in syntheses where partial effect sizes are directly extracted from primary studies, if studies vary in the covariates that are partialled out (say, in regression models or partial *r*s), it may be difficult or impossible to extract similar partial effect indices. Aloe and Becker (2012) follow a suggestion by DuMouchel (1994) and suggest that meta-analysts use meta-regression to account for the absence of key components such as important covariates in the primary-study analyses. The absence of key covariates (such as age or socioeconomic status) in each primary study would be coded using dummy variables, thus, the intercept in the meta-regression model can be interpreted as the partial effect predicted from a model containing all key covariates. The slope coefficients for the dummy variables will indicate the magnitude of the bias introduced by omitting each key covariate. A different approach to this problem is given in Sections 9.7.2 and 21.3.

Because estimation of the partial effect sizes of interest occurs at a later stage with the multivariate matrix approach, the variance of the "synthetic partial effect sizes" computed based on a mean correlation matrix is not estimated in the typical fashion. The estimates can be obtained using mean correlation matrices computed under either common-effect or random-effects assumptions, but separate between-studies and within-study variance components will not be available in the metric of the synthetic index computed from the mean *r* matrix.

The two approaches differ as well in their treatment of moderators. Syntheses of partial effect sizes extracted directly from regression models can easily include typical analyses of moderators (i.e., effect modifiers) such as meta-regressions or analysis-of-variance-like approaches. On the other hand, with models and other partial effect sizes obtained from correlation matrices, the meta-analyst accounts for moderators at the average-correlation-matrix stage. This can be done by grouping correlations from subsets of studies and analyzing them separately using the methods described above, as done in Whiteside and Becker (2000). Using models that parameterize different population correlations for, say, K subsets accomplishes essentially the same thing, for example, by defining $\boldsymbol{\rho}$ as a KP^* vector $(\rho_{112}, \rho_{113}, \ldots, \rho_{212}, \ldots, \rho_{K12}, \rho_{K13}, \ldots, \rho_{K(P-1)P})$ then computing K path models, partial indices, and so on. Differences due to study characteristics then can be evaluated in the metric of the partial effects, though as noted above the variance of the synthetic partial effect size estimated from a mean matrix cannot be partitioned in the typical fashion. That is, between-studies variation and covariation are estimated in the correlation metric then transformed in line with the function of interest, and are not separable components of the variances of functions of the mean matrix, unlike in the partial-effect approach.

For example, assume that one wants to evaluate whether the relationship between self-confidence and sport performance, adjusted for the effect of cognitive and somatic anxiety, is stronger in individual sports than in team sports (Craft et al., 2000). Under a univariate framework, the researcher would directly extract partial effect sizes from the primary studies and compare those from team sports and individual sports. On the other hand, under a multivariate framework, the researcher would extract correlation matrices and sport type from the studies, then separate the team and individual sport studies to compute mean *r* matrices and path models or partial effect sizes for each subset.

16.7 Summary

Regardless of the approach taken, meta-analysts have an array of synthesis methods that enable the exploration of complex relationships. These techniques, when properly employed, can illuminate current understandings and identify avenues for further inquiry. A key advantage of the synthesis of matrices is its capacity to examine indirect effects. Consider Whiteside and Becker's (2000) findings about the frequency of father-child visitation in divorced families. This variable had been dismissed as a predictor of child outcomes based on narrative reviews and analyses of bivariate correlations. In contrast, in their model-based analysis, Whiteside and Becker showed that this predictor, along with pre-divorce father-child involvement, played indirect roles via the quality of the father-child relationship. More frequent father visits were associated with positive child outcomes when the father-child relationship was strong.

Finding that a set of studies has not considered (or not studied thoroughly) a particular promising relationship can highlight the importance of including it in future primary studies. Brown et al. (2015) noted that few studies had examined the role of adherence to keeping doctor or clinic appointments for patients with diabetes. Also, when it was studied, appointment-keeping did not appear with other key predictors of metabolic control such as anxiety. Appointments can be crucial in keeping patients on track with their diabetes management, and existing studies supported its importance (Brown et al., 2016). The fact that patient appointment-keeping behaviors were overlooked in most research studies suggests an important potential predictor for future studies. In a typical meta-analysis of one predictor at a time, appointment-keeping could be overlooked or eliminated due to the small number of studies that have examined it.

These examples show that a focus on complex models, and the correlation matrices underlying them, can lead to better understandings of interrelations among variables than are available through analyses of simple bivariate correlations. Analyses of partial effects play a valuable role as well when full systems of variables are not of interest. Model-based analyses of correlation matrices can bring potential avenues for further investigation into sharper focus.

References

Aloe AM, 2014. An empirical investigation of partial effect sizes for meta-analysis of correlational data. *The Journal of General Psychology* **141**(1): 47–64.

Aloe AM and Becker BJ, 2012. A partial effect size for regression models. *Journal of Educational and Behavioral Statistics* **37**(2): 278–297.

Aloe AM and Thompson CG, 2013. The synthesis of partial effect sizes. *Journal of the Society for Social Work and Research* **4**(4): 390–405.

Aloe AM and Toro Rodriguez R, under review. Synthesis of multivariate and univariate partial effect sizes.

Becker BJ, 1992. Using results from replicated studies to estimate linear models. *Journal of Educational Statistics* **17**(4): 341–362.

Becker BJ, 1995. Using results from replicated studies to estimate linear models: Correction. *Journal of Educational and Behavioral Statistics* **20**(1): 100–102.

Becker BJ, 2000. Multivariate meta-analysis. In Tinsley HEA and Brown S (Eds). *Handbook of Applied and Multivariate Statistics and Mathematical Modeling*. San Diego: Academic Press, 499–525.

Becker BJ, 2009. Model based meta-analysis. In Cooper HM, Hedges LV and Valentine JC (Eds). *The Handbook of Research Synthesis and Meta-Analysis*. 2nd Edition. New York: Russell Sage Foundation, 377–395.

Becker BJ and Aloe AM, 2019. Model-based meta-analysis. In Cooper HM, Hedges LV and Valentine JC (Eds). *The Handbook of Research Synthesis and Meta-Analysis*. 3rd Edition. New York: Russell Sage Foundation, 339–366.

Becker BJ and Fahrbach KR, 1994. A comparison of approaches to the synthesis of correlation matrices. Paper presented to the annual meeting of the American Educational Research Association, at New Orleans, LA.

Becker BJ and Schram CM, 1994. Examining explanatory models through research synthesis. In Cooper HM and Hedges LV (Eds). *The Handbook of Research Synthesis*. 1st Edition. New York: Russell Sage Foundation, 357–381.

Becker BJ and Wu M-J, 2007. The synthesis of regression slopes in meta-analysis. *Statistical Science* **22**(3): 414–429.

Bowman NA, 2010. College diversity experiences and cognitive development: A meta-analysis. *Review of Educational Research* **80**(1): 4–33.

Brown SA, Becker BJ, Garcia AA, Brown A and Ramirez G, 2015. Model-driven meta-analyses for informing health care: A diabetes meta-analysis as an exemplar. *Western Journal of Nursing Research* **37**(4): 517–535.

Brown SA, Garcia AA, Brown A, Becker, BJ, Conn VS, Ramírez G, Winter MA, Sumlin LL, Garcia TJ and Cuevas HE, 2016. Biobehavioral determinants of glycemic control in type 2 diabetes: A systematic review and meta-analysis. *Patient Education and Counseling* **99**(10): 1558–1567.

Brown SA and Hedges LV, 1994. Predicting metabolic control in diabetes: A pilot study using meta-analysis to estimate a linear model. *Nursing Research* **43**(6): 362–369.

Burke DL, Bujkiewicz S and Riley RD, 2018. Bayesian bivariate meta-analysis of correlated effects: Impact of the prior distributions on the between-study correlation, borrowing of strength, and joint inferences. *Statistical Methods in Medical Research* **27**(2): 428–450.

Cheung, MW-L and Chan W, 2004 Testing dependent correlation coefficients via structural equation modeling. *Organizational Research Methods* **7**(2): 206–223.

Cheung, MW-L and Chan W, 2005. Meta-analytic structural equation modeling: A two-stage approach. *Psychological Methods* **10**(1): 40–64.

Craft LL, Magyar TM, Becker BJ and Feltz DL, 2000. The relationship between the CSAI-2 and athletic performance: A meta-analysis. *Journal of Sport and Exercise Psychology* **25**(1): 44–65.

DuMouchel W, 1994. Hierarchical Bayes linear models for meta-analysis. Technical Report Number 27. National Institute of Statistical Sciences. Retrieved July 15, 2010, from http://nisla05.niss.org/technicalreports/tr27.pdf.

Field AP, 2001. Meta-analysis of correlation coefficients: A Monte Carlo comparison of fixed- and random-effects methods. *Psychological Methods* **6**(2): 161–180.

Field AP, 2005. Is the meta-analysis of correlation coefficients accurate when population correlations vary? *Psychological Methods* **10**(4): 444–467.

Fisher RA, 1932. *Statistical Methods for Research Workers*. Edinburgh: Oliver and Boyd.

Furlow CF and Beretvas SN, 2005. Meta-analytic methods of pooling correlation matrices for structural equation modeling under different patterns of missing data. *Psychological Methods* **10**(2): 227–254.

Glass GV, 1976. Primary, secondary, and meta-analysis of research. *Educational Researcher* **5**(10): 3–8.

Gozutok AS, Alghamdi A and Becker BJ, under revision. Worrisome news: Authors of regression studies still do not follow the reporting standards. Manuscript submitted to *Educational Researcher*.
Griffin HD, 1932. On the coefficient of part correlation. *Journal of the American Statistical Association* **27**(179): 298–301.
Gustafson R, 1961. Partial correlations in regression computations. *Journal of the American Statistical Association* **56**(294): 363–366.
Hedges LV and Olkin I, 1985. *Statistical Methods for Meta-Analysis*. Orlando, FL: Academic Press.
Hedges LV and Vevea JL, 1998. Fixed- and random-effects models in meta-analysis. *Psychological Methods* **3**(4): 486–504.
Higgins JPT, Thompson SG, Deeks JJ and Altman DG, 2003. Measuring inconsistency in meta-analyses. *British Medical Journal* **327**(7414): 557–560.
Hoaglin DC, 2016. Misunderstandings about Q and 'Cochran's Q test' in meta-analysis. *Research Synthesis Methods* **35**(4): 485–495.
Knapp G and Hartung J, 2003. Improved tests for a random effects meta-regression with a single covariate. *Statistics in Medicine* **22**(17): 2693–2710.
Jackson D, White IR and Riley RD, 2012. Quantifying the impact of between-study heterogeneity in multivariate meta-analyses. *Statistics in Medicine* **31**(29): 3805–3820.
Keef SP and Roberts LA, 2004. The meta-analysis of partial effect sizes. *British Journal of Mathematical and Statistical Psychology* **57**(1): 97–129.
Martens R, Vealey RS and Burton D, 1990. *Competitive Anxiety in Sport*. Champaign, IL: Human Kinetics.
Olkin I and Siotani M, 1976. Asymptotic distribution of functions of a correlation matrix. In Ikeda S (Ed). *Essays in Probability and Statistics*. Tokyo: Shinko Tsusho, 235–251.
Pearson K, 1915. On the correlation ratio. *Proceedings of the Royal Society Series A* **91**: 492–498.
Prevost AT, Mason D, Griffin S, Kinmonth A-L, Sutton S and Spiegelhalter D, 2007. Allowing for correlations between correlations in random-effects meta-analysis of correlation matrices. *Psychological Methods* **12**(4): 434–450.
Raudenbush SW, Becker BJ and Kalaian S, 1988. Modeling multivariate effect sizes. *Psychological Bulletin* **102**: 111–120.
Riley RD, 2009. Multivariate meta-analysis: The effect of ignoring within-study correlation. *Journal of the Royal Statistical Society: Series A* **172**(4): 789–811.
Riley RD, Abrams KR, Lambert PC, Sutton AJ and Thompson JR, 2007. Bivariate random effects meta-analysis and the estimation of between-study correlation. *BMC Medical Research Methodology* **7**(3): 1–15.
Viechtbauer W, 2005. Bias and efficiency of meta-analytic variance estimators in the random-effects model. *Journal of Educational and Behavioral Statistics* **30**(3): 261–293.
Viechtbauer W, 2007. Hypothesis tests for population heterogeneity in meta-analysis. *British Journal of Mathematical and Statistical Psychology* **60**(1): 29–60.
Viechtbauer W, 2010. Conducting meta-analyses in R with the metafor package. *Journal of Statistical Software* **36**(3): 1–48. Available from http://www.jstatsoft.org/v36/i03/.
Whiteside MF and Becker BJ, 2000. The young child's post-divorce adjustment: A meta-analysis with implications for parenting arrangements. *Journal of Family Psychology* **14**(1): 1–22.

17

The Meta-Analysis of Genetic Studies

Cosetta Minelli and John Thompson

CONTENTS

Human deoxyribonucleic acid (DNA) is a chain of just over three billion chemical bases and most cells of the body contain two copies of this DNA, one inherited from our mother and one from our father. Very occasionally, mutations occur during the copying of the DNA, which get passed on to future generations and eventually these genetic variants can become established in a proportion of the population. Genetic epidemiological studies investigate the relationship between such variants and measures of health (Attia et al., 2009).

The number of meta-analyses of these genetic studies published each year has increased rapidly over the last three decades while the nature of the meta-analyses has changed to reflect developments in technology and primary study design. Early primary studies were often family based and aimed to identify stretches of DNA passed between family members affected by the same disease (linkage studies, see Section 17.3.1), while recent studies are more often population-based and recruit unrelated individuals in order to measure the association between the presence of a particular genetic variant and a disease (Burton et al., 2005).

Until ten years ago, the study of genetic association was dominated by candidate-gene studies designed to investigate the association between a few candidate genetic variants and a specific disease or trait, with the candidate variants selected based on *a priori* knowledge of their biological function (Kwon and Goate, 2000). Then, around 2007, the nature of genetic investigation changed when new technologies made it possible to perform genome-wide association studies (GWAS), in which very large numbers of variants from across the genome are studied simultaneously in a hypothesis-free design, that is with no hypothesized link with the outcome for any specific variant. One of the first GWAS was performed by the Wellcome Trust Case Control Consortium who investigated the associations between 500,000 genetic variants and seven diseases (Burton et al., 2007). Similar studies quickly followed and now it is routine to assess millions of variants on each participant.

The revolution in genetic technology has had a major impact on evidence synthesis, both in terms of the way that it is carried out and its scientific importance. According to HuGE Watch (Yu et al., 2008b), the number of publications on genetic meta-analysis rose from 259 in 2004 to 7068 in 2014; over the same period, the number of genes studied in genetic meta-analyses rose from 252 to 6857 and the number of diseases from 257 to 2400.

In this chapter, we will focus on the meta-analysis of genetic association studies, including candidate-gene studies (Section 17.1) and GWAS (Section 17.2), with a briefer coverage of the meta-analysis of other types of genetic studies (Sections 17.3 and 17.4). We mainly assume that aggregate data are available, because individual participant data are rarely available on genetic studies. We discuss the aims and ideas rather than the mathematical models: these in most cases are the same as those discussed in Chapter 4 and are usually fitted in a frequentist framework, although we note situations where Bayesian analyses are helpful.

17.1 The Meta-Analysis of Candidate-Gene Studies

Of gene-disease associations identified in single candidate-gene studies, the proportion that have been replicated in further studies is disappointingly low (Colhoun et al., 2003). Although, in theory, genuine population diversity might partly explain this, in practice, the reasons for lack of replication are often methodological, including small sample sizes and bias (Ioannidis et al., 2001). Among the most important biases are publication bias and reporting bias (Chapter 13), whereby authors only publish a paper if they obtain statistically significant findings or only report those genetic associations which reach statistical significance. One consequence is that the first study published on a gene-disease association may suggest a genetic effect that is not found, or is found with much smaller magnitude, in subsequent studies, a problem referred to as the "first-study effect" (Trikalinos et al., 2004). More generally, the investigation of between-study heterogeneity and possible reasons for it represents a crucial aspect of the meta-analysis of candidate-gene studies.

Practical guidelines for the conduct and reporting of systematic reviews and meta-analyses of candidate-gene studies have been developed by HuGENet and summarized in their *HuGE Review Handbook* (Little and Higgins, 2006). HuGENet is a global collaboration launched in 1998 to create an information exchange network on the findings and implications of gene discovery in public health (Khoury et al., 2004). Since 2001, HuGENet has maintained a searchable online database of published studies investigating the association of genetic variants with disease traits called the HuGE Navigator (Yu et al., 2008a). This database provides a very useful overview of candidate-gene studies and GWAS, with a set of searching tools available to refine the query by disease trait or gene of interest and to provide additional information on genetic variation, for example prevalence estimates, with direct link to the UCSC Genome Browser (genome browser hosted by the University of California in Santa Cruz, UCSC, and available at https://phgkb.cdc.gov/PHGKB/hNHome.action).

A review of the quality of 120 randomly selected meta-analyses of candidate-gene studies published between 2005 and 2007 (Minelli et al., 2009) showed that the quality of conduct and reporting had improved compared with similar meta-analyses published before 2000 (Attia et al., 2003). The average quality of meta-analyses in this field is similar to that observed in other fields of medicine, with general concerns including lack of explicit reporting of inclusion and exclusion criteria, failure to explore possible sources of heterogeneity, and absence of an investigation of publication bias. The quality of the handling of specific genetic issues, such as the checking of Hardy–Weinberg equilibrium (HWE) and the choice of a genetic model (see Sections 17.1.3 and 17.1.4), is particularly low and has not shown much improvement over time. These issues are especially neglected in those meta-analyses in specialist medical journals that accompany a report on a primary study.

In the following sections, we discuss the key issues of the meta-analysis of candidate-gene association studies and provide some practical advice on how to address them. Some of the methodological aspects presented in these sections will be developed further in the sections on the meta-analysis of GWAS, where we highlight their implications within a hypothesis-free context.

17.1.1 Data Extraction from Genetic Association Studies

The range of options open to the meta-analyst is dictated by the quality of the reporting of the primary studies. The typical candidate-gene study reports on the genotypic effects of biallelic single nucleotide polymorphisms (SNPs), where the genetic variant can take only two possible forms (alleles) and the variation consists of a replacement of a single base with another (Attia et al., 2009). If a SNP can take two forms, A and a, then a participant who inherits a separate copy of their DNA from each parent can have one of three genotypes AA, Aa, or aa. When the outcome under study is binary, such as disease status, then the data from the study can be summarized as counts of participants in a 2×3 table and when this information is available, the meta-analysis is equivalent to an analysis of individual participant data (IPD), as discussed in Chapter 8. For a continuous outcome, such as the measurement of a biomarker, the number, mean, and standard deviation of the measurements in each cell of the 2×3 table, would, under a Gaussian model, be equivalent to having IPD (see Chapter 5).

Unfortunately, published articles may instead only give an effect estimate, such as a log odds ratio or a regression coefficient, and its standard error, confidence interval, or p-value. The difficulty with these estimates is that they are always obtained based on strong assumptions; those assumptions may vary between studies, and without the IPD, they cannot be checked (Ioannidis et al., 2002). In particular, the most important decisions that the researchers of the primary studies will have had to make are the choice of genetic model and whether or not to adjust for covariates. Since a SNP has three genotypes, different assumptions can be made about the effect of the heterozygous Aa genotype in relation to AA and aa; this assumed pattern is called the genetic model. If Aa has the same effect on the outcome as the aa genotype then the A allele is called recessive; if Aa has the same effect as AA then A is dominant; if Aa has an effect midway between aa and AA (on some scale) then A is said to be additive. There are other possibilities, such as the overdominant genetic model in which the Aa genotype shows the largest effect, although they are far less common. Reports of associations under different genetic models cannot be meta-analyzed using standard methods, and a measure under one model cannot usually be used to deduce the association under another.

The impact of covariate adjustment on a meta-analysis is less clear cut. Since a participant's genotype is assigned before birth, adjustment for subsequent factors, such as smoking, will not alter the expected size of the association unless either the factor lies on the causal pathway from gene to outcome, or it interacts with the gene. When the meta-analyst can confidently exclude these possibilities, it would be valid to meta-analyze adjusted and unadjusted effect estimates together, but of course such detailed knowledge of the biology of a SNP is often not available. Methods for combining adjusted and unadjusted effect estimates are also discussed in Chapter 21.

17.1.2 Assessment of Study Quality

The assessment of the quality of the primary studies is an important aspect of any meta-analysis and for candidate-gene studies the main tool is the STREGA (STrengthening the REporting of Genetic Association studies) statement, proposed in 2009 by Little et al. (Little et al., 2009). In addition to general aspects of the meta-analyses of observational studies, STREGA covers specific methodological issues, including assessment of genotyping error, evaluation of HWE, justification of the selection of genes under study, and investigation of population stratification due to the mixture of participants from heterogeneous genetic backgrounds. A more recent tool is Q-Genie, which provides an overall quality score for the genetic association studies (Sohani et al., 2015). The GRIPS statement also extends STREGA but by addressing issues concerned with the quality of genetic risk predictive studies (Janssens et al., 2011).

17.1.3 Choice of the Genetic Model

When the published data are equivalent to IPD, the meta-analyst will need to decide which genetic model is most biologically plausible and the conclusions may well be sensitive to this assumption (Minelli et al., 2005b). The comparison between the three genotypes can be summarized by either (i) ignoring the Aa, and comparing aa with AA; (ii) performing separate pairwise comparisons; (iii) assuming a *recessive model* to justify combining the aa and Aa genotypes and comparing aa + Aa with AA; (iv) assuming a *dominant model* and comparing aa with Aa + AA; or (v) assuming an *additive model* and regressing the outcome on the number of A alleles.

The assumption of a genetic model in options (iii) to (v) increases statistical power, but there should be *a priori* knowledge to support the choice. In practice, the meta-analysis often provides only weak evidence to aid selection of the genetic model, especially when the allele frequency is low. Assuming a "wrong" genetic model is a potential source of bias in the pooled estimate of the genetic effect (Minelli et al., 2005b).

When the underlying genetic model is unknown, it is better to use pairwise comparisons of the three genotypes, and the loss of power may be reduced through the use of a bivariate meta-analysis (Minelli et al., 2005a) (Chapter 9). In particular, the data can be utilized more efficiently using a "genetic model-free" approach, which estimates the genetic model along with the genetic effect size (Minelli et al., 2005a, 2005b).

17.1.4 Evaluation of HWE

The Hardy–Weinberg Equilibrium law states that if two alleles, A and a, with frequencies p and $q=1-p$, are in equilibrium in a population, then the proportion of people with genotypes AA, Aa, and aa will be p^2, $2pq$, and q^2, respectively (Sham, 2001). HWE describes the steady state for a population, where there are no selective forces (e.g., mutation, inbreeding, selective survival) acting on a particular locus or gene. Genuine departures from HWE in populations do occur but their overall impact is usually slight. On the other hand, departures from HWE in a sample may reflect methodological problems with that study, such as genotyping error, population stratification, or selection bias (Khoury et al., 2004), so that checking whether the allele frequencies at a particular SNP observe HWE proportions can be used as a quality measure.

When the equivalent of IPD is available, the meta-analyst should check for HWE separately in each study, with a view to identifying problematic studies that depart from HWE. For a cohort study, HWE should be tested in the whole study population, and in a case-control study, it should be tested in the controls, since these are more representative of the study population. The cases may be out of HWE because of the impact of the genetic variant on disease risk, so a small departure from HWE in the cases is not necessarily diagnostic of a problem.

Different tests have been proposed to evaluate departures from HWE (Wang and Shete, 2012), with the most commonly used being the chi-squared and the exact tests (Minelli et al., 2008), but all of them suffer from low statistical power. Therefore, it is advisable to measure the magnitude of the deviation from HWE as well as its significance, but this is very rarely done (Minelli et al., 2008). Adjustment of the final pooled estimate of the genetic effect for the magnitude of the deviation, or a more drastic approach consisting in excluding studies where the genetic variant investigated is not in HWE, has been recommended (Trikalinos et al., 2006). We suggest that lack of HWE should be treated as a reason for further investigation of a primary study rather than as grounds for its exclusion. An empirical investigation of departures from HWE in published meta-analyses showed that the chance of an extreme deviation from HWE decreases with increasing sample size (Minelli et al., 2008). The recommended exclusion of studies based on hypothesis testing of HWE at the level of 0.05 is questionable as larger studies will produce smaller p-values for the same sized departure from HWE. Excluding them (or adjusting their results for the magnitude of the departure) even when there is no evidence of underlying methodological problems may be inappropriate and just as likely to be detrimental as beneficial.

When the meta-analysis is based on summary statistics that do not enable HWE to be assessed, it must rely on reported HWE assessments from the published reports.

17.1.5 Assessment of Between-Study Heterogeneity

Hypothesis testing alone is insufficient to assess between-study heterogeneity because the tests have low power and do not provide information on its magnitude, which can be directly measured by the between-study variance (τ^2) or quantified using I^2 (Chapter 4) (Higgins et al., 2003). I^2 is much more used than τ^2 and yet its limitations are rarely acknowledged (Higgins, 2008). I^2 describes the proportion of the total variance that is due to heterogeneity, not the magnitude of the heterogeneity itself (Higgins and Thompson, 2002). Also, I^2 is influenced by the sample size of the studies included in the meta-analyses, since it depends on within-study error as well as between-study variation; if the studies are large, as it is often the case for candidate-gene studies and the rule for GWAS, I^2 could be large even in the presence of a small τ^2. Our suggestion is always to report both τ^2 and I^2.

The priority in the presence of heterogeneity is to investigate its causes, which include differences in study quality and methodology, but also population-specific gene-environment interactions. Investigation of heterogeneity due to study-level factors can be performed using meta-regression or subgroup analysis (see Chapter 7) accompanied by formal interaction testing, although statistical significance should be interpreted in the light of the low power of interaction tests (Brookes et al., 2004). In the absence of IPD, meta-regression based on average participant-level characteristics is not recommended, as it is difficult to interpret (Berlin et al., 2002) and has very little power to explain heterogeneity (Lambert et al., 2002). It is important to note that any proposed explanation for heterogeneity should be treated cautiously even if it shows statistical significance in a formal test, as the hypothesis will have been formulated and tested based on the same data.

Heterogeneity is often seen as a nuisance because most genetic association studies are designed to measure average genetic effects over a population and not to uncover heterogeneity due to gene-gene and gene-environment interactions. However, as more genetic research is conducted in different parts of the world and more genetic data are collected from studies where participants have undergone thorough assessment of physiological and clinical characteristics as well as lifestyle and environmental exposures, the exploration of gene-environment interaction through meta-analysis will become a real possibility. The power to detect gene-environment interactions within an individual study will be limited, but meta-analysis of gene-environment interactions could lead to the discovery of the reasons behind some of the between-study heterogeneity and provide insights into the physiopathological mechanism to disease. We return to this issue in Section 17.3.2.

17.1.6 Use of Random-Effects versus Common-Effect Models

Reporting heterogeneity and understanding its causes is very important, since a meta-analysis that ignores it might misleadingly suggest a lack of genetic effect. When heterogeneity between studies cannot be explained, a decision needs to be made on whether to continue with the analysis using a random-effects model or to refrain from pooling. It is important to remember that the summary result from a random-effects meta-analysis represents an estimate of the *average* effect and not of a common effect, and the use of prediction intervals has been proposed as a better way to convey uncertainty than confidence intervals in random-effects meta-analyses (Higgins et al., 2009; Riley et al., 2011). There are no clear guidelines on whether and when it is sensible to refrain from pooling, but as a rough guide, we suggest only using a random-effects model in genetic meta-analysis if the between-study standard deviation is less than 25% of the pooled effect size (Minelli et al., 2009).

17.1.7 Evaluation of Publication Bias

Publication bias has often been neglected in meta-analyses of candidate-gene studies (Minelli et al., 2009), although methods for identifying and addressing it would be the same as those presented in Chapter 13 for a general meta-analysis.

17.2 The Meta-Analysis of Genome-Wide Association Studies

A genome-wide association study is an epidemiological design in which hundreds of thousands of genetic variants are measured in order to discover which of them are associated with a particular disease or trait. The participants are usually recruited in a case-control or a cohort design and the measured genetic variants are usually SNPs.

Most GWAS are analyzed by taking the SNPs one at a time and performing a hypothesis test of association with a single trait or disease outcome, using an additive genetic model. This is similar to what is done in candidate-gene studies, and yet there are specific issues in GWAS. The most important is the problem of multiple testing; to guard against false positives, very stringent significance thresholds are used with the current consensus being that $p < 5 \times 10^{-8}$ is the appropriate level for declaring genome-wide significance. Such a low threshold means that very large samples are required to provide the power to detect the small genetic associations that are found with most traits. For instance, to have 90% power at genome-wide significance to detect an association between a binary trait with prevalence of 20% and a SNP that has a true odds ratio of 1.1 per allele and an allele frequency of 10%, we would require about 55,000 cases and 55,000 controls (calculation performed using QUANTO sample size calculator by USC Biostats, available at http://biostats.usc.edu/software).

The cost of the SNP array for measuring all of the genetic variants plus its processing has fallen considerably since large-scale GWAS were first performed in around 2007, and by 2018, the total cost varied between about $100 and $400 per participant depending on the number of variants. This cost is low enough that GWAS have become very popular, but it is still high enough to ensure that studies only recruit a few thousand participants. The result is that most studies are under-powered by themselves and researchers have been forced to form international consortia to meta-analyze studies of the same trait or disease outcome (Austin et al., 2012). Concern over confidentiality and a reluctance to lose control of the raw data have meant that most research groups will not release the raw SNP measurements and the consortia are forced to meta-analyze summary data.

Selected GWAS data could be incorporated into a meta-analysis of a biologically determined set of SNPs, in which case they can be handled in the same way as any other candidate-gene meta-analysis. However, when groups of primary investigators collaborate in a consortium with a view to pooling summary data, the main aim of the meta-analysis is usually to provide increased power to identify new genetic variants rather than to summarize evidence on those that have been previously identified. The meta-analysis methods used by such consortia are the same as those already discussed for candidate-gene studies, but there are additional practical considerations that we cover in the following sections.

There is a long-standing debate in genetic epidemiology over whether most common diseases are caused by a small number of common genetic variants; the so-called common-disease common-variant hypothesis (Smith and Lusis, 2002). It was originally thought that a few common variants with relatively large effects would be responsible for most of the genetic heritability of common disease, but the reality uncovered by GWAS has proved to be much more complicated. For example, investigations of the genetics of height have shown that larger and larger studies find more and more variants with smaller and smaller effects (Wood et al., 2014). Improved technology will soon make it economic to sequence the entire genome of every participant so that we will be able to search for rare variants, which would never be included on a GWAS chip. Emphasis on rare variants will make meta-analysis even more important (see Section 17.2.7).

17.2.1 Inclusion of Studies in a GWAS Meta-Analysis Consortium

Ideally, the studies in a GWAS meta-analysis should all be based on exactly the same outcome, but in reality, this is rarely possible. Most diseases are themselves amalgamations of many subtypes and the inclusion criteria can vary greatly between studies. For example, the CARDIoGRAM meta-analysis of coronary artery disease included studies with different proportions of people with angina, or who had bypass surgery, or who survived a heart attack (Preuss et al., 2010). Some genes might associate with all of these subcategories but others could be more specific, leading to heterogeneity between the studies.

A second issue is whether or not to include studies from different regions of the world: for example, is it appropriate to include a study from China in a GWAS meta-analysis that is predominantly based on people of European origin? The frequencies of most genetic variants show wide variation between ethnicities (Gibbs et al., 2003) but this is not the main concern. The key question is whether it is reasonable to assume that the effect of the genetic variant on the outcome will be the same in all populations (Marigorta and Navarro, 2013). Shared biology would suggest that the same genes will be important in all populations, but gene-environment interactions (Hunter, 2005) might introduce substantial heterogeneity in average effect size. Gene-gene interactions might also introduce heterogeneity when allele frequencies differ across populations (Battle et al., 2007).

The final inclusion issue concerns the set of genetic variants. It is very likely that different GWAS arrays will have been used in the different studies, and different arrays include different variants. It is now usual for all studies to impute a larger number of SNPs than they actually measure, using a reference panel such as the participants from the 1000 genomes project (Howie et al., 2009; Li et al., 2010). By standardizing the imputation, the meta-analysis will have comparable information on millions of variants, although any particular variant might be measured in some studies and imputed in others. Ideally, we would take the uncertainty in the imputation into account in the meta-analysis, but in practice a threshold for acceptable imputation is usually chosen and all imputed variants that pass this threshold are included as if they were measured (de Bakker et al., 2008).

In a consortium of primary investigators, the set of studies defines itself. When this is not the case, the meta-analyst might search web-based catalogs of results from GWAS such as the NHGRI catalog (Welter et al., 2014) or the HuGE navigator (Yu et al., 2008a). These are useful resources which provide an overview of all published GWAS studies relating to the phenotype of interest, so that the primary research group can contact other groups either to invite them to join the consortium or to ask for a replication look-up, that is summary results for the SNPs of interest to confirm their findings.

17.2.2 Data Quality Control

When the meta-analysis is performed by a consortium, it should be possible to standardize the methods of quality control and primary analysis used to create the summary statistics. However, it is often the case that each study will have performed an analysis of its own data prior to joining the consortium and they may be reluctant to re-analyze as this might produce different effect estimates calculated from the same data. Alternatively, they may have good reasons for choosing a method of primary analysis that suits the particular nature of their own study design. As a result, it is very common to find small differences in the primary analysis methods. Experience suggests that such differences are not of great importance, but the researchers need to be on the look-out for systematic differences in association that might relate to the primary analysis of the data, perhaps by performing subgroup analyses.

Study-level quality control typically involves excluding SNPs or participants that have a high level of missing values and excluding SNPs that deviate a long way from HWE. SNPs where the frequency of the less common allele (minor allele frequency, MAF) is below some agreed threshold are also usually excluded, not only because of lack of power, but also because of increased misclassification due to genotyping error. Finally, since genome-wide data are available, they can be used to measure the degree of relatedness between participants making it possible to remove duplicates, participants whose gender does not match that reported, or participants who are unexpectedly closely related (Laurie et al., 2010).

DNA comes in the form of a double strand for each of the two (maternal and paternal) chromosomes, with the bases arranged in pairs so that cytosine (C) always pairs with guanine (G) and adenine (A) always pairs with thymine (T) (Attia et al., 2009). Reading the DNA along one strand might produce ...AGT... while reading along the other would give ...TCA... Ideally, all of the studies will use the same strand to describe their SNPs, but this cannot be taken for granted and careful strand checks need to be undertaken before embarking on the meta-analysis. A similar problem can arise over the definition of the risk allele in the primary analysis. Suppose that a SNP has two alleles, C and T, on the agreed strand. Some studies might measure the association between the number of copies of C and the outcome, while others measure the association with the number of copies of T. Even agreement to measure the effect of the minor allele might not solve the problem, as it is possible that C might be less common in one study population while T is less common in another; this is most likely to happen for allele frequencies close to 50%.

Unlike a candidate-gene meta-analysis, the sheer scale of GWAS meta-analysis makes it difficult to spot unexpected features in the data and so it is much harder to maintain data quality. There is, however, one big advantage that GWAS meta-analyses have over candidate-gene meta-analysis. Since all genetic variants are available to the meta-analysis, there is no selection problem leading to publication or reporting bias.

17.2.3 Assessment and Adjustment for Population Stratification

Population stratification is a type of confounding typical of genetic epidemiology, and it is caused by a mix of different ethnic groups within a single study population when the frequency of the genetic variant and the incidence of disease both vary across ethnicities within a single study (Cardon and Palmer, 2003). Population stratification is a theoretically important confounder of genetic associations, although its practical impact in real studies has been debated (Thomas and Witte, 2002; Wacholder et al., 2002). Investigation and correction for population stratification should be performed at the individual study-level prior to any meta-analysis of genetic association studies.

While the risk of confounding by population stratification is present in both candidate-gene studies and GWAS, it is only when genome-wide data are available that the problem can be properly investigated and corrected for. In GWAS, the two most widely used approaches to this problem are genomic control and structured association methods. Genomic control (Devlin and Roeder, 1999) is a method for controlling the inflation of test statistics, first developed for binary traits and then generalized for quantitative traits (Bacanu et al., 2002). It uses markers that are not linked with the disease investigated to correct for any exaggeration of statistical significance that is caused by population stratification. Often, QQ-plots are used to illustrate the inflation in the test statistics. We describe these in Section 17.2.5.

Structured association methods use genetic information to estimate and control for population stratification. These methods rely on the use of genome-wide data to categorize the participants on the basis of their ethnic origin, and typically involve some type of multidimensional scaling or principal component analysis (Miclaus et al., 2009; Price et al., 2006). We can remove the confounding due to population stratification by adjusting the analysis for the major components. The exact nature of such data-based adjustments, for instance, the exact method or the number of components, will usually be study-specific.

17.2.4 Pooling Genome-Wide Data across Studies

In a standard GWAS meta-analysis, each study provides summary data on the measured or imputed genetic variants in the form of an estimate and its standard error (e.g., log odds ratios for a binary outcome and regression coefficients for a continuous outcome). Once these summary data have been thoroughly checked and aligned, the meta-analysis proper can begin. The most commonly used approach is a common-effect meta-analysis (see Chapter 4) based on inverse-variance weighting applied separately to each variant (Gogele et al., 2012), assuming no between-study heterogeneity in the genetic effects.

As already noted, there are many potential ways in which heterogeneity can be introduced into the effect size, and adjustment for it would require a random-effects meta-analysis. However, heterogeneity will not be uniform across the whole genome, and a criterion is required for deciding which SNPs need a common-effect meta-analysis and which need a random-effects meta-analysis. There is no consensus as to the best approach. Some researchers have advocated universal common-effect analysis, others have advocated universal random-effects analysis despite the resulting loss of power in the absence of heterogeneity (Thompson et al., 2011). In practice, common-effect meta-analysis tends to be used unless the heterogeneity exceeds some predefined threshold, but this ignores the theoretical implications of this conditioning on the final p-value for association. Moreover, investigation of possible causes of heterogeneity is rarely done in GWAS meta-analyses (Gogele et al., 2012). It is worth remembering that heterogeneity which is widespread across SNPs may indicate that the dissimilarity between studies is so great that a GWAS meta-analysis is not appropriate.

As the number of GWAS has grown so it has become practical to consider meta-analyses that distinguish closely associated phenotypes or different ethnicities. We can ask whether, for example, a genetic variant is associated with some types of asthma but not others, or whether genes that affect heart disease risk are the same in Europe and South Asia.

17.2.5 Visualization of GWAS Meta-Analysis Findings

The whole set of meta-analysis results gives us a way of assessing whether or not our meta-analysis model was appropriate. The analysis was conducted to identify a relatively small number of genetic variants that are associated with the disease, but it is assumed that the over-whelming mass of variants will show no association. If this is the case, most variants will have p-values that will be randomly distributed between zero and one. Results of both primary GWAS and meta-analyses of GWAS are often graphically presented using a Manhattan plot, where association ($-\log_{10}$) p-values for all variants are plotted against their location in the genome (Figure 17.1). The strict level for genome-wide significance ensures that very few of these null variants are identified as true associations ("hits"). Further visualization of the results can be provided using a regional association plot (Figure 17.2), which shows ($-\log_{10}$) p-values for SNPs within a specific region against their genomic position, together with gene annotations, estimated recombination rates, i.e., rates at which linkage is broken (Attia et al., 2009), and linkage disequilibrium (pairwise correlation) between the top hit and the surrounding SNPs.

Most genetic variants will have no impact on the disease under study, so if a histogram were plotted of all of the p-values from a primary study or from a GWAS meta-analysis, it should take the form of a uniform distribution with a small spike close to zero due to the few true associations. If we do not see this pattern, then something is wrong. An alternative graphical representation of the full set of significance tests is the quantile-quantile

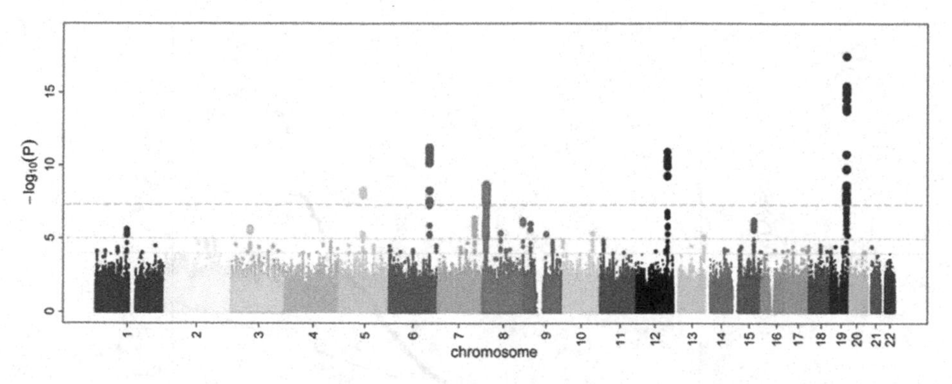

FIGURE 17.1
Manhattan plot. Manhattan plot depicting genetic variants strongly associated retinal venular caliber in a meta-analysis of four GWAS. Each dot represents a SNP, with the X-axis showing genomic location and Y-axis showing association level. [From: Ikram MK et al. Four novel loci (19q13, 6q24, 12q24, and 5q14) influence the microcirculation in vivo. *PLoS Genet* 2010; 6(10):e1001184.]

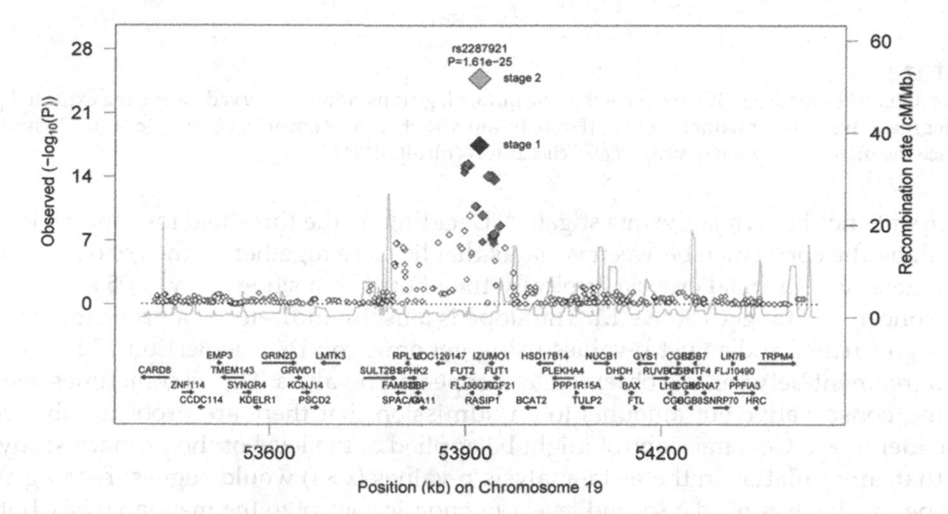

FIGURE 17.2
Regional association plot. Chromosome 19q13 regional association plot showing Stage 1 (discovery phase) p-value (large black diamond) and Stage 2 (replication phase) p-value (large gray diamond) for the top SNP. P-values from Stage 1 for additional SNPs at each locus are shaded according to their linkage disequilibrium with the top SNP, with darker shades corresponding to greater linkage disequilibrium. [From: Ikram MK et al. Four novel loci (19q13, 6q24, 12q24, and 5q14) influence the microcirculation in vivo. *PLoS Genet* 2010; 6(10):e1001184.]

(QQ) plot, in which the sorted test statistics are plotted against their expected values under the null hypothesis. Almost equivalent to this is a plot of $-\log_{10}$ of the sorted p-values against the expected values under a uniform distribution, as illustrated in Figure 17.3. In such a plot, each point represents a SNP and the majority of points will cluster around a straight line representing equality between observed and expected values. When the slope of the points, ignoring the tail caused by the few true associations, deviates from 1, then we have a problem, which needs to be investigated. The slope is usually referred to as inflation factor or lambda (λ). The most likely causes of a problem in a single study are population stratification, genotyping error, or cryptic relatedness (relatedness among

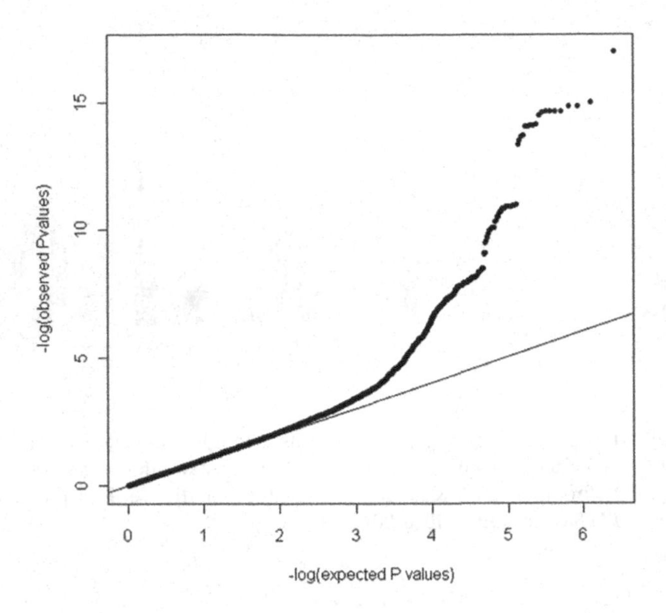

FIGURE 17.3
QQ plot. Quantile-quantile (QQ)-plot showing the minus log-transformed observed versus the expected p-values after meta-analysis for retinal venular. [From: Ikram MK et al. Four novel loci (19q13, 6q24, 12q24, and 5q14) influence the microcirculation in vivo. *PLoS Genet* 2010; 6(10):e1001184.]

participants not known to the investigator). Deciding on the threshold for concern is made difficult by the correlation between variants that lie close together on the genome. So most researchers use informal criteria, typically they hope for a slope below 1.05 and get seriously concerned if it goes above 1.1. The slope is a useful indicator of a problem, but some people go further and adjust p-values using genomic control (see Section 17.2.3) so as to force agreement between the observed and expected p-values. This is sometimes justified as being conservative but amounts to an admission that there are problems that could not be identified. Genomic control might be applied at the level of the primary study, but after that, any inflation in the meta-analysis p-values ($\lambda > 1$) would suggest heterogeneity. Some people have applied a second level of genomic control to the meta-analysis, but this is likely to result in an unnecessary loss of power. A much better approach would be to investigate the heterogeneity and to correct for it through the use of random-effects models when necessary (Thompson et al., 2011).

17.2.6 Replication of GWAS Meta-Analysis Findings

Candidate-gene studies gained a poor reputation during the years before GWAS because of the number of reported associations that turned out to be false positives (Colhoun et al., 2003; Ioannidis et al., 2001; Trikalinos et al., 2004). The National Cancer Institute (NCI) and National Human Genome Research Institute (NHGRI) set up a working group on replication in genetic association studies (Chanock et al., 2007) and many leading journals changed their publication policy to insist on replication or some other supporting evidence before the study could be accepted for publication, and these concerns have carried over into the GWAS era. Large consortia sometimes include most of the available studies in their GWAS meta-analysis, so that replication based on the existing results of other GWAS ("*in silico*" replication) can be difficult. In these cases, a replication might be needed which involves measuring the top genetic variants in an existing cohort that has not yet had a genome-wide scan (*"de novo"* replication). Another option is to support the finding by obtaining gene expression or other genetic data that link the variant to the disease.

When a set of top variants are sent to replication, the appropriate test is the significance of the variants in the replication sample adjusted for the number of variants sent for replication. For this reason, the number of SNPs that can be replicated is limited by the power of the replication study, or replication meta-analysis if replication is based on multiple studies. It is quite common to meta-analyze the results from the original GWAS (discovery data) with the replication results. However, the discovery meta-analysis will be affected by an upward bias referred to as "winner's curse" (Ioannidis et al., 2001), while the replication will give unbiased estimates of the genetic effect size (Gogele et al., 2012; Thompson et al., 2011). For this reason, it is better to declare replication based on the statistical significance in the replication sample alone.

GWAS meta-analyses are based on multiple hypothesis tests and typically seek to measure the strength of evidence for association (p-value) separately for each SNP. However, they could be viewed as a screen to identify the genes most worthy of future study. In reality, this is often what happens, with the most significant variants followed-up in replication studies or in functional studies investigating their mechanism of action. In this screening paradigm, the false discovery rate (FDR) among SNPs that are followed-up is more relevant than their p-value; although the SNPs would be ranked in the same order, using a FDR of 5% or 10% would lead to many more discoveries than a p-value threshold of 5×10^{-8}.

The selection of SNPs for replication is usually based solely on p-values from the discovery meta-analysis. An alternative more powerful approach is SNP prioritization that incorporates external biological information. This approach is rarely used (Gogele et al., 2012), and a practical limitation is in identifying biological data that are informative enough to have a substantial impact. With the growth in biological knowledge and development of new technologies, novel types of genetic data are becoming widely available. Flexible Bayesian methods have been proposed to combine all of this information (Pereira et al., 2017; Thompson et al., 2013; Yang et al., 2017), although their routine implementation on a genome-wide scale is limited by the computing power.

17.2.7 Rare Variants

Using family studies, it is possible to estimate the proportion of the variation in a trait that is due to genetics (the heritability) and the proportion that is due to the environment. The failure of currently identified common genetic variants to explain the anticipated heritability has led to speculation that common diseases might be the result of rare DNA variants (Eichler et al., 2010). Perhaps there are critical genes that can be affected by a number of different variants, each might be rare, but many people will carry one of these variants and so have an affected gene. The huge sample sizes needed to detect individual rare variants are unlikely to be obtained even in a meta-analysis and so we are forced to rethink our overall approach.

GWAS studies have targeted selected common variants, typically with MAF higher than 5%, while exome arrays that concentrate on SNPs in protein coding regions of the genome have targeted low-frequency (MAF between 0.5% and 5%) and rare (MAF lower than 0.5%) variants. With next-generation sequencing (NGS), we are now just at the point where DNA sequencing, i.e., the process of determining the accurate order of all nucleotides along the genome, will become economically feasible across large genomic regions or even across the whole genome. In this way, NGS will identify new rare variants and perhaps variants that are specific to a very small group of related participants.

In the attempt to increase statistical power, methods have been developed that combine rare variants within a defined region to provide a risk score that can be associated with the trait or disease. These include a simple burden score that counts the number of rare variants in a region, with variations grouping the variants by their MAF, the sequence kernel association test (SKAT) and its variations, which estimate the direction of effect and incorporate that in the score, and other methods reviewed by Lee et al. (Lee et al., 2014).

The meta-analysis of regional scores is not problematic if all studies follow exactly the same protocol, but this is not always the case as the tests can be used in many different ways; the researcher must choose the extent of the region that will be covered, which rare variants from within the region will be included in the score, and how each rare variant will be weighted. Unless the studies are willing to co-operate, so as to ensure comparability of methods, then meta-analysis will be dangerous.

Often, rare variant methods are described in terms of a statistical test. When the scores are derived in exactly the same way, score statistics weighted using their variances are combined using common- or random-effects meta-analysis. Tang and Lin (Tang and Lin, 2015) provide a review of methods and compare their own meta-analysis program, MASS (Tang and Lin, 2013), with the stand-alone RAREMETAL program (Feng et al., 2014), and two R packages, MetaSKAT (Lee et al., 2013) and seqMeta (Lumley et al., 2012). The programs make slightly different assumptions and offer different options, for example, to analyze continuous, binary, or survival phenotypes, and to calculate Monte Carlo p-values.

17.3 The Meta-Analysis of Other Types of Genetic Studies

In the following sections, we will briefly mention the main methodological issues surrounding the meta-analysis of other types of genetic studies, providing references for further reading to the interested reader.

17.3.1 Meta-Analysis of Genome-Wide Linkage Studies

Linkage analysis of family data works by identifying stretches of DNA present in family members with disease and absent in those without disease, and relies on the tendency for parts of the genome to be inherited together as a consequence of their physical proximity (Attia et al., 2009). A measured genetic marker passed down through a family in a way that consistently accompanies disease therefore suggests the presence of a nearby gene with a functional effect (Dawn Teare and Barrett, 2005). Linkage studies can find chromosomal regions containing genes associated with disease but cannot point toward individual genetic variants, which can only be identified by further genetic association or functional studies.

Linkage analysis has played a major role in the study of monogenic disorders since the late 1980s (Botstein and Risch, 2003). In complex diseases, genome-wide linkage studies (or "genome scans") have been used to identify chromosomal regions harboring susceptibility genes in an hypothesis-free approach, but they have limited power to detect small to moderate effects compared with GWAS. Meta-analysis of published genome scans can be used to increase power, but is characterized by methodological difficulties due to variability in study designs and methods, including differences in family structures, genetic makers, statistical methods, genetic models, and reporting of results (Dempfle and Loesgen, 2004).

The classical Fisher's method to combine p-values (Fisher, 1932) (Chapter 3), as well as derived methods proposed to improve performance in this setting (Badner and Golding, 1999; Province, 2001; Zaykin et al., 2002), can accommodate variability in study designs. An alternative non-parametric approach that can be used with any linkage test statistic reported in published genome scans is the genome search meta-analysis (GSMA) method by Wise et al. (Wise et al., 1999). GSMA ranks the test statistics of regions of approximately equal length ("bins") across the genome within each study, and for each region it sums ranks across scans. Since ranking considers only the relative significance within a study and ignores differences in sample sizes, ranks can be weighted by the number of cases or families included, with significance thresholds for the GSMA statistics assessed empirically through permutation. This may require extensive simulations and unweighted analyses are therefore often performed in practice (Dempfle and Loesgen, 2004).

While these methods only allow hypothesis testing, Gu et al. first proposed a method that provides a pooled estimate of the genetic effect based on the proportion of alleles shared identical-by-descent (IBD) (Gu et al., 1998), although the requirements of similarity in markers genotyped and availability of IBD sharing statistics from the published genome scans may limit its applicability in practice. Other approaches have been subsequently proposed to combine effect estimates from published genome scans (Kippola and Santorico, 2010). It is important to note that even the meta-analysis of genome scans using raw data, often considered as the gold standard, is problematic when different marker panels are used, since this introduces a systematic difference in IBD estimates between studies (Dempfle and Loesgen, 2004).

17.3.2 Meta-Analysis of Studies on Genetic Interactions

Genetic effects on a trait may be modified by environmental factors and vice versa. The interplay between genes and environment has been recognized as influencing the risk of most common complex diseases and can explain some true heterogeneity observed in genetic association studies (see Section 17.1.5). Moreover, gene-environment (G-E) interactions might be able to identify genetic variants with weak main effects, i.e., limited effect in the absence of the environmental exposure, but strong interaction effects, which would not be detected by marginal genetic analysis that ignores the environmental exposure (Hein et al., 2008).

G-E interactions might be important to identify participants at particular risk of disease, improve understanding of pathophysiological mechanisms, and target pharmacological interventions to subgroups of patients. However, investigation of interactions is difficult due to limited statistical power. As a rule of thumb for a case-control study, detecting an interaction effect will require a sample size at least four times larger than that required to detect a main effect of the same magnitude (Smith and Day, 1984). Moreover, the power of G-E interaction studies can be heavily reduced by measurement error in the environmental exposure, which tends to be much larger than genotyping error (Burton et al., 2009).

Meta-analysis across several studies may therefore be the only way to achieve adequate power. Although limited knowledge is currently available to guide the choice of the most suitable analytical strategy for the meta-analysis of G-E interactions, particularly in the context of genome-wide investigations, this represents a very active field of methodological research.

There is often biological knowledge to motivate the investigation of interactions of an environmental factor with genes involved in a specific pathophysiological mechanism. An example is the interaction between oxidative exposures, such as smoking and air pollution, and antioxidant genes (Minelli and Gogele, 2011). In the context of GWAS, G-E interactions can be investigated across the whole genome in a genetic hypothesis-free approach (Eichler et al., 2010). With a large number of GWAS studies currently available for many disease traits, international consortia are increasingly looking at the presence of G-E interactions in their datasets with the main aim of identifying further susceptibility loci undetected by the original main effect analysis. Although hypothesis-driven and hypothesis-free investigations of G-E interactions are usually viewed as opposite approaches, the two could be combined using modeling strategies that exploit available biological knowledge in mining genome-wide data (Thomas, 2010).

Although IPD meta-analysis would be useful for studying G-E interactions (Chapter 8), this is rarely done in practice because people are reluctant to share genetic data, and meta-analysis based on aggregate data is commonly used. The simplest meta-analysis approach to the investigation of G-E interactions is the meta-analysis of the interaction alone, i.e., a meta-analysis of interaction regression coefficients across studies. However, given the low power of interaction tests, particularly in the context of genome-wide analyses due to the need to correct for multiple testing, a simple two-stage strategy can be used where a meta-analysis of the interaction effect is performed only for SNPs which show a marginal effect at some pre-specified significance threshold (Kooperberg and Leblanc, 2008). However, this strategy works only under the assumption that SNPs with interaction effects will also show marginal effects. If you are unwilling to make this assumption, you can use methods that jointly test main and interaction effects. One approach is the meta-analysis of the two degree-of-freedom test. Briefly, a two degree-of-freedom chi-square test of both main and interaction effects (with the null hypothesis being that both effects are 0) is performed in each study (Kraft et al., 2007), and the p-values from this joint test can then be meta-analyzed using Fisher's method (Manning et al., 2011). This approach shares the limitations of Fisher's method, i.e., it does not provide estimates of the magnitude of main and interaction effects and does not consider the direction of the effects. To overcome this, Manning et al. (Manning et al., 2011) proposed an alternative joint meta-analysis method that simultaneously summarizes main and interaction effects regression coefficients from fitted interaction models. This method, which is based on multivariate meta-analysis (see Chapter 9), provides estimates and confidence intervals for both coefficients, a joint test of significance and a test of heterogeneity. Through simulation work, the authors showed better performance of their method compared with the meta-analysis of the two degree-of-freedom test (Manning et al., 2011). Compared with the meta-analysis of the interaction alone and the two-stage strategy, the joint meta-analysis showed the highest power when both main genetic and interaction effects were present, while in the presence of an interaction effect but no main genetic effect, the meta-analysis of the interaction alone was the most powerful. Similarly, a classical meta-analysis of genetic association had the highest power in the presence of a main genetic effect but no interaction effect (Manning et al., 2011). The method has been implemented in METAL for use in GWAS (http://csg.sph.umich.edu/abecasis/Metal). Additional methods, including case-only and empirical Bayes approaches, are described in a recent review (Gauderman et al., 2017).

The effects of a participant's genotype on the response to a drug, in terms of both efficacy and toxicity, commonly referred to as pharmacogenetics, is a special case of G-E interaction where the "environmental factor" is the drug. Investigators embarking on a meta-analysis of pharmacogenetic studies may face additional methodological issues specific to this type of investigation, including the problems of compliance to treatment and choice and definition of outcome (Jorgensen and Williamson, 2008). Some extent of non-compliance is estimated to occur in 30–40% of treated participants, but adjustments for non-compliance are very rarely performed. The intention-to-treat principle commonly used in the analysis of randomized controlled trials pragmatically accounts for non-compliance and thus provides results that reflect what is likely to happen in real life. In a pharmacogenetic study, however, where the aim is to understand the influence of genetic variation on a patient's response to the drug and thus the approach is more explanatory than pragmatic, non-compliance will lead to an underestimate of the effects of interest and adjustment for non-compliance should be made. In pharmacogenetic studies choosing and defining "patient response" to the drug is often difficult, and this can lead to heterogeneity in outcome definition and increases the possibility of outcome reporting bias.

The issues in the analysis of gene-gene (G-G) interactions are similar to those for G-E interactions, but when G-G interactions are investigated across the whole genome, with millions of genetic variants tested for interaction with each other, the penalty due to correction for multiple testing becomes too high even for large-scale meta-analyses. Two-stage strategies are therefore usually employed; the first filtering stage selects SNPs with marginal effects at a pre-specified significance threshold, which are then tested for interactions in the second stage. Again, this approach is limited by its ability to detect interactions only between SNPs which show marginal effects. Alternative two-stage strategies have been proposed where the filtering process in the first stage is based on biological knowledge of the relationships between genes (e.g., the genes belonging to the same pathways or networks) rather than statistical evidence of marginal effects (Wei et al., 2014). Little methodological work has investigated strategies to meta-analyze evidence on G-G interactions across different GWAS.

17.4 Genes and Causal Inference: The Meta-Analysis of Mendelian Randomization Studies

Mendelian randomization is a method for estimating the causal effect of a risk factor on a disease or trait using genetic variants as instrumental variables (IVs) (Smith and Ebrahim, 2003). This approach has increasingly been used in epidemiology to distinguish causality from correlation.

IVs are widely used in econometrics but are limited by the strong assumptions that they require. Suppose we wish to decide whether X causes Y but for ethical or practical reasons we cannot perform a randomized experiment. If we can find a factor G such that (i) G is associated with X, (ii) G is independent of any confounders between X and Y, and (iii) G is independent of Y conditional on X and the confounders, then G is a valid instrument and by measuring X, Y, and G, we can deduce whether X is a cause of Y (Lawlor et al., 2008).

About 15 years ago, it was noticed that genetic variants would be strong candidates for use as instruments, and the use of genes in IV analysis has become known as Mendelian randomization (MR). Assumption (i) is satisfied by the choice of the gene. Assumption (ii) is likely to hold because genes are fixed at conception and cannot be altered by lifestyle or environment factors that typically confound epidemiological studies. The main problem is with assumption (iii), which needs to be justified (Lawlor et al., 2008); in particular, it could be violated by pleiotropy, which means that the gene influences Y only through X and not through any other pathway. Unfortunately, most individual genes only have a small effect on biological risk factors, with very low power to test for the causal effect of X on Y and low precision of the estimate of the causal effect. It is natural therefore to look to meta-analysis as a solution to this problem.

In MR, meta-analysis could be used in one of two ways: either the estimate of the effects of G on X and G on Y could come from different meta-analyses and be combined in a single MR (Davey Smith and Hemani, 2014), or a series of MRs could be conducted in different study populations and the resulting causal estimates could be combined in a meta-analysis. The second of these possibilities would be straightforward but is not used because of the lack of primary MR studies.

It is a feature of MR that the estimates of the effects of G on X and of G on Y do not need to come from the same study (Burgess et al., 2013), although if they do not, one is forced to make a strong assumption of equivalence between the studies, such that the expected estimate of the effect of G on X is the same as it would have been had it been measured in the G-Y studies. As more and more genetic meta-analyses are published so more information is available on the association of genes, G with risk factors, X, and traits or disease outcomes, Y. Performing a MR analysis has therefore been reduced to little more than a process of looking up results from published genetic meta-analyses, and as such, it has been very widely used and possibly abused, with insufficient attention given to the underlying assumptions.

Genetic meta-analyses performed by consortia provide huge sample sizes and therefore provide the highly precise estimates of the G-X and G-Y effects that are needed for MR. When the G-X association cannot be convincingly shown to be non-zero, G is described as a weak instrument and a weak instrument will not only lead to imprecise causal estimates, but it can also produce large biases.

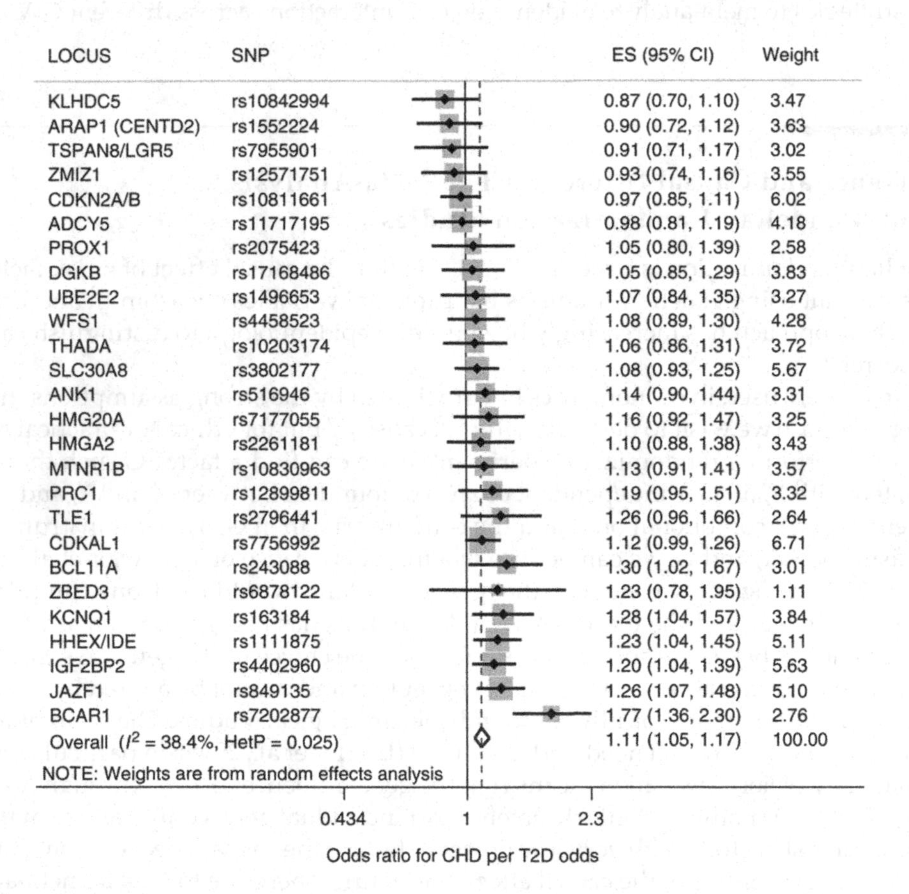

FIGURE 17.4

Forest plot of MR estimates. Forest plot of the MR estimates of the effect of type 2 diabetes on the risk of coronary heart disease from each of the 26 SNPs used as instruments. Shown for each SNP is the 95% confidence interval of the MR estimate and the inverse-variance weight in the random-effects meta-analysis. [From: Ahmad et al. A Mendelian randomization study of the effect of type-2 diabetes on coronary heart disease. *Nat Commun* 2015; 6:7060. Licensed under CC BY 4.0 (https://creativecommons.org/licenses/by/4.0/legalcode).]

A common way of gaining power and precision is to use multiple instruments, in this case, multiple genetic variants that affect the same risk factor (Palmer et al., 2012). These can be combined into a genetic risk score that replaces the individual variant G so that the stronger effect of G on X leads to increased power. Alternatively, the variants can be used separately to derive a set of MR estimates of the causal effect of X on Y and then these estimates can be combined in a meta-analysis over the different SNPs. The latter approach has the advantage of highlighting any heterogeneity in the MR estimates that might be due to violations of the strong IV assumptions.

The idea of meta-analyzing MR estimates over SNPs (Figure 17.4) has been used to create tests of heterogeneity that might be indicative of pleiotropy (Greco et al., 2015), and also to correct for pleiotropy under the assumption of independence between the pleiotropy and the G-X effect, using an adaptation of Egger regression for small study bias (see Chapter 13) to test for bias from pleiotropy and provide a causal effect estimate adjusted for it (Bowden et al., 2015).

17.5 Other Future Directions

The meta-analysis of genetic associations has undergone massive changes in the last 20 years, driven by advances in technology. This trend is certain to continue with the availability of whole genome sequence data and information on related biological processes, such as epigenetics, gene expression, proteomics, and metabolomics. Not only will it be important to analyze these new data sources separately, but researchers will want to know how they relate to one another so that they can follow the biological process from gene, via gene regulation, to gene product, and on to action of that product in the body. These datasets will be huge, but most sample sizes will be limited by cost, and meta-analysis will remain important.

References

Attia J, Thakkinstian A and D'Este C, 2003. Meta-analyses of molecular association studies: Methodologic lessons for genetic epidemiology. *Journal of Clinical Epidemiology* **56**(4): 297–303.

Attia J, Ioannidis JP, Thakkinstian A, McEvoy M, Scott RJ, Minelli C, Thompson J, Infante-Rivard C and Guyatt G, 2009. How to use an article about genetic association: A: Background concepts. *JAMA* **301**(1): 74–81.

Austin MA, Hair MS and Fullerton SM, 2012. Research guidelines in the era of large-scale collaborations: An analysis of Genome-wide Association Study Consortia. *American Journal of Epidemiology* **175**(9): 962–969.

Bacanu SA, Devlin B and Roeder K, 2002. Association studies for quantitative traits in structured populations. *Genetic Epidemiology* **22**(1): 78–93.

Badner JA and Goldin LR, 1999. Meta-analysis of linkage studies. *Genetic Epidemiology* **17**(Suppl 1): S485–S490.

Battle NC, Choudhry S, Tsai H-J, Eng C, Kumar G, Beckman KB, Naqvi M, Meade K, Watson HG, LeNoir M and Burchard EG, 2007. Ethnicity-specific gene–gene interaction between IL-13 and IL-4Rα among African Americans with asthma. *American Journal of Respiratory and Critical Care Medicine* **175**(9): 881–887.

Berlin JA, Santanna J, Schmid CH, Szczech LA and Feldman HI, 2002. Individual patient- versus group-level data meta-regressions for the investigation of treatment effect modifiers: Ecological bias rears its ugly head. *Statistics in Medicine* **21**(3): 371–387.

Botstein D and Risch N, 2003. Discovering genotypes underlying human phenotypes: Past successes for mendelian disease, future approaches for complex disease. *Nature Genetics* **33**(Suppl): 228–237.

Bowden J, Davey Smith G and Burgess S, 2015. Mendelian randomization with invalid instruments: Effect estimation and bias detection through Egger regression. *International Journal of Epidemiology* **44**(2): 512–525.

Brookes ST, Whitely E, Egger M, Smith GD, Mulheran PA and Peters TJ, 2004. Subgroup analyses in randomized trials: Risks of subgroup-specific analyses; power and sample size for the interaction test. *Journal of Clinical Epidemiology* **57**(3): 229–236.

Burgess S, Butterworth A and Thompson SG, 2013. Mendelian randomization analysis with multiple genetic variants using summarized data. *Genetic Epidemiology* **37**(7): 658–665.

Burton PR, Tobin MD and Hopper JL, 2005. Key concepts in genetic epidemiology. *Lancet* **366**(9489): 941–951 (London, England).

Burton PR, Clayton DG, Cardon LR, Craddock N, Deloukas P, Duncanson A, Kwiatkowski DP, McCarthy MI, Ouwehand WH, Samani NJ and Todd JA, 2007. Association scan of 14,500 nonsynonymous SNPs in four diseases identifies autoimmunity variants. *Nature Genetics* **39**(11): 1329–1337.

Burton PR, Hansell AL, Fortier I, Manolio TA, Khoury MJ, Little J and Elliott P, 2009. Size matters: Just how big is BIG?: Quantifying realistic sample size requirements for human genome epidemiology. *International Journal of Epidemiology* **38**(1): 263–273.

Cardon LR and Palmer LJ, 2003. Population stratification and spurious allelic association. *Lancet* **361**(9357): 598–604 (London, England).

Chanock SJ, Manolio T, Boehnke M, Boerwinkle E, Hunter DJ, Thomas G, Hirschhorn JN, Abecasis G, Altshuler D, Bailey-Wilson JE and Brooks LD, 2007. Replicating genotype-phenotype associations. *Nature* **447**(7145): 655–660.

Colhoun HM, McKeigue PM and Davey Smith G, 2003. Problems of reporting genetic associations with complex outcomes. *Lancet* **361**(9360): 865–872 (London, England).

Davey Smith G and Hemani G, 2014. Mendelian randomization: Genetic anchors for causal inference in epidemiological studies. *Human Molecular Genetics* **23**(R1): R89–R98.

Dawn Teare M and Barrett JH, 2005. Genetic linkage studies. *Lancet* **366**(9490): 1036–1044 (London, England).

de Bakker PI, Ferreira MA, Jia X, Neale BM, Raychaudhuri S and Voight BF, 2008. Practical aspects of imputation-driven meta-analysis of genome-wide association studies. *Human Molecular Genetics* **17**(R2): R122–R128.

Dempfle A and Loesgen S, 2004. Meta-analysis of linkage studies for complex diseases: An overview of methods and a simulation study. *Annals of Human Genetics* **68**: 69–83.

Devlin B and Roeder K, 1999. Genomic control for association studies. *Biometrics* **55**(4): 997–1004.

Eichler EE, Flint J, Gibson G, Kong A, Leal SM, Moore JH and Nadeau JH, 2010. Missing heritability and strategies for finding the underlying causes of complex disease. *Nature Reviews Genetics* **11**(6): 446–450.

Feng S, Liu D, Zhan X, Wing MK and Abecasis GR, 2014. RAREMETAL: Fast and powerful meta-analysis for rare variants. *Bioinformatics* **30**(19): 2828–2829 (Oxford, England).

Fisher RA, 1932. *Statistical Methods for Research Workers*. London: Oliver and Boyd.

Gauderman WJ, Mukherjee B, Aschard H, Hsu L, Lewinger JP, Patel CJ, Witte JS, Amos C, Tai CG and Conti D, 2017. Update on the state of the science for analytical methods for gene-environment interactions. *American Journal of Epidemiology* **186**(7): 762–770.

Gibbs RA, Belmont JW, Hardenbol P, Willis TD, Yu F, Yang H, Ch'ang LY, Huang W, Liu B, Shen Y and Tam PK, 2003. The international HapMap project. *Nature* **426**(6968): 789–796.

Gogele M, Minelli C, Thakkinstian A, Yurkiewich A, Pattaro C, Pramstaller PP, Little J, Attia J and Thompson JR, 2012. Methods for meta-analyses of genome-wide association studies: Critical assessment of empirical evidence. *American Journal of Epidemiology* **175**(8): 739–749.

Greco MF, Minelli C, Sheehan NA and Thompson JR, 2015. Detecting pleiotropy in Mendelian randomisation studies with summary data and a continuous outcome. *Statistics in Medicine* **34**(21): 2926–2940.

Gu C, Province M, Todorov A and Rao DC, 1998. Meta-analysis methodology for combining nonparametric sibpair linkage results: Genetic homogeneity and identical markers. *Genetic Epidemiology* **15**(6): 609–626.

Hein R, Beckmann L and Chang-Claude J, 2008. Sample size requirements for indirect association studies of gene-environment interactions (G × E). *Genetic Epidemiology* **32**(3): 235–245.

Higgins JP and Thompson SG, 2002. Quantifying heterogeneity in a meta-analysis. *Statistics in Medicine* **21**(11): 1539–1558.

Higgins JP, Thompson SG, Deeks JJ and Altman DG, 2003. Measuring inconsistency in meta-analyses. *BMJ* (Clinical Research ed) **327**(7414): 557–560.

Higgins JP, Thompson SG and Spiegelhalter DJ, 2009. A re-evaluation of random-effects meta-analysis. *Journal of the Royal Statistical Society: Series A* **172**(1): 137–159.

Higgins JP, 2008. Commentary: Heterogeneity in meta-analysis should be expected and appropriately quantified. *International Journal of Epidemiology* **37**(5): 1158–1160.

Howie BN, Donnelly P and Marchini J, 2009. A flexible and accurate genotype imputation method for the next generation of genome-wide association studies. *PLoS Genetics* **5**(6): e1000529.

Hunter DJ, 2005. Gene–environment interactions in human diseases. *Nature Reviews Genetics* **6**(4): 287–298.

Ioannidis JP, Ntzani EE, Trikalinos TA and Contopoulos-Ioannidis DG, 2001. Replication validity of genetic association studies. *Nature Genetics* **29**(3): 306–309.

Ioannidis JP, Rosenberg PS, Goedert JJ and O'Brien TR, 2002. Commentary: Meta-analysis of individual participants' data in genetic epidemiology. *American Journal of Epidemiology* **156**(3): 204–210.

Janssens AC, Ioannidis JP, van Duijn CM, Little J and Khoury MJ, 2011. Strengthening the reporting of Genetic RIsk Prediction Studies: The GRIPS Statement. *PLoS Medicine* **8**(3): e1000420.

Jorgensen AL and Williamson PR, 2008. Methodological quality of pharmacogenetic studies: Issues of concern. *Statistics in Medicine* **27**(30): 6547–6569.

Khoury MJ, Little J and Burke W, 2004. *Human Genome Epidemiology: A Scientific Foundation for Using Genetic Information to Improve Health and Prevent Disease*. New York: Oxford University Press.

Kippola TA and Santorico SA, 2010. Methods for combining multiple genome-wide linkage studies. *Methods in Molecular Biology* **620**: 541–560 (Clifton, NJ).

Kooperberg C and Leblanc M, 2008. Increasing the power of identifying gene × gene interactions in genome-wide association studies. *Genetic Epidemiology* **32**(3): 255–263.

Kraft P, Yen YC, Stram DO, Morrison J and Gauderman WJ, 2007. Exploiting gene-environment interaction to detect genetic associations. *Human Heredity* **63**(2): 111–119.

Kwon JM and Goate AM, 2000. The candidate gene approach. *Alcohol Research & Health* **24**(3): 164–168.

Lambert PC, Sutton AJ, Abrams KR and Jones DR, 2002. A comparison of summary patient-level covariates in meta-regression with individual patient data meta-analysis. *Journal of Clinical Epidemiology* **55**(1): 86–94.

Laurie CC, Doheny KF, Mirel DB, Pugh EW, Bierut LJ, Bhangale T, Boehm F, Caporaso NE, Cornelis MC, Edenberg HJ and Gabriel SB, 2010. Quality control and quality assurance in genotypic data for genome-wide association studies. *Genetic Epidemiology* **34**(6): 591–602.

Lawlor DA, Harbord RM, Sterne JA, Timpson N and Davey Smith G, 2008. Mendelian randomization: Using genes as instruments for making causal inferences in epidemiology. *Statistics in Medicine* **27**(8): 1133–1163.

Lee S, Teslovich TM, Boehnke M and Lin X, 2013. General framework for meta-analysis of rare variants in sequencing association studies. *American Journal of Human Genetics* **93**(1): 42–53.

Lee S, Abecasis GR, Boehnke M and Lin X, 2014. Rare-variant association analysis: Study designs and statistical tests. *American Journal of Human Genetics* **95**(1): 5–23.

Li Y, Willer CJ, Ding J, Scheet P and Abecasis GR, 2010. MaCH: Using sequence and genotype data to estimate haplotypes and unobserved genotypes. *Genetic Epidemiology* **34**(8): 816–834.

Little J and Higgins JPT, 2006. *The HuGENet HuGE Review Handbook*, Version 1.0. Ottawa, Ontario, Canada: Department of Epidemiology and Community Medicine, Faculty of Medicine, University of Ottawa. Available from http://www.hugenet.ca.

Little J, Higgins JP, Ioannidis JP, Moher D, Gagnon F, von Elm E, Khoury MJ, Cohen B, Davey-Smith G, Grimshaw J and Scheet P, 2009. Strengthening the reporting of genetic association studies (STREGA): An extension of the strengthening the reporting of observational studies in epidemiology (STROBE) statement. *Journal of Clinical Epidemiology* **62**(6): 597–608.e4.

Lumley T, Brody J, Dupuis J and Cupples A, 2012. *Meta-Analysis of a Rare-Variant Association Test*. University of Auckland Available from http://stattech.wordpress.fos.auckland.ac.nz/files/2012/11/skat-meta-paper.pdf.

Manning AK, LaValley M, Liu CT, Rice K, An P, Liu Y, et al., 2011. Meta-analysis of gene-environment interaction: Joint estimation of SNP and SNP × environment regression coefficients. *Genetic Epidemiology* **35**(1): 11–18.

Marigorta UM and Navarro A, 2013. High trans-ethnic replicability of GWAS results implies common causal variants. *PLoS Genetics* **9**(6): e1003566.

Miclaus K, Wolfinger R and Czika W, 2009. SNP selection and multidimensional scaling to quantify population structure. *Genetic Epidemiology* **33**(6): 488–496.

Minelli C and Gogele M, 2011. The role of antioxidant gene polymorphisms in modifying the health effects of environmental exposures causing oxidative stress: A public health perspective. *Free Radical Biology & Medicine* **51**(5): 925–930.

Minelli C, Thompson JR, Abrams KR and Lambert PC, 2005a. Bayesian implementation of a genetic model-free approach to the meta-analysis of genetic association studies. *Statistics in Medicine* **24**(24): 3845–3861.

Minelli C, Thompson JR, Abrams KR, Thakkinstian A and Attia J, 2005b. The choice of a genetic model in the meta-analysis of molecular association studies. *International Journal of Epidemiology* **34**(6): 1319–1328.

Minelli C, Thompson JR, Abrams KR, Thakkinstian A and Attia J, 2008. How should we use information about HWE in the meta-analyses of genetic association studies? *International Journal of Epidemiology* **37**(1): 136–146.

Minelli C, Thompson JR, Abrams KR, Thakkinstian A and Attia J, 2009. The quality of meta-analyses of genetic association studies: A review with recommendations. *American Journal of Epidemiology* **170**(11): 1333–1343.

Palmer TM, Lawlor DA, Harbord RM, Sheehan NA, Tobias JH, Timpson NJ, et al., 2012. Using multiple genetic variants as instrumental variables for modifiable risk factors. *Statistical Methods in Medical Research* **21**(3): 223–242.

Pereira M, Thompson JR, Weichenberger CX, Thomas DC and Minelli C, 2017. Inclusion of biological knowledge in a Bayesian shrinkage model for joint estimation of SNP effects. *Genetic Epidemiology* **41**(4): 320–331.

Preuss M, König IR, Thompson JR, Erdmann J, Absher D, Assimes TL, et al., 2010. Design of the Coronary ARtery DIsease Genome-Wide Replication And Meta-Analysis (CARDIoGRAM) Study: A genome-wide association meta-analysis involving more than 22 000 cases and 60 000 controls. *Circulation: Cardiovascular Genetics* **3**(5): 475–483.

Price AL, Patterson NJ, Plenge RM, Weinblatt ME, Shadick NA and Reich D, 2006. Principal components analysis corrects for stratification in genome-wide association studies. *Nature Genetics* **38**(8): 904–909.

Province MA, 2001. The significance of not finding a gene. *American Journal of Human Genetics* **69**(3): 660–663.

Riley RD, Higgins JP and Deeks JJ, 2011. Interpretation of random effects meta-analyses. *BMJ* (Clinical Research ed) **342**: d549.

Sham P, 2001. *Statistics in Human Genetics*. London, UK: Arnold Publishers.

Smith PG and Day NE, 1984. The design of case-control studies: The influence of confounding and interaction effects. *International Journal of Epidemiology* **13**(3): 356–365.

Smith GD and Ebrahim S, 2003. 'Mendelian randomization': Can genetic epidemiology contribute to understanding environmental determinants of disease? *International Journal of Epidemiology* **32**(1): 1–22.

Smith DJ and Lusis AJ, 2002. The allelic structure of common disease. *Human Molecular Genetics* **11**(20): 2455–2461.

Sohani ZN, Meyre D, de Souza RJ, Joseph PG, Gandhi M, Dennis BB, et al., 2015. Assessing the quality of published genetic association studies in meta-analyses: The quality of genetic studies (Q-Genie) tool. *BMC Genetics* 16: 50.

Tang ZZ and Lin DY, 2013. MASS: Meta-analysis of score statistics for sequencing studies. *Bioinformatics* **29**(14): 1803–1805 (Oxford, England).

Tang ZZ and Lin DY, 2015. Meta-analysis for discovering rare-variant associations: Statistical methods and software programs. *American Journal of Human Genetics* **97**(1): 35–53.

Thomas DC and Witte JS, 2002. Point: Population stratification: A problem for case-control studies of candidate-gene associations? *Cancer Epidemiology, Biomarkers & Prevention* **11**(6): 505–512.

Thomas D, 2010. Methods for investigating gene-environment interactions in candidate pathway and genome-wide association studies. *Annual Review of Public Health* 31: 21–36.

Thompson JR, Attia J and Minelli C, 2011. The meta-analysis of genome-wide association studies. *Briefings in Bioinformatics* **12**(3): 259–269.

Thompson JR, Gogele M, Weichenberger CX, Modenese M, Attia J, Barrett JH, et al., 2013. SNP prioritization using a Bayesian probability of association. *Genetic Epidemiology* **37**(2): 214–221.

Trikalinos TA, Ntzani EE, Contopoulos-Ioannidis DG and Ioannidis JP, 2004. Establishment of genetic associations for complex diseases is independent of early study findings. *European Journal of Human Genetics: EJHG* **12**(9): 762–769.

Trikalinos TA, Salanti G, Khoury MJ and Ioannidis JP, 2006. Impact of violations and deviations in Hardy-Weinberg equilibrium on postulated gene-disease associations. *American Journal of Epidemiology* **163**(4): 300–309.

Wacholder S, Rothman N and Caporaso N, 2002. Counterpoint: Bias from population stratification is not a major threat to the validity of conclusions from epidemiological studies of common polymorphisms and cancer. *Cancer Epidemiology, Biomarkers & Prevention* **11**(6): 513–520.

Wang J and Shete S, 2012. Testing departure from Hardy-Weinberg proportions. *Methods in Molecular Biology* 850: 77–102 (Clifton, NJ).

Wei WH, Hemani G and Haley CS, 2014. Detecting epistasis in human complex traits. *Nature Reviews Genetics* **15**(11): 722–733.

Welter D, MacArthur J, Morales J, Burdett T, Hall P, Junkins H, et al., 2014. The NHGRI GWAS Catalog, a curated resource of SNP-trait associations. *Nucleic Acids Research* **42**(D1): D1001–D1006.

Wise LH, Lanchbury JS and Lewis CM, 1999. Meta-analysis of genome searches. *Annals of Human Genetics* **63**(Pt 3): 263–272.

Wood AR, Esko T, Yang J, Vedantam S, Pers TH, Gustafsson S, et al., 2014. Defining the role of common variation in the genomic and biological architecture of adult human height. *Nature Genetics* **46**(11): 1173–1186.

Yang J, Fritsche LG, Zhou X and Abecasis G, 2017. A scalable Bayesian method for integrating functional information in genome-wide association studies. *American Journal of Human Genetics* **101**(3): 404–416.

Yu W, Gwinn M, Clyne M, Yesupriya A and Khoury MJ, 2008. A navigator for human genome epidemiology. *Nature Genetics* **40**(2): 124–125.

Yu W, Wulf A, Yesupriya A, Clyne M, Khoury MJ and Gwinn M, 2008. HuGE Watch: Tracking trends and patterns of published studies of genetic association and human genome epidemiology in near-real time. *European Journal of Human Genetics: EJHG* **16**(9): 1155–1158. Available from www.hugenavigator.net/HuGENavigator/startPageWatch.do.

Zaykin DV, Zhivotovsky LA, Westfall PH and Weir BS, 2002. Truncated product method for combining P-values. *Genetic Epidemiology* **22**(2): 170–185.

18

Meta-Analysis of Dose-Response Relationships

Nicola Orsini and Donna Spiegelman

CONTENTS

18.1 Introduction

A dose-response analysis describes the changes of a response across levels of a quantitative factor. The quantitative factor could be an administered drug or an exposure. A meta-analysis of dose-response, or exposure-disease, relations aims at identifying the summary trend emerging from multiple studies trying to answer the same research question. Examples of questions usually asked are: Is there any association at all? Is the response changing approximately at a constant rate throughout the observed exposure range?

Is there any substantial change in the outcome beyond a certain exposure level? What is the optimal dose associated with most of the effect? These questions arise in the presence of a single study or multiple studies, individual or summarized data, experimental or observational data.

Two-stage methods for investigating the relationship between quantitative exposures and disease outcomes based on individual participant data from multiple studies have been proposed and evaluated (Sauerbrei and Royston, 2011; White et al., 2019). The main focus of this chapter is two-stage dose-response meta-analysis based on summarized data. This approach requires the quantitative exposure to be categorized and modeled with indicator variables using one exposure level as referent (Turner et al., 2010). A categorical approach is widely used because it greatly facilitates the presentation of crude data (cases, totals, person-years) and measures of association in a tabular form. Greenland and Longnecker (1992) explained how to efficiently estimate a trend based on a set of adjusted published relative risks using a weighted linear regression model. Subsequently, methodological articles appeared on flexible modeling of dose-response associations (Orsini et al., 2012; Bagnardi et al., 2004; Liu et al., 2009; Rota et al., 2010), how to deal with covariances of correlated outcomes (Hamling et al., 2008; Berrington and Cox, 2003), how to assess publication bias (Shi and Copas, 2004), how to assign a typical dose to an exposure interval (Takahashi and Tango, 2010), how to assess the goodness-of-fit (Discacciati et al., 2017), how to limit extrapolations (Crippa et al., 2018), and how to avoid exclusion of studies (Crippa et al., 2019). Alternative meta-analytic methods for combining dose-response data have been evaluated in the field of nutritional epidemiology (Yu et al., 2013).

The data required for dose-response meta-analysis are presented in a tabular form. As an example, consider Table 18.1 showing summarized incidence rate data investigating the association between alcohol intake, as measured in grams/day, in relation to colorectal cancer rates arising from a large cohort study of men (Health Professionals Follow-up Study (Giovannucci et al., 1995)). The individual time-to-event outcomes were summarized using incidence rates (sum of cases/sum of person-time). During 439,244 person-years, a total of 408 cases of colorectal cancer were identified during the follow-up period 1986–1996.

This study will be analyzed in detail in this chapter separately and together with seven other studies included in a pooled analysis (Cho et al., 2004). Table 18.1 is study 2 in Table 18.4. Alcohol intake was categorized into six levels using the lowest as referent. To understand the steps involved in a dose-response meta-analysis, we start by providing a brief sketch of the commonly used two-stage approach.

In Section 18.2, we concisely describe a possible framework for dose-response meta-analysis. In Section 18.3, we specify linear and non-linear dose-response models, explain how to conduct statistical inference, and present algorithms to reconstruct the missing covariance for a single summarized study. We then apply the methods to a table of published data arising from a large cohort study (Section 18.4). In Section 18.5, we explain how to combine dose-response functions across multiple summarized studies, test for heterogeneity, test hypotheses about the parameters, and estimate and present the summary dose-response relationship. An application is then shown using summarized data from eight published prospective cohort studies of alcohol intake and colorectal cancer rate (Section 18.6). In Section 18.7, we compare a dose-response analysis of nine prospective studies based on individual participant data to that of the summarized data. In Section 18.8, we conclude discussing some advantages and disadvantages of a two-stage dose-response meta-analysis based on summarized data, indicating recent methodological developments, and suggesting the most updated statistical packages.

TABLE 18.1

Incidence Rate of Colorectal Cancer in Relation to Alcohol Intake (Grams/Day) in the Health Professionals Follow-up Study (Cho et al., 2004)

Alcohol intake	Median grams/day	No. of cases	Person-years	Rate per 10,000	Crude rate ratio (95% CI)	Adjusted rate ratio (95% CI)
0	0	100	103,002	9.7	1.00 Ref.	1.00 Ref.
[0,5)	2.1	65	106,826	6.1	0.63 (0.46, 0.86)	0.66 (0.48, 0.90)
[5, 15)	9.5	104	119,846	8.7	0.89 (0.68, 1.18)	0.91 (0.69, 1.20)
[15, 30)	18.8	63	58,034	10.9	1.12 (0.82, 1.53)	1.10 (0.79, 1.52)
[30, 45)	36.7	46	33,081	13.9	1.43 (1.01, 2.03)	1.23 (0.85, 1.76)
≥45	59.4	30	18,455	16.3	1.67 (1.11, 2.52)	1.41 (0.92, 2.17)

Rate ratios were adjusted for age, energy intake (kcal/day), multivitamin use, family history of colorectal cancer, current smoking, past smoking, red meat intake, total milk intake, and dietary folate intake.

18.2 A Sketch of Dose-Response Meta-Analysis

Our goal is to investigate the dose-response relationship that might exist across the available studies, published, and, when possible, unpublished. Different dose-response functions can be defined according to the available data and research questions (see Box 18.1). A simple example (Figure 18.1) can clarify the distinction made between the curves μ_t, μ_{f_i}, and μ_{g_i} (e.g., linear predictors of suitable generalized linear models or survival models). μ_t is the true dose-response curve that we aim to estimate. μ_{f_i} is the piecewise constant function that is estimated by the study author on individual data, and that forms the summarized data used by the meta-analyst. μ_{g_i} is the smooth dose-response function estimated by the meta-analyst on the summarized data. The function μ_g is the summary dose-response curve estimated by averaging the regression coefficients defining μ_{g_i} across studies.

BOX 18.1 SKETCH OF DOSE-RESPONSE META-ANALYSIS

1. Unknown dose-response function

$$\mu_t(x) = t(x, \beta)$$

2. Study author piecewise constant function on the ith individual data

$$\mu_{f_i} \mid x_i = f_i(x_i, \gamma_i)$$

3. Meta-analyst smooth function on the ith summarized data

$$\mu_{g_i} \mid x_i = g(x_i, \theta_i)$$

4. Between-studies model

$$\hat{\theta} \text{ as summary of } \hat{\theta}_i \text{ with } i = 1, \ldots, I \text{ studies}$$

5. Summary dose-response function

$$\mu_g \mid x = g(x, \hat{\theta})$$

Suppose there is a U-shaped relationship between body mass index and hyponatremia risk in the population of non-elite marathon runners; body mass index extremes are associated with higher hyponatremia risks. The true population dose-response curve is named μ_t and it is represented by the quadratic mid-gray curve. A prospective study including 488 Boston marathon runners is then conducted (Almond et al., 2005). Data were analyzed with a logistic regression model and the investigators found that relative to runners in the middle range of body mass index 20–25 kg/m^2, runners with a body mass index

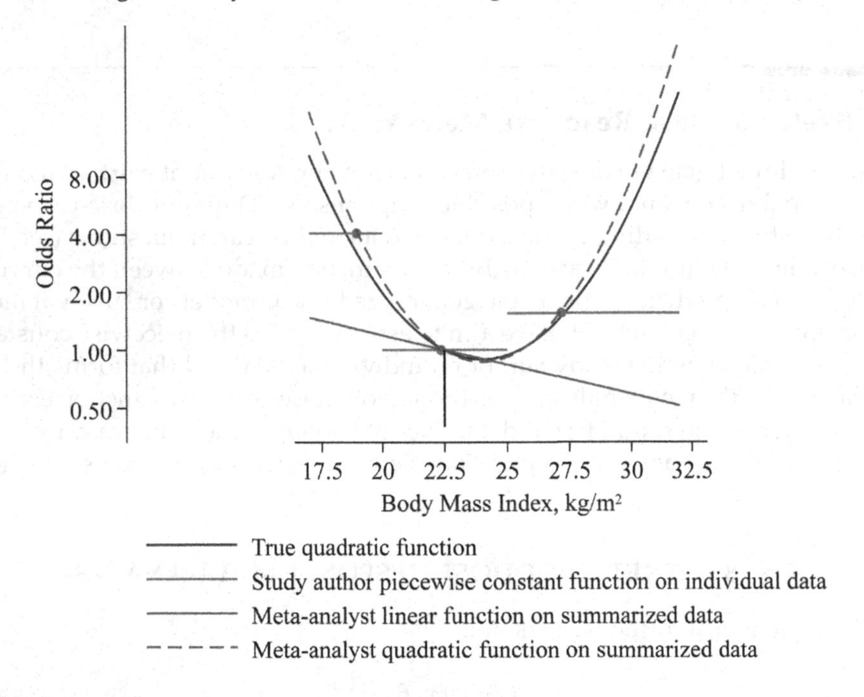

FIGURE 18.1
Graphical presentation of the (a) true dose-response function (μ_t quadratic mid-gray curve) between body mass index and hyponatremia odds using 22.5 kg/m^2 as referent; (b) study author piecewise constant function estimated from individual data (μ_{f_i} dark gray horizontal segments with mean dose values represented by dots) using the interval 20–25 kg/m^2 as referent; (c) meta-analyst linear function estimated from summarized data (μ_{g_i} light gray solid line) using 22.5 kg/m^2 as referent; and (d) meta-analyst quadratic function estimated from summarized data (μ_{g_i} light gray dashed curve) using 22.5 kg/m^2 as referent.

below 20 kg/m^2 had a 4-fold higher odds of hyponatremia and runners above 25 kg/m^2 had a 55% higher odds of hyponatremia, respectively. In this specific study, the authors divided the exposure range into three intervals using two cutpoints at 20 and 25 kg/m^2 (Table 18.3). Given the categorical coding scheme of the body mass index, it follows that μ_{f_i} is the empirical odds ratio represented by the dark gray piecewise horizontal line. To exploit the contribution of this study in forming a judgment about the true dose-response relationship the meta-analyst needs to specify a dose-response function to be estimated from this study as well as all comparable studies. This common summary dose-response function estimated from a specific study is named μ_{g_i}. The expected dose-response relationships described by μ_{f_i} and μ_{g_i} have in common that they are estimated on the same study and that the investigator has to decide how the quantitative exposure relates to the response. The main difference is that the function f_i is estimated by the study author using individual data while the function g_i is estimated by the meta-analyst using aggregated data obtained with f_i. The other difference is that f_i is likely to vary across studies carried out in different parts of the world (different numbers of categories, cutpoints, and reference intervals) whereas g_i (straight line, quadratic curve) is typically estimated from all the studies. Figure 18.1 suggests that the linear function (μ_{g_i} is the light gray solid line) may not be the best strategy to investigate a truly non-linear relationship. Not surprisingly, a quadratic function (μ_{g_i} is the light gray quadratic dashed curve) computed exclusively using the three summary points (gray dots) with coordinates [log(4.05), 18.95], [log(1), 22.36], and [log(1.55), 27.19] would provide a better approximation of the U-shaped association between body mass index and hyponatremia risk present in the population from which this sample of 488 runners was taken.

We can now describe the synthesis of dose-response relationships using a two-stage approach in more general terms. The true functional relationship t relating the dose x to μ, the expected value of the outcome random variable, is unknown. It can probably be approximated by a linear combination of data and parameters $t(x,\beta)$.

Data on dose x_i and outcome y_i are collected by $i = 1,\ldots,I$ studies. Observational studies typically collect information on possible confounders, denoted z_i, of the x_i-y_i association. In each study, a model $\mu_{f_i} \mid x_i = f_i(x_i, \gamma_i)$, eventually adjusted for a set of confounders z_i, is used to make inference on the unknown dose-response relation. Quantitative predictors in epidemiological studies are commonly categorized and modeled using indicator variables, so f_i is usually a step-function with jumps equal to γ_i from a reference group at specified cutpoints of the distribution of x_i. In multiplicative models (i.e., generalized linear models for the binomial and Poisson family, Cox regression models), the values of γ_i represent the changes (on natural log scale) of the response (i.e., odds, risk, rate, hazard). The parameters of the functions $\mu_{f_i} \mid x_i$ estimated in each study constitute the input data for a dose-response meta-analysis based on summarized data. For example, in Table 18.1, the function f_i used to describe the change in colorectal cancer rate according to alcohol intake is a piecewise constant with five cutpoints (5, 10, 15, 30, 45 grams/day) and modeled using five indicator variables with 0 grams/day serving as referent.

Often times, the first choice for the dose-response function g_i is a simple straight line. Considering the data from the Health Professional Follow-up Study in Table 18.1, the question might be what is the change in colorectal cancer rate associated with every increment of 12 grams, or one glass of wine, per day of alcohol intake. This is not the only study investigating this specific association (Table 18.4). So within each available study, the dose-response relation $\mu_{g_i} \mid x_i$ is estimated as a function of the assigned doses x_i (e.g., mean/median exposure within each interval) to obtain parameter $\hat{\theta}_i$. The dose-response function

g_i, estimated from the summarized studies, is defined according to the complexity of the research questions and in light of the available knowledge.

Assuming linearity for each of the eight studies included in the meta-analysis of alcohol intake and colorectal cancer leads to one regression coefficient for each study. Across the I studies, standard methods for meta-analysis are used to combine the study specific estimates $\hat{\theta}_i$ into a summary estimate $\hat{\theta}$ incorporating statistical heterogeneity. Heterogeneity could cause the pooled function to be a poor approximation of study-specific truths. Nevertheless, if each of the previous steps is done carefully and model assumptions are holding, the function $g(x,\hat{\theta})$ is expected to be a good approximation of the unknown dose-response relation $t(x,\beta)$.

18.3 Dose-Response Model within Each Study

We consider I studies reporting the results obtained on individual data according to a J_i-step dose-response function $f_i(x_i,\gamma_i)$. The set of J_i-1 cutpoints used to model the exposure x_i in $\mu_{f_i} \mid x_i$, the number of exposure intervals $j = 1,\ldots, J_i$, and the referent $x_{i,\text{ref}}$ may change across studies. The notation for summarized dose-response data for the i-th study is presented in Table 18.2.

Depending on the study design and regression model used to analyze individual data, γ_i may represent a difference between log risks (cases/total), log odds (cases/controls), or log rates (cases/person-time) estimated using the function f_i on individual participant data. Suppose that the only information available to the investigator is that provided in Table 18.2 and the question is what might have been the smooth change of the response across the dose or exposure range in the i-th study. We now describe how to specify and estimate a dose-response model $\mu_{g_i} \mid x_i = g(x_i,\theta_i)$ using a smooth function g other than f using the available summarized data. In a two-stage approach, the function g estimated within each study is the same ($g_1 = g_2 = \cdots = g_I = g$). A case where the function g is allowed to vary

TABLE 18.2

Notation for Summarized Dose-Response Data for the i-th Study

Category	Dose	Cases	Total	Response	Response difference	Variance
j	x_{ij}	c_{ij}	t_{ij}[a]	$\hat{\mu}_{ij}$[b]	$\hat{\gamma}_{ij} = \hat{\mu}_{ij} - \hat{\mu}_{i,ref}$[c]	$\hat{\sigma}_{ij} = var(\hat{\gamma}_{ij})$[d]
1	x_{i1}	c_{i1}	t_{i1}	$\hat{\mu}_{i1}$	$\hat{\gamma}_{i1}$	$\hat{\sigma}_{i1}$
⋮	⋮	⋮	⋮	⋮	⋮	⋮
ref	$x_{i,\text{ref}}$	$c_{i,\text{ref}}$	$t_{i,\text{ref}}$	$\hat{\mu}_{i,\text{ref}}$	$\hat{\gamma}_{i,\text{ref}} = 0$	$\hat{\sigma}_{i,\text{ref}} = 0$
⋮	⋮	⋮	⋮	⋮	⋮	⋮
J_i	x_{iJ_i}	c_{iJ_i}	t_{iJ_i}	$\hat{\mu}_{iJ_i}$	$\hat{\gamma}_{iJ_i}$	$\hat{\sigma}_{iJ_i}$

[a] Depending on the study design, this column reports total person-time, total number of subjects, or total number of non-cases.

[b] The response $\mu_{ij} = \log(c_{ij}/t_{ij})$ can be a log rate, log risk, or log odds.

[c] Measures of association $\hat{\gamma}_{ij}$ (log rate ratio, log risk ratio, log odds ratio) are estimated using J_i-1 indicator variables with multiplicative regression models.

[d] Estimated variance of $\hat{\gamma}_{ij}$.

across studies can be found elsewhere (Crippa et al., 2018). In general, the choice of the function g depends on the research hypothesis.

18.3.1 Linear Dose-Response Model

Assuming a linear dose-response association, one is interested in finding the magnitude of the slope of the dose-response relationship associated with a given increment in the exposure. Given its straightforward interpretation, a very popular dose-response function $\mu_{g_i} \mid x_i$ estimated from summarized data is a straight line.

A linear dose-response model for the i-th study can be defined as follows

$$\hat{\gamma}_i = \mathbf{X}_i \theta_i + \epsilon_i \tag{18.1}$$

$$\hat{\gamma}_i = \begin{bmatrix} \hat{\mu}_{ij} - \hat{\mu}_{i,ref} \\ \vdots \\ \hat{\mu}_{i(J_i-1)} - \hat{\mu}_{i,ref} \end{bmatrix} \quad \mathbf{X}_i = \begin{bmatrix} x_{i1} - x_{i,ref} \\ \vdots \\ x_{i(J_i-1)} - x_{i,ref} \end{bmatrix} \quad \epsilon_i = \begin{bmatrix} \varepsilon_{i1} \\ \vdots \\ \varepsilon_{i(J_i-1)} \end{bmatrix}$$

where the dependent variable $\hat{\gamma}_i$ is a $(J_i - 1)$ vector $(\hat{\mu}_{ij} - \hat{\mu}_{i,\mathrm{ref}})$ of differences in responses about the referent; the design matrix is $\mathbf{X}_i$ is a $(J_i - 1)$ vector of exposure values centered around the referent; and θ_i, a scalar, is the parameter to be estimated. Of note, the row of data in Table 18.2 corresponding to the referent is not included. A distinctive feature of the model is that there is no intercept and the fitted trend is forced to go through the origin. The regression coefficient θ_i represents the change in the response $\mu_{g_i} \mid x_i$ associated with every one unit increase of the exposure x_{ij} in the i-th study. The errors ε_i form a $(J_i - 1)$ vector, with expected value $E(\varepsilon_i) = 0$ and variance-covariance matrix $\mathrm{Cov}(\varepsilon_i) = E(\varepsilon_i \varepsilon_i')$ equal to the following symmetric matrix given by:

$$\mathrm{Cov}(\varepsilon_i) = \mathbf{C}_i = \begin{bmatrix} \sigma_{i11} & & & & \\ \vdots & \ddots & & & \\ \sigma_{ij1} & & \sigma_{ijk} & & \\ \vdots & & & \ddots & \\ \sigma_{i(J_i-1)1} & \cdots & \sigma_{i(J_i-1)k} & \cdots & \sigma_{i(J_i-1)(J_i-1)} \end{bmatrix}. \tag{18.2}$$

Thus, the response variable $\hat{\gamma}_i$ has expected value $E(\hat{\gamma}_i) = \mathbf{X}_i \theta_i$ and asymptotic variance/covariance matrix $\mathrm{Cov}(\hat{\gamma}_i) = \mathbf{C}_i$. The matrix $\mathbf{C}_i$, arising from the study author piecewise constant function $\mu_{f_i} \mid x_i$ (not from fitting $\mu_{g_i} \mid x_i$), can be estimated as the inverse Fisher information. The off-diagonal elements of $\mathbf{C}_i$, however, are not typically published and procedures have been developed to estimate them from the available information (Section 18.3.6). Inference on the study-specific dose-response relationship $\mu_{g_i} \mid x_i$ can be made using generalized least squares, maximum likelihood methods, or Bayesian methods (Section 18.3.3).

18.3.2 Flexible Dose-Response Model

We can extend the linear regression model specified above (equation 18.1) to estimate more complicated functions (e.g., U-shape, J-shape, V-shape, threshold, spike at zero, steps). This is achieved by using p transformations of the dose x_i. The rate of change of the response $\mu_{g_i} \mid x_i$ depends upon the actual values of x_i. Given the small number of non-reference values available from each study, where J_i typically ranges from three to five, we limit to functional forms that can be expressed in terms of maximum $p=2$ transformations of the dose. The flexible functional relationship on summarized data $\mu_{g_i} \mid x_i = g(x_i, \theta_i)$ can be parametrized in terms of $\boldsymbol{\theta}_i$ and the dose as follows

$$\hat{\gamma}_i = g(x_i, \theta_i) + \varepsilon_i, \quad \varepsilon_i \sim N(0, \mathbf{C}_i), \quad i = 1, \ldots, I \tag{18.3}$$

where $\boldsymbol{\theta}_i$ is the $p \times 1$ vector of regression coefficients to be estimated within each study. Using two transformations, the regression function can be written as

$$g(x_i, \theta_i) = \left[g_1(x_{ij}) - g_1(x_{i,\text{ref}})\right]\theta_{1i} + \left[g_2(x_{ij}) - g_2(x_{i,\text{ref}})\right]\theta_{2i} \tag{18.4}$$

where, for example, $g_1(x_{ij})$ is the value of the first transformation for the j-th exposure level in the i-th study, and $g_1(x_{i,\text{ref}})$ is the value of the first transformation at the referent in the i-th study. The design matrix $\mathbf{X}_i$ is equal to

$$\mathbf{X}_i = \begin{bmatrix} g_1(x_{i1}) - g_1(x_{i,ref}) & g_2(x_{i1}) - g_2(x_{i,ref}) \\ \vdots & \vdots \\ g_1(x_{i(J_i-1)}) - g_1(x_{i,ref}) & g_2(x_{i(J_i-1)}) - g_2(x_{i,ref}) \end{bmatrix}. \tag{18.5}$$

We now specify some examples of functional forms that can expressed in terms of two transformations. A flexible tool able to answer a variety of scientific questions arising in dose-response analysis is represented by regression splines.

A single spline transformation of the exposure x is defined by the location of a knot k and a certain degree d as follows $I(x > k)(x - k)^d$ with $d = 0, 1, 2, 3$. The indicator function $I(x > k)$ takes value 1 if $x > k$ and 0 if $x \le k$. A more compact way of writing a spline is $(x - k)_+^d$. According to the degree of the polynomial and the number of knots, we can estimate piecewise constant, linear, quadratic, and cubic splines.

18.3.2.1 Piecewise Constant Splines

The question of interest is whether there is any change in level of the response at certain values of the exposure. Considering the set of two knots $\mathbf{k} = (k_1, k_2)$, the splines are generated as follows

$$\begin{aligned} g_1(x_{ij}) &= (x_{ij} - k_1)_+^0 = I(x_{ij} > k_1) \\ g_2(x_{ij}) &= (x_{ij} - k_2)_+^0 = I(x_{ij} > k_2) \end{aligned} \tag{18.6}$$

Upon definition of the correct design matrix $\mathbf{X}_i$ as explained above, the parameter θ_{1i} quantifies the shift in response, or log relative risk, comparing the interval $k_1 < x_{ij} \le k_2$ relative to the interval $x_{ij} \le k_1$. The parameter θ_{2i} represents the shift in response comparing the interval $x_{ij} > k_2$ relative to the preceding interval $k_1 < x_{ij} \le k_2$. Figure 18.2a shows the dose-response relationship with the relative risks ($e^{\theta_{1i}}$ and $e^{\theta_{2i}}$) on the y-axis.

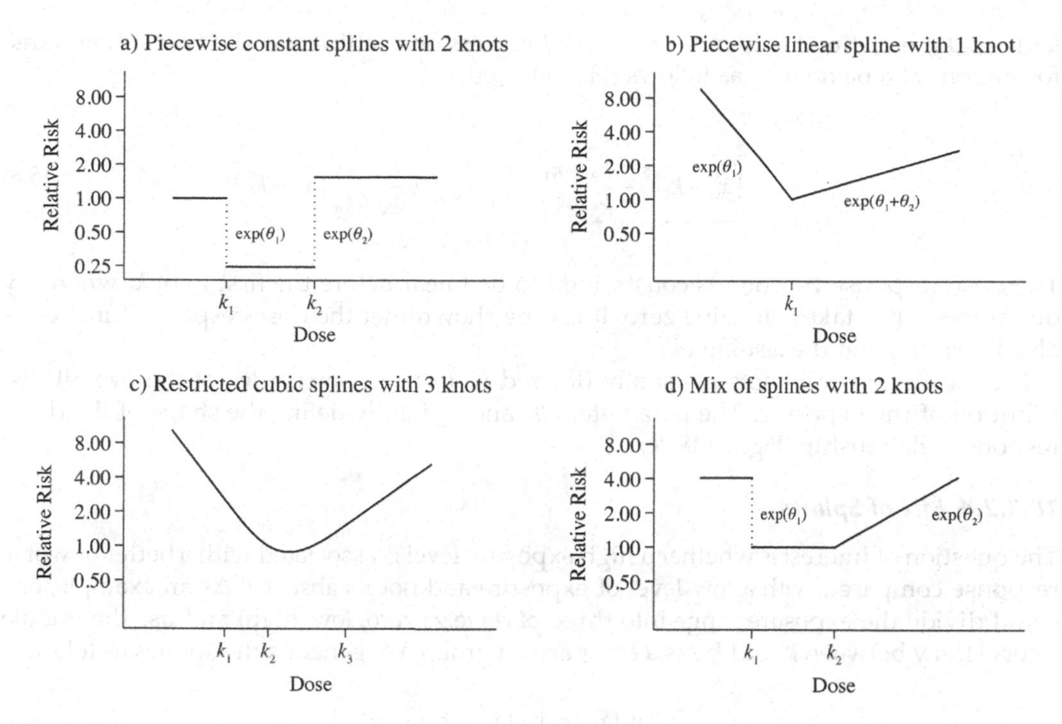

FIGURE 18.2
Illustration of dose-response relationships using (a) piecewise constant splines with two knots at k_1 and k_2; (b) piecewise linear spline with one knot at k_1; (c) restricted cubic splines with three knots at k_1, k_2, and k_3; and (d) mix of piecewise constant and linear splines with knots at k_1 and k_2.

18.3.2.2 Piecewise Linear Splines

The question of interest is whether there is any change in the slope of the response at a certain value of the exposure (i.e., V-shape, threshold). Given one knot $k=k_1$, the transformations to use are

$$g_1(x_{ij}) = x_{ij}$$
$$g_2(x_{ij}) = (x_{ij} - k_1)_+^1 \tag{18.7}$$

where $g_1(x_{ij})$ is simply the identity function representing the linear polynomial piece of the spline. Therefore, the parameter θ_{1i} quantifies the linear trend (log relative risk associated with every one unit increase of the dose) in the interval $x_{ij} \le k_1$. The parameter θ_{2i} is the change in linear trend when moving from the interval $x_{ij} \le k_1$ to the interval $x_{ij} > k_1$. The sum of the two parameters $(\theta_{1i}+\theta_{2i})$ quantifies the linear trend when $x_{ij} > k_1$. Figure 18.2b shows the dose-response relationship with the relative risks ($e^{\theta_{1i}}$ and $e^{\theta_{1i}+\theta_{2i}}$) on the y-axis.

18.3.2.3 Restricted Cubic Splines

The question of interest is whether the dose-response function is curvilinear (i.e., U-shape, J-shape, threshold). The response is smoothly and continuously varying throughout the exposure range with few constraints on the possible shape to limit the possible instability in the tails of the exposure distribution. A restricted cubic spline model with three knots

$\mathbf{k}=(k_1,k_2,k_3)$ is defined only in terms of $p=2$ regression coefficients. The two spline transformations can be defined as follows (Harrell, 2001):

$$g_1(x_{ij}) = x_{ij}$$
$$g_2(x_{ij}) = \frac{(x_{ij}-k_1)_+^3 - \frac{k_3-k_1}{k_3-k_2}(x_{ij}-k_2)_+^3 + \frac{k_2-k_1}{k_3-k_2}(x_{ij}-k_3)_+^3}{(k_3-k_1)^2} \tag{18.8}$$

The dose-response function is constrained to be linear before the first knot k_1 where, by definition, $g_2(x_{ij})$ takes on value zero. It can be shown that the dose-response function is also linear beyond the last knot k_3.

The location of the knots is usually derived from fixed percentiles of the overall distribution of the exposure. The parameters θ_{1i} and θ_{2i} jointly define the shape of the dose-response relationship (Figure 18.2c).

18.3.2.4 Mix of Splines

The question of interest is whether a high exposure level is associated with a better or worse response compared with a low level of exposure and not its absence. As an example, one could divide the exposure range into three parts (e.g., zero, low, high) and use the middle interval, say between k_1 and k_2, as a comparison group. We generate the splines as follows

$$g_1(x_{ij}) = 1-(x_{ij} > k_1)_+^0$$
$$g_2(x_{ij}) = (x_{ij}-k_2)_+^1 \tag{18.9}$$

where $g_1(x_{ij})$, complement of the degree-0 spline with a knot at k_1, is just an indicator variable taking on value 1 if $x_{ij} \le k_1$ and 0 otherwise.

The parameter θ_{1i} quantifies the shift in response (log relative risk) comparing the zero-exposure group ($x_{ij} \le k_1$) relative to the middle interval ($k_1 < x_{ij} \le k_2$). The parameter θ_{2i} represents the linear trend (log relative risk associated with every one unit increase of the dose) in the highly exposed group defined as $x_{ij} > k_2$. An illustration of the dose-response function is provided in Figure 18.2d.

18.3.3 Within-Study Estimation

Once the functional form $\mu_{g_i} \mid x_i$ has been specified and the appropriate design matrix $\mathbf{X}_i$ created, generalized least squares (GLS) can be used to efficiently estimate the regression coefficients $\hat{\theta}_i$ and the corresponding variance-covariance matrix $\mathbf{S}_i$ by minimizing

$$\left(\hat{\gamma}_i - g(\mathbf{X}_i,\boldsymbol{\theta}_i)\right)' \mathbf{C}_i^{-1}\left(\hat{\gamma}_i - g(\mathbf{X}_i,\boldsymbol{\theta}_i)\right) \tag{18.10}$$

generally using a numerical optimization algorithm. If g is a linear combination of the parameters $\boldsymbol{\theta}_i$, as in the parametrizations given in Section 18.3.2, the closed form solution can be written as

$$\hat{\theta}_i = \left(\mathbf{X}_i'\mathbf{C}_i^{-1}\mathbf{X}_i\right)^{-1}\mathbf{X}_i'\mathbf{C}_i^{-1}\hat{\gamma}_i$$
$$\mathbf{S}_i = \mathrm{Var}\left(\hat{\theta}_i\right) = \left(\mathbf{X}_i'\mathbf{C}_i^{-1}\mathbf{X}_i\right)^{-1} \tag{18.11}$$

GLS estimation does not impose any distributional assumption for the random errors ε_i, whereas maximum likelihood (ML) estimation assumes a distribution. Under the assumption that random errors are normally distributed with zero mean and variance-covariance matrix $\mathbf{C}_i$, i.e., $\varepsilon_i \sim N(\mathbf{0}, \mathbf{C}_i)$, the log-likelihood function can be written as

$$l_i(\theta_i; \mathbf{X}_i, \mathbf{C}_i) = -\frac{(J_i - 1)}{2}\log(2\pi) - \frac{1}{2}\log|C_i| - \frac{1}{2}\left[(\gamma_i - \mathbf{X}_i\theta_i)' C_i^{-1}(\gamma_i - \mathbf{X}_i\theta_i)\right]. \quad (18.12)$$

Maximizing equation (18.12) with respect to θ_i is equivalent to solving $\partial l_i / \partial \theta_i = 0$. The solution is the ML estimator of θ_i, which under the normality assumption turns out to be the same as the GLS estimator given by equation (18.11).

18.3.4 Hypotheses Tests

Although the primary interest is in the pooled dose-response relationship (Section 18.5.4), here we briefly explain how to conduct large-sample inference for the i-th study using a Wald-type test or likelihood ratio test. Based on the assumed functional form $\mu_{g_i} \mid x_i$, the hypothesis of overall no exposure-response association can be addressed by testing H_0: $\boldsymbol{\theta}_i = 0$.

A test of the null hypothesis, H_0: $\boldsymbol{\theta}_i = 0$ versus H_A: $\boldsymbol{\theta}_i \neq 0$, can be based on the following Wald-type statistic

$$W_i = \hat{\theta}_i \mathbf{S}_i^{-1} \hat{\theta}_i' \quad (18.13)$$

and a p-value obtained by comparison with a χ^2 distribution with p degrees of freedom. In case of $p = 1$, the Wald-type test for one regression coefficient is the ratio between the estimated parameter and its standard error

$$z_i = \hat{\theta}_i / \sqrt{\mathbf{S}_i} \quad (18.14)$$

which, under the null hypothesis, can be compared with a standard normal distribution.

A likelihood-based method for testing hypotheses consists in estimating a model under the null hypothesis and a larger model with a higher number of parameters. The likelihood ratio test comparing the two alternative models is obtained as

$$\Lambda_i = -2(l_0 - l_1) \quad (18.15)$$

where l_1 and l_0 are the two maximized log-likelihoods (equation 18.12) of the models containing p_1 and p_0 parameters. Under the null hypothesis, Λ_i is compared with a χ^2 distribution with $p_1 - p_0$ degrees of freedom.

18.3.5 Within-Study Confidence Intervals

The study-specific $100(1-\alpha)\%$ Wald-type confidence interval for $\boldsymbol{\theta}_i$ can be computed using the large-sample approximation

$$\hat{\theta}_i \pm z_{\alpha/2}\sqrt{\mathbf{S}_i} \quad (18.16)$$

where $z_{\alpha/2}$ denotes the $(1-\alpha/2)$-quantile of a standard normal distribution (e.g., ±1.96 for 95% confidence interval).

Let x_i^* be the design matrix obtained by applying the function g to a number of reasonable exposure values for the i-th study, and let x_0^* be the same design matrix evaluated at the chosen reference level. The vector of predicted responses is given by

$$(x_i^* - x_0^*)\hat{\theta}_i \tag{18.17}$$

with pointwise $100(1-\alpha)\%$ confidence intervals obtained as

$$(x_i^* - x_0^*)\hat{\theta}_i \pm z_{\alpha/2}\text{diag}\left[(x_i^* - x_0^*)\mathbf{S}_i(x_i^* - x_0^*)'\right]^{1/2}. \tag{18.18}$$

A smooth graph of the study-specific dose-response curve typically requires a grid of dose values whose size exceeds the available J_i dose values. Therefore, the number of rows of x_i^* needed to produce a publication-quality plot is likely to be higher than the number of rows of $\mathbf{X}_i$ used to fit the dose-response model. The reference value x_0^* used to present the predicted study-specific trend may not be the same as the reference value $x_{i,\text{ref}}$ used to present the summarized data. The choice of x_0^* should facilitate the comparison of the estimated dose-response trend with the one arising from other studies.

18.3.6 On the Covariance of Summarized Dose-Response Data

The set of study-specific contrasts, $\hat{\gamma}_i$, has a certain asymptotic covariance structure, $\mathbf{C}_i$, that we assumed as known in fitting the dose-response model. Information on $\mathbf{C}_i$ is needed for valid inference about study-specific trends $\mu_{g_i} \mid x_i$ and eventually to change the reference group of empirical contrasts for later meta-analysis.

We first show how $\mathbf{C}_i$ can be derived directly from the cell counts of a contingency table when the estimated parameters $\hat{\gamma}_i$ have not been adjusted for other covariates. Since all comparisons are made to the reference level, quantifying the covariance involves just the counts of the reference level. Assuming multinomial or Poisson distributions for counts and using the delta-method for large sample inference, the form of $\mathbf{C}_i$ depends on the type of response. In a case-control study where $\hat{\gamma}_{ij}$ are log odds ratios, the elements of $\mathbf{C}_i$ are

$$\hat{\sigma}_{ijk} = \begin{cases} 1/c_{i,\text{ref}} + 1/t_{i,\text{ref}} & \text{if } j \neq k \\ 1/c_{ij} + 1/t_{ij} + 1/c_{i,\text{ref}} + 1/t_{i,\text{ref}} & \text{if } j = k \end{cases}$$

with c indicating the number of cases and t the number of controls. In a cumulative-incidence study where $\hat{\gamma}_{ij}$ are log risk ratios, the elements of $\mathbf{C}_i$ are

$$\hat{\sigma}_{ijk} = \begin{cases} 1/c_{i,\text{ref}} - 1/t_{i,\text{ref}} & \text{if } j \neq k \\ 1/c_{ij} - 1/t_{ij} + 1/c_{i,\text{ref}} - 1/t_{i,\text{ref}} & \text{if } j = k \end{cases}$$

with c and t indicating the number of cases and number of individuals respectively. In a prospective cohort study where $\hat{\gamma}_{ij}$ are log rate ratios, the elements of $\mathbf{C}_i$ are

$$\hat{\sigma}_{ijk} = \begin{cases} 1/c_{i,\text{ref}} & \text{if } j \neq k \\ 1/c_{ij} + 1/c_{i,\text{ref}} & \text{if } j = k \end{cases}$$

with c indicating the number of cases.

When $\hat{\gamma}_{ij}$ have been adjusted for other covariates, the diagonal elements of $\mathbf{C}_i$ are typically derived from reported confidence limits (l,u) using normal-theory $[(u-l)/(2\times z_{\alpha/2})]^2$. The off-diagonal elements of $\mathbf{C}_i$, however, are typically not provided in publications and can be approximated using the Greenland and Longnecker (G&L) method (Greenland and Longnecker, 1992, Orsini et al., 2006) or the Hamling method (Hamling et al., 2008). Both approaches are presented below and require the solving of a system of equations. An empirical evaluation of the approximations and a comparison of assumptions is described elsewhere (Orsini et al., 2012). The authors found no major differences between the results of meta-analyses using G&L or Hamling methods even in the presence of confounding factors.

The application of such algorithms would be unnecessary if the analyst could access the variance/covariance matrix estimated by the multivariable model $\mu_{f_i} \mid x_i$ on the individual participant data. The matrices $\hat{\theta}_i$ and $\mathbf{C}_i$ can be directly available in consortia projects sharing the results of standardized analyses on harmonized data. Sometimes, the average covariance is directly published when $\hat{\theta}_i$ are presented using the floating absolute risk approach (Orsini, 2010).

18.3.6.1 Greenland and Longnecker Method

This method assumes that the correlations between the unadjusted $\hat{\gamma}_{ij}$ are similar to the correlations of the adjusted $\hat{\gamma}_{ij}$ (Greenland and Longnecker, 1992).

The off-diagonal elements of $\mathbf{C}_i$ can be estimated using the following three-step procedure, where formulas used for step 1 and 2 change according to the study type.

For incidence-rate data, where the $\hat{\gamma}_{ij}$ are log rate ratios, the off-diagonal elements $\hat{\sigma}_{ijk}$ are computed as follows:

1. Fit pseudo counts c_{ij}^* and t_{ij}^* to the interior of the contingency table such that

$$\frac{(c_{ij}^* / t_{ij}^*)}{(c_{i,\text{ref}}^* / t_{i,\text{ref}}^*)} = \exp(\gamma_{ij}) \tag{18.19}$$

   where c_{ij}^* is the pseudo number of cases and t_{ij}^* is the pseudo number of person-time at each exposure level corresponding to the adjusted $\hat{\gamma}_{ij}$. The values $c_{i,\text{ref}}^*$ and $t_{i,\text{ref}}^*$ are the pseudo counts corresponding to the reference group ($\hat{\gamma}_{ij}=0$). An iterative algorithm for solving a system of nonlinear equations is given in Appendix 2 of the article by G&L (Greenland and Longnecker, 1992).

2. For $j \neq k$, estimate the asymptotic correlation, $\hat{r}_{ijk}$, of $\hat{\gamma}_{ij}$ and $\hat{\gamma}_{ik}$ by

$$\hat{r}_{ijk} = \hat{s}_{ijk} / (\hat{s}_{ij}\hat{s}_{ik})^{1/2} \tag{18.20}$$

   where $\hat{s}_{ijk} = 1/c_{i,\text{ref}}^*$ (covariance based on pseudo counts) and $\hat{s}_{ij} = (1/c_i^* + 1/c_{i,\text{ref}}^*)$ (variance based on pseudo counts).

3. Estimate the off-diagonal elements, $\hat{\sigma}_{ijk}$, of the asymptotic covariance matrix $\mathbf{C}_i$ by

$$\hat{\sigma}_{ijk} = \hat{r}_{ijk} \times (\hat{\sigma}_{ij}\hat{\sigma}_{ik})^{1/2} \tag{18.21}$$

   where $\hat{\sigma}_{ij}$ and $\hat{\sigma}_{ik}$ are the variances of $\hat{\gamma}_{ij}$ and $\hat{\gamma}_{ik}$ calculated from confidence limits.

TABLE 18.3

Summarized Data Including Estimates of Crude and Adjusted Odds Ratios (95% Confidence Intervals) of Hyponatremia Risk According to Categories of Body Mass Index in a Closed Cohort of Boston Marathon Runners (Almond et al., 2005)

Body mass index, kg/m²	Mean	No. of cases	No. of non-cases	Crude odds ratio (95% CI)	Adjusted odds ratio (95% CI)
<20	18.95	15	34	4.05 (2.00, 8.23)	3.10 (1.35, 7.16)
(20, 25)	22.36	32	294	1.00	1.00
>25	27.19	13	77	1.55 (0.78, 3.10)	1.01 (0.47, 2.18)

Data were fitted with a logistic regression model adjusting for an indicator for weight gain (post-race weight > prerace weight) and race duration (<3:30, 3:30–4:00, >4:00).

The above procedure can be easily extended to the analysis of case-control and cumulative incidence data, upon redefinition of terms in equations (18.19) and (18.20).

As an illustrative example, consider again the association between body mass index and hyponatremia risk presented in Table 18.3. Body mass index has been categorized into three levels (<20, 20–25, >25) with the interval (20–25) serving as referent. Consider the vector of adjusted log odds ratios of hyponatremia [log(3.10), log(1.01)] with corresponding variances [0.1818, 0.1523] derived from confidence limits. The association between body mass index and hyponatremia risk was adjusted for weight change and race duration using a logistic regression model. The algorithm described above allows finding the covariance of these two adjusted log odds ratios. The vectors of cases and non-cases (pseudo counts) corresponding to the adjusted log odds ratios are

$$c_i^* = \begin{bmatrix} 13.69 \\ 36.21 \\ 10.11 \end{bmatrix} \quad t_i^* = \begin{bmatrix} 35.31 \\ 289.79 \\ 79.89 \end{bmatrix}$$

and the asymptotic correlation matrix of the adjusted non-referent log odds ratios based on the above pseudo counts is

$$\hat{r}_{ijk} = \begin{bmatrix} 1 & \\ 0.2261 & 1 \end{bmatrix}$$

with the variance/covariance matrix equal to

$$C_i = \begin{bmatrix} 0.1818 & \\ 0.0376 & 0.1523 \end{bmatrix}.$$

The calculation of the pseudo cases is done so that the margins remained fixed at their observed values. In the G&L method, the sums of the pseudo cases (13.69 + 36.21 + 10.11) and pseudo non-cases (35.31 + 289.79 + 79.89) are equal to the sums of the observed cases

(15 + 32+13) and observed non-cases (34 + 294+77), respectively. Cases and non-cases have just been redistributed across exposure levels in order to match the adjusted log odds ratios.

18.3.6.2 Hamling Method

Hamling and colleagues originally proposed a method to manipulate non-independent $\hat{\gamma}_{ij}$ in order to derive contrasts different from the ones provided in the original publication or to collapse exposure intervals (Hamling et al., 2008). The idea of finding the cell counts corresponding to a set of multivariable-adjusted regression coefficients as well as their variances can be used to back estimate the covariances of $\hat{\gamma}_{ij}$. To estimate the $2 \times J_i$ pseudo counts c_{ij}^* and t_{ij}^* $(j = 1,\dots, J_i)$, an equal number of equations needs to be specified.

The formulas for the log relative risks and related confidence limits for the exposed categories provide $2 \times (J_i - 1)$ equations. To obtain a solution, two more equations capturing basic structure of the summarized data are defined:

- $p_i = t_{i,\text{ref}} / \sum_{j=1}^{J_i} t_{ij}$, ratio between unexposed non-cases and the total number of non-cases
- $z_i = \sum_{j=1}^{J_i} t_{ij} / \sum_{j=1}^{J_i} c_{ij}$, the ratio between total number of non-cases and total number of cases

For incidence-rate data, for example, the system of $2 \times J_i$ equations for c_{ij}^* and t_{ij}^* to be solved can be written as

$$\begin{cases} \exp(\gamma_{ij}) = \dfrac{(c_{ij}^* / t_{ij}^*)}{(c_{i,\text{ref}}^* / t_{i,\text{ref}}^*)} \\ \sigma_{ijk} = (1/c_{ij}^* + 1/c_{i,\text{ref}}^*) \quad j = k \\ p_i^* = t_{i,\text{ref}}^* / \sum_{j=1}^{J_i} t_{ij}^* \\ z_i^* = \sum_{j=1}^{J_i} t_{ij}^* / \sum_{j=1}^{J_i} c_{ij}^* \end{cases}$$

where a solution can be obtained by minimizing the function $(p_i - p_i^*)^2 + (z_i - z_i^*)^2$ through a standard optimization procedure with the Nelder–Mead technique.

Considering the data in Table 18.3, the ratio between unexposed non-cases and the total number of non-cases is p_i is 294/(34 + 294 + 77) = 294/405 = 0.73 while the ratio between total number of non-cases and total number of cases z_i is (34 + 294 + 77)/(15 + 32 + 13) = (405/65) = 6.75. Using the Hamling method, the vectors of pseudo cases and non-cases corresponding to the adjusted log odds ratios are

$$c_i^* = \begin{bmatrix} 9.49 \\ 33.14 \\ 9.57 \end{bmatrix} \quad t_i^* = \begin{bmatrix} 23.61 \\ 255.77 \\ 72.96 \end{bmatrix}$$

where, as expected, $p_i^* = 255.77 / 352.33 = 0.73$ and $z_i^* = 352.33 / 52.20 = 6.75$.

The asymptotic correlation matrix of the adjusted non-referent log odds ratios based on the above pseudo counts is

$$\hat{r}_{ijk} = \begin{bmatrix} 1 & \\ 0.2049 & 1 \end{bmatrix}$$

while the variance/covariance matrix to be used in the within-study estimation procedure is

$$C_i = \begin{bmatrix} 0.1818 & \\ 0.0341 & 0.1523 \end{bmatrix}.$$

It is interesting to note that the sum of the pseudo cases is lower than the sum of observed cases because of the adjustment effect. Cases and non-cases are not only redistributed across exposure levels in order to match the adjusted log odds ratios but their sums are reduced to match the increased variances of the log odds ratios due to adjustment for additional covariates. In this example, the G&L and Hamling methods provide a similar estimate of the covariance (0.0376 vs. 0.0341) between the two log odds ratios.

18.4 Application to a Single Prospective Study

Let us consider the summarized incidence-rate data investigating the association between alcohol intake (grams/day) and colorectal cancer rate arising from a large population-based study of men shown in Table 18.1. The function f_i applied on individual participant data to summarize the dose-response relation was a standard step-function.

Rather than directly using a dose-response regression model for (log) colorectal cancer rate ratios as described in Section 18.3, we start the analysis of Table 18.1 using a traditional Poisson regression model. Therefore, we focus on the second (dose), third (cases), and fourth (person-years) columns of Table 18.1

$$x_{ij} = \begin{bmatrix} 0 \\ 2.1 \\ 9.5 \\ 18.8 \\ 36.7 \\ 59.4 \end{bmatrix} \quad c_{ij} = \begin{bmatrix} 100 \\ 65 \\ 104 \\ 63 \\ 46 \\ 30 \end{bmatrix} \quad t_{ij} = \begin{bmatrix} 103,002 \\ 106,826 \\ 119,846 \\ 58,034 \\ 33,081 \\ 18,455 \end{bmatrix}$$

The Poisson model provides estimates that can then be compared with the introduced linear dose-response model (equation 18.1). Modeling rates rather than rate ratios facilitate the comparison of alternative parametrizations of the dose-response relationship, particularly $\mu_{g_i} \mid x_i$ and $\mu_{f_i} \mid x_i$.

By hypothesizing that colorectal cancer incidence rates, on the natural log scale, are changing linearly with alcohol intake and a Poisson distribution for the number of cases with person-years as offset

$$\xi_{ij} = \beta_0 + \beta_1 x_{ij} + \ln(t_{ij})$$

$$f(c_{ij}) = \frac{e^{-\exp(\xi_{ij})} e^{\xi_{ij} c_{ij}}}{c_{ij}!}$$

the following log-likelihood is maximized

$$l(\beta; c_{ij}, t_{ij}, x_{ij}) = \sum_{j=1}^{6} \{-\exp(\xi_{ij}) + \xi_{ij} c_{ij} - \ln(c_{ij}!)\}$$

using data from Table 18.1. Of note, the Poisson model has an intercept because it is specified in terms of absolute rates rather than rate ratios. We obtain a maximum likelihood estimate of $\hat{\beta}_1 = 0.0134$ with an estimated standard error $SE(\hat{\beta}_1) = 0.0029$. Expressing the linear trend for an increment of 12 grams/day and exponentiating we have that $e^{\hat{\beta}_1 \times 12} = e^{0.1611} = 1.17$. Every 12 grams/day increment of alcohol intake is associated with a 17% higher colorectal cancer rate. A 95% confidence interval for the rate ratio can be obtained by relying on asymptotic normality of the estimator $e^{\hat{\beta}_1 \times 12 \pm 1.96 \times SE(\hat{\beta}_1) \times 12} = (1.10, 1.26)$. The estimated intercept is $\hat{\beta}_0 = -7.1481$. Therefore, the predicted colorectal cancer incidence rate among those who reported 0 grams/day of alcohol intake is $e^{-7.1481} \times 10{,}000 = 7.9$ cases every 10,000 person-years (Figure 18.3).

We now repeat the same analysis but starting with data in a slightly different format. We fit the dose-response model (equation 18.1) assuming a (log) linear dose-response relationship between alcohol intake and colorectal cancer rate using the second (log rate ratios) and sixth (dose) columns of Table 18.1. The vectors $\hat{\gamma}_i$ and $\mathbf{X}_i$ (recall that 0 grams/day is the referent) are

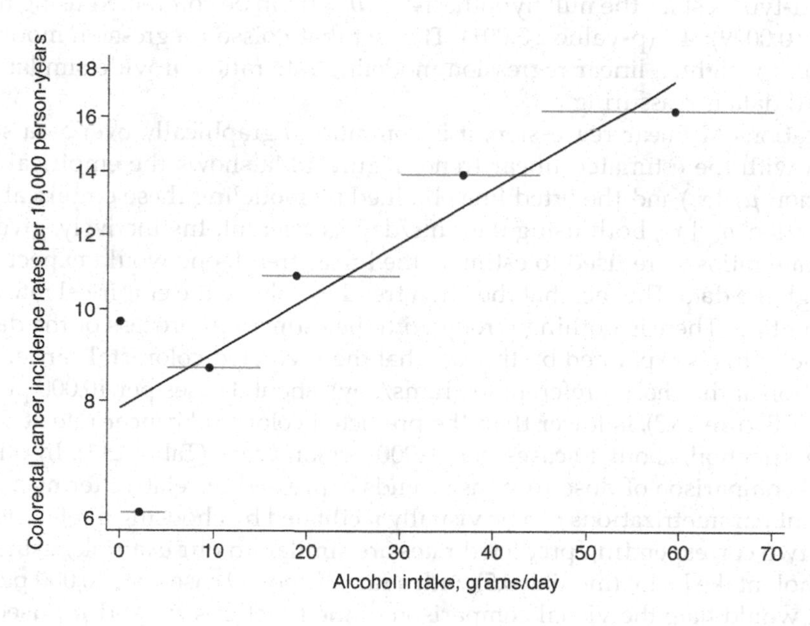

FIGURE 18.3
Alcohol intake in relation to colorectal cancer incidence rates in the Health Professionals Follow-up Study (Giovannucci et al., 1995). Summarized data (dots) were fitted with a Poisson regression model assuming a linear trend (solid line). Horizontal gray lines represent the width of the exposure interval.

$$\hat{\gamma}_i = \log \begin{bmatrix} 0.63 \\ 0.89 \\ 1.12 \\ 1.43 \\ 1.67 \end{bmatrix} \quad \mathbf{X}_i = \begin{bmatrix} 2.1 \\ 9.5 \\ 18.8 \\ 36.7 \\ 59.4 \end{bmatrix}$$

The variance/covariance structure of this set of unadjusted log rate ratios is inversely related to the number of cases within each exposure interval. For example, the asymptotic variance of the log rate ratio comparing the groups (30,45) grams/day and 0 grams/day is (1/46 + 1/100) = 0.0317. The asymptotic covariance between any pair of unadjusted log rate ratios is the inverse of the number of cases that occurred in the reference group, that is 1/100 = 0.01 (Section 18.3.6).

$$C_i = \begin{bmatrix} 0.0254 & & & & \\ 0.01 & 0.0196 & & & \\ 0.01 & 0.01 & 0.0259 & & \\ 0.01 & 0.01 & 0.01 & 0.0317 & \\ 0.01 & 0.01 & 0.01 & 0.01 & 0.0433 \end{bmatrix}$$

The GLS estimate of the linear trend is $\hat{\theta}_i = 0.0129$ with a standard error $SE(\hat{\theta}_i) = 0.0029$. Of note, these estimates are similar to the ones obtained from the Poisson model ($\hat{\beta}_1 = 0.0134$, $SE(\hat{\beta}_1) = 0.0029$). Based on the log-linear dose-response model, every 12 grams/day increment of alcohol intake is associated with a 17% (95% CI = 1.09, 1.25) higher colorectal cancer rate. A Wald-type test for the null hypothesis H_0: $\theta_i = 0$ can be conducted using the statistic z = (0.0129/0.0029) = 4.4 (p-value < 0.001). The fact that Poisson regression modeling absolute rates and weighted linear regression modeling rate ratios provide similar results on summarized data is reassuring.

In applications of linear regression, it is common to graphically overlay a scatter plot of the data with the estimated linear trend. Figure 18.4a shows the empirical rate ratios (step-function $\mu_{f_i} \mid x_i$) and the fitted line obtained by modeling these empirical rate ratios (linear function $\mu_{g_i} \mid x_i$), both using 0 grams/day as referent. Instinctively, given that the empirical rate ratios were used to estimate the linear trend, one would expect the line to pass through the data. The fact that the fitted trend lies above the empirical rate ratios may be disappointing. There is nothing wrong with the estimation process or the data used to fit the model. This is explained by the fact that the predicted colorectal cancer rate using the g_i function at the chosen referent (0 grams/day), about 8 cases per 10,000 person-years (intercept of Figure 18.3), is lower than the predicted colorectal cancer rate at the referent using the f_i function, about 10 cases per 10,000 person-years (Table 18.1). In other words, a graphical comparison of dose-response trends expressed in relative terms arising from two different parametrizations can be visually facilitated by choosing a reference group at which the two corresponding predicted rates are similar. In our example, using the interval of alcohol intake [5, 15) (median 9.5) with a rate of about 9 cases per 10,000 person-years as referent would help the visual comparison of the functions μ_{f_i} and μ_{g_i} used to model the association between alcohol intake and colorectal cancer rate (Figure 18.4b).

The covariance among log rate ratios in this study, 1/100 = 0.01, is not that far from 0. However, it can be interesting to evaluate the impact of ignoring such covariance on the

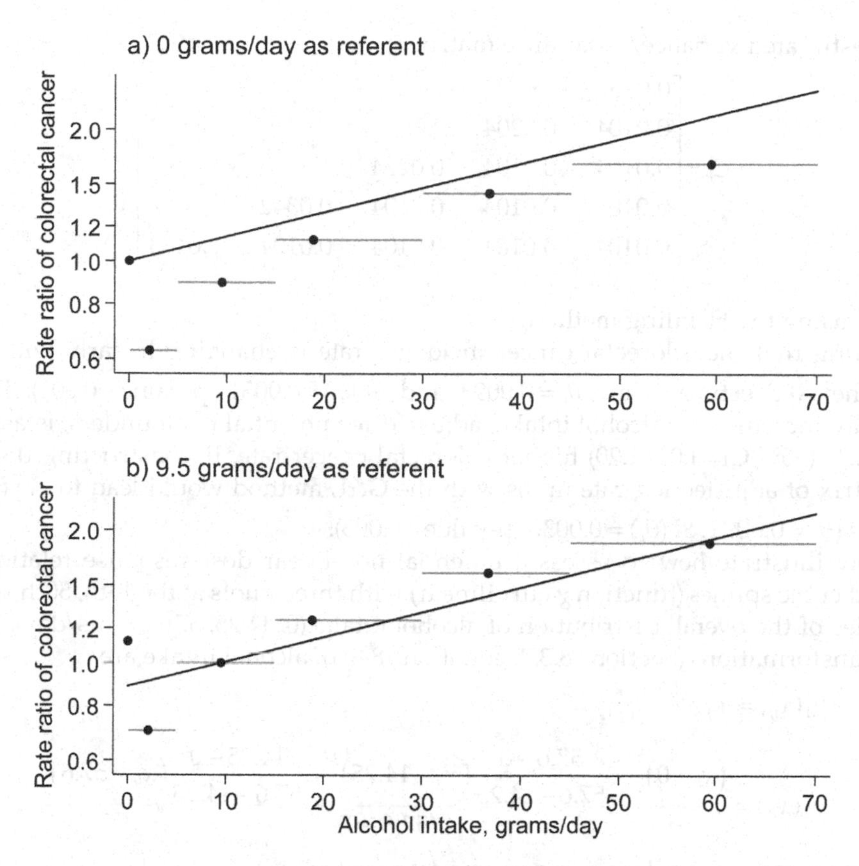

FIGURE 18.4
Alcohol intake and colorectal cancer rate ratios in the Health Professionals Follow-up Study using (a) 0 grams/day and (b) 9.5 grams/day as referent. Summarized data (dots) were fitted with a dose-response model assuming a linear trend (solid line). Horizontal gray lines represent the width of the exposure interval.

estimated linear trend for colorectal cancer rate. We would set the off-diagonal elements of $\mathbf{C}_i$ to zero and re-estimate the model. The estimated linear trend would be 40% lower ($\hat{\theta}_i = 0.0078$). Every 12 grams/day increment of alcohol intake would be associated with a 10% higher colorectal cancer rate.

Given the observational nature of this study, one should assess the dose-response relationship arising from the set of colorectal cancer rate ratios adjusted for potential confounders: age, energy intake (kcal/day), multivitamin use, family history of colorectal cancer, current smoking, past smoking, red meat intake, total milk intake, and dietary folate intake.

Using the last column of Table 18.1, the vector of estimated adjusted log rate ratios is

$$\hat{\gamma}_i = \log \begin{bmatrix} 0.66 \\ 0.91 \\ 1.10 \\ 1.23 \\ 1.41 \end{bmatrix}$$

with an estimated variance/covariance matrix

$$C_i = \begin{bmatrix} 0.0257 & & & & \\ 0.0104 & 0.0204 & & & \\ 0.0104 & 0.0104 & 0.0274 & & \\ 0.0104 & 0.0104 & 0.0104 & 0.0342 & \\ 0.0104 & 0.0104 & 0.0104 & 0.0104 & 0.0483 \end{bmatrix}$$

obtained using the Hamling method.

Assuming that the colorectal cancer incidence rate is changing linearly with alcohol intake, the GLS estimates are $\hat{\theta}_i = 0.0093$ and $SE(\theta_i) = 0.0031$ (p-value = 0.002). Every 12 grams/day increment of alcohol intake, adjusted for potential confounders, is associated with a 12% (95% CI = 1.04, 1.20) higher colorectal cancer rate. Reconstructing the covariance matrix of adjusted log rate ratios with the G&L method would lead to very similar estimates ($\hat{\theta}_i = 0.0092$, $SE(\hat{\theta}_i) = 0.0030$, p-value = 0.003).

We now illustrate how to assess a potential non-linear dose-response relation using restricted cubic splines (function g curvilinear) with three knots at the 10th, 50th, and 90th percentiles of the overall distribution of alcohol intake (0, 14.25, 57.6 grams/day). The two spline transformations (Section 18.3.2, equation 18.8) of alcohol intake are

$$g_1(x_{ij}) = x_{ij}$$

$$g_2(x_{ij}) = \frac{(x_{ij}-0)_+^3 - \frac{57.6-0}{57.6-14.25}(x_{ij}-14.25)_+^3 + \frac{14.25-0}{57.6-14.25}(x_{ij}-57.6)_+^3}{(57.6-0)^2}$$

and design matrix in the dose-response model equal to

$$\mathbf{X}_i = \begin{bmatrix} 2.1 & 0.0028 \\ 9.5 & 0.2584 \\ 18.8 & 1.9650 \\ 36.7 & 10.367 \\ 59.4 & 26.310 \end{bmatrix}$$

The GLS estimates are $(\hat{\theta}_1, \hat{\theta}_2) = (0.0112, -0.0048)$ with a variance/covariance matrix equal to

$$\mathbf{S} = \begin{bmatrix} 0.00008 & 0.00045 \\ -0.00018 & \end{bmatrix}$$

The colorectal cancer incidence rate, using 9.5 grams/day as referent, is estimated to change with alcohol intake according to the function $e^{0.0112\times[g_1(x^*)-9.5]-0.0048\times[g_2(x^*)-0.2584]}$. The Wald-type statistic ($\chi_2^2 = 9.26$, p-value = 0.010) suggests rejection of the null hypothesis $H_0\colon \theta_1 = \theta_2 = 0$ of overall no association between alcohol intake and colorectal cancer incidence rate in this specific study. Furthermore, summarized data presented in Table 18.1 seem compatible with a simpler linear dose-response relationship ($H_0\colon \theta_2 = 0$, $\chi_1^2 = 0.05$, p-value = 0.820).

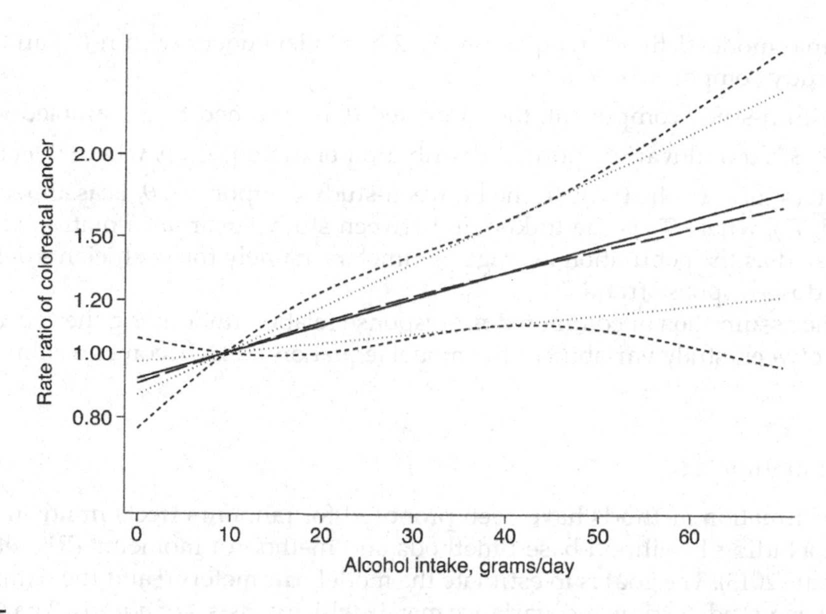

FIGURE 18.5
Adjusted dose-response relationship between alcohol intake and colorectal cancer incidence rate estimated assuming linearity (solid line) with 95% confidence intervals (dotted lines). Data were fitted using a restricted cubic spline model (long dashed line) with three knots at 0, 14.25, and 57.6 grams/day. Short dashed lines represent 95% confidence intervals for the fitted spline model. The value of 9.5 grams/day served as referent. Health Professionals Follow-up Study.

Figure 18.5 presents in a single plot the analysis based on summarized data of this large prospective cohort study by presenting the dose-response relation emerging from the two different parametrizations: linear trend and restricted cubic splines.

18.5 Pooling Dose-Response Association across Studies

The study-specific trends $\mu_{g_i} \mid x_i$ are estimated on each table of summarized data. The functional form g applies equally to all the studies and we then synthesize the distribution of the study-specific parameters $\hat{\theta}_i$ taking into account their weights $\mathbf{S}_i^{-1}$. Statistical models for multivariate meta-analysis can be used to obtain a summary dose-response relationship (Berkey et al., 1998; Ritz et al., 2008; van Houwelingen et al., 2002; Jackson et al., 2011; Gasparrini et al., 2012). The following sections summarize material presented in Chapter 9.

18.5.1 Meta-Analysis

The dose-response analysis within each study provides a total of I vectors of parameter estimates $\hat{\theta}_i$ of length p and the related $p \times p$ covariance matrices $\mathbf{S}_i$. The estimates $\hat{\theta}_i$ obtained in the dose-response analysis within each study are then used as the dependent variables in a random-effects multivariate meta-analysis model

$$\hat{\theta}_i \sim N_p(\theta, \mathbf{S}_i + \mathcal{T}) \qquad (18.22)$$

The marginal model defined in equation 18.22 has independent within-($\mathbf{S}_i$) and common between-study components ($\mathcal{T}$).

In the within-study component, the estimated $\hat{\theta}_i$ is assumed to be sampled with error from $N_p(\boldsymbol{\theta}_i, \mathbf{S}_i)$, a multivariate normal distribution of dimension p, where $\hat{\theta}_i$ is the vector of parameters for the i-th study. In the between-study component, $\hat{\theta}_i$ is assumed sampled from $N_p(\boldsymbol{\theta}, \mathcal{T})$, where $\mathcal{T}$ is the unknown between-study covariance matrix. Here, $\boldsymbol{\theta}$ can be interpreted as the population-average parameters, namely the coefficients defining the summary dose-response trend.

Under the assumption of common dose-response relation underlying the collected studies (zero-between-study variability), the model (equation 18.22) becomes a common-effect model.

18.5.2 Estimation

Different estimation methods have been proposed for random-effects multivariate meta-analysis, including likelihood-based methods and method of moments (Ritz et al., 2008; Jackson et al., 2013). The goal is to estimate the model parameters $\boldsymbol{\theta}$ and the parameters $\mathcal{T}$ of the between-study variance-covariance matrix which consists of $p(p+1)/2$ parameters if no structure for $\mathcal{T}$ is assumed.

ML estimates of $\boldsymbol{\theta}$ and $\mathcal{T}$ can be obtained simultaneously by numerically maximizing the log-likelihood function of the marginal model (18.22), subject to the constraint that $\mathcal{T}$ is positive semi-definite (White, 2011). Under the common assumption that the I studies are independent, the log-likelihood function is proportional to the logarithm of the product of I p-variate normal densities

$$l(\boldsymbol{\theta}, \mathcal{T}; \hat{\theta}_i, \mathbf{S}_i) = -\frac{1}{2}\sum_{i=1}^{I} \log|\mathbf{S}_i + \mathcal{T}| - \frac{1}{2}\sum_{i=1}^{I}\left[\left(\hat{\theta}_i - \boldsymbol{\theta}\right)(\mathbf{S}_i + \mathcal{T})^{-1}\left(\hat{\theta}_i - \boldsymbol{\theta}\right)\right]. \tag{18.23}$$

The parameter vector $\boldsymbol{\theta}$ and its accompanying variance-covariance matrix $\mathbf{V}(\boldsymbol{\theta})$ can be estimated by maximizing (18.23) using the equations (see also Chapter 9)

$$\hat{\theta} = \left(\sum_{i=1}^{I} \mathbf{S}_i^{-1}\right)^{-1} \sum_{i=1}^{I} (\mathbf{S}_i + \mathcal{T})^{-1}\hat{\theta}_i \tag{18.24}$$

$$\mathbf{V}(\hat{\theta}) = \left(\sum_{i=1}^{I} (\mathbf{S}_i + \mathcal{T})^{-1}\right)^{-1}. \tag{18.25}$$

To avoid the downward bias of ML estimates, the parameters of the between-study variance-covariance matrix can be estimated using restricted maximum likelihood (REML) (White, 2011). This bias occurs because ML does not account for the loss of degrees of freedom from the estimation of $\boldsymbol{\theta}$. REML estimation is carried out by iteratively maximizing the following restricted log-likelihood, which is a function of $\mathcal{T}$ only and is proportional to:

$$l_{\text{REML}}(\mathcal{T};\hat{\theta}_i,\mathbf{S}_i,\hat{\theta}) \propto -\frac{1}{2}\sum_{i=1}^{I}\log\left|(\mathbf{S}_i+\mathcal{T})^{-1}\right| - \frac{1}{2}\log\left|\sum_{i=1}^{I}(\mathbf{S}_i+\mathcal{T})^{-1}\right|$$

$$-\frac{1}{2}\sum_{i=1}^{I}\left[\left(\hat{\theta}_i-\hat{\theta}\right)(\mathbf{S}_i+\mathcal{T})^{-1}\left(\hat{\theta}_i-\hat{\theta}\right)\right],$$

where $\hat{\theta}$ is obtained by using equation 18.24.

18.5.3 Between-Study Heterogeneity

Meta-analysis can help identify sources of variation in the dose-response trend $\mu_{g_i} \mid x_i = g(x_i,\theta_i)$ across studies. The hypothesis of no heterogeneity between studies beyond that explained by sampling variability alone can be tested by means of the multivariate extension of the Cochran Q-test (Ritz et al., 2008; Jackson et al., 2012). Formally, the null hypothesis is $H_0\colon \mathcal{T} = \mathbf{0}$ and the test statistic is defined as

$$Q = \sum_{i=1}^{I}\left[\left(\hat{\theta}_i - \hat{\theta}_{CE}\right)\mathbf{S}_i^{-1}\left(\hat{\theta}_i - \hat{\theta}_{CE}\right)\right], \tag{18.26}$$

where $\hat{\theta}_{CE}$ is estimated using a common-effect model. Under the null hypothesis, Q asymptotically follows a χ^2 distribution with $(I-p)$ degrees of freedom. This statistic reduces to the classic Q-statistic if $p=1$ (Cochran, 1954).

18.5.4 Testing Hypotheses

The shape of the summary dose-response curve $\mu_g \mid x = g(x,\theta)$ depends on the vector of estimated coefficients $\hat{\theta}$ under either a common-effect ($\hat{\theta}_{CE}$) or a random-effects model ($\hat{\theta}_{RE}$). The hypothesis of overall no association can be addressed by testing $H_0\colon \boldsymbol{\theta} = \mathbf{0}$. Depending on the functional form g specified, testing one specific regression coefficient may detect specific characteristics of the shape (i.e., non-linearity, shift in level, change in slope). This can be done by testing $H_0^*\colon \boldsymbol{\theta}^* = \mathbf{0}$, where $\boldsymbol{\theta}^*$ refers to the part of the vector of coefficients defining those characteristics. For example, if the function g consists of restricted cubic spline models with three knots (two splines), a p-value detecting departure from linearity can be obtained by testing the coefficient of the second spline equal to zero. Similarly to within-study testing (Section 18.3.4), the Wald-type tests of hypothesis for combined parameters $\boldsymbol{\theta}$ and $\boldsymbol{\theta}^*$ can be based on the estimated $\hat{\theta}$ and its variance/covariance matrix $\mathbf{V}(\hat{\theta})$.

18.5.5 Prediction of the Summary Dose-Response Relationship

The prediction of the summary, average across studies, dose-response association $\mu_g \mid x = g(x,\hat{\theta})$ is an important step to display the results in tabular and graphical form. In particular, the goal is to present how the summary response varies according to levels of the quantitative exposure, choosing a particular value of interest as referent.

Let x^* be the $(v \times p)$ design matrix obtained by applying the function g to a number v of reasonable exposure values, and let x_0^* be the same design matrix evaluated at the chosen reference level. The $(v \times 1)$ vector of predicted summary responses is given by

$$(x^* - x_0^*)\hat{\theta} \tag{18.27}$$

with pointwise $100(1-\alpha)\%$ confidence intervals obtained as follows

$$(x^* - x_0^*)\hat{\theta} \pm z_{\alpha/2}\text{diag}\left[(x^* - x_0^*)\mathbf{V}(\hat{\theta})(x^* - x_0^*)'\right]^{1/2}. \tag{18.28}$$

The values of $\mu_g | x$ are summary log relative risks (difference in log risks, log odds, or log rates). To obtain summary relative risks one needs to exponentiate the above point and interval estimates.

18.6 Application to Multiple Prospective Studies

We combine the dose-response relation between alcohol intake and colorectal cancer rate arising from eight prospective cohort studies including 489,979 women and men participating in the Pooling Project of Prospective Studies of Diet and Cancer (Cho et al., 2004). A total of 3646 cases and 2,511,424 person-years are included in the analysis (Orsini et al., 2012). Each study categorized alcohol intake into six intervals using the same cutpoints. So, an alternative to dose-response meta-analysis in this specific case could be a multivariate meta-analysis of the estimated contrasts between categories. In general, however, cutpoints are likely to vary across studies. Basic information about cases, person-years, incidence rate ratios, and their confidence intervals is summarized into $6 \times 8 = 48$ rows of data ($I=8$, $J_i=6$) (Table 18.4). Rate ratios were adjusted for age, energy intake (kcal/day), multivitamin use, family history of colorectal cancer, current smoking, past smoking, red meat intake, total milk intake, and dietary folate intake. The Hamling method was used to estimate the covariances of multivariable-adjusted rate ratios.

Our first choice to approximate the relation between alcohol intake and colorectal cancer rate arising from these eight prospective studies is the linear function ($\mu_{g_i} | x_i = x_i\theta_i$). Study-specific GLS estimates of the linear trends can be presented either as a forest plot (Figure 18.6) or a dose-response plot (Figure 18.7a). The Q-test does not reject the null hypothesis of equal linear trends across studies ($Q=4.68$, df$=7$, p-value$=0.698$). Under a random-effects model, the average adjusted rate ratio of colorectal cancer associated with every 12 grams/day increment of alcohol intake is $e^{0.077}=1.08$ (95% CI$=1.05$–1.11).

Suppose the association between alcohol intake and colorectal cancer rate might be J-shaped. We can investigate this possibility with a flexible dose-response model using restricted cubic splines of alcohol intake with three knots at the 10th, 50th, and 90th percentiles of the distribution of alcohol intake (0, 14.25, 57.6 grams/day). By doing so, the study-specific trends are no longer forced to be linear and each study is allowed to have a different curvilinear pattern (Figure 18.7b).

The restricted cubic spline model is first fitted within each study using GLS. Study-specific estimated regression coefficients are then pooled using a multivariate random-effects

TABLE 18.4

Summarized Data on Total Alcohol Intake (Grams/Day) and Colorectal Cancer Rate in Eight Prospective Cohort Studies Participating in the Pooling Project of Prospective Studies of Diet and Cancer (Cho et al., 2004, Orsini et al., 2012)

Study (*i*)	Median, grams/day	No. of cases	Person-years	Rate per 10,000	Crude rate ratio (95% CI)	Adjusted rate ratio (95% CI)
1	0.0	28	22,186	12.6	1.00 Ref.	1.00 Ref.
1	1.8	38	43,031	8.8	0.70 (0.43, 1.14)	0.66 (0.40, 1.08)
1	9.2	43	53,089	8.1	0.64 (0.40, 1.03)	0.67 (0.42, 1.09)
1	22.9	32	45,348	7.1	0.56 (0.34, 0.93)	0.61 (0.37, 1.03)
1	35.7	16	19,791	8.1	0.64 (0.35, 1.18)	0.76 (0.40, 1.42)
1	58.4	27	19,920	13.6	1.07 (0.63, 1.82)	1.22 (0.70, 2.15)
2	0.0	100	103,002	9.7	1.00 Ref.	1.00 Ref.
2	2.1	65	106,826	6.1	0.63 (0.46, 0.86)	0.66 (0.48, 0.90)
2	9.5	104	119,846	8.7	0.89 (0.68, 1.18)	0.91 (0.68, 1.20)
2	18.8	63	58,034	10.9	1.12 (0.82, 1.53)	1.10 (0.79, 1.52)
2	36.7	46	33,081	13.9	1.43 (1.01, 2.03)	1.23 (0.85, 1.76)
2	59.4	30	18,455	16.3	1.67 (1.11, 2.52)	1.41 (0.92, 2.17)
3	0.0	446	224,905	19.8	1.00 Ref.	1.00 Ref.
3	1.7	199	105,884	18.8	0.95 (0.80, 1.12)	1.00 (0.84, 1.19)
3	8.6	91	47,328	19.2	0.97 (0.77, 1.21)	1.03 (0.82, 1.30)
3	17.1	29	16,578	17.5	0.88 (0.61, 1.28)	0.95 (0.65, 1.40)
3	37.8	25	10,208	24.5	1.23 (0.83, 1.85)	1.26 (0.83, 1.91)
3	60.6	6	3542	16.9	0.85 (0.38, 1.91)	0.90 (0.40, 2.04)
4	0.0	80	180,925	4.4	1.00 Ref.	1.00 Ref.
4	1.8	70	189,462	3.7	0.84 (0.61, 1.15)	0.93 (0.67, 1.28)
4	10.2	50	125,766	4.0	0.90 (0.63, 1.28)	0.95 (0.66, 1.36)
4	19.5	8	39,169	2.0	0.46 (0.22, 0.96)	0.48 (0.23, 1.00)
4	35.8	8	21,873	3.7	0.83 (0.40, 1.71)	0.82 (0.39, 1.73)
4	58.4	4	6735	5.9	1.34 (0.49, 3.67)	1.36 (0.49, 3.82)
5	0.0	151	234,830	6.4	1.00 (1.00, 1.00)	1.00 Ref.
5	1.8	136	213,956	6.4	0.99 (0.78, 1.25)	1.04 (0.82, 1.32)
5	9.5	72	132,975	5.4	0.84 (0.64, 1.11)	0.85 (0.63, 1.13)
5	19.5	27	44,705	6.0	0.94 (0.62, 1.41)	0.88 (0.58, 1.34)
5	35.9	23	24,416	9.4	1.46 (0.94, 2.27)	1.22 (0.78, 1.93)
5	57.6	11	8613	12.8	1.99 (1.08, 3.66)	1.66 (0.89, 3.12)
6	0.0	168	3480	482.7	1.00 Ref.	1.00 Ref.
6	1.5	155	4099	378.1	0.78 (0.63, 0.97)	0.83 (0.64, 1.09)
6	9.3	93	2033	457.5	0.95 (0.74, 1.22)	0.97 (0.70, 1.36)
6	21.0	44	967	455.1	0.94 (0.68, 1.31)	0.97 (0.64, 1.48)
6	33.3	15	295	509.1	1.05 (0.62, 1.79)	1.51 (0.77, 2.96)
6	50.3	9	100	899.8	1.86 (0.95, 3.64)	2.41 (0.94, 6.17)
7	0.0	87	1599	544.2	1.00 Ref.	1.00 Ref.
7	2.2	147	2145	685.4	1.26 (0.97, 1.64)	1.30 (0.92, 1.82)
7	9.4	138	2984	462.4	0.85 (0.65, 1.11)	0.85 (0.61, 1.18)
7	22.3	151	2445	617.6	1.13 (0.87, 1.48)	1.14 (0.82, 1.59)
7	36.3	75	936	801.6	1.47 (1.08, 2.01)	1.42 (0.94, 2.16)

(*Continued*)

TABLE 18.4 (CONTINUED)

Summarized Data on Total Alcohol Intake (Grams/Day) and Colorectal Cancer Rate in Eight Prospective Cohort Studies Participating in the Pooling Project of Prospective Studies of Diet and Cancer (Cho et al., 2004, Orsini et al., 2012)

Study (*i*)	Median, grams/day	No. of cases	Person-years	Rate per 10,000	Crude rate ratio (95% CI)	Adjusted rate ratio (95% CI)
7	56.2	44	553	795.9	1.46 (1.02, 2.10)	1.54 (0.94, 2.51)
8	0.0	62	22,624	27.4	1.00 Ref.	1.00 Ref.
8	1.0	205	97,700	21.0	0.77 (0.58, 1.02)	0.93 (0.70, 1.24)
8	11.4	109	46,456	23.5	0.86 (0.63, 1.17)	1.09 (0.79, 1.49)
8	22.8	69	27,468	25.1	0.92 (0.65, 1.29)	1.19 (0.84, 1.70)
8	34.2	18	10,279	17.5	0.64 (0.38, 1.08)	0.87 (0.51, 1.48)
8	57.0	29	10,784	26.9	0.98 (0.63, 1.53)	1.43 (0.91, 2.26)

Rate ratios were adjusted for age, energy intake (kcal/day), multivitamin use, family history of colorectal cancer, current smoking, past smoking, red meat intake, total milk intake, and dietary folate intake. A total of 3646 cases and 2,511,424 person-years were included in the analysis.

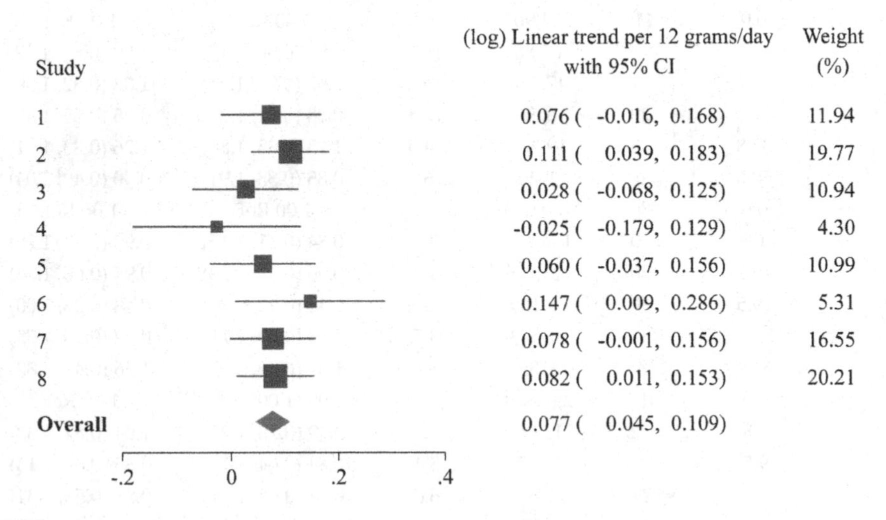

FIGURE 18.6

Forest plot of adjusted (log) linear trends (per 12 grams/day) with 95% confidence interval (CI) for the association between total alcohol intake and colorectal cancer rate for eight prospective cohort studies participating in the Pooling Project of Prospective Studies of Diet and Cancer (Cho et al., 2004, Orsini et al., 2012). The Hamling method was used to estimate the covariances of multivariable-adjusted rate ratios. The combined dose-response trend is derived under the hypothesis of a random-effects model using restricted maximum likelihood for estimation and inference.

meta-analysis. Figure 18.7b presents the study-specific curvilinear predicted trends together with the average dose-response relation given by $e^{-0.0012[g_1(x^*)-g_1(0)]+0.0208[g_2(x^*)-g_2(0)]}$. The large value of the Wald-type statistic $\chi_2^2=26.17$ (p-value <0.001) leads to rejection of the null hypothesis of overall no association between alcohol intake and colorectal cancer rates with a suggestion of departure from linearity ($\chi_1^2=3.75$, p-value $=0.053$). A visual inspection of the summary dose-response relation shown in Figure 18.7b (or Figure 18.8a) indicates an elevated colorectal cancer rate limited to alcohol intake of 30 grams/day or more.

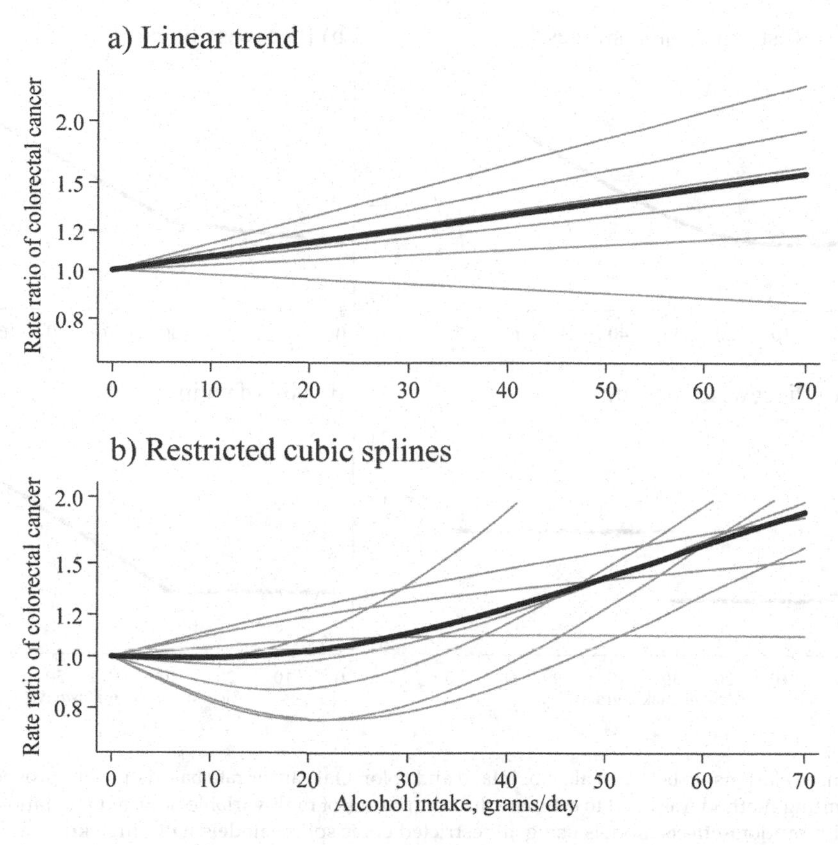

FIGURE 18.7
Dose-response relations between alcohol intake and colorectal cancer rate based on eight prospective studies. Data were fitted with random-effects models using (a) linear trend and (b) restricted cubic splines with three knots at 0, 14.25, and 57.6 grams/day. The Hamling method was used to estimate the covariances of multivariable-adjusted rate ratios. Gray lines are study specific curves and the solid thick line represents the overall trend. The value of 0 grams/day served as referent.

To address the question of whether there is any change in slope relating alcohol intake to colorectal cancer rate at 30 grams/day, we can specify a piecewise linear random-effects dose-response model (Figure 18.8b). Below 30 grams/day, every 12 grams/day increase is associated with a 3% higher colorectal cancer rate (95% CI: 0.97, 1.10). Above 30 grams/day, every 12 grams/day increase is associated with a 17% higher colorectal cancer rate (95% CI: 1.06, 1.30).

The treatment of non-drinkers as a separate group in investigating the alcohol-colorectal cancer rate relationship may inform the specification of alternative dose-response models (Section 18.3.2). A possible functional form is piecewise constant splines with two knots at 1 and 30 grams/day of alcohol intake (Figure 18.8c). Compared with non-drinkers, the colorectal cancer rate among those individuals with an alcohol intake 1 to 30 grams/day is 3% lower (95% CI: 0.87, 1.07). Compared with those drinking 1 to 30 grams/day, those individuals with an alcohol intake higher than 30 grams/day was 31% higher (95% CI: 1.14, 1.50).

Another functional form consists of a mix of different types of splines. For example, a constant at zero (also known as spike at zero) detecting the difference in colorectal cancer rate between non-drinkers and an interval 1 to 30 grams/day that serves as referent, and

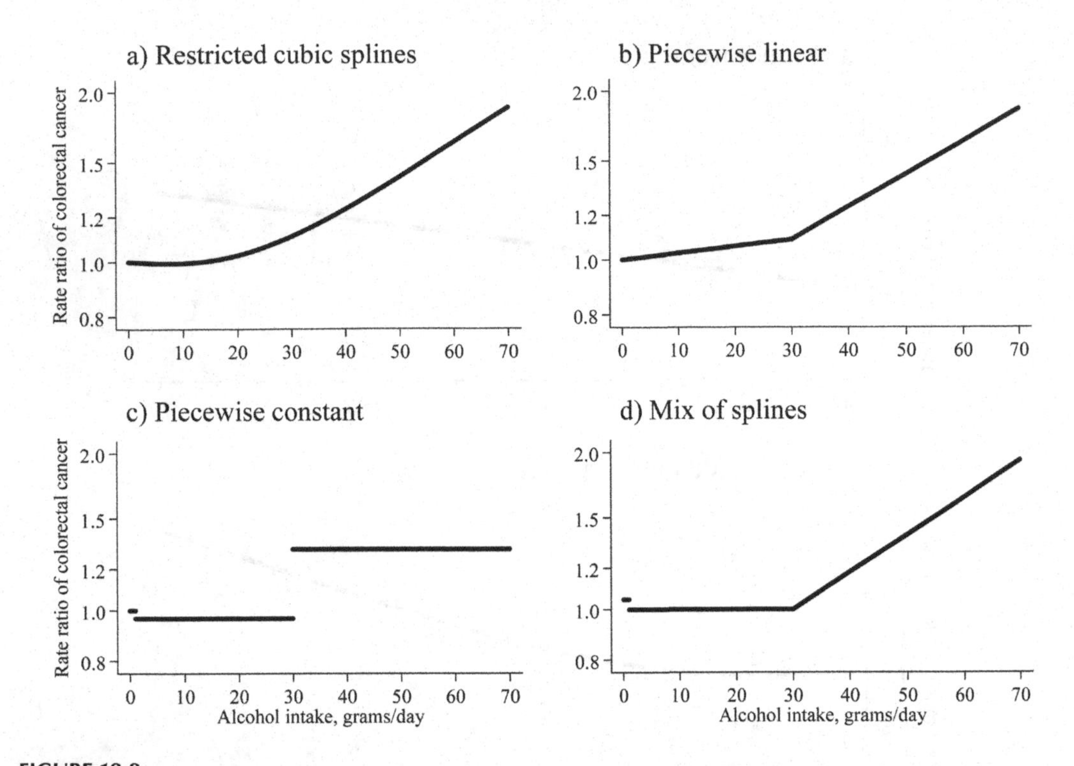

FIGURE 18.8
Dose-response relationship between alcohol intake and colorectal cancer rate based on eight prospective studies. The Hamling method was used to estimate the covariances of multivariable-adjusted rate ratios. Data were modeled with random-effects models using (a) restricted cubic spline models with three knots at 0, 14.25, and 57.6 grams/day; (b) piecewise linear with a knot at 30 grams/day; (c) piecewise constant with knots at 1 and 30 grams/day; and (d) mix of piecewise constant and linear splines with knots at 1 and 30 grams/day.

then a linear trend above 30 grams/day (Figure 18.8d). Compared with an alcohol intake between 1 and 30 grams/day, non-drinkers had a 5% (95% CI: 0.96, 1.14) higher colorectal cancer rate, whereas those drinking above 30 grams/day had a 22% higher (95% CI: 1.13, 1.32) colorectal cancer rate for every 12 grams/day increment.

Figure 18.8 presents the average dose-response relationship between alcohol intake and colorectal cancer rate according to the four different models of dose-response.

18.7 Pooling Individual versus Summarized Data

Here, we compare a dose-response analysis of multiple prospective studies based on individual participant data with that of the summarized data. For illustrative purposes, we use data from the Surveillance, Epidemiology, and End Results (SEER) program of the National Cancer Institute. The SEER program provides data about cancer statistics from several population-based registries in the USA (http://seer.cancer.gov) from San Francisco-Oakland, Connecticut, Metropolitan Detroit, Hawaii, Iowa, New Mexico, Seattle, Utah, and Metropolitan Atlanta that here are considered as different studies. Analysis are based on nine studies of prognostic factors for breast cancer survival including a total of 84,404 women. Year of breast cancer diagnosis ranged from 1988 to 1997. During 554,812 person-years, 8520 women died from breast cancer. Median follow-up time was about six years in all of the studies. The same source of data has been used to illustrate dose-response meta-analysis of individual participant data using fractional polynomials (Sauerbrei and Royston, 2011).

We examine the dose-response relationship between age, a weak prognostic factor, and mortality rates from breast cancer. We categorize age into five intervals (<45, 45–54, 55–64, 65–74, ≥75 years) using the mid-group (55–64 years) as referent. The association between age and mortality rates from breast cancer is adjusted for race, histology, marital status, tumor size, grade, ER, PR, and number of nodes. The adjusted differences in (log) mortality rates from breast cancer along the observed range of age are represented by the standard step-function f_i modeled through indicator variables

$$\mu_{f_i} \mid x_i = \gamma_{i1}I(x \leq 45) + \gamma_{i2}I(45 \leq x < 54) + \gamma_{i3}I(65 \leq x < 74) + \gamma_{i4}I(\geq 75)$$

The values of $\hat{\gamma}_{ij}$, describing a step up or down from the referent, estimated from individual data, constitute the input data of the dose-response meta-analysis based on summarized data. The size of the dataset reduces from 84,404 rows to $(9 \times 5) = 45$ rows. In creating the design matrix for the analysis for summarized data, given the non-zero dose at referent (55–64 years), it is important to center the assigned values of x_i around the assigned referent within each study.

As shown in Table 18.5, the estimated linear trends and their standard errors were similar. Assuming a random-effects dose-response model, every ten years increase of age is associated with a 2% ($e^{0.0228} = 1.02$) higher mortality from breast cancer using summarized data. This estimate, when rounded to two decimal places, is practically undistinguishable from the one estimated on individual data ($e^{0.0227} = 1.02$). The Q-statistic for heterogeneity of the linear trends across studies is small relative to its degrees of freedom ($\chi^2_8 = 13.75$, p-value = 0.089).

TABLE 18.5

Adjusted Linear Trends of Mortality Rates from Breast Cancer per Every Increment of Ten Years of Age within Each Study Based on Summarized Data ($\hat{\theta}_i$) and Individual Participant Data ($\hat{\gamma}_i$)

		Summarized data		Individual data	
Number	**Study**	$\hat{\theta}_i$	$SE(\hat{\theta}_i)$	$\hat{\gamma}_i$	$SE(\hat{\gamma}_i)$
1	San Francisco-Oakland	−0.0066	0.0203	−0.0079	0.0201
2	Connecticut	0.0469	0.0221	0.0404	0.0224
3	Metropolitan Detroit	0.0378	0.0197	0.0307	0.0194
4	Hawaii	0.0725	0.0458	0.0502	0.0457
5	Iowa	0.0100	0.0205	0.0111	0.0216
6	New Mexico	0.0435	0.0362	0.0318	0.0357
7	Seattle (Puget Sound)	0.0263	0.0223	0.0240	0.0226
8	Utah	0.0829	0.0419	0.0786	0.0412
9	Metropolitan Atlanta	−0.0427	0.0269	−0.0537	0.0279
	Random-effects	0.0228	0.0113	0.0227	0.0125

The Hamling method was used to estimate the covariances of multivariable-adjusted rate ratios. Under a random-effects model, estimates were obtained with the restricted maximum likelihood method.

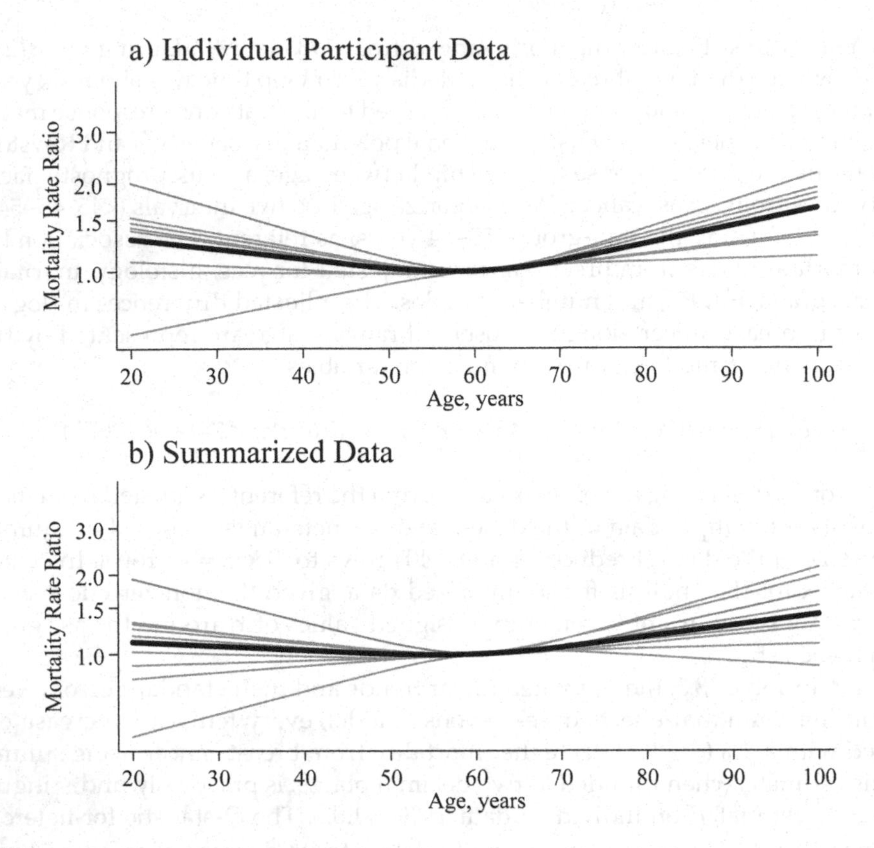

FIGURE 18.9
Mortality rate ratios from breast cancer according to age adjusted for race, histology, marital status, tumor size, grade, ER, PR, and number of nodes. Individual participant data (a) and summarized data (b) were fitted using restricted cubic spline models within each study (gray solid lines) and the summary dose-response functions (thick black line) were estimated using restricted maximum likelihood in a multivariate random-effects model. The value of 60 years served as referent.

We assess the possible non-linear dose-response relationship (elevated mortality rates at the extremes of the age distribution), while taking into account confounders, by using restricted cubic splines with three knots at 42, 61, and 78 years (10th, 50th, and 90th percentile of the overall age distribution). Figure 18.9 shows similar average dose-response shapes estimated on individual and summarized data. Statistical tests for the overall association between age and mortality (individual data: $\chi_2^2 = 38.6$, p-value < 0.001; summarized data: $\chi_2^2 = 19.14$, p-value < 0.001) and departure from linearity (individual data: $\chi_1^2 = 8.21$, $P = 0.004$; summarized data: $\chi_1^2 = 5.65$, p-value $= 0.018$) are consistent under the two approaches.

18.8 Conclusion

This chapter introduced statistical methods that can be useful to describe the change of a response according to the level of quantitative exposure based on tables of summarized data. We explained what constitute the input data and how to specify, interpret, and visualize a dose-response model. We illustrated how to specify a functional form in relation to the research questions and available data in the analysis of single and multiple studies. Through the use of simple exposure transformations such as regression splines, we have shown how to describe common dose-response patterns (i.e., U-shaped, J-shaped, V-shaped, threshold, spike at zero, steps). Graphical presentationsl of predicted study-specific and summary dose-response relationships were illustrated in detail.

Given the increasing number of quantitative reviews and collaborative projects including only observational and prospective studies, the examples chosen to illustrate the methods consist of large cohort studies where the individual time-to-event outcomes are described using event rates. The methods can be readily estimated from other study designs and measures of event occurrence such as odds and risks in case-control and cumulative-incidence data. The extension to differences in means or standardized means of quantitative outcomes in randomized trials has been recently proposed (Crippa and Orsini, 2016a).

There is no doubt that having individual data is better than having summarized data since data are harmonized, checked, and consistently re-analyzed (Riley et al., 2010). Imagine how much information about the dose-response relationship is lost in the SEER example by collapsing 84,404 rows of individual data on outcome and exposure into 45 rows of summarized data. Nevertheless, our empirical comparison of dose-response meta-analysis based on both sources of data indicated a relatively good agreement for both linear and curvilinear relationships. A possible explanation is that summarized data were generated similarly among the combined studies (similar age distribution, cutpoints, and mean age within each interval). If that is the case, sharing standardized analyses on harmonized datasets may represent a possible cheap, quick, and reasonably accurate way of combining results of large studies that is sustainable in the long term.

Having less data does not imply that one needs fewer statistical skills. On the contrary, when modeling quantitative predictors, our experience is that comparing dose-response relations using different parameterizations is challenging regardless of whether one has individual or summarized data. Although the methods for dose-response meta-analysis are not complicated and are implemented in major statistical packages (`drmeta` command in Stata (Orsini, 2018), `metadose` macro in SAS (Li and Spiegelman, 2010), and `dosresmeta` package in R (Crippa and Orsini, 2016b)), it can be difficult to graphically present a dose–response relationship without adequate programming and statistical expertise. It should be noted that finding graphical presentations of uncertainty in dose-response meta-analysis is a challenge because overlaying too many lines makes figures difficult to read.

At the moment, the programs implementing the most recent developments for dose-response meta-analysis are the `dosresmeta` and `drmeta` procedures offering a one-stage approach (Crippa et al., 2019), extension to mean differences and standardized mean differences (Crippa and Orsini, 2016a), and measures of goodness of fit (Discacciati et al., 2017). The one-stage approach, formalized as a linear mixed-effects model, is able to answer more advanced research questions including all the available studies. The mixed-effects framework is particularly suitable for inferential procedures, marginal and conditional predictions, quantification of heterogeneity, goodness-of-fit, and model comparison (Crippa et al., 2019).

Using contrasts in responses obtained from one functional form (i.e., μ_{f_i} consists of steps) to derive contrasts in responses according to another functional form (i.e., μ_{g_i} is a linear trend) is not straightforward and represents a major source of confusion in the application of the method. The example in Section 18.4 showed that, somewhat counterintuitively, the dose-response curve estimated from summarized data, μ_{g_i}, may not go through the data points provided by μ_{f_i}. This is likely to occur when findings based on individual data are presented using an extreme reference group with a very different response compared with the rest of the exposure intervals. In such cases, the apparent dose-response shape suggested by the categorical model (inverse U-shaped) may be completely different from what a linear dose-response model suggests (decreasing linear trend). In this respect, individual data simplify the comparison of alternative functional forms because one can estimate absolute responses, and not just changes in responses relative to an arbitrary referent.

Conducting dose-response meta-analysis requires subject-matter knowledge and careful evaluation of alternative explanations for the estimated summary dose-response relationship. Sources of variation and uncertainty about the exposure (type of assessment, distribution, cutpoint, assignment), the response (different measures of event occurrence, length of follow-up in cohort data, baseline response), and other factors can influence the dose-response relationship (geographical area, characteristics of the population, degree of adjustment in multivariable models). Stratification, meta-regression, and sensitivity analysis may be used to examine how the dose-response shape is changing according to relevant scientific or design factors.

References

Almond CS, Shin AY, Fortescue EB, Mannix RC, Wypij D, Binstadt BA, Duncan CN, Olson DP, Salerno AE, Newburger JW and Greenes D, 2005. Hyponatremia among runners in the Boston Marathon. *The New England Journal of Medicine* **352**(15): 1550–1556.

Bagnardi V, Zambon A, Quatto P and Corrao G, 2004. Flexible meta-regression functions for modeling aggregate dose-response data, with an application to alcohol and mortality. *American Journal of Epidemiology* **159**(11): 1077–1086.

Berkey CS, Hoaglin DC, Antczak-Bouckoms A, Mosteller F and Colditz GA, 1998. Meta-analysis of multiple outcomes by regression with random effects. *Statistics in Medicine* **17**(22): 2537–2550.

Berrington A and Cox D, 2003. Generalized least squares for the synthesis of correlated information. *Biostatistics* **4**(3): 423–431.

Cho E, Smith-Warner SA, Ritz J, van den Brandt PA, Colditz GA, Folsom AR, Freudenheim JL, Giovannucci E, Goldbohm RA, Graham S, Holmberg L, Kim D-H, Malila N, Miller A, Pietinen P, Rohan TE, Sellers T, Speizer F, Willett WC, Wolk A and Hunter DJ, 2004. Alcohol intake and colorectal cancer: A pooled analysis of 8 cohort studies. *Annals of Internal Medicine* **140**(8): 603–613.

Cochran WG, 1954. The combination of estimates from different experiments. *Biometrics* **10**(1): 101–129.

Crippa A and Orsini N, 2016a. Dose-response meta-analysis of differences in means. *BMC Medical Research Methodology* **16**(1): 91.

Crippa A and Orsini N, 2016b. Multivariate dose-response meta-analysis: The dosresmeta R package. *Journal of Statistical Software* **72**(1): 1–15.

Crippa A, Thomas I and Orsini N, 2018. A pointwise approach to dose-response meta-analysis of aggregated data. *International Journal of Statistics in Medical Research* **7**(2): 25–32.

Crippa A, Discacciati A, Bottai M, Spiegelman D and Orsini N, 2019. One-stage dose-response meta-analysis for aggregated data. *Statistical Methods in Medical Research* **28**(5): 1579–1596.

Discacciati A, Crippa A and Orsini N, 2017. Goodness of fit tools for dose-response metaanalysis of binary outcomes. *Research Synthesis Methods* **8**(2): 149–160.

Gasparrini A, Armstrong B and Kenward MG, 2012. Multivariate meta-analysis for nonlinear and other multi-parameter associations. *Statistics in Medicine* **31**(29): 3821–3839.

Giovannucci E, Rimm E, Ascherio A, Stampfer M, Colditz G and Willett W, 1995. Alcohol, low-methionine-low-folate diets, and risk of colon cancer in men. *Journal of the National Cancer Institute* **87**(4): 265–273.

Greenland S and Longnecker MP, 1992. Methods for trend estimation from summarized dose-response data, with applications to meta-analysis. *American Journal of Epidemiology* **135**(11): 1301–1309.

Hamling J, Lee P, Weitkunat R and Ambühl M, 2008. Facilitating meta-analyses by deriving relative effect and precision estimates for alternative comparisons from a set of estimates presented by exposure level or disease category. *Statistics in Medicine* **27**(7): 954–970.

Harrell FE, 2001. *Regression Modeling Strategies: With Applications to Linear Models, Logistic Regression, and Survival Analysis*. New York: Springer.

Jackson D, Riley R and White IR, 2011. Multivariate meta-analysis: Potential and promise. *Statistics in Medicine* **30**(20): 2481–2498.

Jackson D, White IR and Riley RD, 2012. Quantifying the impact of between-study heterogeneity in multivariate meta-analyses. *Statistics in Medicine* **31**(29): 3805–3820.

Jackson D, White IR and Riley RD, 2013. A matrix-based method of moments for fitting the multivariate random effects model for meta-analysis and meta-regression. *Biometrical Journal* **55**(2): 231–245.

Li R and Spiegelman D, 2010. The SAS `metadose` macro. Available from https://www.hsph.harvard.edu/donna-spiegelman/software/metadose/.

Liu Q, Cook NR, Bergström A and Hsieh C-C, 2009. A two-stage hierarchical regression model for meta-analysis of epidemiologic nonlinear dose-response data. *Computational Statistics and Data Analysis* **53**(12): 4157–4167.

Orsini N, 2010. From floated to conventional confidence intervals for the relative risks based on published dose-response data. *Computer Methods and Programs in Biomedicine* **98**(1): 90–93.

Orsini N, 2018. *DRMETA: Stata Module for Dose-Response Meta-Analysis S458546*. Statistical Software Components, Boston College Department of Economics. Available from https://ideas.repec.org/c/boc/bocode/s458546.html.

Orsini N, Bellocco R and Greenland S, 2006. Generalized least squares for trend estimation of summarized dose-response data. *Stata Journal* **6**(1): 40.

Orsini N, Li R, Wolk A, Khudyakov P and Spiegelman D, 2012. Meta-analysis for linear and nonlinear dose-response relations: Examples, an evaluation of approximations, and software. *American Journal of Epidemiology* **175**(1): 66–73.

Riley RD, Lambert PC and Abo-Zaid G, 2010. Meta-analysis of individual participant data: Rationale, conduct, and reporting. *British Medical Journal* **340**: c221.

Ritz J, Demidenko E and Spiegelman D, 2008. Multivariate meta-analysis for data consortia, individual patient meta-analysis, and pooling projects. *Journal of Statistical Planning and Inference* **138**(7): 1919–1933.

Rota M, Bellocco R, Scotti L, Tramacere I, Jenab M, Corrao G, La Vecchia C, Boffetta P and Bagnardi V, 2010. Random-effects meta-regression models for studying nonlinear dose-response relationship, with an application to alcohol and esophageal squamous cell carcinoma. *Statistics in Medicine* **29**(26): 2679–2687.

Sauerbrei W and Royston P, 2011. A new strategy for meta-analysis of continuous covariates in observational studies. *Statistics in Medicine* **30**(28): 3341–3360.

Shi JQ and Copas J, 2004. Meta-analysis for trend estimation. *Statistics in Medicine* **23**(1): 3–19.

Takahashi K and Tango T, 2010. Assignment of grouped exposure levels for trend estimation in a regression analysis of summarized data. *Statistics in Medicine* **29**(25): 2605–2616.

Turner EL, Dobson JE and Pocock SJ, 2010. Categorisation of continuous risk factors in epidemiological publications: A survey of current practice. *Epidemiologic Perspectives and Innovations* **7**(1): 9.

van Houwelingen HC, Arends LR and Stijnen T, 2002. Advanced methods in meta-analysis: Multivariate approach and meta-regression. *Statistics in Medicine* **21**(4): 589–624. ISSN 0277-6715.

White IR, 2011. Multivariate random-effects meta-regression: Updates to `mvmeta`. *STATA Journal* 11(2): 255–270.

White IR, Kaptoge S, Royston P, Sauerbrei W and E. R. F. Collaboration, 2019. Meta-analysis of non-linear exposure-outcome relationships using individual participant data: A comparison of two methods. *Statistics in Medicine* **38**(3): 326–338.

Yu WW, Schmid CH, Lichtenstein AH, Lau J and Trikalinos TA, 2013. Empirical evaluation of meta-analytic approaches for nutrient and health outcome dose-response data. *Research Synthesis Methods* **4**(3): 256–268.

19

Meta-Analysis of Diagnostic Tests

Yulun Liu, Xiaoye Ma, Yong Chen, Theo Stijnen and Haitao Chu

CONTENTS

19.1 Introduction to Meta-Analysis of Diagnostic Tests

Accurate diagnosis of a disease is often the first step toward its treatment and prevention. The performance of a binary test of interest (candidate or index test) is commonly compared with a reference test (preferably a gold standard test) and characterized by accuracy indices. For example, ultrasound is the candidate or index test in diagnosing rotator cuff tears, and arthroscopy serves as a gold standard (Singisetti and Hinsch, 2011).

TABLE 19.1

A 2×2 Table for a Toy Example in a Diagnostic Test for Rotator Cuff Tears

	Arthroscopy		
Ultrasound	**+ (Diseased)**	**– (Non-diseased)**	**Total**
+	32	20	52
–	8	180	188
Total	40	200	240

A pair of commonly used indices are sensitivity (Se) and specificity (Sp). Sensitivity is defined as the probability of testing positive given a person being diseased, and specificity is the probability of testing negative given a person being disease-free (Pepe, 2003). In practice, they can be used to address questions on how well the test reflects the true disease status. For example, suppose that, in one study, 240 participants are tested by both ultrasound and arthroscopy (treated as a gold standard) for rotator cuff tears, and test results are compared in a cross-tabulated 2×2 table (Table 19.1). In this study, Se is estimated as $P($Ultrasound = + | Diseased$)$ = 32/40 = 0.8 and Sp is estimated as $P($Ultrasound = – | Non-diseased$)$ = 180/200 = 0.9. Ideally, a gold standard test has a Se and Sp of 1. Other frequently used indices include positive and negative predictive values (PPV and NPV), and positive and negative diagnostic likelihood ratios (LR+ and LR–). PPV is the probability of diseased given a positive test result and NPV is the probability of disease-free given a negative test result. In this example, PPV is estimated as $P($Diseased | Ultrasound = +$)$ = 32/52 = 0.62 and NPV is estimated as $P($Non-diseased | Ultrasound = –$)$ = 180/188 = 0.96.

Meta-analysis of diagnostic tests (MA-DT) is useful for combining evidence on diagnostic accuracy measures from multiple studies. Compared with conventional meta-analyses of controlled clinical trials, it has several additional statistical challenges that have been extensively studied in the literature, such as possible correlation between test accuracy indices. Such a correlation might, for instance, arise when there is an underlying continuous diagnostic test statistic which is dichotomized: above a certain cut-off value the test is positive and below it is negative. Different cut-off values between studies will induce a negative correlation between sensitivity and specificity, and this correlation has to be accounted for in the analysis method. Other important issues in MA-DT are the lack of a gold standard, possible dependence of the test characteristics on the prevalence of the disease, lack of full verification of the true disease status of all individuals (partial verification), and the need to combine data from mixed study designs. In this chapter, we introduce MA-DT when a gold standard is available in Section 19.2, MA-DT without a gold standard in Section 19.3, and other issues such as combining case-control studies and cohort studies, in which data from the cohort studies allow estimation of the disease prevalence, and partial verification bias in Section 19.4. Throughout this chapter, we assume a binary diagnostic test (either positive or negative) and a binary disease status (either present or absent).

19.2 Methods When There Is a Gold Standard

In meta-analysis, performance of diagnostic tests can be heterogeneous across studies, due to variation in study populations, laboratory methods, and so on. Also, test accuracy indices are likely to be correlated, and studies have found that they can potentially depend on the disease prevalence in a study population (Chu et al., 2009). When a reference test can be considered as a gold standard, several meta-analysis methods are available to account for such heterogeneity and correlations (Rutter and Gatsonis, 2001; Van Houwelingen et al., 2002; Macaskill, 2004; Reitsma et al., 2005; Chu and Cole, 2006; Harbord et al., 2007; Arends et al., 2008; Zwinderman and Bossuyt, 2008). Specifically, random-effects models including the hierarchical summary receiver operating characteristic (HSROC) model (Rutter and Gatsonis, 2001) and bivariate random-effects meta-analysis on sensitivities and specificities (Van Houwelingen et al., 2002; Chu and Cole, 2006; Reitsma et al., 2005) have been recommended (Harbord et al., 2007; Zwinderman and Bossuyt, 2008; Chu and Guo, 2009). Interestingly, in the absence of covariates, these two models turn out to be the same up to a reparameterization (Harbord et al., 2007). In general, generalized linear mixed models (GLMM), which use the exact binomial within-study likelihood, often perform better than the linear mixed models (LMM), which use a normal approximation (Chu and Cole, 2006; Hamza et al., 2008). In addition, a trivariate GLMM has been proposed to jointly model the disease prevalences, sensitivities, and specificities (Chu et al., 2009). In this section, we will outline these methods.

Not discussed in this chapter is meta-analysis of a single series of sensitivities or a single series of specificities to get an overall sensitivity or specificity. This can be done using the methods of Chapters 4 and 5. Furthermore, we will not cover meta-analysis of the diagnostic odds ratio (DOR), which is defined as the odds of a positive test result in the diseased divided by the odds of a positive test result in the non-diseased. Meta-analysis of odds ratios is also extensively discussed in Chapters 4 and 5.

For notation, let n_{idt} denote the number of subjects with index test result t and disease status d for study i $(i=1,2,\ldots,I)$, where $t=1$ for a positive index test and $t=0$ for a negative one, and $d=1$ if diseased and $d=0$ if non-diseased. When the reference test can be considered as a gold standard, n_{i11}, n_{i00}, n_{i01}, and n_{i10} are the number of true positives, true negatives, false positives, and false negatives. See Table 19.2 for the 2×2 table for study i. Let $n_{i1+}=n_{i11}+n_{i10}$ and $n_{i0+}=n_{i01}+n_{i00}$ be the study-specific numbers of diseased and disease-free subjects. Then the study-specific sensitivity and specificity can be estimated as $\widehat{Se}_i=n_{i11}/n_{i1+}$ and $\widehat{Sp}_i=n_{i00}/n_{i0+}$.

19.2.1 Summary Receiver Operating Characteristic Method

When the diagnostic test may be considered as a dichotomization of an underlying continuous test, i.e., the test is positive above a chosen threshold, and negative below the threshold, the receiver operating characteristic (ROC) curve is well established as a method for assessing accuracy of a diagnostic test in a single study (Beck and Schultz, 1986; Metz

TABLE 19.2

A 2×2 Table for ith Study

	Reference test		
Index test	**Positive (+)**	**Negative (−)**	**Total**
Positive(+)	n_{i11}	n_{i01}	
Negative(−)	n_{i10}	n_{i00}	
Total	n_{i1+}	n_{i0+}	n_{i++}

et al., 1998). Specifically, it is a graphical representation of the relationship between Se (i.e., true positive proportion) and (1-Sp) (i.e., the false positive proportion) of a diagnostic test for all possible thresholds. This is equivalent to the curve that plots Se against Sp on the reversed scale for Sp (from 1 to 0 instead of from 0 to 1). A typical ROC curve starts at the origin (0,0) and increases monotonically to (1,1). In general, the closer the curve to the upper left-hand corner of the unit square, the better the performance of the diagnostic test. A summary ROC curve is conceptually similar to the ROC but in the context of a meta-analysis. Moses et al. (1993) were the first to propose a method to construct a summary ROC reflecting the trade-off between sensitivity and specificity across studies. This method is often referred to as the SROC method. To calculate the SROC curve, first, define D and S as the difference and sum of the logit transformed sensitivity and 1-specificity: $D_i = \text{logit}(\widehat{Se}_i) - \text{logit}(1-\widehat{Sp}_i)$ and $S_i = \text{logit}(\widehat{Se}_i) + \text{logit}(1-\widehat{Sp}_i)$ respectively, where $\text{logit}(p) = \log\{p/(1-p)\}$ with $0<p<1$. D_i is interpreted as a measure of test accuracy in discriminating diseased from disease-free subjects, as $D_i = \log(\widehat{OR}_i) = \log\left(\frac{n_{i11}}{n_{i10}} \Big/ \frac{n_{i01}}{n_{i00}}\right)$ is the logarithm of the diagnostic odds ratio for the ith study. When some studies have one or more cell numbers $n_{idt}=0$, an ad hoc continuity correction can be applied by adding a small constant, usually 0.5, to each of the four cells in Table 19.2 of such studies. S_i is interpreted as a function of the test threshold. When $S_i=0$, sensitivity is equal to specificity; S_i is positive when sensitivity is higher than specificity; and it is negative when specificity is higher (Walter, 2002; Reitsma et al., 2005).

Moses et al. (1993) showed that if the transformation that transforms the distribution of the underlying continuous test into a logistic distribution in the non-diseased also leads to a logistic distribution in the diseased, then the plot of $\text{logit}(Se_i)$ against $\text{logit}(1-Sp_i)$ will be exactly a straight line. This can be seen as follows. Let T be the transformed underlying continuous test such that it has a logistic distribution with cumulative distribution $\Pr(T<t|\text{non-disease})=\exp(t)/(1+\exp(t))$. Then the cumulative distribution in the diseased will be $\Pr(T<t|\text{disease})=\exp(-c+dt)/(1+\exp(-c+dt))$ for some values c and d. If t_0 denotes the threshold value of T above which the diagnostic test is declared positive, we have $\text{logit}(1-Sp)=-t_0$ and $\text{logit}(Se)=c-dt_0$. This implies the following linear relationship between the logits of the true positive and the false positive proportion $\text{logit}(Se)=c+d\ \text{logit}(1-Sp)$. Thus, D and S will have a straight line relation as well, say $D=a+bS$. One then can fit a linear regression model, $D=a+bS$, to the data points (S_i,D_i), and the parameters can be easily estimated by ordinary least squares or weighted least squares regression. Inverse within-study variances can be used as weights in the weighted least squares regression: $\text{var}(\log(\widehat{OR}_i))^{-1}$, where $\text{var}(\log(\widehat{OR}_i)) = 1/n_{i11}+1/n_{i10}+1/n_{i01}+1/n_{i00}$ (Van Houwelingen et al., 2002). This SROC method is one possible (imperfect) method, and others will be discussed later.

Once the model parameters, a and b, are estimated, the relationship between Se and Sp can be backtransformed to ROC space as

$$Se = \left\{1+e^{-\hat{a}/(1-\hat{b})} \times (Sp/(1-Sp))^{(1+\hat{b})/(1-\hat{b})}\right\}^{-1}, \tag{19.1}$$

with Se on the y-axis and $(1-Sp)$ on the x-axis. As shown in equation (19.1), the corresponding sensitivity can be derived to any given value of specificity, thereby giving the entire SROC curve. When comparing the performance of multiple diagnostic tests, the SROC curve may be useful to describe the trade-off of sensitivity against specificity (Bossuyt et al., 2013). For example, given the same value of specificity, the corresponding sensitivities of each diagnostic test can be calculated, and their differences can be reported at specified values of the specificity. In addition, summary values for the diagnostic odds ratio can be computed and their corresponding confidence intervals as well.

After constructing the SROC curve, test performance can be summarized either by the area under the curve (AUC) (Walter et al., 1999), or by the Q^* statistic, which measures the value of sensitivity at the point where the curve intersects the diagonal line $x+y=1$ (Moses et al., 1993; Gatsonis and Paliwal, 2006). The larger the AUC or Q^*, the closer the ROC is to the upper left corner of the unit square, and the better the diagnostic performance of the test. The value 1 for AUC or Q^* corresponds with a perfect test.

However, the SROC method has some potential limitations. First, because $D_i = \log(\widehat{OR}_i)$, each study contributes only one outcome measure (the DOR) which depends not only on sensitivity but also on specificity. In other words, the same DOR could be obtained by different combinations of sensitivity and specificity (Bossuyt et al., 2013). Therefore, separate inferences of sensitivity and specificity are unavailable (Reitsma et al., 2005). Second, as it is a common-effect method, it does not well explain between-study heterogeneity (Walter, 2002). Third, an arbitrary continuity correction is needed to handle zero cells. Last, the estimate of the regression coefficient b and its corresponding standard error are affected by the presence of sampling error in the independent variable in the regression, S_i. This measurement error leads to bias toward zero in the estimation ("regression dilution bias", see also Chapter 14) of the parameter b, leading to misleading conclusions drawn from the SROC curve (Gatsonis and Paliwal, 2006). Also, a minor problem is that the measurement errors of D_i and S_i may be somewhat correlated, which is not accounted for in the regression.

In practice, the SROC method is used with the logit transformation for the sensitivity and specificity. However, other link functions might be used instead, such as the probit link (= the inverse cumulative standard normal distribution function). Under the assumption that the transformation which makes the distribution of the underlying continuous test normal for the non-diseased also makes it normal for the diseased, the relation between the probit transformed Se_i and $(1-Sp_i)$ is linear, and the SROC method can be carried through analogously. However, D_i loses its nice log odds ratio interpretation.

19.2.2 Bivariate Linear Mixed Models

In addition to obtaining a SROC curve showing the trade-off between Se and Sp, in practice, researchers are also interested in summarizing overall performance through summary statistics (e.g., mean or median) of sensitivity, specificity, and their between-studies variances. We introduce a bivariate LMM proposed by Reitsma et al. (2005), which assumes a bivariate normal distribution for the true logit transformed sensitivity and specificity pairs:

$$\begin{pmatrix} \text{logit}(Se_i) \\ \text{logit}(Sp_i) \end{pmatrix} \sim N\left(\begin{pmatrix} \alpha \\ \beta \end{pmatrix}, \Sigma\right), \quad \Sigma = \begin{pmatrix} \sigma_\alpha^2 & \rho_{\alpha\beta}\sigma_\alpha\sigma_\beta \\ \rho_{\alpha\beta}\sigma_\alpha\sigma_\beta & \sigma_\beta^2 \end{pmatrix},$$

where α and β are parameters describing mean logit sensitivity and specificity, σ_α^2 and σ_β^2 measure between-study variation, and $\rho_{\alpha\beta}$ measures the correlation between logit sensitivity and specificity. This is the bivariate special case of the general multivariate meta-analysis model discussed in Chapter 9. Note that $\rho_{\alpha\beta}$ denotes between-studies correlation. For example, if it is negative, it means that studies with a larger true sensitivity tend to have a smaller true specificity. Though in practice almost always the logit transformation is used, it might be replaced by any other link function.

To account for the sampling variation, the estimated logit sensitivity and specificity are assumed to be normally distributed given the true sensitivity and specificity of a study:

$$\begin{pmatrix} \text{logit}(\widehat{Se}_i) \\ \text{logit}(\widehat{Sp}_i) \end{pmatrix} \sim N\left(\begin{pmatrix} \text{logit}(Se_i) \\ \text{logit}(Sp_i) \end{pmatrix}, C_i\right),\ C_i = \begin{pmatrix} \text{var}(\text{logit}(\widehat{Se}_i)) & 0 \\ 0 & \text{var}(\text{logit}(\widehat{Sp}_i)) \end{pmatrix},$$

where $\text{var}(\text{logit}(\widehat{Se}_i)) = 1/n_{i11} + 1/n_{i10}$ and $\text{var}(\text{logit}(\widehat{Sp}_i)) = 1/n_{i01} + 1/n_{i00}$. Note that the within-study correlation is zero, since the estimates of sensitivity and specificity are based on different groups of patients.

Consequently, we have

$$\begin{pmatrix} \text{logit}(\widehat{Se}_i) \\ \text{logit}(\widehat{Sp}_i) \end{pmatrix} \sim N\left(\begin{pmatrix} \alpha \\ \beta \end{pmatrix}, \Sigma + C_i\right). \tag{19.2}$$

In fitting this model, the C_i are assumed to be fixed and known. Overall estimates for sensitivity and specificity are obtained as $1/(1+\exp(-\hat{\alpha}))$ and $1/(1+\exp(-\hat{\beta}))$. Note that these are estimates of the *median* sensitivity and specificity. Overall estimates of *mean* sensitivity and specificity might be calculated as

$$\widehat{Se} = \int_{-\infty}^{\infty} \text{logit}^{-1}(\hat{\alpha} + x) f_{\hat{\alpha}}(x)dx \quad \text{and} \quad \widehat{Sp} = \int_{-\infty}^{\infty} \text{logit}^{-1}(\hat{\beta} + x) f_{\hat{\beta}}(x)dx,$$

where $\text{logit}^{-1}(\cdot)$ is the inverse logit function $1/1(1+\exp(-x))$, and $f_\alpha(x)$ and $f_\beta(x)$ are normal density functions with mean 0 and variances $\sigma_{\hat{\alpha}}^2$ and $\sigma_{\hat{\beta}}^2$, respectively. The standard errors of $\widehat{Se}$ and $\widehat{Sp}$ are obtained by the delta method.

The LMM can easily be extended with study-level covariates $\mathbf{Z}$ by replacing α by $\alpha + \gamma\mathbf{Z}_i$ and β by $\beta + \lambda\mathbf{Z}_i$, where γ, λ are the corresponding regression coefficients.

The bivariate method in fact estimates the bivariate normal distribution of the true logit(sensitivity) and logit(specificity) pairs. Arends et al. [2] argue that a summary ROC curve is just the transformation to the ROC space of a straight line characterization of this bivariate normal distribution. There are many reasonable characterizations, for instance, the regression of logit(*Se*) on logit(*Sp*), given by

$$\text{logit}(Se) = \hat{\alpha} + \frac{\hat{\rho}_{\alpha\beta}\hat{\sigma}_\alpha}{\hat{\sigma}_\beta}\left(\text{logit}(Sp) - \hat{\beta}\right). \tag{19.3}$$

This can lead to a quite different curve compared with the one resulting from the SROC method described in the previous section. Alternatively, one can use the regression of logit(*Sp*) on logit(*Se*), which again can lead to a quite different summary ROC curve. Calculation of the regression line of *D* on *S* would be another possibility, which leads to an improved version of the SROC curve of the previous subsection. Arends et al. (2008) described several other reasonable methods to calculate summary ROC curves based on equation (19.2). Chappell et al. (2009) also discussed the choice of an appropriate summary ROC based on the bivariate model.

The summary ROC curves are based on only one reported point for each study specific ROC curve. Therefore, the study specific ROC curves are not identifiable, unless extra assumptions are imposed on their shapes. This means that the summary ROC curves cannot be interpreted as representative for the ROC curves of the individual studies. Only under very strong model assumptions, for instance, on the way the points are selected, can the summary ROC curves, including the ones discussed in the next subsections, be interpreted as estimates of the median, mean, or any other characteristic of the population distribution of study-specific ROC curves. In principle, the study specific ROC curves can have a shape completely different from the summary ROC curve. Therefore, a much less ambitious interpretation applies. For instance, equation (19.3) is interpreted as estimating the median of the reported sensitivities within subpopulations of studies that report the same specificity.

19.2.3 Bivariate Generalized Linear Mixed Models

The LMM discussed in the previous subsection assumes approximate normal distributions for the estimated logit(*Se*) and logit(*Sp*). These approximations can be inaccurate if the study sizes are small, or sensitivity or specificity are close to 1. Also (arbitrary) continuity corrections are needed when there are zero cells. Furthermore, the LMM implicitly assumes that the estimates of logit sensitivity and specificity are not related to their estimated standard errors, which is not true and leads to bias. See Chapter 5 for more on this. Chu and Cole (2006) addressed these shortcomings of the LMM by replacing the approximate normal distributions of logit($\widehat{Se}$) and logit($\widehat{Sp}$) by the exact binomial distributions. Later on, they extended this model from using logit link functions to general link functions (Chu et al., 2010). This model is as follows:

$$n_{i11} \sim \text{Bin}(n_{i11} \mid n_{i1+}; Se_i),\ n_{i00} \sim \text{Bin}(n_{i00} \mid n_{i0+}; Sp_i),\ g(Se_i) = \alpha_i,\ g(Sp_i) = \beta_i, \tag{19.4}$$

where Bin($\cdot\mid\cdot;\cdot$) denotes the binomial distribution. As in the LMM of the previous subsection, α_i and β_i are study-level random effects following a bivariate normal distribution

$$\begin{pmatrix}\alpha_i\\ \beta_i\end{pmatrix} \sim N\left(\begin{pmatrix}\alpha\\ \beta\end{pmatrix}, \begin{pmatrix}\sigma_\alpha^2 & \rho_{\alpha\beta}\sigma_\alpha\sigma_\beta\\ \rho_{\alpha\beta}\sigma_\alpha\sigma_\beta & \sigma_\beta^2\end{pmatrix}\right),$$

and $g(\cdot)$ is a link function such as the logit, probit, or complimentary log-log link. The logit link is widely used, while it is argued that, for some meta-analyses, the choice of link functions may affect model fit and inference (Chu et al., 2010). The parameters σ_α^2 and σ_β^2 represent the between-study variances and $\rho_{\alpha\beta}$ allows for the possible correlation. This bivariate GLMM model is also widely used in other meta-analysis applications. We refer to, for instance, Chapters 5 and 14, where it is used with the logit link. Inference is performed by straightforward likelihood using a GLMM program.

Median *Se* and *Sp* can be estimated as $g^{-1}(\hat{\alpha})$ and $g^{-1}(\hat{\beta})$. As in the LMM, one can easily adjust for study-level covariates. See Hamza et al. (2009b) for a nice case study in which this meta-regression model is applied.

The regression line of $g(Se)$ on $g(Sp)$, $g(Se) = \hat{\alpha} + \hat{\rho}_{\alpha\beta}\dfrac{\hat{\sigma}_\alpha}{\hat{\sigma}_\beta}[g(Sp) - \hat{\beta}]$, plots a summary ROC curve after transforming $g(Se)$ and $g(Sp)$ to the *Se* and *Sp* scale. As discussed in Section 19.2.2, this is only one of several possible summary ROC curves.

Hamza et al. (2008) conducted extensive simulation experiments to compare the behaviors of the bivariate LMM and the bivariate GLMM and found that the bivariate GLMM performed better.

19.2.4 Hierarchical SROC Approach

In this subsection, we present the method proposed by Rutter and Gatsonis (2001). Though all random-effect models are hierarchical, this method is generally called the "hierarchical summary ROC (HSROC)" method. Like the GLMM of the previous subsection, it also uses binomial distributions. At the first level, as for the GLMM model, the HSROC approach assumes binomial distributions for the number of positive outcomes in the ith study:

$$n_{i11} \sim \text{Bin}(n_{i11} \mid n_{i1+}; Se_i), \quad n_{i01} \sim \text{Bin}(n_{i11} \mid n_{i0+}; 1 - Sp_i).$$

Se_i and Sp_i are modeled as:

$$\text{logit}(Se_i) = (\theta_i + 0.5\lambda_i)e^{-0.5\xi}; \ \text{logit}(Sp_i) = -(\theta_i - 0.5\lambda_i)e^{0.5\xi}. \tag{19.5}$$

Note that if θ_i increases, sensitivity increases and specificity decreases. Thus, θ_i models the trade-off between sensitivity and specificity and is associated with the threshold for the underlying continuous test. Therefore, θ_i is called the "cutpoint parameter" or "positivity criterion". When $\xi = 0$, then $\lambda_i = \text{logit}(Se_i) + \text{logit}(Sp_i) = \text{logit}(Se_i) - \text{logit}(1 - Sp_i) = \log(OR_i)$ independent from θ_i. So λ_i is associated with the diagnostic odds ratio and is called the "accuracy parameter". The parameter ξ determines the extent of asymmetry of the summary ROC curve, as can be seen below.

At the second level, θ_i and λ_i are assumed to vary across studies according to independent normal distributions:

$$\begin{pmatrix} \theta_i \\ \lambda_i \end{pmatrix} \sim N\left(\begin{pmatrix} \Theta \\ \Lambda \end{pmatrix}, \begin{pmatrix} \sigma_\Theta^2 & 0 \\ 0 & \sigma_\Lambda^2 \end{pmatrix} \right). \tag{19.6}$$

Rutter and Gatsonis (2001) motivate the independence assumption as: "The assumed conditional independence of θ_i and λ_i reflects assumptions implicit in ROC analysis. In the context of ROC analysis, positivity threshold and accuracy are independent test characteristics that together impose correlation between a test's sensitivity and specificity". By eliminating θ_i from (19.5), it is seen that the underlying pair of true sensitivity and specificity fulfills the following relationship:

$$\text{logit}(Se_i) = \lambda_i e^{-\xi/2} - e^{-\xi}\text{logit}(Sp_i).$$

If one is willing to assume that this is not only true for the underlying pair of true sensitivity and specificity for which we have observed data, but is true for all possible pairs of sensitivity and specificity, a study specific ROC curve is given by

$$\text{logit}(y) = \Lambda_i e^{-\xi/2} - e^{-\xi}\text{logit}(x). \tag{19.7}$$

Note that if $\xi = 0$, the study specific ROC curves are parallel lines with slope one in the logit-logit space. In the Se–Sp space they are symmetric around the line $x + y = 1$. If ξ is not zero, the curves become asymmetric. Also note that Rutter and Gatsonis (2001) make the study specific ROC curves identifiable by assuming that there is no between-study variation in the slopes in the logit-logit space. Given (19.7), it is logical to take the expected value as summary ROC curve:

$$\text{logit}(y) = \Lambda e^{-\xi/2} - e^{-\xi}\text{logit}(x). \tag{19.8}$$

It is estimated by plugging in estimates for Λ and ξ, and the result is known as the HSROC summary curve. It has a straightforward interpretation as the mean curve in the logit-logit space in the population of studies, and as the median ROC curve in the x–y space. However, this interpretation only holds if one believes that (19.7) is true, which is an assumption that cannot be verified by the data and lacks a clear rationale. If one is not sure that (19.7) holds, the interpretation of the HSROC summary curve is not obvious. An alternative summary ROC curve can be constructed by calculating the expectation of logit(Se) given logit(Sp) according to the bivariate normal distribution of θ_i and λ_i in equation (19.6) (Chu and Guo, 2009):

$$\text{logit}(Se) = e^{-0.5\xi}(0.5\hat{\Lambda} + \hat{\Theta}) + \frac{0.25\hat{\sigma}_\Lambda^2 - \hat{\sigma}_\Theta^2}{0.25\hat{\sigma}_\Lambda^2 + \hat{\sigma}_\Theta^2} \times e^{-\xi}\{\text{logit}(Sp) - e^{-0.5\xi}(0.5\hat{\Lambda} - \hat{\Theta})\}. \tag{19.9}$$

Without assuming anything about the shape of the study specific ROC curves, this SROC curve is interpreted as estimating the mean of the reported logit sensitivities in a subpopulation of studies reporting the same specificity.

Rutter and Gatsonis (2001) estimated the model parameters by taking a fully Bayesian approach, while Macaskill (2004) showed how to estimate the parameters through straightforward maximum likelihood using a GLMM program, which is in practice more convenient.

19.2.5 Equivalence between the HSROC and Bivariate GLMM Approaches

As the HSROC and bivariate GLMM approaches seem to be very different, practical investigators may face the question of choosing between these two models. Interestingly, Harbord et al. (2007) showed that, in the absence of study-level covariates, these two approaches are equivalent up to a reparameterization and the parameters are related through the following equations:

$$\xi = \log(\sigma_\alpha / \sigma_\beta),$$

$$\Theta = \frac{1}{2}\{(\sigma_\alpha/\sigma_\beta)^{1/2}\alpha - (\sigma_\beta/\sigma_\alpha)^{1/2}\beta\},$$

$$\Lambda = (\sigma_\alpha/\sigma_\beta)^{1/2}\alpha + (\sigma_\beta/\sigma_\alpha)^{1/2}\beta,$$

$$\sigma_\theta^2 = \frac{1}{2}(\sigma_\alpha\sigma_\beta - \sigma_{\alpha\beta}), \sigma_\Lambda^2 = 2(\sigma_\alpha\sigma_\beta + \sigma_{\alpha\beta}).$$

Harbord et al. (2007) highlighted that the shape parameter ξ of the HSROC is determined alone by the ratio of the variances of logit sensitivity and logit specificity in the bivariate GLMM model, and the correlation between logit sensitivity and logit specificity is not required. It is remarkable that this implies that the HSROC curve could also be estimated simply by standard univariate random-effects meta-analysis (Chapter 5) on sensitivities and specificities separately. (Of course, for further inference such as standard errors and confidence intervals the simultaneous analysis of sensitivities and specificities is still needed.)

19.2.6 A Composite Likelihood Method

In practice, fitting the bivariate GLMM of Section 19.2.3 may lead to non-convergence problems and unstable results due to double integrals in the likelihood function, even when modern computational methods such as Laplace or adaptive Gaussian quadrature approximations are used. Such a problem is non-trivial when the number of studies is relatively small. Bayesian approaches have many fewer problems with convergence. However, especially when the number of studies is small, the choice of the prior distributions can then be crucial. Different reasonable vague priors can show a marked variation in results (Lambert et al., 2005). Chen et al. (2017b) proposed a composite likelihood approach to address the convergence issue.

Denote $\theta_1 = (\alpha, \sigma_\alpha^2)$, $\theta_2 = (\beta, \sigma_\beta^2)$ and $\theta = (\theta_1, \theta_2)$. The log likelihood function of $(\theta_1, \theta_2, \rho_{\alpha\beta})$ is given by

$$\begin{aligned}
&\log L(\theta_1, \theta_2, \rho_{\alpha\beta}) \\
&= \sum_{i=1}^{N} \log \Pr(n_{i00}, n_{i11} \mid n_{i0+}, n_{i1+}, \theta_1, \theta_2, \rho_{\alpha\beta}) \\
&= \sum_{i=1}^{N} \log \int\int \text{Bin}(n_{i00} \mid n_{i0+}; Sp_i) \\
&\quad \times \text{Bin}(n_{i11} \mid n_{i1+}; Se_i) \phi(Se_i, Sp_i; \theta_1, \theta_2, \rho_{\alpha\beta}) dSe_i dSp_i,
\end{aligned} \tag{19.10}$$

where $\phi(\cdot, \cdot; \theta_1, \theta_2, \rho_{\alpha\beta})$ is the bivariate logit normal distribution indexed by $(\theta_1, \theta_2, \rho_{\alpha\beta})$ and $\text{Bin}(\cdot \mid \cdot; \cdot)$ is the binomial distribution.

The motivation for the composite likelihood approach is that the commonly used performance measures of a diagnostic test (e.g., overall sensitivity and specificity, diagnostic likelihood ratios, and diagnostic odds ratio) are all functions of α and β only, and do not involve the correlation parameter $\rho_{\alpha\beta}$. To circumvent the aforementioned computational issue associated with double integrals in the likelihood function, Chen et al. (2017) proposed to construct a pseudo likelihood by using an independent working assumption. Specifically, setting $\rho_{\alpha\beta} = 0$ in equation (19.10), the pseudo likelihood is given by

$$\log L_p(\theta_1, \theta_2) = \log L_1(\theta_1) + \log L_2(\theta_2), \tag{19.11}$$

where

$$\log L_1(\theta_1) = \sum_{i=1}^{N} \log \Pr(n_{i11} \mid n_{i1+};\theta_1)$$

$$= \sum_{i=1}^{N} \left\{ \log \int \text{Binomial}(n_{i11} \mid n_{i1+}, Se_i)\phi(Se_i;\theta_1)dSe_i \right\},$$

$$\log L_2(\theta_2) = \sum_{i=1}^{N} \log \Pr(n_{i00} \mid n_{i0+};\theta_2)$$

$$= \sum_{i=1}^{N} \left\{ \log \int \text{Binomial}(n_{i00} \mid n_{i0+}, Sp_i)\phi(Sp_i;\theta_2)dSp_i \right\},$$

and $\phi(\cdot;\theta_j)$ is the logit normal distribution indexed by θ_j ($j=1,2$). Note that only one-dimensional integrals are involved in the pseudolikelihood. Hence, the approximation errors can be reduced. In addition, the non-convergence or non-positive definite covariance matrix problem is alleviated since there is no correlation parameter involved in the pseudolikelihood. More importantly, in contrast to the bivariate normality assumption made by the standard likelihood method, the pseudolikelihood relies on the marginal normality of logit sensitivity and logit specificity. Hence, the pseudolikelihood based inference may be more robust than the standard likelihood inference to misspecification of the joint distribution assumption. The parameters are estimated by maximizing the pseudolikelihood, robust standard errors are calculated with the sandwich method, and confidence intervals are obtained by Wald's method.

Note that the parameters θ_1 and θ_2, and thereby the above-mentioned descriptive accuracy measures, could also be estimated from two separate simple random-effects meta-analyses. However, to get confidence intervals for measures involving both θ_1 and θ_2, such as the log diagnostic odds ratio, the correlation between the estimates of α and β is needed, and these are provided by the composite likelihood approach. Note that, interestingly, the HSROC curve can be estimated from the composite likelihood method, while the other SROC curves need the correlation parameter.

19.2.7 Trivariate GLMM

The previous bivariate approaches involving only sensitivities and specificities are appropriate if most studies use case-control designs not enabling estimation of the disease prevalence. When cohort study designs are used based on a random sample from the total population, then the disease prevalence is estimable, and we can derive other clinically meaningful indices such as PPV and NPV. In this case, the potential dependence of test performance on the disease prevalence, which is known as "spectrum bias" (Ransohoff and Feinstein, 1978), may be investigated. This is of most concern when the bivariate diagnostic outcome is based on a continuous trait with a threshold. It may lead to a high risk of misclassification, particularly for the subjects with true values around the cutpoint. Furthermore, the misclassification rate may vary between populations; in other words, it depends on the distribution of the true levels of the underlying trait in the population

TABLE 19.3

2×2 Table for Study *i* Accounting for Disease Prevalence

	Reference test		
Index test	**Positive (+)**	**Negative (−)**	**Total**
Positive(+)	n_{i11}	n_{i01}	
	$\pi_i Se_i$	$(1-\pi_i)(1-Sp_i)$	
Negative(−)	n_{i10}	n_{i00}	
	$\pi_i(1-Se_i)$	$(1-\pi)Sp_i$	
Total	n_{i1+}	n_{i0+}	n_{i++}
	π_i	$1-\pi_i$	1

relative to the cutpoint (Brenner and Gefeller, 1997). The diagnostic misclassification and disease prevalence are related because the disease prevalence in the population is determined by this distribution (Brenner and Gefeller, 1997). To account for this potential dependence, we encourage the use of a trivariate GLMM, which is a generalization of the bivariate GLMM, to jointly model the disease prevalence, sensitivity, and specificity (Chu et al., 2009; Ma et al., 2014, 2016b).

We denote the disease prevalence in the *i*th study by π_i and extend the 2×2 table to Table 19.3 incorporating the disease prevalence. The first level of the trivariate GLMM assumes binomial distributions:

$$n_{i1+} \sim \text{Bin}(n_{i++}, \pi_i), n_{i11} \sim \text{Bin}(n_{i1+}, Se_i), n_{i00} \sim \text{Bin}(n_{i0+}, Sp_i). \tag{19.12}$$

Similar to the bivariate GLMM, parameters are modeled via link functions and study level random effects η_i, α_i and β_i: $g(\pi_i)=\eta_i$, $g(Se_i)=\alpha_i$ and $g(Sp_i)=\beta_i$. The random effects are assumed to be trivariate normally distributed: $(\eta_i, \alpha_i, \beta_i)^T \sim MVN((\eta, \alpha, \beta)^T, \Sigma)$, where Σ has diagonal parameters σ_η^2, σ_α^2 and σ_β^2 capturing the between-study variances of the disease prevalence, sensitivity, and specificity, and off-diagonal elements $\sigma_\pi\sigma_\alpha\rho_{\pi\alpha}$, $\sigma_\pi\sigma_\beta\rho_{\pi\beta}$ and $\sigma_\alpha\sigma_\beta\rho_{\alpha\beta}$ depending on the correlations. The medians of interest are derived as $\hat{\pi}_M = g^{-1}(\hat{\eta})$, $\widehat{Se}_M = g^{-1}(\hat{\alpha})$, and $\widehat{Sp}_M = g^{-1}(\hat{\beta})$. An additional merit of the trivariate GLMM is that PPV and NPV can also be estimated:

$$\widehat{PPV} = \frac{g^{-1}(\hat{\eta})g^{-1}(\hat{\alpha})}{g^{-1}(\hat{\eta})g^{-1}(\hat{\alpha}) + \{1-g^{-1}(\hat{\eta})\}\{1-g^{-1}(\hat{\beta})\}},$$
$$\widehat{NPV} = \frac{\{1-g^{-1}(\hat{\eta})g^{-1}(\hat{\beta})}{\{1-g^{-1}(\hat{\eta})g^{-1}(\hat{\beta}) + g^{-1}(\hat{\eta})\{1-g^{-1}(\hat{\alpha})\}}. \tag{19.13}$$

19.2.8 A Case Study: A Meta-Analysis of Ultrasound for Diagnosis of Rotator Cuff Tears

19.2.8.1 Dataset

In this section, we illustrate the application of the introduced methods by a meta-analysis of 23 studies evaluating accuracy of ultrasound in diagnosing rotator cuff tears (with either arthroscopy or open surgical findings as gold standard test) in adults, performed by Smith

TABLE 19.4

Data of Meta-Analysis of Ultrasound for Diagnosis of Rotator Cuff Tears

ID	Study	n_{11}	n_{01}	n_{10}	n_{00}
1	Al-Shawi, 2008	65	12	1	65
2	Alasaarela, 1998	1	0	0	19
3	Breeneke and Morgan, 1992	11	8	14	45
4	Cullen, 2007	11	2	3	21
5	Ferrari, 2002	8	1	10	25
6	Friedman, 1993	2	0	2	0
7	Hedtmann and Fett, 1995	121	0	12	0
8	Iannotti, 2005	26	7	2	16
9	Kang, 2009	2	5	2	5
10	Kayser, 2005	41	16	11	171
11	Labanauskaite, 2002	11	3	2	9
12	Milosavljevic, 2005	17	0	7	6
13	Naqvi, 2009	4	2	0	11
14	Read et al., 1998	6	1	7	28
15	Roberts et al., 2001	5	0	2	7
16	Rutten et al., 2010	8	12	0	24
17	Takagishi, 1996	10	7	10	57
18	Teefey, 2000	10	3	5	17
19	Teefey, 2005	13	4	2	52
20	van Holsbeeck et al., 1995	14	3	1	47
21	Vlychou et al., 2009	44	2	3	7
22	Wiener and Seitz, 1993	64	4	3	71
23	Yen et al., 2004	9	1	1	9

et al. (2011). The data are shown in Table 19.4. For easy comparison of the trivariate GLMM with other approaches, all studies are considered as cohort studies in the sense that the total group is randomly sampled. Otherwise, the trivariate analysis does not make sense. Rotator cuff tear is a common reason for shoulder pain and the incidence of partial rotator cuff tears is reported to be 13% to 32% in cadaveric studies. Ultrasound is non-invasive, less expensive, and can provide dynamic evaluations. However, it has lower sensitivity and specificity in detecting the disease than open surgery or arthroscopic evaluation.

19.2.8.2 Results

The above-mentioned models can be easily fitted by using (G)LMM programs of standard statistical packages. For this example, we used the MIXED and NLMIXED procedures of SAS. For complex models such as the trivariate GLMM, model selection criteria such as Akaike information criterion (AIC) can be used to select an appropriate model for the covariance matrix Σ. To help comparisons across models, we apply logit links for all GLMMs.

The SROC method of Moses et al. (1993) estimates an AUC of 0.91 with parameter estimates $\hat{a} = 3.39$, $\hat{b} = 0.13$ by ordinary least squares and $\hat{a} = 3.57$, $\hat{b} = 0.40$ by weighted regression. The SAS MIXED procedure is used to fit the bivariate LMM, which returns median estimates of *Se* and *Sp* as 0.79 (95% CI = 0.70 to 0.87) and 0.87 (95% CI = 0.82 to 0.91),

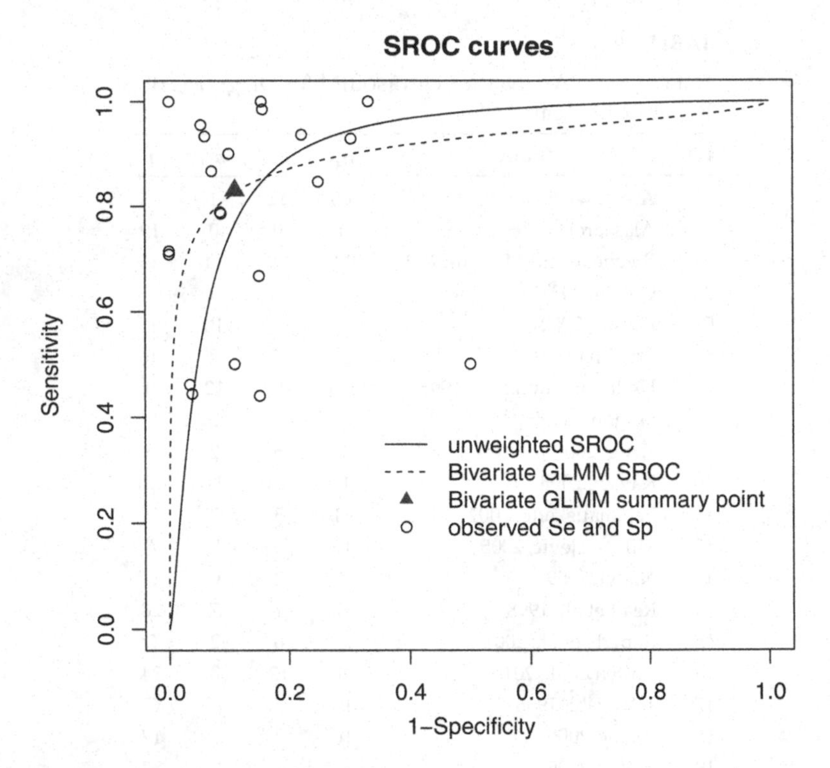

FIGURE 19.1
SROC curves from unweighted SROC, HSROC, and bivariate GLMM methods using logit links.

respectively, and a correlation of −0.18 between logit(*Se*) and logit(*Sp*). Both HSROC and bivariate GLMM can be fitted using the maximum likelihood approach implemented in the SAS NLMIXED procedure. The HSROC approach estimates the median *Se* as 0.83 (95% CI = 0.74 to 0.91) and *Sp* as 0.89 (95% CI = 0.85 to 0.93), with parameter estimates $\hat{\theta} = -0.74$, $\hat{\lambda} = 3.89$, and $\hat{\xi} = -0.52$. As it should, the bivariate GLMM gives identical median estimates to the HSROC approach. A negative correlation of 0.3 is estimated between logit(*Se*) and logit(*Sp*) and AUC is 0.91. When using the composite likelihood approach to fit the bivariate GLMM, median *Se* and *Sp* estimates are 0.83 (95% CI = 0.74 to 0.91) and 0.89 (95% CI = 0.85 to 0.93), respectively. The results are same as for the standard maximum likelihood approach. The best trivariate GLMM (selected by AIC) has the same median estimates of *Se* and *Sp* as the bivariate GLMM, and prevalence is estimated as 0.45 (95% CI = 0.30 to 0.60). Figure 19.1 plots the estimated SROC curves.

19.3 Methods without a Gold Standard

Compared with the well-developed meta-analysis models assuming a gold standard, limited tools are available when the reference test is imperfect. Latent class random-effects models are most commonly used (Walter et al., 1999; Sadatsafavi et al., 2010; Chu et al., 2009), but a Bayesian approach which extends the HSROC model has also been developed

TABLE 19.5

2×2 Table When the Reference Test Is Not a Gold Standard, Where n_{ijk} Are the Observed Counts and p_{ijk} Are the Joint Probabilities

	Reference test		
Index test	**Positive (+)**	**Negative (–)**	**Total**
Positive(+)	n_{i11}	n_{i01}	
	$p_{i11} = \pi_i Se_{i1} Se_{i2} + (1-\pi_i)(1-Sp_{i1})(1-Sp_{i2})$	$p_{i01} = \pi_i Se_{i1}(1-Se_{i2}) + (1-\pi_i)(1-Sp_{i1})Sp_{i2}$	
Negative(–)	n_{i10}	n_{i00}	
	$p_{i10} = \pi_i(1-Se_{i1})Se_{i2} + (1-\pi_i)Sp_{i1}(1-Sp_{i2})$	$p_{i00} = \pi_i(1-Se_{i1})(1-Se_{i2}) + (1-\pi_i)Sp_{i1}Sp_{i2}$	
Total	n_{i1+}	n_{i0+}	n_{i++}
	$p_{i1+} = \pi_i Se_{i2} + (1-\pi_i)(1-Sp_{i2})$	$p_{i0+} = \pi_i(1-Se_{i2}) + (1-\pi_i)Sp_{i2}$	1

(Dendukuri et al., 2012). In this section, we introduce the latent class random-effects model by Chu et al. (2009), the HSROC approach by Dendukuri et al. (2012), and their relationships.

19.3.1 Latent Class Random-Effects Model (a Multivariate GLMM)

Suppose that in all studies we have a random sample of individuals for which the results of two diagnostic tests, an index and a reference test, are available, but a gold standard is absent. Let $T_j, j=1,2$, denote the index test and the reference test, respectively, and (Se_{ij}, Sp_{ij}) denote sensitivity and specificity for T_j in study i. Assuming independence of the two tests given the disease status, we can use these pairs of parameters and the prevalence, π_i, to construct the 2×2 table, as shown in Table 19.5.

The log likelihood can then be written as:

$$\log L = \sum_{i}^{I} \{n_{i11}\log(p_{i11}) + n_{i10}\log(p_{i10}) + n_{i01}\log(p_{i01}) + n_{i00}\log(p_{i00})\}, \qquad (19.14)$$

where the n_{ijk} are the observed counts, and p_{ijk} the probabilities in terms of prevalence, sensitivity, and specificity as shown in Table 19.5. We note that conditional independence is assumed in this model, i.e., the two test results are conditionally independent given the true disease status. If there are doubts about this assumption, a sensitivity analysis can be conducted by incorporating a correlation parameter (Chu et al., 2009). We refer to Section 19.5 for more methods under conditional dependence. Under the assumption of conditional independence, there are five unknown parameters. They are non-identifiable from one 2×2 table with three unconstrained cells. When a second study is sampled with a different prevalence, assuming the diagnostic tests have the same accuracy in both populations, there are six unconstrained cells with sufficient degrees of freedom to estimate six parameters, including two prevalences, two sensitivities, and two specificities.

Random effects are used to model between-studies heterogeneity and potential correlations. It is assumed that

$$g(\pi_i) = \eta_i;\ g(Se_{ij}) = \alpha_{ij};\ g(Sp_{ij}) = \beta_{ij}, \quad j = 1,2,$$

follow a multivariate normal distribution: $N((\eta, \alpha_1, \beta_1, \alpha_2, \beta_2)^T, \Sigma)$ with variance-covariance matrix

$$\Sigma = \begin{pmatrix} \sigma_\eta^2 & \rho_{\eta\alpha_1}\sigma_\eta\sigma_{\alpha_1} & \rho_{\eta\beta_1}\sigma_\eta\sigma_{\beta_1} & \rho_{\eta\alpha_2}\sigma_\eta\sigma_{\alpha_2} & \rho_{\eta\beta_2}\sigma_\eta\sigma_{\beta_2} \\ & \sigma_{\alpha_1}^2 & \rho_{\alpha_1\beta_1}\sigma_{\alpha_1}\sigma_{\beta_1} & \rho_{\alpha_1\alpha_2}\sigma_{\alpha_1}\sigma_{\alpha_2} & \rho_{\alpha_1\beta_2}\sigma_{\alpha_1}\sigma_{\beta_2} \\ & & \sigma_{\beta_1}^2 & \rho_{\beta_1\alpha_2}\sigma_{\beta_1}\sigma_{\alpha_2} & \rho_{\beta_1\beta_2}\sigma_{\beta_1}\sigma_{\beta_2} \\ & & & \sigma_{\alpha_2}^2 & \rho_{\alpha_2\beta_2}\sigma_{\alpha_2}\sigma_{\beta_2} \\ & & & & \sigma_{\beta_2}^2 \end{pmatrix}.$$

The variance parameters describe the between-study variation in the prevalence and accuracy measures, and the correlation parameters describe the dependence among them.

Median estimates of prevalence, sensitivities, and specificities are given by $\hat{\pi}_M = g^{-1}(\hat{\eta})$, $\widehat{Se}_{jM} = g^{-1}(\widehat{\alpha_j})$, and $\widehat{Sp}_{jM} = g^{-1}(\widehat{\beta}_j)$, $j=1,2$. Variance and correlation parameter estimates can be derived from $\hat{\Upsilon}$. Covariates $\mathbf{Z}$ can be adjusted for through linear regression terms added to the mean vectors, for instance, $g(\pi_i) = \eta_i + \gamma \mathbf{Z}_i$.

19.3.2 Extended Hierarchical SROC Model

Along the line of work by Dendukuri et al. (2012), Liu et al. (2015) provided an extended HSROC framework by allowing study-specific cut-off and accuracy values for the reference test. They assume that the two diagnostic tests T_j, $j=1,2$, are based on latent variables Z_{ij} and a study-specific "cut-off" value θ_{ij}, where T_j is positive if $Z_{ij} \geq \theta_{ij}$ and is negative otherwise. Also, the Z_{ij} are assumed to follow independent location-scale distributions $h(z \mid \mu, \sigma) = \sigma^{-1} h((z-\mu)/\sigma)$ with the location and scale parameters $-\lambda_{ij}/2$ and $\exp(-\beta_j/2)$, respectively, when disease status is negative, and $\lambda_{ij}/2$ and $\exp(\beta_j/2)$ when disease status is positive. The λ_{ij} and θ_{ij} follow a $N(\Lambda_j, \sigma_{\Lambda_j}^2)$ and $N(\Theta_j, \sigma_{\Theta_j}^2)$ distribution, respectively, assumed independent for model identifiability. Based on these assumptions, the study-specific disease prevalence, sensitivity, and specificity of T_j in the ith study, on the transformed scale, can be calculated as

$$g^{-1}(\pi_i) = -\theta_{i0}, g^{-1}(Se_{ij}) = -(\theta_{ij} - \lambda_{ij}/2)\exp(-\beta_j/2),$$

$$g^{-1}(Sp_{ij}) = (\theta_{ij} + \lambda_{ij}/2)\exp(\beta_j/2),$$

where g^{-1} is the inverse of the cumulative distribution function of $h(\cdot|0,1)$. (Note that, compared with the original HSROC method, Liu et al. (2015) changed the sign of the threshold parameter θ_{ij}, assuming that higher values of the underlying continuous diagnostic test correspond with the disease.)

It is assumed that the study-specific values $(\theta_{i0}, \theta_{i1}, \lambda_{i1}, \theta_{i2}, \lambda_{i2})^T$ jointly follow a multivariate normal distribution with mean $\mathcal{H} = (\Theta_0, \Theta_1, \Lambda_1, \Theta_2, \Lambda_2)^T$ and variance Ω, where

$$\Omega = \begin{pmatrix} \sigma_{\Theta_0}^2 & \sigma_{\Theta_0\Theta_1} & \sigma_{\Theta_0\Lambda_1} & \sigma_{\Theta_0\Theta_2} & \sigma_{\Theta_0\Lambda_2} \\ & \sigma_{\Theta_1}^2 & 0 & \sigma_{\Theta_1\Theta_2} & \sigma_{\Theta_1\Lambda_2} \\ & & \sigma_{\Lambda_1}^2 & \sigma_{\Lambda_1\Theta_2} & \sigma_{\Lambda_1\Lambda_2} \\ & & & \sigma_{\Theta_2}^2 & 0 \\ & & & & \sigma_{\Lambda_2}^2 \end{pmatrix}.$$

Here the parameters Θ_0, Θ_j, and Λ_j are the overall means, respectively, of the transformed prevalence, the cut-off values, and the transformed accuracy values for tests T_1 and T_2. The variance parameters $\sigma_{\Theta_0}^2$, $\sigma_{\Theta_1}^2$, $\sigma_{\Lambda_1}^2$, $\sigma_{\Theta_2}^2$, and $\sigma_{\Lambda_2}^2$ describe the between-study variation in the prevalences, the cut-off, and accuracy values. The covariance parameters $\sigma_{\Theta_0\Theta_1}, \sigma_{\Theta_0\Lambda_1}, \sigma_{\Theta_0\Theta_2}$, and $\sigma_{\Theta_0\Lambda_2}$ describe the dependence between the prevalence cut-off value θ_{i0} and test characteristic parameters $(\theta_{i1}, \lambda_{i1}, \theta_{i2}, \lambda_{i2})$, and the covariance parameters $\sigma_{\Theta_1\Theta_2}$, $\sigma_{\Theta_1\Lambda_2}$, $\sigma_{\Lambda_1\Theta_2}$, and $\sigma_{\Lambda_1\Lambda_2}$ describe the dependence between the two tests. This model can be fitted by standard likelihood methods in GLMM programs, such as SAS PROC NLMIXED.

19.3.3 Relations between the Multivariate GLMM and Extended HSROC Models

The two models described in Sections 19.3.1 and 19.3.2, although seemingly very different, are closely related and some of their submodels are equivalent. Liu et al. (2015) provided the exact relations between the parameters of these two models and assumptions under which the two models can be reduced to equivalent submodels, and also showed that some submodels of one framework do not have corresponding equivalent submodels in the other framework. Their results generalize the important relations between the bivariate GLMM and HSROC models discussed in Section 19.2.5 as established by Harbord et al. (2007) when the reference test is a gold standard. For more details, we refer to Liu et al. (2015).

19.3.4 A Case Study: A Meta-Analysis of the Papanicolaou Test (Pap Smear) for Diagnosis of Cervical Neoplasia

19.3.4.1 Dataset

We use a meta-analysis of accuracy of the Pap test for the diagnosis of cervical cancer collected by Fahey et al. (1995) to illustrate the models under the setting without a gold standard. The Pap test is a quick, non-invasive, and relatively inexpensive way to detect cervical cancer. Fifty-nine studies were included in this meta-analysis using the Pap test as the diagnostic test and the histology test as the gold standard. However, previous studies argued that the histology test has sensitivity of 0.97 and specificity of 0.62, thus it cannot be treated as a gold standard (Walter et al., 1999) The data is listed in Table 19.6.

19.3.4.2 Results

The latent class random-effects model is fitted using the SAS NLMIXED procedure. As more random effects are included now than in the previous models, there are convergence problems. The number of parameters in the covariance matrix of the random-effects matrix is as many as 13, and, in order to obtain convergence, we decrease that number. First, we assume the disease prevalence to be independent of sensitivities and specificities and next we apply a forward-selection procedure to select the random effects, based on AIC. The resulting final model contains disease prevalence, both the sensitivity and specificity of the Pap test and the specificity of the histology test as random effects. The Pap test is estimated to have a median sensitivity of 0.66 (95% CI = 0.57 to 0.73) and a median specificity of 0.84 (95% CI = 0.77 to 0.90). In addition, median prevalence is estimated as 0.64 (95% CI = 0.54 to 0.73) and the histology test has an estimated median sensitivity of 0.90 (95% CI = 0.88 to 0.92) and specificity of 0.99 (95% CI = 0.96 to 1.00). The extended HSROC method is also fitted by using the NLMIXED procedure, yielding estimates of median *Se* as 0.65 (95% CI = 0.57 to 0.73) and *Sp* as 0.83 (95% CI = 0.77 to

TABLE 19.6

Data of Meta-Analysis of the Pap Smear Test for Diagnosis of Cervical Neoplasia

ID	Study	n_{11}	n_{10}	n_{00}	n_{01}
1	Alloub et al.	8	23	84	3
2	Alons-van Kordelaar and Boon	31	43	14	3
3	Anderson et al.	70	121	25	12
4	Anderson et al.	65	6	6	10
5	Anderson et al.	20	19	4	3
6	Andrews et al.	35	20	156	92
7	August	39	111	271	7
8	Bigrigg et al.	567	140	157	117
9	Bolger and Lewis	26	12	18	37
10	Byme et al.	38	17	37	28
11	Chomet	45	15	48	35
12	Engineer and Misra	71	10	306	87
13	Fletcher et al.	4	36	5	0
14	Frisch et al.	2	3	21	2
15	Giles et al.	5	3	182	9
16	Giles et al.	38	7	62	21
17	Gunderson et al.	4	16	31	2
18	Haddad et al.	87	12	9	13
19	Hellberg et al.	15	65	15	3
20	Helmerhorst et al.	41	61	29	1
21	Hirschowitz et al.	76	11	12	12
22	Jones DED et al.	10	48	174	4
23	Jones MH et al.	28	28	77	11
24	Kashimura et al.	3	5	1	0
25	Kealy	79	13	182	26
26	Koonlng et al.	61	27	35	20
27	Koonlng et al.	62	16	49	20
28	Kwikkel et al.	284	68	68	31
29	Lozowski et al.	66	20	44	25
30	Maggi et al.	40	12	47	43
31	Morrison EAB et al.	11	1	2	1
32	Morrison BW et al.	23	10	44	50
33	Nyirjesy	65	42	13	13
34	Okagaki and Zelterman	1270	263	1085	927
35	Oyer and Hanjanl	223	74	83	22
36	Parker	154	20	237	30
37	Pearlstone et al.	6	12	81	2
38	Ramlrez et al.	7	3	4	4
39	Reld et al.	12	11	60	5
40	Robertson et al.	348	212	103	41
41	Schauberger et al.	8	11	34	4
42	Shaw	12	6	0	2
43	Singh et al.	95	2	1	9

(Continued)

TABLE 19.6 (CONTINUED)

Data of Meta-Analysis of the Pap Smear Test for Diagnosis of Cervical Neoplasia

ID	Study	n_{11}	n_{10}	n_{00}	n_{01}
44	Skehan et al.	40	20	19	18
45	Smith et al.	71	20	18	13
46	Soost et al.	1205	454	241	186
47	Soutter et al.	5	52	27	20
48	Soutter et al.	35	12	12	9
49	Spitzer et al.	10	5	32	31
50	Stafi	3	3	15	5
51	Syrjanen et al.	118	44	183	40
52	Szarewski	13	82	17	3
53	Tait et al.	38	13	62	14
54	Tawa et al.	14	67	291	25
55	Tay et al.	12	6	12	14
56	Upadhyay et al.	238	2	16	52
57	Walker et al.	111	20	39	44
58	Wetrich	491	250	702	164
59	Wheelock and Kaminski	49	39	31	16

0.90) for the Pap test, median *Se* of 0.90 (95% CI = 0.88 to 0.92) and *Sp* of 0.99 (95% CI = 0.96 to 1.00) for the histology test, and median 0.64 (95% CI = 0.54 to 0.73) for prevalence. This model converges without the need to reduce the number of covariance parameters. These results confirm that the two approaches are very similar, if not equivalent, in the absence of covariates (Liu et al., 2015).

19.4 Mixture of Case-Control and Cohort Studies and Partial Verification Bias

In practice, meta-analyses of diagnostic tests often contain both cohort and case-control study designs. The trivariate GLMM in Section 19.2.7 can be applied to cohort studies to obtain prevalence estimates, and thus further to obtain PPV and NPV estimates. However, it cannot be directly applied to case-control studies. In this section, we will present a hybrid model that can combine cohort and case-control studies in a meta-analysis, and, at the same time, corrects for partial verification bias (Ma et al., 2016a). Partial verification is an important source of bias that occurs when the selection of samples to be tested by a reference test is affected by the results of a diagnostic test (Ransohoff and Feinstein, 1978; De Groot et al., 2011). We use the example described in Section 19.1 again, assuming that 81% of ultrasound positive patients are verified by arthroscopy while only 20% of ultrasound negative patients are verified. Test results can be presented in a 2×3 table (Table 19.7), extended from the 2×2 table (Table 19.1). If partial verification is ignored and only fully observed results are used, the Se and Sp of ultrasound will be estimated as $26/28 = 0.93$ and $36/52 = 0.7$, which are biased estimates.

TABLE 19.7

A 2×3 Table for a Toy Example: With Partial Verification

	Arthroscopy			
Ultrasound	**+ (Diseased)**	**– (Non-diseased)**	**Missing**	**Total**
+	26	16	10	52
–	2	36	150	188
Total	28	52	160	240

19.4.1 A Hybrid Hierarchical Model Combining Case-Control and Cohort Studies and Adjusting for Partial Verification

In this section, we assume that the reference test is the gold standard. We index the I studies such that the first I_1 studies are cohort studies and the remaining $I-I_1$ are case-control studies. For the case-control studies, we do not consider missing values because subjects with unverified disease status generally do not exist and subjects with missing test results usually are ignored. Ignoring missing diagnostic test results in case-control studies is valid under the assumption that they are random given the disease status. For the cohort studies, we allow that there are missing values in D (e.g., caused by partial verification of disease status) or T (e.g., caused by partial testing). We assume that the missing values are missing at random (MAR), meaning that $P(D \text{ missing} \mid D,T)$ does not depend on the disease status D, and $P(T \text{ missing} \mid D,T)$ does not depend on the test result T. As before, we use n_{idt} to denote the number of subjects with disease status $D=d(d=0,1)$ and test $T=t(t=0,1)$. Let $\omega_{imt} = P(D \text{ missing} \mid T=t)$ and $\omega_{idm} = P(T \text{ missing} \mid D=d)$. In cohort studies, subjects with both D and T missing are usually ignored, which is valid under the assumption that the probability that both are missing does not depend on the true disease status and test result. Therefore, we assume zero probability for both D and T missing.

Table 19.6 presents the data structure for the ith study using either study design. The right panel is for a case-control study. We a assume a simple design: the cases and controls are randomly sampled with a fixed sample size from the populations of cases and controls, respectively. Thus, the right panel is the typical 2×2 table as seen before. The left panel, for a cohort study, extends the standard 2×2 table to a 3×3 table allowing for missing values in T and D. The numbers of subjects with missing disease status and observed test result t $(t=0,1)$ are denoted by n_{imt}, and the numbers of subjects with missing test result and observed disease status d $(d=0,1)$ are denoted by n_{idm}. The cell probabilities are derived as follows (we suppress the index i):

$$\begin{aligned} p_{11} &= P(T=1, T \text{ obs}, D=1, D \text{ obs}) \\ &= P(T \text{ obs}, D \text{ obs} \mid T=1, D=1)P(T=1, D=1), \\ &= \{1 - P(T \text{ mis } \textit{or } D \text{ mis} \mid T=1, D=1)\}\pi Se. \end{aligned}$$

As T and D cannot be both missing, we have

$$p_{11} = \{1 - P(T \text{ mis} \mid T=1, D=1) - P(D \text{ mis} \mid T=1, D=1)\}\pi Se.$$

Because of the MAR assumption we get

$$p_{11} = \{1 - P(T \text{ mis} \mid D=1) - P(D \text{ mis} \mid T=1)\}\pi Se = (1 - \omega_{1m} - \omega_{m1})\pi Se.$$

Likewise, the cell probabilities p_{10}, p_{01} and p_{00} are derived. As T missing implies that D is observed, we have:

$$p_{1m} = P(T \text{ missing}, D = 1, D \text{ observed}) = P(T \text{ missing}, D = 1).$$

Thus we get

$$\begin{aligned} p_{1m} &= P(T \text{ missing}, T = 1, D = 1) + P(T \text{ missing}, T = 0, D = 1) \\ &= \omega_{1m}\pi Se + \omega_{1m}\pi(1 - Se) = \omega_{1m}\pi. \end{aligned}$$

Likewise the cell probabilities p_{0m}, p_{m1} and p_{m0} are derived.

Write $\boldsymbol{\omega} = \{\boldsymbol{\omega}_i\}$ and $\boldsymbol{\theta} = \{\boldsymbol{\theta}_i\}$, where $\omega_i = (\omega_{i0m}, \omega_{i1m}, \omega_{im0}, \omega_{im1})$ and $\theta_i = (\pi_i, Se_i, Sp_i)$ for study i. The likelihood can be obtained by assuming independence of subjects conditional on $\boldsymbol{\theta}_i$ and $\boldsymbol{\omega}_i$, and using a multinomial distribution for cohort studies and binomial distributions for case-control studies (following Table 19.8). Because all terms involving $\boldsymbol{\omega}$ occur multiplicatively in the cell probabilities, the likelihood factors out such that $L(\theta, \boldsymbol{\omega} \mid \text{Data}) \propto L(\theta \mid \text{Data}) \times L(\omega \mid \text{Data})$, where

$$\begin{aligned} L(\theta \mid \text{Data}) &\propto \prod_{i=1}^{I} Se_i^{n_{i11}} (1 - Sp_i)^{n_{i01}} (1 - Se_i)^{n_{i10}} Sp_i^{n_{i00}} \\ &\prod_{i=1}^{I_1} \pi_i^{\sum_j n_{i1j}} (1 - \pi_i)^{\sum_j n_{i0j}} h_{i1}^{n_{i1m}} h_{i0}^{n_{i0m}}, \end{aligned} \tag{19.15}$$

where $h_{i1} = \pi_i Se_i + (1 - \pi_i)(1 - Sp_i)$, $h_{i0} = \pi_i(1 - Se_i) + (1 - \pi_i)Sp_i$ and $j = 0, 1, m$.

We assume the same trivariate GLMM for $\boldsymbol{\theta}_i = (\pi_i, Se_i, Sp_i)$ as in Section 19.2.7. For some chosen link function g, $g(\pi_i) = \eta_i$, $g(Se_i) = \alpha_i$, and $g(Sp_i) = \beta_i$ are assumed to follow a trivariate normal distribution: $(\eta_i, \alpha_i, \beta_i)^T \sim MVN((\eta, \alpha, \beta)^T, \Sigma)$, where $\boldsymbol{\Sigma}$ has diagonal parameters σ_η^2, σ_α^2 and σ_β^2 capturing the between-study variances of the disease prevalence, sensitivity, and specificity, and off-diagonal elements involving the correlations.

Though the model can be fitted by maximum likelihood, we follow the authors (Ma et al., 2016a) in fitting the model by a fully Bayesian approach, using Markov chain Monte Carlo (MCMC) methods for parameter estimation (Chapter 6). We use non-informative normal priors on fixed effects η, α, β (e.g., $N(0,10^2)$) and a Wishart prior on the precision matrix $\boldsymbol{\Sigma}^{-1}$: $\Sigma^{-1} \sim W(\mathbf{R}, v)$, where $\mathbf{R}$ is a 3 × 3 matrix, and a small number is chosen as the degrees of freedom v ($v \geq 3$). Using the MCMC samples of η, α, and β, the posterior median π, Se and Sp can be sampled by $g^{-1}(\eta)$, $g^{-1}(\alpha)$, and $g^{-1}(\beta)$, respectively. PPV and NPV can also be estimated as in equation (19.13).

Parallel to the Bayesian sampling approach of Ma et al. (2016a), Chen et al. (2015) proposed a likelihood approach to fit the hybrid GLMM. To overcome the potential computational difficulties, such as non-convergence and singular information matrices in the standard full likelihood inference of the hybrid GLMM model, Chen et al. (2015) proposed an alternative inference procedure based on composite likelihood, analogous to what we did in Section 19.2.6.

TABLE 19.8

Data Display for the ith Study When It Is a Cohort Study and When It Is a Case-Control Study

Index test	Gold standard				
	Cohort ($i=1,\ldots,I_1$)			Case-control ($i=I_1+1,\ldots,I$)	
	+	−	Missing	+	−
+	n_{i11}	n_{i01}	n_{i1m}	n_{i11}	n_{i01}
	$(1-\omega_{im1}-\omega_{i1m})\pi_i Se_i$	$(1-\omega_{im1}-\omega_{i0m})(1-\pi_i)(1-Sp_i)$	$\omega_{im1}\{\pi_i Se_i+(1-\pi_i)(1-Sp_i)\}$	Se_i	$1-Sp_i$
−	n_{i10}	n_{i00}	n_{i0m}	n_{i10}	n_{i00}
	$(1-\omega_{im0}-\omega_{i1m})\pi_i(1-Se_i)$	$(1-\omega_{im0}-\omega_{i0m})(1-\pi_i)Sp_i$	$\omega_{im0}\{\pi_i(1-Se_i)+(1-\pi_i)Sp_i\}$	$1-Se_i$	Sp_i
Missing	n_{i1m}	n_{i0m}			
	$\omega_{i1m}\pi_i$	$\omega_{i0m}(1-\pi_i)$			

n_{idt} are cell counts and p_{idt} are the corresponding probabilities.

19.4.2 A Case Study: A Meta-Analysis of Gadolinium-Enhanced Magnetic Resonance Imaging in Detecting Lymph Node Metastases

We use an example to illustrate the model introduced in this section (Klerkx et al., 2010). Thirty-two studies were reported assessing diagnostic accuracy of gadolinium-enhanced magnetic resonance imaging (MRI) in detecting lymph node metastases, with the histopathology test as the reference gold standard test. By the QUADAS criterion (Whiting et al., 2003), this data contains six case-control studies and seven of the cohort studies are subject to partial verification. The data are shown in Table 19.9.

TABLE 19.9

Data for the Meta-Analysis of Gadolinium-Enhanced MRI in Detecting Lymph Node Metastases

ID	Study	n_{11}	n_{01}	n_{10}	n_{00}	n_{1m}	n_{0m}
Cohort Studies							
1	Bley, 2005	7	3	3	6	0	0
2	Drew, 1999	7	5	5	12	0	0
3	Gaa, 1999	18	6	3	19	0	0
4	Hasegawa, 2003	11	1	4	14	0	0
5	Kang, 2000	9	5	3	29	0	0
6	Kaza, 2006[a]	3	2	1	9	1	7
7	Krupski, 2002	6	0	2	7	0	0
8	Kvistad, 2000	15	9	8	33	0	0
9	Low, 2003	15	7	1	25	0	0
10	Wallengren, 1996[ab]	2	1	0	7	2	0
11	Einspieler, 1991[a]	5	1	2	3	9	4
12	Hawighorst, 1998	13	6	3	11	0	0
13	Hallscheidt, 1998[a]	7	1	1	23	NA	NA
14	Luciani, 2004	7	1	1	7	0	0
15	Sheu, 2001[ab]	9	2	4	26	38	0
16	Manfredi, 2004[ab]	1	1	1	18	16	0
17	Murray, 2002	10	0	17	20	0	0
18	Okizuka, 1996	10	5	3	14	0	0
19	Oellinger, 2000	5	8	2	17	0	0
20	Rockall, 2007[a]	4	5	1	40	23	23
21	Barentsz, 1996	12	2	2	41	0	0
22	Ramsay, 2004[ab]	2	5	2	7	9	0
23	Hunerbein, 2000	3	1	1	22	0	0
24	Matsuoka, 2003	5	2	0	12	0	0
25	Thurnher, 1991[a]	6	3	1	11	NA	NA
26	Mumtaz, 1997	36	4	6	29	0	0
Case-control studies							
27	Heuck, 1997	16	2	2	22	NA	NA
28	Kim, 2000	91	16	65	45	NA	NA
29	Vorreuther, 1990	4	0	1	31	NA	NA
30	Tempany, 2000	5	8	25	133	NA	NA
31	Matsuoka, 2004	18	6	8	22	NA	NA
32	Medl, 1995	6	6	1	16	NA	NA

[a] The study has partial verification.
[b] The numbers of n_{1m} and n_{0m} are arbitrarily assigned such that $n_{0m}=0$.

We fit the data using the hybrid GLMM model with logit link function using WinBUGS (Lunn et al., 2000). We chose the degree of freedom v in the Wishart prior on $\mathbf{\Sigma}^{-1}$ as $v=4$, as a smaller v indicates weaker prior for the correlation coefficient parameters (Barnard et al., 2000; Tokuda et al., 2001). After 100,000 burn-in samples, 1,000,000 posterior samples are collected. The median estimates and 95% credible intervals (CrI) are: prevalence is 0.39 (95% CrI = 0.34 to 0.45), sensitivity is 0.77 (95% CrI = 0.70 to 0.82), and specificity is 0.84 (95% CrI = 0.79 to 0.89). NPV is estimated as 0.85 (95% CrI = 0.80 to 0.89), PPV is 0.76 (95% CrI = 0.69 to 0.83), LR+ is 3.25 (95% CrI = 2.35 to 4.63), and LR– is 0.31 (95% CrI = 0.22 to 0.43).

19.5 Future Research Topics

There are many remaining interesting topics in MA-DT not discussed in this chapter. First, when a gold standard is lacking, conditional dependence can exist between the index and reference tests (e.g., when they are based on a similar biological phenomenon). Several attempts have been made to account for this dependence through a correlation parameter (Chu et al., 2009), an additional latent class random effect (Qu et al., 1996), and multivariate probit models (Xu and Craig, 2009).

Second, in addition to partial verification bias, other potential sources of bias such as publication bias also threaten accurate evaluation of test performance. Simple methods such as funnel plots (Song et al., 2002) and the trim-and-fill method (Duval and Tweedy, 2000) have been used for detection of and correction for publication bias. However, as pointed out in Chapter 13, these are inappropriate diagnostics for detecting publication bias. Moreover, separate funnel plots for sensitivity and specificity are difficult to interpret and test simultaneously. To address publication bias in diagnostic test meta-analysis, further research efforts are needed.

Third, other than gaining efficiency by combining cohort and case-control studies, a great potential gain could be found in extending MA-DT to network MA-DT, where two or more candidate tests can be evaluated simultaneously and all studies reporting any of the candidate tests can be combined in one analysis (Trikalinos et al., 2014). Network meta-analysis has been discussed in Chapter 10. Ma et al. (2018) and Lian et al. (2019a) have developed Bayesian hierarchical models for network meta-analysis of multiple diagnostic tests. Further research in this area is needed.

Fourth, in this chapter, we restricted to meta-analyses where a pair of sensitivity and specificity is available for only one specific cut-off point, which may differ between studies. This makes the interpretation of a summary ROC curve problematic. It cannot be interpreted as a kind of mean or median of the study-specific ROC curves. Sometimes, sensitivities and specificities are given for more than one cut-off point, giving two or more pairs of sensitivity and specificity per study. In principle, this makes the distribution of the study-specific ROC curves identifiable and allows the estimation of summary ROC curves that have a clear interpretation as the median ROC curve (Hamza et al., 2009; Dukic and Gatsonis, 2003). Also, more research in this area is needed.

Last, meta-analysis of diagnostic tests methods can be integrated in the research synthesis that combines two (or more) meta-analyses to correct exposure or outcome misclassification. Lian et al. (2019b) present a Bayesian approach on synthesizing two separate sets of meta-analyses to correct misclassification bias: one meta-analysis on the association between a misclassified exposure and an outcome (main studies), and the other on the association between the misclassified exposure and the true exposure (validation studies). More methodological developments and applications in this area are needed.

19.6 Software for MA-DT

Programs for fitting bivariate/trivariate GLMMs, HSROC, and the advanced statistical models for MA-DT are available in many statistical packages, including SAS, STATA, and R. In SAS, the PROC NLMIXED, PROC MIXED, and PROC GLIMMIX procedures are commonly used to fit these models, see SAS codes in the appendix of the tutorial paper by Ma et al. (2013). In STATA, both the `midas` (Dwamena, 2009) and `metandi` (Harbord and Whiting, 2009) commands are useful to fit the bivariate GLMM (Reitsma et al., 2005) and the HSROC models (Rutter and Gatsonis, 2001). In addition, there are many user-written packages for conducting MA-DT in R. Among them, the `mada` (Doebler and Holling, 2015) package which provides the established standard models and some current approaches to diagnostic meta-analysis. The `hsroc` (Schiller and Dendukuri, 2015) package performs the joint meta-analysis of sensitivity and specificity with or without a gold standard reference test. For advanced methodology, Chen et al. (2017) have developed the R package `xmeta` for multivariate meta-analysis of continuous or binary outcomes and the composite likelihood methods as described in Section 19.2.6.

References

Arends LR, Hamza TH, van Houwelingen JC, Heijenbrok-Kal MH, Hunink MGM and Stijnen T, 2008. Bivariate random effects meta-analysis of ROC curves. *Medical Decision Making* **28**(5): 621–638.

Barnard J, McCulloch R and Meng XL, 2000. Modeling covariance matrices in terms of standard deviations and correlations, with application to shrinkage. *Statistica Sinica* **10**(4): 1281–1312.

Beck JR and Shultz EK, 1986. The use of relative operating characteristic (ROC) curves in test performance evaluation. *Archives of Pathology & Laboratory Medicine* **110**(1): 13–20.

Bossuyt P, Davenport C, Deeks J, Hyde C, Leeflang M and Scholten R, 2013. Chapter 11: Interpreting results and drawing conclusions. In *Cochrane Handbook for Systematic Reviews of Diagnostic Test Accuracy Version 0.9*. The Cochrane Collaboration.

Brenner H and Gefeller O, 1997. Variation of sensitivity, specificity, likelihood ratios and predictive values with disease prevalence. *Statistics in Medicine* **16**(9): 981–991.

Chappell FM, Raab GM and Wardlaw JM, 2009. When are summary ROC curves appropriate for diagnostic meta-analyses? *Statistics in Medicine* **28**(21): 2653–2668.

Chen Y, Hong C, Chu H and Liu Y, 2017a. R-package xmeta. Available from https://cran.r-project.org/package=xmeta.

Chen Y, Liu Y, Ning J, Cormier J and Chu H, 2015. A hybrid model for combining case-control and cohort studies in systematic reviews of diagnostic tests. *Journal of the Royal Statistical Society: Series C* **64**(3): 469–489.

Chen Y, Liu Y, Ning J, Nie L, Zhu H and Chu H, 2017b. A composite likelihood method for bivariate meta-analysis in diagnostic systematic reviews. *Statistical Methods in Medical Research* **26**(2): 914–930.

Chu H, Chen S and Louis TA, 2009. Random effects models in a meta-analysis of the accuracy of two diagnostic tests without a gold standard. *Journal of the American Statistical Association* **104**(486): 512–523.

Chu H and Cole SR, 2006. Bivariate meta-analysis of sensitivity and specificity with sparse data: A generalized linear mixed model approach. *Journal of Clinical Epidemiology* **59**(12): 1331–1332.

Chu H and Guo H, 2009. Letter to the editor: A unification of models for meta-analysis of diagnostic accuracy studies. *Biostatistics* **10**(1): 201–203.

Chu H, Guo H and Zhou Y, 2010. Bivariate random effects meta-analysis of diagnostic studies using generalized linear mixed models. *Medical Decision Making* **30**(4): 499–508.

Chu H, Nie L, Cole SR and Poole C, 2009. Meta-analysis of diagnostic accuracy studies accounting for disease prevalence: Alternative parameterizations and model selection. *Statistics in Medicine* **28**(18): 2384–2399.

De Groot JAH, Bossuyt PMM, Reitsma JB, Rutjes AWS, Dendukuri N, Janssen KJM and Moons KGM, 2011. Verification problems in diagnostic accuracy studies: Consequences and solutions. *British Medical Journal* **343**: d4770.

Dendukuri N, Schiller I, Joseph L and Pai M, 2012. Bayesian meta-analysis of the accuracy of a test for tuberculous pleuritis in the absence of a gold standard reference. *Biometrics* **68**(4): 1285–1293.

Doebler P and Holling H, 2015. Meta-analysis of diagnostic accuracy with mada. Available from https://cran.rproject.org/web/packages/mada/vignettes/mada.pdf.

Dukic V and Gatsonis C, 2003. Meta-analysis of diagnostic test accuracy assessment studies with varying number of thresholds. *Biometrics* **59**(4): 936–946.

Duval S and Tweedie R, 2000. A nonparametric trim and fill method of accounting for publication bias in meta-analysis. *Journal of the American Statistical Association* **95**(449): 89–98.

Dwamena B, 2009. Midas: Stata module for meta-analytical integration of diagnostic test accuracy studies. *Statistical Software Components*.

Fahey MT, Irwig L and Macaskill P, 1995. Meta-analysis of pap test accuracy. *American Journal of Epidemiology* **141**(7): 680–689.

Gatsonis CA and Paliwal P, 2006. Meta-analysis of diagnostic and screening test accuracy evaluations: Methodologic primer. *American Journal of Roentgenology* **187**(2): 271–281.

Hamza TH, Arends LR, Van Houwelingen HC and Stijnen T, 2009a. Multivariate random effects meta-analysis of diagnostic tests with multiple thresholds. *BMC Medical Research Methodology* **9**(1): 73.

Hamza TH, Van Houwelingen HC, Heijenbrok-Kal MH and Stijnen T, 2009b. Associating explanatory variables with summary receiver operating characteristic curves in diagnostic meta-analysis. *Journal of Clinical Epidemiology* **62**(12): 1284–1291.

Hamza TH, Van Houwelingen HC and Stijnen T, 2008. The binomial distribution of meta-analysis was preferred to model within-study variability. *Journal of Clinical Epidemiology* **61**(1): 41–51.

Harbord RM, Deeks JJ, Egger M, Whiting P and Sterne JAC, 2007. A unification of models for meta-analysis of diagnostic accuracy studies. *Biostatistics* **8**(2): 239–251.

Harbord RM and Whiting P, 2009. Metandi: Meta-analysis of diagnostic accuracy using hierarchical logistic regression. *Stata Journal* **9**(2): 211.

Klerkx WM, Bax L, Veldhuis WB, Heintz APM, Mali WPTM, Peeters PHM and Moons KGM, 2010. Detection of lymph node metastases by gadolinium enhanced magnetic resonance imaging: Systematic review and meta-analysis. *Journal of the National Cancer Institute* **102**(4): 244–253.

Lambert PC, Sutton AJ, Burton PR, Abrams KR and Jones DR, 2005. How vague is vague? A simulation study of the impact of the use of vague prior distributions in MCMC using WinBUGS. *Statistics in Medicine* **24**(15): 2401–2428.

Lian Q, Hodges JS and Chu H, 2019a. A Bayesian hierarchical summary receiver operating characteristic model for network meta-analysis of diagnostic tests. *Journal of the American Statistical Association* **114**(527): 949–961.

Lian Q, Hodges JS, MacLehose R and Chu H, 2019b. A Bayesian approach for correcting exposure misclassification in meta-analysis. *Statistics in Medicine* **38**(1): 115–130.

Liu Y, Chen Y and Chu H, 2015. A unification of models for meta-analysis of diagnostic accuracy studies without a gold standard. *Biometrics* **71**(2): 538–547.

Lunn DJ, Thomas A, Best N and Spiegelhalter D, 2000. Winbugs-a Bayesian modelling framework: Concepts, structure, and extensibility. *Statistics and Computing* **10**(4): 325–337.

Ma X, Chen Y, Cole SR and Chu H, 2016a. A hybrid Bayesian hierarchical model combining cohort and case-control studies for meta-analysis of diagnostic tests: Accounting for partial verification bias. *Statistical Methods in Medical Research* **25**(6): 3015–3037.

Ma X, Lian Q, Chu H, Ibrahim JG and Chen Y, 2018. A Bayesian hierarchical model for network meta-analysis of multiple diagnostic tests. *Biostatistics* **19**(1): 87–102.

Ma X, Nie L, Cole SR and Chu H, 2016b. Statistical methods for multivariate meta-analysis of diagnostic tests: An overview and tutorial. *Statistical Methods in Medical Research* **25**(4): 1596–1619.

Ma X, Suri MFK and Chu H, 2014. A trivariate meta-analysis of diagnostic studies accounting for prevalence and non-evaluable subjects: Re-evaluation of the meta-analysis of coronary CT angiography studies. *BMC Medical Research Methodology* **14**(1): 128.

Macaskill P, 2004. Empirical Bayes estimates generated in a hierarchical summary ROC analysis agreed closely with those of a full Bayesian analysis. *Journal of Clinical Epidemiology* **57**(9): 925–932.

Metz CE, Herman BA and Shen JH, 1998. Maximum likelihood estimation of receiver operating characteristic (ROC) curves from continuously-distributed data. *Statistics in Medicine* **17**(9): 1033–1053.

Moses LE, Shapiro D and Littenberg B, 1993. Combining independent studies of a diagnostic test into a summary ROC curve: Data-analytic approaches and some additional considerations. *Statistics in Medicine* **12**(14): 1293–1316.

Pepe MS, 2003. *The Statistical Evaluation of Medical Tests for Classification and Prediction*, Chapter 2. Oxford: Oxford University Press.

Qu Y, Tan M and Kutner MH, 1996. Random effects models in latent class analysis for evaluating accuracy of diagnostic tests. *Biometrics* **52**(3): 797–810.

Ransohoff DF and Feinstein AR, 1978. Problems of spectrum and bias in evaluating the efficacy of diagnostic tests. *New England Journal of Medicine* **299**(17): 926–930.

Reitsma JB, Glas AS, Rutjes AWS, Scholten RJPM, Bossuyt PM and Zwinderman AH, 2005. Bivariate analysis of sensitivity and specificity produces informative summary measures in diagnostic reviews. *Journal of Clinical Epidemiology* **58**(10): 982–990.

Rutter CM and Gatsonis CA, 2001. A hierarchical regression approach to meta-analysis of diagnostic test accuracy evaluations. *Statistics in Medicine* **20**(19): 2865–2884.

Sadatsafavi M, Shahidi N, Marra F, FitzGerald MJ, Elwood KR, Guo N and Marra CA, 2010. A statistical method was used for the meta-analysis of tests for latent TB in the absence of a gold standard, combining random-effect and latent-class methods to estimate test accuracy. *Journal of Clinical Epidemiology* **63**(3): 257–269.

Schiller I and Dendukuri N, 2015. HSROC: An R package for Bayesian meta-analysis of diagnostic test accuracy. Available from https://core.ac.uk/download/pdf/23797204.pdf

Singisetti K and Hinsche A, 2011. Shoulder ultrasonography versus arthroscopy for the detection of rotator cuff tears: Analysis of errors. *Journal of Orthopaedic Surgery* **19**(1): 76–79.

Smith TO, Back T, Toms AP and Hing CB, 2011. Diagnostic accuracy of ultrasound for rotator cuff tears in adults: A systematic review and meta-analysis. *Clinical Radiology* **66**(11): 1036–1048.

Song F, Khan KS, Dinnes J and Sutton AJ, 2002. Asymmetric funnel plots and publication bias in meta-analyses of diagnostic accuracy. *International Journal of Epidemiology* **31**(1): 88–95.

Tokuda T, Goodrich B, Van Mechelen I, Gelman A and Tuerlinckx F, 2001. Visualizing distributions of covariance matrices. Technical report, University of Leuwen, Belgium and Columbia University, USA. Available from http://www.stat.Columbia/edu/gelman/research/unpublished/Visualization.pdf (Cited on pages 114, 116, 117 and 119).

Trikalinos TA, Hoaglin DC, Small KM, Terrin N and Schmid CH, 2014. Methods for the joint meta-analysis of multiple tests. *Research Synthesis Methods* **5**: 294–312.

Van Houwelingen HC, Arends LR and Stijnen T, 2002. Advanced methods in meta-analysis: Multivariate approach and meta-regression. *Statistics in Medicine* **21**(4): 589–624.
Walter SD, 2002. Properties of the summary receiver operating characteristic (SROC) curve for diagnostic test data. *Statistics in Medicine* **21**(9): 1237–1256.
Walter SD, Irwig L and Glasziou PP, 1999. Meta-analysis of diagnostic tests with imperfect reference standards. *Journal of Clinical Epidemiology* **52**(10): 943–951.
Whiting P, Rutjes AW, Reitsma JB, Bossuyt PMM and Kleijnen J, 2003. The development of QUADAS: A tool for the quality assessment of studies of diagnostic accuracy included in systematic reviews. *BMC Medical Research Methodology* **3**: 25.
Xu H and Craig BA, 2009. A probit latent class model with general correlation structures for evaluating accuracy of diagnostic tests. *Biometrics* **65**(4): 1145–1155.
Zwinderman AH and Bossuyt PM, 2008. We should not pool diagnostic likelihood ratios in systematic reviews. *Statistics in Medicine* **27**(5): 687–697.

20

Meta-Analytic Approach to Evaluation of Surrogate Endpoints

Tomasz Burzykowski, Marc Buyse, Geert Molenberghs, Ariel Alonso, Wim Van der Elst, and Ziv Shkedy

CONTENTS

20.1 Introduction

The development of new drugs is facing unprecedented challenges today, with more molecules than ever potentially available for clinical testing and a better targeting of the populations likely to respond, but a slow, costly, and inefficient clinical development process. A very important factor influencing the duration and complexity of this process is the choice of endpoint(s) used to assess drug efficacy. Often, the most clinically relevant endpoint, which we will term the *true endpoint* and denote by T, is difficult to use in a trial. This happens if the measurement of this clinical endpoint (1) is costly to measure (e.g., cachexia, a condition associated with malnutrition and involving loss of muscle and fat tissue, is assessed using expensive equipment that measures the levels of nitrogen, potassium, and water in the patient's body); (2) is difficult to measure (e.g., quality-of-life assessments involve multi-dimensional, poorly validated instruments); (3) requires a large sample size because of low incidence of the event of interest (e.g., cytotoxic drugs may have rare but serious side effects, such as leukemias induced by topoisomerase inhibitors); or (4) requires a long follow-up time (e.g., survival in early stage cancers). A potential strategy in these cases is to look for a *surrogate endpoint* or a *surrogate biomarker*, which we will denote by S, that can be measured more cheaply, more conveniently, more frequently, or earlier than the true clinical endpoint T of interest. The intention of the strategy is to estimate the effect of treatment in any new trial using S instead of T.

From a regulatory perspective, an endpoint or a biomarker is considered acceptable for efficacy determination only after its establishment as a valid indicator of clinical benefit, i.e., after its evaluation as a surrogate marker for the true endpoint of interest (Burzykowski et al., 2005). Note that we prefer the term "evaluation" to "validation" or "qualification" of potential surrogate endpoints (Schatzkin and Gail, 2002). Evaluation involves a host of considerations, ranging from statistical conditions to clinical and biological evidence. For a biomarker to be used as a "valid" surrogate, a number of conditions must be fulfilled. The ICH Guidelines on Statistical Principles for Clinical Trials state that:

> In practice, the strength of the evidence for surrogacy depends upon (i) the biological plausibility of the relationship, (ii) the demonstration in epidemiological studies of the prognostic value of the surrogate for the clinical outcome, and (iii) evidence from clinical trials that treatment effects on the surrogate correspond to effects on the clinical outcome. (International Conference on Harmonization of Technical Requirements for Registration of Pharmaceuticals for Human Use, 1998)

In the remainder of this chapter, we will focus on the latter two conditions.

These considerations naturally point to the need of proper definitions. Historically, first definitions were proposed under the assumption that the evaluation of a surrogate is conducted by using data from a single clinical trial. However, it turned out that the single-trial setting is too limiting and that a meaningful evaluation requires data from multiple trials, i.e., a meta-analytic framework.

20.1.1 Single-Trial Setting

Prentice (1989) proposed to define a surrogate endpoint as "a response variable for which a test of the null hypothesis of no relationship to the treatment groups under comparison is also a valid test of the corresponding null hypothesis based on the true endpoint" (Prentice, 1989, p. 432). This definition cannot be operationalized if data are available from a single trial only, since it would require multiple estimates of treatment effects on both endpoints to be available, preferably from a large number of experiments. In the case of a single trial, Prentice suggested to use the following operational criteria:

1. Treatment has a significant impact on the surrogate endpoint.
2. Treatment has a significant impact on the true endpoint.
3. The surrogate endpoint has a significant impact on the true endpoint.
4. The full effect of treatment upon the true endpoint is captured by the surrogate.

Prentice's fourth criterion implies that the true endpoint T does not depend on treatment Z given the surrogate endpoint S. The criterion is very appealing at first sight, because it seems to provide a mathematical way of testing that there exists a biological mechanism through which the surrogate fully captures the effect of treatment on the true endpoint. In cancer trials, for example, if the surrogate (say, tumor progression) were the unique mechanism in the causal chain leading from treatment exposure to the final endpoint (say, death), then the effect of a new treatment on survival would be *causally* (and therefore completely) explained by the effect of this treatment on the surrogate. Unfortunately, fulfillment of Prentice's fourth criterion does not support any such claim of causality (Buyse et al., 2015). This is because the criterion requires the strong assumption that there are no "unmeasured confounders". The assumption cannot be tested and, hence, the criterion is never guaranteed to provide a valid test of surrogacy (Buyse et al., 2015).

Freedman, Graubard, and Schatzkin (1992) argued that Prentice's fourth criterion raises a conceptual difficulty since it requires the statistical test for treatment effect on the true endpoint to be *non*-significant after adjustment for the surrogate. The non-significance of this test does not prove that the effect of treatment upon the true endpoint is *fully* captured by the surrogate, and therefore, they proposed to estimate the proportion of the treatment effect mediated by the surrogate (*proportion explained, PE*). In this paradigm, a valid surrogate would be one for which *PE* is equal to one. In practice, a surrogate would be deemed acceptable if the lower limit of the confidence interval of *PE* was "sufficiently" close to one.

Difficulties surrounding *PE* have been discussed in the literature (Buyse and Molenberghs, 1998; Choi et al., 1993; Daniels and Hughes, 1997; Flandre and Saidi, 1999; Lin et al., 1997; Volberding et al., 1990). *PE* will tend to be unstable when the effect of treatment on the true endpoint is close to zero, a situation that is likely to occur in practice. As Freedman, Graubard, and Schatzkin (1992) themselves acknowledged, the confidence limits of *PE* will tend to be rather wide (and sometimes even unbounded if Fieller's confidence intervals are used), unless large sample sizes are available and/or a very strong treatment effect is observed on the true endpoint. Note that strong treatment effects are uncommon, but large sample sizes can be available in meta-analyses of randomized clinical trials. Another, more serious issue, arises when there is an interaction between the treatment and surrogate endpoint. In that case, even the definition of *PE* becomes problematic, because there is no single effect of treatment, as the effect changes with the values of the surrogate endpoint (Molenberghs et al., 2005).

Buyse and Molenberghs (1998) proposed two other quantities for the evaluation of a surrogate endpoint: the relative effect (*RE*), which is the ratio of the effects of treatment upon the final and the surrogate endpoint, and the treatment-adjusted association between the surrogate and the true endpoint, ρ_Z. Molenberghs et al. (2002) showed that a simple relationship can be derived between *PE*, *RE*, and ρ_Z. From this relationship, it follows that *PE* is, in fact, not a proportion and can assume any real value, which makes the interpretation of *PE* problematic. Begg and Leung (2000) used a similar argument to warn against the use of *PE*.

The two measures proposed by Buyse and Molenberghs (1998) allow getting more insight into the properties of a surrogate than *PE*. *RE* is a useful quantity to predict the effect of treatment upon the true endpoint, having observed the effect of treatment upon the surrogate endpoint. If both *RE* and treatment effect on the surrogate are estimated precisely, then the predicted effect upon the true endpoint will, in turn, be precise enough to be useful. Thus, *RE* can be seen as a measure related to condition "(iii) evidence from clinical trials that treatment effects on the surrogate correspond to effects on the clinical outcome" in the ICH Guidelines (International Conference on Harmonization of Technical Requirements for Registration of Pharmaceuticals for Human Use, 1998).

Additionally, one would expect a good surrogate to have a strong association with the true endpoint within individuals, hopefully reflecting some biological pathway from the surrogate endpoint to the true endpoint. Such an association could be captured by ρ_Z. A large value of ρ_Z would provide indirect evidence that the surrogate is plausible on biological grounds, since the true endpoint would then be largely determined by the surrogate endpoint regardless of any treatment effect. Thus, the measure can be seen as related to condition "(ii) the demonstration in epidemiological studies of the prognostic value of the surrogate for the clinical outcome" in the ICH Guidelines (International Conference on Harmonization of Technical Requirements for Registration of Pharmaceuticals for Human Use, 1998). Of course, such evidence should ideally be supplemented with genuine biological evidence.

In practice, the use of RE and ρ_Z to evaluate surrogate endpoints is also complicated by a few problems. As noted by Buyse and Molenberghs (1998), the confidence intervals for RE can be wide. This difficulty can be overcome by sufficiently large sample sizes, though. More importantly, however, if we want to use RE, estimated using data from a single trial, for predicting the treatment effect on the true endpoint for a new trial (given the effect on the surrogate), it is necessary to assume that the relationship between the treatment effects on the surrogate and the true endpoints is multiplicative (Buyse and Molenberghs, 1998; Buyse et al., 2000; Molenberghs et al., 2002). This assumption may be untenable in practice, and it cannot be checked using data from a single trial. To verify the assumption, Buyse and Molenberghs (1998) suggested the use of data from multiple randomized trials. Bearing in mind the aforementioned difficulties encountered when trying to evaluate a surrogate using data from a single randomized trial, several other authors have also proposed a meta-analytic approach when data can be collected from multiple randomized clinical trials (Buyse et al., 2000; Daniels and Hughes, 1997; Gail et al., 2000).

20.1.2 Multiple-Trials Setting

The suggestion of Buyse and Molenberghs (1998) to evaluate surrogates by using data from multiple randomized trials coincided with the proposal of The Biomarker Definitions Working Group (2001; Ellenberg and Hamilton, 1989). The definitions formulated by the latter group have since been widely adopted. A clinical endpoint is considered the most credible indicator of drug response and defined as a characteristic or variable that reflects how a patient feels, functions, or survives. In clinical trials aimed at establishing the worth of new therapies, clinically relevant endpoints should be used, unless a biomarker or other endpoint is available that has risen to the status of surrogate endpoint. A biomarker is defined as a characteristic that can be objectively measured as an indicator of healthy or pathological biological processes, or pharmacological responses to therapeutic intervention. A surrogate endpoint is a biomarker that is intended for substituting a clinical endpoint. A surrogate endpoint is expected to predict clinical benefit, harm, or lack of these. Toward this aim, a prediction model is needed. Such a model can be built using data from multiple randomized trials.

In this chapter, we provide an outline of a meta-analytic approach that was originally formulated for two continuous, normally distributed (Gaussian) outcomes (Buyse et al., 2000). We will also discuss extensions that allow applying the approach to endpoints of different types. A more thorough review of the methods for evaluation of surrogate endpoints can be found in recently published texts (Alonso et al., 2017; Buyse et al., 2015).

20.2 A Meta-Analytic Approach for Two Normally Distributed Outcomes

This section outlines a meta-analytic approach that was originally formulated for two continuous, normally distributed (Gaussian) outcomes (Buyse et al., 2000). Even though the models in this case are relatively straightforward, they pose considerable computational challenges. Simplified modeling approaches will be discussed in Section 20.2.2. The models can be extended to other outcome types, ranging from continuous, binary, ordinal, time-to-event, and longitudinally measured outcomes (Burzykowski et al., 2005). These will be taken up in Section 20.3.

20.2.1 The Hierarchical Model

We use a hierarchical two-level model for the situation of a surrogate and a true endpoint that are jointly normally distributed. Let T_{ij} and S_{ij} be the random variables denoting the true and surrogate endpoints for the jth subject in the ith trial, respectively, and let Z_{ij} be the indicator variable for treatment. At the first stage, consider the following fixed-effects models:

$$S_{ij} = \mu_{Si} + \alpha_i Z_{ij} + \varepsilon_{Sij}, \tag{20.1}$$

$$T_{ij} = \mu_{Ti} + \beta_i Z_{ij} + \varepsilon_{Tij}, \tag{20.2}$$

where μ_{Si} and μ_{Ti} are trial-specific intercepts, α_i and β_i are trial-specific effects of treatment Z_{ij} on the endpoints in trial i, and ε_{Sij} and ε_{Tij} are correlated error terms, assumed to be zero-mean normally distributed with the variance-covariance matrix

$$\Sigma = \begin{pmatrix} \sigma_{SS} & \sigma_{ST} \\ & \sigma_{TT} \end{pmatrix}. \tag{20.3}$$

At the second stage, we decompose

$$\begin{pmatrix} \mu_{Si} \\ \mu_{Ti} \\ \alpha_i \\ \beta_i \end{pmatrix} = \begin{pmatrix} \mu_S \\ \mu_T \\ \alpha \\ \beta \end{pmatrix} + \begin{pmatrix} m_{Si} \\ m_{Ti} \\ a_i \\ b_i \end{pmatrix}, \tag{20.4}$$

where the second term on the right-hand side of (20.4) is assumed to follow a zero-mean normal distribution with variance-covariance matrix

$$D = \begin{pmatrix} d_{SS} & d_{ST} & d_{Sa} & d_{Sb} \\ & d_{TT} & d_{Ta} & d_{Tb} \\ & & d_{aa} & d_{ab} \\ & & & d_{bb} \end{pmatrix}. \tag{20.5}$$

Combination of the above two steps leads to a classical hierarchical, random-effects model:

$$S_{ij} = \mu_S + m_{Si} + \alpha Z_{ij} + a_i Z_{ij} + \varepsilon_{Sij}, \tag{20.6}$$

$$T_{ij} = \mu_T + m_{Ti} + \beta Z_{ij} + b_i Z_{ij} + \varepsilon_{Tij}. \tag{20.7}$$

Here, μ_S and μ_T are fixed intercepts, α and β are fixed treatment effects, m_{Si} and m_{Ti} are random intercepts, and a_i and b_i are random treatment effects in trial i for the surrogate and true endpoints, respectively. The vector of random effects $(m_{Si}, m_{Ti}, a_i, b_i)$ is assumed to be zero-mean normally distributed with the variance-covariance matrix (20.5). The error terms ε_{Sij} and ε_{Tij} follow a zero-mean bivariate normal distribution with the variance-covariance matrix (20.3).

After fitting model (20.6)–(20.7), surrogacy is captured by means of two quantities: the "individual-level" and "trial-level" coefficients of determination, denoted respectively by R^2_{indiv} and R^2_{trial}. The former measures the association between S and T at the level of the individual patient, after adjustment for Z and trial effects, while the latter quantifies the association between the treatment effects on S and T at the trial level (adjusting for the trial-specific level of S).

R^2_{indiv} is based on (20.3) and takes the following form:

$$R^2_{\text{indiv}} = R^2_{\varepsilon_{Tij}|\varepsilon_{Sij}} = \frac{\sigma^2_{ST}}{\sigma_{SS}\sigma_{TT}}. \tag{20.8}$$

R^2_{trial} is given by

$$R^2_{\text{trial}} = R^2_{b_i|m_{Si},a_i} = \frac{\begin{pmatrix} d_{Sb} \\ d_{ab} \end{pmatrix}^T \begin{pmatrix} d_{SS} & d_{Sa} \\ d_{Sa} & d_{aa} \end{pmatrix}^{-1} \begin{pmatrix} d_{Sb} \\ d_{ab} \end{pmatrix}}{d_{bb}}. \tag{20.9}$$

The above quantities are unitless and, if the corresponding variance-covariance matrix is positive definite, lie within the unit interval.

A surrogate could be adopted when R^2_{indiv} and R^2_{trial} are both sufficiently close to one. Some authors (Lassere et al., 2007) and health authorities (e.g., the German Institute for Quality and Efficiency in Health Care, IQWiG, 2011) have suggested thresholds that have to be met by these measures of association before a surrogate is considered acceptable. While such thresholds provide useful guidance, there will always be clinical and other judgments involved in the decision process.

20.2.2 Simplified Modeling Strategies

Fitting the random-effects model (20.6)–(20.7) to the data is a numerically complex task and often poses a considerable computational challenge. When such computational difficulties are encountered, the model can be simplified. For instance, it is possible to simplify the random-effects model by removing the random trial-specific intercepts m_{Si} and m_{Ti}. In this case, $(\alpha_i, \beta_i)^T$ follows a bivariate normal distribution with mean $(\alpha, \beta)^T$ and variance-covariance matrix

$$\widetilde{D} = \begin{pmatrix} d_{aa} & d_{ab} \\ d_{ab} & d_{bb} \end{pmatrix} \tag{20.10}$$

and the R^2_{trial} for the reduced model simplifies to

$$R^2_{\text{trial}(r)} = R^2_{b_i|a_i} = \frac{d^2_{ab}}{d_{aa}d_{bb}}. \tag{20.11}$$

Alternatively, one can consider a two-stage approach (Buyse et al., 2000; Tibaldi et al., 2003), in which the trial-specific effects are treated as fixed. Effectively, the approach follows the two-stage construction presented in Section 20.2.1. In particular, the first-stage model takes the form (20.1)–(20.2). At the second stage, following the spirit of model (20.4) without considering μ_{Ti}, the estimated treatment effect on the true endpoint is regressed on the treatment effect on the surrogate and the intercept associated with the surrogate endpoint:

$$\hat{\beta}_i = \lambda_0 + \lambda_1 \hat{\mu}_{Si} + \lambda_2 \hat{\alpha}_i + \varepsilon_i. \tag{20.12}$$

An estimate of the trial-level R^2_{trial} is obtained from the regression model (20.12). Empirical evidence (Buyse et al., 2000; Burzykowski et al., 2004) suggests that, in fact, using the intercept associated with the surrogate endpoint does not improve the prediction. Hence, in practice, one often considers regressing $\hat{\beta}_i$ on $\hat{\alpha}_i$ only and estimating $R^2_{\text{trial}(r)}$.

The two-stage approach is also useful in the case of evaluation of non-Gaussian endpoints, as will be seen in Section 20.3.

Another major simplification consists of fitting separate models for the true and surrogate endpoints (Tibaldi et al., 2003). If the trial-specific effects are considered fixed, models (20.1)–(20.2) are fitted separately, i.e., the corresponding error terms in the two models are assumed independent. Similarly, if the trial-specific effects are considered random, models (20.6)–(20.7) are fitted separately. Such univariate models do not provide a direct estimate of R^2_{indiv}, but interest focuses here on R^2_{trial}. In addition, R^2_{indiv} can be estimated by making use of the correlation between the residuals from two separate univariate models.

When the univariate approach and/or the two-stage fixed-effects approach is chosen, there is a need to adjust for the estimation error in trial-specific contributions used in the linear regression (20.12). This is because ignoring the error leads to biased estimates of R^2_{trial} (Burzykowski and Cortiñas Abrahantes, 2005). One simple way of doing so in the case of normally distributed endpoints is to use weighting according to trial size. In other situations, for example, when S and T are time-to-event endpoints, such an approach is problematic, as the precision of the estimation is related to the number of events, which may be different for the two endpoints; a single weight will not reflect the difference. A more general approach, based on measurement-error models and applicable to different types of endpoints, was considered by Burzykowski and Cortiñas Abrahantes (2005). In particular, assume that the estimated treatment effects $\hat{\alpha}_i$ and $\hat{\beta}_i$ follow the model

$$\begin{pmatrix} \hat{\alpha}_i \\ \hat{\beta}_i \end{pmatrix} = \begin{pmatrix} \alpha_i \\ \beta_i \end{pmatrix} + \begin{pmatrix} \varepsilon_{ai} \\ \varepsilon_{bi} \end{pmatrix}, \tag{20.13}$$

where the estimation errors ε_{ai} and ε_{bi} are zero-mean normally distributed with variance-covariance matrix

$$\Omega_i = \begin{pmatrix} \omega_{aa,i} & \omega_{ab,i} \\ \omega_{ab,i} & \omega_{bb,i} \end{pmatrix} \tag{20.14}$$

and $(\alpha_i,\beta_i)^T$ follows a bivariate normal distribution with mean $(\alpha,\beta)^T$ and variance-covariance matrix $\widetilde{D}$ given in (20.10). Consequently, $(\hat{\alpha}_i, \hat{\beta}_i)^T$ follows a bivariate normal distribution with mean $(\alpha,\beta)^T$ and variance-covariance matrix $\widetilde{D}+\Omega_i$. One can then fit model (20.13)–(20.14) while fixing matrices Ω_i at their values estimated in the first-stage model and obtain an estimate of $\widetilde{D}$ (see Chapter 5 in Alonso et al. (2017) and also Chapter 9 in this book). The latter is used to estimate $R^2_{\text{trial}(r)}$ defined in (20.11).

Renfro et al. (2012) have taken a completely different route to address the computational challenges of the second stage of hierarchical modeling. They propose performing this second stage, trial-level evaluation within a Bayesian framework. They assume a vague multivariate normal prior for the mean treatment effects with a vague Wishart prior for the precision of these effects. They show, through simulations and real case studies, that the Bayesian approach yields estimates of R^2_{trial} when the likelihood-based approach does not (Renfro et al., 2012). Even though the choice of priors remains a key issue, it may be preferable to use a Bayesian approach to compute R^2_{trial} with due allowance for the variance-covariance of the estimated treatment effects, rather than to use an unadjusted R^2_{trial} that ignores the estimation error. Shkedy and Torres Barbosa (2005) have also investigated the use of Bayesian methodology and conclude that even a relatively non-informative prior has a strongly beneficial impact on the algorithms' performance.

20.2.3 Other Units of Analysis

We have assumed above that several trials were available, in which case the meta-analytic method uses trial as the unit of analysis. If a single multicenter trial of sufficient size is available, it may also be reasonable to "split" this trial into smaller units of analysis, such as country, center, or investigator (Burzykowski et al., 2001; Buyse et al., 2003, 2011; Laporte et al., 2013; Scher et al., 2014). This choice may depend on practical considerations, such as the information available in each unit, experts' considerations about the most suitable unit for a specific problem, the number of potential units, and the number of patients per unit. From a technical point of view, the most desirable situation is when the number of units and the number of patients per unit are both sufficiently large (Cortiñas Abrahantes et al., 2004). When such is not the case, one can resort to simplified strategies, acknowledging the limitations of the results obtained. Clearly, an analysis based on a simplified strategy that shows good performance of a potential surrogate may support efforts to obtain more data for further analysis. Conversely, an analysis based on a simplified strategy that shows poor performance of a potential surrogate may be misleading. For instance, most published papers evaluating surrogacy that use treatment effects extracted from the literature ignore estimation error and simply fit a naïve regression line through the observed treatment effects. This approach is likely to yield biased estimates of R^2_{trial} and, as such, it may be quite misleading.

20.3 A Meta-Analytic Approach for Non-Gaussian Endpoints

Section 20.2 considered the joint distribution of the random variables governing the surrogate and true endpoints in the situation where both were Gaussian random variables. In practice, the endpoints are often non-Gaussian variables; they can be binary (e.g., pathological complete response), categorical (e.g., tumor response category: complete response, partial response, stable disease, progressive disease), censored continuous (e.g., time to tumor recurrence or overall survival), longitudinal (e.g., tumor measurements over time), or multivariate longitudinal (e.g., joint measurements of circulating tumor cells and prostate-specific antigen over time) endpoints. In those cases, the two-stage approach, outlined in Section 20.2.2, offers a solution. The models used at the first stage depend on the type of variables observed in the problem at hand. In this section, we discuss some of those models.

20.3.1 Binary Endpoints

In this case, a single-stage approach is still possible. In particular, Renard et al. (2002a) have shown that extension of the random-effects model (20.6)–(20.7) to this situation is easily done using a latent variable formulation. That is, one posits the existence of a pair of continuously distributed latent variable responses $(\tilde{S}_{ij}, \tilde{T}_{ij})$ that produce the actual values of (S_{ij}, T_{ij}). These unobserved variables are assumed to have a joint normal distribution and the realized values follow by double dichotomization. On the latent-variable scale, we obtain a model similar to (20.1)–(20.2); in the matrix (20.3), the variances are set equal to unity in order to ensure identifiability. This leads to the following model:

$$\begin{cases} \Phi^{-1}(P[S_{ij}=1 \mid Z_{ij}, m_{S_i}, a_i, m_{T_i}, b_i]) = \mu_S + m_{S_i} + (\alpha + a_i)Z_{ij}, \\ \Phi^{-1}(P[T_{ij}=1 \mid Z_{ij}, m_{S_i}, a_i, m_{T_i}, b_i]) = \mu_T + m_{T_i} + (\beta + b_i)Z_{ij}, \end{cases}$$

where Φ denotes the standard normal cumulative distribution function. Renard et al. (2002a) used pseudo-likelihood methods to estimate the model parameters. Similar ideas have been used in cases where one of the endpoints is continuous, with the other one binary or categorical (Geys, 2005).

20.3.2 Failure-Time Endpoints

Assume now that S_{ij} and T_{ij} are failure-time endpoints. In this case, the two-stage approach can be applied. In particular, model (20.1)–(20.2) is replaced by a model for two correlated failure-time random variables. Burzykowski et al. (2004) used copulas toward this end (Clayton, 1978; Dale, 1986; Hougaard, 1986). One then assumes that the joint survivor function of (S_{ij}, T_{ij}) in trial i can be written as

$$F(s,t) = P(S_{ij} \geq s, T_{ij} \geq t) = C_\delta\{F_{Sij}(s), F_{Tij}(t)\}, \quad s,t \geq 0, \tag{20.15}$$

where (F_{Sij}, F_{Tij}) denote marginal survivor functions and C_δ is a copula, i.e., a distribution function on $[0,1]^2$ with $\delta \in R^1$.

When the hazard functions are specified, estimates of the parameters for the joint model can be obtained using maximum likelihood. Shih and Louis (1995) discuss alternative estimation methods. Different copulas may be used, depending on assumptions made about the nature of the association between the surrogate and the true endpoint; such assumptions are generally unavailable, in which case the best fitting copula may be chosen (Clayton, 1978; Dale, 1986; Hougaard, 1986). The association parameter is generally hard to interpret. However, it can be shown that there is a link with Kendall's concordance-coefficient τ and Spearman's rank-correlation coefficient ρ (Genest and McKay, 1986). These easier-to-interpret coefficients can be used as a measure of association at the individual level (Burzykowski et al., 2004).

At the second stage, $R^2_{\text{trial}(r)}$ is computed based on the pairs of treatment effects estimated at the first stage (preferably, taking into account the estimation error; see Section 20.2.2).

20.3.3 An Ordered Categorical Surrogate and a Failure-Time True Endpoint

Let S_{ij} be an (ordered) categorical endpoint and T_{ij} is a failure-time endpoint. Ideas similar to those presented in Section 20.3.2 can be used (Burzykowski et al., 2004). In this case, one marginal distribution is a proportional odds logistic regression (or ordinary logistic regression if the surrogate is a binary response), while the other is a proportional hazards model. If the Plackett copula (Dale, 1986) is chosen to capture the association between both endpoints, the global odds ratio is a relatively easy to interpret measure of individual-level association (Burzykowski et al., 2004). At the second stage, $R^2_{\text{trial}(r)}$ is computed based on the pairs of treatment effects estimated at the first stage (see Section 20.2.2).

20.3.4 Other Cases

For the case of a binary surrogate and normally distributed true endpoint, one could apply the two-stage fixed-effects modeling outlined in Section 20.2.2. In the first stage, a bivariate normal model corresponding to (20.1)–(20.2) for $\tilde{S}_{ij}$ (the latent variable underlying the surrogate S_{ij}) and T_{ij} can be formulated (Molenberghs et al., 2001). At the second stage, decomposition (20.4) is used. Measures to assess the quality of the surrogate both at the trial- and individual level are then obtained. This case has received full attention in Assam et al. (2007).

Frequently, a surrogate endpoint is based on longitudinal measurements of some outcome of interest or of some relevant biomarker over time, while the true endpoint is a failure-type endpoint. In most forms of advanced cancer, for instance, the size of the tumor is measured repeatedly over time, with tumor shrinkage indicating treatment benefit, and tumor growth lack of responsiveness to treatment. In prostate cancer, the prostate-specific antigen (PSA) is repeatedly measured over time, with a maintenance at low levels indicating treatment benefit, and a rise predicting tumor recurrence, which may call for intensified treatment regimens.

Renard et al. (2002b) and Alonso et al. (2005) showed that moving from a univariate setting to a multivariate setting for normally distributed measurements is challenging. The R^2 measures proposed in Buyse et al. (2000) are no longer applicable. If treatment effect can be assumed constant over time, then (20.9) can still be useful to evaluate surrogacy at the trial level. However, the situation is different at the individual level: R^2_{indiv} is no longer applicable, and new concepts are needed. They have been discussed in Alonso et al. (2003).

Extensions to repeated measurements of non-Gaussian endpoints are difficult. Alonso et al. (2005) introduced a new measure to evaluate surrogacy at the individual level when both responses are measured over time or in general when multivariate or repeated measures are available. Alonso and Molenberghs (2007) have pushed the search for universal measures of surrogacy a step further and proposed a unifying approach to surrogate evaluation based on information theory. Their proposal avoids the need for a joint, hierarchical model, and allows for unification across different types of endpoints. The approach is elegant and universal but is beyond the scope of this chapter; the reader is referred to reference (Alonso and Molenberghs, 2007) for details.

20.4 Surrogate Threshold Effect

The key motivation for evaluating a surrogate endpoint is the ability to predict the effect of treatment on the true endpoint based on the observed effect of treatment on the surrogate endpoint. Suppose that we have fitted model (20.1)–(20.2) to data from a meta-analysis of N trials. Suppose further that a new trial has been conducted for which data are available on the surrogate endpoint but not on the true endpoint. It is essential to explore the quality of the prediction of the effect of Z on T in the new trial $i=0$, based on the information contained in the trials $i=1,\ldots,N$ used in the evaluation process, and the estimate of the effect of Z on S in the new trial. We can fit the following linear model to the surrogate outcomes S_{0j}:

$$S_{0j} = \mu_{S0} + \alpha_0 Z_{0j} + \varepsilon_{S0j}. \tag{20.16}$$

We are interested in an estimate of the effect $\beta+b_0$ of Z on T, given the effect of Z on S. Toward this end, one can observe that $(\beta + b_0 \mid m_{S0}, a_0)$, where m_{S0} and a_0 are, respectively, the surrogate-specific random intercept and the treatment effect in the new trial, follows a normal distribution with mean linear in μ_{S0}, μ_S, α_0, and α, and variance

$$\text{Var}(\beta + b_0 \mid m_{S0}, a_0) = (1 - R^2_{\text{trial}})\text{Var}(b_0). \tag{20.17}$$

Here, $\text{Var}(b_0)$ denotes the unconditional variance of the trial-specific random effect of Z on T. The smaller the conditional variance given by (20.17), the better the precision of the prediction.

In practice, we must use the estimated form of model (20.1)–(20.2) and estimates of m_{S0} and a_0. Denote by $\boldsymbol{\vartheta}$ the fixed-effects parameters and variance components related to the model, with $\hat{\vartheta}$ denoting the corresponding estimates. Fitting the linear model (20.16) to data on the surrogate endpoint from the new trial provides estimates for m_{S0} and a_0. The prediction variance can be written as follows:

$$\begin{aligned}\text{Var}(\beta + b_0 \mid \hat{\mu}_{S0}, \hat{\alpha}_0, \hat{\vartheta}) \approx\ & f\{\text{Var}(\hat{\mu}_{S0}, \hat{\alpha}_0)\} + f\{\text{Var}(\hat{\vartheta})\} \\ & + \left(1 - R^2_{\text{trial}}\right)\text{Var}(b_0),\end{aligned} \tag{20.18}$$

where $f\{\text{Var}(\hat{\mu}_{S0}, \hat{\alpha}_0)\}$ and $f\{\text{Var}(\hat{\vartheta})\}$ are functions of the asymptotic variance-covariance matrices of $(\hat{\mu}_{S0}, \hat{\alpha}_0)^T$ and $\hat{\vartheta}$, respectively. The third term on the right-hand side of (20.18), which is equivalent to (20.17), describes the prediction's variability if μ_{S0}, α_0, and $\boldsymbol{\vartheta}$ were known. The first two terms describe the contributions to the variability due to the use of the estimates of these parameters, respectively in the new trial (estimation of μ_{S0} and of α_0) and in the meta-analysis (estimation of $\boldsymbol{\vartheta}$).

In reality, the parameters of models (20.1)–(20.2) and (20.16) have always to be estimated, in which case the prediction variance is given by the three terms on the right-hand side of (20.18). It is useful however to consider two theoretical situations:

1. **No estimation error.** If the parameters of the mixed-effects model (20.1)–(20.2) and the single-trial regression model (20.16) were known, the prediction variance for $\beta+b_0$ would be expressed by (20.17) and the precision of the prediction would be

driven entirely by the value R^2_{trial}. While this situation is of theoretical relevance only, as it would require an infinite number of trials and infinite sample sizes for the estimation in the meta-analysis and in the new trial, it provides important insights about the intrinsic quality of the surrogate, and shows that R^2_{trial} measures the "potential" validity of a surrogate endpoint at the trial-level.

2. **Estimation error only in the meta-analysis.** This scenario is again possible only in theory, as it would require an infinite sample size in the new trial. But it can provide information of practical interest since, with an infinite sample size, the parameters of the single-trial regression model (20.16) would be known and the first term on the right-hand side of (20.18), $f\{\text{Var}(\hat{\mu}_{S0}, \hat{\alpha}_0)\}$, would vanish. In this case, (20.18) would provide the minimum variance of the prediction of β+b_0 that is achievable when the size of the meta-analysis is finite. Gail et al. (2000) used this fact to point out that the use of a surrogate validated through a meta-analytic approach will be less efficient than the direct use of the true endpoint. Even so, a surrogate can be of great use in terms of reduced sample size, shortened trial duration, or both.

The second scenario presented above indicates that R^2_{trial} may not be fully capturing the validity of a surrogate, defined in terms of the reduction of the conditional variance of β+b_0. Based on this observation, Burzykowski and Buyse (2006) proposed the "surrogate threshold effect" (STE) as a useful measure of surrogacy when interest focuses on predicting the treatment effect on the true endpoint, having observed the treatment effect on the surrogate. Assume the prediction of β+b_0 can be made independently of μ_{S0}. Under this assumption, the conditional mean of β+b_0 is a linear function of α_0, the treatment effect on the surrogate. Assume further that α_0 is estimated without error. The conditional variance of β+b_0 can be written as

$$\text{Var}(\beta + b_0 \mid \alpha_0, \hat{\vartheta}) \approx f\{\text{Var}(\hat{\vartheta})\} + (1 - R^2_{\text{trial}(r)})\text{Var}(b_0). \tag{20.19}$$

Since in linear mixed-effects models the maximum likelihood estimates of the covariance parameters are asymptotically independent of the fixed-effects parameters (Verbeke and Molenberghs, 2000), one can show that the prediction variance (20.19) can be expressed approximately as a quadratic function of α_0.

Consider a $(1-\gamma)100\%$ prediction interval for β+b_0:

$$E(\beta + b_0 \mid \alpha_0, \vartheta) \pm z_{1-\gamma/2}\sqrt{\text{Var}(\beta + b_0 \mid \alpha_0, \vartheta)}, \tag{20.20}$$

where $z_{1-\gamma/2}$ is the $(1-\gamma/2)$ quantile of the standard normal distribution. The limits of the interval (20.20) are functions of α_0. Denote by $l(\alpha_0)$ and $u(\alpha_0)$, respectively, the lower and upper prediction limit functions of α_0, corresponding to (20.20). One can then compute a value of α_0 such that, for example,

$$l(\alpha_0) = 0. \tag{20.21}$$

This value is called the *surrogate threshold effect*. The STE is the smallest treatment effect on the surrogate necessary to be observed to predict a significant treatment effect on the true endpoint. The STE depends on the variance of the prediction. The larger the variance, the larger the absolute value of the STE. In practical terms, one would hope to get a value of the STE that can realistically be achieved, given the range of treatment effects on the surrogate observed in previous clinical trials. If the STE were too large to be achievable, the surrogate would not be useful for the purposes of predicting a treatment effect on the true endpoint. In such a case, the use of the surrogate would not be reasonable, even if the surrogate were "potentially" valid, i.e., with $R^2_{\text{trial}(r)} \simeq 1$. The STE thus provides important information about the usefulness of a surrogate in practice. It can also be used to design a clinical trial aimed at showing an effect on the surrogate that exceeds the STE, which is anticipated to predict a significant effect on the true endpoint.

20.5 Case Studies

In this section, we illustrate the concepts and difficulties inherent in the evaluation of surrogate endpoints using the data from two meta-analyses of randomized clinical trials. In both case studies, progression-free survival (PFS) is evaluated as a surrogate for overall survival (OS). These two examples have been chosen because they illustrate how a quantitative evaluation can inform the use of surrogate endpoints: even though the same endpoints are considered, we will come to different conclusions about the use of PFS as a surrogate for OS. Based on the measures of surrogacy introduced in Sections 20.2.1 and 20.4, we will argue that PFS can be used as a reasonable surrogate for OS in advanced colorectal cancer, but not in advanced gastric cancer.

20.5.1 Advanced Colorectal Cancer

This analysis used data for 3089 patients from ten historical trials comparing fluorouracil (FU) with FU + leucovorin (LV) or FU + LV with raltitrexed (Buyse et al., 2007). As all the trials included FU + LV, a natural choice was to use this treatment as the common "reference" (control) when defining treatment contrasts.

The first-stage model was defined by using the Hougaard copula and Weibull marginal distributions for PFS and OS. The individual-level association, quantified by Spearman's rank correlation coefficient, was equal to 0.82 (95% confidence interval, CI: [0.81, 0.84]), indicating substantial correlation between PFS and OS for a given patient.

At the trial level, there was a tight association between the treatment effects on PFS and on OS (circles in Figure 20.1). Note that HRs were defined with the hazard function for FU + LV in the numerator; hence, HR < 1 implied that FU + LV was better than the comparator. Without adjustment for the estimation error in treatment effects, $R^2_{\text{trial}(r)}$ was equal to 0.926 (95% CI: [0.837, 1.014]). (Note that the CI extends beyond 1 due to the use of a normal approximation; here and later, we choose not to truncate the CIs to the [0,1] interval.) After adjusting for the estimation error, $R^2_{\text{trial}(r)}$ was estimated to be equal to 0.978 (95% CI: [0.883, 1.074]). The latter estimate was obtained from an estimated variance-covariance matrix of the random trial-specific treatment effects with a condition number equal to 188.5, which is large, but not extreme. The linear regression model adjusted for estimation errors was

$$\log(HR_{OS}) = 0.003 + 0.807 \times \log(HR_{PFS}), \tag{20.22}$$

where the standard errors of the intercept and slope were estimated to be equal to 0.042 and 0.109, respectively. This regression line is labeled "Predicted" in Figure 20.1. Each trial

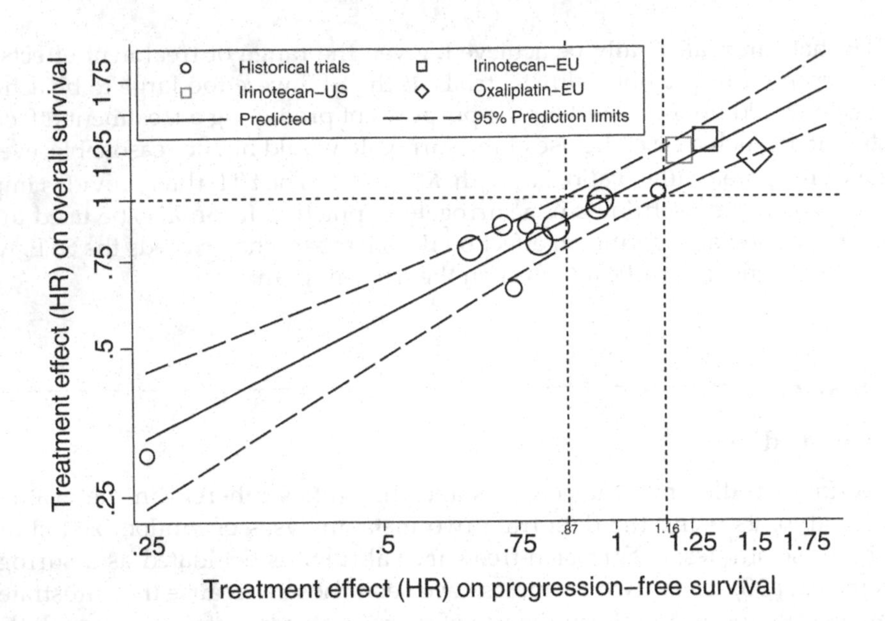

FIGURE 20.1
Trial-level association between treatment effects on PFS and OS in advanced colorectal cancer (both axes on a log scale).

is represented by a circle with a size proportional to the trial sample size. The 95% prediction limits indicate the range of effects on OS that can be expected for a given effect on PFS.

STE (indicated by the vertical dashed line in Figure 20.1) was equal to 0.87 (based on the upper prediction limit) or 1.16 (based on the lower prediction limit). Hence, in a future trial using similar treatment modalities as in the set of trials in the meta-analysis, a HR_{PFS} smaller than 0.87 (or larger than 1.16) would predict, with 95% confidence, a HR_{OS} significantly smaller (or larger) than 1.

One of the trials (see the bottom-left corner of the plot in Figure 20.1) exhibited extreme treatment benefits in terms of both PFS and OS. Exclusion of the trial resulted in a much weaker association between the treatment effects. Without adjustment for the estimation error in treatment effects, $R^2_{\text{trial}(r)}$ was equal to 0.738 (95% CI: [0.441, 1.035]). Analysis with an adjustment for the estimation error encountered numerical instabilities, leading to an estimate of the variance-covariance matrix of the random trial-specific treatment effects with a condition number of the order of 10^7. The issue is that the exclusion of the single trial reduces the between-trial variability of the treatment effects, which makes it a smaller part of the total variability. This results in problems with the estimation of the variance-covariance matrix.

The results of the surrogate evaluation could be externally validated using three trials not included in the meta-analysis. Two of these trials (one in the USA and one in Europe, indicated by squares in Figure 20.1) compared FU + LV with FU + LV + irinotecan, and one European trial (indicated by the diamond in Figure 20.1) compared FU + LV with FU + LV + oxaliplatin. Note that the trials suggested that the modified FU + LV treatments were better than FU + LV (in the plot, the HRs are larger than 1, because the hazard of FU + LV is in the numerator). One can observe that the predicted effects agreed extremely well with the observed effects in trials testing irinotecan, but less well (though still within the prediction limits) in the trial testing oxaliplatin, in which the predicted effect overestimated the observed effect. The latter may be due to the effect of second-line treatments (e.g., a cross-over of FU+LV patients to the oxaliplatin arm upon progression) (Buyse et al., 2007).

It is interesting to note that in the three validation trials, the prediction intervals for the treatment effect on OS were narrower than the CI of the observed effect. In particular, the 95% prediction intervals were equal to [1.00, 1.55], [0.96, 1.43], and [1.12, 1.75] for the irinotecan-US, irinotecan-EU, and the oxaliplatin trial, respectively, while the corresponding 95% CIs were equal to [1.02, 1.67], [1.00, 1.53], and [0.94, 1.55], respectively. This underscores the potential gain arising from the use of a surrogate endpoint for which more events are available than for the true clinical endpoint.

Taken together, these findings suggest that in advanced colorectal cancer, PFS can be used as a reliable surrogate for OS, at least for the type of 5FU-based chemotherapies considered in the analysis.

20.5.2 Advanced Gastric Cancer

The analysis included individual data on 4069 patients with advanced or recurrent gastric cancer randomized in 20 randomized trials with documented OS and PFS (Paoletti et al., 2013).

Unlike in the advanced colorectal cancer example (Section 20.5.1), no common "reference" (control) treatment could be used for the purpose of defining treatment contrasts. Thus, a pragmatic choice was made (Paoletti et al., 2013): the treatment that contained the larger number of drugs (e.g., triple combinations vs. double combinations) or, in the case of an equal number of drugs, the treatment that included the newer agent was selected as the "experimental" one. The results presented below are conditional on this choice; adopting other definitions of treatment contrasts might change the results.

The first-stage model was defined by using the Hougaard copula and Weibull marginal distributions for PFS and OS. The individual-level association, quantified by Spearman's rank correlation coefficient, was equal to 0.853 (95% CI: [0.852, 0.854]), indicating substantial correlation between PFS and OS for a given patient.

At the trial level, the association between the treatment effects on PFS and on OS was only moderate, as shown in Figure 20.2. Without adjustment for the estimation error in treatment effects, $R^2_{\text{trial}(r)}$ was equal to 0.621 (95% CI: [0.359, 0.9541]). With the adjustment, $R^2_{\text{trial}(r)}$ was equal to 0.606 (95% CI: [0.041, 1.170]). The large confidence interval reflects the uncertainty around this estimate, due in part to the small sample sizes of some of the trials included in the meta-analysis. The linear regression model adjusted for estimation errors was

$$\log(HR_{OS}) = 0.042 + 0.779 \times \log(HR_{PFS}), \tag{20.23}$$

where the standard errors of the intercept and slope were estimated to be equal to 0.079 and 0.295, respectively. This regression line is labeled "Predicted" in Figure 20.2. Each trial is represented by a circle of a size proportional to the trial sample size. The 95% prediction limits indicate the range of effects on OS that can be expected for a given effect on PFS.

The moderate correlation at the trial level is reflected by STE equal to 0.56 (indicated by the vertical dashed line in Figure 20.2). Hence, one should observe a HR_{PFS} smaller than 0.56 in order to predict, with 95% confidence, a HR_{OS} significantly smaller than 1.

The results of the surrogate evaluation could be externally validated using 12 trials not included in the meta-analysis, using treatment effects extracted from reports published in the literature after the meta-analysis was completed. Figure 20.3 shows the same

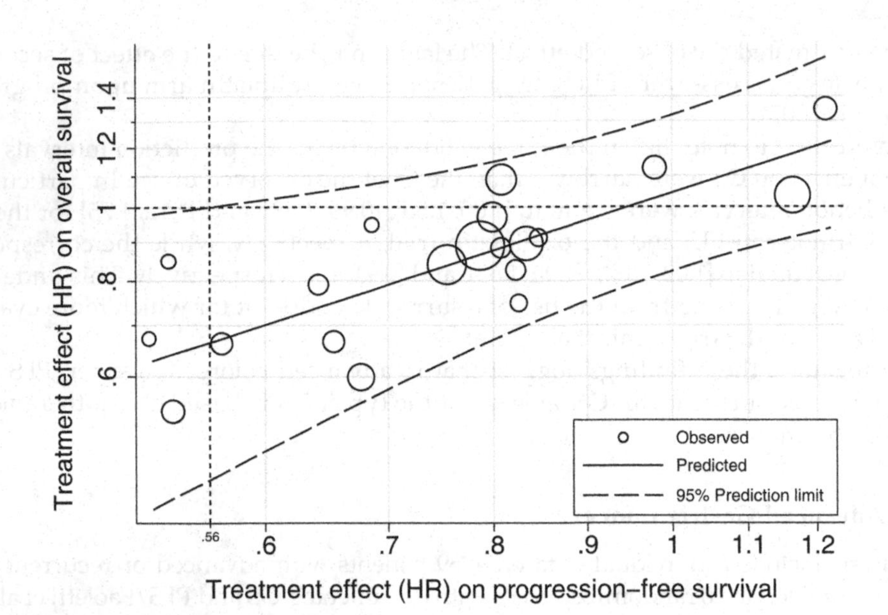

FIGURE 20.2
Trial-level association between treatment effects on PFS and OS in advanced gastric cancer (both axes on a log scale).

regression line as on Figure 20.2. The observed treatment effects on survival (HR_{OS}) are shown (as squares) for these 12 trials, as well as the effects predicted from the effects on the surrogate (HR_{PFS}) in these trials, along with their 95% predictive intervals. As can be seen, the prediction intervals were wide and included one in all trials, which means that the observed effects on PFS would not have allowed to predict a significant effect on OS

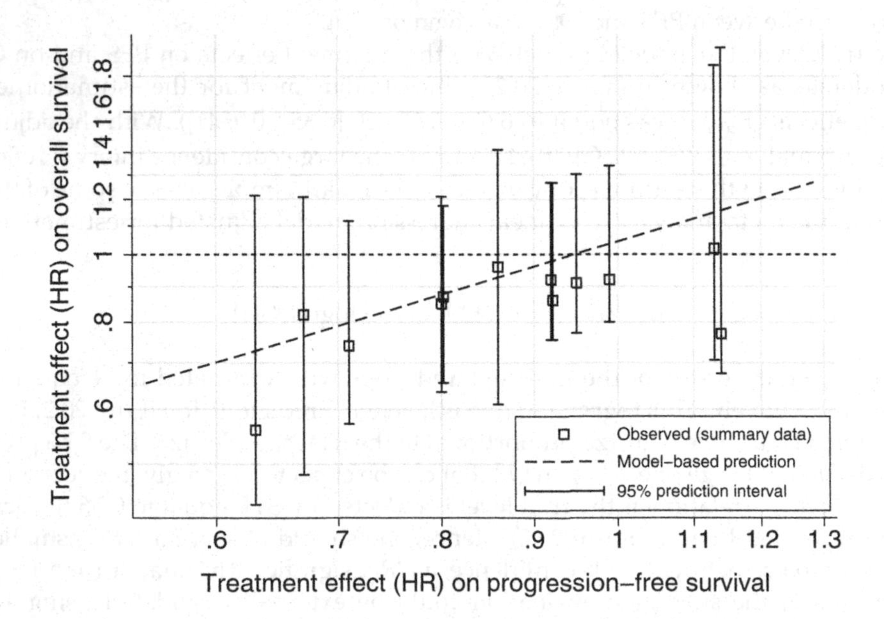

FIGURE 20.3
Observed (squares) versus predicted treatment effect on OS in validation trials (both axes on a log scale).

in any of the 12 trials. Yet, three of the 12 trials showed a statistically significant effect of treatment on survival (Paoletti et al., 2013).

All in all, PFS does not seem to be a useful surrogate for OS in advanced gastric cancer. These findings parallel those in advanced breast cancer (Burzykowski et al., 2008). They are at variance with those in advanced colorectal cancer and advanced ovarian cancer, where PFS seemed a good surrogate for OS (Burzykowski et al., 2001; Buyse et al., 2007). In advanced lung cancer, the value of PFS as a surrogate for OS is questionable (Laporte et al., 2013). It seems difficult to draw general conclusions about the surrogacy of PFS as a surrogate for OS in advanced forms of cancer.

20.6 Concluding Remarks

Over the years, a variety of surrogate marker evaluation strategies have been proposed. The initial attempts to evaluate a surrogate endpoint using data from a single trial have been shown inadequate theoretically, and indeed none of these attempts has, to the best of our knowledge, successfully identified an acceptable surrogate endpoint for use in the clinic. Today, much of the attention has shifted to meta-analytic situations in which data are available from several trials. The approach has been formally adopted by, for example, IQWiG (Institut für Qualität und Wirtschaftlichkeit im Gesundheitswesen, 2015).

Note that, within the meta-analytic approach, the evaluation of the validity of a surrogate at both the individual-patient and trial level requires availability of individual-patient data. If only the validity at the trial level is of interest, one could consider conducting the analyses by using only estimates of treatment effects (or summary data that allow computing the estimates). A simple approach would then be to use model (20.12) and the resulting estimate of the trial-level coefficient of determination R^2_{trial}. However, as argued in Section 20.2.2, when fitting model (20.12), one should adjust for the estimation error that is present in the estimated treatment effects. Toward this aim, one could use model (20.13)–(20.14) or methods developed for measurement-error models with an error in the equation (Burzykowski and Cortiñas Abrahantes, 2005). It should be noted, however, that both methods require, in particular, estimates of the covariances between estimation errors present in the estimated treatment effects on the surrogate and true endpoints. Such estimates are usually unavailable in published reports of randomized clinical trials. Thus, in practice, the estimation-error-adjusted analysis is rarely possible when using only published summary data.

When considering the meta-analytic approach, Gail et al. (2000) pointed to several issues that might undermine the utility of the methodology:

1. The difficulty in defining the set of historical trials (and drugs), data of which could be used to evaluate a surrogate for development of new, future drugs.
2. Potential lack of historical studies that could be available for analysis.
3. The need for availability of individual data for both endpoints.
4. The loss of precision associated with the prediction of the treatment effect on the true endpoint instead of its direct estimation.
5. The use of realistic models for the treatment effects.

6. The need for extending the hierarchical normal model to other types of endpoints.
7. The problem of unanticipated delayed toxicity that can be missed if shorter studies based on surrogate endpoints are conducted.

The first and the last of these issues (1, 7) can be seen as related to the concept of the use of surrogate endpoints in general, and not particularly to the meta-analytic approach *per se*. The selection of the relevant set of historical studies is not trivial. A surrogate valid for a treatment with a particular mode of action may not be necessarily valid for an intervention based on a different mode of action. Recognizing this fact, IQWiG (Institut für Qualität und Wirtschaftlichkeit im Gesundheitswesen, 2015) specified, somewhat generally, that the "studies on which the validation was based must therefore have been conducted in patient populations and interventions that allow conclusions on the therapeutic indication investigated in the benefit assessment as well as on the test intervention and comparator intervention".

Points 5–6 have been addressed by the methodological developments that have been summarized in this chapter. The lack of historical studies (2) and/or individual-patient data (3) is still an important limitation. However, the current initiatives aimed at granting public access to clinical trial data may help in this respect. Finally, the loss of precision due to prediction of the treatment effect on the true endpoint may not always take place, as illustrated by the colorectal cancer case study (Section 20.5.1). And even if there is a loss of precision, it may be acceptable as the price to pay for the gain in shortened duration of the trials using the surrogate endpoint.

From a practical implementation point of view, the use of the meta-analytic approach to evaluation of surrogate endpoints requires, in general, dedicated software. Only for the case of two continuous endpoints (see Section 20.2) can standard programs implementing linear mixed-effects models be used. More information about software available for other cases can be found in Alonso et al. (2017).

Finally, it is worth noting that the meta-analytic approach to evaluation of surrogate endpoints importantly differs from the "classical", common-effect meta-analysis in several aspects:

- It considers the bivariate outcome and focuses on estimation of the association between the treatment effects rather than estimation of the pooled treatment effect on a single endpoint.
- It is based on the assumption of random trial-specific treatment effects (see Chapter 5).
- It benefits, rather than suffers, from the between-trial heterogeneity, as the latter is necessary for reliable estimation of the association between treatment effects on the surrogate and true endpoints, as illustrated by the colorectal cancer case study (Section 20.5.1).

References

Alonso A, Geys H, Molenberghs G and Vangeneugden T, 2003. Validation of surrogate markers in multiple randomized clinical trials with repeated measurements. *Biometrical Journal* **45**(8): 931–945.

Alonso A, Molenberghs G, Geys H and Buyse M, 2005. A unifying approach for surrogate marker validation based on Prentice's criteria. *Statistics in Medicine* **25**(2): 205–211.

Alonso A and Molenberghs G, 2007. Surrogate marker evaluation from an information theoretic perspective. *Biometrics* **63**(1): 180–186.

Alonso A, Bigirumurame T, Burzykowski T, et al. 2017. *Applied Surrogate Endpoint Evaluation Methods with SAS and R*. Boca Raton, FL: Chapman & Hall/CRC.

Assam P, Tilahun A, Alonso A and Molenberghs G, 2007. Information-theory based surrogate marker evaluation from several randomized clinical trials with continuous true and binary surrogate endpoints. *Clinical Trials* **4**(6): 587–597.

Begg C and Leung D, 2000. On the use of surrogate endpoints in randomized trials. *Journal of the Royal Statistical Society: Series A* **163**(1): 26–27.

Biomarkers Definition Working Group, 2001. Biomarkers and surrogate endpoints: Preferred definitions and conceptual framework. *Clinical Pharmacological Therapy* **69**(3): 89–95.

Burzykowski T, Molenberghs G, Buyse M, Geys H and Renard D, 2001. Validation of surrogate endpoints in multiple randomized clinical trials with failure-time endpoints. *Journal of the Royal Statistical Society: Series C* **50**(4): 405–422.

Burzykowski T, Molenberghs G and Buyse M, 2004. The validation of surrogate endpoints using data from randomized clinical trials: A case-study in advanced colorectal cancer. *Journal of the Royal Statistical Society: Series A* **167**(1): 103–124.

Burzykowski T and Cortiñas Abrahantes J, 2005. Validation in the case of two failure-time endpoints. In Burzykowski T, Molenberghs G and Buyse M (Eds). *The Evaluation of Surrogate Endpoints*. New York: Springer.

Burzykowski T, Molenberghs G and Buyse M, 2005. *The Evaluation of Surrogate Endpoints*. New York: Springer.

Burzykowski T and Buyse M, 2006. Surrogate threshold effect: An alternative measure for meta-analytic surrogate endpoint validation. *Pharmaceutical Statistics* **5**(3): 173–186.

Burzykowski T, Buyse M, Piccart-Gebhart MJ, Sledge G, Carmichael J, Luück HJ, Mackey JR, Nabholtz JM, Paridaens R, Biganzoli L and Jassem J, 2008. Evaluation of tumor response, disease control, progression-free survival, and time to progression as potential surrogate endpoints in metastatic breast cancer. *Journal of Clinical Oncology* **26**(12): 1987–1992.

Buyse M and Molenberghs G, 1998. Criteria for the validation of surrogate end-points in randomized experiments. *Biometrics* **54**(3): 1014–1029.

Buyse M, Molenberghs G, Burzykowski T, Renard D and Geys H, 2000. The validation of surrogate endpoints in meta-analyses of randomized experiments. *Biostatistics* **1**(1): 49–68.

Buyse M, Vangeneugden T, Bijnens L, et al. 2003. Validation of biomarkers as surrogates for clinical endpoints. In Bloom JC and Dean RA (Eds). *Biomarkers in Clinical Drug Development*. New York: Marcel Dekker.

Buyse M, Burzykowski T, Carroll K, Michiels S, Sargent DJ, Miller LL, Elfring GL, Pignon JP and Piedbois P, 2007. Progression-free survival is a surrogate for survival in advanced colorectal cancer. *Journal of Clinical Oncology* **25**(33): 5218–5224.

Buyse M, Michiels S, Squifflet P, Lucchesi KJ, Hellstrand K, Brune ML, Castaigne S and Rowe JM, 2011. Leukemia-free survival as a surrogate endpoint for overall survival in the evaluation of maintenance therapy for patients with acute myeloid leukemia in complete remission. *Haematologica* **96**(8): 1106–1112.

Buyse M, Molenberghs G, Paoletti X, et al. 2015. Statistical evaluation of surrogate endpoints with examples from cancer clinical trials. *Biometrical Journal* **58**(1): 104–132.

Choi S, Lagakos S, Schooley RT and Volberding PA, 1993. CD4+ lymphocytes are an incomplete surrogate marker for clinical progression in persons with asymptomatic HIV infection taking zidovudine. *Annals of Internal Medicine* **118**(9): 674–680.

Clayton DG, 1978. A model for association in bivariate life tables and its application in epidemiological studies of familial tendency in chronic disease incidence. *Biometrika* **65**(1): 141–151.

Cortiñas Abrahantes J, Molenberghs G, Burzykowski T, et al. 2004. Choice of units of analysis and modeling strategies in multilevel hierarchical models. *Computational Statistics and Data Analysis* **47**(3): 537–563.

Dale JR, 1986. Global cross ratio models for bivariate, discrete, ordered responses. *Biometrics* **42**(4): 909–917.
Daniels MJ and Hughes MD, 1997. Meta-analysis for the evaluation of potential surrogate markers. *Statistics in Medicine* **16**(17): 1515–1527.
Ellenberg SS and Hamilton JM, 1989. Surrogate endpoints in clinical trials: Cancer. *Statistics in Medicine* **8**(4): 405–413.
Flandre P and Saidi Y, 1999. Letter to the editor: Estimating the proportion of treatment effect explained by a surrogate marker. *Statistics in Medicine* **18**(1): 107–115.
Freedman LS, Graubard BI and Schatzkin A, 1992. Statistical validation of intermediate endpoints for chronic diseases. *Statistics in Medicine* **11**(2): 167–178.
Gail MH, Pfeiffer R, van Houwelingen HC and Carroll RJ, 2000. On meta-analytic assessment of surrogate outcomes. *Biostatistics* **1**(3): 231–246.
Genest C and McKay J, 1986. The joy of copulas: Bivariate distributions with uniform marginals. *American Statistician* **40**(4): 280–283.
Geys H, 2005. Validation using single-trial data: Mixed binary and continuous outcomes. In Burzykowski T, Molenberghs G and Buyse M (Eds). *The Evaluation of Surrogate Endpoints*. New York: Springer.
Hougaard P, 1986. Survival models for heterogeneous populations derived from stable distributions. *Biometrika* **73**(2): 387–396.
Institut für Qualität und Wirtschaftlichkeit im Gesundheitswesen, 2011. Validity of surrogate endpoints in oncology. IQWiG Reports – Commission, No. A10–05. Available from https://www.iqwig.de/download/A10-05_Executive_Summary_v1-1_Surrogate_endpoints_in_oncology.pdf.
Institut für Qualität und Wirtschaftlichkeit im Gesundheitswesen, 2015. *General Methods* (version 4.2). Available from https://www.iqwig.de/en/methods/methods-paper/general-methods-previous-versions.3021.html.
International Conference on Harmonization of Technical Requirements for Registration of Pharmaceuticals for Human Use, 1998. ICH Harmonised Tripartite Guideline. Statistical principles for clinical trials. *Federal Register* **63**(179): 49583. Available from http://www.ich.org/pdfICH/e9.pdf.
Laporte S, Squifflet P, Baroux N, et al. 2013. Prediction of survival benefits from progression-free survival benefits in advanced non small cell lung cancer: Evidence from a pooled analysis of 2,334 patients randomized in 5 trials. *BMJ Open* **3**(3): 3.03.
Lassere M, Johnson K, Boers M, Tugwell P, Brooks P, Simon L, Strand V, Conaghan PG, Ostergaard M, Maksymowych WP and Landewe R, 2007. Definitions and validation criteria for biomarkers and surrogate endpoints: Development and testing of a quantitative hierarchical levels of evidence schema. *Journal of Rheumatology* **34**(3): 607–615.
Lin DY, Fleming TR and DeGruttola V, 1997. Estimating the proportion of treatment effect explained by a surrogate marker. *Statistics in Medicine* **16**(13): 1515–1527.
Molenberghs G, Buyse M and Burzykowski T, 2005. The history of surrogate endpoint validation. In Burzykowski T, Molenberghs G and Buyse M (Eds). *The Evaluation of Surrogate Endpoints*. New York: Springer.
Molenberghs G, Buyse M, Geys H, Renard D, Burzykowski T and Alonso A, 2002. Statistical challenges in the evaluation of surrogate endpoints in randomized trials. *Controlled Clinical Trials* **23**(6): 607–625.
Molenberghs G, Geys H and Buyse M, 2001. Evaluation of surrogate end-points in randomized experiments with mixed discrete and continuous outcomes. *Statistics in Medicine* **20**(20): 3023–3038.
Paoletti X, Oba K, Bang YJ, et al. on behalf of the GASTRIC Group, 2013. Progression-free survival as a surrogate for overall survival in patients with advanced/recurrent gastric cancer: A meta-analysis. *Journal of the National Cancer Institute* **105**(21): 1608–1612.
Prentice RL, 1989. Surrogate endpoints in clinical trials: Definitions and operational criteria. *Statistics in Medicine* **8**(4): 431–440.

Ray ME, Bae K, Hussain MH, Hanks GE, Shipley WU and Sandler HM, 2009. Potential surrogate endpoints for prostate cancer survival: Analysis of a phase III randomized trial. *Journal of the National Cancer Institute* **101**(4): 228–236.

Renard D, Geys H, Molenberghs G, Burzykowski T and Buyse M, 2002a. Validation of surrogate endpoints in multiple randomized clinical trials with discrete outcomes. *Biometrical Journal* **44**(8): 1–15.

Renard D, Geys H, Molenberghs G, et al. 2002b. Validation of a longitudinally measured surrogate marker for a time-to-event endpoint. *Journal of Applied Statistics* **30**(2): 235–247.

Renfro LA, Shi Q, Sargent DJ and Carlin BP, 2012. Bayesian adjusted R^2 for the meta-analytic evaluation of surrogate time-to-event endpoints in clinical trials. *Statistics in Medicine* **31**(8): 743–761.

Schatzkin A and Gail M, 2002. The promise and peril of surrogate end points in cancer research. *Nature Reviews Cancer* **2**(1): 19–27.

Scher HI, Heller G, Molina A, et al. 2014. Circulating tumor cell biomarker panel as an individual-level surrogate for survival in metastatic castration-resistant prostate cancer. *Jounral of Clinical Oncology* **33**(12): 1348–1355.

Shih JH and Louis TA, 1995. Inferences on association parameter in copula models for bivariate survival data. *Biometrics* **51**(4): 1384–1399.

Shkedy Z and Barbosa FT, 2005. Bayesian evaluation of surrogate endpoints. In Burzykowski T, Molenberghs G and Buyse M (Eds). *The Evaluation of Surrogate Endpoints*. New York: Springer.

Tibaldi FS, Cortiñas Abrahantes J, Molenberghs G, Renard D, Burzykowski T, Buyse M, Parmar M, Stijnen T and Wolfinger R, 2003. Simplified hierarchical linear models for the evaluation of surrogate endpoints. *Journal of Statistical Computation and Simulation* **73**(9): 643–658.

Verbeke G and Molenberghs G, 2000. *Linear Mixed Models for Longitudinal Data*. New York: Springer.

Volberding PA, Lagakos SW, Koch MA, Pettinelli C, Myers MW, Booth DK, Balfour Jr HH, Reichman RC, Bartlett JA, Hirsch MS and Murphy RL, 1990. Zidovudine in asymptomtic human immunodeficiency virus infection: A controlled trial in persons with fewer than 500 CD4-positive cells per cubic millimeter. *New England Journal of Medicine* **322**(14): 941–949.

21

Meta-Analysis of Epidemiological Data, with a Focus on Individual Participant Data

Angela Wood, Stephen Kaptoge, Michael Sweeting, and Clare Oliver-Williams

CONTENTS

21.1 Introduction

Increasing numbers of meta-analyses of observational data are being conducted to achieve greater precision, reduced overfitting, and increased generalizability in epidemiological studies. The application of individual participant data (IPD, Chapter 8) or aggregate data (Chapter 4) meta-analysis to observational data presents different challenges from meta-analysis of randomized controlled trials. First, how to combine information across multiple observational studies with different study designs (e.g., cohort, nested case-control, case-cohort) that have different principal analysis approaches and estimated measures of association (e.g., hazard ratios and odds ratios). In addition, studies can be prospective or retrospective and the latter usually induce selection and information biases. Second, how to adjust for known or potential confounders which may or may not be observed for all studies. Third, how to assess effect modification, which can be assessed within studies but also between studies. Fourth, how to account for measurement error in observational data. Fifth, how to deal with missing data.

Chapter 8 described the advantages of meta-analysis of IPD from multiple randomized trials compared with meta-analysis of aggregated published data, including harmonization of disease outcomes; ability to update follow-up information; ability to analyze unpublished data, hence reducing the likelihood of publication bias (selective outcome reporting); and determination of how effects depend on age, sex, and other potential effect modifiers. Meta-analysis of IPD from multiple epidemiological studies has all these advantages and more, including harmonization of risk markers; consistent approaches to adjustment for confounding; characterization of the shape of exposure-risk relationships; greater ability to correct for regression dilution bias; and detailed characterization of risk prediction models.

In this chapter, we address and illustrate the challenges of performing IPD meta-analysis. We use IPD from the Fibrinogen Studies Collaboration (FSC) and the Emerging Risk Factors Collaboration (ERFC). The FSC conducted a meta-analysis of the association of plasma fibrinogen, a major coagulation protein in circulating blood, and risk of coronary heart disease (CHD) based on individual-level data from 154,211 participants from 31 prospective cohort studies (Fibrinogen Studies Collaboration, 2004, 2005). The ERFC largely subsumed the FSC, extending IPD meta-analyses to other emerging and established risk factors, including lipid, inflammation, adiposity, and glycemic markers, with data now available in up to 2.5 million participants from 130 prospective cohort studies (Danesh et al., 2007; Di Angelantonio et al., 2009; Kaptoge et al., 2010). Our examples use survival models on time-to-event outcomes, but the methods can be adapted for other outcome types. We use two-stage analysis methods, since one-stage analysis methods for time-to-event outcomes can be complex; one-stage analysis methods for IPD are discussed in Chapter 8.

21.2 Combining Estimates from Different Study Designs

21.2.1 Handling Different Study Designs

A main aim of IPD meta-analysis is to quantify exposure-disease associations using individual-level data from relevant population samples. Depending on context, this may often involve data from different study designs, necessitating careful planning of analyses to ensure that appropriate statistical models are used for inference.

Prospective cohort studies sample participants from a relevant population who, at baseline, are assessed for exposures and other covariates of interest, and are subsequently tracked for the occurrence of incident outcomes. In addition, prospective cohort studies provide the relevant sampling frame for the design of case-cohort and nested case-control studies, which may be preferable designs in certain circumstances, including studying associations with costly to measure exposures and rare outcomes. The case-cohort design selects a random sample of the original cohort as a representative referent set (known as a subcohort) and then includes all recorded incident cases of the relevant disease outcome, irrespective of subcohort membership. The nested case-control design on the other hand selects incident cases recorded in a prospective cohort and matches them to suitable controls from the same cohort (e.g., by sex and age). The controls may either be sampled from concurrent outcome-free participants (or survivors at the end of a defined study period) to each closely match the characteristics of an index case (i.e., individually matched) or simply selected from a predefined pool of suitable controls to match the characteristics of recorded cases (i.e., frequency matched). In other variations, disease cases and appropriate controls (e.g., of similar sex and age) are identified contemporaneously from the same setting (e.g., hospital, registry, or other health records). Inclusion of clinical trials in meta-analysis of observational studies often involves the non-intervention arms of the trial or stratification of the analyses by trial arm.

A key consideration when planning a meta-analysis is the choice of an appropriate measure of association as discussed in Chapter 3. Prospective studies allow the estimation of hazard ratios or incidence rate ratios, whereas case-control studies allow the estimation of odds ratios. The extent to which odds ratios may be regarded as approximating the same measure of relative risk as hazard ratios or rate ratios needs to be considered in light of the plausibility of the rare disease assumption (Greenland and Thomas, 1982).

A two-stage approach, where study-specific estimates are first obtained in stage 1 then combined in stage 2, is convenient for handling different study designs in meta-analysis and can allow the inclusion of studies with only aggregate data. For analysis of time-to-event data based on the Cox proportional hazards (PH) model (Cox, 1972), the hazard at time t after baseline for each of $i=1,\ldots,N$ studies, with strata $s=1,\ldots,S_i$ (e.g., $S_i=2$ for stratification by sex) and individuals $k=1,\ldots,K_i$ with exposure of interest E_{ik} and other covariates $\mathbf{X}_{ik}$ is:

$$\log h_{isk}(t \mid E_{ik}, \mathbf{X}_{ik}) = \log h_{0is}(t) + \beta_i E_{ik} + \gamma_i \mathbf{X}_{ik} \tag{21.1}$$

The evolution of risk over time is thus modeled independently for each stratum in each study, as represented by the non-parametric baseline hazards $h_{0is}(t)$. The β_i are the parameters of interest, being the log hazard ratios (HRs) per unit increase in the exposure in study i, adjusted for the confounding effects of the covariates $\mathbf{X}_{ik}$. The estimated log HRs are combined over studies using the random-effects meta-analysis model described in Chapter 5 (Higgins et al., 2009).

Inclusion of other study designs in the meta-analysis would involve specifying an appropriate association model instead of (21.1) above, such as weighted Cox regression models for case-cohort studies (Barlow et al., 1999; Prentice, 1986), and logistic regression models for case-control studies (Breslow et al., 1980):

$$\log\left(\frac{p_{ik}}{1-p_{ik}}\right) = \beta_{0i} + \beta_i E_{ik} + \gamma_i \mathbf{X}_{ik}. \tag{21.2}$$

Differences in results of different study designs can be further assessed by meta-regression as described in Section 21.4.

21.2.2 Checking Some Key Modeling Assumptions

The above specification of the meta-analysis model assumes the existence of a log-linear relationship between exposure and outcome, which should be initially checked by flexibly assessing the shape of the dose-response relationship, for example using methods described in Chapter 18. Figure 21.1 shows assessments of the dose-response association between C-reactive protein (CRP) and CHD outcome in the ERFC meta-analysis (Kaptoge et al., 2010). Adjusted study-specific log HRs by deciles were combined using multivariate random-effects meta-analysis and plotted against similarly combined geometric mean CRP levels.

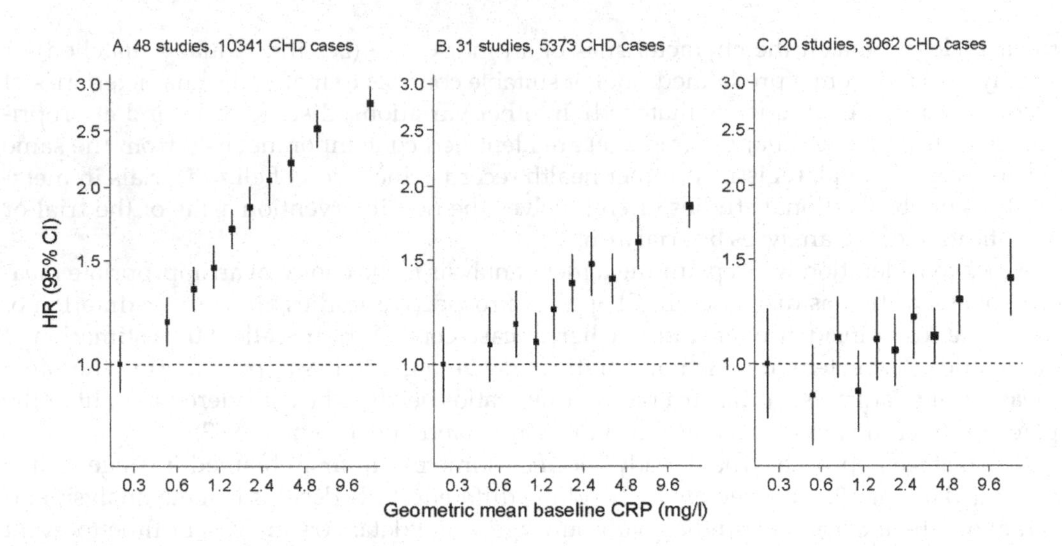

FIGURE 21.1
ERFC meta-analysis results on shape of dose-response association of baseline $\log_e$ CRP concentration and incident CHD with adjustment for (A) age, sex, and study; (B) age, sex, study, systolic blood pressure, smoking, history of diabetes, body mass index, concentrations of $\log_e$ triglycerides, non-HDL cholesterol, and HDL cholesterol, and alcohol consumption; and (C) preceding plus fibrinogen.

The results in Figure 21.1 provide good evidence for the existence of a log-linear association between $\log_e$ CRP and CHD outcome, both in minimally adjusted analyses and in further adjusted analyses. Furthermore, application of methods more appropriate for meta-analysis of continuous dose-response relationships using fractional polynomials (White et al., 2019) provides further evidence in favor of a log-linear relationship (Figure 21.2). The exact magnitude of association can therefore be quantified based on the meta-analysis of continuous log-linear associations with adjustment for confounding factors.

21.2.3 Adjusted Associations in the Presence of Systematically Missing Confounders

Exposure-disease associations estimated from observational data are subject to confounding by measured and unmeasured risk factors. Statistical adjustment of confounding factors measured *a priori* may help mitigate confounding bias to some extent. However, except for some well-recognized confounding factors in some circumstances (e.g., age, sex, smoking, diabetes, blood pressure, adiposity, and lipids for cardiovascular disease), the confounding factors available or considered by researchers would often vary by study (i.e., systematically missing).

IPD meta-analysis can better address the problem of systematically missing confounders across studies than literature-based meta-analysis. Analyses can be restricted to subsets of studies with complete sets of confounders measured and results compared (subset restriction, as illustrated in Table 21.1) (Fibrinogen Studies Collaboration, 2005; Kaptoge et al., 2010). Alternatively, modeling approaches may be utilized to borrow information across models with correlated parameters as further discussed in Section 21.4 (Fibrinogen Studies Collaboration, Jackson et al., 2009).

Table 21.1 summarizes the ERFC CRP-CHD meta-analysis results, with progressive adjustment for confounders. The pooled estimate has been scaled to correspond to a 1 SD higher exposure value, here 1.11 higher $\log_e$ CRP or approximately a 3-fold higher CRP. To

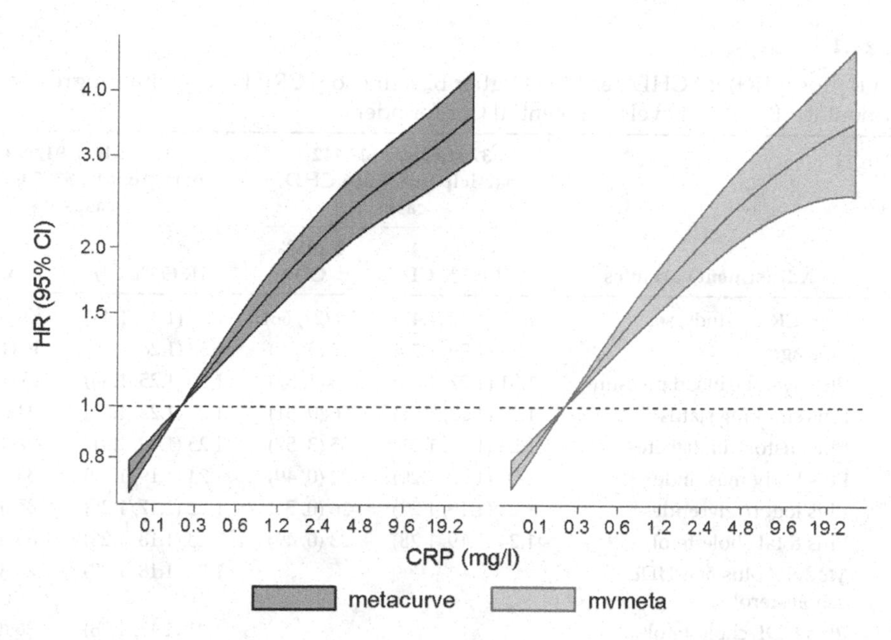

FIGURE 21.2
Random-effects meta-analysis of CRP-CHD association using fractional polynomials with adjustment for age, sex, and study. For details of the metacurve and mvmeta methods see White et al. (2019).

simply address systematically missing confounders, two sets of analyses have been conducted for comparison; one restricted to 37 studies contributing complete data for models 1 through 8 and another comprising data from 31 studies contributing complete data for models 1 through 11 (Table 21.1). Repeating the progressive adjustment in the restricted set of studies provides a basis for judging whether inferences are broadly comparable. In this particular example, the similarly adjusted HRs per 1 SD higher baseline $\log_e$ CRP were very similar in the 31 studies as in the 37 studies. Adjustment for high-density lipoprotein (HDL) and non-HDL cholesterol (instead of total cholesterol) plus alcohol consumption only slightly further attenuates the HR. Between-study heterogeneity, quantified by the I^2 statistic, was moderate and decreased with adjustment for confounders (Table 21.1).

21.2.4 Proportional Hazards

For valid inference using the Cox regression model, it is essential that the proportional hazards (PH) assumption is met, meaning that the regression coefficients in model (21.1) do not change with time since baseline measurement. While the effect of any covariate measured at baseline may plausibly decrease over time, the prime interest is whether the PH assumption is appropriate for the exposure of interest. This can be evaluated in each study separately by including an interaction between the exposure and time, or by the commonly used diagnostic based on Schoenfeld residuals (Collett, 1994; Thompson et al., 2010). The independent χ_1^2 statistics can be summed across the N studies, yielding a χ_N^2 statistic testing the hypothesis that PH holds in each study. This approach is, however, not a powerful test against the plausible alternative hypothesis that HRs tend to decline over time in all studies. A better method is to meta-analyze the interaction terms between the exposure and time over studies, analogous to investigating effect modification as further described below (Thompson et al., 2010).

TABLE 21.1

Hazard Ratios (HRs) for CHD per 1 SD Higher Baseline $\log_e$ CRP Levels with Progressive Adjustment for Baseline Levels of Potential Confounders

		37 studies, 109,742 participants, 8056 CHD cases		31 studies, 91,990 participants, 5373 CHD cases	
Model	**Adjustment variables**	**HR (95% CI)**	**I^2 (95% CI)**	**HR (95% CI)**	**I^2 (95% CI)**
1	$\log_e$ CRP | study, sex	1.37 (1.32, 1.43)	50 (27, 66)	1.37 (1.31, 1.44)	53 (30, 69)
2	Plus age	1.34 (1.29, 1.39)	39 (9, 59)	1.33 (1.28, 1.39)	46 (18, 65)
3	Plus systolic blood pressure	1.31 (1.27, 1.36)	36 (4, 57)	1.30 (1.25, 1.36)	43 (12, 63)
4	Plus smoking status	1.27 (1.23, 1.32)	26 (0, 51)	1.27 (1.22, 1.32)	34 (0, 58)
5	Plus history of diabetes	1.26 (1.22, 1.31)	35 (3, 57)	1.25 (1.20, 1.31)	43 (13, 63)
6	Plus body mass index	1.24 (1.20, 1.29)	22 (0, 49)	1.24 (1.19, 1.29)	33 (0, 57)
7	Plus $\log_e$ triglycerides	1.23 (1.19, 1.27)	26 (0, 51)	1.22 (1.17, 1.28)	35 (0, 58)
8	Plus total cholesterol	1.24 (1.19, 1.28)	23 (0, 49)	1.23 (1.18, 1.28)	33 (0, 57)
9	Model 7 plus non-HDL cholesterol	–	–	1.22 (1.18, 1.27)	26 (0, 53)
10	Plus HDL cholesterol	–	–	1.21 (1.17, 1.26)	26 (0, 52)
11	Plus alcohol consumption	–	–	1.21 (1.16, 1.26)	30 (0, 55)

21.3 Investigating Effect Modification

An important advantage of IPD is that it provides the opportunity for systematic investigation of the exposure-risk relationship at different levels of other variables, just as exploration of treatment effect modification in randomized trials was shown to be an advantage of IPD in Chapter 8. This evaluation of factors that modify the overall log HRs estimated above involves assessing their interactions with the exposure of interest, which can be performed using either within-study information (e.g., for individual level effect modifiers such as age) or between-study information (e.g., for study level effect modifiers such as study population or laboratory methods). Using between-study evidence for assessing individual level effect modifiers is subject to aggregation bias and often gives different results from the within-studies evidence, as discussed in Chapter 8.

To assess individual-level effect modifiers using within-study information (Kaptoge et al., 2010; Thompson et al., 2010) a two-step procedure is again adopted, first estimating the interaction in each study separately. For example, for a single potential effect modifier X_{ik}, the model in study i is:

$$\log h_{isk}(t \mid E_{ik}, X_{ik}) = \log h_{0is}(t) + \beta_i E_{ik} + \gamma_i X_{ik} + \delta_{Wi} E_{ik} X_{ik} \tag{21.3}$$

For testing the PH assumption, X_{ik} is replaced by time t or a function such as log(t). The estimates of the interaction terms δ_{Wi} are combined using random-effects meta-analysis as described in Chapter 4. The overall interaction, δ_W say, is then based on only within-study information (Chapter 8). Model (21.3) can be extended by including adjustments for other confounders, and indeed their interactions with the exposure of interest; this enables investigation of whether, as is possible, a particular interaction is confounded by other main effects or interactions.

To assess study-level effect modifiers, the evidence relies entirely on between-study comparisons using random-effects meta-regression (Chapter 7) (Thompson and Sharp, 1999). Using the estimates of β_i from (21.1), the meta-analysis model is extended to include a study level covariate X_i by writing:

$$\begin{aligned} \hat{\beta}_i &= \beta_i + \varepsilon_i; \qquad \text{where } \varepsilon_i \sim N\left(0, s_i^2\right) \\ \beta_i &= \beta + \delta_B X_i + \eta_i; \quad \text{where } \eta_i \sim N\left(0, \tau^2\right) \end{aligned} \tag{21.4}$$

and where δ_B is the between-study interaction term, with statistical significance assessed allowing for the residual between-study heterogeneity τ^2.

A few variables, notably sex and ethnic group, have potential interactions for which both within-study and between-study information may be important. For example, studies involving both men and women provide within-study information on sex interactions, while studies involving members of one sex alone can only be used to assess interactions across studies. In this case, the within-study interaction δ_w is estimated as in model (21.3) based on studies of both sexes, and the between-study interaction δ_B is estimated using model (21.4) in which X_i is the proportion of women in each study. Provided they are similar, these two asymptotically independent estimates of interaction can themselves be combined. As between-study information on interactions is prone to numerous potential sources of aggregation bias (Thompson and Higgins, 2002), there is a trade-off between increased precision and possible bias in choosing whether to use between-study information in addition to within-study information (Riley et al., 2008a; Simmonds and Higgins, 2007; Jackson et al., 2006).

Presenting interactions is not easy. For a binary variable identifying two subgroups, the exponent of the interaction term is a ratio of HRs, but it is simpler to present two separate meta-analyses, one in each subgroup. However, because within-studies and between-studies evidence may disagree, the (multivariate) meta-analytic weighting of study-specific subgroup estimates is different from the weighting of study-specific interactions. So neither the estimates nor the confidence intervals of the subgroup-specific estimates are necessarily compatible with the estimate and confidence interval of the interaction term. In practice, this problem is not usually severe. For categorical variables, consistency can better be achieved by conducting multivariate meta-analysis of the main effect term(s) $\boldsymbol{\beta}_s$ and interaction term(s) $\boldsymbol{\delta}_s$ and then deriving subgroup-specific estimates as linear combinations of relevant meta-analysis coefficients, although this approach also uses between-studies evidence. For example, for a binary variable such as sex with males coded as reference, the exponent of β will be the HR in males and exponent of $\beta + \delta$ will be the HR in females. For continuous variables, the exponent of the interaction term is a ratio of HRs per unit increase in the effect modifier. Similarly, for presentation, it is easier to present the HR estimates according to study-specific quantile groups (e.g., thirds or fifths) of the effect modifier distribution with use of multivariate meta-analysis to obtain the subgroup-specific HRs.

Figure 21.3 shows HRs for CHD per 1 SD higher baseline $\log_e$ CRP according to subgroups of characteristics that are largely individual-level (smoking, age, BMI), study-level (study design, sample type), or both (sex). For the continuous variables, age and BMI, the HRs have been estimated with categorization into three groups, but the p-value for interaction comes from meta-analysis of their continuous interactions with $\log_e$ CRP. The subgroup-specific estimates have been estimated as a linear combination of relevant coefficients following multivariate meta-analysis of the main effect and interaction terms as described above. For categorical variables such as sex and smoking, the p-value for interaction derives from a 1

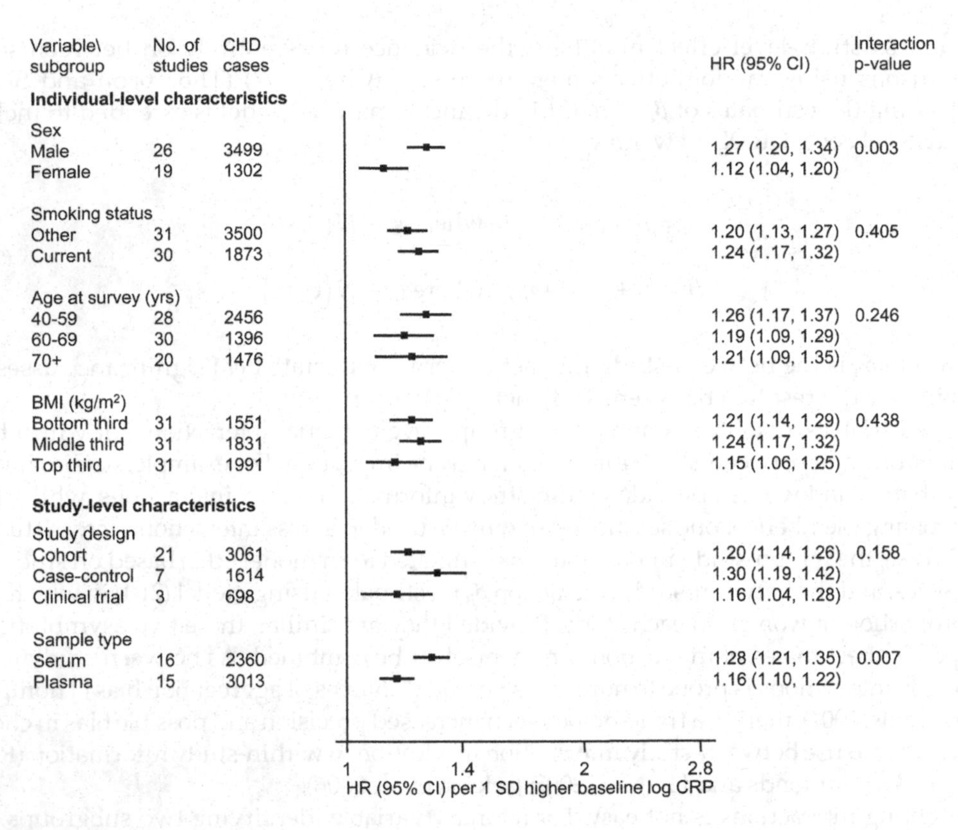

Variable\ subgroup
No. of studies
CHD cases
HR (95% CI)
Interaction p-value
Individual-level characteristics
Sex
Male 26 3499 1.27 (1.20, 1.34) 0.003
Female 19 1302 1.12 (1.04, 1.20)
Smoking status
Other 31 3500 1.20 (1.13, 1.27) 0.405
Current 30 1873 1.24 (1.17, 1.32)
Age at survey (yrs)
40-59 28 2456 1.26 (1.17, 1.37) 0.246
60-69 30 1396 1.19 (1.09, 1.29)
70+ 20 1476 1.21 (1.09, 1.35)
BMI (kg/m2)
Bottom third 31 1551 1.21 (1.14, 1.29) 0.438
Middle third 31 1831 1.24 (1.17, 1.32)
Top third 31 1991 1.15 (1.06, 1.25)
Study-level characteristics
Study design
Cohort 21 3061 1.20 (1.14, 1.26) 0.158
Case-control 7 1614 1.30 (1.19, 1.42)
Clinical trial 3 698 1.16 (1.04, 1.28)
Sample type
Serum 16 2360 1.28 (1.21, 1.35) 0.007
Plasma 15 3013 1.16 (1.10, 1.22)
1
1.4
2
2.8
HR (95% CI) per 1 SD higher baseline log CRP

FIGURE 21.3
Hazard ratios for CHD per 1 SD higher baseline $\log_e$ CRP according to subgroups of largely individual (sex, smoking, age, BMI) and study-level (study design, sample type) characteristics.

df Wald test of the pooled interaction term. For the variable sex, which is also a study-level characteristic, the multivariate meta-analysis approach gave similar results as the alternative approach described above that involves first deriving a between-study interaction estimate by meta-regression of study-specific log HRs on the proportion of females (Figure 21.4), followed by meta-analysis to combine with estimated within-study interactions from studies involving both sexes (Figure 21.5). In this particular example, there was consistency between the two types of interactions (both indicating about 12% lower HR in females per 1 SD higher baseline $\log_e$ CRP), with the within-study interactions contributing much of the evidence in the random-effects meta-analysis (i.e., 82% total weight) (Figure 21.6). For study-level characteristics such as study design and sample type, the inferences have been based on random-effects meta-regression model, relying solely on between-study differences.

21.4 Handling Studies with Different Sets of Confounders by Modeling Approaches

The subset restriction approach to handling systematically missing confounders as in Table 21.1, while simple, has limitations, including data-driven selection of subsets of confounder

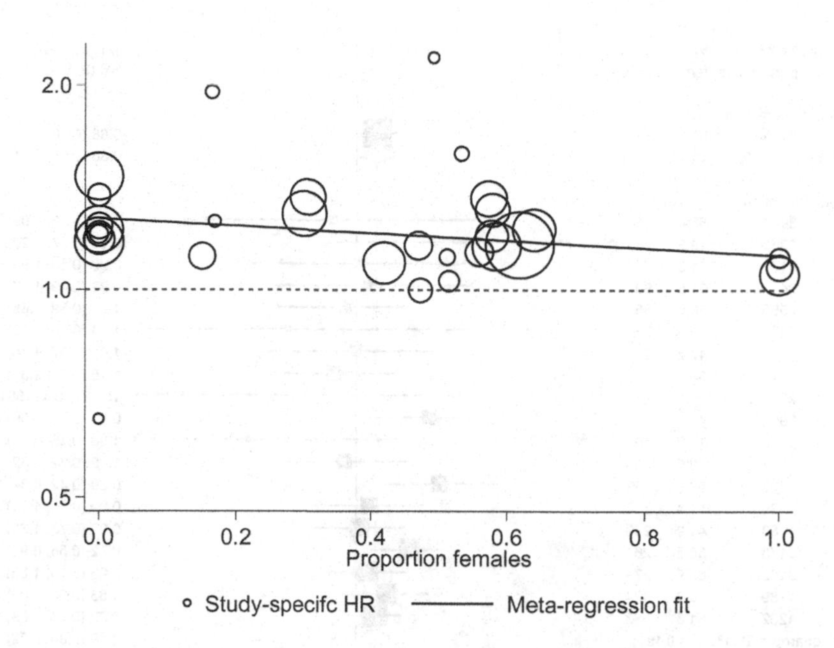

FIGURE 21.4
Inference of between-study interaction for sex by meta-regression of study-specific log HR estimates on proportion of females within studies.

models, and potential selection biases, should reasons for the systematically missing data be related to quality of the studies. Jackson et al. (2009) proposed a more formal modeling framework (discussed in Chapter 9 from a multivariate meta-analysis perspective) to address the problem of systematically missing confounders in the IPD meta-analysis context. Observing that studies provide either partial or full information on confounders, Jackson et al. proposed use of standard bivariate random-effects meta-analysis as a joint model for analysis of the fully and partially adjusted coefficients. Studies that provide full details of all potential confounders can be used to estimate both fully and partially adjusted coefficients, and ascertain the nature of the association between the estimates, while those that provide only a subset of confounders can be used only to estimate partially adjusted coefficients. The bivariate model enables simultaneous combination of both partially and fully adjusted coefficients, while to some extent "borrowing of strength" between them (Riley et al., 2007) depending on the degree of within- and between-study correlations. In particular, the combined fully adjusted coefficient borrows strength from the studies that only provide partially adjusted coefficients.

A two-stage approach is adopted under the proportional hazards modeling framework. At the first stage, partially and (where possible) fully adjusted coefficients are estimated from each study, together with their standard errors; a key issue is estimating the "within-study" correlation of the two estimates. At the second stage, the results are combined in a bivariate meta-analysis. For a particular study, denoting the complete covariates as $\boldsymbol{X}_{ik}$ and the systematically missing covariates as $\boldsymbol{Z}_{ik}$, with other notation as in (21.1), two proportional hazards regression models can be fitted to estimate fully (superscript f) and partially adjusted (superscript p) coefficients:

$$\log h^f_{isk}(t \mid E_{ik}, \boldsymbol{X}_{ik}, \boldsymbol{Z}_{ik}) = \log h^f_{0is}(t) + \beta^f_i E_{ik} + \gamma^f_i \boldsymbol{X}_{ik} + \varphi^f_i \boldsymbol{Z}_{ik} \tag{21.5}$$

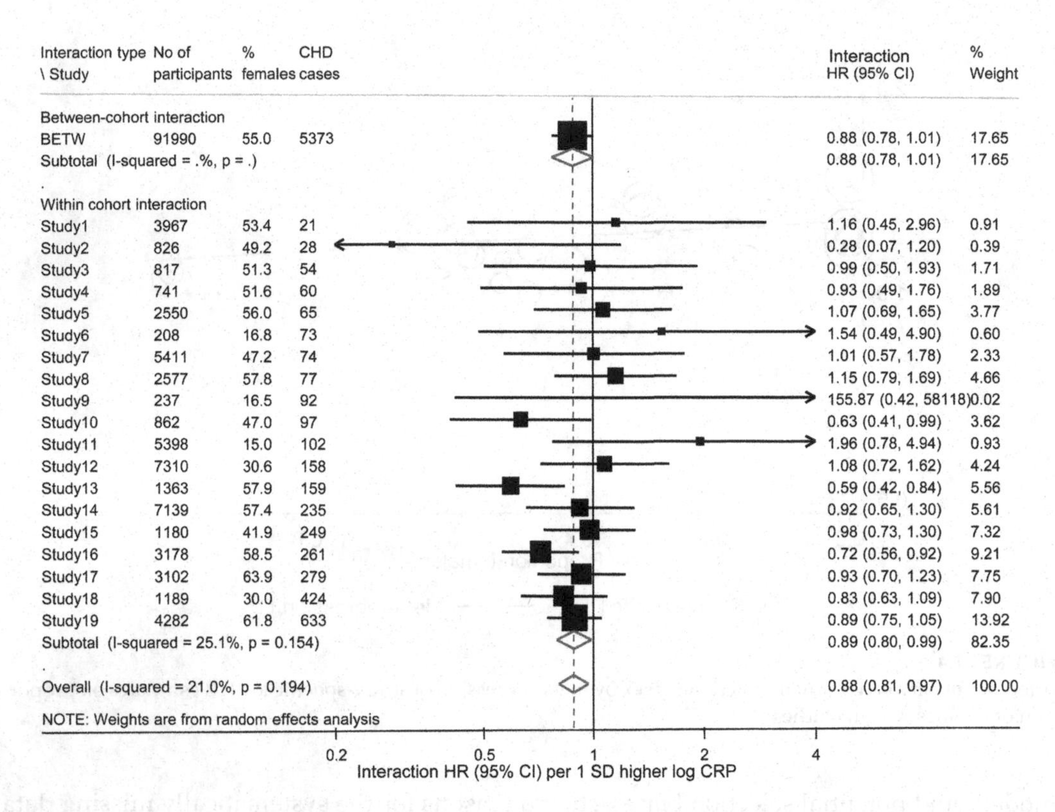

Interaction type \ Study
No of participants
% females
CHD cases
Interaction HR (95% CI)
% Weight
Between-cohort interaction
BETW 91990 55.0 5373 0.88 (0.78, 1.01) 17.65
Subtotal (I-squared = .%, p = .) 0.88 (0.78, 1.01) 17.65
Within cohort interaction
Study1 3967 53.4 21 1.16 (0.45, 2.96) 0.91
Study2 826 49.2 28 0.28 (0.07, 1.20) 0.39
Study3 817 51.3 54 0.99 (0.50, 1.93) 1.71
Study4 741 51.6 60 0.93 (0.49, 1.76) 1.89
Study5 2550 56.0 65 1.07 (0.69, 1.65) 3.77
Study6 208 16.8 73 1.54 (0.49, 4.90) 0.60
Study7 5411 47.2 74 1.01 (0.57, 1.78) 2.33
Study8 2577 57.8 77 1.15 (0.79, 1.69) 4.66
Study9 237 16.5 92 155.87 (0.42, 58118)0.02
Study10 862 47.0 97 0.63 (0.41, 0.99) 3.62
Study11 5398 15.0 102 1.96 (0.78, 4.94) 0.93
Study12 7310 30.6 158 1.08 (0.72, 1.62) 4.24
Study13 1363 57.9 159 0.59 (0.42, 0.84) 5.56
Study14 7139 57.4 235 0.92 (0.65, 1.30) 5.61
Study15 1180 41.9 249 0.98 (0.73, 1.30) 7.32
Study16 3178 58.5 261 0.72 (0.56, 0.92) 9.21
Study17 3102 63.9 279 0.93 (0.70, 1.23) 7.75
Study18 1189 30.0 424 0.83 (0.63, 1.09) 7.90
Study19 4282 61.8 633 0.89 (0.75, 1.05) 13.92
Subtotal (I-squared = 25.1%, p = 0.154) 0.89 (0.80, 0.99) 82.35
Overall (I-squared = 21.0%, p = 0.194) 0.88 (0.81, 0.97) 100.00
NOTE: Weights are from random effects analysis
0.2
0.5
1
2
4
Interaction HR (95% CI) per 1 SD higher log CRP

FIGURE 21.5
Meta-analysis of between- and within-study interaction estimates for sex.

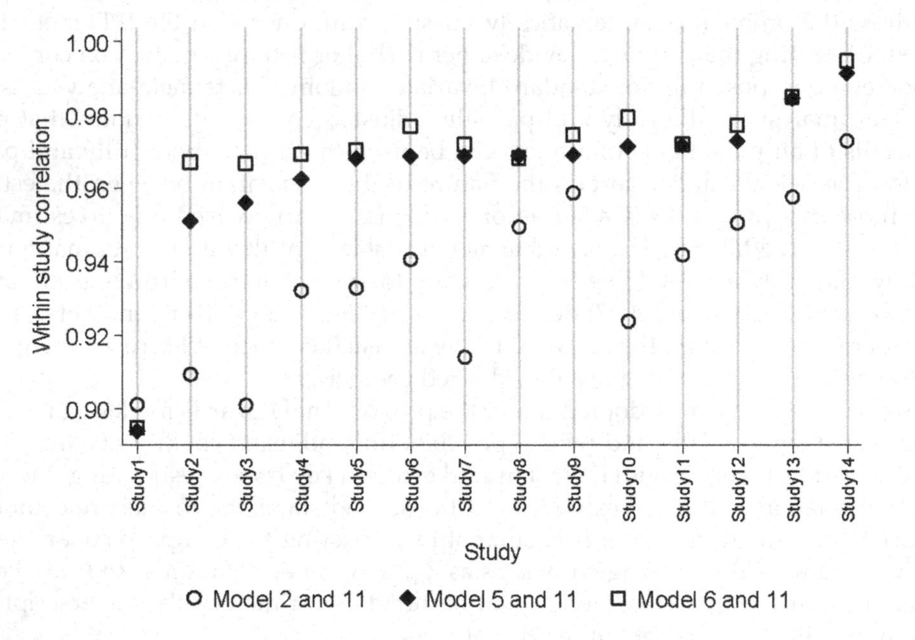

FIGURE 21.6
Estimated within-study correlations of log hazard ratios per 1 g/L higher baseline fibrinogen in pairs of adjusted models. The adjustment model index numbers are as listed in Table 21.1 and the correlations were estimated using the record stacking method.

$$\log h_{isk}^{p}(t \mid E_{ik}, \boldsymbol{X}_{ik}) = \log h_{0is}^{p}(t) + \beta_i^p E_{ik} + \gamma_i^p \boldsymbol{X}_{ik} \tag{21.6}$$

Of particular interest is the scalar regression coefficient for the primary exposure variable E_{ik} in the fully and partially adjusted model, $\hat{\beta}_i^f$ and $\hat{\beta}_i^p$, respectively; the other coefficients representing confounder associations. The standard bivariate random-effects meta-analysis model can be used for inference (Chapter 9), where the two outcomes are the fully and partially adjusted regression coefficients of the exposure of interest.

The key challenge is obtaining estimates of the within-study correlations of the fully and partially adjusted association estimates. Chapter 9 discussed alternative approaches, including non-parametric bootstrap and the alternative bivariate random-effects meta-analysis model that does not require knowledge of the within-study correlations (Riley et al., 2008b). An additional approach for estimating the within-study correlations is the "record stacking" approach. The pairwise correlations are estimated from the observed data by stacking the relevant model variables, such that each participant contributing to both the partially and fully adjusted models has two records in the stacked dataset, and then contextual variables coding for associations within each stack are generated (i.e., the contextual variables equal the original variable values within a stack and are zero elsewhere). The proportional hazards regression model is then fitted to the stacked data for each study using the contextual variables as the independent variables, and additionally stratifying by the variable indicating the stacked records (i.e., adjustment subset). A robust covariance matrix is estimated to account for non-independence of the stacked records (Lin and Wei, 1989).

In the 31 studies contributing data to the FSC fibrinogen-CHD meta-analysis (Table 21.2) (Fibrinogen Studies Collaboration, 2005; Kaptoge et al., 2007), data were complete for

TABLE 21.2

Summary of Confounder-Adjustment Models Considered in the Assessment of Fibrinogen-CHD Associations in the FSC

Model No.	Adjustment variables	No of studies	No of participants[a]	CHD cases	% missing data within study, median (5th, 95th) percentiles
1	Study, sex	31	154,012	7110	0 (0, 0)
2	Plus age	31	154,012	7110	0 (0, 0)
3	Plus smoking status	31	154,012	7110	0 (0, 0)
4	Plus systolic blood pressure	31	154,012	7110	0 (0, 0)
5	Plus body mass index	31	154,012	7110	0 (0, 0)
6	Plus total cholesterol	31	154,012	7110	0 (0, 0)
7	Model 5 plus HDL cholesterol	24	113,808	4073	0 (0, 31)
8	Plus LDL cholesterol	20	96,267	3327	1 (0, 5)
9	Plus alcohol consumption	20	95,123	3262	3 (0, 18)
10	Plus $\log_e$ triglycerides	16	81,116	2639	3 (0, 20)
11	Plus history of diabetes	14	71,918	2528	4 (0, 38)

[a] One hundred and ninety-nine participants from the original 154,211 withdrew their consent and have been removed from this analysis.

fibrinogen and some conventional cardiovascular risk factors (age, sex, smoking, systolic blood pressure, body mass index [BMI], total cholesterol) and were mostly studywise missing for other variables of interest (e.g., high- and low- density lipoprotein cholesterol, alcohol consumption, and history of diabetes) (Table 21.2). Note that the 11 models considered here differ slightly from those described in Table 21.1.

The FSC analyses used two subset restrictions; one comprising data from 31 studies contributing complete data for models 1 through 6 and another comprising data from 14 studies contributing complete data for models 1 through 11, with broadly similar results obtained for similarly adjusted models (Table 21.3).

For illustrative purposes, we consider models 2, 5, and 6 as partially adjusted models that may be used to inform estimation of the fully adjusted coefficients through bivariate models. Figure 21.6 shows within-study correlations estimated from the record stacking method in the 14 studies with complete data, suggesting high correlations that decrease with dissimilarity of models (i.e., adjusted covariates). The magnitude of correlations was very similar across the studies, as depicted in Figure 21.7 for model 5 and model 11 with study-specific 95% confidence regions shown.

Results of the bivariate random-effects meta-analysis model for systematically missing confounders, using partially adjusted model 5 to inform estimation of fully adjusted model 11 coefficients, are summarized in Table 21.4. Inferences from the alternative methods for estimating within-study correlations were broadly similar in this particular example.

Different choices of the partially adjusted model, and extensions using two or three partially adjusted models in multivariate random-effects meta-analysis, are compared with a simple meta-analysis of the 14 fully adjusted coefficients in Table 21.5. Any choice of a partially adjusted model borrows strength and improves precision, but different choices gave similar benefits. The use of directly estimated study-specific correlations seemed to offer a somewhat greater degree of borrowing of strength than use of the Riley overall correlation model, not requiring knowledge of within-study correlations (Table 21.5).

TABLE 21.3

Hazard Ratios (HRs) for CHD per 1 g/L Higher Baseline Fibrinogen Levels When Using Subset Restriction Approach to Address Systematically Missing Confounders

		31 studies, 154,012 participants, 7110 CHD cases		14 studies, 71,918 participants, 2528 CHD cases	
Model	**Adjustment variables**	**HR (95% CI)**	**I^2 (95% CI)**	**HR (95% CI)**	**I^2 (95% CI)**
1	Study, sex	1.72 (1.60, 1.86)	73 (62, 81)	1.66 (1.45, 1.90)	80 (67, 88)
2	Plus age	1.57 (1.47, 1.67)	65 (49, 76)	1.55 (1.38, 1.74)	72 (52, 84)
3	Plus smoking status	1.46 (1.38, 1.56)	57 (36, 71)	1.47 (1.32, 1.63)	64 (36, 80)
4	Plus systolic blood pressure	1.43 (1.35, 1.51)	50 (24, 67)	1.44 (1.30, 1.59)	58 (24, 77)
5	Plus body mass index	1.41 (1.33, 1.48)	42 (12, 62)	1.42 (1.30, 1.57)	54 (16, 75)
6	Plus total cholesterol	1.38 (1.31, 1.45)	36 (2, 59)	1.40 (1.28, 1.52)	45 (0, 70)
7	Model 5 plus HDL cholesterol	–	–	1.37 (1.25, 1.49)	47 (1, 71)
8	Plus LDL cholesterol	–	–	1.33 (1.23, 1.43)	28 (0, 62)
9	Plus alcohol consumption	–	–	1.32 (1.23, 1.42)	21 (0, 58)
10	Plus $\log_e$ triglycerides	–	–	1.32 (1.22, 1.43)	31 (0, 64)
11	Plus history of diabetes	–	–	1.31 (1.22, 1.41)	26 (0, 61)

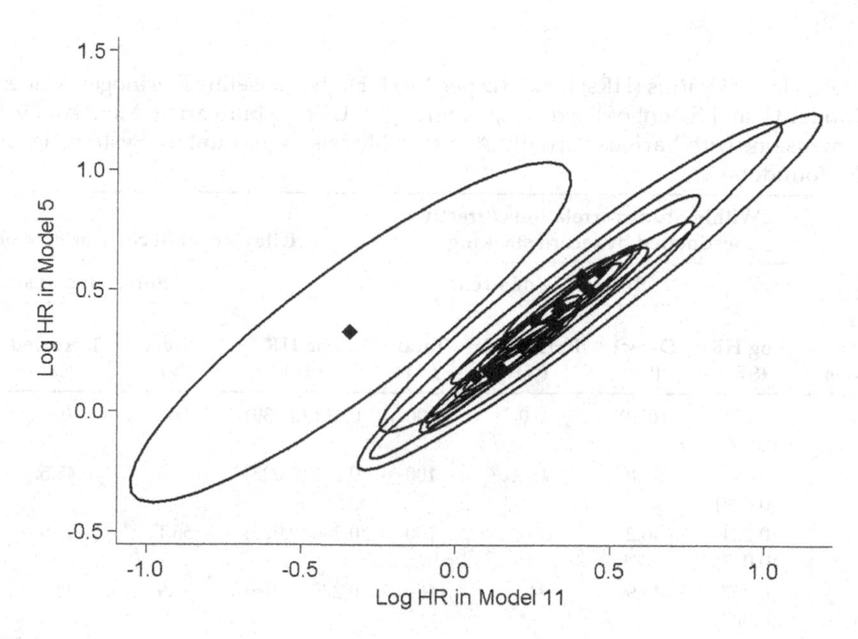

FIGURE 21.7
Scatter plots of study-specific log hazard ratios per 1 g/L higher baseline fibrinogen in models 5 and 11, with 95% confidence regions.

TABLE 21.4

Estimated Log Hazard Ratios for CHD per 1 g/L Higher Baseline Fibrinogen Values and between Study Variances in Fully Adjusted Model 11 (β_f and τ_f^2) and Partially Adjusted Model 5 (β_p and τ_p^2), Using Bivariate-Meta-Analysis to Account for Systematically Missing Confounders

Correlation method	β_f (SE)	τ_f^2 (SE)	β_p (SE)	τ_p^2 (SE)
Bootstrap	0.271 (0.026)	0.005 (0.004)	0.346 (0.030)	0.011 (0.006)
Analytic	0.275 (0.027)	0.006 (0.004)	0.358 (0.031)	0.013 (0.006)
Modified analytic	0.272 (0.027)	0.005 (0.004)	0.350 (0.030)	0.011 (0.006)
Record stacking	0.274 (0.027)	0.006 (0.004)	0.349 (0.029)	0.010 (0.006)
Riley overall ρ	0.266 (0.026)	0.002 (N/A)	0.344 (0.026)	0.006 (N/A)

21.5 Dealing with Measurement Error in Meta-Analysis of Observational Studies

21.5.1 Measurement Error in Epidemiological Analyses

In epidemiological studies, exposures and/or covariates are often measured with technical error and/or are subject to fluctuations within individuals, leading to biased estimates of associations in analyses using only single observed measurements. We class sources of variability together as "within-person variability" which is often quantified by the regression dilution ratio (RDR) (Carroll et al., 2006; Rosner et al., 1990, 1992). Various methods have been proposed to correct for the effect of within-person variability in single studies,

TABLE 21.5

Estimated Log Hazard Ratios (HRs) for CHD per 1 g/L Higher Baseline Fibrinogen Values in Fully Adjusted Model 11, and Extent of Borrowing of Strength, Using Multivariate Meta-Analysis of Varying Dimensions with Various Partially Adjusted Models to Account for Systematically Missing Confounders

Partially adjusted model coefficient(s) included in multivariate meta-analysis	Within-study correlations directly estimated by record stacking				Riley's overall correlation model			
		Borrowing of strength				Borrowing of strength		
	Log HR (SE)	Direct (%)	Borrowed (%)	Total (%)	Log HR (SE)	Direct (%)	Borrowed (%)	Total (%)
None	0.271 (0.039)	100.0	0.0	100	0.271 (0.039)	100.0	0.0	100
2	0.284 (0.028)	51.8	48.2	100	0.282 (0.025)	54.5	45.5	100
5	0.274 (0.027)	46.2	53.8	100	0.266 (0.026)	56.1	43.9	100
6	0.267 (0.027)	45.9	54.1	100	0.275 (0.016)	56.9	43.1	100
2 5	0.277 (0.027)	45.3	54.7	100	0.279 (0.027)	52.4	47.6	100
2 6	0.272 (0.027)	45.5	54.5	100	0.282 (0.027)	52.5	47.5	100
5 6	0.274 (0.027)	45.9	54.1	100	0.278 (0.029)	55.0	45.0	100
2 5 6	0.277 (0.028)	45.2	54.8	100	0.285 (0.029)	51.8	48.2	100

and increasingly in the context of IPD meta-analysis. All methods require additional information about the error-prone risk factors, such as repeat measures on all or some individuals or true measures on some individuals. Ideally, this information comes from individuals internal to the study, although if necessary, it can come from external data. In the context of meta-analysis, the information is likely to come from a subset of individuals in a subset of studies. Early work (Lewington et al., 2002; MacMahon et al., 1990) focused on methods for correcting for univariate measurement error in meta-analyses. The Prospective Studies Collaboration estimated a usual risk factor level from a non-linear time-dependent regression calibration model that pooled all studies (Lewington et al., 2002), whereas the FSC estimated study- and time-specific RDRs, which were then combined allowing for within- and between-study heterogeneity (Fibrinogen Studies Collaboration, Wood et al., 2006). Both ignored error in confounders. More recently, a general multivariate regression calibration technique was extended to a meta-analysis framework (Wood et al., 2009) and is described below.

21.5.2 Regression Calibration with Application to the Cox Model

We will assume a Cox proportional hazards disease model as described by equations (21.1) and (21.2) but will denote $\boldsymbol{E}_{ik} = (E_{ik1}, E_{ik2}, \ldots, E_{ikP})$ as an unobserved vector of *usual (long-term average) levels* of P risk factors. We denote $\boldsymbol{W}_{ik0}$ as a vector of *observed* measurements of the corresponding baseline risk factors $\boldsymbol{E}_{ik}$, available for all individuals in all studies and

$W_{ik1},\ldots, W_{ikR,}$ to be a set of vectors containing repeat measures r = 1,…,R of the *observed* risk factors available for some or all individuals in some or all studies. Assuming (i) model 21.1 is correct, (ii) E_{ik} is not available for all individuals, (iii) W_{ik0} and W_{ikr} are unbiased measures of the true E_{ik}, and (iv) the errors in W_{ik0} and W_{ikr} are independent of each other and of the true value (i.e., non-differential errors), then approximately unbiased estimators of β_i may be obtained from a Cox regression using conditional expectations $E[W_{ikr} \mid W_{ik0}]$ in place of E_{ik}. These conditional expectations are estimated from a regression calibration model.

21.5.3 Regression Calibration Model for Meta-Analysis of Multiple Studies

To construct a regression calibration model for estimating $E[W_{ikr} \mid W_{ik0}]$, we assume data are available from some or all of the N studies, with study i having all K_i individuals providing baseline measurements and up to K_i individuals providing up to R_i repeat measures on the P risk factors. For simplicity, we assume the error distribution in the risk factors is time-independent, although the models may be extended to incorporate time trends (Lewington et al., 2002). Let the baseline and the rth repeat measurement for the pth risk factor in study i be denoted by W_{ik0p} and W_{ikrp} for (k = 1,…, K_i, r = 1,…, R_i), respectively. A general linear multivariate RC model, initially ignoring non-error-prone covariates for ease of exposition, takes the form:

$$W_{ikrp} = a_{irp} + b_{ip1}W_{ik01} + b_{ip2}W_{ik02} + \cdots + b_{ipP}W_{ik0P} + z_{ikp} + e_{ikrp}$$
$$\text{for } i = 1,\ldots,N,\ K = 1,\ldots,K_i,\ r = 1,\ldots,R_i,\ p = 1,\ldots,P \tag{21.7}$$

where $(e_{ikr1},\ldots,e_{ikrP})^T \sim MVN(\mathbf{0}, \boldsymbol{\Sigma})$ and $(z_{ik1},\ldots,z_{ikP})^T \sim MVN(\mathbf{0},\boldsymbol{\Phi})$ for all r = 1,…, R_i, and $\boldsymbol{\Sigma}$ and $\boldsymbol{\Phi}$ are $(P \times P)$ variance-covariance matrices denoting residual error and individual-specific variation, respectively. Non-error-prone confounders, such as baseline age and sex , can easily be incorporated in equation (21.7). For example, if Age_{ik0} denotes baseline age for individual k in study i, the term $\gamma_p Age_{ik0}$ or $\gamma_{ip} Age_{ik0}$ can be added to the right-hand side of equation (21.7). Note that repeat measurements are regressed on the baseline measurements rather than vice-versa to ensure applicability to individuals and studies with only baseline measurements. If true measures rather than repeat measures were available for some studies or some individuals, then W_{ikrp} could be replaced by E_{ikp} in model (21.7).

If sufficiently many true or repeat measures are available in all studies, then $\boldsymbol{\Sigma}$ and $\boldsymbol{\Phi}$ may be replaced by study-specific matrices $\boldsymbol{\Sigma}_i$ and $\boldsymbol{\Phi}_i$, reducing model (21.7) to N independent multivariate models. Regression calibration techniques can then be applied as standard at the study level. However, usually only a subset of studies has data available on repeat measurements and it is necessary to transfer the information to the other studies. Two techniques are available to combine the regression coefficients across studies: (i) averaged regression coefficients and (ii) empirical Bayes regression coefficients, both described briefly below.

(i) *Averaged regression coefficients*

An averaged set of coefficients from equation (21.7) can be estimated using the further set of meta-analysis models;

$$\hat{b}_{ipq} = b_{pq} + u_{ipq} + v_{irpq} \text{ for } p,\quad q = 1,\ldots,P \tag{21.8}$$

where the study random-effect terms are $u_i=(u_{i11}, u_{i12}, \ldots, u_{i1P}, u_{i21}, \ldots, u_{iPP}) \sim$ MVN($\mathbf{0}$,$\boldsymbol{\Psi}$) and $\boldsymbol{\Psi}$ is a $P^2 \times P^2$ variance-covariance matrix, and where arranging the random terms as a vector $\mathbf{v}_i=(v_{i11}, v_{i12}, \ldots, v_{i1P}, v_{i21}, \ldots, v_{iPP})$, then $\mathbf{v}_i \sim \mathrm{N}(\mathbf{0},\boldsymbol{\Omega})$ where $\boldsymbol{\Omega}$ is a $P^2 \times P^2$ variance-covariance matrix. The estimates of b_{pq} are then the averaged regression coefficients, and b_{pp} is the pooled RDR for the pth error-prone risk factor. When considerable heterogeneity in the regression coefficients (b_{ipq}) exists, averaged regression coefficients may not provide an appropriate correction (although they are likely to be far better than no correction).

(ii) *Empirical Bayes regression coefficients*

The fully empirical Bayes regression coefficients which incorporate the individual and study random effects can be extracted from (21.7) and (21.8) as:

$$\hat{b}_{pq} + \hat{u}_{ipq} + \hat{v}_{irpq} \quad \text{for } p,q = 1,\ldots,P$$

where estimates $\hat{u}_{ipq}$ and $\hat{v}_{irpq}$ are best linear unbiased predictors of the random effects obtained as described in Goldstein (2003). These estimates are "optimally" weighted averages that combine information derived from the individual and/or study with the mean for all similar groups. The advantage of these estimates is that they reflect the between-individual and/or between-study heterogeneity for individuals and studies with repeat measures and provide averaged estimates for individuals and studies without repeat measures.

The study-specific, averaged or empirical Bayes regression coefficients can be used to construct the correction matrix (CM) with the diagonal component $\hat{b}_{pp}$ representing the estimated adjusted RDR for the pth risk factor (adjusted for all other risk factors included in the RC model). Thus, denoting $\hat{\boldsymbol{\beta}}^{un}$ as the vector of estimated uncorrected risk associations (e.g., log HRs) with estimated variance-covariance matrix $\hat{v}^{un} = \mathrm{var}\ (\hat{\boldsymbol{\beta}}^{un})$ then corrected risk associations and corresponding variance-covariance matrix are estimated as $\hat{\boldsymbol{\beta}} = \hat{\boldsymbol{\beta}}^{un}$ $\mathbf{CM}^{-1}$ and $\hat{v} = \mathbf{CM}^{-1}\hat{v}^{un}\mathbf{CM}^{-\mathrm{T}}$, respectively. Alternatively, the study-specific, averaged, or empirical Bayes regression coefficients can be used to estimate the conditional expectations of the usual risk factors (i.e., $\mathrm{E}[\boldsymbol{W}_{ikr} \mid \boldsymbol{W}_{ik0}]$, or $\mathrm{E}[\boldsymbol{E}_{ik} \mid \boldsymbol{W}_{ik0}]$) which can be directly used in the disease model to provide the estimates $\boldsymbol{\beta}_s$ in equation (21.1), before meta-analysis as in equation (21.2).

21.5.4 Example: Emerging Risk Factors Collaboration

The ERFC (Kaptoge et al., 2010) collected baseline and repeat information on CRP, systolic blood pressure, BMI, $\log_e$ triglycerides, and total cholesterol from 160,309 participants from 54 studies (Table 21.6). Repeat information on CRP was available for a subset of 22,124 participants from 14 studies. Participants who provided repeat measurements tended to be younger ($p = 0.002$), were less likely to be smokers ($p = 0.02$) or be diagnosed with diabetes ($p = 0.002$), and had lower systolic blood pressure ($p = 0.009$) than participants without repeat data in the same studies.

The extent of within-person variability in $\log_e$ CRP, systolic blood pressure, BMI, $\log_e$ triglycerides, and total cholesterol was quantified by the adjusted regression dilution ratio pooled over studies. The pooled regression dilution ratio of $\log_e$ CRP was 0.58 (0.52–0.63), indicating a year-to-year consistency that was broadly similar to those for

TABLE 21.6

Summary of Repeat Information and Adjusted Regression Dilutions Ratios for $\log_e$ CRP, Systolic Blood Pressure, BMI, $\log_e$ Triglycerides and Total Cholesterol among 160,309 Participants from 54 Studies

Variable	No. of studies with repeats	No. of individuals with repeats	No. of repeats	Pooled RDR (adjusted for age, sex, $\log_e$ CRP, systolic blood pressure, smoking status, BMI, $\log_e$ triglycerides, and total cholesterol)
Log_e CRP	14	22,124	24,222	0.58 (0.52, 0.63)
Systolic blood pressure	22	40,976	108,712	0.54 (0.51, 0.57)
BMI	19	35,875	102,699	0.97 (0.95, 0.98)
Log_e triglycerides	20	43,055	89,801	0.65 (0.61, 0.69)
Total cholesterol	20	44,405	95,145	0.60 (0.56, 0.64)

systolic blood pressure (0.54, 0.51–0.57) and total cholesterol (0.60, 0.56–0.64) (Table 21.7). Study- and resurvey-specific RDRs were also examined and are shown for CRP in Figure 21.8.

Cox proportional hazards models were used to calculate HRs in relation to $\log_e$ CRP, adjusted for age, sex, systolic blood pressure, smoking status, BMI, $\log_e$ triglycerides, and total cholesterol. Analyses involved a two-stage approach with estimates of association calculated separately within each study before pooling across studies by random-effects meta-analysis. HRs were calculated on baseline levels and error-corrected usual levels of CRP and confounders by use of conditional expectations predicted from regression calibration models (equation (21.7)).

Without correction for regression dilution bias, the age- and sex-adjusted HR (95% CI) of CHD per 1 SD higher $\log_e$ CRP levels was 1.34 (1.29, 1.39) and became 1.24 (1.19, 1.28) after further adjustment for systolic blood pressure, smoking, history of diabetes, body mass index, triglycerides, and total cholesterol (Table 21.7). In analyses that corrected for within-person variability in CRP only, the fully adjusted HR was 1.46 (1.37, 1.56), whereas in analyses that further corrected for within-person variability in all error-prone confounders, the fully adjusted HRs were attenuated to 1.42 (1.33, 1.51). This is because residual confounding caused by within-person variability in baseline confounders has been corrected for.

TABLE 21.7

Hazard Ratios (HRs) for CHD per 1 SD Higher Baseline and Usual $\log_e$ CRP Levels (~1.11 $\log_e$ mg/l) with Adjustment for Baseline and Usual Levels of Potential Confounders

Adjustment variables	HR (95% CI) Baseline $\log_e$ CRP levels Baseline confounders	HR (95% CI) Usual $\log_e$ CRP levels Baseline confounders	HR (95% CI) Usual $\log_e$ CRP levels Usual confounders
37 studies, 109,742 participants, 8056 CHD cases			
Age, sex, study	1.34 (1.29, 1.39)	1.64 (1.54, 1.75)	1.64 (1.54, 1.75)
Plus systolic blood pressure	1.31 (1.27, 1.36)	1.59 (1.49, 1.69)	1.57 (1.47, 1.67)
Plus smoking, history of diabetes	1.26 (1.22, 1.31)	1.49 (1.40, 1.59)	1.45 (1.36, 1.55)
Plus BMI, $\log_e$ triglycerides, total cholesterol	1.24 (1.19, 1.28)	1.46 (1.37, 1.56)	1.42 (1.33, 1.51)

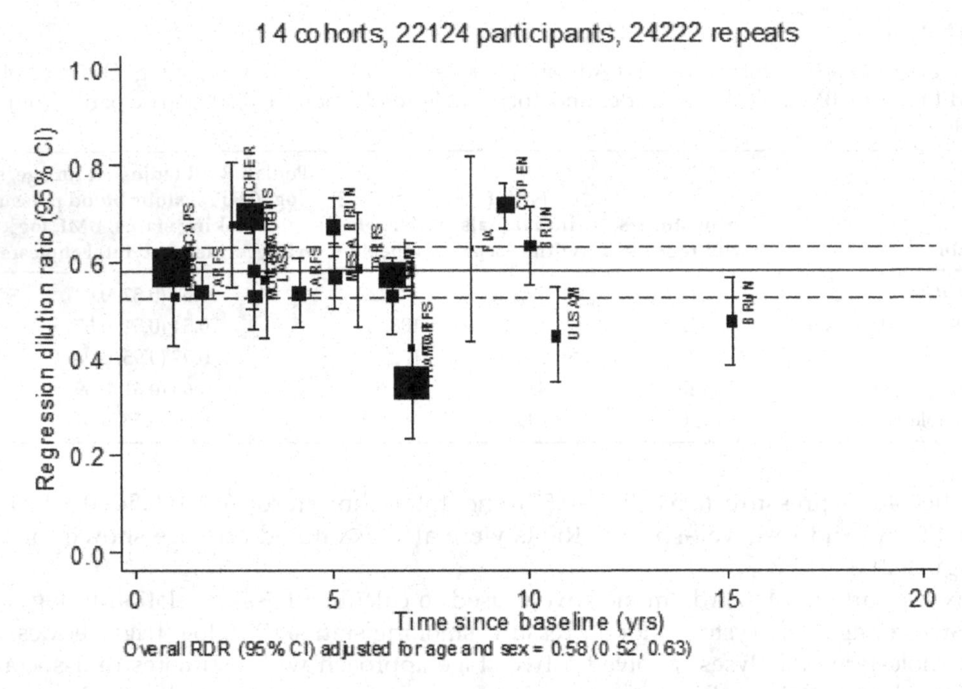

FIGURE 21.8
Regression dilution ratios for $\log_e$ CRP concentration plotted against time since baseline measurement by study. The sizes of the markers are inversely proportional to the variance of the regression dilution ratio. CI indicates confidence interval. This figure was published in *The Lancet*, 375, The Emerging Risk Factors Collaboration, C-reactive protein concentration and risk of coronary heart disease, stroke, and mortality: an individual participant meta-analysis, WebAppendix page 16, Copyright Elsevier (2010).

21.6 Dealing with Missing Data in Meta-Analysis of Observational Studies

21.6.1 Problems Caused by Missing Data in Multiple Epidemiological Studies

Missing data present specific challenges in a meta-analysis context. Specifically, the degree and nature of missing data may differ between studies, and this extra heterogeneity requires special consideration. For example, a complete-case analysis can be unbiased in a single study under the missing at random assumption, but different degrees of missing data between several studies will lead to changes in study weights contributing to an inverse-variance weighted meta-analysis (Section 21.4) and hence different (but not necessarily biased) meta-analysis estimates in comparison to meta-analyzing the unobserved full data.

Missing data may occur for a number of different reasons both within and outside the control of investigators. Factors within their control tend to lead to systematic patterns of missing data, such as data on a variable being missing for an entire study (Section 21.3). Factors outside their control tend to lead to sporadically missing data, where data on a variable are missing for a few individuals with no clear pattern to the missingness. In this section, we focus on sporadic missing data in IPD meta-analyses. Specifically, we focus on the multiple imputation strategy under a MAR assumption, which is increasingly used in applied research due to recent software development. We consider two main issues: whether imputation should be within or across studies, and the order for applying the multiple imputation rules and inverse-variance weighted meta-analysis.

21.6.2 Multiple Imputation and Meta-Analysis

In multiple imputation, missing values are imputed several times by drawing random values from the conditional distribution of the missing values according to a specified imputation model using observed data values in order to form a completed dataset. The parameter estimates and standard errors from each of these imputed datasets are combined using Rubin's rules (Schafer, 1997). Under the MAR assumption, a multiple imputation analysis can be efficient (include all relevant information) and give unbiased estimates. For clarity, we consider imputation of a single incomplete variable, using a regression model as the imputation model; similar ideas apply with multiple incomplete variables.

It is always important to consider the compatibility of the imputation model and the meta-analysis model. Imputation and analysis models are compatible if a joint model exists under which both models are conditionals (Arnold et al., 2001). In practice, to achieve compatibility, it is recommended that the imputation model should not have stronger assumptions than the analysis model (Schafer, 1997). In IPD meta-analysis data, a separate imputation model could be fitted in each study. Here, Rubin's rules should be applied within each study before meta-analyzing the estimates across studies using a common-effect or random-effects meta-analysis model. Alternatively, an imputation model could be fitted across all studies simultaneously. In this case, the study intercepts could be fixed effects or follow a random-effects model, and the covariate effects could be common across studies (a common-covariate-effect imputation model) or follow a random-effects model (a random-covariate-effects imputation model). A random-covariate-effects imputation model is compatible with both common-effect and random-effects meta-analysis, whereas a common-covariate-effect imputation model is only compatible with a common-effect meta-analysis. Here, meta-analysis should be applied to each imputed dataset before applying Rubin's rules on the pooled estimates (Burgess et al., 2013). Table 21.8 summarizes the properties of a multiple imputation meta-analysis in common settings.

21.6.3 Example

We illustrate a multiple imputation meta-analysis using data from the ERFC on 53,723 participants from ten studies to explore the association of low-density lipoprotein cholesterol (LDL-C, units mmol/L) with systolic blood pressure (SBP, units mmHg) using body mass index (BMI, units kg/m^2) as a covariate. Participants are those with complete data on SBP and BMI in studies with data on LDL-C. Missing data on LDL-C was introduced for 20% of participants in each study by discarding observations completely at random. We report complete-data, complete-case, and imputation analyses, using a common-covariate-effects imputation model and using within-study imputation. A random-covariate-effects imputation model was not considered as it cannot easily be implemented in currently available software packages. Common-effect and random-effects meta-analysis methods were used. For each imputation model, 50 imputed datasets were generated.

Results are given in Table 21.9. There was considerable heterogeneity between the studies, with $I^2=77\%$ (95% CI: 58%, 87%) in the complete-data meta-analysis. Compared with the complete-case analysis, we would expect the imputation analyses to bring the point estimate closer to the complete-data estimate and reduce its standard error (but not lower

TABLE 21.8

Summary of Desirable Properties of a Multiple Imputation Meta-Analysis for Common Combinations of Imputation Models and Inverse-Variance Weighted Meta-Analyses

Quality	Separate imputation model fitted in each study		Imputation model fitted simultaneously across all studies			
			Common-covariate-effects imputation model		Random-covariate-effects imputation model	
	Possess quality	Comment	Possess quality	Comments	Possess quality	Comments
Compatible with random-effects meta-analysis	✓	Rubin's rules should be applied before meta-analysis.	×	Stronger assumptions in imputation model relative to analysis models.	✓	Meta-analysis should be applied before Rubin's rules.
Compatible with common-effects meta-analysis	✓	Rubin's rules should be applied before meta-analysis. Weaker assumption in imputation model relative to analysis model.	✓	Meta-analysis should be applied before Rubin's rules.	✓	Meta-analysis should be applied before Rubin's rules. Weaker assumptions in imputation model relative to analysis model.
Can handle sporadic missing data	✓		✓		✓	
Can handle systematic missing data	×		✓		✓	
Can handle multivariable missing data	✓		✓		✓	Difficult in practice
Maintains inter-variable correlations between studies	✓		×		✓	
Maintains between-study heterogeneity	×	May induce between-study heterogeneity (even when there is no heterogeneity in the original data)	×	Induces a dependence between imputed values in different studies. May weaken between-study heterogeneity	✓	
Efficient	×	Does not maximize the potential of the multiple-study framework, and may be disadvantageous for smaller studies.	✓		✓	
Easy to implement in available software	✓		✓		×	

TABLE 21.9

Regression Coefficients for the Association of Low-Density Lipoprotein Cholesterol (mmol/L) with Systol ic Blood Pressure (mmHg) Adjusting for Body Mass Index (kg/m^2) from Complete-Data, Complete-Case, and Multiple Imputation Analyses with Common-Covariate-Effects and Separate Study Imputation Models Using Common-Effect and Random-Effects Meta-Analysis Models

Analysis method	Common-effect meta-analysis	Random-effects meta-analysis
Complete-data	1.084 (0.069)	1.189 (0.225)
Complete-case	1.105 (0.078)	1.166 (0.231)
Multiple imputation		
Common-covariate-effects imputation model	1.093 (0.081)	1.278 (0.220)
Separate imputation model in each study	1.110 (0.078)	1.177 (0.226)

Estimates with standard error (in brackets). With the common-covariate-effects imputation model, meta-analysis of imputed datasets was performed prior to combining estimates using Rubin's rules. With separate imputation models, Rubin's rules were applied within each study prior to meta-analysis.

than that from the complete-data analysis). This occurs only for the random-effects meta-analysis using the separate imputation models fitted in each study. Concerningly, using the common-covariate-effects imputation model and the random-effects analysis model, the apparent precision of the multiple imputation analysis is greater than that of the complete-data analysis. This is because the common-covariate-effects imputation model has ignored the between-study heterogeneity when imputing the values and thus weakened the between-study heterogeneity.

21.7 Discussion

With the increasing number of multi-study collaborations and consortia in epidemiological research, combining observational results across studies using meta-analysis is becoming ever more common. This brings with it new challenges regarding the often disparate designs and data available in each study as well as computational challenges in IPD meta-analysis with large datasets.

A number of approaches aimed at addressing these challenges have been discussed in this chapter. Each comes with its own underlying assumptions that researchers should be aware of and fully disclose when reporting results. To help in this matter, guidelines on the transparent presentation of Meta-analysis of Observational Studies in Epidemiology (MOOSE) (Stroup et al., 2000) have been drawn up and these should be followed along with more detailed reporting of statistical methods in non-standard analyses.

Software for implementation of IPD meta-analyses is now readily available. For example, in Stata, the commands metan, metaan, and ipdmetan can be used to perform two-stage inverse-variance IPD univariate meta-analysis (Fisher, 2015), while multivariate IPD meta-analysis can be performed using the commands mvmeta_make and mvmeta. Stata programs developed in the course of ERFC/FSC meta-analyses (e.g., mvmetaipd, mvshape) have been made available online (Kaptoge S, www.phpc.cam.ac.uk/ceu/research/erfc/stata/) to handle studies with a number of different designs (i.e., cohort, case-cohort, or case-control) and systematically missing confounders. The latter commands can also automatically restrict data to participants with complete data on a maximum adjustment model, thus allowing the same individuals and studies to be used in a set of model comparisons and handling measurement error. Several user-written software packages are now available in R for performing multilevel multiple imputation (e.g., mitml, micemd, pan (Grund et al., 2016)), although it remains difficult to implement these within a two-stage meta-analysis framework.

We have primarily considered the application of frequentist methods in the meta-analysis of observational data, due to their popularity and availability of software. Meta-analysis models can also be formulated and estimated using Bayesian methods. In the context of large-scale IPD meta-analysis of observational studies such as ERFC with multiple confounding factors, the specification of a full Bayesian hierarchical model is a formidable task, and estimation via Markov chain Monte Carlo (MCMC) methods is also likely to be computationally expensive.

We thank investigators and participants of the several studies that contributed data to the ERFC. The ERFC coordinating center was underpinned by program grants from the British Heart Foundation and the UK Medical Research Council (MRC), with project-specific support received from the National Institute for Health Research (NIHR; Cambridge Biomedical Research Centre at the Cambridge University Hospitals NHS Foundation Trust), British United Provident Association UK Foundation, and an unrestricted educational grant from GlaxoSmithKline. A variety of funding sources have supported recruitment, follow-up, and laboratory measurements in the studies contributing data to the ERFC, which are listed on the ERFC website (www.phpc.cam.ac.uk/ceu/erfc/list-of-studies/). The views expressed are those of the authors and not necessarily those of the NHS, the NIHR, or the Department of Health and Social Care.

References

Arnold BC, Castillo E and Sarabia JM, 2001. Conditionally specified distributions: An introduction. *Statistical Science* **16**(3): 249–265.

Barlow WE, Ichikawa L, Rosner D and Izumi S, 1999. Analysis of case-cohort designs. *Journal of Clinical Epidemiology* **52**(12): 1165–1172.

Breslow NE, Day NE and International Agency for Research on Cancer, 1980. Statistical methods in cancer research. Volume I: The analysis of case-control studies. *Statistical Methods in Cancer Research* **1**: 346.

Burgess S, White IR, Resche-Rigon M and Wood AM, 2013. Combining multiple imputation and meta-analysis with individual participant data. *Statistics in Medicine* **32**(26): 4499–4514.

Carroll RJ, Ruppert D, Stefanski LA and Crainiceanu CM, 2006. *Measurement Error in Nonlinear Models: A Modern Perspective*. Boca Raton, FL: CRC Press.

Collett D, 1994. *Modelling Survival Data in Medical Research*. London: Chapman & Hall.

Cox DR, 1972. Regression models and life tables. *Journal of the Royal Statistical Society: Series B* **34**(2): 187–220.

Danesh J, Erqou S, Walker M, et al., 2007. The Emerging Risk Factors Collaboration: Analysis of individual data on lipid, inflammatory and other markers in over 1.1 million participants in 104 prospective studies of cardiovascular diseases. *European Journal of Epidemiology* **22**(12): 839–869.

Di Angelantonio E, Sarwar N, Perry P, et al., 2009. Major lipids, apolipoproteins, and risk of vascular disease. *JAMA* **302**(18): 1993–2000.

Fibrinogen Studies Collaboration, 2004. Collaborative meta-analysis of prospective studies of plasma fibrinogen and cardiovascular disease. *The European Journal of Cardiovascular Prevention & Rehabilitation* **11**(1): 9–17.

Fibrinogen Studies Collaboration, Danesh J, Lewington S, et al., 2005. Plasma fibrinogen level and the risk of major cardiovascular diseases and nonvascular mortality: An individual participant meta-analysis. *JAMA* 294(14): 1799–1809.

Fibrinogen Studies Collaboration, Jackson D, White I, et al., 2009. Systematically missing confounders in individual participant data meta-analysis of observational cohort studies. *Statistics in Medicine* **28**(8): 1218–1237.

Fibrinogen Studies Collaboration, Wood AM, White I, Thompson SG, Lewington S and Danesh J, 2006. Regression dilution methods for meta-analysis: Assessing long-term variability in plasma fibrinogen among 27,247 adults in 15 prospective studies. *International Journal of Epidemiology* **35**(6): 1570–1578.

Fisher DJ, 2015. Two-stage individual participant data meta-analysis and generalized forest plots. *Stata Journal* **15**(2): 369–396.

Goldstein H, 2003. *Multilevel Statistical Models*. 3rd Edition. London: Edward Arnold.

Greenland S and Thomas DC, 1982. On the need for the rare disease assumption in case-control studies. *American Journal of Epidemiology* **116**(3): 547–553.

Grund S, Lüdtke O and Robitzsch A, 2016. Multiple imputation of multilevel missing data: An introduction to the R package pan. *SAGE Open* **6**(4): 1–17.

Higgins JPT, Thompson SG and Spiegelhalter DJ, 2009. A re-evaluation of random-effects meta-analysis. *Journal of the Royal Statistical Society. Series A* **172**(1): 137–159.

Jackson CH, Best N and Richardson S, 2006. Improving ecological inference using individual-level data. *Statistics in Medicine* 25(12): 2136–2159.

Kaptoge S, White IR, Thompson SG, et al., 2007. Associations of plasma fibrinogen levels with established cardiovascular disease risk factors, inflammatory markers, and other characteristics: Individual participant meta-analysis of 154,211 adults in 31 prospective studies. *American Journal of Epidemiology* **166**(8): 867–879.

Kaptoge S, Di Angelantonio E, Lowe G, et al., 2010. C-reactive protein concentration and risk of coronary heart disease, stroke, and mortality: An individual participant meta-analysis. *Lancet* **375**(9709): 132–140.

Lewington S, Clarke R, Qizilbash N, Peto R, Collins R and Prospective Studies Collaboration, 2002. Age-specific relevance of usual blood pressure to vascular mortality: A meta-analysis of individual data for one million adults in 61 prospective studies. *Lancet* **360**(9349): 1903–1913.

Lin DY and Wei LJ, 1989. The robust inference for the cox proportional hazards model. *Journal of the American Statistical Association* **84**(408): 1074–1078.

MacMahon S, Peto R, Collins R, et al., 1990. Blood pressure, stroke, and coronary heart disease. Part 1, prolonged differences in blood pressure: Prospective observational studies corrected for the regression dilution bias. *Lancet* **335**(8692): 765–774.

Prentice RL, 1986. A case-cohort design for epidemiologic cohort studies and disease prevention trials. *Biometrika* **73**(1): 1–11.

Riley RD, Abrams KR, Sutton AJ, Lambert PC and Thompson JR, 2007. Bivariate random-effects meta-analysis and the estimation of between-study correlation. *BMC Medical Research Methodology* **7**(1): 3.

Riley RD, Lambert PC, Staessen JA, et al., 2008a. Meta-analysis of continuous outcomes combining individual patient data and aggregate data. *Statistics in Medicine* **27**(11): 1870–1893.

Riley RD, Thompson JR and Abrams KR, 2008b. An alternative model for bivariate random-effects meta-analysis when the within-study correlations are unknown. *Biostatistics* **9**(1): 172–186.

Rosner B, Spiegelman D and Willett WC, 1990. Correction of logistic regression relative risk estimates and confidence internals for measurement error: The case of multiple ocvariates measured with error. *American Journal of Epidemiology* **132**(4): 734–745.

Rosner B, Spiegelman D and Willett WC, 1992. Correction of logistic regression relative risk estimates and confidence intervals for random within-person measurement error. *American Journal of Epidemiology* **136**(11): 1400–1413.

Schafer J, 1997. *Analysis of Incomplete Multivariate Data*. Chapman & Hall/CRC.
Simmonds MC and Higgins JPT, 2007. Covariate heterogeneity in meta-analysis: Criteria for deciding between meta-regression and individual patient data. *Statistics in Medicine* **26**(15): 2982–2999.
Stroup DF, Berlin JA, Morton SC, et al., 2000. Meta-analysis of observational studies in epidemiology: A proposal for reporting. Meta-analysis Of Observational Studies in Epidemiology (MOOSE) group. *JAMA* **283**(15): 2008–2012.
Thompson SG and Higgins JPT, 2002. How should meta-regression analyses be undertaken and interpreted? *Statistics in Medicine* **21**(11): 1559–1573.
Thompson SG and Sharp SJ, 1999. Explaining heterogeneity in meta-analysis: A comparison of methods. *Statistics in Medicine* **18**(20): 2693–2708.
Thompson S, Kaptoge S, White I et al., 2010. Statistical methods for the time-to-event analysis of individual participant data from multiple epidemiological studies. *International Journal of Epidemiology* **39**(5): 1345–1359.
White I, Kaptoge S, Royston P, Sauerbrei W and The Emerging Risk Factors Collaboration (ERFC), 2019. Meta-analysis of non-linear exposure-outcome relationships using individual participant data: A comparison of two methods. *Statistics in Medicine* **38**(3): 326–338.
Wood AM, Thompson SG and Kostis JB, 2009. Correcting for multivariate measurement error by regression calibration in meta-analyses of epidemiological studies. *Statistics in Medicine* **28**(7): 1067–1092.

22

Meta-Analysis of Prediction Models

Ewout Steyerberg, Daan Nieboer, Thomas Debray, and Hans van Houwelingen

CONTENTS

22.1 Introduction

Prediction models aim to provide estimates of absolute risk of an outcome (or endpoint). In medicine, common outcomes are the presence of a disease (establishing a diagnosis according to a reference standard; Chapter 19) or occurrence of a future event (prognosis, e.g., mortality within 30 days, at six months, or longer follow-up).

Prediction models are increasingly common in the medical literature. These models may inform patients and physicians by providing individualized risk predictions and support decision making. Predictions might be improved by combining information from different data sources or published models using meta-analysis techniques.

In this introduction, we describe commonly used prediction models for binary and survival outcomes, discuss considerations on improving prediction models to have broad validity, and give a set-up of this chapter.

22.1.1 Prediction Models

We focus on prediction models for binary outcomes. Let X denote the P-dimensional column vector of the observed predictor values (e.g., patient characteristics) and Y the observed outcome (0 = no event, 1 = event). A common approach to model the relation between X and Y is logistic regression. The corresponding probabilities, $p(x) = P(Y = 1 \mid X = x)$ are then related to x by adopting a logit link function:

$$\text{logit}\, p(x) = \alpha + x'\beta$$

where α is the model intercept (or constant) and β is a column vector of regression coefficients.

We also consider the Cox regression model for survival outcomes, where the hazard function is modeled as

$$h(t \mid x) = h_0(t)\exp(x'\beta).$$

The survival probability at time t is calculated as

$$S(t \mid x) = P(T > t \mid x) = \exp\{-H_0(t)\exp(x'\beta)\}$$

where $H_0(t) = \int_0^t h_0(s)ds$. We notice that the logistic regression model contains a constant α while the Cox model contains a baseline hazard $h_0(t)$ that plays the role of a generalized constant. Both α and $h_0(t)$ reflect the baseline risk in a prediction model. The regression coefficients β reflect the relative effects of the predictors x. The relative effect of the pth predictor is commonly expressed as $\exp(\beta_p)$, representing an adjusted odds ratio in logistic regression models, and an adjusted hazard ratio in Cox regression models (Harrell, 2001). For example, we may consider the strength of association for males versus females (a binary predictor) with an outcome, or the possibly non-linear shape of the association for age (a continuous predictor) with the outcome.

As estimates of α and β are approximately normally distributed within studies, the relative effect estimates from multiple studies may be studied with traditional meta-analysis (MA) approaches (see Chapter 21 and elsewhere). For prognostic models, however, the focus is on absolute risk estimates: the probability of an event $P(Y=1|x)$ or the survival probability $S(t|x)$. These depend on the joint effect of multiple predictors as well as the baseline risk per study. Hence, more advanced MA techniques are needed to apply evidence synthesis in the context of prediction models.

22.1.2 Evaluating Prediction Models

Prediction models are commonly developed in relatively small samples (the "development data") from a specific setting, for example, a single hospital. This often leads to decreased performance of the prediction model in new settings due to statistical overfitting at development (Van Houwelingen and Le Cessie, 1990). Common dimensions of performance include discrimination and calibration. Discrimination refers to the model's ability to distinguish between patients who do and do not develop the event of interest. It is usually measured using the concordance statistic, which ranges from 0.5 (no discrimination) to 1 (perfect discrimination) for sensible models. Calibration refers to the correspondence of model predictions, on average, to observed outcome frequencies. This can be quantified using several measures (Steyerberg, 2009).

Performance of a prediction model should ideally be determined using independent data, giving an unbiased assessment. We may then distinguish between the model's reproducibility and transportability (Justice et al., 1999; Debray et al., 2015). Reproducibility refers to the model's performance in the setting in which it was developed. Reproducibility can usually be assessed in the development data by applying cross-validation or bootstrapping methods. When the reproducibility is poor (e.g., substantial overfitting), improvements are possible by applying penalization techniques in the model development (Harrell, 2001; Steyerberg, 2009).

Whereas reproducibility relates to the model's performance in the original setting or population, transportability refers to the model's performance in populations different from the development population (Justice et al., 1999). Assessment of transportability requires consideration of data or results from different patient populations. Between-setting heterogeneity is then the focus of the evaluation. Improvement of the transportability of a prediction model requires knowledge of the medical context and data on the sources of between-setting heterogeneity.

22.1.3 Meta-Analysis for a Global Prediction Model

Improvements in model reproducibility are possible by considering meta-analysis techniques to increase the effective sample size and hence give better summary estimates of relative effects and absolute risks.

A complicating factor is that multiple different models may have been published for the same type of patients for similar outcomes, varying in the predictors that are included to make predictions (Mushkudiani et al., 2008). Differences in the categorizations of predictors may also occur, and different functional forms of continuous predictors may have been used in published prediction models. Hence, meta-analysis of such published models for relative or absolute risks is difficult.

Access to individual patient data (IPD) allows for consideration of common model specifications, with summary estimates and between-setting heterogeneity estimates for relative and absolute risks. MA techniques are essential here to assess the presence of between-study heterogeneity in baseline risk and predictor effects. If sufficiently homogeneous, data from various studies may be combined to provide a "global prediction model".

In the following, we start by supposing that we have access to IPD from multiple studies and consider whether we can develop a global prediction model with validity across multiple settings. We assume complete availability of the outcome Y and predictors X. A previously proposed framework is shown in Figure 22.1. We start by estimating study-specific predictor effects β_i in study $i=1,\ldots,I$. These may be pooled to provide common predictor effects β while allowing a study-specific baseline risk α_i. We may subsequently assess performance by study, and pool study-specific performance estimates ("validation", see Figure 22.1).

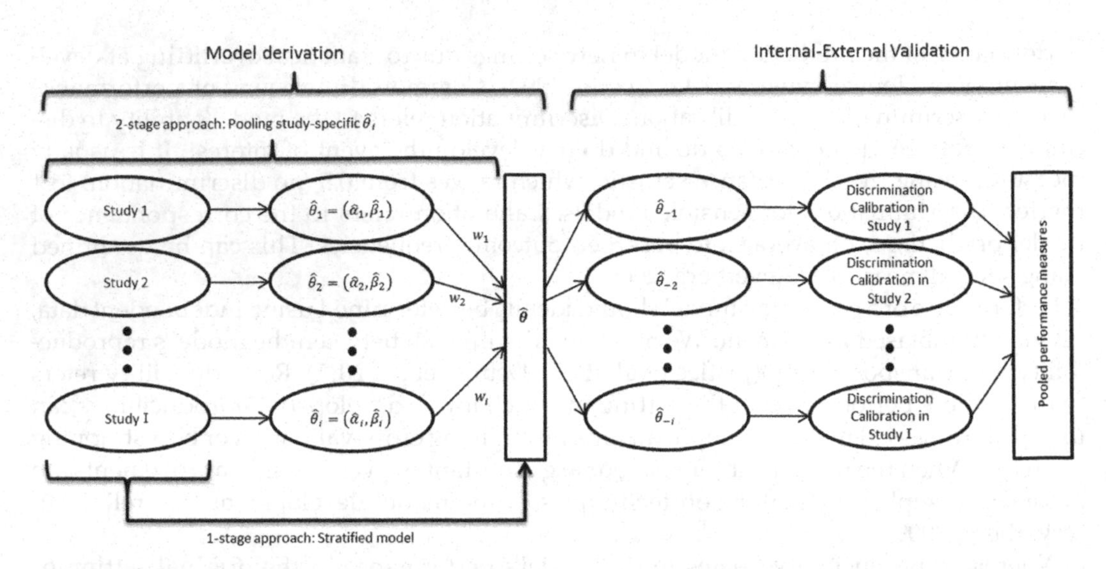

FIGURE 22.1
Schematic representation of development of a global prediction model ("model derivation") and validation ("assessment of predictive ability"). Model derivation considers stratified estimates of predictor effects β_i. A one-stage or two-stage approach may be followed to pool common effects for a global model. The performance of this global model with pooled estimated for β may be assessed in each study, with subsequent pooling of performance estimates θ_i if heterogeneity is considered acceptable. Figure based on Pennells et al. (2014).

22.1.4 Case Study

For illustrative purposes, we consider $I=15$ cohort studies of patients suffering from traumatic brain injury (TBI). These studies were included in the IMPACT project, where a total of 25 prognostic factors were considered for prediction of a six-month outcome (Murray et al., 2007). We here focus on mortality, which occurred in 20 to 40% of the included patients. Increasingly complex prediction models were developed, with three main predictors in a base model: age, motor score, and pupillary reactivity (Steyerberg et al., 2008). This base model was subsequently extended with results from computerized tomography (CT) scans and laboratory values, so that other predictors were hypoxia, hypotension, CT scan class, and traumatic subarachnoid hemorrhage (tSAH). Descriptive statistics of these studies are shown in Table 22.1. The membership C-statistic will be discussed in Section 22.2.2.

22.1.5 Chapter Set-Up

We first focus on the development of a *"global model" defined for the population of all studies*, by considering IPD from different studies, with stratification by study. We discuss preliminary steps, dealing with missing values using imputation techniques and data inspection issues (Section 22.2). We describe three models with various options to study heterogeneity in the joint effect of predictors as well as heterogeneity in the baseline risk (Section 22.3). We then describe methods using published data (Section 22.4). If the results from studies are reasonably homogeneous, the MA can produce a global prediction model that is valid across the population of studies. If the MA suggests that a global model is reasonable, validation in a specific setting may nevertheless be advised, with possible updating of the baseline risk and/or predictor effects. Strictly speaking, such validation and updating approaches are beyond the scope of the meta-analysis itself and will not be discussed here.

TABLE 22.1

Description of 15 IMPACT Datasets of 11,022 Patients with Traumatic Brain Injury

No.	Name	Recruitment	Type	Sample size	Mortality	Membership C-statistic
1	TINT	1991–1994	RCT	1118	25%	0.62
2	TIUS	1991–1994	RCT	1041	22%	0.65
3	SLIN	1994–1996	RCT	409	23%	0.76
4	SAP	1995–1997	RCT	919	23%	0.60
5	PEG	1993–1995	RCT	1510	24%	0.67
6	HIT I	1987–1989	RCT	350	28%	0.68
7	UK4	1986–1988	Survey	791	45%	0.64
8	TCDB	1984–1987	Survey	603	44%	0.67
9	SKB	1996–1996	RCT	126	27%	0.75
10	EBIC	1995–1995	Survey	822	34%	0.63
11	HIT II	1989–1991	RCT	819	23%	0.63
12	NABIS	1994–1998	RCT	385	26%	0.65
13	CSTAT	1996–1997	RCT	517	22%	0.61
14	PHARMO	2001–2004	RCT	856	17%	0.68
15	APOE	1996–1999	Survey	756	15%	0.73

Most of the models we describe may be fitted in either the frequentist or the Bayesian framework, and are usually fitted in the frequentist framework. We indicate where Bayesian methods are specifically needed to use informative priors.

22.2 Preliminary Steps

Before fitting prediction models, we first consider how we may deal with missing values within and between studies. We then consider how we may assess between-study heterogeneity with respect to predictor and outcome distribution, predictor effects, absolute risk estimates, and model performance.

22.2.1 Imputation of Missing Values

Patients with missing predictor or outcome values are often discarded during the development and validation of prediction models, leading to a complete case analysis. This approach is usually suboptimal as information is discarded and becomes particularly problematic when missing values occur for multiple predictors. When individual datasets are affected by missing values, several approaches can be followed (Chapter 21).

When dealing with IPD-MA, special attention is required to the potential for between-study heterogeneity in the correlation structure. If studies are quite homogeneous, a simple overall imputation model may suffice, with study as a main effect in the imputation model. The imputation model then only adjusts for heterogeneity in the levels and prevalence of missing predictors. A more refined method is to fully stratify by study, or to adopt a multilevel imputation model allowing between-study heterogeneity for all (or most) coefficients of the imputation model (Audigier et al., 2018; Burgess et al., 2013; Jolani et al., 2015).

A challenging situation occurs when a predictor is systematically missing in one or more studies, for example, when it has not been measured. It is then no longer feasible to fully stratify imputation models by study, whereas multilevel imputation models may suffer from estimation problems as there is no simple estimation of within-study variability. Advanced multilevel imputation models have recently been proposed to overcome technical estimation problems (Jolani, 2018). We note that such imputation of systematically missing values may only be reasonable if a global prediction model holds.

22.2.2 Between-Study Heterogeneity in Predictors and Outcomes

Before undertaking any formal meta-analysis, it is important to evaluate similarities and differences between the available datasets. This assessment gives insight into whether we are studying reproducibility or transportability of a prediction model (Debray et al., 2015). Differences in "case-mix" (the distribution of the predictors within a study) typically arise from differences in design and setting. For instance, for the 15 studies on TBI, four were performed in relatively unselected populations, with broad inclusion criteria (Type="Survey" in Table 22.1). Inclusion criteria were stricter for the 11 randomized controlled trials (Type="RCT"). The studies also vary in the calendar time of enrolment of patients. In line with these differences, we note substantial differences in the incidence of six-month mortality: this was 17% in the most recent trial (study 14) versus 42% in a survey (study 7).

The distributions of the predictors showed substantial variability, both in mean levels of predictors and distributions of the predictors within studies (Figure 22.2). We note broader distributions in the surveys compared with the trials, in line with their broader inclusion criteria.

Another summary measure of similarity between studies can be obtained from a membership model, where we aim to quantify how well we can distinguish which patient belongs to which study. The membership model is obtained by fitting a multinomial prediction model with membership of study as the outcome variable and including relevant predictors. For Table 22.1, we included only baseline characteristics to represent case-mix differences, while we recognize that the outcome Y could also be included. Differences between studies can be quantified by the membership (C-)statistic (Van Calster et al., 2012). Here, we repeatedly compare the model's ability to predict the probability of belonging with a specific study (Table 22.1). A high membership C-statistic for a study means that the baseline characteristics in that study are systematically different from the baseline characteristics in other studies. We note that studies 3, 9, and 15 seem to be somewhat different from the other studies (Table 22.1, C-statistics > 0.7).

22.3 Development of Prediction Models from IPD

22.3.1 Modeling the Data

After imputing missing values (Section 22.2.1) and describing between-study heterogeneity (Section 22.2.2), the next step is to explore predictor-outcome associations across the

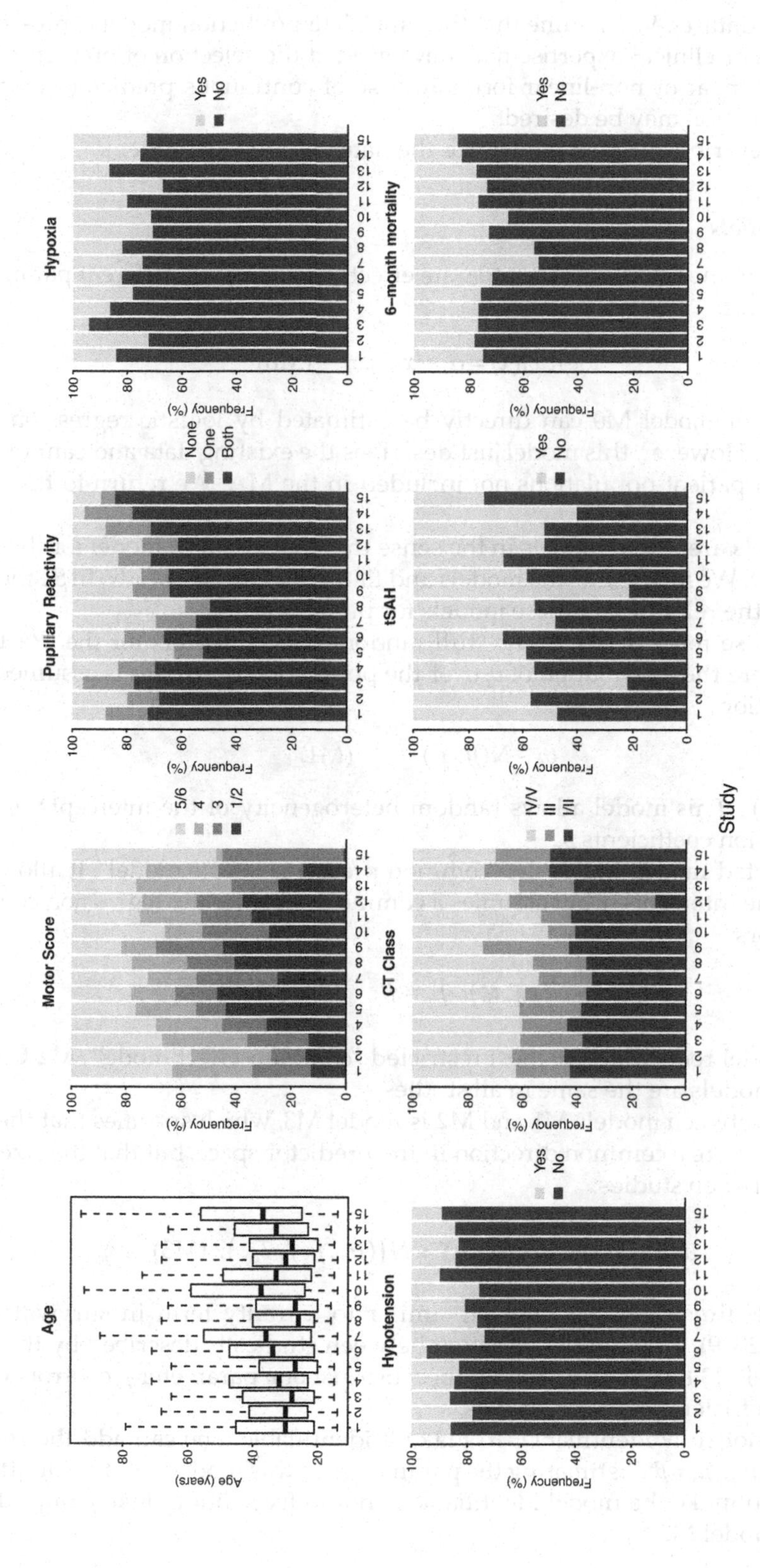

FIGURE 22.2
Distribution of eight patient characteristics and six-month mortality in 15 studies with 11,022 TBI patients.

pooled IPD-MA dataset. We assume that the candidate prediction model is pre-specified. Prior knowledge or clinical expertise may have guided the selection of predictors as well as the choice of linear or non-linear forms in case of continuous predictors. In practice, some form of selection may be desired.

We will first describe different models for the data and how to fit them.

22.3.1.1 The Models

The starting point, model M0, is the "separate effects model" with different parameters in each study $i=1,\ldots,I$:

$$\text{logit } p_i(x) = \alpha_i + x'\beta_i. \qquad \text{(M0)}$$

The parameters of model M0 can directly be estimated by logistic regression in each study separately. However, this model just describes the existing data and cannot be used for prediction in patient populations not included in the MA. We return to this issue in Section 22.3.2.

The next models are global models in the sense that they define a model for the population of all studies. We first define the models and fit them to the TBI study. In Section 22.3.2 we will discuss the relevance of these models for prediction.

The first of these models, M1, is the "full random-effects model" for the $(P+1)$-vector $\theta_i = (\alpha_i, \beta_{i'})'$, where the distribution of θ_i over the population of studies is assumed to be a normal distribution

$$\theta_i \sim N(\mu, \mathcal{T}) \qquad \text{(M1)}$$

with $\mu = (\mu_\alpha, \mu'_\beta)'$. This model allows random heterogeneity of the intercepts α_i and the vector of regression coefficients β_i.

A more restricted model, M2, is the "common predictor effects model". It allows a random effect for the intercepts α_i but assumes a common effect for the regression coefficients β of the predictors.

$$\alpha_i \sim N(\mu_\alpha, \tau_\alpha^2), \;\; \beta_i = \beta. \qquad \text{(M2)}$$

If $\tau_\alpha^2 = 0$, the model reduces to the most restricted "common-effect model", M2-CE, where the prediction models are the same in all studies.

Intermediate between models M1 and M2 is model M3, which specifies that the regression coefficients share a common direction in the predictor space, but that the size of their effects varies between studies:

$$\beta_i = \gamma_i \, \beta^* \text{ where } (\alpha_i, \gamma_i)' \sim N\left((\mu_\alpha, \mu_\gamma)', \mathcal{T}_{\alpha\gamma}\right). \;\; \text{(M3)}$$

Here γ_i is a one-dimensional parameter, similar to a frailty term in survival analysis (Legrand et al., 2009). The random variation between studies is described by the pair (α_i, γ_i). We call model M3 the "rank=1 model" because just one parameter γ_i controls variation in regression coefficients between studies.

Model M3 is not fully identifiable. To make it identifiable, one can add the restriction that μ_γ=1, and hence $\mu_\beta=\beta^*$. Estimating the parameters in this model is not straightforward. An easier way to make the model identifiable is not to fix μ_γ but to take $\beta=\beta_{M2}$, the point estimate from model M2.

22.3.1.2 Fitting the Different Models

Before fitting models M0–M3, we center the predictors with respect to the pooled means in the whole dataset. These means are given in the first row of Table 22.2 for the sake of completeness. The resulting study-specific estimates from fitting model M0 to the $I = 15$ studies are depicted in a forest plot (Figure 22.3) and shown in the top part of Table 22.2.

Model M1 can be fitted by standard two-stage multivariate meta-analysis (van Houwelingen et al., 2002) (Chapter 9) and/or one-stage generalized linear mixed modeling (GLMM) software. Several methods are available (Riley et al., 2015). To avoid identifiability problems, we require $I > P + 1$. The two-stage approach uses the parameter estimates θ_i and the within-study covariance matrices S_i from fitting model M0. A key benefit of having access to IPD is that we can readily estimate the whole of S_i, whereas the correlations in S_i are typically not available from published studies. The parameters of the whole model can be estimated from the marginal distribution of the estimated parameters

TABLE 22.2

Results for Models M0, M1, M2, and M3

Study	Intercept	Age	Motor score	Pupillary reactivity	Hypoxia	Hypo tension	CT class	tSAH
Pooled means								
		3.49	3.59	1.54	2.98	0.215	0.170	0.454
M0: separate effects								
1	−1.22 (0.09)	0.20 (0.05)	−0.39 (0.08)	0.41 (0.11)	0.36 (0.20)	1.03 (0.21)	0.56 (0.10)	1.01 (0.17)
2	−1.40 (0.10)	0.21 (0.07)	−0.40 (0.08)	0.36 (0.11)	0.46 (0.18)	0.75 (0.19)	0.34 (0.10)	0.74 (0.17)
3	−1.35 (0.22)	0.28 (0.09)	−0.28 (0.12)	0.71 (0.23)	−0.36 (0.58)	0.97 (0.35)	0.47 (0.15)	0.70 (0.37)
4	−1.34 (0.09)	0.20 (0.06)	−0.14 (0.07)	0.74 (0.11)	0.68 (0.24)	0.22 (0.23)	0.33 (0.10)	0.82 (0.18)
5	−1.73 (0.10)	0.21 (0.05)	−0.52 (0.06)	0.52 (0.08)	0.33 (0.16)	0.77 (0.17)	0.38 (0.08)	0.54 (0.14)
6	−1.41 (0.19)	0.30 (0.09)	−0.45 (0.13)	0.82 (0.17)	0.00 (0.38)	−0.60 (0.63)	0.38 (0.08)	0.95 (0.29)
7	−0.93 (0.11)	0.43 (0.05)	−0.30 (0.09)	1.01 (0.12)	0.07 (0.21)	1.21 (0.22)	0.36 (0.11)	0.70 (0.19)
8	−0.73 (0.12)	0.47 (0.07)	−0.42 (0.10)	0.57 (0.12)	0.36 (0.27)	1.31 (0.25)	0.43 (0.12)	0.63 (0.21)
9	−1.28 (0.35)	0.38 (0.16)	−0.23 (0.22)	0.34 (0.26)	−0.40 (0.54)	0.71 (0.59)	0.64 (0.28)	0.72 (0.63)
10	−1.41 (0.12)	0.40 (0.05)	−0.45 (0.09)	0.80 (0.12)	0.54 (0.23)	0.73 (0.24)	0.31 (0.11)	0.81 (0.19)
11	−1.44 (0.11)	0.22 (0.06)	−0.40 (0.09)	0.43 (0.11)	0.21 (0.23)	0.34 (0.30)	0.38 (0.11)	0.97 (0.19)
12	−1.49 (0.17)	0.24 (0.10)	−0.39 (0.11)	0.68 (0.14)	0.35 (0.28)	0.83 (0.34)	0.36 (0.21)	0.60 (0.26)
13	−1.43 (0.14)	0.22 (0.09)	−0.42 (0.11)	0.68 (0.14)	−0.04 (0.34)	0.26 (0.30)	0.52 (0.14)	0.76 (0.24)
14	−1.61 (0.11)	0.17 (0.07)	−0.34 (0.09)	0.29 (0.16)	0.06 (0.23)	0.46 (0.26)	0.53 (0.11)	0.42 (0.21)
15	−2.07 (0.18)	0.52 (0.07)	−0.59 (0.15)	0.91 (0.16)	0.33 (0.28)	0.54 (0.37)	0.29 (0.14)	0.47 (0.26)
M1: full random effects								
$\hat{\mu}_p(se)$	−1.35 (0.07)	0.28 (0.03)	−0.38 (0.03)	0.61 (0.06)	0.27 (0.07)	0.71 (0.10)	0.40 (0.03)	0.72 (0.06)
$\hat{\tau}_p$	0.25	0.09	0.07	0.17	0.08	0.27	0.06	0.08
% PI	[−1.92, −0.78]	[0.08, 0.48]	[-0.55, −0.20]	[0.21, 1.01]	[0.05, 0.50]	[0.08, 1.34]	[0.25, 0.55]	[0.50, 0.93]
M2: common predictor effects/M3: rank = 1								
$\hat{\mu}_\alpha, \hat{\beta}_p$ (se)	−1.34 (0.08)	0.30 (0.02)	−0.39 (0.02)	0.61 (0.03)	0.28 (0.06)	0.72 (0.05)	0.39 (0.03)	0.72 (0.05)
M2: $\hat{\tau}_\alpha = 0.29$ /M3: $\hat{\tau}_\alpha = 0.24, \hat{\tau}_\gamma = 0.14$								

PI = prediction interval.

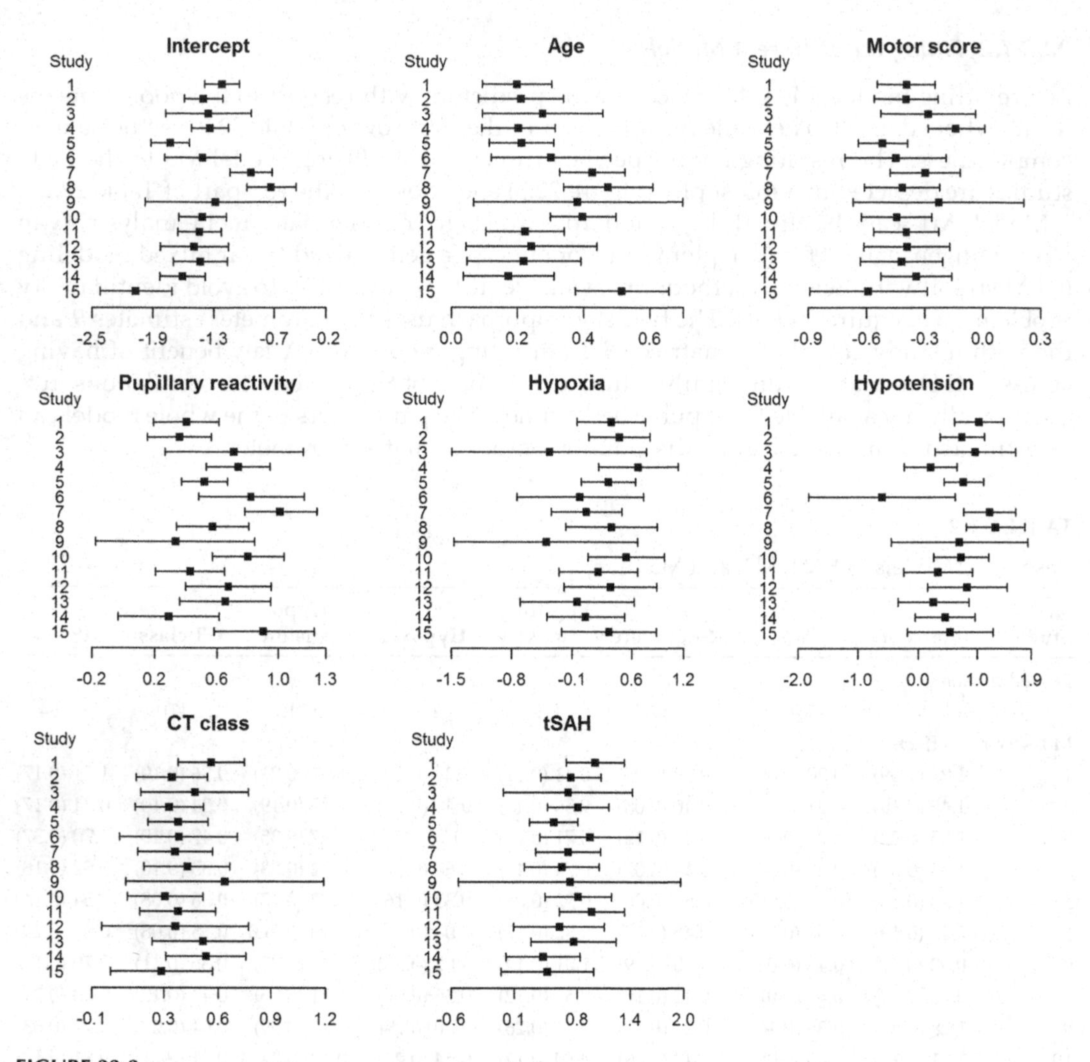

FIGURE 22.3
Forest plots showing estimated regression coefficients and associated 95% confidence intervals from fitting model M0 to all studies.

$$\theta_i \sim N(\mu, \mathcal{T} + S_i).$$

In our analysis, the GLMM approach is used. The results are given in Table 22.2 under the heading M1. The heterogeneity for each predictor p is shown in Figure 22.3 and quantified in Table 22.2 by the estimated standard deviation $\hat{\tau}_p$ of the random effect and the 95% prediction intervals, $\hat{\mu}_p \pm 1.96\hat{\tau}_p$. The values of $\hat{\tau}_p$ cannot be used to compare the heterogeneity of the predictors, because $\hat{\tau}_p$ is not scale-independent. A simple scale-independent measure can be obtained from the ratio of $\hat{\tau}_p$ to the standard error of $\hat{\mu}_p$. The largest values are seen for intercept, age, pupillary reactivity, and hypotension. The analysis also produces an estimate of the correlation of the predictor effects between studies (results not shown). There are some large correlations, but there is not much of a pattern.

Next, we fit model M2, also using GLMM software. Results can be found in Table 22.2. Notice that the common β is very close to μ_β of model M1 and that $\hat{\tau}_\alpha$ is a bit larger.

TABLE 22.3

Summary of Aspects of Between-Study Heterogeneity

No.	Name	Type	Mortality	Membership C	α_i from M2-FE	γ_i from M3-FE
1	TINT	RCT	25%	0.62	−1.22	1.02
2	TIUS	RCT	22%	0.65	−1.40	0.88
3	SLIN	RCT	23%	0.76	−1.35	1.00
4	SAP	RCT	23%	0.60	−1.34	0.87
5	PEG	RCT	24%	0.67	−1.73	0.92
6	HIT I	RCT	28%	0.68	−1.41	1.03
7	UK4	OBS	45%	0.64	−0.93	1.20
8	TCDB	OBS	44%	0.67	−0.73	1.14
9	SKB	RCT	27%	0.75	−1.28	0.90
10	EBIC	OBS	34%	0.63	−1.41	1.10
11	HIT II	RCT	23%	0.63	−1.44	0.90
12	NABIS	RCT	26%	0.65	−1.49	0.97
13	CSTAT	RCT	22%	0.61	−1.43	0.97
14	PHARMOS	RCT	17%	0.68	−1.61	0.81
15	APOE	OBS	15%	0.73	−2.07	1.03

Columns 1–5 are copied from Table 22.1 for convenience.

In fitting model M3, we use the simplified model with $\beta^* = \beta_{M2}$. The only difference with M2 is the extra random effect γ. The last line in Table 22.2 shows that part of the random variation shifts from α to γ ($\hat{\tau}_\alpha$ drops from 0.29 in M2 to 0.24 in M3).

The heterogeneity between the studies can be judged from fitting alternative versions of models M2 and M3, M2-FE and M3-FE, where α_i and γ_i are treated as fixed instead of random effects (Table 22.3). We note that studies 7 and 8 had a higher observed mortality than others. The high mortality in these observational studies was not explained by predictor differences (γ_i), and so was reflected in higher than average model intercepts (α_i). Study 14 had somewhat lower mortality than could be expected from the estimated α_i. The predictor effects were somewhat stronger than average in study 7, and weakest in study 14.

22.3.2 Using the Global Models for Prediction

Now we want to use the models to make predictions in a new patient population. The global models M1–M3 presented in Section 22.3.1.1 are of the form

$$\text{logit } p_i(x) = \mu(x) + v_i(x)$$

where $\mu(x)$ is a common term and $v_i(x)$ is a study-specific random term with mean zero, as in the table:

Model	$\mu(x)$	$v_i(x)$
M1	$\mu_\alpha + x'\mu_\beta$	$e_{\alpha i} + x'e_{\beta i}$
M2	$\mu_\alpha + x'\beta$	$e_{\alpha i}$
M3	$\mu_\alpha + x'\beta\mu_\gamma$	$e_{\alpha i} + x'\ \beta e_{\gamma i}$

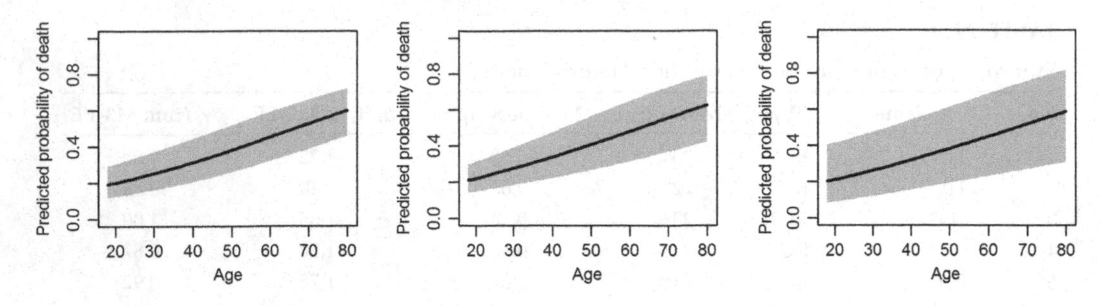

FIGURE 22.4
95% ranges for predicted probability of death at different ages of having TBI across studies for models M2, M3, and M1, respectively. Other patient characteristics were motor score=3; both pupils reacting; no hypoxia; no hypotension; CT Class=5; and a tSAH.

where $e_{\alpha i} = \alpha_i - \mu_\alpha$, $e_{\beta i} = \beta_i - \mu_\beta$, and $e_{\gamma i} = \gamma_i - \mu_\gamma$. The distribution of these random terms is specified for each model. So far, only multivariate normal models have been considered, but other models are feasible as well.

The global models could be used in different ways. First, any of the three global models could be used to obtain an "average model" by computing the average of $p_i(x)$ over the population of studies. This would be close to the "common-effect" model, M2-CE. (Note that the $\mu(x)$ above are poor approximations to an average model because $E\left[p(x)\right] \neq p(E\left[x\right].)$

However, care is needed when a prediction is required for a specific patient population not represented in the MA. The people of that population will share the same random factors and the optimal model for that population might be far from the "average model". *It is necessary to explore the variability of the predicted probabilities before deciding to use the "average model"*. This can be done for any individual with predictor value x by computing both $\mu(x)$ and the standard deviation of $v(x)$, $sd(x) = \sqrt{\text{var}(v(x))}$. The 95% range, for example, of the model probabilities can then be obtained from $\text{expit}\left(\mu(x) \pm 1.96sd(x)\right)$, where $\text{expit}(\mu) = \exp(\mu)/(1+\exp(\mu))$ is the inverse of the logit link function. This is close to the 95% prediction interval, but not quite, because the 95% range defined here does not take account of the uncertainties in the estimated model parameters. The situation is quite simple in model M2, where the random part does not depend on x and $sd(x)$ is equal to the standard deviation τ_α of α. For the TBI data, $\hat{\tau}_\alpha = 0.29$ (Table 22.2) and the 95% range is $\text{expit}\left(\mu(x) \pm 0.57\right)$.

We illustrate these approaches in Figure 22.4 by plotting the 95% range across different ages for otherwise average risk patients in the TBI case study. As expected, the range is smallest in the common predictor effects model M2, where we assume common predictor effects across studies. The variation is larger when we allow for between-study differences in the predictor effects in the rank=1 model M3, and largest in the full random-effects model M1.

An alternative visual approach to assessing the performance of an average model is to assess calibration and discrimination for cross-validated performance. We may develop any model M1–M3 for all studies, hold out one, and (as a poor but quick approximation) use $\mu(x)$ for prediction. We test each of these models in the individual held-out studies. This procedure has been referred to as internal-external validation (Royston et al., 2004; Steyerberg and Harrell, 2016). It can be combined with meta-analysis to assess summary performance and heterogeneity (Debray et al., 2013). In our case study, we assessed calibration using "calibration plots" of observed outcome probability against predicted outcome probability. We noted some variability in the calibration slopes (Figure 22.5). We assessed discrimination by the C-statistic. The pooled C-statistic was 0.79, with a 95% confidence interval from 0.76 to 0.81 and a 95% prediction interval from 0.69 to 0.90 (Figure 22.6). This indicates that model discrimination substantially varies across different settings and populations, possibly due to differences in case-mix distribution.

The different ways of assessing the performance of the average model depend very much on the clinical context. Clinical input is needed and there is no simple statistical criterion for deciding on the acceptability of a prediction model. If the random variation in the model of choice is judged to be too large, the average model can still serve as a tool to get improved predictions in existing datasets with the same predictors by using any of the models M1–M3 as a Bayesian prior. This will be helpful if the existing dataset is small, while in large datasets, the posterior model might appear be the same as the observed model in the data. This is an example of empirical Bayes methodology. However, empirical Bayes estimates are very sensitive to the model for the random effect and they should be handled with care (Van Houwelingen, 2014). Another possibility is some form of calibration that fits a logistic regression model on the one-dimensional predictor $z=x'\beta$, for example, logit $p(z)=\alpha+z$ as in M2 or logit $p(z)=\alpha+\gamma z$ as in M3. If no data are yet available in the new population, then the global model may be of no help. A solution is to go back to model M0 and select the study that is closest to the own population.

22.3.3 Extension to Survival Data

If we have individual survival data and assume proportional hazards for predictor effects, we can estimate the regression parameters β (from the partial likelihood) and the baseline hazard $h_0(t)$ (by the Breslow estimator) in each study separately (Vittinghoff, 2005). This will give an estimate of cumulative survival probability $S(t_0|x)$ at any time t_0 of interest. It is also possible to obtain the covariance matrix of $(\ln \hat{H}_0(t_0),\hat{\beta})$. The implication is that a prognostic meta-analysis for fixed t_0 can be carried out along the lines as described above for model M1 to derive a pooled model for $S(t_0|x)$. The structure is identical, only the link function is different. Note that missing outcome data (censoring) are implicitly handled by standard Cox regression analysis. Common-effect models like M2-CE can also be readily obtained. Adding a stratified baseline hazard to a Cox regression model allows the baseline hazard to vary freely between studies. Including study as a predictor or random effect is also possible; it requires the additional assumption that the baseline hazard varies proportionally between studies.

The rank=1 model M3 can be obtained in a similar two-stage procedure as in the case of logistic regression models. The fixed-effects version with study-specific α_i and γ_i can easily be implemented. The random version with two random effects is harder to implement.

This analysis can be extended to a set of fixed time points $(t_1,t_2,...,t_l)$. In analogy with the above we use the notation $(\alpha_1,...,\alpha_l)$ for $(\ln(H_0(t_1)),\ldots,\ln(H_0(t_l)))$. A model for $(\alpha_1,...,\alpha_l)$ can be based on models for time-varying frailties (Fiocco et al., 2009; Putter and van Houwelingen, 2015). Such a meta-model provides insight into a prognostic survival model and the between-study heterogeneity under the assumption of proportional hazards within each study, considering multiple time points rather than a single time point t_0. Such a pooled model can also be used to obtain conditional survival probabilities $P(T>t+w\,|\,T\geq t,x)$ under proportional hazards (see Chapter 15). Conditional prognostic models that relax the proportional hazards assumption can be obtained by applying the landmark approach (Putter and van Houwelingen, 2017). Using a fixed prediction window,

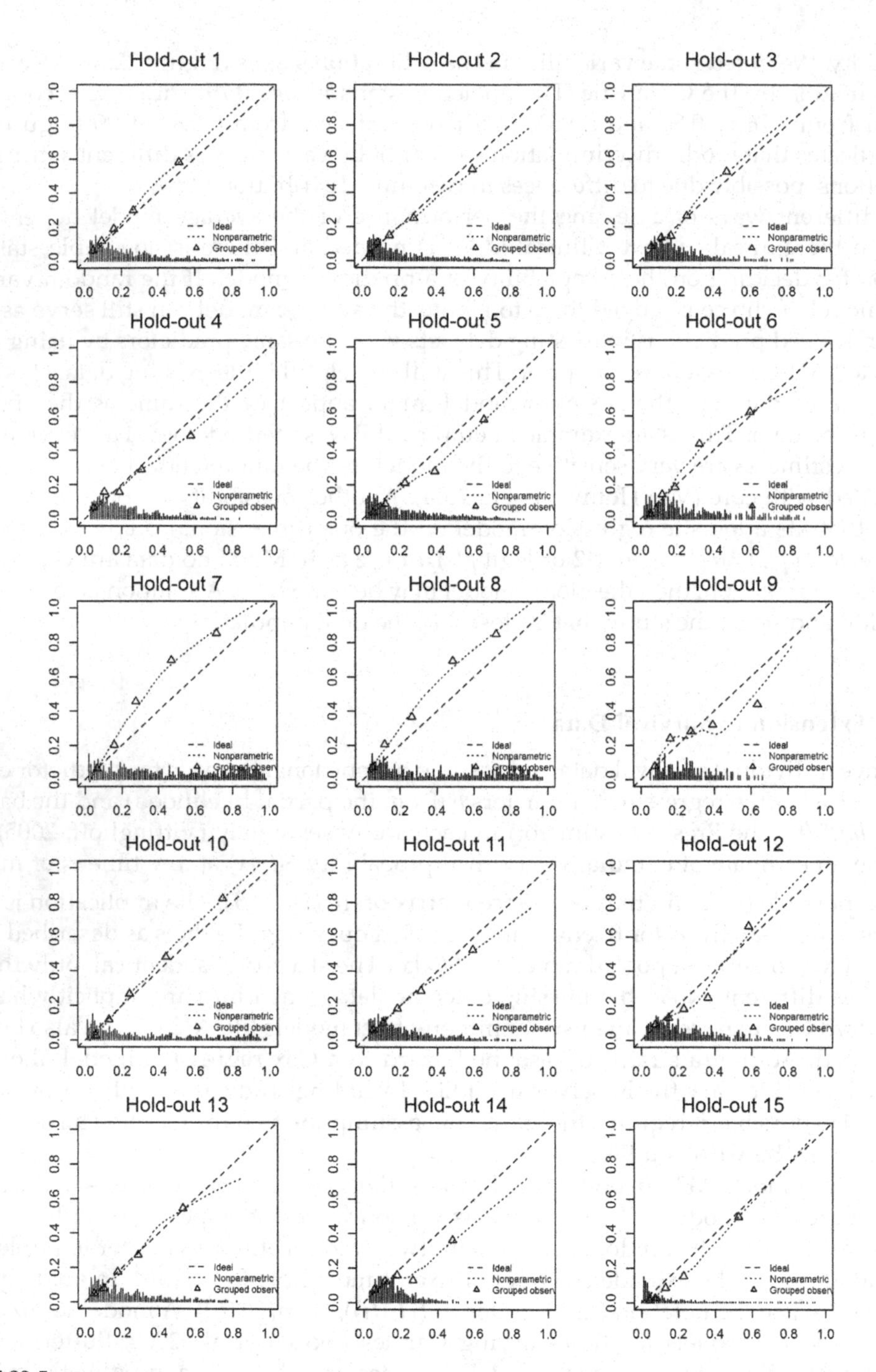

FIGURE 22.5
Calibration plots of the common-effect model (M2) from internal-external validation of a global TBI model developed without each of the 15 studies at a time evaluated for each of the "hold-out" studies. Plot shows observed outcome probability against predicted outcome probability. The dashed lines are the ideal lines, the dotted lines are the loess smoother, and the triangles are the means in five quintile groups.

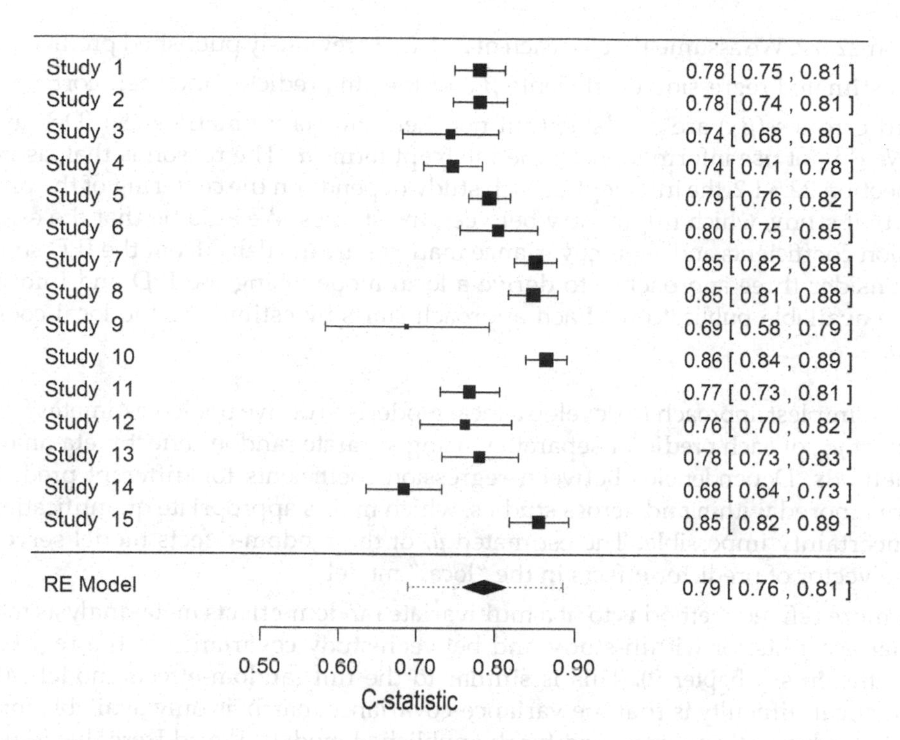

FIGURE 22.6
Estimated C-statistic of the global model for each of the 15 studies in the TBI case study, using internal-external cross-validation. Dotted interval is a 95% prediction interval.

the models are similar to the model with fixed t_0 described above and can be analyzed in the same way. Moving the landmark point of prediction gives insights into the development of heterogeneity over time.

22.4 Development of Prediction Models from Published Studies

Often, there is no access to IPD from multiple studies to estimate a global prediction model. In this section, we explore how summary data from published studies can be used to derive a "local model" using the published data and a single IPD dataset, which will improve the prediction in the population from which the IPD are drawn and might be of interest outside the setting of that population as well.

22.4.1 Multiple Studies, Same Multivariable Predictors

We first consider the situation where several studies have published a prediction model for a certain outcome using the same (or a very similar) set of predictors. (This is an idealized situation; the more realistic situation where published models differ between studies is discussed in Section 22.4.2.)

We will describe how the additional information can be used to improve the prediction in the single IPD set. This has some similarity with the approaches for IPD data discussed in Section 22.3.2. We assume that for each of $i = 1,...,I$ previously published prediction models, the estimated regression coefficients $\hat{\beta}_{ip}$ for the pth predictor and their corresponding standard errors $se(\hat{\beta}_{ip})$ are available (but not their covariance matrices S_i) (Debray et al., 2012). We do not use information on the intercept terms α_i. The reason is that, as pointed out in Section 22.3.1.2, the intercept in each study depends on the centering of the predictor data in that study, which might vary between the studies. We assume that the estimated regression coefficients and their covariance matrices are available from the IPD study.

We consider three approaches to derive a local model using the IPD and information from the available publications. Each approach starts by estimating the local coefficient vector β.

(1) The simplest approach to develop a local model is to derive pooled estimates for the effect $\mu_{\beta p}$ of each predictor separately using separate random-effect meta-analysis methods. Dependencies between regression coefficients for different predictors are ignored within and across studies, which makes appropriate quantification of uncertainty impossible. The estimated μ_β of the random-effects model serves as the vector of predictor effects in the "local" model.

(2) A more refined method is to fit a multivariate random-effects meta-analysis model that accounts for within-study and between-study covariance of the regression coefficients (Chapter 9). This is similar to the full random-effects model M1. A practical difficulty is that the variance-covariance matrix is only available for the IPD, and usually not reported for the published models. We address this issue by assuming the correlations between predictor effects in the published studies are the same as the correlations observed in the IPD study. As in (1), the estimated μ_β of the random-effect model serves as the vector of predictor effects in the "local" model.

(3) Finally, we may apply a Bayesian approach where a summary of the previously published regression coefficients, produced by method (2) above, serves as prior for the regression coefficients in the IPD. This prior distribution does not include estimates from the IPD, so unlike methods (1) and (2), this method follows good practice in not using the IPD twice. The posterior mean of the β serves as predictor of the new model. This third approach is motivated by the potential heterogeneity between settings, where the local model may not work optimally for the setting where the IPD originates from.

In all three approaches, we then estimate the model intercept in the IPD. This ensures that the derived model is on average well calibrated for the setting of the IPD, while the coefficients are based on a broader set of evidence. This may be especially advantageous if the IPD is relatively small. The model intercept is estimated by fitting a regression model in the IPD, using the linear predictor that is calculated from the pooled regression coefficients as an offset (Debray et al., 2012).

In an empirical evaluation, all aggregation approaches performed similarly, and yielded better outcome predictions than models that were based solely on the IPD (Debray et al., 2012; Martin et al., 2017, 2018). This was especially the case if the IPD set was relatively small. Assessing the degree of heterogeneity between IPD and literature findings is important. The more homogeneity of predictor effects, the better the meta-analysis of previous studies allows for the development of a local model that provides better predictions for the setting where the IPD dataset originates from.

22.4.2 Multiple Studies, Different Multivariable Predictors

The situation where published studies all considered the same prediction model is relatively rare. A more common situation is that multiple studies provide models with different but overlapping sets of predictors. We consider various approaches for aggregating previously published prediction models for one IPD dataset containing all predictors from the published models. As in Section 22.4.1, the goal is to obtain a better prediction for the setting of the IPD dataset.

A first step is to assess the statistical validity of published models for the setting where we have IPD available. Only models with reasonable performance are further considered for the development of a local model.

Next, we may consider various degrees of recalibration of each published model to the IPD setting, and then provide a local model using a linear weighting scheme (model averaging [Debray et al., 2014]). A more attractive approach is a variant of ensemble learning, commonly known as "stacked generalizations" (Martin et al., 2018). The predictions of each published model are then treated as the predictor variables of the "meta-model" and subsequently create a linear combination of M model predictions:

$$\text{logit}(\pi_i) = \alpha_0 + \sum_{m=1}^{M} \alpha_m LP_{i,m}.$$

The predictions $LP_{i,m}$ (on the linear predictor scale) of each model are weighted by an independent parameter α_m that emphasizes good prediction overall and penalizes models with poor performance or extreme predictions (similar to logistic calibration). We apply the constraint $\alpha_m \geq 0$ to ensure that models do not make a negative contribution to the meta-model. The intercept α_0 ensures that the baseline risk of the synthesis model is optimal for the IPD sample (similar to intercept updating) (Vergouwe et al., 2017). The unknown parameters $\alpha_0, \alpha_1, \ldots, \alpha_M$ can be estimated using maximum likelihood. This approach works well if the models are not strongly collinear; if there is strong collinearity, dropping a correlated but poorer performing model might be considered (Martin et al., 2018; Debray et al., 2014).

22.4.3 Multiple Studies, Only Univariable Predictor Effect Estimates

Finally, we consider meta-analytic approaches to develop a local model if multiple studies have only reported the univariable relations of predictors to an outcome, rather than multivariable effects. This situation is common in medical research.

Considering prior information from published studies is particularly relevant when the IPD sample of interest is relatively small (Debray et al., 2012). This was the motivation for such an approach in a case study of aortic aneurysm patients (Steyerberg et al., 1995; Steyerberg et al., 2000).

The basic approach is that we first pool the published estimates of univariable regression coefficients for each predictor, yielding a meta-analysis average $\hat{\mu}^{uv}_{\beta p}$. Common- or random-effect meta-analysis techniques can be used. Second, we estimate the change from univariable ($\hat{\mu}^{uv,IPD}_{\beta p}$) to multivariable coefficient ($\hat{\mu}^{mv,IPD}_{\beta p}$) of the predictor in the IPD, yielding $\hat{\delta}_p = \hat{\mu}^{mv,IPD}_{\beta p} - \hat{\mu}^{uv,IPD}_{\beta p}$. This change is then used as an adaptation for $\mu_{\beta p}$, giving a literature-informed estimate for the predictor's multivariable coefficient as $\hat{\mu}^{uv}_{\beta p} + \hat{\delta}_p$. Further improvements are possible by incorporating weakly informative prior distributions (Debray et al., 2012).

This approach is simpler than that discussed in Chapters 9 and 21 and illustrated with the Fibrinogen Studies Collaboration (Jackson et al., 2009), because it assumes that the unadjusted and adjusted coefficients have a constant difference, rather than a joint distribution. The assumption is that the correlation structure in the IPD is similar to that in the literature studies.

22.4.4 MA of Published Studies, No IPD

Alternative approaches have been proposed which do not even require the availability of IPD (Sheng et al., 2014). A local model is then developed by using information from published associations between predictors of interest and from correlations between each pair of the predictors of interest. Here it is assumed that predictor values from the overall population follow a multivariate normal distribution. No specific care is taken to account for potential between-study heterogeneity. This approach is also known as synthesis analysis and is loosely related to meta-analysis.

22.5 Concluding Remarks

In this chapter, we focused on the development and validation of prediction models across multiple studies. Meta-analytical techniques have an essential role in this context. The "internal-external validation" approach is useful to study validity by study. IPD from multiple studies provide the best opportunity to investigate, quantify, and report any heterogeneity in the baseline risks and predictor effects of a prediction model. The performance of a global model may vary substantially by setting, in terms of calibration and discrimination. In our experience, differences in baseline risk are most common, leading to poor calibration of predictions. Correcting calibration is hence the first focus in validation of a model for a specific setting.

A second point of attention is heterogeneity in predictor effects. If we notice large heterogeneity for one or more predictor effects, we may consider dropping these predictor(s) from the global model, and this may be sensible when these predictors do not offer great improvement in discriminative ability. This ensures that the global model includes predictors with more or less homogeneous effects across studies (Debray et al., 2013). However, strong heterogeneity may also be caused by missed non-linear effects or interactions between predictors. If the pooled estimate indicates that the predictor has a strong but heterogeneous predictive effect, we may decide to keep the predictor in the global model, since the predictor overall contains important predictive information. For example, we found substantial heterogeneity for the effect of pupillary reactivity in our case study. The overall effect was relatively large, arguing against dropping this predictor from a global model. This heterogeneity implies that the predictive performance of the global model will vary across the different studies. Hence, local adaptations may be required, not only for the baseline risk, but also for the effect of specific predictors.

A specific interest may lie in quantifying the value of new predictors, such as new markers to predict cardiovascular disease, beyond what is possible with traditional predictors (Kaptoge et al., 2012). The incremental value of a marker may then be studied per study, with MA approaches applied to performance measures such as the increment in the C-statistic. Large sample size and the availability of multiple data sources are important to strengthen conclusions on such a marker.

References

Audigier V, White IR, Jolani S, Debray TPA, Quartagno M, Carpenter JR, et al., 2018. Multiple imputation for multilevel data with continuous and binary variables. *Statistical Science* **33**(2): 160–183.

Burgess S, White IR, Resche-Rigon M, Wood AM, 2013. Combining multiple imputation and meta-analysis with individual participant data. *Statistics in Medicine* **32**(26): 4499–4514.

Debray TP, Koffijberg H, Vergouwe Y, Moons KG and Steyerberg EW, 2012a. Aggregating published prediction models with individual participant data: A comparison of different approaches. *Statistics in Medicine* **31**(23): 2697–2712.

Debray TP, Koffijberg H, Lu D, Vergouwe Y, Steyerberg EW and Moons KG, 2012b. Incorporating published univariable associations in diagnostic and prognostic modeling. *BMC Medical Research Methodology* **12**(121).

Debray TP, Moons KG, Ahmed I, Koffijberg H and Riley RD, 2013. A framework for developing, implementing, and evaluating clinical prediction models in an individual participant data meta-analysis. *Statistics in Medicine* **32**(18): 3158–3180.

Debray TP, Koffijberg H, Nieboer D, Vergouwe Y, Steyerberg EW and Moons KG, 2014. Meta-analysis and aggregation of multiple published prediction models. *Statistics in Medicine* **33**(14): 2341–2362.

Debray TP, Vergouwe Y, Koffijberg H, Nieboer D, Steyerberg EW and Moons KG, 2015. A new framework to enhance the interpretation of external validation studies of clinical prediction models. *Journal of Clinical Epidemiology* **68**(3): 279–289.

Fiocco M, Putter H and van Houwelingen JC, 2009. Meta-analysis of pairs of survival curves under heterogeneity: A Poisson correlated gamma-frailty approach. *Statistics in Medicine* **28**(30): 3782–3797.

Harrell FE, 2001. *Regression Modeling Strategies: With Applications to Linear Models, Logistic Regression, and Survival Analysis*. New York: Springer.

Jackson D, White I, Kostis JB, Wilson AC, Folsom AR, Wu K, Chambless L, Benderly M, Goldbourt U, Willeit J, et al., 2009. Systematically missing confounders in individual participant data meta-analysis of observational cohort studies. *Statistics in Medicine* **28**(8): 1218–1237.

Jolani S, Debray TP, Koffijberg H, van Buuren S and Moons KG, 2015. Imputation of systematically missing predictors in an individual participant data meta-analysis: A generalized approach using MICE. *Statistics in Medicine* **34**(11): 1841–1863.

Jolani S, 2018. Hierarchical imputation of systematically and sporadically missing data: An approximate Bayesian approach using chained equations. *Biometrical Journal. Biometrische Zeitschrift* **60**(2): 333–351.

Justice AC, Covinsky KE and Berlin JA, 1999. Assessing the generalizability of prognostic information. *Annals of Internal Medicine* **130**(6): 515–524.

Kaptoge S, Di Angelantonio E, Pennells L, Wood AM, White IR, Gao P, Walker M, Thompson A, Sarwar N, Caslake M, et al., 2012. C-reactive protein, fibrinogen, and cardiovascular disease prediction. *The New England Journal of Medicine* **367**(14): 1310–1320.

Legrand C, Duchateau L, Janssen P, Ducrocq V and Sylvester R, 2009. Validation of prognostic indices using the frailty model. *Lifetime Data Analysis* **15**(1): 59–78.

Martin GP, Mamas MA, Peek N, Buchan I and Sperrin M, 2017. Clinical prediction in defined populations: A simulation study investigating when and how to aggregate existing models. *BMC Medical Research Methodology* **17**(1): 1.

Martin GP, Mamas MA, Peek N, Buchan I and Sperrin M, 2018. A multiple-model generalisation of updating clinical prediction models. *Statistics in Medicine* **37**(8): 1343–1358.

Murray GD, Butcher I, McHugh GS, Lu J, Mushkudiani NA, Maas AI, Marmarou A and Steyerberg EW, 2007. Multivariable prognostic analysis in traumatic brain injury: Results from the IMPACT study. *Journal of Neurotrauma* **24**(2): 329–337.

Mushkudiani NA, Hukkelhoven CW, Hernandez AV, Murray GD, Choi SC, Maas AI and Steyerberg EW, 2008. A systematic review finds methodological improvements necessary for prognostic models in determining traumatic brain injury outcomes. *Journal of Clinical Epidemiology* **61**(4): 331–343.

Pennells L, Kaptoge S, White IR, Thompson SG, Wood AM and Emerging Risk Factors C, 2014. Assessing risk prediction models using individual participant data from multiple studies. *American Journal of Epidemiology* **179**(5): 621–632.

Putter H and van Houwelingen HC, 2015. Dynamic frailty models based on compound birth-death processes. *Biostatistics* **16**(3): 550–564.

Putter H and van Houwelingen HC, 2017. Understanding landmarking and its relation with time-dependent cox regression. *Statistics in Biosciences* **9**(2): 489–503.

Riley RD, Price MJ, Jackson D, Wardle M, Gueyffier F, Wang J, Staessen JA and White IR, 2015. Multivariate meta-analysis using individual participant data. *Research Synthesis Methods* **6**(2): 157–174.

Royston P, Parmar MK and Sylvester R, 2004. Construction and validation of a prognostic model across several studies, with an application in superficial bladder cancer. *Statistics in Medicine* **23**(6): 907–926.

Sheng E, Zhou XH, Chen H, Hu G and Duncan A, 2014. A new synthesis analysis method for building logistic regression prediction models. *Statistics in Medicine* **33**(15): 2567–2576.

Steyerberg EW and Harrell FE Jr., 2016. Prediction models need appropriate internal, internal-external, and external validation. *Journal of Clinical Epidemiology* **69**: 245–247.

Steyerberg EW, Kievit J, de Mol Van Otterloo JC, van Bockel JH, Eijkemans MJ and Habbema JD, 1995. Perioperative mortality of elective abdominal aortic aneurysm surgery. A clinical prediction rule based on literature and individual patient data. *Archives of Internal Medicine* **155**(18): 1998–2004.

Steyerberg EW, Eijkemans MJ, van Houwelingen JC, Lee KL and Habbema JD, 2000. Prognostic models based on literature and individual patient data in logistic regression analysis. *Statistics in Medicine* **19**(2): 141–160.

Steyerberg EW, Mushkudiani N, Perel P, Butcher I, Lu J, McHugh GS, Murray GD, Marmarou A, Roberts I, Habbema JD and Maas AI, 2008. Predicting outcome after traumatic brain injury: Development and international validation of prognostic scores based on admission characteristics. *PLOS Medicine* **5**(8): e165; discussion e.

Steyerberg EW, 2009. *Clinical Prediction Models: A Practical Approach to Development, Validation, and Updating*. New York: Springer.

Van Calster B, Vergouwe Y, Looman CW, Van Belle V, Timmerman D and Steyerberg EW, 2012. Assessing the discriminative ability of risk models for more than two outcome categories. *European Journal of Epidemiology* **27**(10): 761–770.

Van Houwelingen JC and Le Cessie S, 1990. Predictive value of statistical models. *Statistics in Medicine* **9**(11): 1303–1325.

Van Houwelingen HC, Arends LR and Stijnen T, 2002. Advanced methods in meta-analysis: Multivariate approach and meta-regression. *Statistics in Medicine* **21**(4): 589–624.

Van Houwelingen HC, 2014. The role of empirical Bayes methodology as a leading principle in modern medical statistics. *Biometrical Journal Biometrische Zeitschrift* **56**(6): 919–932.

Vergouwe Y, Nieboer D, Oostenbrink R, Debray TPA, Murray GD, Kattan MW, Koffijberg H, Moons KGM and Steyerberg EW, 2017. A closed testing procedure to select an appropriate method for updating prediction models. *Statistics in Medicine* **36**(28): 4529–4539.

Vittinghoff E, 2005. *Regression Methods in Biostatistics: Linear, Logistic, Survival, and Repeated Measures Models*. New York: Springer.

23

Using Meta-Analysis to Plan Further Research

Claire Rothery, Susan Griffin, Hendrik Koffijberg, and Karl Claxton

CONTENTS

Meta-analysis provides a central estimate of the relative effectiveness of an intervention within a range of plausible values, for example, a confidence or credible interval around the central estimate of effect. When the range of plausible values does not support clear superiority, or inferiority, compared with alternative interventions, this uncertainty has consequences on health outcomes. If the intervention is chosen when the balance of evidence suggests it is expected to be effective, there is a risk that one of the alternative interventions is more effective and would improve health outcomes to a greater extent (Claxton et al., 2016). Uncertainty, therefore, creates the potential for adverse health consequences as the expected benefits of the intervention may not be realized and the resources committed by the use of the intervention wasted. Similarly, if an intervention is not expected to perform better than the available alternatives rejecting its use may risk failing to provide access to a valuable intervention if the health outcomes would have been greater than expected (McKenna et al., 2015). The scale of the health consequences of this uncertainty is determined by the chance that the intervention is not the most effective, how much less effective it is likely to be, and the size of the patient population facing the uncertain treatment choice. When the scale of the consequences is large, further research may be valuable since it can resolve uncertainty and ultimately lead to better treatment decisions for patients and better use of finite resources (Claxton et al., 2016).

Health outcomes can also be improved by ensuring that the findings of existing evidence (e.g., based on the results of a meta-analysis) are fully implemented into practice. In fact, the potential improvements in health outcomes by encouraging the implementation of what existing evidence suggests is the most effective intervention may well exceed the potential improvements in health outcomes through conducting further research (Claxton et al., 2013). However, early implementation of an intervention when existing evidence is still inconclusive may not be appropriate if it means that the type of research required to obtain conclusive evidence becomes impossible (e.g., unable or unethical to complete recruitment) or too costly to conduct. Implementation efforts can also be difficult and costly to reverse if the results of subsequent research find that the intervention is not as effective as the previous evidence suggested. Therefore, the question of when the evidence from a meta-analysis is sufficient to justify implementation requires a quantitative assessment of both the value of additional evidence and the value of implementation efforts to improve health outcomes.

Value of information analysis provides a conceptual framework to assess the need for additional evidence, the consequences of uncertainty, and the relative priority of alternative topics for research prioritization decisions (Claxton, 1999; Claxton et al., 2012; Claxton and Sculpher, 2006; Eckermann and Willan, 2007; Fleurence and Meltzer, 2013). It can be used to establish: (i) whether the evidence currently available is sufficient to support the use of the intervention in practice, i.e., whether on balance the evidence from the meta-analysis suggests that the intervention achieves better outcomes than the available alternatives; (ii) whether additional evidence might be needed to resolve uncertainty in the range of plausible values; (iii) what type and design of research might resolve this uncertainty; (iv) whether it would be better to withhold the intervention until additional evidence becomes available; and (v) whether any proposed research is considered a priority relative to other topics competing for the same resources. In this chapter, we demonstrate how value of information analysis can be applied to standard results of meta-analysis to determine the need for and value of additional research.

Throughout this chapter, we use the results of a meta-analysis of corticosteroids following traumatic brain injury to provide a worked example. Despite 19 randomized controlled trials conducted between the years of 1972 and 1995, the effect of corticosteroids on death and disability following traumatic brain injury was unclear (Alderson and Roberts, 1997, 1999). The CRASH (Corticosteroid Randomisation After Significant Head injury) trial, which first reported in the year 2004, was stopped early after enrolling 10,008 adults with brain injury (CRASH Trial Collaborators, 2004, 2005). It reported a higher risk of death or severe disability associated with the use of corticosteroids compared with no treatment. The global value of the CRASH trial appears, with hindsight, self-evident. However, the prevention of thousands of iatrogenic deaths worldwide hinges on the fact that this trial took place. We conduct a retrospective analysis of the evidence available before CRASH to show how the results of meta-analysis could have been used to determine the need for and value of further research (McKenna et al., 2015).

23.1 The Value of Additional Evidence

The principles of value of information analysis have a firm foundation in statistical decision theory with diverse applications in areas of research such as business decisions, engineering, environmental risk analysis, economics, and finance. Over recent years, these methods have been extended to research priority setting in the evaluation of healthcare interventions (Claxton and Sculpher, 2006; Basu and Meltzer, 2007; Claxton et al., 2004, 2005; Colbourn et al., 2007; Hassan et al., 2009). Most commonly, they have been applied in the context of probabilistic decision analytic models used to estimate expected cost-effectiveness of alternative interventions. In this chapter, we show that the same type of analysis can be applied to standard results of meta-analysis without the necessity to implement a full decision theoretic model or undertake a cost-effectiveness analysis (Claxton et al., 2016).

Standard results of meta-analysis provide a relative measure of effect (e.g., odds ratio) for an intervention compared with another, expressed in terms of its central value and associated uncertainty (confidence or credible interval) (Sutton and Abrams, 2001; Whitehead, 2002). When this uncertainty is combined with information about baseline risk and number of patients facing the uncertain treatment choice (e.g., expected incidence in the population whose treatment choice can be informed by the decision), the absolute effect of the uncertainty on health outcomes is obtained. Value of information analysis determines an estimate of the potential health benefits that could be gained if this uncertainty about treatment choice were resolved completely. Although further research cannot entirely eliminate uncertainty, it reduces the associated consequences by rendering them increasingly unlikely. For this reason, the estimate of the expected consequences of uncertainty represents an expected upper bound on the potential value of further research.

We now consider how this quantitative estimate of the expected health benefits of additional research is obtained. Figure 23.1 shows the forest plot of the evidence available before the CRASH trial comparing the use of corticosteroids with control (placebo or no treatment) for the outcome of mortality following traumatic brain injury (McKenna et al., 2015). The relative treatment effect is obtained from a Bayesian random-effects meta-analysis of the trials before CRASH, where it is assumed that the study-specific true relative effects can be described as coming from a common random-effects distribution. The relative effect that is reported in Figure 23.1 is the mean of this distribution calculated on the log odds scale with an odds ratio of 0.93 in favor of the use of corticosteroids and a

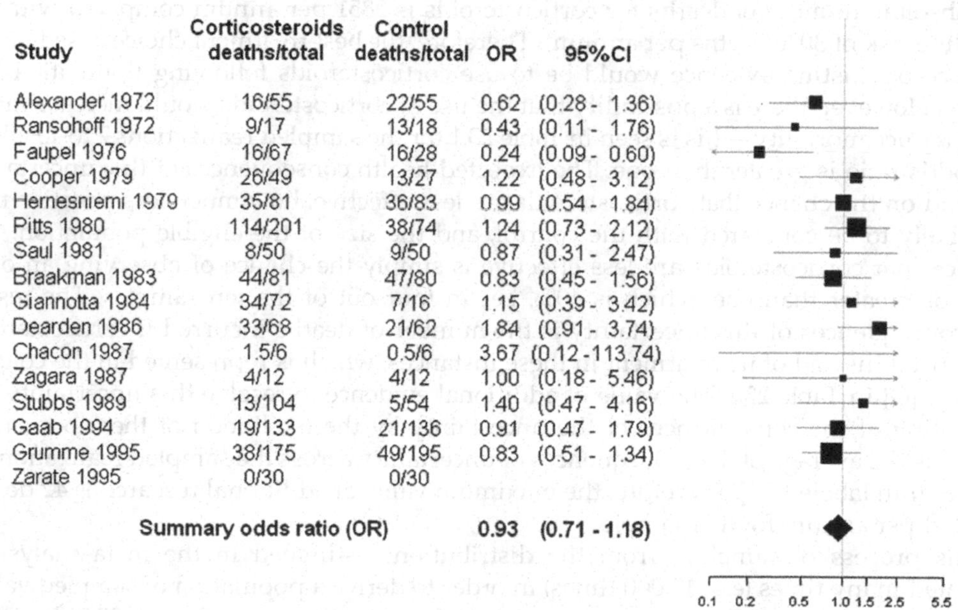

Study	Corticosteroids deaths/total	Control deaths/total	OR	95%CI
Alexander 1972	16/55	22/55	0.62	(0.28 - 1.36)
Ransohoff 1972	9/17	13/18	0.43	(0.11 - 1.76)
Faupel 1976	16/67	16/28	0.24	(0.09 - 0.60)
Cooper 1979	26/49	13/27	1.22	(0.48 - 3.12)
Hernesniemi 1979	35/81	36/83	0.99	(0.54 - 1.84)
Pitts 1980	114/201	38/74	1.24	(0.73 - 2.12)
Saul 1981	8/50	9/50	0.87	(0.31 - 2.47)
Braakman 1983	44/81	47/80	0.83	(0.45 - 1.56)
Giannotta 1984	34/72	7/16	1.15	(0.39 - 3.42)
Dearden 1986	33/68	21/62	1.84	(0.91 - 3.74)
Chacon 1987	1.5/6	0.5/6	3.67	(0.12 - 113.74)
Zagara 1987	4/12	4/12	1.00	(0.18 - 5.46)
Stubbs 1989	13/104	5/54	1.40	(0.47 - 4.16)
Gaab 1994	19/133	21/136	0.91	(0.47 - 1.79)
Grumme 1995	38/175	49/195	0.83	(0.51 - 1.34)
Zarate 1995	0/30	0/30		
Summary odds ratio (OR)			**0.93**	**(0.71 - 1.18)**

FIGURE 23.1
Random-effects meta-analysis of the effect of corticosteroids on mortality including the evidence available before the CRASH trial. Adapted from McKenna, C., et al., Methods to place a value on additional evidence are illustrated using a case study of corticosteroids after traumatic brain injury, *J Clin Epidemiol.* 2015 Sep 18. doi:10.1016/j.jclinepi.2015.09.011 Copyright © 2015.

95% credible interval of 0.71 to 1.18. This effect estimate and credible interval describe the distribution of possible values for the relative treatment effect compared with the control. The baseline risk for the control can be derived from a pooled analysis of the control arms of the trials or, alternatively, from an external source relevant to the target population. The baseline odds of death for traumatic brain injury is 0.54 with 95% credible interval 0.33 to 0.88, obtained from a pooled random-effects analysis of the control arms in Figure 23.1.

The two quantities of baseline risk and relative effect determine the absolute effect of corticosteroids on mortality. The impact of uncertainty on number of deaths is obtained by sampling from the distributions of relative effect and baseline risk by preserving the structural correlation between these two quantities and then multiplying by the number of individuals whose treatment choice is to be informed by the decision. This is illustrated in Table 23.1 for ten random (equally likely) samples taken from independent distributions of relative effect and baseline risk for ease of demonstrating the principles. Note that correlation in outcomes should be preserved where possible. For example, a multivariate or bivariate meta-analysis may be more appropriate to account for the dependence between multiple and possibly correlated outcomes in a single analysis (Riley, 2009). The absolute number of deaths is estimated for an eligible population of 8800 patients per annum (estimated as the approximate annual incidence of traumatic brain injury in the UK). Each sampled value from the distributions is interpreted as one possible realization of uncertainty, i.e., one possible "true" value of how patient outcomes might turn out, as supported by the evidence. The last row of Table 23.1 represents the average across the ten sampled values corresponding to the central estimate of effect.

Across the ten sampled values, the balance of evidence indicates that corticosteroids is expected to reduce the number of deaths per annum by 179 compared with control (i.e., the absolute number of deaths for corticosteroids is 2851 per annum compared with the baseline risk of 3030 deaths per annum). Therefore, the best treatment choice based on the balance of existing evidence would be to use corticosteroids following traumatic brain injury. However, there is a possibility that the use of corticosteroids would increase rather than reduce mortality—this is seen in Table 23.1 for the sampled realizations 7 to 10 where the odds ratio is greater than one. The expected health consequences of this uncertainty depend on the chance that corticosteroids are less effective, how much less effective they are likely to be compared with the control, and the size of the eligible population. The chance that corticosteroids are less effective is simply the chance of observing an odds ratio of greater than one, which is 40%, i.e., in four out of the ten samples. The resulting consequences of this uncertainty is the number of deaths incurred if corticosteroids were used instead of no treatment in these instances, which is represented in the column labeled [G] in Table 23.1. The value of additional evidence to resolve this uncertainty is a weighting of the consequences of the uncertainty by the likelihood of them occurring, which is the average of the consequences of uncertainty across the sampled realizations in the column labeled [G]. Therefore, the maximum value of additional research is 42 deaths averted per annum for this sample.

This process of sampling from the distributions estimated in the meta-analysis is repeated many times (e.g., 10,000 times) in order to derive a population of sampled values for the consequences of uncertainty in number of deaths per annum. Figure 23.2 shows the resulting distribution for the use of corticosteroids following traumatic brain injury, based on the evidence available before CRASH. The balance of evidence favored corticosteroids with a 72% chance that corticosteroids were effective and reduced mortality. However, there was a 28% chance that corticosteroids were not effective, and mortality would have been lower if corticosteroids had not been used. Examining this 28% chance that

TABLE 23.1

Calculating the Absolute Health Impact of an Intervention (Corticosteroids) on Mortality and the Value of Additional Evidence

				Deaths per annum for an incidence of 8800 eligible patients			Value of additional evidence		
Sampled realization of uncertainty	Odds ratio for death (corticosteroids vs. control) [A]	Baseline odds of death for control [B]	Odds of death for corticosteroids (=A × B)[a] [C]	Control (=8800 × B/ (1+B)) [D]	Corticosteroids (=8800 × C/ (1+C)) [E]	Absolute effect for corticosteroids (=E-D) [F]	Consequences of uncertainty for corticosteroids (=F if A>1) [G]	Perfect information (=minimum of D and E) [H]	Consequences of uncertainty for corticosteroids for MCD=50 (G≥50) [I]
1	0.71	0.88	0.62	4119	3384	–735	0	3384	0
2	0.77	0.74	0.57	3743	3194	–548	0	3194	0
3	0.80	0.67	0.54	3531	3071	–460	0	3071	0
4	0.85	0.60	0.51	3300	2972	–328	0	2972	0
5	0.95	0.54	0.51	3086	2984	–102	0	2984	0
6	0.98	0.49	0.48	2894	2855	–39	0	2855	0
7	1.01	0.43	0.43	2646	2665	18	18	2646	0
8	1.02	0.40	0.41	2514	2550	36	36	2514	0
9	1.05	0.35	0.37	2281	2365	83	83	2281	83
10	1.18	0.33	0.39	2183	2466	283	283	2183	283
Average	0.93	0.54	0.48	3030	2851	–179	42	2809	37

[a] Distributions for baseline odds and odds ratio are assumed independent for ease of demonstrating the principles to the reader.
MCD = minimum clinical difference.

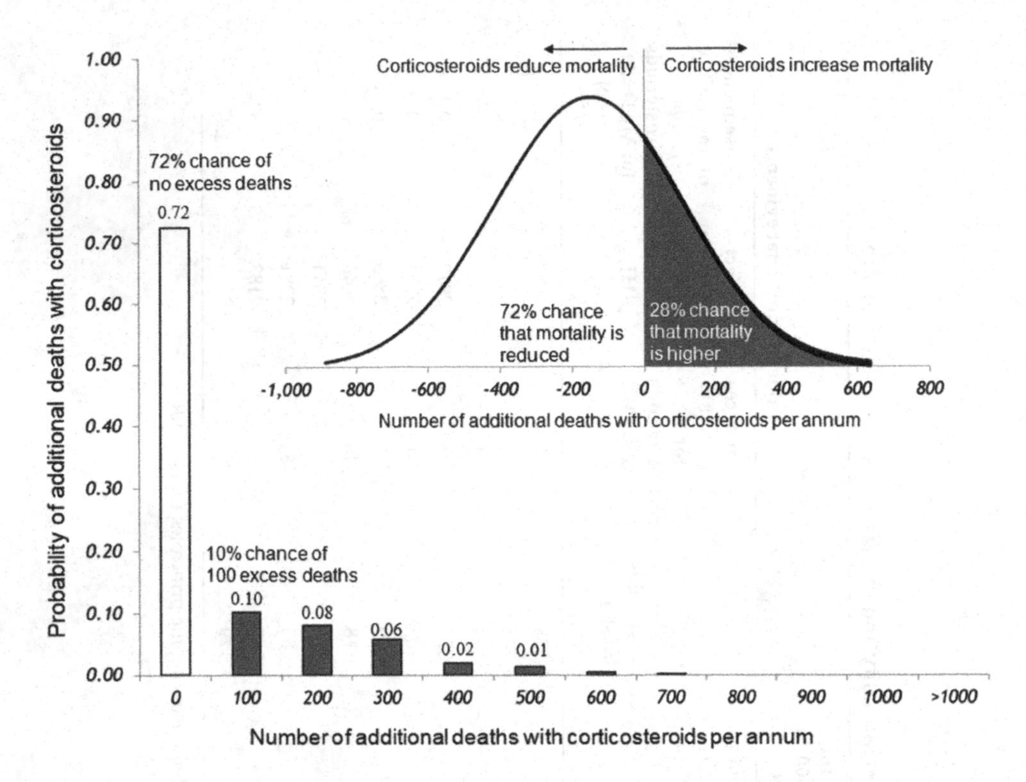

FIGURE 23.2
Distribution of the consequences of uncertainty in number of deaths per annum for corticosteroids following traumatic brain injury.

corticosteroids were not effective reveals that the adverse health consequences were not uniform. For example, there was a greater chance of more limited adverse consequences (10% chance of 100 additional deaths per annum) compared with a smaller chance of greater adverse consequences (1% chance of 500 additional deaths per annum). The mean of this distribution provides an estimate of the expected adverse health consequences of uncertainty and, by implication, the maximum value of conducting further research (42 deaths averted per year).

Note that the methods described above assume that the target setting for the decision to conduct further research is exactly equal to the average setting from the studies included in the meta-analysis. This may not be the case in practice, and it may be more relevant to consider the prediction interval for the effect in a new study. The predictive distribution has been proposed as a more realistic way to characterize the uncertainty in the treatment effect in a new future study by reflecting the between-study heterogeneity in the meta-analysis of effects (Welton et al., 2015).

23.2 The Value of Implementing Research Findings

Further research may not be the only way to improve health outcomes (Claxton et al., 2016). Examining the expected benefits of an intervention based on the findings of existing research from a meta-analysis indicates the value of implementation efforts to get results of research into practice. The expected absolute benefits of implementing current research findings is obtained by combining the central estimate of relative effect with the estimate of baseline risk. For example, in Table 23.1, the expected outcomes from implementing corticosteroids are 2851 deaths per annum compared with a baseline risk of 3030 deaths per annum. Therefore, the balance of evidence favors corticosteroids, which means that on average the benefits (odds ratio < 1) exceed the harms (odds ratio > 1) so a reduction of 179 deaths per annum would be expected by ensuring that corticosteroids are widely implemented. However, due to uncertainty in the effects of corticosteroids, the current level of utilization or uptake is more likely to represent some proportion of the population receiving corticosteroids. For example, prior to CRASH, clinical practice in the UK did not reflect the balance of evidence that favored the use of corticosteroids on the endpoint of mortality, with approximately 12% of patients receiving them (McKeating et al., 1998). For the sampled values in Table 23.1, the expected outcome with 12% utilization of corticosteroids is 3008 (=0.12*2851 + 0.88*3030) deaths per annum. Therefore, if clinical practice were changed to reflect the balance of existing evidence, we would expect to observe 158 (=3008 − 2851) fewer deaths per annum, which represents the expected value of implementation efforts.

Table 23.2 shows the value of implementation and the value of further research expressed as a simple 2×2 table based on the level of information available for corticosteroids and the level of implementation in practice (Fenwick et al., 2008). This table allows a direct comparison to be made between the benefits of conducting further research and the benefits of implementing current evidence findings. In this example, the potential improvements in health outcomes by encouraging the implementation of what existing evidence suggests is the most effective intervention exceed the potential improvements in health outcomes through conducting further research. This distinction between the value of implementation and the value of additional evidence is important as additional research may not be the only or most effective way to change practice. Continuing to conduct research to

TABLE 23.2
A 2×2 Table for the Expected Value of Implementation and the Expected Value of Further Research for the Data in Table 23.1

	Deaths per annum	
	Information	
Implementation	**Current (existing evidence)**	**Perfect (uncertainty resolved)**
Current level	A = 3008	B = 3008[a]
Perfect level	C = 2851	D = 2809

Cell A, expected value of a decision made on the basis of current evidence and current level of utilization
Cell B, expected value of a decision made with perfect information and current level of utilization
Cell C, expected value of a decision made on the basis of current evidence and perfect utilization
Cell D, expected value of a decision made with perfect information and perfect utilization
[a]B = A under the assumption that information alone has no effect on current level of utilization
Expected value of implementation efforts = C − A = 158 deaths averted per annum[a]
Expected value of further research = D − C = 42 deaths averted per annum[a]
Expected value of further research and implementation efforts = D − A = 200 deaths averted per annum[a]

[a] Numbers may not add up due to rounding.

influence implementation rather than because there is real value in acquiring additional evidence itself is inappropriate for two main reasons: (i) the limited resources to fund research could be used elsewhere to acquire additional evidence in areas where it is genuinely needed and offers greater potential health benefits; and (ii) there are negative health effects to patients enrolled in research and allocated to interventions which are expected to be less effective. Therefore, in situations where expected benefits of interventions are substantial, but not yet demonstrated to be statistically significant, waiting to implement research findings until traditional rules of statistical significance are achieved may come at a considerable cost to patient outcomes (Claxton et al., 2016).

However, decisions about implementing research findings must also be weighed against the potential opportunity costs, which can include the value of research forgone as a consequence of earlier implementation (Claxton et al., 2012). For example, if implementation makes it impossible or more difficult to conduct the type of research required to reduce uncertainty then the value of further research is forgone by implementation. Efforts to implement research findings may also prove to be as costly as the costs of conducting additional research. It may also be difficult (and costly) to reverse treatment decisions if subsequent research demonstrates that the intervention implemented is not effective. Therefore, in many circumstances, it may be better to delay the use of the intervention until additional research is undertaken (Claxton et al., 2012).

23.3 Minimum Clinical Difference in Outcomes

For the results of research to have an impact on practice, a substantial improvement in effectiveness may need to be demonstrated, for example, clinical practice may be unlikely to change without observing highly statistically significant results. In these cases, a minimum clinical difference in outcomes between alternative interventions may need to be observed to change practice (Claxton et al., 2013). Although any improvement in outcome is valuable, there may be circumstances when larger differences are required. For example, larger differences may be required when there are other aspects of outcome not captured in the primary endpoint of the meta-analysis (e.g., adverse events or quality of life impacts) or resource implications associated with widespread implementation. If evidence must demonstrate larger differences in effect, this will tend to reduce the value of additional research because larger differences are less likely to be observed than smaller ones.

For the data presented in Table 23.1, a minimum clinical difference in the outcome of mortality means that clinical practice will only change from corticosteroids (reflecting the balance of existing evidence) to the control if the control reduces the odds of death by a pre-specified minimum difference when compared with the odds of death for corticosteroids. This minimum clinical difference in the odds of death can also be expressed as a minimum reduction in the absolute number of deaths that would need to be observed with additional evidence. The last column of Table 23.1 (column labeled [I]) shows the value of additional evidence for a minimum clinical difference of 50 deaths averted per annum. The chance that the intervention is less effective than the control occurs in realizations 7 to 10 (i.e., odds ratio > 1) but if further research must demonstrate a difference of 50 deaths averted, then this only occurs in realizations 9 and 10. The need to demonstrate a reduction of 50 additional deaths per annum reduces the expected value of additional research from 42 to 37 deaths averted per year.

23.4 Taking Account of Other Aspects of Outcome

For the endpoint of mortality, the benefits of further research and implementation efforts to change practice are expressed in number of deaths averted. However, in many contexts, other aspects of outcome associated with morbidity may need to be considered. The benefits of research and implementation can be expressed as the number of events avoided for a harmful outcome or gained for a benefit outcome, while for a continuous scale of outcome, it may be expressed as the improvement per patient or the number of patients experiencing a clinically important improvement (Claxton et al., 2016). For corticosteroids following traumatic brain injury, mortality is only one aspect of outcome because the impact on disability and subsequent survival is also important. Although the outcome reported in 16 of the trials before CRASH was number of deaths (Alexander, 1972; Braakman et al., 1983; Chacon, 1987; Cooper et al., 1979; Dearden et al., 1986; Faupel et al., 1976; Gaab et al., 1994; Giannotta et al., 1984; Grumme et al., 1995; Hernesniemi and Troupp, 1979; Pitts and Kaktis, 1980; Ransohoff, 1972; Saul et al., 1981; Stubbs et al., 1989; Zagara et al., 1987; Zarate and Guerrero, 1995), the Glasgow Outcome Scale (Jennett and Bond, 1975), which categorizes individuals into one of five health states: (i) death; (ii) persistent vegetative; (iii) severe disability; (iv) moderate disability; and (v) recovery, was also used to assess neurological outcomes in seven of the trials (Braakman et al., 1983; Cooper et al., 1979; Dearden et al., 1986; Faupel et al., 1976; Gaab et al., 1994; Giannotta et al., 1984; Grumme et al., 1995). A further two trials reported the combined number of individuals who were dead, vegetative, or severely disabled at the end of the study (Hernesniemi and Troupp, 1979; Pitts and Kaktis, 1980). The meta-analysis can be extended to include the effects of corticosteroids on both survival and disability. A scenario analysis based on the meta-analysis of trials that only report the composite endpoint of death, vegetative, or severe disability was conducted to assess whether these other aspects of outcome might change the potential value of further research and implementation efforts. Note that a bivariate meta-analysis (Chapter 9) on both endpoints (death and disability) could be conducted using all 16 trials. A Bayesian random-effects meta-analysis for the composite endpoint changes both the estimate of effectiveness of corticosteroids and the uncertainty associated with it. Based on the random-effects analysis, the existing evidence no longer favors the use of corticosteroids, with an odds ratio of 1.10 and 95% credible interval of 0.81 to 1.53. This is because the evidence from the trials suggested that the use of corticosteroids was also associated with a greater proportion of patients surviving in a vegetative or severely disabled state, i.e., on balance corticosteroids appeared to increase the risk of severe disability. The chance that corticosteroids are now effective is only 30% and the value of further research to resolve this uncertainty is 65 deaths and severe disabilities averted per annum. The value of implementing current evidence findings (i.e., control, which is no use of corticosteroids following traumatic brain injury) is 23 deaths and severe disabilities averted per annum.

Mortality and morbidity endpoints reported in clinical trials and commonly used as the endpoint in a meta-analysis often reflect only limited aspects of outcome. Furthermore, use of these endpoints leads to the difficulty of interpreting the value of further research and implementation across diverse clinical areas. For example, knowing the number of deaths and severe disabilities averted in traumatic brain injury is not sufficient to make a comparison with the number of events averted (e.g., thromboembolic events) in another clinical area. Linking these clinical endpoints to a more comprehensive measure of health outcome could be informed by a meta-analysis of both primary and secondary endpoints and by using other sources of evidence on subsequent survival and quality of life, while

TABLE 23.3

Random-Effects Meta-Analysis of the Effect of Corticosteroids on Glasgow Outcome Scale Outcomes

	Proportion of individuals (95% CrI) by treatment		Health-related quality of life weight (SE)[a]	Years lived in full health (QALY)[b]
Outcome	Corticosteroids	Control		
Death	33.5 (22.8, 45.2)	35.3 (24.8, 46.9)	0.00	0.00
Vegetative	4.8 (2.8, 7.5)	3.8 (2.4, 5.9)	0.08 (0.16)	0.56
Severe disability	13.5 (8.3, 20.1)	10.7 (7.1, 15.8)	0.26 (0.25)	3.24
Moderate disability	11.6 (8.6, 14.8)	12.1 (9.2, 15.1)	0.63 (0.27)	10.51
Good recovery	36.5 (28.1, 44.8)	38.0 (30.1, 45.6)	0.85 (0.19)	15.39

Adapted from McKenna, C., et al., Methods to place a value on additional evidence are illustrated using a case study of corticosteroids after traumatic brain injury, J Clin Epidemiol. 2015 Sep 18. doi:10.1016/j.jclinepi.2015.09.011 Copyright © 2015.

CrI = credible interval; SE = standard error.

[a] The conventional scale for health-related quality of life weights is between 0 (death) and 1 (perfect health). A weight of 0.5 is equivalent to half a year in full health.

[b] Average age of 50 years and outcomes discounted at a rate of 3.5% per annum.

reflecting any additional uncertainty that might be introduced. For example, evidence of expected survival and quality of life post traumatic brain injury can be used to translate mortality consequences into life years and quality-adjusted life years (QALYs) (Weinstein et al., 2009). By translating health effects into a more generic measure of outcome, the health benefits of further research and implementation efforts can be compared across diverse clinical areas.

Table 23.3 shows the proportion of individuals expected to be in each of the Glasgow Outcome Scale states by treatment group based on a Bayesian random-effects meta-analysis of the effect of corticosteroids following traumatic brain injury (McKenna et al., 2015). These clinical endpoints can now be linked to other evidence about life expectancies given survival in a particular health state, as well as estimates of the quality of life associated with different levels of disability described by these states (Table 23.3). For example, Shavelle et al. (2007) estimated the expected life expectancy of an individual following traumatic brain injury by age and severity of disability, while taking account of changes in health status over time (Shavelle et al., 2007). This is then combined with health-related quality of life weights, which are used to weight survival in worse health states lower than survival in full health, in order to quantify an individual's remaining life expectancy in terms of years lived in full health (QALYs) (Aoki et al., 1998). This allows the different aspects of outcome from the meta-analysis to be combined into a generic measure of equivalent years lived in full health, i.e., the effects on mortality and morbidity due to the intervention are adjusted for the quality in which the life years are likely to be lived.

In the same way as before, a distribution of the overall health consequences of uncertainty in number of QALYs is derived by sampling from the distributions of uncertainty in Glasgow Outcome Scale outcomes and combining these with uncertainty in the external evidence. This analysis of the effects of corticosteroids on a more comprehensive measure of health outcome suggests that on balance corticosteroids were not expected to be effective and its effects were more uncertain compared with the endpoint of mortality alone. If the use of corticosteroids had been based on this evidence then there would be a 63% chance that not using corticosteroids was the best decision, i.e., a 63% chance of no avoidable lost years of full health. However, there was a 37% chance that corticosteroids could have improved health outcomes, including a 23% chance of between 0 and 3000 years of full health gained per year and 14% chance of more than 3000 years of full health gained per year (derived from the distribution of the health consequences of uncertainty in number of QALYs). The resulting expected consequences of this uncertainty or the potential gains through conducting further research is 1067 years of full health gained each year in the UK. Prior to CRASH, 12% of patients with traumatic brain injury received corticosteroids in the UK (McKeating et al., 1998). Therefore, the value of both implementing the uncertain findings of existing evidence and acquiring additional research that would resolve this uncertainty is greater at 1264 years of full health gained per annum.

23.5 The Type of Evidence Required and the Design of Research

Once a single endpoint is linked to other endpoints and external evidence, the same type of analysis can be extended to identify which sources of uncertainty are most important and what type of evidence is likely to be most valuable (Claxton et al., 2012). This can start to indicate the type of research design that is likely to be required and identify the most important endpoints to include in the design. It can also be used to consider whether there are other types of research that could be conducted relatively quickly and cheaply for which the results might confirm whether or not more lengthy and expensive research is really needed, i.e., it can indicate the sequence in which different studies might be conducted (Griffin et al., 2010).

The expected benefits of resolving uncertainty in different sources of evidence are established using the same principles as above. For example, if the health outcome of alternative interventions depends on two groups of uncertain parameters, the decision based on the balance of existing evidence follows as before where the intervention that is expected to offer the best outcome is chosen. If the uncertainty associated with one of these groups of parameters could be resolved (group 1), the decision maker would know which values that parameter would take before choosing between the alternative interventions. The values of the other group of parameters (group 2) remain uncertain so the best that the decision maker can do is to select the intervention which provides the best outcomes for each value of group 1 (by averaging over outcomes in group 2). However, which particular value group 1 will take is unknown before the research is conducted so the expected benefits when uncertainty associated with group 1 is fully resolved is the average of these outcomes over all the possible values of group 1. Table 23.4 provides an illustration of the value of additional evidence for the relative effect of corticosteroids on mortality (corresponding to data in Table 23.1) and assuming that the research does not provide information on baseline risk of death (i.e., uncertainty remains in this parameter). A single sampled value is first drawn from the distribution of values for the odds ratio of death and held fixed. A sample of values is then drawn from the uncertain distribution of baseline risk and used to calculate the corresponding number of deaths for the alternative interventions, conditional on the fixed value of the odds ratio. However, the odds ratio is unknown before the research is conducted so we need to repeat this process until a population of sampled results is derived for different values of the odds ratio. The resulting expected value of further research about the relative effectiveness parameter is the average of the

TABLE 23.4

Calculating the Value of Additional Evidence for the Relative Effect (Odds Ratio) of Corticosteroids on Mortality

				Deaths per annum for an incidence of 8,800 eligible patients			Value of additional evidence for odds ratio	
Sampled realization of uncertainty	Odds ratio for death (corticosteroids vs. control) [A]	Baseline odds of death for control [B]	Odds of death for corticosteroids[a] (=A × B) [C]	Control (= 8800 × B/(1+B)) [D]	Corticosteroids (=8800 × C/(1+C)) [E]	Absolute effect for corticosteroids (=E–D) [F]	Consequences of uncertainty for corticosteroids (=F if A>1 else 0) [G]	Perfect information (=minimum of D and E) [H]
		0.88	0.57	3743	3194	–548		
1	0.77	0.54	0.42	3086	2584	–501		
		0.33	0.25	2183	1783	–400		
Average realization outcome				3004	2520	–481	0	2520
		0.74	0.73	3743	3699	–43		
2	0.98	0.60	0.59	3300	3258	–42		
		0.33	0.34	2281	2248	–34		
Average realization outcome				3108	3068	–40	0	3068
		0.67	0.70	3531	3634	104		
3	1.05	0.54	0.57	3086	3184	98		
		0.35	0.35	2183	2265	81		
Average realization outcome				2933	3028	94	94	2933
Overall average	0.93	0.54	0.50	3015	2872	–143	31	2841

[a] Distributions for baseline odds and odds ratio are assumed independent for ease of demonstrating the principles to the reader.

consequences of uncertainty across the sampled results (column labeled [G]), which is 31 deaths averted per annum for this sample.

For the case of corticosteroids following traumatic brain injury the different sources of uncertainty include: (i) the effect of corticosteroids on the risk of death; (ii) the effect of corticosteroids on the risk of disability; and (iii) uncertainty in the health-related quality of life weights. Separating apart these different sources of uncertainty (results not shown) suggests that further evidence about the combined effect of corticosteroids on the risk of death and severe disability is of most value. Future research which includes mortality as an endpoint alone is of limited value. There is also very little additional value to research which includes the less severe disability outcomes or quality of life weights.

This type of analysis can also be extended to consider the sequence in which different types of studies might be conducted (Griffin et al., 2010). For example, whether research on the parameters of group 1 should come first, followed by research on the parameters of group 2 conditional on the results of group 1 (or vice versa), or whether research about different groups of parameters should be conducted simultaneously. The sequential design of research allows for the abandonment of research on a second parameter if there is no longer value to additional evidence on the basis of additional information obtained for the first parameter.

The analysis can be extended even further to consider the optimal sample size of a future clinical trial. The expected value of sample information (EVSI) provides the value of a decision based on having additional sample information (Ades et al., 2004; McKenna and Claxton, 2011). This is achieved by predicting possible sample results that would be obtained from a new study with a sample size of n. These predicted results are then combined with the evidence available (prior information) to determine the weight of evidence that could exist following a new study. For example, for a new study that informs the relative treatment effect, a value is sampled from the prior distribution of the range of possible values for the odds ratio. This is then used to generate a possible sample result for a new study of sample size n through a likelihood function, which represents the probability of observing the new sample result conditional on the value of the prior parameter. This predicted sample result is then combined with the prior to form a predicted posterior value, which in turn is used to calculate the corresponding health outcomes for the intervention and control, i.e., using the same methods as above but with the posterior value of the distribution in the analysis instead of the prior. This process is repeated until a population of sampled posterior results is derived for each new sample size n. If this sample result was known in advance the decision maker would choose the intervention which offers the best outcomes. However, the actual results of the sample are not known before the research is conducted and therefore the expected value of a decision taken with sample information is found by averaging the best outcomes over the distribution of possible sample results. The EVSI is the difference between the expected value of the decision with sample information and the expected value of decision without sample information. It is worth noting that the methods described here are for a linear relationship between the sampled parameters and difference in outcomes between the alternative interventions. If a non-linear relationship holds, the EVSI calculations can become computationally intensive (see Ades et al. (2004) for further details).

To establish the optimal sample size for a particular type of study these calculations need to be repeated for a range of sample sizes. The difference between the EVSI and the costs of acquiring the sample information is the expected payoff to research. The optimal sample size is simply the value of n that generates the maximum payoff in health outcomes. The same type of analysis can also be used to evaluate a range of different dimensions of the research design such as the length of follow-up of the study and which endpoints and interventions to include (Conti and Claxton, 2009). It should be recognized that the costs of research not only include the resources consumed in conducting it but also the opportunity costs falling on those patients enrolled in the research (e.g., patients allocated to the treatment arm which is expected to be less effective). Therefore, optimal research design depends on a number of factors, including whether patients have access to the intervention while the research is being conducted, how long the research will take to report (determined by the length of follow-up and recruitment rates), and also the likelihood that the results of research will be implemented into clinical practice, for example, if clinical practice depends on a trial reporting a statistically significant result for a particular effect size, this will influence the optimal sample size. This analysis applied to corticosteroids following traumatic brain injury suggests that the original sample size for CRASH (10,000 patients for corticosteroids, 10,000 for control) may have been too large and that most of the uncertainty could have been resolved quicker and cheaper with a smaller trial. However, if the implementation of the results of research depends on the trial reporting a statistically significant result for the effect size specified in the original research proposal (a 2% reduction in the risk of the unfavorable outcome of death, persistent vegetative, and severe disability), the sample size for CRASH would appear appropriate.

23.6 Assessing the Value and Priority of Proposed Research

The analyses described above provide an estimate of the health benefits of further research and the value of implementing current evidence findings. The question that now remains is whether the value of the research is sufficient to justify it being prioritized and commissioned. This requires an assessment of whether the research is potentially worthwhile and whether it should be prioritized over other research topics that could be commissioned with the same resources. These assessments require some consideration of the period of time over which the additional evidence generated by research is likely to be relevant to clinical practice, as well as the likely time required for research to complete and report. For example, the patient population will not benefit from the results of the research until it becomes available and, if treatment decisions cannot be reversed, then it is only those patient incidents after the research reports that will realize any of the potential benefits. Also, the information generated by research will not be valuable indefinitely because other changes occur over time (Claxton et al., 2012). For example, new and more effective interventions may become available, making the current intervention and comparators obsolete and possibly rendering information about their effectiveness irrelevant to future clinical practice. Also, other evaluative research may already have been commissioned which may resolve much of the uncertainty. Therefore, the actual time horizon for evidence generated by research is unknown since it is a proxy for a complex and uncertain process of future changes (Philips et al., 2008). However, some judgment about the time horizon is required in order to make decisions about research priorities. This assessment might be possible based on historical evidence and judgments about whether a particular area of research is likely to experience these changes and the likelihood that the intervention would be approved and diffused into clinical practice.

The CRASH trial was proposed to the UK Medical Research Council in the year 2000. A time horizon of 15 years may have been a reasonable but conservative judgment at that time since there were no other trials underway and previously few major innovations that had transformed the treatment of traumatic brain injury. CRASH was not expected to report before the year 2004. Therefore, the overall (undiscounted) expected health benefits from this trial were 13,904 additional years of full health (1264 years of full health gained per annum × 11 years from 2004 to 2015). In the UK, both health benefits and costs are discounted at the same rate of 3.5% per annum specified by the UK Treasury (to reflect the fact that resources committed today could be invested at a real rate of return to provide more resource in the future that is used to generate more health). Therefore, the benefits of research are discounted over this time horizon of 15 years and it is the discounted value of 10,266 additional years of full health which represents the expected value of the CRASH trial. The question that now remains is whether these expected benefits were sufficient to justify the expected costs of the CRASH trial (£2.2 million) and whether it represented a priority compared with the other research topics that could have been commissioned using the same resources. One way to address this question is to ask whether similar expected benefits (or greater) could be generated elsewhere for this cost, or equivalently whether the costs of the research would generate more health benefits if the resources were made available elsewhere (Claxton et al., 2013).

A recent study in the UK has estimated the relationship between changes in National Health Service (NHS) expenditure and health outcomes. This study suggests that it costs the NHS £114,000 to avert one death, £25,000 to gain one life year, and £13,000 to gain one QALY (Claxton et al., 2015). These figures can be used to assess whether the opportunity costs of research are less than the expected benefits and, if so, it would suggest that the proposed research is potentially worthwhile. Using these estimates, the costs of CRASH could have been used to avoid 19 deaths (£2.2 million/£114,000) and generate 169 QALYs elsewhere, which are substantially less than the expected health benefits of CRASH. Alternatively, the NHS would have had to spend an additional £133 million (10,266 QALYs*£13,000 per QALY) between 2004 and 2015 to generate expected health benefits similar to those offered by CRASH. However, most research funders have limited resources (with constraints relevant to a specific budgetary period) and cannot draw directly on other (or future) resources within the NHS. Therefore, a similar analysis would need to be conducted for all research proposals competing for the same limited research resources in order to identify those which are likely to be worthwhile. Once a shortlist is drawn up, then it becomes possible to select from this shortlist the research that is likely to offer the greatest value. This type of analysis therefore allows the value and priority of proposed research to be assessed.

It is useful to reconsider the analysis set out above once the results of CRASH became available by updating the meta-analysis and the estimates of the expected benefits of further research. Although the trials prior to CRASH in Figure 23.1 were synthesized using a Bayesian random-effects meta-analysis, the results of CRASH, which was designed and commissioned to be of high quality and directly relevant to the target patient population and clinical practice, are entered as a fixed effect, i.e., the meta-analysis is updated by including the results from CRASH as a fixed effect and using the evidence pre-dating CRASH as a random-effects prior in the synthesis. Other methods to synthesize the evidence from CRASH may be appropriate. For example, it may be more appropriate to analyze CRASH on its own or base inferences on shrunken study level estimates from meta-analysis (Welton et al., 2015; Welton and Ades, 2012).

The first results of CRASH were reported in 2004 showing the effect of corticosteroids on death within 14 days in 10,008 randomized adults with significant head injury (CRASH Trial Collaborators, 2004). The relative risk of death compared with placebo was 1.18 (95%

CI 1.09–1.27; p = 0.0001). The final results of CRASH were reported in 2005 for outcomes on the Glasgow Outcome Scale at six months after injury (CRASH Trial Collaborators, 2005). The risk of death at six months was higher in the corticosteroid group than in the placebo group (1248 [25.7%] versus 1075 [22.3%] deaths; relative risk 1.15, 95% CI 1.07–1.24; p = 0.0001). Similarly, the risk of death or severe disability was higher in the corticosteroid group (1828 [38.1%] versus 1728 [36.3%] dead or severely disabled; relative risk 1.05, 95% CI 0.99–1.10; p = 0·079). When the results of CRASH are included in the meta-analysis, the chance that corticosteroids reduce mortality is zero (<0.0001), while the chance that corticosteroids improve survival and quality of life is effectively zero (probability of 0.005). Therefore, when the analysis of the potential value of further research is updated, there are no expected benefits of acquiring additional evidence after CRASH. In this sense, CRASH was a definitive trial which resolved all the decision uncertainty about the effects of corticosteroids on mortality and disability following traumatic brain injury.

23.7 Other Considerations

The analyses above demonstrate how methods of value of information analysis can be integrated with the results of meta-analysis to inform the value and priority of future research. The application of value of information analysis to the results from a standard meta-analysis is not technically challenging; however, in some contexts, more sophisticated forms of meta-analysis may be required (e.g., Bayesian meta-analysis to link multiple endpoints), while in other contexts, careful consideration of the relevance and quality of evidence is required (e.g., different weights may be assigned to different studies in the meta-analysis to reflect alternative judgments about the relevance of the evidence to the target population and the potential for bias, as in Chapter 12) (Claxton et al., 2013). In these cases, the same considerations that would be made before conducting the most appropriate meta-analysis must also be made before assessing the value of further research. For example, if there are important sources of heterogeneity between the studies, a common-effect meta-analysis is unlikely to be appropriate due to the assumption that each of the studies are estimating the same underlying effect; however, a random-effects meta-analysis may also not be appropriate to reflect heterogeneity if there are reasonably held judgments about the differences across studies and the relevance of particular studies to a target population. In these circumstances, expert elicitation may be used to provide explicit weights to reflect judgments about the relevance of particular studies (Cooke, 1991; O'Hagan et al., 2006). These judgments about the quality and relevance of evidence not only change the estimate of effectiveness but also the uncertainty associated with it and, therefore, the value of additional evidence and the need for further research (Claxton et al., 2013).

The analyses above have also focused on using the results of standard meta-analysis comparing two alternative interventions to estimate the value of further research. Commonly, however, there are a number of alternative interventions available, only one of which can be selected as the appropriate treatment for a patient. Over recent years, methods to extend meta-analysis to a more generalized form of evidence synthesis have developed rapidly (see Chapter 10). Conceptually, value of information analysis based on network meta-analysis (NMA) is not different from standard meta-analysis comparing two alternative interventions. However, NMA may require more advanced statistical methods, and choices made for the NMA will likely influence subsequent estimates of the

TABLE 23.5

Network of Evidence for Second-Line Treatments for Advanced Ovarian Cancer

	Median weeks survival (number of patients)			
Trial ID	**Paclitaxel (Pac)**	**Topotecan (Top)**	**PLDH**	**Hazard ratio (95% CI)**
039	53.0 (n = 114)	63.0 (n = 112)	–	0.914[a] (0.680, 1.226)
30–49	–	59.7 (n = 235)	62.7 (n = 239)	1.216[a] (1.000, 1.478)
30–57	56.3 (n = 108)	–	46.6 (n = 108)	0.931[b] (0.702, 1.234)

PLDH = pegylated liposomal doxorubicin hydrochloride; CI = confidence interval.

[a] Hazard ratio for topotecan versus comparator.

[b] Hazard ratio for paclitaxel versus PLDH.

value of further research. Importantly, failing to conduct analysis based on an appropriate and complete NMA will lead to biased estimates of the value of further research.

For example, consider the network of evidence illustrated in Table 23.5 for three alternative interventions of paclitaxel (Pac), topotecan (Top), and pegylated liposomal doxorubicin hydrochloride (PLDH) for second-line treatment of advanced ovarian cancer (Griffin et al., 2006; Main et al., 2006). The results of the clinical trials in Table 23.5 present an inconsistent picture of the relative effectiveness of the interventions, for example, Top is more effective than Pac (trial 039) and PLDH is more effective than Top (trial 30–49) but the evidence from trial 30–57 suggests that Pac is more effective than PLDH. By restricting the analyses to the three separate pairwise comparisons made in the existing trials, three different estimates of effectiveness and uncertainty would be generated with each associated with ignoring one of the three trials. Alternatively, analysis could be restricted to comparisons based on a common comparator (indirect treatment comparison, ITC). In this case, there is no single comparator that is common to all three trials (only two out of the three trials have a common comparator). Therefore, restricting the evidence synthesis to a method that requires a single common comparator would be inappropriate as one trial would be excluded and, insofar as this excluded trial provides some information, this would bias the estimate of effectiveness and overestimate the uncertainty associated with it.

A more appropriate NMA would exploit the full network of evidence, making use of common comparators and the indirect comparisons from combinations of studies. For example, when comparing PLDH to Top, the information from trial 30–49 (PLDH vs. Top) provides direct evidence. However, the combination of trials 039 (Top vs. Pac) and 30–57 (PLDH vs. Pac) tells us something about PLDH vs. Top based on the common comparator of Pac. In addition, both trial 039 (Top vs. Pac) in combination with trial 30–49 (PLDH vs. Top) and trial 30–57 (PLDH vs. Pac) in combinations trial 30–49 (PLDH vs. Top) tells us something about PLDH compared with Top. An analysis of the expected health benefits of further research based on this NMA suggests that research could potentially result in 227 deaths averted per annum for the target population in the UK. This contrasts sharply with an estimate of 766 deaths averted per annum from an incomplete NMA based only on common comparators (ITC) and 379 deaths averted per annum by summing the expected health benefits of additional evidence across each pairwise comparison (Claxton et al., 2013). This indicates the importance of an appropriate NMA because the potential benefits of further research will be estimated with bias using an incomplete NMA restricted to a common comparator or the sum of potential benefits over pairwise comparisons.

Although the meta-analyses described here are Bayesian and use non-informative priors, they could also be performed using a frequentist framework. However, Bayesian approaches may be preferred for random-effects meta-analysis, as they allow for greater uncertainty than the frequentist approach by enabling estimation of both the overall population effect and the between-study effect by the data (Cooper et al., 2004; Smith et al., 1995). In addition, when cumulative meta-analyses are used to assess the value of further research, repeated testing of updated meta-analyses affects assessments of statistical significance in the frequentist framework, increasing the risk of satisfying a predefined significance level by chance, whereas Bayesian approaches would not suffer from this problem.

23.8 Conclusions

Studies often conclude with the recommendation that further research is needed. However, this statement is often not supported by any quantitative assessment of the need or consequences if the research is not undertaken. In this chapter, we have shown that a simple extension of standard results of meta-analysis can be used to provide an indication of the scale of the health benefits of further research. These estimates can be used to assess whether the proposed research is potentially worthwhile compared with the costs of conducting the research and relative to the benefits and costs of other research proposals competing for the same limited resources. This can start to inform the questions posed in research prioritization and commissioning decisions by identifying where and what type of research would be most valuable. It can also start to inform other important policy questions including whether the widespread use of an intervention should be encouraged based on the balance of existing evidence and whether sufficient resources are being devoted to evaluative research.

References

Ades AE, Lu G and Claxton K, 2004. Expected value of sample information calculations in medical decision modeling. *Medical Decision Making* **24**(2): 207–227.

Alderson P and Roberts I, 1997. Corticosteroids in acute traumatic brain injury: Systematic review of randomised controlled trials. *British Medical Journal* **314**(7098): 1855–1859.

Alderson P and Roberts I, 1999. *Corticosteroids for Acute Traumatic Brain Injury (Cochrane Review).* Oxford: Update Software: The Cochrane Library.

Alexander E, 1972. Medical management of closed head injuries. *Clinical Neurosurgery* **19**: 210–250.

Aoki N, Kitahara T, Fukui T, Beck JR, Soma K, Yamamoto W, et al., 1998. Management of unruptured intracranial aneurysm in Japan: A Markovian decision analysis with utility measurements based on the Glasgow Outcome Scale. *Medical Decision Making* **18**(4): 357–364.

Basu A and Meltzer D, 2007. Value of information on preference heterogeneity and individualized care. *Medical Decision Making* **27**(2): 112–127.

Braakman R, Schouten HJA, Blaauw-van Dishoeck M and Minderhoud JM, 1983. Megadose steroids in severe head injury. *Journal of Neurological Surgery* **58**(3): 326–330.

Chacon L, 1987. Brain edema in severe head injury on children treated with and without dexametasone. *Med Crit Venez* **2**: 75–79.

Claxton K, 1999. The irrelevance of inference: A decision-making approach to the stochastic evaluation of health care technologies. *Journal of Health Economics* **18**(3): 341–364.

Claxton KP and Sculpher MJ, 2006. Using value of information analysis to prioritise health research. Some lessons from recent UK experience. *Pharmacoeconomics* **24**(11): 1055–1068.

Claxton K, Eggington S, Ginnelly L, Griffin S, McCabe C, Philips Z, et al., 2005. A pilot study of value of information analysis to support research recommendations for the National Institute for Health and Clinical Excellence: Centre for Health Economics. Research Paper 4. York: Centre for Health Economics, University of York.

Claxton K, Ginnelly L, Sculpher MJ, Philips Z and Palmer S, 2004. A pilot study on the use of decision theory and value of information analysis as part of the NHS Health Technology Assessment Programme. *Health Technology Assessment* **8**(31): 1–103.

Claxton K, Palmer S, Longworth L, Bojke L, Griffin S, McKenna C, et al., 2012. Informing a decision framework for when NICE should recommend the use of health technologies only in the context of an appropriately designed programme of evidence development. *Health Technology Assessment* **16**(46): 1–33.

Claxton K, Griffin S, Koffijberg H and McKenna C, 2013. Expected health benefits of additional evidence: Principles, methods and applications. CHE Research Paper 83. Centre for Health Economics, University of York.

Claxton KP, Martin S, Soares MO, Rice N, Spackman E, Hinde S, et al., 2015. Methods for the estimation of the NICE cost effectiveness threshold. *Health Technology Assessment* **19**(14): 1–503.

Claxton K, Griffin S, Koffijberg H and McKenna C, 2015. How to estimate the health benefits of additional research and changing clinical practice. *BMJ* 2015;351:h5987.

Colbourn TE, Asseburg C, Bojke L, Philips Z, Welton NJ, Claxton K, et al., 2007. Preventive strategies for group B streptococcal and other bacterial infections in early infancy: Cost effectiveness and value of information analyses. *British Medical Journal* **335**(7621): 655.

Conti S and Claxton K, 2009. Dimensions of design space: A decision-theoretic approach to optimal research design. *Medical Decision Making* **29**(6): 643–660.

Cooke RM, 1991. *Experts in Uncertainty: Opinion and Subjective Probability in Science*. New York: Oxford University Press.

Cooper PR, Moody S, Clark WK, Kirkpatrick J, Maravilla K, Gould AL and Drane W, 1979. Dexamethasone and severe head injury. *Journal of Neurological Surgery* **51**(3): 307–316.

Cooper NJ, Sutton AJ, Abrams KR, Turner D and Wailoo A, 2004. Comprehensive decision analytical modelling in economic evaluation: A Bayesian approach. *Health Economics* **13**(3): 203–226.

CRASH Trial Collaborators, 2004. Effect of intravenous corticosteroids on death within 14 days in 10008 adults with clinically significant head injury (MRC CRASH trial): randomised placebo-controlled trial. *Lancet* **364**: 1321–1328.

CRASH Trial Collaborators, 2005. Final results of MRC CRASH, a randomised placebo-controlled trial of intravenous corticosteroid in adults with head injury – Outcomes at 6 months. *Lancet* **365**(9475): 1957–1959.

Dearden NM, Gibson JS, McDowall DG, Gibson RM and Cameron MM, 1986. Effect of high dose dexamethasone on outcome from severe head injury. *Journal of Neurological Surgery* **64**(1): 81–88.

Eckermann S and Willan AR, 2007. Expected value of information and decision making in HTA. *Health Economics* **16**(2): 195–209.

Faupel G, Reulen HJ, Muller D and Schurmann K, 1976. Double-blind study on the effects of steroids on severe closed head injury. In Pappius MM and Feindel W (Eds). *Dynamics of Brain Edema*. Berlin: Springer-Verlag.

Fenwick E, Claxton K and Sculpher M, 2008. The value of implementation and the value of information: Combined and uneven development. *Medical Decision Making* **28**(1): 21–32.

Fleurence RL and Meltzer DO, 2013. Toward a science of research prioritization? The use of value of information by multidisciplinary stakeholder groups. *Medical Decision Making* **33**(4): 460–462.

Gaab MR, Trost HA, Alcantara A, Karimi-Nejad A, Moskopp D, Schultheiss R, et al., 1994. "Ultrahigh" dexamethasone in acute brain injury. *Zentralblatt für Neurochirurgie* **55**(3): 135–143.
Giannotta SL, Weiss MH, Apuzzo MLJ and Martin E, 1984. High dose glucocorticoids in the management of severe head injury. *Neurosurgery* **15**(4): 497–501.
Griffin S, Bojke L, Main C and Palmer S, 2006. Incorporating direct and indirect evidence using bayesian methods: An applied case study in ovarian cancer. *Value in Health* **9**(2): 123–131.
Griffin S, Welton NJ and Claxton K, 2010. Exploring the research decision space: The expected value of information for sequential research designs. *Medical Decision Making* **30**(2): 155–162.
Grumme T, Baethmann A, Kolodziejczyk D, Krimmer J, Fischer M, von Eisenhart Rothe B, et al., 1995. Treatment of patients with severe head injury by triamcinolone: A prospective, controlled multicenter trial of 396 cases. *Research in Experimental Medicine* **195**(4): 217–229.
Hassan C, Hunink MG, Laghi A, Pickhardt PJ, Zullo A, Kim DH, et al., 2009. Value-of-information analysis to guide future research in colorectal cancer screening. *Radiology* **253**(3): 745–752.
Hernesniemi J and Troupp H, 1979. A clinical retrospective and a double blind study of betamethasone in severe closed brain injuries. *Acta Neurochrurgica* **28**(2): 499.
Jennett B and Bond M, 1975. Assessment of outcome after severe brain damage. *Lancet* **1**(7905): 480–484.
Main C, Bojke L, Griffin S, Norman G, Barbieri M, Mather L, et al., 2006. Topotecan, pegylated liposomal doxorubicin hydrochloride and paclitaxel for second-line or subsequent treatment of advanced ovarian cancer: A systematic review and economic evaluation. *Health Technology Assessment* **10**(9): 1–132.
McKeating EG, Andrews PJ, Tocher JI and Menon DK, 1998. The intensive care of severe head injury: A survey of non-neurosurgical centres in the United Kingdom. *British Journal of Neurosurgery* **12**(1): 7–14.
McKenna C and Claxton K, 2011. Addressing adoption and research design decisions simultaneously: The role of value of sample information analysis. *Medical Decision Making* **31**(6): 853–865.
McKenna C, Griffin S, Koffijberg H and Claxton K, 2015. Methods to place a value on additional evidence are illustrated using a case study of corticosteroids after traumatic brain injury. *Journal of Clinical Epidemiology* **70**: 183–190.
O'Hagan A, Buck CE, Daneshkhah A, Eiser JR, Garthwaite PH, Jenkinson DJ, et al., 2006 *Uncertain Judgements: Eliciting Experts' Probabilities*. Chichester: Wiley.
Philips Z, Claxton K and Palmer S, 2008. The half-life of truth: What are appropriate time horizons for research decisions? *Medical Decision Making* **28**(3): 287–299.
Pitts LH and Kaktis JV, 1980. Effect of megadose steroids on ICP in traumatic coma. In Shulman K, Marmarou A, Millar JD, et al. (Eds). *Intracranial Pressure IV*. Berlin: Springer-Verlag.
Ransohoff J. The effects of steroids on brain edema in man. In: Reulen HJ, Schurmann K (editors), 1972. *Steroids and Brain Edema*. New York: Springer-Verlag, 211–213.
Riley RD, 2009. Multivariate meta-analysis: The effect of ignoring within-study correlation. *Journal of the Royal Statistical Society: Series A* **172**(4): 789–811.
Saul TG, Ducker TB, Salcman M and Carro E, 1981. Steroids in severe head injury. *Journal of Neurological Surgery* **54**(5): 596–600.
Shavelle RM, Strauss DJ, Day SM and Ojdana KA, 2007. Life expectancy. In: Zasler ND, Katz DI and Zafonte RD (editors). *Brain Injury Medicine: Principles and Practice*. New York: Demos Medical Publishing.
Smith TC, Spiegelhalter DJ and Thomas A, 1995. Bayesian approaches to random-effects meta-analysis: A comparative study. *Statistics in Medicine* **14**(24): 2685–2699.
Stubbs DF, Stiger TR and Harris WR. Multinational controlled trial of high-dose methylprednisolone in moderately severe head injury. In Capildeo R (Ed). *Steroids in Diseases of the Central Nervous System*. Chichester: John Wiley & Sons, 163–168.
Sutton AJ and Abrams KR, 2001. Bayesian methods in meta-analysis and evidence synthesis. *Statistical Methods in Medical Research* **10**(4): 277–303.
Weinstein MC, Torrance G and McGuire A, 2009. QALYs: The basics. *Value in Health* **12**: S5–S9.

Welton N and Ades AE, 2012. Research decisions in the face of heterogeneity: What can a new study tell us? *Health Economics* **21**(10): 1196–1200.

Welton NJ, Soares MO, Palmer S, Ades AE, Harrison D, Shankar-Hari M and Rowan KM, 2015. Accounting for heterogeneity in relative treatment effects for use in cost-effectiveness models and value-of-information analyses. *Medical Decision Making* **35**(5): 608–621.

Whitehead A, 2002. *Meta-Analysis of Controlled Clinical Trials*. West Sussex, UK: John Wiley & Sons, Ltd.

Zagara G, Scaravilli P, Carmen Belluci M, Seveso M, 1987. Effect of dexamethasone on nitrogen metabolism in brain-injured patients. *Journal of Neurological Surgery* **31**: 207–212.

Zarate IO and Guerrero JG, 1995. Corticosteroids in paediatric patients with severe head injury. *Practica Pediatrica* **4**: 7–14.

Index

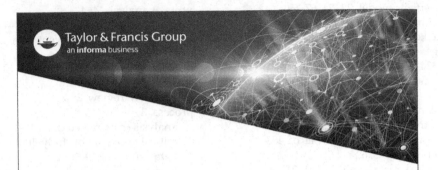

Printed in the United States
By Bookmasters